William Miller

A Dictionary of English names of plants applied in England and among English-speaking people

William Miller

A dictionary of English names of plants applied in England and among English-speaking people

ISBN/EAN: 9783337225179

Printed in Europe, USA, Canada, Australia, Japan

Cover: Foto ©Paul-Georg Meister /pixelio.de

More available books at **www.hansebooks.com**

A DICTIONARY

OF

ENGLISH NAMES OF PLANTS

APPLIED IN

ENGLAND AND AMONG ENGLISH-SPEAKING PEOPLE

TO

CULTIVATED AND WILD PLANTS, TREES, AND SHRUBS.

By WILLIAM MILLER.

IN TWO PARTS. ENGLISH–LATIN AND LATIN–ENGLISH.

LONDON: JOHN MURRAY, ALBEMARLE STREET.
1884.

No one who has had experience of the progress of Botany as a science can doubt that it has been more impeded in this country by the repulsive appearance of the names which it employs than by any other cause whatever, and that, in fact, this has proved an invincible obstacle to its becoming the serious occupation of those who are unacquainted with the learned languages, or who, being acquainted with them, are fastidious about euphony and Greek or Latin purity.—DR. LINDLEY.

Botany has this great practical advantage over all other sciences as a means of universal culture, that the materials are the most generally accessible of any scientific material in the world. . . . . What is needed is that its terminology should be popularised. . . . . Historically almost the first of sciences, botany is naturally and eductionally first in order to the enquiring mind. Its objects are near our homes, awakening to our minds, and inviting to our touch. Botany is adapted to be the universal preparatory science, the science to infuse the scientific sense. Why should we allow a pile of heterogeneous names to stand as a barrier between our people and the fairest gate of knowledge? These strange names are all but barren of interest in themselves; what interest they possess springs wholly out of the objects they represent. The objects and their mutual relations might be learnt quite as effectually through congenial names, if only one-thousandth part of the labour that has been expended on those were bestowed on these.—Prof. EARLE, " English Plant Names," p. cix.

PREFACE.

The compilation of this Dictionary was undertaken at the request of Mr. W. Robinson, of THE GARDEN newspaper, at whose expense the work is published, he having advocated in his journal a more general use of English names for the plants, trees, and shrubs which are commonly grown in our gardens and pleasure-grounds, and who wished the horticultural public to have at command a list of all such names now applied to these as well as to all other cultivated and useful plants including our native flora and the native plants and trees of America and the colonies.

It is an undeniable fact that the vast majority of people of all classes who take an interest in horticultural pursuits consists of those who, never having received any classical or botanical training, find it difficult to learn and remember, and impossible to understand, the Latin or scientific names by which plants are spoken of and described by botanists. These names, however useful and even necessary they may be as technical terms to the systematic botanist, become a senseless jargon in the vain attempt to fix them amongst our "household words," and most of us are keenly alive to the inconsistency of employing words from a foreign and even dead language to name such familiar everyday objects as the flowers and shrubs which are grown in our gardens and woods. Notwithstanding the copious use of Latin, it would be a grievous mistake to suppose that English names do not exist for most of our cultivated plants, the fact being that such names do exist, and abundantly, many of them dating back to the days of Spenser, Shakespeare, Gerard, and Parkinson—nearly 300 years ago—although they have now fallen into disuse, and are only to be met with in books, in consequence of what the Rev. John Earle, in his excellent little volume on "English Plant Names," terms "the gratuitous rejection of good native names in favour of some Latin name, through mere contempt for homely things and affectation of novelty." No farther back, indeed, than the commencement of the present century, it would appear that, even amongst gardeners, it was the ordinary custom to speak of plants by their English names, as we find the poet Crabbe thus describing an exceptional case :—

> High-sounding words our worthy gardener gets,
> And at his club to wondering swains repeats ;
> He there of *Rhus* and *Rhododendron* speaks,
> And *Allium* calls his Onions and his Leeks.
> * * * * * *
> There *Arums*, there *Leontodons* we view,
> And *Artemisia* grows where Wormwood grew.
> (Crabbe's "Parish Register," Part I., Baptisms.)

and there can be little doubt that many good old English plant-names which, happily, are still preserved to us in books, have been gradually ousted from popular use and sacrificed for Latin terms, not from any conviction that these were better or more appropriate, but simply through the spread of the craze for "high-sounding words." To quote Mr. Earle further, "The adoption of classical words was in deference to the prestige of the classical languages at first, then it became a piece of scholastic pedantry which, spreading ever wider and wider, became at length a fashion because it was a flag of social pretension."

A botanist writes in THE GARDEN (vol. xxiii., p. 403) :—"But what do we see in popular naming ? . . . . The whole business breeds nothing but confusion, as if there was not enough already in the same direction." Such a remark comes with a peculiarly bad grace from a scientific botanist, and may be regarded as a stone thrown by one who lives in a glass house of rather extensive dimensions, when we consider the deplorable condition of his own pet nomenclature in this respect. There is, in fact, no greater stumbling-block and no more torturing embarrassment in the way of the botanical student than the swarms of synonyms which beset him at almost every step and, like the aliases of a culprit who is "wanted," serve rather to conceal than to point out the subjects to which they are applied. The whole family of the Coniferæ, for instance, is almost smothered in this way, as anyone may see who chooses to look into the last edition of Gordon's "Pinetum," where he will find that nearly all the trees there described have a greater or less number of synonyms applied to them, several of them as many as half-a-dozen or more apiece ! The practical results of this extreme plurality of scientific synonyms are well exemplified by an instance which occurred last year, when a correspondent of THE GARDEN wrote to the effect that "The Bluebell of Scotland is Agraphis nutans," and that "the English Bluebell is Hyacinthus non-scriptus" (THE GARDEN for June 9, 1883, p. 523), in evident ignorance that the two names are synonymous for the same plant, which has yet the two other synonyms of Scilla nutans and Hyacinthus anglicus.

Many of the new names which have appeared in THE GARDEN were absolutely needed for plants which previously had no popular

English names, and others are decided improvements on older names. As an instance in which it is a clear gain to "ring in the new, ring out the old," I may mention "Torch Lily," which I should think few would hesitate to adopt instead of "Red-hot-poker Plant." In giving the popular and the scientific names of plants together, there is little left for the scientists to complain of.

"One very simple view of the subject has been apparently overlooked, perhaps from its very simplicity. Why should plants and flowers be *the only things* that we are to have no means of speaking of in our own language? The utility and necessity of the botanical names no one denies—a noble and simple invention, a "lingua franca" for the learned of all nations, though grievously overburdened with synonyms and masses of cumbersome "uncrystallised" matter. But why must people who love flowers know them by these names only—names that to many of them convey no sort of meaning—for all people who cultivate or enjoy flowers have not such a knowledge of the dead languages as to make the names intelligible? And why in any case speak in a dead language only of things so essentially living and affecting our daily use and happiness? Why should a piece of pedantic tyranny be imposed on us in this matter, *and this only?* Animals, birds, and insects also have their necessary scientific names, but no one reproves us for talking of a horse, or a sparrow, or a dragon-fly. Diseases have their universal names derived from Latin and Greek, used in scientific treatises and among members of the medical profession, and yet we commonly talk of gout, and small-pox, and scarlet fever. The bones and muscles of our bodies are all known in anatomy by such technical names, and yet in our every-day talk we may speak of rib, thigh-bone, and shoulder-blade. Why, then, should flowers only, of all the subjects that need a common language for purposes of classification and scientific research, have their purely technical appellations imposed on us, to the exclusion of such simple words in our own language as we use in other absolutely analogous cases? . . . . Does it not come to this, that both kinds of names are necessary, each for its proper purpose; the scientific name for classification, for study, for international research and correspondence, for business, for all rather dry and hard purposes; but for daily life among flowers, in poetry and popular books, for common use among the many people whose enjoyment of flowers does not approach any scientific purpose, the familiar names in our own tongue? Let me ask our learned men, who possess the dead languages, and therefore do not feel the need of the simpler means of expression, to descend in imagination to the level of those to whom Day Lily has a distinct meaning, while Hemerocallis is a jumble of senseless syllables. Let them think how absurd it would be if some arbitrary tyranny obliged us to call other things of common utility or enjoyment by long Latin names. . . . . Why are plants, and plants only, to be banished to this philological limbo, a place of weariness and lifelessness, that those who love flowers for their beauty's sake do not care to have to explore in order to find names by which to know their treasures? Will not our kindly *savants* rather help us to the supply of the living want and give us well-made English names in place of the perhaps ill-constructed ones that we should find for ourselves?" (THE GARDEN, vol. xxiv., p. 59.)

The above from a lady correspondent of THE GARDEN puts the case fairly for English names, and I shall conclude my quotations with Mr. Ruskin's remarks (in "Proserpina") on botany as now taught:—

"Yesterday evening I was looking over the first book in which I studied botany—"Curtis's Magazine," published in 1795 at No. 3, St. George's Crescent, Blackfriars Road, and sold by the principal booksellers in Great Britain and Ireland. Its plates are excellent, so that I am always glad to find in it the picture of a flower I know. And I came yesterday upon what I suppose to be a variety of a favourite flower of mine, called, in Curtis, 'the St. Bruno's Lily.' I am obliged to say " what I suppose to be a variety," because my pet Lily is branched,* while this is drawn as unbranched, and especially stated to be so. And the page of text in which this statement is made is so characteristic of botanical books and botanical science, not to say all science as hitherto taught for the blessing of mankind, and of the difficulties thereby accompanying its communication, that I extract the page entire, printing it as nearly as possible in facsimile.

[318]

ANTHERICUM LILIASTRUM. SAVOY ANTHERICUM,
or ST. BRUNO'S LILY.

Class and Order.
HEXANDRIA MONOGYNIA.
Generic Character.
Cor. 6-petala, patens. Caps. ovata.
Specific Character and Synonyms.
ANTHERICUM *Liliastrum* foliis planis, scapo simplicissimo, corollis campanulatis, staminibus declinatis. *Linn. Syst. Vegetab. ed.* 14. *Murr. p.* 330. *Ait. Kew. v.* 1. *p.* 449.
HEMEROCALLIS floribus patulis fecundis. *Hall. Hist. n.* 1230.
PHALANGIUM magno flore. *Bauh. Pin.* 29.
PHALANGIUM Allobrogicum majus. *Clus. cur. app. alt.*
PHALANGIUM Allobrogicum. The Savoye Spider-wort. *Park. Parad. p.* 150. *tab.* 151. *f.* 1.

Botanists are divided in their opinions respecting the genus of this plant; Linnæus considers it as an Anthericum, Haller and Miller make it an Hemerocallis.

It is a native of Switzerland, where, Haller informs us, it grows abundantly in the Alpine meadows, and even on the summits of the mountains; with us it flowers in May and June.

It is a plant of great elegance, producing on an unbranched stem, about a foot and a half high, numerous flowers of a delicate white colour, much smaller than, but resembling in form, those of the common white lily, possessing a considerable degree of fragrance. Their beauty is heightened by the rich orange colour of their antheræ; unfortunately they are but of short duration.

Miller describes two varieties of it differing merely in size.

A loamy soil, a situation moderately moist, with an eastern or western exposure, suits this plant best ; so situated, it will increase by its roots, though not very fast, and by parting these in the autumn it is usually propagated.

Parkinson describes and figures it in his *Parad. Terrest.*, observing that "divers allured by the beauty of its flowers, had brought it into these parts."

---

* At least, it throws off its flowers on each side in a bewilderingly pretty way; a real Lily can't branch, I believe; but, if not, what is the use of the botanical books saying "on an unbranched stem?"

Now you observe, in this instructive page, that you have in the first place eight names given you for one flower; and that, among these eight names, you are not even at liberty to make your choice, because the united authority of Haller and Miller may be considered as an accurate balance to the single authority of Linnæus; and you ought therefore for the present to remain, yourself, balanced between the sides. You may be farther embarrassed by finding that the Anthericum of Savoy is only described as growing in Switzerland. And farther still, by finding that Mr. Miller describes two varieties of it, which differ only in size, while you are left to conjecture whether the one here figured is the larger or smaller, and how great the difference is.

Farther, if you wish to know anything of the habits of the plant, as well as its eight names, you are informed that it grows both at the bottoms of the mountains and the tops; and that, with us, it flowers in May and June,—but you are not told when in its native country.

The four lines of the last clause but one may indeed be useful to gardeners; but —although I know my good father and mother did the best they could for me in buying this beautiful book; and though the admirable plates of it did their work and taught me much—I cannot wonder that neither my infantine nor boyish mind was irresistibly attracted by the text, of which this page is one of the most favourable specimens; nor, in consequence, that my botanical studies were—when I had attained the age of fifty—no farther advanced than the reader will find them in the opening chapter of this book.

Which said book was therefore undertaken to put, if it might be, some elements of the science of botany into a form more tenable by ordinary human and childish faculties; or—for I can scarcely say I have yet any tenure of it myself—to make the paths of approach to it more pleasant. In fact, I only know of it the pleasant distant effects which it bears to simple eyes; and some pretty mists and mysteries, which I invite my young readers to pierce, as they may, for themselves,—my power of guiding them being only for a little way.

Pretty mysteries, I say, as opposed to the vulgar and ugly mysteries of the so-called science of botany,—exemplified sufficiently in this chosen page. Respecting which, please observe farther: Nobody—I can say this very boldly—loves Latin more dearly than I; but, precisely because I do love it (as well as for other reasons), I have always insisted that books, whether scientific or not, ought to be written either in Latin or English, and not in a doggish mixture of the refuse of both."

It may not be out of place to observe that our leading nurserymen might, if so disposed, render valuable assistance to the work of disseminating a knowledge of the English names of plants, if, in their catalogues, they made it a practice to give the vernacular names along with the botanical ones. This is very largely done by American nurserymen, and, although it may seem invidious to single out any one establishment, it may be useful to mention the catalogue of hardy perennial plants issued by Messrs. Woolson & Co., Passaic, New Jersey, as suggestive in this respect.

With regard to the present volume, it has been carefully compiled from all available sources of information in our standard botanical works, British and Colonial Floras, the leading horticultural journals, and the catalogues of British, American, and Australian nurserymen. Being simply a dictionary of names (of which it contains over 15,000), it formed no part of the plan to introduce any matter descriptive of the plants, trees, and shrubs which it enumerates, and this must be looked for in botanical works devoted to the purpose. The Latin or botanical names given are those which are most commonly employed and best known, and synonyms are rarely noted, the few that are mentioned being chiefly in those cases where two botanical names are pretty equally in use for the same plant, as, Centaurea—Amberboa—moschata (Sweet Sultan), Calla—Richardia—æthiopica (Lily-of-the-Nile), &c. Familiar generic names, like Azalea, Crocus, Fuchsia, Iris, Phlox, &c., which have, to all intents and purposes, become English names, are retained as such. In the case of our native British plants, I have purposely omitted trivial local names, such as "Dog-chowps" and "Cuddy's-lugs," which savour of a chaw-baconism verging close on barbarism, and I have also left unnoticed the coarse and often grossly indelicate names which occur in some of the old writers on plants, as no useful purpose could be served by the reproduction of such names.

The book is probably far from being exhaustive or complete, as new names will be constantly arising from time to time, and it must be regarded as only a first step in the direction which it takes, but it is hoped that it will prove a handy and useful volume of reference, and a means of making the study of plants less technical and difficult to English-speaking people.

September, 1884.   W. M.

A DICTIONARY

OF

ENGLISH PLANT NAMES.

ABBREVIATIONS—*Sp.*, *species*; *var.*, *variety*; *fl.-pl.*, *flore-pleno.*

Aar. *Alnus glutinosa*
Aaron. *Arum maculatum*
Aaron's Beard. *Hypericum calycinum*
Aaron's Rod. *Verbascum Thapsus*
Abbey. *Populus alba*
Abele Tree. *Populus alba*
Absinth. *Artemisia Absinthium*
Abuta-root, or Butua-root. *Cissampelos Pareira*
Acacia, Clammy. *Robinia viscosa*
 Large-leafleted Rose. *Robinia hispida macrophylla*
 Parasol. *Robinia umbraculifera*
 Rice's. *Acacia Illecana*
 Rose. *Robinia hispida*
 Siris. See Siris-Acacia
 Smooth Tree. *Acacia Julibrissin*
 Three-thorned. *Gleditschia triacanthos*
 Two-spiked. *Acacia lophantha*
 Weeping. *Gleditschia Bugoti pendula*
Acajou-wood. The timber of *Cedrela brasiliensis*
Ach-root. *Morinda tinctoria*
Ach-weed or Ash-weed. *Ægopodium Podagraria*
Ache. *Apium graveolens*
Achocon Tree, of Peru. *Leonia glycycarpa*
Aconite. The genus *Aconitum*
 Common. *Aconitum Napellus*
 Indian or Nepaul. *Aconitum ferox*
 Winter. *Eranthis hyemalis*
Acorn, Sweet. The fruit of *Quercus Ballota*
Adam and Eve. *Corallorrhiza odontorrhiza*. Applied also to the tubers of various native Orchises, and to the common Arum
 N. American. *Aplectrum hyemale*
Adam's Flannel. *Verbascum Thapsus*
Adam's Needle. The genus *Yucca*; also *Scandix Pecten*
 Aloe-leaved. *Yucca aloifolia*
 Channelled-leaved. *Yucca canaliculata*
 Common. *Yucca gloriosa*
 Conspicuous. *Yucca conspicua*
 Drooping-leaved. *Yucca aloifolia var. pendula* and *Yucca draconis*
 Flaccid. *Yucca flaccida*
 Glaucous. *Yucca gloriosa var. glaucescens*
 Grass-leaved. *Yucca graminifolia*

Adam's Needle, Hollow-leaved. *Yucca concava*
 Narrow-leaved. *Yucca angustifolia*
 Oblique-leaved. *Yucca obliqua*
 Pointed-flowered. *Yucca acuminata*
 Recurved-leaved. *Yucca recurva*
 Reddish-edged. *Yucca rufo-cincta*
 Scalloped-leaved. *Yucca crenulata*
 Silvery. *Yucca nivea*
 Slender-leaved. *Yucca tenuifolia*
 Superb. *Yucca superba*
 Thready. *Yucca filamentosa*
 Upright. *Yucca stricta*
 Wavy-leaved. *Yucca undulata*
Adam's Needle and Thread. *Yucca filamentosa*
Adder-spit. *Pteris aquilina*
Adder-wort. *Polygonum Bistorta*
Adder's Fern. *Polypodium vulgare*
Adder's Flower. *Lychnis diurna*
Adder's Grass. *Orchis mascula*
Adder's Meat. *Arum maculatum*
Adder's Spear. *Ophioglossum vulgatum*
Adder's Tongue. *Ophioglossum vulgatum*
 Yellow. *Erythronium americanum*
Adder's Violet. See Violet
Adonis Flower. *Adonis autumnalis*
 Pyrenean. *Adonis pyrenaica*
Affadil. *Narcissus Pseudo-Narcissus*
African Rubber Tree. See Rubber
Agarics. A genus of Fungi
Agila-wood. *Aquilaria ovata* and *A. Agallochum*
Ag-leaf. *Verbascum Thapsus*
Agnes's (St.) Flower. The genus *Erinosma*
Agrimony, Bastard. *Ageratum conyzoides*
 Common. *Agrimonia Eupatoria*
 Creeping. *Agrimonia repens*
 Hemp. *Eupatorium cannabinum*
 Sweet-scented. *Agrimonia odorata*
 Three-leaved. *Agrimonia agrimonioides*
 Water. *Eupatorium cannabinum*
 Water Hemp. *Bidens cernua* and *B. tripartita*
Ague-root. *Aletris farinosa*
Ague Tree. *Laurus Sassafras*
Ague-weed, Indian. *Eupatorium perfoliatum*
Agworm Flower. *Stellaria Holostea*

B

English Names of Cultivated, Native,

Aikraw. *Stictina scrobiculata*
Ailanto. *Ailantus glandulosa*
Ail-weed. *Cuscuta Trifolii*
Air-flower, or Air-plant. *Aëranthes grandiflorum*
Fragrant. *Aërides odorata*
Ajowan, Ajouan, or Javanee, plant. *Ptychotis Ajowan*
Akaroa, or Cotton Tree of New Zealand *Plagianthus betulina* and *P. urticina*
Aka Tree. *Metrosideros scandens*
Akee Tree. *Blighia (Cupania) sapida*
Alaternus, Blotched-leaved. *Rhamnus Alaternus maculatus*
Broad-leaved. *Rhamnus Alaternus latifolius*
Common. *Rhamnus Alaternus*
Hybrid. *Rhamnus hybridus*
Narrow-leaved. *Rhamnus Alaternus angustifolius*
Silver-edged. *Rhamnus Alaternus argenteus*
Alder, American Black. *Alnus incana* and *Prinos verticillatus*
Berry-bearing. *Rhamnus Frangula*
Black. *Rhamnus Frangula.*
Californian. *Alnus rhombifolia*
Californian Red. *Alnus rubra*
Common. *Alnus glutinosa*
Common White. *Clethra alnifolia*
Fern-leaved. *Alnus imperialis asplenifolia*
Green or Mountain. *Alnus viridis*
Heart-leaved. *Alnus cordifolia*
Hoary-leaved. *Alnus incana*
Michaux's White. *Clethra Michauxi*
Mountain. See Alder, Green
Panicled White. *Clethra paniculata*
Pointed-leaved White. *Clethra acuminata*
Red. *Cunonia capensis*
Sea-side. *Alnus maritima*
Smooth. *Alnus serrulata*
S. African White. *Platylophus trifoliatus*
Speckled. *Alnus incana*
Spotted. *Hamamelis virginica*
Tag. *Alnus rubra*
Turkey. *Alnus oblongata*
Witch. *Fothergilla alnifolia*
Woolly White. *Clethra tomentosa*
Alder-Buckthorn. *Rhamnus Frangula*
Alecost, or Alecoast. *Tanacetum Balsamita*
Ale-hoof. *Nepeta Glechoma*
Aleppo Millet Grass. See Grass
Alerse Tree. *Libocedrus tetragona*
Alexanders, or Alisanders. *Smyrnium Olusatrum*
Candy. *Smyrnium apiifolium*
Golden. *Zizia integerrima*
Alexander's Foot. *Anacyclus Pyrethrum*
Alfa, or Alpha, Grass. See Grass
Alfalfa. *Medicago sativa*
Alga marina. *Chondrus (Fucus) crispus*
Algaroba. *Prosopis juliflora*
Algoborillo. *Cæsalpinia brevifolia*
Alisanders. See Alexanders
Aliways. *Aloës*
Alk Gum Tree. *Pistacia Terebinthus*
Alkanet. The genus *Anchusa*
American. *Lithospermum canescens*
Bastard. *Lithospermum arvense*
Cape. *Anchusa capensis*
Evergreen, or Green. *Anchusa sempervirens*

Alkanet, Hybrid. *Anchusa hybrida*
Italian. *Anchusa italica*
Mountain. *Arnica montana*
Alkanna. *Lithospermum hirtum*
Alkekeng. *Physalis Alkekengi*
All-bone. *Stellaria Holostea*
Alleluia. *Oxalis Acetosella*
Aller Tree. *Alnus glutinosa*
All-good. *Chenopodium Bonus-Henricus*
All-heal. *Prunella vulgaris* and *Valeriana officinalis*
Allicampane. See Elecampane
Alligator-wood. *Guarea grandifolia*
Alison. See Alysson
All seed. *Radiola Millegrana*, *Chenopodium polyspermum*, and the genus *Polycarpon*
Four-leaved. *Polycarpon tetraphyllum*
Allspice, Carolina. *Calycanthus floridus*
Jamaica. *Eugenia Pimenta*
Japan. *Chimonanthus fragrans (Calycanthus præcox)*
Wild American. *Lindera Benzoin (Benzoin odoriferum)*
Alme. *Ulmus campestris*
Almendor, or Almond of Brazil. *Geoffroya superba*
Almond Tree, African. *Brabejum stellatifolium*
Brazil. See Almendor
Bitter. *Amygdalus communis var. amara*
Common. *Amygdalus communis*
Country. *Terminalia Catappa*
Double-flowered Dwarf. *Amygdalus pumila*
Dwarf. *Amygdalus nana*
Earth. *Cyperus esculentus*
Java. *Canarium commune*
Long-fruited. *Amygdalus communis rar. macrocarpa*
Malabar. *Terminalia Catappa*
Prostrate. *Amygdalus prostrata*
Siberian. *Amygdalus sibirica*
Silvery-leaved. *Amygdalus orientalis (A. argentea)*
S. American. *Geoffroya superba*
Weeping. *Amygdalus communis var. pendula*
W. Indian wild. *Hippocratea comosa*
Wild, Cape of Good Hope. *Brabejum stellatifolium*
Woolly. *Amygdalus incana*
Aloe, American. *Agave americana*
Cape. *Aloë ferox*
False. *Agave virginica*
Fetid. *Agave fœtida*
Partridge-breast. *Aloë rariegata*
Pearl. *Aloë margaritifera*
Proliferous. *Aloë prolifera*
Utah. *Agave utahensis*
Vera Cruz. *Agave lurida*
Water. See Water-Aloe
Yellow-flowered. *Aloë vulgaris*
Aloes. The inspissated juice of the leaves of various species of *Aloë*
Barbadoes. From *Aloë vulgaris*
Bombay. See Aloes, Socotrine
Cape. From *Aloë ferox*, *A. perfoliata*, and *A. linguæformis*

Aloes, Curaçao. From *Aloë vulgaris*
E. Indian. See Aloes, Socotrine
Meka. From *Aloë socotrina*
Socotrine, Bombay, E. Indian, or Zanzibar.
From *Aloë socotrina*
Zanzibar. See Aloes, Socotrine
Aloes-wood. *Aloexylon Agallochum*
Alpam Root, of Malabar. The root of *Bragantia Wallichi*
Alphabet Plant. *Spilanthes Acmella*
Alpine (a corruption of "Orpine"). *Sedum Telephium*
Alsike. *Trifolium hybridum*
Alstonia Bark Tree. *Alstonia (Echites) scholaris*
Althæa-Frutex, Common. *Hibiscus syriacus*
Double. *Hibiscus syriacus elegantissimus*
Painted Lady. *Hibiscus syriacus oculatus*
Purple. *Hibiscus syriacus purpureus*
Alum. *Symphytum officinale*
Alum Root. *Heuchera americana;* also the root of *Geranium maculatum*
Currant-leaved. *Heuchera ribifolia*
Downy. *Heuchera pubescens*
Small-flowered. *Heuchera micrantha*
Smooth. *Heuchera glabra*
Alysson, Purple. *Aubrietia purpurea*
Small Yellow. *Alyssum calycinum*
Sweet. *Alyssum maritimum*
White. *Arabis alpina*
Amadou, Punk, Spunk, Touchwood, German or Vegetable Tinder. *Polyporus (Boletus) fomentarius* and *Polyporus igniarius*
Hard. *Polyporus igniarius*
S. American. The wood of *Hernandia guianensis*
Amaranth, Edible. *Amarantus oleraceus*
Globe. *Gomphrena globosa*
Golden. *Amarantus salicifolius*
Showy. *Amarantus speciosus*
Thorny. *Amarantus spinosus*
Three-coloured. *Amarantus tricolor*
Amaranth Feathers. *Humea elegans*
Amber. *Hypericum perforatum*
Sweet. *Hypericum Androsæmum*
Amber Tree. *Anthospermum æthiopicum*
Amboyna - wood. *Pterospermum indicum*
Amel Corn. *Triticum amyleum*
Amelanchier, Common. *Amelanchier vulgaris*
Flowery. *Amelanchier florida*
Oval-leaved. *Amelanchier ovalis*
Red-branched. *Amelanchier sanguinea*
Ameos. An old name for *Ammi majus*
American Cress. *Barbarea præcox*
Plants. A garden term, applied to *Rhododendrons, Andromedas, Azaleas,* and some other plants which require to be grown in moist peaty soil
Water Weed. *Anacharis Alsinastrum*
Ammoniacum Gum Plant. *Dorema ammoniacum*
Amole, or Indian Soap Plant. *Chlorogalum pomeridianum*
Ananbeam. *Euonymus europæus*
Andrew's (St.) Cross. *Ascyrum Crux-Andreæ*

Andromeda, Acute - leaved. *Andromeda acuminata*
Box-leaved. *Andromeda buxifolia*
Branching. *Andromeda racemosa*
Bundle-flowered. *Andromeda fasciculata*
Cluster-flowered. *Andromeda racemosa*
Free-flowering. *Andromeda floribunda*
Himalayan. *Andromeda (Cassiope) fastigiata*
Jamaica. *Andromeda jamaicensis*
Japanese. *Andromeda japonica*
Large-flowered. *Andromeda speciosa*
Mossy. *Andromeda (Cassiope) hypnoides*
Oval-leaved. *Andromeda ovalifolia*
Spike-flowered. *Andromeda spicata*
Spiny-leaved. *Andromeda Catesbæi*
Square-stemmed. *Andromeda (Cassiope) tetragona*
Thick-leaved. *Andromeda coriacea*
Andurion. *Eupatorium cannabinum*
Anemone. (See also Wind-flower.)
Cyclamen-leaved. *Anemone palmata*
Garden varieties of. *Anemone coronaria*
Poppy. *Anemone coronaria*
Rue. *Thalictrum anemonoides*
Snowdrop. *Anemone sylvestris*
Water. *Ranunculus aquatilis*
Wood. *Anemone nemorosa*
Yellow Wood. *Anemone ranunculoides*
Anet. *Anethum graveolens*
Angelica, Garden. *Archangelica officinalis*
Great. *Archangelica atropurpurea*
Wild. *Angelica sylvestris*
Angelica Tree. Various species of *Fraxinus*
Virginian. *Aralia spinosa*
Angelico. *Ligusticum actæifolium*
Angelim. *Tipuana heteroptera* and several species of *Andira*
Angel's Eyes. *Veronica Chamædrys*
Angel's Trumpets. The flowers of *Brugmansia suaveolens*
Angle Berries. *Lathyrus pratensis*
Angle Pod. The genus *Gonolobus*
Angola - weed. *Roccella fuciformis* and *Ramulina furfuracea*
Angostura Bark, Carony Bark, or Cusparia Bark Tree. *Galipea Cusparia*
Anise, or Aniseed. *Pimpinella Anisum* and *Myrrhis odorata*
Orinoco. *Ocotea cymbarum*
Star. *Illicium anisatum*
Aniseed Plant, of commerce. *Pimpinella Anisum*
Aniseed Tree, Red-flowered Florida. *Illicium Floridanum*
Star. *Illicium anisatum*
Anny. *Pimpinella Anisum*
Anthony Nut. *Staphylea pinnata*
Antidote Cocoon. *Feuillæa cordifolia*
Ant-hill Grass. *Festuca sylvatica*
Ant Tree. *Triplaris Bonplandiana*
Apache Plume. *Fallucia paradoxa*
Apes-on-horseback. *Bellis hortensis prolifera*
Apple, Adam's. *Citrus Limetta*
Alligator. *Anona palustris*
Argyle. *Eucalyptus pulverulenta*
Astrachan. *Pyrus Malus astracanica*
Australian. *Angophora lanceolata*

Apple, Australian Brush or Bush. *Achras australis*
Balsam. *Momordica Balsamina*
Beef. *Sapota rugosa*
Bitter. *Cucumis (Citrullus) Colocynthis*
Cane. *Arbutus Unedo*
Carthaginian. *Punica Granatum*
Cherry. *Pyrus baccata*
Chinese. *Pyrus (Malus) spectabilis*
Common Cultivated. Vars. of *Pyrus Malus*
Coral-flowered. *Pyrus Malus var. floribunda*
Deadsea, or "of Sodom." The fruit of *Solanum sodomeum*; also applied to the galls of *Quercus infectoria*
Devil's. *Mandragora officinalis*
Diœcious. *Pyrus Malus dioica*
Double-flowered Chinese. *Pyrus spectabilis fl.-pl.*
Elephant's. *Feronia elephantum*
Fir. The cones of *Pinus Abies*
Free-flowering Chinese. *Pyrus spectabilis floribunda*
Golden. *Ægle Marmelos* and *Spondias lutea*
Hen. *Pyrus Aria*
Jew's. *Solanum Melongena*
Kai, Kau, or Kei. *Aberia Caffra*
Kangaroo. *Solanum laciniatum*
Kau. See Apple, Kai
Kei. See Apple, Kai
Love. *Solanum Lycopersicum*
Mad. *Solanum Melongena*
Malay. *Jambosa (Eugenia) malaccensis*
Mammee. *Mammea americana*
Marvellous. See Marvellous Apples
Mess. *Karstenia quinquenervia*
Mexican. *Casimiroa edulis*
Monkey. *Clusia flava*
Monkey, of Sierra Leone. *Anisophyllum laurinum*
Monkey, of the West Indies. *Anona palustris*
New S. Wales. *Angophora subvelutina*
N. American Crab. *Pyrus coronaria*
N. American Narrow-leaved Crab. *Pyrus angustifolia*
Oak. A gall produced by insects, on the leaves and twigs of the Oak
Of Jerusalem. *Momordica Balsamina*
Of Peru. *Nicandra physaloides*
Of Scripture. Probably the Apricot (*Prunus armeniaca*) or the Quince (*Pyrus Cydonia*)
Of Sodom. See Apple, Dead Sea
Of the Earth. An old name for *Aristolochia rotunda* and the genus *Cyclamen*
Oregon Crab. *Pyrus rivularis*
Otahcite. *Spondius (Poupartia) dulcis*
Paradise. *Pyrus Malus præcox*
Prairie. *Psoralea esculenta*
Rose. *Eugenia malaccensis* and other species
Sea. *Manicaria Plukeneti*
Star. *Chrysophyllum Cainito*
Sugar. *Anona squamosa* and *Rollinia Sieberi*
Thorn. *Datura Stramonium* and other species
Victoria. *Angophora lanceolata*
Wild Balsam. *Echinocystis lobata*
Wild Star. *Chrysophyllum olivæforme*

Apple, Wood. *Feronia elephantum*
Apple-berry, Australian. The genus *Billardiera*
Climbing. *Billardiera scandens*
Colour-changing. *Billardiera mutabilis*
Long-flowered. *Billardiera longiflora*
Oval-leaved. *Billardiera ovalis*
Apple Haw. *Cratægus æstivalis*
Apple Mint. *Mentha rotundifolia*
Apple Moss. See Moss
Apple-peru. *Datura Stramonium*
Apple Pie. *Epilobium hirsutum*
Apples, Cedar. See Cedar Apples
Apricock. An old name for Apricot
Apricot Tree, Briançon. *Armeniaca briantiaca*
Common. *Armeniaca vulgaris* (*Prunus armeniaca*)
Oval-leaved. *Armeniaca vulgaris var. ovalifolia*
Siberian. *Armeniaca sibirica*
Thick-fruited. *Armeniaca dasycarpa*
Variegated-leaved. *Armeniaca vulgaris foliis variegatis*
Wild. *Mammea americana*
Aps (Aspen). *Populus tremula*
Apron, Tanner's. *Primula Auricula*
Arbor-vitæ, American. *Thuja occidentalis*
American Tom Thumb. *Thuja occidentalis var. ericoides*
Bagshot Park. *Thuja occidentalis var. densa* or *compacta*
Belgian Variegated. *Thuja occidentalis var. Verraeneana*
Broad-leaved. The genus *Thujopsis*
Bush. *Thuja dumosa*
Chilian. *Libocedrus chilensis*
Chinese. *Biota (Thuja) orientalis*
Dwarf Chinese. *Biota orientalis elegantissima*
Ever-golden. *Thuja semper-aurea*
Giant. *Thuja gigantea*
Golden. *Thuja aurea*
Golden Chinese. *Biota orientalis aurea*
Hatchet-leaved. *Thujopsis dolabrata*
Japanese. *Biota orientalis Sieboldi (Thuja japonica)*
Jointed. *Callitris quadrivalvis*
New Zealand. *Libocedrus Doniana*
Nootka Sound. *Thuja plicata*
Pyramidal Chinese. *Biota orientalis pyramidalis*
Sickle-spine-coned Chinese. *Biota falcata*
Silvery-leaved. *Thuja occidentalis argentea*
Western. *Thuja gigantea*
Arbute Tree. *Arbutus Unedo*
Andrachne or Oriental. *Arbutus Andrachne*
Densely-flowered. *Arbutus densiflora*
Oriental. See Arbute Tree, Andrachne
Tall. *Arbutus procera*
Trailing. *Epigæa repens*
Woolly-branched. *Arbutus tomentosa*
Archall or Orchil. *Roccella tinctoria*
Archangel. *Archangelica officinalis*, *Stachys sylvatica*, and various species of *Lamium*
White. *Lamium album*
Yellow. *Lamium Galeobdolon*
Archer, Water. *Sagittaria sagittæfolia*
Argan Tree. *Argania Sideroxylon*

Argel, or Arghel. *Salenostemma Argel*
Argemone. *Potentilla anserina*
Argentine, or Argentina. *Onopordum Acanthium* and *Potentilla anserina*
Arghel. See Argel
Armstrong. *Polygonum aviculare*
Arnotta, or Anotta. *Bixa orellana*
Arn Berries. The fruit of *Rubus Idæus*
Arn Tree. *Sambucus nigra*
Ar-Nut. *Bunium flexuosum*
Aronia, Arbutus-leaved. *Pyrus arbutifolia*
  Black-fruited. *Pyrus melanocarpa*
  Downy-branched. *Pyrus pubescens*
  Large-leaved. *Pyrus grandifolia*
  Many-flowered. *Pyrus floribunda*
Arracacha. *Arracacia esculenta* (*Conium Arracacha*)
Arrow Grass. *Triglochin palustre*
Arrow-Head, or Arrow Leaf, Chinese. *Sagittaria chinensis*
  Chilian. *Sagittaria montevidensis*
  Common. *Sagittaria sagittæfolia*
  Double. *Sagittaria sagittæfolia fl.-pl.*
  Various-leaved. *Sagittaria variabilis*
  Yellow-flowered. *Sagittaria simplex*
Arrow-Leaf. See Arrow-Head
Arrow-Poison, Gaboon. See Arrow Poison, Tropical African
  Guiana, Curari, or Ourali. *Strychnos toxifera*
  Javanese. *Strychnos Tieute*
  Malay. *Antiaris toxicaria*
  Tropical African, or Gaboon. *Strophanthus Kombe*, (*S. hispidus*)
Arrow-root Plant, Bermuda. *Maranta arundinacea*
  Chinese. *Nelumbium speciosum*
  E. Indian. *Curcuma angustifolia* and other species
  English. *Solanum tuberosum*
  Mexican. *Dion edule*
  Oswego. Indian Corn (*Zea Mays*)
  Portland. *Arum maculatum*
  South Sea and Sandwich Islands. *Tacca pinnatifida* (*Tacca oceanica*)
Arrow-wood, Californian. *Viburnum ellipticum*
  Downy. *Viburnum pubescens*
  Maple-leaved. *Viburnum acerifolium*
  N. American. *Viburnum dentatum*
Artichoke, French or Globe. *Cynara Scolymus*
  Jerusalem. *Helianthus tuberosus*
  Prickly. *Cynara Cardunculus*
Artichoke Gall. See Gall
Artillery Plant. *Pilea serpyllifolia* and *Pilea herniariæfolia*
Arum, Arrow. *Peltandra virginica*
  Bog. *Calla palustris*
  Dragon. *Arum Dracunculus*
  Giant. *Conophallus Titanum*
  Italian. *Arum italicum*
  Water. *Calla palustris*
Arum Lily, Common. *Calla* (*Richardia*) *æthiopica*
  Spotted-leaved. *Richardia maculata*
  Yellow. *Richardia hastata*
Asafœtida Plant. *Narthex Asafœtida* and *Scorodosma fœtidum*
  Persian. *Ferula persica*

Asarabacca. *Asarum europæum*
  Sweet-scented. *Asarum virginicum*
Ash, American Black or Water. *Fraxinus sambucifolia*
  American Flowering. *Ornus americana*
  Aucuba-leaved. *Fraxinus americana var. aucubæfolia*
  Bitter. *Picræna* (*Simaruba*) *excelsa*
  Black Mountain. *Eucalyptus Leucoxylon*
  Blue. *Fraxinus quadrangulata*
  Bosc's. *Fraxinus americana var. Bosci*
  Brown-branched. *Fraxinus americana var. fusca*
  Cape. *Ekebergia capensis*
  Carolina Water. *Fraxinus platycarpa*
  Chinese. *Fraxinus chinensis*
  Cloth-leaved. *Fraxinus americana var. pannosa*
  Common. *Fraxinus excelsior*
  Dwarf. *Fraxinus excelsior var. nana* (*Fraxinus excelsior humilis*)
  Elliptic-leaved. *Fraxinus americana var. elliptica*
  Flowering. *Fraxinus Ornus*
  Fungous-barked. *Fraxinus excelsior var. fungosa*
  Golden-barked. *Fraxinus excelsior var. aurea*
  Gray. *Fraxinus americana var. cinerea*
  Green. *Fraxinus viridis*
  Ground. *Ægopodium Podagraria* and *Angelica sylvestris*
  Hoop. *Celtis crassifolia*
  Horizontal-branched. *Fraxinus excelsior var. horizontalis*
  Jerusalem. *Isatis tinctoria*, or *Reseda Luteola*
  Kincairney. *Fraxinus excelsior var. Kincairniæ*
  Lentiscus-leaved. *Fraxinus lentiscifolia*
  Long-leaved. *Fraxinus americana var. longifolia*
  Manna. *Fraxinus Ornus var. rotundifolia*
  Many-flowered Flowering. *Fraxinus Ornus var. floribunda*
  Mixed. *Fraxinus americana var. mixta*
  Mountain, or Wild. *Pyrus Aucuparia*
  Mountain, of Australia. *Panax dendroides*
  Mountain, of New South Wales. *Eucalyptus hæmastoma* and *Eucalyptus virgata*
  Mountain, of N. America. *Pyrus americana*
  Narrow-leaved. *Fraxinus excelsior var. angustifolia*
  Nepaul. *Fraxinus floribunda*
  Northern Prickly. *Xanthoxylon americanum*
  Oregon. *Fraxinus oregana*
  Ovate-leaved. *Fraxinus americana var. ovata*
  Pale-barked. *Fraxinus excelsior var. pallida*
  Poison. *Chionanthus virginica* and *Rhus venenatum*
  Polemonium-leaved. *Fraxinus americana var. polemonifolia*
  Powdery. *Fraxinus americana var. pulverulenta*
  Prickly. *Xanthoxylon fraxineum*
  Purple-barked. *Fraxinus excelsior var. purpurascens*

Ash, Red. *Alphitonia excelsa*
Red American. *Fraxinus pubescens*
Reddish-veined. *Fraxinus americana var. rubicunda*
Richard's. *Fraxinus americana var. Richardi*
Rim. *Celtis occidentalis*
Round-leafleted Flowering. *Fraxinus Ornus var. rotundifolia*
Rufous-haired. *Fraxinus americana var. rufa*
Schiede's. *Fraxinus Schiediana*
Sharp-fruited. *Fraxinus excelsior var. parifolia oxycarpa*
Silver-striped-leaved. *Fraxinus excelsior var. argentea*
Small-leaved. *Fraxinus excelsior var. parvifolia*
Southern Prickly. *Xanthoxylon carolinianum* (*Xanthoxylon Clava-Herculis*)
Striped-barked. *Fraxinus excelsior var. jaspidea*
Striped-barked Flowering. *Fraxinus Ornus var. striata*
Variegated Weeping. *Fraxinus aurea pendula*
Various-leaved. *Fraxinus heterophylla*
Wafer. *Ptelea trifoliata*
Warted-barked. *Fraxinus excelsior var. verrucosa*
Water. *Fraxinus sambucifolia*
Weeping. *Fraxinus excelsior var. pendula*
Weeping Mountain. *Pyrus Aucuparia pendula*
Western Mountain. *Pyrus sambucifolia*
White. *Fraxinus americana var. alba*
Whorled-leaved. *Fraxinus excelsior var. verticillata*
Yellow-edged-leafleted. *Fraxinus excelsior var. lutea*
Ash-Barberry. The genus *Mahonia*
Ash Keys, or Ash Candles. The fruit of *Fraxinus excelsior*
Ash-weed, or Ach-weed. *Ægopodium Podagraria*
Asparagus, Bath or French. *Ornithogalum pyrenaicum*
Common. *Asparagus officinalis*
Decumbent. *Asparagus decumbens*
Ethiopian. *Asparagus æthiopicus*
Feathery. *Asparagus plumosus*
French. See Asparagus, Bath
Garden Hedge, of Madeira. *Asparagus albus*
Giant. *Asparagus Broussoneti*
Racemose. *Asparagus racemosus*
Sickle-branched. *Asparagus falcatus*
Slender-leaved. *Asparagus tenuifolius*
S. African. *Asparagus laricinus*
S. European. *Asparagus acutifolius*
Asp, or Aspen. *Populus tremula*
American Quaking. *Populus tremuloides*
White. *Populus alba*
American Large-toothed. *Populus grandidentata*
Soft. See Poplar, Soft
Weeping. *Populus tremula var. pendula*
Asphodel, Bog or Lancashire. *Narthecium ossifragum*
Bog, American. *Narthecium americanum*
Branching. *Asphodelus ramosus*

Asphodel, False. The genus *Tofieldia*
Lancashire. See Asphodel, Bog
Onion. *Asphodelus fistulosus*
Scotch. *Tofieldia palustris*
Sub-alpine. *Asphodelus sub-alpinus*
White-flowered. *Asphodelus albus*
Yellow-flowered. *Asphodelus luteus*
Assagay, Assegay, or Hassagay Tree. *Curtisia faginea*
Ass's Foot. *Tussilago Farfara*
Aster. (See also Star-wort)
Cape. *Agathæa amelloides*
China. *Callistemma hortense* (*Callistephus hortensis*)
Double-bristled. The genus *Diplopappus*
Dwarf Amellus. *Aster bessarabicus*
Golden. The genus *Chrysopsis*
New Zealand. Various species of *Celmisia*
Purple Mexican. *Cosmos bipinnatus*
Stokes's. *Stokesia cyanea*
White-topped. The genus *Seriocarpus*
Asthma Plant. *Nonatelia officinalis*
Weed. *Lobelia inflata*
Atlee Gall. A Gall found on *Tamarix orientalis*
Augers. *Salix viminalis*
Aul. *Alnus glutinosa*
Aum Tree. *Ulmus campestris*
Auricula, Common. *Primula Auricula*
Yellow Alpine. *Auricula alpina*
Autumn Bell-flower. *Gentiana Pneumonanthe*
Crocus. *Colchicum autumnale*
Ava, or Kava, Shrub. *Macropiper methysticum*
Ave Grace. *Ruta graveolens*
Avellano Nut. See Nut
Avens, Chiloe. *Geum chiloënse*
Common. *Geum urbanum*
Creeping. *Geum reptans*
Double Scarlet. *Geum coccineum fl.-pl.*
Drooping. *Geum rivale*
Drummond's. *Dryas Drummondi*
Golden. *Geum aureum*
Labrador Mountain. *Dryas tenella*
Large-flowered Scarlet. *Geum coccineum grandiflorum*
Mountain. *Dryas octopetala*
Pyrenean. *Geum pyrenaicum*
Three-flowered. *Geum triflorum*
Scarlet-flowered. *Geum coccineum*
Water or Drooping. *Geum rivale*
White. *Geum virginianum*
Wood. *Geum sylvaticum*
Yellow-flowered Mountain. *Geum montanum*
Averill. *Narcissus Pseudo-Narcissus*
Avignon Berries. The berries of *Rhamnus infectorius*
Awl Tree. *Morinda citrifolia*
Awl-wort. *Subularia aquatica*
Axe, Flower of the. *Lobelia urens*
Ax-weed. *Ægopodium Podagraria*
Ax-wort, or Ax-sitch. *Securigera Coronilla*
Ayapana. *Eupatorium triplinerve*
Azalea, Alpine. *Loiseleuria* (*Azalea*) *procumbens*
Blunt-leaved. *Azalea obtusa*
Bright-flowered. *Azalea amœna*
Chinese. *Azalea sinensis*
Clammy. *Azalea viscosa*

and Foreign Plants, Trees, and Shrubs.

Azalea, Common Yellow. *Azalea pontica*
Crisp-flowered. *Azalea crispiflora*
Daniels's. *Azalea Danielsiana*
Dwarf Glaucous. *Azalea glauca*
Fine Golden. *Azalea calendulacea var. chrysolecta*
Flame-coloured. *Azalea calendulacea*
Hoary. *Azalea canescens*
Indian. *Azalea indica*
Ledum-leaved. *Azalea ledifolia*
Marigold. *Azalea calendulacea*
Naked-flowered. *Azalea nudiflora*
Ovate-leaved. *Azalea ovata*
Pontic. *Azalea pontica*
Proliferous. *Azalea nudiflora var. prolifera*
Scaly. *Azalea squamata*
Shining-leaved. *Azalea nitida*
Showy. *Azalea speciosa*
Smooth. *Azalea arborescens*
Soft-leaved. *Azalea mollis*
Sweet-scented. *Azalea viscosa var. odorata*
Tall Glaucous. *Azalea hispida*
Tree. *Azalea arborescens*
Two-coloured. *Azalea bicolor*
White-flowered. *Azalea ovata var. alba*
Azaleas, Ghent. Varieties of *A. calendulacea, A. nudiflora, A. speciosa,* and *A. viscosa*
Azarole Thorn. *Crataegus Azarolus*
Azzy Tree. *Crataegus Oxyacantha*

Babington's Curse. *Anacharis Alsinastrum* (*Elodea canadensis*)
Bacaba Palm. See Palm
Baccobolts. *Typha latifolia*
Bachelor's Buttons. A name applied to various button-shaped flowers
Back-wort. *Symphytum officinale*
Bacon-weed. *Chenopodium album*
Badderlocks. *Alaria esculenta*
Badger's-Bane. *Aconitum Meloctonum*
Bad-jong. *Acacia microbotrya*
Badmoney or Baldmoney. *Meum athamanticum* and various species of *Gentiana*
Bael, or Bhel, Fruit. The fruit of *Ægle Marmelos*
Bag Nut. Another name for Bladder Nut
Bairn-wort. A northern name for the Daisy
Bajree Millet. See Millet
Baldmoney. See Badmoney
Baldwein. An old name for Gentian
Bale-worth. *Papaver somniferum*
Ballata Tree. *Bumelia retusa*
Ball Thistle. Another name for Globe Thistle
Balm, Bastard. *Melittis Melissophyllum*
Common. *Melissa officinalis*
Bee. *Monarda didyma*
Field. *Calamintha Nepeta*
Heart-leaved. *Cedronella cordata*
Indian. *Trillium pendulum*
Moldavian. *Dracocephalum moldavicum*
Molucca. The genus *Moluccella*
Ox. *Collinsonia canadensis*
Balm of Gilead. *Cedronella triphylla* (*Dracocephalum canariense*)
Hoary. *Cedronella cana*
S. American. *Icica Carana*
Balm of Gilead Tree. *Balsamodendron* (*Amyris*) *Gileadense* and *Populus balsamifera var. candicans*

Balm of Heaven. *Oreodaphne californica*
Balmony. *Chelone glabra*
Balsam, Broad-leaved. *Sciadophyllum capitatum*
Common Yellow. *Impatiens Noli-me-tangere*
Common Garden. Varieties of *Impatiens Balsamina*
Hardy Indian. *Impatiens glanduligera*
Kentish. *Mercurialis perennis*
Orange-flowered. *Impatiens fulva*
Seaside. *Croton balsamiferum*
Sonsonate. An American name for the Balsam of Peru
Water. *Hydrocera* (*Tytonia*) *natans*
W. Indian Garden. *Justicia pectoralis*
White. *Gnaphalium polycephalum*
Yellow. *Croton flavens*
Zanzibar. *Impatiens Sultani*
Balsam Apple. *Momordica Balsamina*
Balsam Bog. *Azorella* (*Bolax*) *glebaria*
Balsam Herb. *Dianthera repens*
Jamaica. *Justicia comata*
Balsam of Peru Plant. *Myrospermum Peruiferum*
Balsam of Tolu Plant. *Myrospermum Toluiferum*
Balsam Poplar. *Populus balsamifera*
Balsam Root, Californian. The genus *Balsamorrhiza*
Balsam Tree, Acouchi. *Icica Aracouchini*
Bayee. *Balsamodendron pubescens*
Carpathian. *Pinus Cembra*
Canada. *Pinus* (*Abies*) *balsamea* and *Pinus Fraseri*
Copalm. *Liquidambar styraciflua*
Florida. *Amyris Floridana*
Gogul. *Balsamodendron Roxburghi*
Hungarian. *Pinus Pumilio*
Jamaica. *Clusia flava*
Maria. *Verticillaria acuminata*
Mecca or Roghen. *Balsamodendron Gileadense*
Old Field. *Gnaphalium polycephalum*
Quinquino or White. *Myrospermum pubescens*
Roghen. See Balsam, Mecca
St. Thomas. *Sorindeia trimera*
Tamacoari. *Carapa fascicultata*
Umiri. *Humirium floribundum*
W. Indian. The genus *Clusia*
White. See Balsam, Quinquino
Bamboo. The genus *Bambusa*
Australian. *Poa ramigera*
Berry-bearing. *Melocanna bambusoides*
Blow-pipe. *Arthrostylidium Schomburgki*
Common. *Bambusa arundinacea*
Dark-stemmed. *Bambusa nigra*
Fortune's. *Bambusa Fortunei*
Grayish. *Bambusa viridi-glaucescens*
Hardy. *Bambusa falcata*
Male. *Dendrocalamus strictus*
Metake. *Bambusa japonica* (*B. Metake*)
Netted-veined. *Bambusa reticulata*
Ningala. *Arundinaria falcata*
Orinoco. *Bambusa latifolia*
Praong. *Arundinaria Hookeriana*
Sacred. *Nandina domestica*
Simmonds's. *Bambusa Simmondsi*
Slender. *Bambusa gracilis*

Bamboo, Striped. *Bambusa striata*
 Yellow-stemmed. *Bambusa aurea*
Bamboo Briar. A species of Smilax
Banana Tree. *Musa sapientum*
 Abyssinian. *Musa Ensete*
 Channelled-leaved. *Musa vittata*
 Dwarf Chinese. *Musa Cavendishi*
 Livingstone's. *Musa Livingstoniana*
 Scarlet-bracted. *Musa coccinea*
 Striped-leaved. *Musa zebrina*
 Sumatran. *Musa Sumatrana*
Band Plant. *Vinca major*
Bandolier Fruit. The berries of *Zanonia indica*
Bane-berry, Common. *Actæa spicata*
 Red. *Acæta spicata var. rubra*
 White. *Actæa alba*
Bane-wort. *Atropa Belladonna* and some species of *Ranunculus*
Bank Cress. *Barbarea præcox* and *Sisymbrium officinale*
Banner Plant. The genus *Anthurium*
Ban-nut. The fruit of *Juglans regia*
Ban-wort, or Ban-wood. *Bellis perennis*
Banyan Tree. *Ficus indica*
 Lord Howe's Island. *Ficus columnaris*
Baobab Tree. *Adansonia digitata*
Barbadoes Pride. *Adenanthera pavonina* and *Poinciana pulcherrima*
Barbara's (St.) Herb. *Barbarea vulgaris*
Barbary Buttons. *Medicago scutellata*
Barber's Brushes. *Dipsacus sylvestris*
Barberry, or Berbery. The genus *Berberis*
 Ash-leaved. The genus *Mahonia*
 Asiatic. *Berberis asiatica*
 Awned-leaved. *Berberis aristata*
 Beale's. *Berberis Bealei*
 Box-leaved. *Berberis fascicularis*
 Bundle-flowered. *Berberidopsis corallina*
 Coral. *Berberidopsis corallina*
 Cratægus-like. *Berberis cratægina*
 Creeping-rooted. *Berberis (Mahonia) repens*
 Crowded-racemed. *Berberis (Mahonia) fascicularis*
 Darwin's. *Berberis Darwini*
 Edging. *Mahonia repens*
 Edible-fruited. *Berberis dulcis*
 Fuegian. *Berberis empetrifolia*
 Golden. *Berberis stenophylla*
 Holly-leaved. *Berberis (Mahonia) Aquifolium*
 Hooker's. *Berberis Hookeri*
 Japanese. *Berberis japonica*
 Many-flowered. *Berberis floribunda*
 Narrow-leaved. *Berberis stenophylla*
 N. American. *Berberis canadensis*
 Notch-petalled. *Berberis emarginata*
 Ribbed-leaved. *Berberis (Mahonia) nervosa*
 Siberian. *Berberis sibirica*
 Various-leaved. *Berberis heterophylla*
 Wallich's. *Berberis Wallichi*
 Whitish-leaved. *Berberis dealbata (B. glauca)*
Barberry Fungus. *Æcidium Berberidis*
Bargeman's Cabbage. *Brassica campestris*

Barilla Plant. *Salsola Kali* and *Salsola Soda*
Bark Plant, Alcornoco or Alcornoque. Various species of *Byrsonima,* also *Bowdichia virgilioides*
 Babul or Babur. *Acacia arabica*
 Barberry. *Berberis aristata, B. asiatica,* and *B. Lycium*
 Bastard Cabbage. *Andira inermis*
 Bonace or Burn-nose. *Daphne tinifolia*
 Bur. *Triumfetta semitriloba*
 Canella or White Wood. *Canella alba*
 Calisaya. *Cinchona Calisaya* and *C. Boliviana*
 Calisaya, of Santa Fé. *Cinchona latifolia*
 Caribbean. *Exostemma caribbæum*
 Cascarilla, or Sweet Wood. *Croton Cascarilla*
 China. *Buena hexandra*
 Clove. *Dicypellium caryophyllatum*
 Conessi. *Wrightia antidysenterica*
 Culilawan. *Cinnamomum Culiliwan*
 Dita. *Alstonia scholaris*
 False Angostura. *Strychnos Nux-vomica*
 French Guiana. *Portlandia hexandra*
 Gray. *Cinchona peruviana*.
 Jesuit's. Various species of *Cinchona*
 Kunro. *Rhizophora mucronata*
 Lodh. *Symplocos racemosa*
 Loxa Crown. *Cinchona officinalis (C. condaminea)*
 Malambo or Matius. *Croton Malamba*
 Matius. See Bark, Malambo
 Monesia. *Chrysophyllum Buranheim*
 Morinda. *Morinda citrifolia, M. tinctoria,* and *M. umbellata*
 Muruxi. *Byrsonima spicata*
 Niepa. *Samadera indica*
 Ordeal, Sassy, or Saucy. *Erythrophleum guineense*
 Panococco. *Swartzia tomentosa*
 Peruvian. Various species of *Cinchona*
 Philadelphia. *Quercus tinctoria*
 Pitayo. *Cinchona Pitayensis*
 Pottery. Various species of *Licania*
 Quercitron. *Quercus tinctoria*
 Quillai. *Quillaia saponaria*
 Red. *Cinchona succirubra*
 Californian. *Daphnidostaphylis glauca*
 Rohun. *Soymida febrifuga*
 Sassafras. *Atherosperma moschata*
 Shag. *Pithecolobium micradenium*
 Sassy. See Bark, Ordeal
 Saucy. See Bark, Ordeal
 Soft Columbian. *Cinchona latifolia*
 Sweet Wood. See Bark Plant, Cascarilla
 Tanekaha. *Phyllocladus trichomanoides*
 Tawai. *Weinmannia racemosa*
 W. Indian. *Exostemma caribbæum*
 White Wood. See Bark Plant, Canella
 Winter's. See Cinnamon, Winter's
 Worm. *Andira inermis*
Bark Tree, Stringy, of Australia. Various species of *Eucalyptus*
 Tasmanian Stringy. *Eucalyptus gigantea*
Barley, Battledore or Sprat. *Hordeum Zeocriton*
 Bear. *Hordeum vulgare var. hexastichum*
 Common. *Hordeum distichum*
 Fan. *Hordeum Zeocriton*

Barley, Mouse, Wall, or Wild. *Hordeum murinum*
Nepaul. *Critho ægiceras* (*Hordeum nepalense, H. trifurcatum*)
Pearl or Pot. The grain of common Barley from which the skin has been removed
Pot. See Barley, Pearl
Putney. *Hordeum Zeocriton*
Red Sea. *Hordeum deficiens*
Sea-side. *Hordeum maritimum*
Six-rowed. *Hordeum hexastichum*
Sprat. See Barley, Battledore
Two-rowed. *Hordeum distichum*
Wall. See Barley, Mouse
Wild. See Barley, Mouse
Barm Fungus. See Fungus, Yeast
Barm-leaf. *Melissa officinalis*
Barfoot. *Helleborus fœtidus*
Barnabas, or St. Barnaby's, Thistle. *Centaurea solstitialis*
Baron's Mercury. Male plant of *Mercurialis annua*
Barrel Tree. *Sterculia* (*Delabechea*) *rupestris*
Barren Ivy. The small-leaved form of *Hedera Helix*
Barren Strawberry. *Potentilla Fragariastrum*
Barren-wort, American. *Vancouveria hexandra*
Common. *Epimedium alpinum*
Large-flowered. *Epimedium macranthum* (*E. grandiflorum*)
Large Yellow-flowered. *Epimedium pinnatum*
Muschi's. *Epimedium Muschianum*
Purple-flowered. *Epimedium purpureum*
Two-leaved. *Epimedium diphyllum* (*Aceranthus diphyllus*)
Violet-flowered. *Epimedium violaceum*
Barricari-seed Plant. *Adenanthera pavonina*
Barrigon Tree. *Carolinia* (*Pachira*) *Barrigon*
Barrow Roses. *Rosa spinosissima*
Barton's Flower, Golden. *Bartonia aurea*
Bar-wood. *Baphia nitida*
Basam, Basom, Basam, Bisom, or Beesom. *Sarothamnus scoparius* and *Calluna vulgaris*
Base Broom. *Genista tinctoria*
Base Rocket. *Reseda Luteola*
Basil, American Wild. The genus *Pycnanthemum*
Bush. *Ocymum minimum*
Chinese. *Plectranthus nudiflorus*
Cow. *Saponaria Vaccaria*
E. Indian. *Ocymum gratissimum*
Field. *Calamintha Clinopodium* and *C. Acinos*
Holy, or Monk's. *Ocymum sanctum*
Purple-stalked. *Ocymum sanctum*
Stone. *Calamintha Clinopodium*
Sweet. *Ocymum Basilicum*
Wild. *Calamintha Clinopodium*
Basil Barm. *Calamintha Acinos*
Basil Thyme. *Calamintha Acinos*
Basil Weed. *Calamintha Clinopodium*
Basket Fern. *Nephrodium Filix-mas*

Basket Wyth. *Tournefortia bicolor*
Bass, or Bast. *Scirpus lacustris* and *Tilia parvifolia*
Bassinet. Applied to *Geranium sylvaticum* and some species of *Ranunculus*
Bass-wood, or White-wood. *Tilia americana*
White. *Tilia heterophylla*
Bast Palm. *Attalea funifera* and *Leopoldinia Piassaba*
Bast Tree, Cuba. *Paritium elatum*
Russian. *Tilia europæa*
Bastard Nigella. An old name for *Agrostemma Githago*
Batter Dock, or Butter Dock. *Petasites vulgaris* and *Rumex obtusifolius*
Baum-leaf. *Melittis Melissophyllum*
Bawchan, or Bawchee, Seeds. The seeds of *Psoralea corylifolia*
Bawme, or Baum. *Melissa officinalis*
Bay, Rose. *Epilobium angustifolium*
Bay Tree, Common. *Laurus nobilis*
Dwarf. *Daphne Laureola* and *D. Mezereum*
E. Indian. *Persea indica*
E. Indian Rose. *Tabernæmontana coronaria*
Loblolly. *Gordonia lasianthus*
Madeira. *Laurus fœtens*
Poison. *Illicium Floridanum*
Red. *Persea Carolinensis*
Rose. *Nerium Oleander*
Sweet. *Laurus nobilis*
White. *Magnolia glauca*
Bay-berry. *Myrica cerifera*
Californian. *Myrica californica*
W. Indian. *Eugenia Pimenta*
Bay-berries. The fruit of *Laurus nobilis*
Bay Laurel. *Laurus nobilis*
Bay Oak. *Quercus sessiliflora*
Bay Willow. *Epilobium angustifolium*
Bays, Willow. *Salix pentandra*
Baziers. *Primula Auricula*
Bdellium Tree, Indian, or Googul. *Balsamodendron Mukul*
Bead Tree. *Melia Azedarach*
Beak Rush. See Rush
Beak Sedge, Brown. *Rhynchospora fusca*
White. *Rhynchospora alba*
Beam Tree. *Pyrus Aria*
Swedish. *Pyrus intermedia*
Bean, Algaroba. *Ceratonia Siliqua*
Algerian Wax. A variety of *Phaseolus vulgaris*
Asparagus. *Dolichos sesquipedalis*
Broad Windsor. *Faba vulgaris var. macrosperma*
Butter. A tender-podded variety of *Phaseolus vulgaris*
Calabar. See Calabar Bean
Cujumary. The seed of *Aydendron Cujumary*
Diverse-leaved Kidney. *Phaseolus diversifolius*
Egyptian. The seed of *Nelumbium speciosum*
Egyptian Kidney. *Dolichos Lab-lab*
Field, Horse, or Tick. *Faba vulgaris var equina*
French, or Kidney. *Phaseolus vulgaris*
Frijol. *Phaseolus Hernandezi*
Garden. *Faba vulgaris var. hortensis*

Bean, Goa. The seeds of *Psophocarpus tetragonolobus*
Green Windsor. *Faba vulgaris var. chlorosperma*
Haricot. The seed of *Phaseolus vulgaris*
Horse. See Bean, Field
Horse, of the W. Indies. *Canavalia gladiata*
Horse-eye. *Dolichos (Mucuna) pruriens*
Indian. *Catalpa bignonioides*
Lima Kidney. *Phaseolus lunatus*
Long-pod. *Faba vulgaris var. ensiformis*
Malacca. The seed of *Semecarpus Anacardium*
Mazagan. *Faba vulgaris var. præcox*
Mezquit. *Prosopis glandulosa*
Molucca. The seed of *Guilandina Bonducella*
Ordeal, of Old Calabar. *Physostigma venenosum*
Ox-eye. *Dolichos (Mucuna) urens*
Pichurim. The seed of *Nectandra Puchury*
Pigeon. A small-seeded variety of the Field Bean
Pythagorean. The seed of *Nelumbium speciosum* (*Nelumbo nucifera*)
Ram's Horn. *Dolichos bicontortus*
Red. *Vigna unguiculata*
Sacred. *Nelumbium speciosum*
St. Ignatius's. *Strychnos Ignatii*
Scarlet Runner. *Phaseolus multiflorus var. coccineus*
Seaside. *Canavalia obtusifolia* and *Vigna luteola*
Screw. *Prosopis pubescens*
Soja, or Soy. *Soja hispida*
Straight. *Vicia Faba*
Sugar. *Phaseolus saccharatus* and *P. lunatus*
Sword. *Canavalia gladiata* and the genus *Entada*
Tick. See Bean, Field
Tonquin. *Dipterix odorata*
Tornillo. *Prosopis pubescens*
Tree. See Bean Tree
Underground. *Arachis hypogæa* and *Voandzeia subterranea*
Yam. *Dolichos tuberosus*
Year. *Phaseolus vulgaris*
Yellow Sacred, or Yellow Water. *Nelumbium luteum*
White Runner. *Phaseolus multiflorus var. albiflorus*
Wild, of N. America. *Apios tuberosa*
Bean Caper. The genus *Zygophyllum*
Common. *Zygophyllum Fabago*
Bean Tree, Australian. *Bauhinia Hookeri* and *Castanospermum australe*
Deccan. *Butea superba*
Red. *Erythrina Corallodendron*
Bean Trefoil. An old name for Laburnum
Bean Vine, Wild. *Amphicarpæa monoica*
Bear, Bere, Beer, Beir, Big, or Bigg **Barley.** *Hordeum vulgare var. hexastichum*
Bear-Bane. *Aconitum arctophonum*
Bear-berry, Alpine or Black. *Arctostaphylos (Arbutus) alpina*
Californian. *Rhamnus Purshianus*
Common. *Arctostaphylos Uva-ursi*
N. W. American. *Rhamnus Purshianus*

Bear-bind. *Convolvulus arvensis, C. sepium, Lonicera Periclymenum, Polygonum Convolvulus,* and the genus *Calystegia*
Bear Grass. The genus *Yucca*
Beard Grass. See Grass
Beard-tongue. The genus *Pentstemon*
Bear's Bed. *Polytrichum juniperinum*
Bear's-breech. The genus *Acanthus*
Armed. *Acanthus spinosissimus*
Common. *Acanthus mollis*
Long-leaved. *Acanthus longifolius*
Soft-leaved. *Acanthus mollis*
Spiny. *Acanthus spinosus*
Stately. *Acanthus latifolius*
Bear's-ear. *Primula Auricula*
Bear's-ear Sanicle. *Cortusa Matthioli*
Bear's-foot. *Helleborus fœtidus, H. viridis, H. niger, Aconitum Napellus,* and *Alchemilla vulgaris*
Bear's Garlick. *Allium ursinum*
Bear's Grass. *Camassia esculenta*
Bear-wort. *Meum athamanticum*
Beast's-Bane. *Aconitum theriophonum*
Beaver Poison. *Cicuta maculata*
Beaver Tree, or Beaver-wood. *Magnolia glauca*
Beck Bean. *Menyanthes trifoliata*
Bedeguar, or Sweet Briar Sponge. A gall found on the Sweet Briar and other Roses
Bede Sedge. *Sparganium ramosum*
Bedewen, or Bedwen. *Betula alba*
Bedlam Cowslip. The Paigle or larger Cowslip and *Pulmonaria officinalis*
Bedstraw. All the species of *Galium,* except *G. Aparine;* also applied to *Desmodium Aparines*
Bloomer's. *Galium Bloomeri*
Corn. *Galium tricorne*
Cross-wort. *Galium Cruciata*
Great. *Galium Mollugo*
Heath. *Galium saxatile*
Hedge. *Galium Mollugo*
Lady's. *Galium rerum*
Marsh. *Galium palustre*
Northern. *Galium boreale*
Rough. *Galium asprellum*
Swamp. *Galium uliginosum*
Wall. *Galium parisiense*
Yellow. *Galium verum*
Bedwind. *Convolvulus sepium*
Beebeeree, or Bibiri, Tree. *Nectandra Rodiæi*
Bee-bread. *Trifolium pratense*
Bee-feed, Californian. *Eriogonum fasciculatum*
Bee Flower. *Ophrys apifera*
Bee Larkspur. *Delphinium grandiflorum* and other cultivated species
Bee, Day, or Deye **Nettle.** *Galeopsis Tetrahit, G. versicolor,* and *Lamium album*
Bee Orchis. See Orchis
Beech, American Purple. *Fagus ferruginea*
Australian. *Tretona australis*
Blue. *Carpinus americana*
Common. *Fagus sylvatica*
Copper-coloured. *Fagus sylvatica var. cuprea*
Crested. *Fagus sylvatica var. cristata*
Cut-leaved. *Fagus sylvatica var. incisa*
Dutch. *Populus alba*
Evergreen. *Fagus betuloides*

Beech, Fern-leaved. *Fagus sylvatica var. asplenifolia*
Horn, Horse, or Hurst. *Carpinus Betulus*
N. S. Wales. *Monotoca elliptica*
New Zealand. *Fagus fusca*
Oblique-leaved. *Fagus obliqua*
Purple. *Fagus sylvatica var. purpurea*
Seaside. *Exostemma caribbæum*
Water. *Carpinus americana* and *Platanus occidentalis*
Weeping. *Fagus sylvatica var. pendula*
White. *Carpinus Betulus*
Beech-drops, False. *Monotropa Hypopitys*
Virginian. *Orobanche virginica*
Beech-fern. *Polypodium Phegopteris*
Beech-fungus. *Cyttaria Darwini*
Beech-wheat. *Polygonum Fagopyrum*
Beef-wood. The genus *Casuarina*
Australian. *Casuarina stricta*
N. S. Wales. *Stenocarpus salignus*
Red. *Ardisia coriacea*
Queensland. *Banksia compar*
White. *Schœpfia chrysophylloides*
Beef-steak Plant. *Begonia Evansiana*
Beef-suet Tree. *Shepherdia argentea*
Beer. See Bear
Bee's Nest. *Daucus Carota*
Beesom. *Sarothamnus scoparius*
Beet, Chard, Leaf, Seakale, or Spinach. *Beta brasiliensis* and *B. Cicla*
Red. *Beta vulgaris*
Seakale. See Beet, Chard
Sea-side or Wild. *Beta maritima*
Sicilian. *Beta Cicla*
Spinach. See Beet, Chard
Sugar. A cultivated variety of *Beta maritima*
White. *Beta Cicla*
Wild Marsh. *Statice Limonium*
Beetle, March or Marish. *Typha latifolia*
Beggar's Basket. *Pulmonaria officinalis*
Blanket. *Verbascum Thapsus*
Buttons. The flower-heads of *Arctium Lappa*
Lice. *Cynoglossum Morisoni*
Needle. *Scandix Pecten*
Beggar Ticks, or Stick-tight. *Bidens frondosa*
Swamp *Bidens connata*
Beggar-weed. *Polygonum aviculare* and various other troublesome weeds
Begonia, Chestnut-leaved. *Begonia castaneæfolia*
Collared. *Begonia manicata*
Coral-flowered. *Begonia corallina*
Dwarf Scarlet. *Begonia Davisi*
Eight-petalled. *Begonia octopetala*
Ever-blooming. *Begonia semperflorens*
Frœbel's. *Begonia Frœbeli*
Fuchsia-like. *Begonia fuchsioides*
Geranium-leaved. *Begonia geranioides*
King. *Begonia Rex*
Many-flowered. *Begonia multiflora*
Peruvian. *Begonia peruviana*
Roezl's. *Begonia Roezli*
Shining. *Begonia lucida*
Socotran. *Begonia socotrana*
Sweet-scented. *Begonia suaveolens*
Tuberous-rooted. *Begonia tuberosa*
Two-coloured. *Begonia discolor*

Begonia, Veitch's. *Begonia Veitchi*
White-and-Scarlet. *Begonia alba coccinea*
Begoon. See Bringal
Beir. See Bear
Belder Root. *Œnanthe crocata*
Belladonna. *Atropa Belladonna*
Belladonna Lily. See Lily
Bell-bind. *Convolvulus sepium*
Bell-bottle. *Scilla nutans*
Belleisle Cress. *Barbarea præcox*
Bell-flower. The genus *Campanula*
Alliaria-leaved. *Campanula alliariæfolia*
Ailioni's. *Campanula Allioni*
Alpine. *Campanula alpina*
American. *Quamoclit vulgaris*
Attic. *Campanula attica*
Australian. *Wahlenbergia capillaris*
Autumn. *Campanula autumnalis (Platycodon autumnale)* and *Gentiana Pneumonanthe*
Barrelier's. *Campanula Barrelicri*
Bearded. *Campanula barbata*
Branching. *Campanula divaricata*
Broad-leaved. *Campanula latifolia*
Candelabrum. *Campanula macrostyla*
Carpathian. *Campanula carpatica*
Chinese. *Platycodon grandiflorum*
Clustered. *Campanula glomerata*
Corn-field. *Campanula hybrida*
Creeping. *Campanula rapunculoides*
Crowded-flowered. *Campanula aggregata*
Dark-coloured Dwarf. *Campanula pulla*
Diamond-leaved. *Campanula rhomboidea*
Diminutive. *Campanula pusilla*
Diminutive White. *Campanula pusilla alba*
Dwarf. *Campanula pumila*
Elatine. *Campanula Erinus*
Elegant. *Campanula elegans*
Equal-leaved. *Campanula isophylla*
Flax-leaved. *Campanula linifolia*
Forked. *Campanula Erinus*
Fragile. *Campanula fragilis*
Gland. The genus *Adenophora*
Great. *Campanula grandis*
Gummy. *Campanula sarmatica (C. gummifera)*
Henderson's. *Campanula Hendersoni*
Hohenhacker's. *Campanula Hohenhackeri*
Host's. *Campanula Hosti*
Ivy-leaved. *Campanula (Wahlenbergia) hederacea*
Langsdorff's. *Campanula Langsdorffiana*
Large-flowered. *Campanula macrantha*
Large-flowered White. *Campanula macrantha alba*
Leutwein's. *Campanula Leutweini*
Ligurian. *Campanula isophylla (C. floribunda)*
Long-flowered. *Campanula nobilis*
Long-leaved. *Campanula longifolia*
Long-styled. *Campanula macrostyla*
Lorey's. *Campanula Loreyi*
Madeira. *Wahlenbergia lobelioides*
Marsh. *Campanula aparinoides*
Michaux's. *Michauxia campanuloides*
Milk-white. *Campanula lactiflora*
Mont Cenis. *Campanula cenisia*
Mount Gargano. *Campanula garganica*
Nettle-leaved. *Campanula urticæfolia, C. celtidifolia,* and *C. Trachelium*

Bell-flower, N. S. Wales. *Wahlenbergia gracilis*
Noble. *Campanula grandiflora* (*Platycodon grandiflorum*)
Peach-leaved. *Campanula persicifolia*
Pendulous. *Campanula* (*Symphyandra*) *pendula*
Portenschlag's. *Campanula Portenschlagiana*
Primrose-leaved. *Campanula primulæfolia*
Pyramidal. *Campanula pyramidalis*
Rainer's. *Campanula Raineri*
Sage-leaved. *Campanula collina*
Shining. *Campanula nitida*
Showy. *Campanula speciosa*
Sibthorp's. *Campanula Sibthorpi*
Silvery-leaved Dwarf. *Edraianthus Pumilio* and other species
Soldanella - flowered. *Campanula soldanellæflora*
Spear-leaved. *Campanula lanceolata*
Spotted. *Campanula punctata*
Spreading. *Campanula patula*
Swamp, of California. *Campanula linnæifolia*
Tall American. *Campanula americana*
Tasmanian. *Wahlenbergia littoralis*
Thompson's. *Campanula Thompsoni*
Tufted. *Campanula cæspitosa* (*C. pumila*)
Turban. *Campanula turbinata*
Van Houtte's. *Campanula Van Houttei*
Vase-flowered. *Campanula turbinata*
Wall. *Campanula muralis*
Wanner's. *Campanula* (*Symphyandra*) *Wanneri*
Woolly-leaved. *Campanula petræa*
Zoys's. *Campanula Zoysi*
Bell Heather, or Bell Heath. *Erica Tetralix*
Bell Rose. *Narcissus Pseudo-Narcissus*
Bell Ware. *Zostera marina*
Bell Woodbind. *Convolvulus sepium*
Bell-wort, Large-flowered. *Uvularia grandiflora*
Sessile-leaved. *Uvularia sessilifolia*
Bell-weed (Ball-weed ?). *Centaurea nigra*
Bells, Candlemas. *Galanthus nivalis*
Canterbury. *Campanula Medium* and *C. Trachelium*
Coventry. *Campanula Medium* and *Anemone Pulsatilla*
Dead Men's. *Digitalis purpurea*
Easter. *Stellaria Holostea*
Hedge. *Convolvulus sepium* and *Convolvulus arvensis*
Peach. *Campanula persicifolia*
Sea. *Convolvulus Soldanella*
Steeple. *Campanula pyramidalis*
Witch's. *Campanula rotundifolia*
Wood. *Scilla nutans*
Belly-ache Bush. *Jatropha gossypiifolia*
Belvedere. *Kochia* (*Chenopodium*) *Scoparia*
Bembil. *Eucalyptus populifolia*
Ben, or White Ben. *Silene inflata*
Ben-dock. *Œnanthe crocata*
Bengal Quince. *Ægle Marmelos*
Root. The root of *Zingiber Casumunar*
Benjamin Bush. *Lindera Benzoin*
Bennels. *Phragmites communis*
Benner-gowan. *Bellis perennis*
Bennet, Herb. *Geum urbanum*

Bent. *Psamma Arenaria;* also applied to the old stalks of various Grasses
Bent Grass. *Agrostis vulgaris, Aira cæspitosa,* and *A. flexuosa*
Bent-wood. *Hedera Helix*
Ben-weed. *Senecio Jacobæa*
Benzoin Bush. See Benjamin
False. *Terminalia mauritiana*
Berberry. See Barberry
Bere. See Bear
Bergamot. *Mentha citrata*
American Wild. *Monarda fistulosa*
Medicinal. *Citrus Bergamia var. vulgaris*
Berries. The fruit of *Ribes Grossularia*
Yellow or Persian. The berries of *Rhamnus infectorius*
Bertram. An old name for *Anacyclus Pyrethrum*
Besom-weed. The genus *Thlaspi*
Betel-nut. See Nut
Betel-nut Palm. See Palm
Beth Root, or Birth Root. *Trillium erectum*
Betle-leaf. See Betle-pepper
Betle-pepper or Betle-leaf. See Pepper
Betony, Brook or Water. *Scrophularia aquatica*
Common. *Stachys Betonica*
Field. *Stachys arvensis*
Large-flowered. *Betonica grandiflora*
Marsh. *Stachys palustris*
St. Paul's. *Veronica serpyllifolia*
Wood. *Stachys Betonica*
Wood, American. *Pedicularis canadensis*
Bhang or Hashish. The dried leaves and small stalks of *Cannabis sativa* mixed with flour and various other additions
Bhel Fruit. See Bael Fruit
Bibiri Tree. See Beebeeree
Bifoil. *Listera ovata*
Big or Bigg. See Bear
Bigold. An old name for *Chrysanthemum segetum*
Big Root or Bitter Root, Californian. *Megarrhiza californica* (*Echinocystis fabacea*)
Big Tree, Californian. *Sequoia gigantea*
Bikh, Bisk, Bish, or Bis, Poison, of Nepaul. *Aconitum ferox*
Bilberry. *Vaccinium Myrtillus*
Bear. *Arctostaphylos Uva-ursi*
Bog. *Vaccinium uliginosum*
Jamaica. *Vaccinium meridionale*
Kamtschatkan. *Vaccinium præstans*
New Zealand. *Gaultheria antipoda*
Whortle. *Vaccinium Myrtillus*
Bilimbi Tree. *Averrhoa Bilimbi*
Bilsted. *Liquidambar styraciflua*
Bind-corn. *Polygonum Convolvulus*
Bind-weed. *Convolvulus arvensis* and several other climbing plants
Blue-flowered. *Ipomœa cœrulea*
Blue Rock. *Convolvulus mauritanicus*
Blushing. *Convolvulus erubescens*
Bryony-leaved. *Convolvulus bryoniæfolius*
Cantabrian. *Convolvulus cantabricus*
Dahurian. *Calystegia dahurica* (*Convolvulus dahuricus*)
Double-blossomed. *Calystegia pubescens fl.-pl.*

Bind-weed, Dwarf. *Calystegia spithamea*
Mallow. *Convolvulus althæoides*
Pigmy. *Convolvulus lineatus*
Riviera. *Convolvulus althæoides*
Rosy-flowered. *Convolvulus sepium var. roseus*
Scarlet-flowered. *Ipomœa coccinea*
Shrubby. *Convolvulus Cneorum*
Silky. *Convolvulus sericeus*
Silvery. *Convolvulus Cneorum*
White Star. *Ipomæa lacunosa (Convolvulus micranthus)*
Bindweed Nightshade. *Circæa Lutetiana*
Bindwith. The genus *Clematis*, especially *C. Vitalba*
Bindwood. *Hedera Helix*
Binweed. *Senecio Jacobæa*
Biophytum. *Biophytum sensitivum (Oxalis sensitiva)*
Bine, or Bines. *Convolvulus arvensis;* also the stalk of the Hop (*Humulus Lupulus*)
Birch, American, White, Gray, or Old Field. *Betula alba var. populifolia*
Antarctic. See Birch, Fuegian
Black. *Betula lenta*
Canoe, or Paper. *Betula papyracea*
Cherry. *Betula lenta*
Common. *Betula alba*
Common Weeping. *Betula alba var. pendula*
Cut-leaved. *Betula populifolia var. laciniata*
Cut-leaved Weeping. *Betula incisa var. pendula*
Dwarf. *Betula nana*
Dwarf American. *Betula glandulosa*
E. Indian. *Betula Bhojputtra*
Fern-leaved. *Betula laciniata*
Fern-leaved Weeping. *Betula laciniata var. pendula*
Fuegian, or Antarctic. *Betula antarctica*
Gray. See Birch, American White, and Birch, Yellow
Hairy Dwarf. *Betula pumila*
Jamaica. *Bursera gummifera*
Kamtschatka Weeping. *Betula tristis*
Low. *Betula pumila*
Mahogany. *Betula lenta*
Old Field. See Birch, American White
Otago. *Fagus Menziesii*
Paper. See Birch, Canoe
Red, or River. *Betula nigra*
River. See Birch, Red
Sweet. *Betula lenta*
W. Indian. *Bursera gummifera*
White New Zealand. *Fagus Solandri* and *F. cliffortioides*
Yellow, or Gray. *Betula lutea*
Bird Briar. *Rosa canina*
Bird Cherry. *Prunus Padus*
Bird Eagles. The fruit of *Cratægus Oxyacantha*
Bird Grass. *Poa trivialis*
Bird Plant, Mexican. *Heterotoma lobelioides*
Bird Thistle. *Carduus lanceolatus*
Bird's Bill. *Trigonella ornithorrhynchus*
Birds'-head Orchid. See Orchid
Bird's Eyes. *Veronica Chamædrys* and many other plants with small bright flowers
Bird's Foot, or Bird's-foot Trefoil. *Lotus corniculatus*

Bird's Foot, Channel Islands. *Ornithopus ebracteatus (Arthrolobium ebracteatum)*
Common. *Ornithopus perpusillus*
Hairy. *Ornithopus compressus*
Bird's Nest. *Daucus Carota*
Yellow. *Monotropa Hypopitys*
Bird's-nest Fern. See Fern
Bird's-nest Orchis. See Orchis
Bird's Tongue. *Senecio paludosus,* the genus *Ornithoglossum, Stellaria Holostea,* and a few other plants, probably from the shape of the leaves
Birk (Birch). *Betula alba*
Birk Apples. Cones of *Pinus sylvestris*
Birth-root. *Trillium erectum*
Birth-wort. The genus *Aristolochia*
Broad-leaved. *Aristolochia Sipho*
Sweet-scented. *Aristolochia odoratissima*
Upright. *Aristolochia Clematitis*
Woolly. *Aristolochia tomentosa*
Bis. See Bikh
Bish. See Bikh
Bishop's Cap. See Mitre-wort
Bishop's Elder. *Ægopodium Podagraria*
Bishop's Hat. *Epimedium alpinum*
Bishop's Leaves. *Scrophularia aquatica*
Bishop's Weed. *Ægopodium Podagraria*
Common. *Ammi majus*
Mock. The genus *Discopleura*
Prickly-seeded. *Ammi copticum*
Toothpick. *Ammi Visnaga*
True. *Ammi copticum*
Bishop's Wort. *Stachys Betonica*
Bisk. See Bikh
Bistort, Great. *Polygonum Bistorta*
Small. *Polygonum viviparum*
Bitch-wood. *Piscidia carthaginensis*
Biting Clematis. *Clematis Vitalba*
Biting Dragon. An old name for Tarragon
Bitter-bark Tree. *Pinckneya pubens*
Bitter-blain. *Vandellia diffusa*
Bitter Cress. The genus *Cardamine.* (See also Cress.)
Bitter King. *Soulamea amara*
Bitter Nut, N. American. *Carya amara*
Bitter Root. *Lewisia rediviva*
Californian. See Big Root
Natal. *Gerrardanthus macrorrhiza*
Bittersgall. The fruit of *Pyrus Malus var. acerba*
Bittersweet. *Solanum Dulcamara* and *Spiræa Ulmaria*
Climbing. *Celastrus scandens*
Bitter Vetch. The genus *Orobus*
Black-rooted. *Orobus niger*
Gmelin's. *Orobus Gmelini*
Hoary. *Orobus canescens*
Spring. *Orobus vernus*
Yellow-flowered. *Orobus aurantius*
Bitter Weed. *Ambrosia artemisiæfolia*
Bitter-wood. The genera *Picramnia* and *Simaruba;* also *Xylopia glabra* and other species
Guiana. *Xylopia frutescens*
Madagascar. *Carissa Xylopicron*
N. American. *Simaruba glauca*
W. Indian. *Xylopia glabra*
White. *Trichilia spondoides*

Bitter-wort. Various species of *Gentiana*
Black Alder. *Rhamnus Frangula*
Black Bead Shrub. *Pithecolobium Unguis-cati*
Black-berry. *Rubus fruticosus, Vaccinium Myrtillus*, and *Ribes nigrum*
Common or High, American. *Rubus villosus*
High. See Blackberry, Common
Low. *Rubus canadensis*
Low Bush. *Rubus trivialis*
Oregon. *Rubus ursinus*
Running Swamp. *Rubus hispidus*
Sand. *Rubus cuneifolius*
Black-boy Tree. The genus *Xanthorrhœa*
Black Bryony. *Tamus communis*
Black-butt. *Eucalyptus pilularis*
Black-cap. *Typha latifolia*
Black Dogwood. *Prunus Padus*
Black-drink Tree. *Ilex Cassine*
Black Grass. *Alopecurus agrestis* and a few other Grasses
Black-heads. *Typha latifolia*
Black-heart. *Vaccinium Myrtillus*; also a variety of Cherry
Black Heath. *Erica cinerea*
Black Hellebore. *Astrantia major*
Black Horehound. *Ballota nigra*
Blacking Plant. *Hibiscus Rosa-sinensis*
Black-knot Fungus. See Fungus
Black Maiden-hair. *Adiantum Capillus-Veneris*
Black Maire, New Zealand. *Olea Cunninghamii*
Black Pot-herb. See Pot-herb
Black Root. *Symphytum officinale* and *Leptandra virginica*
American. *Pterocaulon pycnostachyum*
Black Sampson. The genus *Echinacea*
Black Top. *Centaurea scabiosa*
Blackthorn. *Prunus spinosa*
W. Indian. *Acacia Furnesiana*
Black-wood, Australian. *Acacia Melanoxylon*
E. Indian. *Dalbergia latifolia*
N. S. Wales. *Acacia Melanoxylon*
St. Helena. *Melhania Melanoxylon*
W. Australian. *Acacia penninervis*
Black-wort. *Symphytum officinale*
Bladder Campion. *Silene inflata*
Bladder Fern. The genus *Cystopteris*
Bladder Herb. *Physalis Alkekengi*
Bladder Ketmia. *Hibiscus (Ketmia) Trionum*
Bladder Moss. See Moss
Bladder Nut, African. *Royena lucida*
American. *Staphylea trifoliata*
Common. *Staphylea pinnata*
Ivory-flowered. *Staphylea colchica*
Bladder Pod, American. *Vesicaria Shortii*
Bladder Seed. *Physospermum cornubiense*
Grecian. *Vesicaria græca*
Inflated. *Vesicaria utriculata*
Bladder Senna, Annual. *Colutea herbacea*
Cape. *Sutherlandia frutescens*
Common. *Colutea arborescens*
Nepaul. *Colutea nepalensis*
Pocock's. *Colutea Haleppica (C. Pocockii)*
Red-flowered. *Colutea cruenta*
Smaller. *Colutea media*

Bladder-wort. The genus *Utricularia*
Endres's. *Utricularia Endresi*
Inflated. *Utricularia inflata*
Small. *Utricularia minor*
Blaeberry. *Vaccinium Myrtillus*
Blanket Flower, Blunt-toothed. *Gaillardia amblyodon*
Bristly. *Gaillardia aristata*
Lance-leaved. *Gaillardia lanceolata*
Large-blossomed. *Gaillardia grandiflora*
Loisel's. *Gaillardia Loiseli*
Blanket Leaf. *Verbascum Thapsus*
Blawort. *Campanula rotundifolia* and *Centaurea Cyanus*
Blazing Star. *Helonias dioica, Liatris squarrosa*, and *Chamælirion luteum*
Bleeding Heart. *Aristotelia peduncularis, Cheiranthus Cheiri, Colocasia esculenta*, and *Dicentra formosa*; also a variety of Cherry
Plumy. *Dicentra eximia*
Bleeding Nun. *Cyclamen europæum*
Bleeding Willow. *Orchis Morio*
Blessed Thistle. *Carduus benedictus* and *C. Marianus*
Blewart. *Veronica Chamædrys*
Blewits (Blue Hats). *Agaricus personatus*
Blimbing. *Averrhoa Bilimbi*
Blind Ball. *Lycoperdon Bovista* and other species
Blind Eyes. *Papaver Rhœas*, or *P. dubium*
Blind Nettle. *Lamium album* and various other Labiate plants with leaves resembling those of the Nettle
Blinding Tree. *Excœcaria Agallocha*
Blinks. *Montia fontana*
Blister Plant. *Ranunculus acris* and other species
Blite. *Chenopodium Bonus-Henricus* and various species of *Atriplex* and other Chenopodiaceous plants
Coast, or Sea. *Blitum maritimum (Suæda maritima)*
Shrubby Sea. *Schoberia fruticosa*
Wild. *Amarantus Blitum*
Blood Berry. *Rivina humilis*
Blood Cups. *Peziza coccinea*
Blood Flower. The genus *Hæmanthus*
W. Indian. *Asclepias curassavica*
Blood Hilder (Blood Elder). *Sambucus Ebulus*
Blood Root. *Potentilla Tormentilla* and *Sanguinaria canadensis*
Bloodstrange. *Myosurus minimus*
Blood Tree. *Croton gossypifolium*
Blood-wood, Australian. *Eucalyptus corymbosa* and *E. eximia*
E. Indian. *Lagerstrœmia Reginæ*
Jamaica. *Gordonia Hæmatoxylon*
Norfolk Island. *Baloghia lucida*
Queensland. *Eucalyptus paniculata*
Victoria. *Eucalyptus corymbosa*
Blood Vine. *Epilobium angustifolium*
Blood-wort. *Rumex sanguineus* and a few other plants
Bloody Dock. *Rumex sanguineus*
Bloody Finger. *Digitalis purpurea*
Bloody-man's Finger. *Orchis mascula* and *Arum maculatum*

and Foreign Plants, Trees, and Shrubs. 15

Bloody Warrior. *Cheiranthus Cheiri*
Blooming Sally. *Epilobium angustifolium* and *E. hirsutum*
Blooming Spurge. See Spurge
Blow Ball. *Leontodon Taraxacum*
Blubber Grass. *Bromus mollis* and other species
Blue Ball. *Scabiosa succisa*
Blue Bell. *Scilla nutans* and *Campanula rotundifolia*
Large. *Scilla campanulata*
New Zealand. Various species of *Wahlenbergia*, especially *W. saxicola*
Of Scotland. *Campanula rotundifolia*
Spanish. *Scilla campanulata*
Spreading. *Scilla patula*
Blue Berry, American Common or Swamp. *Vaccinium corymbosum*
British. *Vaccinium Myrtillus*
Canada. *Vaccinium canadense*
Dwarf. *Vaccinium pennsylvanicum*
Low. *Vaccinium vacillans*
Blue Bonnets, or Blue Caps. *Scabiosa succisa, Centaurea Cyanus,* and a few other plants
Blue Bottle. *Scilla nutans, Centaurea Cyanus,* and various other blue flowers
Blue Bush, Mexican. *Ceanothus azureus*
Pimpled. *Ceanothus papillosus*
Tooth-leaved. *Ceanothus dentatus*
Blue Buttons. *Scabiosa succisa* and various other plants with button-like blue flowers
Blue Chamomile. *Aster Tripolium*
Blue Cowslip. *Pulmonaria angustifolia*
Blue Creeper, Tasmanian. *Comesperma volubilis* (*C. gracilis*)
Blue Curls. *Iris dichotoma* and *Trichostemma dichotoma*
Blue Daisy. *Aster Tripolium* and the genus *Globularia*
Bluets. *Houstonia cœrulea*
Blue Eyes. *Veronica Chamædrys*
Blue-eyed Grass. See Grass
Blue Grass. Various species of *Carex*
Blue Hearts. *Buchnera americana*
Blue Innocence. *Houstonia cœrulea*
Blue John. *Taxus stricta*
Blue Kiss. *Scabiosa succisa*
Blue Legs. *Agaricus personatus*
Blue Money. *Anemone Pulsatilla*
Blue Mould of Cheese. See Mould
Blue Rocket. *Aconitum pyramidale* and other species
Blue Runner. *Nepeta Glechoma*
Blue Seggin. *Iris fœtidissima*
Blue Stars. *Veronica Chamædrys*
Blue Tangles. *Vaccinium frondosum*
Blue Venus' Pride. *Houstonia cœrulea*
Blue Weed. *Echium vulgare*
Blue-wood. *Condalia obovata*
Blush-wort. The genus *Erythræa*
Bo Tree. *Ficus indica*
Boar Thistle. *Carduus lanceolatus* and *C. arvensis*
Boar's Ears. *Primula Auricula*
Boar's Foot. *Helleborus viridis*
Bobbin Joan. *Arum maculatum*
Bobbins. *Arum maculatum, Nymphæa alba,* and *Nuphar lutea*

Bockwheat (Buckwheat). *Polygonum Fagopyrum*
Bog Bean, or Buck Bean. *Menyanthes trifoliata*
Bog Berry. *Vaccinium Oxycoccos*
Bog Moss. See Moss
Bog Myrtle. *Myrica Gale*
Bog Nut. *Menyanthes trifoliata*
Bog Pimpernel. *Anagallis tenella*
Bog Rhubarb. *Petasites vulgaris*
Bog Straw-berry. *Comarum palustre*
Bog Trefoil. *Menyanthes trifoliata*
Bog Violet. *Pinguicula vulgaris*
Bolbonac. An old name for *Lunaria biennis*
Boldo, or Boldu, Tree. *Boldoa fragrans*
Boliaun. *Senecio Jacobæa*
Bollas, Bullas, or Bullace. The fruit of *Prunus insititia* and *P. spinosa*
Bolts. An old name for *Trollius europæus*
Bombast. An old name for the genus *Gossypium*
Bonduc Seeds. See Nicker Seeds
Bone Flower. *Bellis perennis*
Bone Seed. The genus *Osteospermum*
Bone-set. *Symphytum officinale*
American. *Eupatorium perfoliatum* and *E. hyssopifolium*
Climbing. *Mikania scandens*
Upland. *Eupatorium sessilifolium*
Bongrace Moss. *Splachnum rubrum*
Bonace, or Burn-nose, Tree. *Daphne tinifolia*
Boodle, or Buddle. *Chrysanthemum segetum*
Boon Tree, or Boor Tree. *Sambucus nigra*
Boor's Mustard. *Thlaspi arvense*
Bootes. An old name for *Caltha palustris*
Borage. *Borago officinalis*
Bell-flowered. *Borago laxiflora*
Early-flowering. *Borago orientalis* (*B. cordifolia*)
Fairy. *Eritrichium nanum*
Spotted Golden. *Arnebia echioides*
Borecole. See Cabbage
Boss Fern. Various species of *Nephrodium*
Bottle Brush. *Equisetum sylvaticum, E. arvense,* and *Hippuris vulgaris*
Bottle-brush Flowers. The flowers of *Beaufortia splendens, Melaleuca hypericifolia, Metrosideros floribunda,* and some species of *Callistemon*
Bottle-cod Root. *Capparis cynophallophora*
Bottle Tree, Australian. *Delabechea* (*Sterculia*) *rupestris*
Victorian. *Delabechea* (*Sterculia*) *diversifolia*
Bouncing Bet. *Saponaria officinalis*
Boutry Tree. *Sambucus nigra*
Bower Plant, Fragrant. *Marsdenia suaveolens*
Bower, Virgin's. *Clematis Vitalba*
Bowman's Root. *Gillenia trifoliata* and *Euphorbia corollata*
Box Berry. *Gaultheria procumbens*
Box Elder. *Acer Negundo* (*Negundo aceroides*)
Californian. *Negundo californicum*
Curled-leaved. *Acer Negundo crispum*
Maple-leaved. *Acer Negundo*
Box Thorn. The genus *Lycium*
Ash-coloured. *Lycium cinereum*

Box Thorn, Barbary. *Lycium barbarum*
Carolina. *Lycium carolinense*
Chinese. *Lycium chinense*
Chili. *Vestia lycioides*
European. *Lycium europæum*
Four-stamened. *Lycium tetrandrum*
Fuchsia-like. *Lycium fuchsioides*
Russian. *Lycium ruthenicum*
Shaw's. *Lycium Shawii*
Slender. *Lycium tenue*
Small-leaved. *Lycium microphyllum*
Spear-leaved. *Lycium lanceolatum*
Stiff. *Lycium rigidum*
Top-shape l-fruited. *Lycium turbinatum*
Trew's. *Lycium Trewianum*
Very Prickly. *Lycium horridum*
Box Tree, American. *Cornus florida*
Asses'. The genus *Lycium*
Australian. Various species of *Eucalyptus*
Bastard. *Eucalyptus Leucoxylon*
Balearic. *Buxus balearica*
Chinese. *Buxus chinensis* and *Murraya exotica*
Common. *Buxus sempervirens*
Dwarf, or Edging. *Buxus sempervirens suffruticosa*
False. *Schæfferia frutescens*
Flowering. *Vaccinium Vitis-Idæa*
Fortune's Round-leaved. *Buxus Fortunei rotundifolia*
Gold-edged. *Buxus sempervirens aurea*
Gray, Queensland. *Lophostemon macrophyllus*
Gray, Victoria. *Eucalyptus dealbata*
Japanese. *Euonymus japonicus var. radicans*
Minorca. *Buxus balearica*
Myrtle-leaved. *Buxus sempervirens myrtifolia*
Narrow-leaved. *Buxus sempervirens angustifolia*
New Zealand. *Veronica buxifolia*
Prickly. *Ruscus aculeatus* and the genus *Lycium*
Pyramidal. *Buxus sempervirens pyramidata*
Red, N. S. Wales. *Lophostemon australis*
Red, S. E. Australian. *Eucalyptus polyanthema*
Rosemary - leaved. *Buxus sempervirens rosmarinifolia*
Round-leaved. *Buxus sempervirens rotundifolia*
Silvery-leaved. *Buxus sempervirens argentea*
Tasmanian. *Bursaria spinosa*
Tree. *Buxus sempervirens arborescens*
White, Australian. *Eucalyptus albens*
White, N. S. Wales. *Pittosporum undulatum*
Yellow, Australian. *Eucalyptus melliodora*
Box-wood, American or False. *Cornus florida*
W. Indian. *Tecoma pentaphylla* and *Vitex umbrosa*
Boy's Love, or Lad's Love. *Artemisia Abrotanum*
Boy's Mercury. *Mercurialis annua*
Brab Tree. *Borassus flabelliformis*
Bracelet-wood Tree. *Jacquinia armillaris*
Bracken, or Brake. *Pteris aquilina;* sometimes applied to other large Ferns
Brake, Buck-horn. *Osmunda regalis*
Brake-root. *Polypodium vulgare*

Brakes, Rock. *Allosorus crispus*
Bramble. *Rubus fruticosus*
Arctic. *Rubus arcticus*
Blue. *Rubus cæsius*
Canadian. *Rubus canadensis*
Cut-leaved. *Rubus fruticosus var. laciniatus*
Dog. *Ribes Cynosbati*
Double-flowered. *Rubus discolor fl.-pl.* and *R. fruticosus fl.-pl.*
Dwarf Crimson-flowered. *Rubus arcticus*
Hawthorn-leaved. *Rubus cratægifolius*
Japanese Climbing. *Rubus phœnicolasius*
Mountain. *Rubus Chamæmorus*
Nepaul. *Rubus nepalensis*
New Zealand. *Rubus australis*
Nootka Sound. *Rubus Nutkanus*
Purple-flowered. *Rubus odoratus*
Rose-leaved. *Rubus rosæfolius*
Rosy-flowered. *Rubus spectabilis*
Showy-flowered. *Rubus spectabilis*
Showy White-flowered. *Rubus deliciosus*
Stone. *Rubus saxatilis*
Sweet-scented. *Rubus odoratus*
Branch Peas. An old name for Garden Peas
Brancorn, or Brawn (the "Smut" in Wheat). *Ustilago segetum*
Brandy Bottle. *Nuphar lutea*
Brandy Mint. *Mentha piperita*
Branke Ursine. An old name for Acanthus
Brank, or Buck-wheat. *Fagopyrum esculentum*
Brasiletto. See Braziletto
Brassica, or Brassock. *Sinapis arvensis*
Brauna-wood Tree. *Melanoxylon Brauna*
Brawlins. *Arctostaphylos Uva-ursi* and *Vaccinium Vitis-Idæa*
Brawn. See Brancorn
Braziletto-wood, or Brazil-wood of commerce. Several species of *Cæsalpinia*
Brazil. *Peltophorum Vogellianum*
Bahama. *Cæsalpinia Crista*
Jamaica. *Peltophorum Linnæi (Cæsalpinia brasiliensis)*
Brazil-nut Tree. *Bertholletia excelsa*
Bread, Cuckoo's. *Oxalis Acetosella* and *Cardamine pratensis*
Australian Native. *Mylitta australis*
Tartar. The root of *Crambe tatarica*
Bread-and-milk. *Cardamine pratensis*
Bread-fruit Tree. *Artocarpus incisa*
African. *Treculia africana*
Nicobar. *Pandanus odoratissimus*
N. Australian. *Gardenia edulis*
Bread Mould. See Mould
Bread-nut Tree. *Brosimum Alicastrum*
Bastard. *Pseudolmedia spuria*
Bread Root, Missouri. *Psoralea esculenta*
Break-axe. *Sloanea jamaicensis*
Break-bones. *Stellaria Holostea*
Break-stone. The genus *Saxifraga*
Breeches Flower, or Dutchman's Breeches. *Dielytra (Dicentra) Cucullaria*
Briar, or Brier. A general name for Wild Roses and Brambles
Austrian. *Rosa lutea var. punicea*
Cat. See Cat Briar
Green. *Smilax rotundifolia*
Sensitive. The genus *Schrankia*
Sweet. *Rosa rubiginosa*

and Foreign Plants, Trees, and Shrubs. 17

Briar Root (a corruption of the French "Bruyère,") of which pipes are made. *Erica arborea*
Briar Rose. *Rosa canina*
Bride's Laces. *Phalaris arundinacea*
Bride-wort. *Spiræa Ulmaria*
Bright Meadow. *Caltha palustris*
Brimstone, Vegetable. The spores of *Lycopodium* and *Selago*, used in France and Germany in the manufacture of fireworks
Brimstone-wort. *Peucedanum palustre*
Bringal, Brinjal, or Begoon. The fruit of *Solanum Melongena*
Bristle Fern. *Trichomanes radicans* and other species
Bristle Moss. See Moss
Broad Leaf, New Zealand. *Griselinia lucida*
Broad Seed. The genus *Ulospermum*
Broccoli. *Brassica oleracea* var. *Botrytis asparagoides*
Bronze Leaf, Rodgers's. *Rodgersia podophylla*
Brook Bean. *Menyanthes trifoliata*
Brook Leek. An old name for *Arum Dracunculus*
Brook Lime. *Veronica Beccabunga*
American. *Veronica americana*
Brook Mint. *Mentha hirsuta*
Brook-weed, Common. *Samolus Valerandi*
Tasmanian. *Samolus littoralis*
Broom, African. The genus *Aspalathus*
Amsanto. *Genista amsantica* (*G. anxantica*)
Arrow-jointed. *Genista sagittalis*
Black-rooted. *Cytisus nigricans*
Butcher's. The genus *Ruscus*; see also Butcher's Broom
Close-branched. *Genista congesta*
Common. *Cytisus* (*Sarothamnus*) *scoparius*
Cypress. *Taxodium capense*
Dyer's. *Genista tinctoria*
Early-flowering. *Genista præcox*
Egyptian. *Genista ægyptiaca*
Flax-leaved. *Genista linifolia*
Flowery. *Genista florida*
German. *Genista germanica*
Gibraltar. *Genista gibraltarica*
Hairy. *Genista pilosa*
He. *Cytisus Laburnum*
Jamaica Mountain. *Baccharis Scoparia*
Jointed. *Genista sagittalis*
Large-spined. *Genista horrida*
Leafless. *Genista aphylla*
Long-twigged. *Genista virgata*
Milk-wort-leaved. *Genista polygalæfolia*
Mt. Etna. *Genista ætnensis*
Mountain. *Baccharis Scoparia*
New Zealand. Various species of *Carmichælia*
Pink-flowered. *Notospartium Carmichæliæ*
Portugal. *Genista lusitanica*
Procumbent. *Genista procumbens*
Prostrate. *Genista prostrata*
Purging. *Genista purgans*
Quadrangular-branched. *Genista tetragona*
Rock. *Genista tinctoria* and other species
Round-podded. *Genista sphærocarpa*

Broom, Rush. The genus *Viminaria*; also applied to *Spartium junceum*
Scorpion. *Genista Scorpius*
Siberian. *Genista sibirica*
Silky-leaved. *Genista sericea*
Small-flowered. *Genista parviflora*
Spanish. *Spartium junceum*
Spanish Dwarf Prickly. *Genista hispanica*
Spreading. *Genista patula*
Swan River. *Comesperma Scoparia*
Sweet. *Scoparia dulcis*
Teneriffe. *Cytisus nubigenus*
Thorn. *Ulex europæus*
Three-spined. *Genista triacanthos*
Trailing. *Genista humifusa*
Triangular-stemmed. *Genista triquetra*
Umbel-flowered. *Genista umbellata*
W. Indian Prickly. *Parkinsonia aculeata*
White-flowered. *Genista monosperma*
White Portugal. *Genista multiflora*
Whitish-leaved. *Genista candicans*
Wood. *Genista sylvestris*
Violet-flowered. *Genista aphylla*
Broom Bush. *Parthenium Hysterophorus*
Broom Corn. *Sorghum vulgare* and *S. saccharatum*
Broom Goose-foot. *Kochia Scoparia*
Broom Grass. *Andropogon scoparius*
Broom-weed. *Corchorus siliquosus*
Broom-rape. The genus *Orobanche*, especially *O. major*
American. *Phelipæa Ludoviciana*
Blue-flowered. *Orobanche cærulea*
Branched. *Orobanche ramosa*
Clove-scented. *Orobanche caryophyllacea*
Great. *Orobanche major*
Naked. The genus *Aphyllon*
Red. *Orobanche rubra*
Small. *Orobanche minor*
Tall. *Orobanche elatior*
Virginian. *Epiphegus virginiana*
Brother-wort. An old name for *Thymus Serpyllum*
Brown Bugle. *Ajuga reptans*
Brown Dragons. *Arum atro-rubens*
Brown-wort. *Scrophularia aquatica*
Bruise-wort. *Bellis perennis*
Bruliaun. *Bromus secalinus*
Brunel. *Prunella vulgaris*
Brush Bush. *Eucryphia pinnata*
Brush Grass. See Grass
Brussels Sprouts. *Brassica oleracea gemmifera* (*B. o. bullata minor*)
Bryony. *Bryonia dioica* and *Tamus communis*
Bastard. *Cissus sicyoides*
Black. *Tamus communis*
Red or White. *Bryonia dioica*
Buaze-fibre-Plant. *Securidaca longipedunculata* (*Lophostylis pallida*)
Bucchu. See Buchu
Bucco. See Buchu
Buck Bean, Common. *Menyanthes trifoliata*
Fringed, or Round-leaved. *Menyanthes* (*Villarsia*) *nymphæoides*
Buchu, Bucchu, Bucco, Bucha, or Buka, leaves. The leaves of *Barosma betulina*, *B. crenulata*, and *B. serratifolia*

c

Buck - eye. Various species of *Pavia* (*Æsculus*)
California. *Pavia* (*Æsculus*) *californica*
Deep-toothed-leaved Red-flowered. *Pavia* (*Æsculus*) *rubra var. sublaciniata*
Dwarf, Red - flowered. *Pavia* (*Æsculus*) *rubra var. humilis*
Dwarf Red-and-yellow-flowered. *Pavia* (*Æsculus*) *discolor*
Fetid, or Ohio. *Pavia* (*Æsculus*) *glabra*
Long-fruited. *Pavia* (*Æsculus*) *macrocarpa*
Long-spiked-flowered. *Pavia* (*Æsculus*) *macrostachya*
Neglected. *Pavia* (*Æsculus*) *neglecta*
Ohio. See Buck-eye, Fetid
Red-flowered. *Pavia* (*Æsculus*) *carnea*
Sharp-toothed-leaved Red-flowered. *Pavia* (*Æsculus*) *rubra var. arguta*
Sweet, or Big. *Pavia* (*Æsculus*) *flava*
Variegated - flowered. *Pavia* (*Æsculus*) *hybrida*
Weeping Red - flowered Dwarf. *Pavia* (*Æsculus*) *rubra var. humilis pendula*
Yellow-flowered. *Pavia* (*Æsculus*) *flava*
Buck-grass. *Lycopodium clavatum*
Buck-horn Moss. *Lycopodium clavatum*
Buckie-berries. The fruit of *Rosa canina*
Buckler Fern. The genus *Lastræa*
Buckler Mustard, Ear-podded. *Biscutella auriculata*
Buck-mast. An old name for the Beech
Buckrams. *Allium ursinum*
Buck's-horn. *Senebiera Coronopus*
Buck's-horn Plantain. *Plantago Coronopus*
Buckthorn, Alder. *Rhamnus Frangula*
Alpine. *Rhamnus alpinus*
American. *Frangula Caroliniana*
Azorean. *Rhamnus latifolius*
Box-leaved. *Rhamnus burifolius*
Broad-leaved. *Rhamnus latifolius*
Canadian Sea. *Hippophaë Canadensis*
Common. *Rhamnus catharticus* and *Prunus spinosa*
Dahurian. *Rhamnus dahuricus*
Downy. *Rhamnus pubescens*
Dwarf. *Rhamnus pumilus*
Dwarf Rock. *Rhamnus rupestris*
Dyer's. *Rhamnus tinctorius*
Frangula-like. *Rhamnus franguloides*
Long-leaved. *Rhamnus longifolius*
Lycium-like. *Rhamnus lycioides*
Olive-leaved. *Rhamnus oleoides*
Red-wooded. *Rhamnus Erythroxylon*
Rock. *Rhamnus saxatilis*
Sea. *Hippophaë rhamnoides*
Slender-branched. *Rhamnus virgatus*
Small. *Rhamnus pusillus*
Southern. *Bumelia lycioides*
Valencia. *Rhamnus valentinus*
Willow-leaved Sea. *Hippophaë salicifolia*
Wulfen's. *Rhamnus Wulfenii*
Yellow-berried. *Rhamnus infectorius*
Buckwheat, Climbing. *Polygonum Convolvulus*
Common. *Polygonum Fagopyrum*
E. Indian. *Polygonum emarginatum*
Perennial. *Fagopyrum cymosum*
W. Indian. *Anredera scandens*

Buckwheat, Wild, Californian. *Eriogonum fasciculatum*
Buckwheat Tree. *Mylocaryum ligustrinum*
Buddle. *Chrysanthemum segetum*
Buffalo Berry, Missouri. *Shepherdia argentea*
Buffalo-grass. See Grass
Buffalo-nut. *Pyrularia* (*Hamiltonia*) *oleifera*
Bufflehorn-wood. The wood of *Burchellia capensis*
Buffles-ball-wood. The wood of *Gardenia Thunbergii*
Bug Agaric. *Agaricus muscarius*
Bug-bane. *Actæa racemosa*
American. *Cimicifuga americana*
False. *Trautvetteria palmata*
Fetid. *Cimicifuga fœtida*
Bug-seed. *Corispermum hyssopifolium*
Bugle, Alpine. *Ajuga alpina*
Common. *Ajuga reptans*
Erect. *Ajuga genevensis*
Musky. *Ajuga Iva*
Pyramidal. *Ajuga pyramidalis*
Variegated-leaved. *Ajuga reptans variegata*
Yellow. *Ajuga Chamæpitys*
Bugle-weed. *Lycopus virginicus*
Bugloss, or Viper's Bugloss. *Echium vulgare.*
Also applied to some other rough-leaved plants
Purple. *Echium violaceum*
Pyrenean. *Echium pyrenaicum*
Red-flowered. *Echium rubrum*
Russian. *Arnebia echioides*
Sea. *Mertensia maritima*
Small. *Anchusa* (*Lycopsis*) *arvensis*
Bugloss Cowslip. *Pulmonaria officinalis*
Buka. See Buchu
Bukkum-wood. *Cæsalpinia Sappan*
Bukul. *Mimusops Elengi*
Bullace. *Prunus insititia*
Double-blossomed. *Prunus insititia florepleno*
Jamaica. *Melicocca bijuga*
Red-fruited. *Prunus insititia fructû rubro*
Yellowish-white-fruited. *Prunus insititia fructû luteo-albo*
Bull Berries. The fruit of *Vaccinium Myrtillus*
Bull Daisy. *Chrysanthemum Leucanthemum*
Bullet Tree, or Bully Tree. *Sapota Sideroxylon* and *Myrsine læta*
Bastard. *Bumelia retusa*
Black. *Bumelia ingens* and *B. nigra*
Guiana. A species of *Mimusops*
Jamaica. *Lucuma mammosa*
Mountain. *Bumelia montana*
White. *Dipholis salicifolia*
Bull Foot. An old name for Colt's-foot (*Tussilago Farfara*)
Bullimong. An old name for Buckwheat
Bullock's-eye. *Sempervivum tectorum*
Bullock's-heart. *Anona reticulata*
Bull-grass. *Bromus mollis*
Bull-hoof. *Murucuja ocellata*
Bull-rush, or Bul-rush. *Scirpus lacustris* and *Typha latifolia*
Great American. *Scirpus validus*

Bull-rush, Lesser. *Typha angustifolia*
Nile. *Papyrus antiquorum* (*Cyperus Papyrus*)
River. *Scirpus fluviatilis*
Bulls and Cows. *Arum maculatum*
Bull Thistle. *Carduus lanceolatus*
Bull-weed. *Centaurea nigra*
Bull-wort. *Ammi majus*
Bully Tree. See Bullet Tree
Bul-rose. *Narcissus Pseudo-Narcissus*
Bul-rush. See Bull-rush
Bumbo, or Bungo, Tree. *Daniellia thurifera*
Bum-wood. *Rhus Metopium*
Bunch Berry. *Cornus canadensis* and the fruit of *Rubus saxatilis*
Bunch Flower. *Melanthium Virginicum*
Bungo. See Bumbo .
Bunkuss Grass. See Grass
Bunt Fungus. See Fungus
Bunya-Bunya Tree. *Araucaria Bidwillii*
Bur, or Burr. *Arctium Lappa* and a few other prickly-fruited plants
Bathurst. *Xanthium spinosum*
Clot. *Arctium Lappa*
New Zealand. The genus *Acæna*
Paroquet. The genus *Triumfetta*
Styptic, or Velvet. *Priva echinata*
Burdock. *Arctium Lappa*
Edible. *Lappa edulis*
Prairie. *Silphium terebinthinaceum*
Small. *Xanthium strumarium*
Burgrass, West Indian. *Cenchrus echinatus*
Bur Marigold. *Bidens tripartita*
Burk Tree. *Betula alba*
Burnet. *Sanguisorba officinalis* and *Poterium Sanguisorba*
Canadian. *Poterium canadense*
Great. *Sanguisorba officinalis*
Salad. *Poterium Sanguisorba*
Burnet Rose. *Rosa spinosissima*
Burnet Saxifrage. *Pimpinella Saxifraga*
Burning Bush. *Dictamnus Fraxinella*
American. *Euonymus atropurpureus*
Burn-nose Tree. See Bonace
Burnt-weed. *Scolopendrium vulgare*
Burn-wood. *Rhus Metopium*
Bur Parsley. *Caucalis daucoides*
Bur Reed, Common. *Sparganium ramosum*
Floating. *Sparganium natans*
Unbranched. *Sparganium simplex*
Burst-wort. Another name for Rupturewort
Bur Thistle. *Carduus lanceolatus*
Bur Tree. *Sambucus nigra*
Bur-weed. *Galium Aparine*
Knotted. *Caucalis* (*Torilis*) *nodosa*
W. Indian. The genus *Triumfetta*
Bush Grass. *Calamagrostis Epigejos*
Bush Lawyer, of New Zealand. *Rubus australis*
Butcher's Broom, Broad-leaved. *Ruscus Hypophyllum var. latifolius*
Climbing. *Ruscus androgynus*
Common. *Ruscus aculeatus*
Double-leaved. *Ruscus Hypoglossum*
Butter-and-Eggs. *Linaria vulgaris*, various species of *Narcissus*, and a few other flowers which are of two shades of yellow

Butter-and-Tallow Tree. *Pentadesma butyracea*
Butter Bean. See Bean
Butter Bur. *Petasites vulgaris*
Butter-cup. Various species of *Ranunculus*
Chervil-leaved. *Ranunculus chærophyllus*
Grass-leaved. *Ranunculus gramineus*
Gowan's. *Ranunculus Gowani*
Mountain. *Ranunculus montanus*
Montpelier. *Ranunculus monspeliacus*
Spiked. *Ranunculus spicatus*
Water. *Caltha palustris*
White. *Ranunculus amplexicaulis*
Butter Daisy. Various species of *Ranunculus*
Butter Dock. *Rumex obtusifolius*
Buttered Haycocks. *Linaria vulgaris*
Butter Flower. Another name for Butter-cup
Butterfly Flower. The genus *Schizanthus*
Butterfly Orchis. See Orchis
Butterfly Plant, E. Indian. *Phalænopsis amabilis*
S. American. *Oncidium Papilio*
W. Indian. *Oncidium Papilio*
Butterfly-weed. *Asclepias tuberosa* and the genus *Calochortus*
or Flag Flower, Virginian. *Vexillaria virginica*
Butter Grass. See Grass
Butter Leaves. The leaves of *Atriplex hortensis*
Butter-nut Tree, Common. *Caryocar nuciferum*
Guiana. *Caryocar tomentosum*
N. American. *Juglans cinerea*
Butter Root. *Pinguicula vulgaris*
Butter Tree, African, or Park's. *Bassia Parkii*
E. Indian, or Nepaul. *Bassia butyracea*
Shea. *Bassia Parkii*
Butter-weed. *Erigeron canadensis* and *Senecio lobatus*
Butter-wort, Common. *Pinguicula vulgaris*
Irish, or Large-flowered. *Pinguicula grandiflora*
Long-leaved. *Pinguicula longifolia*
Long-spurred. *Pinguicula caudata*
Mountain. *Pinguicula alpina*
Pale-flowered. *Pinguicula lusitanica*
Vallisneria-leaved. *Pinguicula vallisneriæfolia*
White-flowered. *Pinguicula alpina*
Yellow-flowered. *Pinguicula lutea*
Button Bush. *Cephalanthus occidentalis*
Button Flower. The genus *Gomphia*
African. *Dais cotinifolia*
Glossy-leaved. *Gomphia nitida*
Button Gall. See Gall
Button Grass. *Avena elatior*
Button-hole. *Scolopendrium vulgare*
Button Snake-root, Dense-spiked. *Liatris pycnostachya*
Hairy-cupped. *Liatris elegans*
Long-spiked. *Liatris spicata*
Button Tree. The genus *Conocarpus*
Button-weed. The genus *Borreria* (*Spermacoce*), *Centaurea nigra*, *Diodia virginica*, and *D. teres*

Button-wood. *Platanus occidentalis*
White. *Laguncularia racemosa*
Buttons. *Tanacetum vulgare*
Thorny. *Medicago muricata*

Cabaret. *Asarum europæum*
Cabbage. *Brassica oleracea*
Arkansas. *Strepthanthus obtusifolius*
Borecole, or Kale. *Brassica oleracea acephala*
Brussels Sprouts. *Brassica oleracea bullata minor*
Chinese (Pak-choi and Pe-tsai). *Brassica chinensis*
Couve Tronchuda. *Brassica oleracea costata*
Dog's. *Thelygonum Cynocrambe*
Drum-head. *Brassica oleracea capitata*
Kale. See Cabbage, *Borecole*
Kerguelen's Land. *Pringlea antiscorbutica*
Kohl-Rabi. *Brassica Caulo-rapa*
Meadow. *Symplocarpus fœtidus*
St. Patrick's. *Saxifraga umbrosa*
Savoy. *Brassica oleracea bullata major*
Sea. *Crambe maritima*
Sea Otter's. *Nereocystis Lutkeana*
Skunk. *Symplocarpus fœtidus*
Turnip. *Brassica Napo-brassica*
Turnip-rooted. *Brassica Caulo-rapa*
Wild Californian. *Caulanthus crassicaulis* and *C. procerus*
Cabbage Palm Tree. *Areca oleracea* and *Chamærops* (*Sabal*) *Palmetto*
Cabbage Tree, Australian. *Corypha australis*
Bastard. *Andira inermis*
Bastard, or Black, of St. Helena. *Melanodendron integrifolium*
Bastard, of S. America. The genus *Geoffroya*
Black. See Cabbage Tree, Bastard
Canary Island. *Cucalia Kleinia*
Small-umbelled. *Commidendron spurium*
W. Indian. *Oreodoxa oleracea* and *Andira inermis*
Cabbage-wood. *Eriodendron anfractuosum*
Cacao Butter Tree. *Theobroma Cacao*
Cacoon, or Cocoon, W. Indian. *Entada scandens* and *Fevillea* (*Feuillæa*) *cordifolia*
Cactus, Cochineal. *Opuntia cochinillifera* and *O. Tuna*
Common Hardy. *Opuntia vulgaris*
Elephant's-tooth. *Mammillaria elephantidens*
Erect. *Mammillaria erecta*
Hardy Dwarf. *Opuntia humilis*
Houllett's Woolly. *Pilocereus Houlletti* (*P. fossulatus*)
Leaf-flowering. The genus *Epiphyllum*
Many-headed Hedgehog. *Echinocactus polycephalus*
Melon. See Melon Cactus
Nipple. The genus *Mammillaria*
Old Man. *Pilocereus senilis*
Pin-pillow. *Opuntia curassavica*
Rat's-tail. *Cactus flagelliformis*
Silvery. *Echinocactus myriostigma*
Simpson's Hardy Hedgehog. *Echinocactus Simpsoni*

Cactus, Turk's-Cap. The genus *Melocactus*
White-spined Hardy. *Opuntia missouriensis var. leucospina*
Winter. The genus *Epiphyllum*
Cactus Dahlia. *Dahlia Juarezi*
Cad-weed. *Heracleum Sphondylium*
Caffre Bread. *Encephalartos Caffer* and several species of *Zamia*
Caffre Butter. *Combretum butyrosum*
Cain-and-Abel. The tubers of *Orchis latifolia*
Cainito, of Australia. *Niemeyera* (*Lucuma*) *prunifera*
Cairn Tangle. *Laminaria digitata*
Cajeput, or Cajuput, Tree. *Melaleuca Leucodendron* and *Oreodaphne* (*Tetranthera*) *californica*
Calaba Tree. *Calophyllum Calaba*
Calabar Bean, or Chop Nut. *Physostigma venenosum*
Calabash, Sweet. The fruit of *Passiflora maliformis*
Calabash Nutmeg. *Monodora Myristica*
Calabash Tree. Various species of *Crescentia*
Calabur Tree. *Muntingia Calabura*
Calalu. A W. Indian name for *Phytolacca* and some other plants
Branched. *Solanum nodiflorum*
Green. *Euxolus viridis*
Prickly. *Amarantus spinosus*
Small-leaved. *Euxolus caudatus*
Spanish. *Phytolacca octandra*
Calamander-wood, or Coromandel-wood. The wood of *Diospyros quæsita* and *D. oppositifolia*
Calambac. *Aloexylon Agallochum*
Calamint, Alpine. *Calamintha alpina*
Common. *Calamintha officinalis*
Field. *Calamintha Acinos* and *C. Nepeta*
Hedge. *Calamintha Clinopodium*
Medicinal. *Calamintha officinalis*
Tom Thumb. *Calamintha glabella*
Small-flowered. *Calamintha Nepeta*
Wood. *Calamintha sylvatica*
Calathian Violet. *Gentiana Pneumonanthe*
Calavanche. *Cicer arietinum*
Cale, or Kale. See Cabbage
Calf's Foot. *Arum maculatum*
Calf' Snout. *Antirrhinum Orontium*
Calico Bush. *Kalmia latifolia*
Calisaya Bark Tree, or Yellow Cinchona Bark Tree. *Cinchona Calisaya*
Callcedra-wood. The timber of *Flindersia australis*
Callemundoo. See Cattemandoo
Calool Tree. *Sterculia quadrifida*
Columba, or Colombo, Root. *Jateorrhiza palmata* (*Cocculus palmatus*)
Calvary Clover. See Clover
Calverkeys. See Culverkeys
Caltrops. *Centaurea Calcitrapa*, *Tribulus terrestris*, and other species
Water. *Potamogeton densus*, *P. crispus*, *Trapa natans*, and other species
Camass, or Camash. See Quamash
Camata, or Camatena. A term applied to the unripe acorns of *Quercus Ægilops*, which are used for tanning and dyeing

Camatena. See Camata
Cambi Resin Plant. *Gardenia lucida*
Cambie Leaf. *Nymphæa alba* and *Nuphar lutea*
Cambric Grass Plant. *Urtica (Bœhmeria) nivea*
Cambridgeshire Oaks. Willows; in allusion to the marshy soil of the county
Cambuca, of Brazil. *Marliera (Rubachia) glomerata*
Camellia. *Camellia japonica* and other species
Anemone-flowered. *Camellia anemonæflora*
Apple-flowered. *Camellia maliflora*
Bright Crimson. *Camellia Beallii* and *C. Matthotiana*
Bright Rose-coloured. *Camellia Dride*
Bright rose, crimson-rayed. *Camellia Optima*
Captain Rawes's. *Camellia reticulata*
Carmine-rose. *Camellia imbricata*
Creamy-white. *Camellia ochroleuca* and *C. candidissima*
Deep Rose, large-flowered. *Camellia Rubens*
Double White. *Camellia alba plena*
Fine Rosy-flowered. *Camellia Valteraredo*
Fine White. *Camellia compacta alba*
Lady Banks's. *Camellia Sasanqua*
Large Crimson. *Camellia Triumphans*
Large White. *Camellia Grunelli*
Matthoti's. *Camellia Matthotiana*
Netted. *Camellia reticulata*
Pure White, Fringed. *Camellia fimbriata*
Pure White, Imbricated. *Camellia Matthotiana alba*
Red, White-ribboned. *Camellia Queen Victoria*
Rosy-pink. *Camellia Saccoi nova*
Small-flowered. *Camellia euryoides*
Snow-white. *Camellia Spinii*
Superb Rosy-flowered. *Camellia Lombarda*
Three-coloured. *Camellia tricolor nova* and *C. tricolor imbricata plena*
Vermilion-red. *Camellia commensa*
White, Crimson-striped. *Camellia Mrs. Cope*
Yellowish-white. *Camellia sulcata*
Camel Tree, or Camelopard's Tree. *Acacia Giraffæ*
Camel's Hay. *Andropogon Schœnanthus*
Camomile, or Chamomile. *Anthemis nobilis*
Blue. *Aster Tripolium*
Corn. *Anthemis arvensis*
Cut-leaved. *Anthemis chia*
Dog's. *Anthemis Cotula*
False. *Boltonia glastifolia*
German. *Matricaria Chamomilla*
Purple. *Aster Tripolium*
Purple-stalked. *Anthemis valentina*
Red. *Adonis autumnalis*
Roman. *Anthemis nobilis*
Scentless. *Anthemis inodora fl.-pl.*
Scotch. *Anthemis nobilis*
Silvery. *Anthemis Aizoon*
Stinking. *Anthemis Cotula*
Wild. *Matricaria Chamomilla*

Camomile, Yellow. *Anthemis tinctoria*
Camooyne, or Camowyne. *Anthemis nobilis* and *A. Cotula*
Campeachy-wood, or Log-wood. *Hæmatoxylon campechianum*
Campernelle. *Narcissus odorus*
Camphire. *Crithmum maritimum*
Of Scripture. Supposed to be *Lawsonia alba*
Camphor Tree. *Laurus (Cinnamomum) Camphora* and *Dryobalanops aromatica*
Camphor-wood. *Callitris Ventenatii*
Campion. Various species of *Lychnis* and *Silene*
Berry-bearing. *Cucubalus bacciferus*
Bunge's. *Lychnis Bungeana*
Bladder. *Silene inflata*
Double Red. *Lychnis dioica fl.-pl.*
Moss. *Silene acaulis*
Constantinople. *Lychnis Chalcedonica*
Red. *Lychnis diurna*
Rose. *Lychnis coronaria.* (See also Rose Campion)
Senno. *Lychnis Senno*
Spanish. *Silene Otites*
Starry. *Silene stellata*
White. *Lychnis vespertina*
Cammock, or Cammick. *Ononis arvensis, O. spinosa,* and a few other plants
Camwood. *Baphia nitida*
Canary Creeper. *Tropæolum aduncum*
Canary Grass. *Phalaris canariensis*
Canary Seed. The seed of *Phalaris canariensis*
Canary-wood. *Persea indica* and *P. canariensis*
Cancer Root. *Conopholis (Orobanche) americana* and *Epiphegus virginiana*
One-flowered. *Aphyllon uniflorum*
Cancer-wort. *Linaria spuria* and *L. Elatine;* also an old name for the genus *Veronica*
Canchalagua, of California. The genus *Erythræa*
Candle-berry Myrtle. *Myrica cerifera* and *M. Gale*
Candle-berry Tree. *Aleurites triloba*
Candlemas Bells. *Galanthus nivalis*
Candle Nut. See Nut
Candle Nut Tree. *Virola sebifera*
Candle Plant. *Cacalia coccinea*
Candle Rush. *Juncus communis*
Candle Tree, Panama. *Parmentiera cerifera*
Society Islands. *Aleurites moluccana*
Candle-wick. *Verbascum Thapsus*
Candle-wood, Californian. *Fouquiera splendens*
Jamaica. *Gomphia guianensis*
S. American. *Sciadophyllum capitatum*
White and Black. *Amyris balsamifera*
Can Dock. *Nuphar lutea* and *Nymphæa alba*
Candy Mustard. An old name for *Æthionema saxatile*
Candytuft, Alpine. *Iberis alpina* and *Iberis stylosa*
Bitter. See Candytuft, Wild
Broad-leaved. *Iberis semperflorens*
Buban's. *Iberis Bubani*

English Names of Cultivated, Native,

Candytuft, Common Annual. *Iberis umbellata*
Coris-leaved. *Iberis corifolia*
Dwarf Alpine. *Iberis petræa*
Evergreen. *Iberis sempervirens*
Flax-leaved. *Iberis linifolia*
Garrex's. *Iberis Garrexiana*
Gibraltar. *Iberis gibraltarica*
Glaucous. *Iberis jucunda (Æthionema coridifolium)*
Late White. *Iberis correæfolia (I. coriacea)*
Lebanon. *Æthionema coridifolium (Iberis jucunda)*
Rock. *Iberis saxatilis*
Rosemary-leaved. *Iberis rosmarinifolia*
Tenore's. *Iberis Tenoreana*
Wild, or Bitter. *Iberis amara*
Winged. *Iberis pinnata*
Cane, Dumb. *Caladium (Dieffenbachia) seguinum*
Chair-bottom. *Calamus Rotang, C. rudentum, C. verus,* and *C. riminalis*
Great Rattan. *Calamus rudentum*
Ground Rattan. *Rhapis flabelliformis*
Imphee. *Sorghum saccharatum*
Large American. *Arundinaria macrosperma*
Malacca. *Calamus Scipionum*
Rattan. *Calamus Draco (C. Rotang)*
Small American. *Arundinaria tecta*
Snake. *Kunthia montana*
Sugar. *Saccharum officinarum*
Sugar, Chinese. *Sorghum saccharatum*
Tobago. *Bactris minor*
Whangee, or Wanghee. *Phyllostachys nigra*
Wild. *Arundo occidentalis* and *A. saccharoides*
Cane Apple. *Arbutus Unedo*
Cane Killer. *Alevtra brasiliensis*
Canella Bark Tree. *Canella alba*
Canker, or Canker Rose. *Rosa canina* and *Papaver Rhæas*
Canker Berry. The fruit of *Rosa canina*; also *Solanum bahamense*
Canker-Lettuce. See Lettuce
Canker-weed. *Senecio Jacobæa*
Canker-wort. *Leontodon Taraxacum*
Canna-down, or Cannach. *Eriophorum vaginatum*
Cannon-ball Tree. *Couroupita (Lecythis) guianensis*
Canoe-wood. *Liriodendron tulipiferum*
Canterbury Bells. *Campanula Medium* and *C. Trachelium*
Caoutchouc Tree. (See also India-rubber)
Bornean. *Urceola elastica*
E. Indian. *Ficus elastica*
Panama. *Castilloa elastica* and *C. Markhamiana*
Pará. *Hevea brasiliensis, H. guianensis,* and other species
Madras. *Euphorbia Cattimandoo*
Cap, Friar's, Soldier's, or Turk's. *Aconitum Napellus*
Cape Asparagus. *Aponogeton distachyon*
Cape Gooseberry. *Physalis peruviana*
Cape Gum Tree. *Acacia horrida*
Cape Pond-weed. *Aponogeton distachyon*

Cape Treasure Flower. *Gazania pavonia*
Cape Weed. *Roccella tinctoria*
Australian. *Hypochæris radicata* and *Cryptostemma calendulacea*
New Zealand. *Hypochæris radicata*
Caper Bean. See Bean Caper
Caper Bush, Common. *Capparis spinosa*
Timbuctoo. *Capparis sodada*
Wild. *Euphorbia Lathyris*
Caper Spurge. *Euphorbia Lathyris*
Caper Tree, N. S. Wales. *Busbeckia arborea*
Capillaire. *Adiantum Capillus-Veneris*
Capivi. See Copaiba
Caprifole. *Lonicera Caprifolium*
Caps. Fungi of all kinds
Capsicum, Winter Cherry. *Solanum Pseudo-Capsicum*
Carageen, Carrageen, or Carrigeen, Moss. *Chondrus crispus*
Caramba, or Carambola, Tree. *Averrhoa Carambola*
Carameile. *Lathyrus macrorrhizus*
Carandas. *Carissa Carandas*
Caravaun-beg. *Prunella vulgaris*
Caraway, or Carrawny, Common. *Carum Carui*
Edible-rooted. *Carum Gairdneri* and *C. Kelloggii*
Tuberous-rooted. *Carum (Bunium) Bulbocastanum*
Whorled-leaved. *Carum (Sison) verticillatum*
Wild. *Anthriscus sylvestris* and *Cacalia atriplicifolia*
Carberry. Another name for the Gooseberry
Carbolic-acid Plant. *Andromeda Leschenaulti*
Carcel Oil Plant. *Brassica Napus oleifera*
Cardamom Plant, Bastard. *Alpinia Cardamomum*
Bastard, Siam. *Amomum xanthoides*
Bengal. *Amomum aromaticum*
Java. *Amomum maximum*
Malabar, or Small. *Elettaria (Alpinia) Cardamomum*
Round, or Clustered. *Amomum Cardamomum*
Small. See Cardamom, Malabar
Wild, of S. Africa. *Xanthoxylon capense*
Carde Thistle. *Dipsacus sylvestris*
Cardinal Flower. *Lobelia cardinalis* and *Cleome cardinalis*
Card-leaf Tree. Various species of *Clusia*
Cardoon. *Cynara Cardunculus*
Care. *Pyrus Aucuparia*
Caribbæan Bark Plant. *Exostemma floribundum*
Caricature Plant. *Graptophyllum hortense*
Carl Hemp. *Cannabis sativa*
Carlin Heather. *Erica vinerea*
Carlin Spurs. *Genista anglica*
Carline Thistle. *Carlina vulgaris*
Acanthus-leaved. *Carlina acanthifolia*
Carlock. *Sinapis arvensis*
Carnation, or Carnadine. *Dianthus Caryophyllus* and vars.
Spanish. *Poinciana pulcherrima*
Carnation Grass. *Carex glauca* and *C. panicea*
Carnation Tree. *Cacalia Kleinia (Kleinia neriifolia)*

Carnivorous Plants. A term lately applied to *Dionœa muscipula*, *Darlingtonia californica*, the *Droseras*, and other insect-catching plants, on the supposition that they feed on the insects which they entrap.
Carob Tree. *Ceratonia Siliqua*
Carolina Allspice, Glaucous-leaved. *Calycanthus glauca*
Smooth-leaved. *Calycanthus lœvigata*
Carony Bark Plant. See Angostura Bark Plant
Carpenter-grass. *Prunella vulgaris*
Carpenter's-herb. *Rivina humilis*
Carpenter's-leaf. *Galax aphylla*
Carpet-plant. *Ionopsidion acaule*; also a general term applied to plants of dwarf dense foliage, which are used to form an ornamental turf or "carpet"
Carpet-weed. The genus *Mollugo*
Carrageen. See Carageen
Carraway. See Caraway
Carrigeen. See Carageen
Carrion Flower. *Cyprosmanthus herbaceus*, *Smilax herbacea*, and the genus *Stapelia*
Carrot, Candy, or Cretan. *Athamanta cretensis*
Common. Cultivated varieties of *Daucus Carota*
Cretan. See Carrot, Candy
Deadly, or Stinking. The genus *Thapsia*
Peruvian. *Arracacha esculenta*
Shining-leaved. *Daucus Gingidium*
Spanish. *Daucus Visnaga*
Stinking. See Carrot, Deadly
Tasmanian. *Geranium parviflorum*
Wild. *Daucus Carota*
Carrot-tree. *Monizia edulis*
Caroy Seeds. The seeds of *Carum Carui*
Cascarilla-bark Tree. *Croton Eleuteria* (*C. Cascarilla*)
False Bahama. *Croton lucidum*
Jamaica. *Croton Sloanei*
Cashaw Tree. *Prosopis dulcis* and *Prosopis juliflora*
Cashew-nut Tree. *Anacardium occidentale*
Wild. *Anacardium Rhinocarpus*
Casque-wort. The genus *Galeandra*
Cassareep-plant. See Cassava
Cassava, Cassareep, or Mandioc, plant. *Jatropha* (*Janipha*) *Manihot* (*Manihot utilissima*)
Bitter. *Manihot utilissima*
Sweet. *Manihot Aipi*
Cassava-wood. *Turpinia occidentalis*
Cassena. *Ilex Cassine*
Cassia, Clove. *Dicypellium caryophyllatum*
Horse. *Cathartocarpus marginatus*
Maryland. *Cassia marilandica*
Poet's. *Osyris alba* (*Cassia poetica*)
Purging. *Cassia Fistula*
Cassia-bark Tree. Several species of *Cinnamomum*
Cassioberry-bush. *Viburnum lœvigatum*
Cassidony. *Lavandula Stœchas*
Golden. *Helichrysum Stœchas*
Cast-me-down. A corruption of *Cassidony*
Castor-bean. *Ricinus communis*
Red. *Ricinus sanguineus*
Castor-oil-plant. *Ricinus communis*

Castor-wood. *Magnolia glauca*
Catamaran - wood Tree. *Gyrocarpus asiaticus*
Catananch, Blue. *Catananche cœrulea* and vars.
Lobed-leaved. *Catananche lobata*
Cat-berries. *Ribes Grossularia*
Cat-briar. The genus *Smilax*
Catch-fly. The genera *Silene* and *Lychnis*
Alpine. *Silene alpestris*
Autumn. *Silene Schafta*
Bolander's. *Silene Bolanderi*
Caucasian. *Silene caucasica* (*S. Zawadskii*)
Clammy. *Lychnis viscosa*
Double-flowered Seaside. *Silene maritima fl.-pl.*
Dover. *Silene paradoxa*
Drooping. *Silene pendula*
Elizabeth's. *Silene Elisabethœ*
Four-cleft. *Silene quadrifida*
Four-toothed. *Silene quadridentata*
German. *Lychnis Viscaria*
Giant. *Silene gigantea*
Green-flowered. *Silene viridiflora*
Greig's. *Silene Greigi*
Hooker's. *Silene Hookeri*
Italian. *Silene pendula*
Large Scarlet. *Silene rotundifolia*
Lobel's. *Silene Armeria*
Night-flowering. *Silene noctiflora*
Nottingham. *Silene nutans*
Orchis-flowered. *Silene orchidea*
Pendulous. *Silene pendula*
Pennsylvanian. *Silene pennsylvanica*
Pigmy. *Silene Pumilio* and *Silene tenella*
Requien's. *Silene Requieni*
Round-leaved. *Silene rotundifolia*
Royal. *Silene regia*
Saxifrage. *Silene Saxifraga*
Seaside. *Silene maritima*
Sickle-leaved. *Silene falcata*
Sleepy, or Snapdragon. *Silene antirrhina*
Small-flowered. *Silene anglica*
Snapdragon. See Catchfly, Sleepy
Striate... *Silene conica*
Twisted-petalled. *Silene saxatilis*
Umbel-flowered. *Silene orientalis*
Cat-chop. *Mesembryanthemum felinum*
Catch-weed. *Galium Aparine*
Catechu Tree, Black, Pegu, Cutch, or Terra japonica. *Acacia* (*Mimosa*) *Catechu* and *A. Suma*
Gambier. See Gambier
Pale. See Gambier
Pegu. See Catechu Tree, Black
Catechu Palm Tree. *Areca Catechu*
Caterpillar Fungus. Various species of *Cordiceps*
Caterpillar Plant. The genus *Scorpiurus*
Common. *Scorpiurus vermiculatus*
Furrowed. *Scorpiurus sulcatus*
Hairy. *Scorpiurus subvillosus*
Prickly. *Scorpiurus muricatus*
Cat-gut. *Tephrosia Virginiana*
Cat-in-Clover. *Lotus corniculatus*
Catjang. *Cajanus indicus*
Cat-mint, Cat-nep, or Cat-nip. *Nepeta Cataria* and *Calamintha officinalis*

Cat-mint, Blue-flowered. *Nepeta carulea*
Common. *Nepeta Cataria*
Large-flowered. *Nepeta macrantha*
Malabar. *Anisomeles malabarica*
Scallop-leaved. *Nepeta Mussini*
Small. *Nepeta Nepetella*
Cat-nip, or Cat-nep. See Cat-mint
Cat-nut. *Bunium flexuosum*
Cat-o'-nine-tails. *Typha latifolia*
Cat-tree, or Cat-wood. *Euonymus europæus*
Cat-whin. *Genista anglica, Rosa canina, Rosa spinosissima,* and *Ulex nanus*
Cat's-claw. *Bignonia Unguis-cati, Dolichos filiformis,* and *Inga Unguis-cati.*
Cat's-ear. *Hypochæris radicata* and other species; also applied to *Gnaphalium dioicum* and the genus *Antennaria*
Cat's-foot. *Nepeta Glechoma*
Mountain. *Antennaria dioica rosea*
Cat's-milk. *Euphorbia Helioscopia*
Cat's-paw Creeper. *Bignonia Unguis-cati*
Cat's-tail. The catkin of the Hazel or Willow; also applied to *Equisetums, Hippuris,* and a few other plants
Cat's-tail Grass. *Phleum pratense*
Cat's-tongue. *Apargia serotina*
Cattimandoo, Cattemundoo, or Callemundoo, Gum Plant. *Euphorbia antiquorum* (*E. Cattimandoo*)
Catteridge Tree. *Cornus sanguinea*
Cattle-poison Plant, W. Australian. *Gastrolobium trilobum, G. obovatum,* and *G. spinosum*
Cauliflower, or Broccoli. *Brassica oleracea var. Botrytis cauliflora*
Causeway Grass. *Poa annua*
Cavan Tree. *Acacia Cavenia*
Cavern Moss. See Moss
Caviuna-wood. The wood of *Dalbergia nigra*
Caxes. See Kexes
Cebadilla, or Cevadilla, Seeds. The fruit of *Asagræa officinalis* (*Veratrum officinale*)
Cedar, Australian. *Melia australis*
Barbadoes. *Juniperus barbadensis*
Bastard, Barbadoes. *Cedrela odorata*
Bastard, Guiana. *Icica altissima*
Bastard, Jamaican. *Guazuma tomentosa*
Bastard, N. S. Wales. *Dysoxylon rufum*
Bastard, W. Indian. *Guazuma ulmifolia*
Bermuda. *Juniperus bermudiana*
Brazilian. *Cedrela brasiliensis*
British Columbia. *Thuja gigantea* and *Juniperus Henryana*
Bussaco. *Cupressus lusitanica var. pendula*
Californian. *Thuja gigantea*
Canary Islands. *Juniperus Cedrus*
Chinese. *Cedrela sinensis*
Dominica. *Bignonia Leucoxylon*
Dwarf Red. *Juniperus virginiana var. humilis*
E. Indian. *Cedrela Toona* and *Cedrus Deodara*
False. *Cedrela odorata*
Goa. *Cupressus lusitanica*
Honduras. *Cedrela odorata*

Cedar, Incense. The genus *Libocedrus*
Jamaica or W. Indian. *Cedrela odorata*
Lebanon. *Cedrus Libani*
Mt. Atlas. *Cedrus atlantica*
Oregon White. *Cupressus* (*Chamæcyparis*) *Lawsoniana*
Pencil-wood. *Juniperus bermudiana*
Prickly. *Juniperus Oxycedrus*
Van Dieman's Land. *Cyathodes Oxycedrus*
Port Orford. *Cupressus* (*Chamæcyparis*) *Lawsoniana*
Queensland. *Pentaceras australis*
Red Australian. *Cedrela australis*
Red Californian. *Libocedrus decurrens*
Russian. *Pinus Cembra*
Sharp. *Acacia Oxycedrus* and *Juniperus Oxycedrus*
Silvery. *Cedrus atlantica*
Silvery-leaved Red. *Juniperus virginiana var. argentea*
Singapore. *Cedrela Toona*
Stinking. *Torreya taxifolia*
Virginia Red. *Juniperus virginiana*
Water. The genus *Chamæcyparis*
Weeping Red. *Juniperus virginiana var. pendula*
W. Indian. See Cedar, Jamaica
White. *Cupressus thyoides, Libocedrus decurrens,* and *Thuja occidentalis*
White-wood. *Tecoma Leucoxylon*
Yellow. *Thujopsis borealis*
Cedar-apples. Gall-like tubercles, produced on the bark of *Juniperus virginiana* by the fungus *Podisoma macropus*
Cedar-wood, Guiana. *Icica altissima*
Cedron. The fruit of *Simaba Cedron*
Ceiba Tree, or God Tree. *Eriodendron anfractuosum* (*Bombax Ceiba*)
Celandine, Greater. *Chelidonium majus*
Japanese. *Chelidonium japonicum* (*C. grandiflorum*)
Small. *Ranunculus Ficaria*
Tree. *Bocconia frutescens*
Celeriac, or Turnip-rooted Celery. *Apium graveolens rapaceum*
Three-coloured. *Apium graveolens rapaceum var. tricolor*
Celery, Australian. *Apium prostratum*
Common. *Apium graveolens*
New Zealand. *Apium australe*
Turnip-rooted. See Celeriac
Water. *Ranunculus sceleratus*
Centaury, American. The genus *Sabbatia*
American Sweet. *Cacalia suaveolens*
Australian. *Centaurea australis*
Babylonian. *Centaurea babylonica*
Black. *Centaurea nigra*
Californian. *Centaurea chironoides*
Common. *Erythræa Centaurium*
Corn. *Centaurea Cyanus*
Erect. *Centaurea stricta*
Greater. *Centaurea Scabiosa*
Guiana. *Eracum guianense*
Jersey. *Centaurea aspera*
Large-headed. *Centaurea macrocephala*
Mealy. *Centaurea dealbata*
Mountain. *Centaurea montana*
One-flowered. *Centaurea uniflora*

Centaury, Phrygian. *Centaurea Phrygia*
Star Thistle. *Centaurea Calcitrapa*
Swiss. *Rhaponticum cynaroides*
Yellow-flowered. *Centaurea solstitialis* and *Chlora perfoliata*
Century Plant. *Agave americana*
Queen Victoria's. *Agave Victoriæ-reginæ*
Cereus, Night-flowering. *Cereus grandiflorus*
Cevadilla. See Cebadilla
Chaca. See Choco
Chafe-weed. *Gnaphalium sylvaticum*
Chaff-flower. *Alternanthera Achyrantha*
Chaff-seed. *Schwalbea americana*
Chaff-weed. *Centunculus minimus*
Chalice-flower. *Narcissus Pseudo-Narcissus*
Chamæleon, Black. *Cardopatium corymbosum*
White. *Carlina gummifera*
Chamiso of California. The genus *Adenostoma*
Champaca-tree. *Michelia Champaca*
Champignon, Fairy-ring. See Mushroom, Fairy-ring
Champillion. *Agaricus arvensis*
Champ-wood. The timber of *Michelia Champaca* and *M. excelsa*
Chandelier-flower. *Brunsvigia (Amaryllis) Josephina*
Chandelier-tree. *Pandanus Candelabrum*
Chanar, or Chanal, tree. *Gourliea decorticans*
Chantarelle. *Cantharellus cibarius*
Chaplet-flower, Madagascar. *Stephanotis floribunda*
Charas, or Churrus, plant. *Cannabis sativa*
Chards. The blanched stalks of Artichokes, Leaf-Beet, &c.
Charity. *Polemonium cæruleum*
Charles's-sceptre, King. *Pedicularis Sceptrum Carolinum*
Charlock. *Sinapis arvensis* and *S. alba* Jointed, or White. *Raphanus Raphanistrum*
Chaste-Tree. *Vitex Agnus-castus*
Chats (Catkins). "Keys" of the Ash and Maple; also cones of Fir-trees
Chaulmugra-seed-plant. *Gynocardia odorata*
Chaw-stick, St. Domingo. *Gouania domingensis*
Chayota-plant. *Sechium edule*
Chay-root, or Che-root. *Oldenlandia (Hedyotis) umbellata*
Cheat, Cheats, or Chess. *Bromus secalinus* and *Lolium temulentum*
American. *Bromus Kalmii*
Cheddar Pink. *Dianthus cæsius*
Cheese-rennet. *Galium verum*
Cheese-room. *Agaricus arvensis*
Cheeses, or Cheese-cakes. The fruit of *Malva sylvestris*
Chequer-berry. *Mitchella repens* and *Gaultheria procumbens*
Chequered Daffodil, Chequered Lily, or Chequered Tulip. *Fritillaria Meleagris*
Chequer-tree. *Pyrus torminalis*
Cherimoyer-fruit. The fruit of *Anona Cherimolia*
Che-root. See Chay-root
Cherries, Dog's. The fruit of the Woodbine
Cherry. *Prunus Cerasus*
American Bird. *Prunus pennsylvanica*

Cherry, American Wild Black. *Prunus serotina*
American Wild Red. *Prunus pennsylvanica*
Australian. *Exocarpus cupressiformis*
Australian Brush. *Trochocarpa laurina*
Barbadoes. *Malpighia glabra, M. punicifolia*, and *Eugenia uniflora*
Bastard. *Cerasus Pseudo-Cerasus*
Bastard, of the W. Indies. *Ehretia tinifolia*
Beech, or Brush. *Trochocarpa laurina*
Bigarreau. *Cerasus duracina var. cordigera*
Bird. *Prunus Padus*
Black. *Atropa Belladonna*
Black Choke. *Cerasus hyemalis*
Black-fruited. *Cerasus nigra*
Broad-leaved, of the W. Indies. *Cordia macrophylla*
Brush. See Cherry, Beech
Californian Wild. *Prunus demissa*
Canadian. *Cerasus pumila*
Capollin Bird. *Cerasus Capollin*
Cayenne. *Eugenia Michelii*
Choke. *Prunus virginiana*
Clammy. *Cordia Collococca*
Common. *Cerasus vulgaris (Prunus Cerasus)*
Cornelian. *Cornus mascula*
Cow-itch. *Malpighia urens*
Cultivated. Varieties, pure and hybrid, raised from *Cerasus caproniana, C. Juliana*, and *C. duracina*
Double-flowered. *Cerasus sylvestris fl.-pl.* and *C. vulgaris fl.-pl.*
Downy. *Cerasus pubescens*
Dwarf. *Cerasus (Prunus) pumila*
Ever-flowering. *Cerasus semperflorens*
Evergreen. *Cerasus caroliniana*
False. *Cerasus Pseudo-Cerasus*
Flemish. A variety of *Cerasus vulgaris*
Gean. *Cerasus Juliana*
Ground. *Cerasus Chamæcerasus* and the genus *Physalis*
Hautbois. *Cerasus caproniana*
Helmet-fruited. *Cerasus Juliana var. Beaumiana*
Horned. *Cerasus cornuta*
Hottentot. *Cassine Maurocenia*
Jamaica. *Ficus pedunculata*
Japan. *Cerasus (Prunus) japonica*
Jerusalem. *Solanum Pseudo-capsicum*
Kentish. A variety of *Cerasus vulgaris*
Mahaleb. *Cerasus Mahaleb*
May Duke. A variety of *Cerasus vulgaris*
Native Australian. *Exocarpus cupressiformis*
Naughty Man's. *Atropa Belladonna*
Nepaul Bird. *Cerasus nepalensis*
N. S. Wales. *Nelitris ingens*
Northern Choke. *Cerasus borealis*
Peach-leaved. *Cerasus persicæfolia*
Perfumed. *Cerasus Mahaleb*
Pigmy. *Cerasus pygmæa*
Red Cornish. *Cerasus Padus rubra*
Red Winter. *Physalis Alkekengi*
Round-fruited. *Cerasus sphærocarpa*
St. Julian's. *Cerasus Juliana*
Sand. *Cerasus (Prunus) pumila*
Saw-leaved. *Cerasus serrulata*
Somniferous Winter. *Physalis somnifera*
Surinam. A species of *Eugenia*

Cherry, Sweet-scented Ground. *Cerasus Chamæcerasus fragrantissima*
W. Indian. *Cerasus occidentalis* and various species of *Malpighia* and *Bunchosia*
White-heart. *Cerasus duracina*
Wild. *Prunus Avium*
Willow-leaved. *Cerasus salicina*
Winter. *Physalis Alkekengi* and various species of *Solanum*
Wooden. The genus *Hakea*
Cherry-Bay. Another name for Portugal Laurel
Teneriffe. *Cerasus Hixa*
Cherry-Crab. *Pyrus Malus baccata*
Cherry-Pepper. *Capsicum cerasiforme*
Cherry-pie-flower. *Heliotropium peruvianum* and *Epilobium hirsutum*
Chervil. *Anthriscus Cerefolium*
Broad Tooth-pick. *Tordylium syriacum*
Bur. *Anthriscus vulgaris*
Great. *Myrrhis odorata*
Mock. *Anthriscus sylvestris*
Parsnip. *Chærophyllum bulbosum (Anthriscus bulbosus)*
Spanish Tooth-pick. *Ammi Visnaga*
Tuberous-rooted, or Turnip-rooted. *Chærophyllum bulbosum*
Wild. *Anthriscus sylvestris* and *Scandix Pecten*
Chess. See Cheat
Chess-apple. The fruit of *Pyrus Aria*
Chestnut, American. *Castanea vesca* var. *americana*
Antilles. *Cupania americana*
Cape. *Brabejum stellatifolium*
Double-flowered Horse. *Æsculus Hippocastanum fl.-pl.*
Double-flowered Red Horse. *Æsculus Hippocastanum rubra fl.-pl.*
Earth. *Bunium flexuosum* and *Conopodium denudatum*
Fern-leaved. *Castanea vesca aspleniifolia*
Fiji. *Inocarpus edulis*
Flesh-coloured-flowered Horse. *Æsculus carnea*
Golden-leaved. *Castanea chrysophylla*
Guiana. *Carolinea princeps*
Horse. *Æsculus Hippocastanum*
Kaffir. *Brabejum stellatifolium*
Large-spiked Horse. *Pavia macrostachya*
Moreton Bay. *Castanospermum australe*
Otaheite. *Inocarpus edulis*
Pale-flowered Horse. *Æsculus pallida*
Red-flowered Horse. *Æsculus rubra* or *rubicunda*
Smooth-fruited Horse. The genus *Pavia*
Spanish, or Sweet. *Castanea vesca*
Variegated. *Castanea vesca variegata*
Water. *Trapa natans*
Yellow. *Quercus Castanea*
Yellow-flowered Horse. *Æsculus (Pavia) flava*
Chestnut-plant, Seaside. *Entada Gigalobium*
Chia-plant of California. *Salvia columbaria*
Chicalote of California. *Argemone hispida*
Chica-plant. *Bignonia Chica*
Chicory. *Cichorium Intybus*

Chicory, Large-rooted Brussels, or "Witloof." A variety of *Cichorium Intybus*
Chicot. *Gymnocladus canadensis*
Chick-weed. Black Winter-green, American. *Trientalis americana, Stellaria media*
Chinese. *Claytonia sibirica*
Forked. *Anychia (Queria) dichotoma*
Golden. *Stellaria graminea aureo-variegata*
Great American. *Stellaria pubera*
Indian. *Mollugo verticillata*
Jagged. *Holosteum umbellatum*
Mouse-ear. The genus *Cerastium,* especially *C. triviale.* (See also Mouse-ear-Chickweed)
Purple. *Arenaria rubra*
Sea. *Arenaria peploides*
Silver. *Paronychia argyrocoma*
Water. *Montia fontana*
W. Indian. *Drymaria cordata*
Winter-green. *Trientalis europæa*
Chillies. See Pepper, Cayenne
Chimney-plant. *Campanula pyramidalis*
Chimney-sweeps. *Plantago lanceolata* and *Luzula campestris*
China-Aster. See Aster
China-grass. *Bæhmeria (Urtica) nivea*
China-root-plant. *Smilax China (S. ferox)*
W. Indian. *Cissus sicyoides*
Chinese-grass. See China-grass
Chink-wort, or Scripture-wort. The genus *Opegrapha*
Chinquapin. *Castanea pumila*
Water. *Nelumbium luteum*
Western. *Castanopsis chrysophylla*
Chip-tree. *Thrinax argentea*
Chirayta, or Chiretta. *Ophelia Chirata (Gentiana Chirayita)*
Chittagong-wood. The timber of *Cedrela Toona, Chickrassia tabularis,* and some other trees
Chives. *Allium Schœnoprasum*
Choco, Choko, or Chaca, plant. *Sechium edule*
Chocolate, Indian. *Geum rivale*
Chocolate-nut-tree. *Theobroma Cacao*
Chocolate-root. *Geum canadense*
Choke-berry. *Aronia (Pyrus) arbutifolia*
Choke-dog. *Gonolobus obliquus*
False. *Gonolobus carolinensis*
Choke-pear. *Pyrus communis*
Choko-plant. See Choco
Chola-plant. *Cicer arietinum*
Choop-tree. *Rosa canina*
Chop-nut. See Calabar Bean
Chowlee-plant. *Vigna (Dolichos) sinensis*
Chriseis of Californian. *Eschscholtzia californica*
Christmas. *Ilex Aquifolium*
Christmas Gambol. *Ipomæa sidæfolia*
Christmas Pride. *Ruellia paniculata*
Christmas Rose, Common. *Helleborus niger*
Great. *Helleborus altifolius (H. niger maximus)*
Green-flowered. *Helleborus abchasicus*
Large-flowered. *Helleborus altifolius (H. niger maximus)*
Madagascar. The genus *Hydnora*

Christmas Rose, Narrow - leaved. *Helleborus angustifolius*
Plum-coloured. *Helleborus colchicus*
Purple. *Helleborus purpurascens*
Purplish - red - flowered. *Helleborus atrorubens*
Rose-coloured. *Helleborus orientalis*
Syrian. *Helleborus vesicarius*
Christmas Tree, Australian. *Ceratopetalum gummiferum*
Tasmanian. *Bursaria spinosa*
Christopher, Herb. *Actæa spicata* and *Osmunda regalis*
Stinking. *Scrophularia aquatica* and *S. nodosa*
Christ's-Eye. *Inula Oculus-Christi*
Christ's-Hair. *Scolopendrium vulgare*
Christ's Thorn. *Cratægus Pyracantha* and *Paliurus aculeatus*
Christ's-wort. *Helleborus niger*
Chrysanthemum, Alpine. *Chrysanthemum alpinum*
Common Garden. Varieties of *Chrysanthemum sinense*
Feverfew. *Chrysanthemum Parthenium*
Mountain. *Chrysanthemum montanum*
Northern. *Chrysanthemum arcticum*
Old Garden. *Chrysanthemum coronarium*, with its varieties *album* and *aureum*
Scarlet. *Chrysanthemum atro-coccineum*
Scentless. *Chrysanthemum inodorum*
Sicilian. *Chrysanthemum coronarium*
Tansy-leaved. *C. millefoliatum*
Three - coloured. *Chrysanthemum tricolor* and *C. carinatum*
Yellow-and-crimson. *Chrysanthemum Burridgianum*
Chufa. *Cyperus esculentus*
Chulan-tree. *Chloranthus inconspicuus* and *Aglaia odorata*
Champaka-tree. *Michelia Champaca*
Chupa-Chupa-tree. See Sapote
Church-brooms. *Dipsacus sylvestris*
Church-wort. An old name for Pennyroyal
Churnstaff. *Euphorbia Helioscopia*
Ciboul. *Allium fistulosum*
Perennial. *Allium lusitanicum*
Cicely, Fool's. *Æthusa Cynapium*
Sweet. *Myrrhis odorata*
Sweet, of N. America. The genus *Osmorrhiza*
Wild. *Anthriscus sylvestris*
Cich, or Ciches. An old name for Chick-pea
Ciderage. An old name for *Polygonum Hydropiper*
Cider-tree, Tasmanian. *Eucalyptus Gunni*
Cigar-box-wood. The wood of *Cedrela odorata* and *C. sinensis*
Cinchona-bark-plant. See Bark-plant, Peruvian
Yellow. See Bark-plant. *Calisaya*
Cineraria, Alpine. *Cineraria alpina*
Great-leaved. *Cineraria (Ligularia) macrophylla*
Silvery-leaved. *Cineraria acanthifolia (C. maritima)*
Cinnamon-tree. *Cinnamomum Zeylanicum*
Bark, Winter's. *Drimys Winteri*

Cinnamon-tree, Bastard. *Laurus Cassia*
Black. *Pimenta acris*
Chinese. *Cinnamomum Cassia*
Isle of France. *Oreodaphne cupularis*
Mountain. *Cinnamodendron corticosum*
Santa Fé. *Nectandra cinnamomoides*
Wild. *Canella alba* and *Myrcia acris*
Cinnamon-root. *Inula Conyza*
Cinnamon-Sedge. *Acorus Calamus*
Cinquefoil, Alpine. *Potentilla alpestris*
Brilliant. *Potentilla splendens*
Calabrian. *Potentilla calabrica*
Clusius's. *Potentilla Clusiana*
Colorado Silvery. *Potentilla Hippiana*
Common. *Potentilla reptans*
Creeping. *Potentilla reptans*
Dark Crimson. *Potentilla atrosanguinea*
Dwarf. *Potentilla nana*
Dwarfest. *Potentilla minima*
Garden varieties of. *Potentilla hybrida*
Goose-grass. *Potentilla anserina*
Hoary. *Potentilla argentea*
Lady's Mantle. *Potentilla alchemilloides*
Marsh. *Comarum palustre*
Norwegian. *Potentilla norvegica*
Pyrenean. *Potentilla pyrenaica*
Rock. *Potentilla rupestris*
Shining. *Potentilla nitida*
Showy. *Potentilla speciosa*
Shrubby. *Potentilla fruticosa*
Silvery. *Potentilla argentea*
Snowy. *Potentilla nivalis*
Spring. *Potentilla verna*
Three-toothed-leaved. *Potentilla tridentata*
Tormentil. *Potentilla Tormentilla*
Tree. *Potentilla fruticosa*
Tufted. *Potentilla cæspitosa*
White-flowered. *Potentilla alba*
Cipper-nut. *Bunium flexuosum*
Cistus, Bog. *Cistus ladaniferus*
Common Gum. *Cistus cyprius*
Ladanum Gum. *Cistus ladaniferus*
Citron. *Citrus medica* var. *Cedra (C. acida)*
Fingered. *Citrus sarcodactylis*
Citronella-Oil-plant. *Andropogon Nardus*
City Avens. *Geum urbanum*
Clary. *Salvia Sclarea*
Meadow. *Salvia pratensis*
Silvery. *Salvia argentea*
Wild. *Heliotropium indicum* and *Salvia Verbenacea*
Clay-weed. *Tussilago Farfara*
Clematis, Alpine. *Atragene (Clematis) alpina*
American. *Atragene (Clematis) americana*
Austrian. *Clematis alpina (Atragene austriaca)*
Bell-flowered. *Clematis campanulata*
Biting. *Clematis Vitalba*
Chinese. *Clematis chinensis*
Curled-sepaled. *Clematis crispa*
Cylindrical-flowered. *Clematis cylindrica*
David's. *Clematis Davidiana*
Entire-leaved. *Clematis integrifolia*
Erect. *Clematis erecta*
Evergreen. *Clematis calycina (C. balearica)* and *C. cirrhosa*

Clematis, Fool's-Parsley-leaved. *Clematis æthusæfolia*
Glaucous-leaved. *Clematis glauca*
Hardy Purple-flowered. *Clematis Viticella var. venosa*
Jackman's. *Clematis Jackmanni* and vars.
Large-flowered. *Clematis florida*
Leathery-flowered. *Clematis Viorna*
Minorca. *Clematis balearica*
Mountain. *Clematis montana*
Net-veined-leaved. *Clematis reticulata*
New Zealand. *Clematis indivisa lobata*
Ochotsk. *Atragene (Clematis) ochotensis*
Open-flowered. *Clematis patens*
Oriental. *Clematis orientalis*
Oval-leaved. *Clematis ovata*
Panicled. *Clematis paniculata*
Scarlet-flowered. *Clematis coccinea*
Siberian. *Atragene (Clematis) sibirica*
Sims's. *Clematis Simsii*
Sweet-scented. *Clematis Flammula*
Tube-flowered. *Clematis tubulosa*
Veined Vine-bower. *Clematis Viticella venosa*
Vine-bower. *Clematis Viticella*
Virginian. *Clematis virginiana*
Western. *Atragene (Clematis) occidentalis*
Winter-flowering. *Clematis calycina (C. balearica)*
Woolly. *Clematis lanuginosa*
Clear-weed. *Pilea pumila*
Cleats. *Petasites vulgaris* and *Tussilago Farfara*
Cleavers. *Galium Aparine*
Small. *Galium tinctorium*
Cliff-pink, or Cleve-pink. *Dianthus cæsius*
Cliff-rose. *Armeria maritima*
Mexican. *Cowania mexicana*
Climber, Sportsman's. *Cissus venatorum*
Cling-stone. A term applied to some varieties of Peaches and Nectarines, the flesh of which adheres to the stone. (See "Freestone")
Clite, or Clithe. An old name for the Burdock
Cloak Fern. See Fern
Clock, Shepherd's. *Anagallis arvensis* and *Tragopogon pratensis*
Clod-weed. *Filago germanica* and *Scabiosa arvensis*
Clog-weed. *Heracleum Sphondylium*
Close Sciences. An old name for *Hesperis matronalis*
Clot, or Clote. *Nuphar lutea*
Clot-bur, or Clod-bur. *Arctium Lappa*
Spiny. *Xanthium spinosum*
Cloud-berry. *Rubus Chamæmorus*
Cloud-grass. See Grass
Clove Gillyflower, or Clove Pink. *Dianthus Caryophyllus*
Clove-scented Creeper. *Lettsomia Bonanox*
Clove-strip. *Jussiæa repens*
Clove-tongue. An old name for *Helleborus niger*
Clove-tree. *Caryophyllus aromaticus*
Wild. *Eugenia (Pimenta) acris*
Clover. The genus *Trifolium*

Clover, Alsike. *Trifolium hybridum*
American Bush. The genus *Lespedeza*
Bastard. *Trifolium hybridum*
Bersin. *Trifolium Alexandrinum*
Bird's-foot. *Lotus corniculatus*
Bladder-podded. *Trifolium spumosum*
Boccone's. *Trifolium Bocconi*
Bokhara, or Cabul. *Melilotus alba*
Brown. *Trifolium spadiceum*
Buffalo. *Trifolium reflexum* and *T. pennsylvanicum*
Cabul. See Clover, Bokhara
Canadian Bush. *Hedysarum canadense*
Calvary. *Medicago Echinus*
Carolina. *Trifolium Carolinianum*
Clustered. *Trifolium glomeratum*
Common. *Trifolium pratense* and *T. repens*
Crimson. *Trifolium incarnatum*
Dutch. *Trifolium repens*
Egyptian. *Trifolium Alexandrinum*
Golden. *Trifolium agrarium*
Hare's-foot. *Trifolium arvense*
Hart's, King's, or Plaster. *Melilotus officinalis*
Hop. *Trifolium procumbens*
Japanese. *Lespedeza striata*
King's. See Clover, Hart's
Knotted. *Trifolium striatum*
Lesser. *Trifolium procumbens*
Maltese. *Hedysarum coronarium*
Mayad. *Trifolium subrotundum*
Oval-headed. *Trifolium alpestre*
Pin. See Pin-clover
Plaster. See Clover, Hart's
Prairie. The genus *Petalostemon*
Purple. *Trifolium pratense*
Red. *Trifolium pratense*
Reversed. *Trifolium resupinatum*
Rough. *Trifolium scabrum*
Running Buffalo. *Trifolium stoloniferum*
St. Mawe's. *Medicago maculata*
Sand. *Trifolium suffocatum*
Sea-egg. *Medicago Echinus*
Sea-side. *Trifolium maritimum*
Slender. *Trifolium filiforme*
Snail. The genus *Medicago*
Soola. *Hedysarum coronarium*
Starry. *Trifolium stellatum*
Stone. *Trifolium arvense*
Strawberry-headed. *Trifolium fragiferum*
Striped-flowered. *Trifolium involucratum*
Subterranean. *Trifolium subterraneum*
Sulphur. *Trifolium ochroleucum*
Sweet. The genus *Melilotus*
Treacle. *Psoralea bituminosa*
Upright. *Trifolium strictum*
White. *Trifolium repens*
Winter. *Mitchella repens*
Yellow. *Medicago lupulina*, *Trifolium procumbens*, and *T. minus*
Yellow Suckling. *Trifolium filiforme*
Zig-zag. *Trifolium medium*
Clover-Trefoil. *Trifolium medium*
Clown's All-heal, or Wound-wort. *Stachys palustris*
Clown's Lungwort. *Verbascum Thapsus*
Clown's Mustard. *Iberis amara*
Clown's Treacle. *Allium sativum*

Club, Shepherd's. *Verbascum Thapsus*
Club-grass. The genus *Scirpus*
Club-moss. The genera *Lycopodium* and *Selaginella*
Alpine. *Lycopodium alpinum*
American Dwarf. *Selaginella apus* and *S. rupestris*
Common. *Lycopodium clavatum*
Creeping. *Selaginella apus*
Fir. *Lycopodium Selago*
Lesser. *Lycopodium selaginoides*
Marsh. *Lycopodium inundatum*
Savin-leaved. *Lycopodium alpinum*
Shining. *Lycopodium lucidulum*
Tree. *Selaginella cæsia arborea*
Club-rush. The genus *Scirpus*; also applied to *Typha latifolia*
Bristly. *Scirpus setaceus*
Clustered. *Scirpus Holoschœnus*
Creeping. *Scirpus palustris*
Few-flowered. *Scirpus pauciflorus*
Floating. *Scirpus fluitans*
Many-stalked. *Scirpus multicaulis*
Needle. *Scirpus acicularis*
Or Bul-rush, Lake. *Scirpus lacustris*
Savi's. *Scirpus Savii*
Sea-side. *Scirpus maritimus*
Sharp. *Scirpus pungens*
Triangular-stemmed. *Scirpus triqueter*
Tufted. *Scirpus cæspitosus*
Wood. *Scirpus sylvaticus*
Cnout-berry. *Rubus Chamæmorus*
Coakum. *Phytolacca decandra*
Cob-nut. *Corylus Avellana* var. *grandis*
Jamaica. The genus *Omphalea*
Coca, Mexican. *Richardsonia scabra*
Coca - leaf - tree, Peruvian. *Erythroxylon Coca*
Cocculus - indicus - plant. *Anamirta (Menispermum) Cocculus*
Cochineal Cactus. *Opuntia cochinillifera*
Cockle, or Corn Cockle. *Lychnis Githago*
Cockle-bur, or Clot-bur. *Agrimonia Eupatoria* and the genus *Xanthium*
Cock-grass. *Bromus mollis* and *B. secalinus*
Cock-rose. *Papaver Rhœas*
Cocks. *Plantago lanceolata*
Cock's-comb. *Celosia cristata*. Applied also to *Rhinanthus Crista-galli*, and locally to a few other native plants in Scotland
Cock's-foot. An old name for *Chelidonium majus* and also for *Columbine*
Cock's-foot Grass. *Dactylis glomerata*
Cock's-head. *Onobrychis Caput-galli*
W. Indies. *Desmodium tortuosum*
Cock's-spur, of the W. Indies. *Pisonia aculeata*
Cock-spur Thorn. See Thorn
Coco. See Cocoa-root
Cocoa-wood. *Inga vera*
Cocoa Nibs. The beans of *Theobroma Cacao* roasted and split or broken
Cocoa-nut, Buddha's. *Sterculia alata*
Cocoa-nut Palm. *Cocos nucifera*
Double, or Sea. *Lodoicea Seychellarum*
Small Prickly. *Cocos guineensis*
Cocoa-plum. *Chrysobalanus Icaco*
Cocoa-root, or Coco. The root of *Colocasia antiquorum*

Cocoa-tree. *Theobroma Cacao*
Cocoon. See Cacoon
Cocus-wood. *Brya Ebenus*
Codlings-and-Cream. *Epilobium hirsutum*
Coffee, Date. Roasted Date-stones
Swedish. The seeds of *Astragalus bœticus*
Coffee-Blight-Fungus. See Fungus
Coffee-Climber. *Periploca Mauritiana*
Coffee-tree, Arabian. *Coffea arabica*
Californian. *Rhamnus californica*
Kentucky. *Gymnocladus canadensis*
Liberian. *Coffea liberica*
Peruvian. *Coffea racemosa*
Wild, of the W. Indies. *Faramea odoratissima, Zuelania latioides*, and *Eugenia disticha*
Coker-nut. See Nut
Cogwood - tree. *Hernandia sonora* and *Laurus Chloroxylon*
Cohosh. *Actæa racemosa*
Black. *Cimicifuga racemosa*
Blue. *Caulophyllum thalictroides*
White. *Actæa alba*
Cole. *Brassica Napus*
Colesat. See Cole-seed
Cole-seed, or Colesat. *Brassica Napus oleifera*
Cole-wort. *Brassica oleracea*
Heart-leaved. *Crambe cordifolia*
Rushy. *Crambe juncea*
Sea. *Crambe maritima*
Shrubby. *Crambe fruticosa*
Colic-root. *Aletris farinosa* and *Dioscorea villosa*
Colic-weed. *Corydalis glauca*
Climbing. *Adlumia cirrhosa*
Collar Moss. See Moss
Collins's-flower, Large-flowered. *Collinsia grandiflora*
Many-coloured. *Collinsia multicolor*
Two-coloured. *Collinsia bicolor*
Various-leaved. *Collinsia heterophylla*
Collinson's-flower. *Collinsonia canadensis*
Colocynth. See Apple, Bitter
Himalyan. *Citrullus (Cucumis) Pseudocolocynthis*
Colombo Root. See Calumba Root
American. *Frasera verticillata*
Coloquintida. An old name for *Colocynth*
Colpoon-tree. *Cassine Colpoon*
Colt's-foot, Alpine. *Tussilago alpina*
Arrow-leaved. *Nardosma (Tussilago) sagittata*
Common. *Tussilago Farfara*
Fragrant. *Tussilago (Nardosma) fragrans*
N. American. *Nardosma palmata*
Spotted. *Farfugium grande*
Variegated. *Tussilago Farfara variegata*
W. Indian. The genus *Pothomorpha*
Colt's-tail. *Erigeron canadensis*
Columbine, Alpine. *Aquilegia alpina*
Altaian. *Aquilegia glandulosa*
Burger's. *Aquilegia Burgeriana*
Californian. *Aquilegia californica*
Canadian. *Aquilegia canadensis*
Common. *Aquilegia vulgaris*
Feathered or Tufted. *Thalictrum aquilegifolium*

Columbine, Golden-flowered. *Aquilegia chrysantha*
Green-flowered. *Aquilegia viridiflora*
Large Orange. *Aquilegia truncata* (*A. californica, A. eximia*)
Long-bracted. *Aquilegia longibracteata*
Longest-spurred. *Aquilegia longissima*
Mauve-coloured. *Aquilegia grata*
Orange-and-yellow. *Aquilegia canadensis*
Pyrenean. *Aquilegia pyrenaica*
Rocky Mountain. *Aquilegia cœrulea*
Scented. *Aquilegia fragrans*
Skinner's. *Aquilegia Skinneri*
White-flowered. *Aquilegia vulgaris alba*
Yellow Long-spurred. *Aquilegia leptoceras lutea*
Columbo, American. The genus *Frasera*
Colza, or Coltza. *Brassica Napus oleifera*
Comb, Lady's, Shepherd's, or Venus's. *Scandix Pecten-Veneris*
Comb-Fern. See Fern
Comb Finger-grass. See Grass
Comet-plant. *Cometes alternifolia*
Comfrey, Bohemian. *Symphytum bohemicum*
Caucasian. *Symphytum caucasicum*
Common. *Symphytum officinale*
Crimean. *Symphytum tauricum*
Forage. See Comfrey, Prickly
Iberian. *Symphytum ibericum*
Prickly, or Forage. *Symphytum asperrimum* (*S. peregrinum*)
Tuberous-rooted. *Symphytum tuberosum*
Wild, American. *Cynoglossum virginicum*
Compass-plant. *Silphium laciniatum*
Cone-flower, Californian. *Rudbeckia californica*
Drummond's. *Rudbeckia Drummondi* (*R. columnaris, Obeliscaria pulcherrima*)
Cut-leaved. *Rudbeckia laciniata*
Glowing. *Rudbeckia fulgida*
Hairy. *Rudbeckia hirta*
Large. *Rudbeckia maxima*
Newman's. *Rudbeckia Newmani*
Purple. The genus *Echinacea*
Shining. *Rudbeckia nitida*
Showy. *Rudbeckia speciosa*
Three-lobed. *Rudbeckia triloba*
Virginian. *Rudbeckia virginiana*
Yellow. The genus *Rudbeckia*
Cone-head. The genus *Strobilanthes*
Conessi-bark-tree. *Wrightia antidysenterica*
Connemon, of Japan. *Cucumis Conomon*
Consound, Comfrey. *Symphytum officinale*
King's. *Delphinium Consolida*
Middle. *Ajuga reptans*
Saracen's. *Senecio Saracenicus*
Contrayerva-root. The root of *Dorstenia Contrayerva*
Convolvulus, Trellis. *Ipomœa tuberosa*. (See also Bind-weed)
Cooch-grass. See Couch-grass
Cool-tankard. *Borago officinalis*
Cool-weed. *Urtica pumila*
Cooper's-wood. *Alphitonia excelsa* and *Pomaderris apetala*
Copaiba, or Copaiva, Balsam-plant. Several species of *Copaifera*
Copai-ye-wood. *Vochysia guianensis*

Copal-gum-tree, Brazilian. *Hymenœa martiana*
E. African. *Trachylobium Hornemannianum*
Sierra Leone. *Copaifera Gibourtiana* (*Gibourtia copallifera*)
Copalchi-bark-plant. *Croton niveum*
Brazilian. *Strychnos Pseudo-quina*
Mexican. *Croton Pseudo-china*
Copra. The commercial name for Cocoa-nut-oil
Cop-rose, or Copper-rose. *Papaver Rhœas*
Copse Laurel. *Daphne Laureola*
Coptis Root. *Coptis Teeta*
Coquilla Nuts. See Nuts
Coral, Garden. *Rochea* (*Crassula*) *coccinea*
Coral-bead-plant. *Abrus precatorius*
Coral-berry. *Symphoricarpus vulgaris*
Coral Creeper. *Kennedya prostrata*
Coral Pea-tree. *Adenanthera pavonina*
Coral-plant. *Jatropha multifida*
Coral-root. *Dentaria bulbifera* and the genus *Corallorrhiza*
Spurless. *Corallorrhiza innata*
Coral-teeth. *Corallorrhiza odontorrhiza*
Coral-tree, East Indian. *Erythrina indica*
West Indian. *Erythrina Corallodendron*
Cord-grass. *Spartina stricta*
Cord-moss. See Moss
Coriander, Common. *Coriandrum sativum*
Cork. *Lecanora tartarea* and *Rocella tinctoria*
Cork-tree, Common. *Quercus Suber*
E. Indian. *Adansonia digitata* and *Bignonia suberosa* (*Millingtonia hortensis*)
New Zealand. *Entelea arborescens*
Siberian. *Phellodendron amurense*
Cork-wood. *Hibiscus tiliaceus*
Marsh. *Anona palustris*
N. S. Wales. *Duboisia myoporoides*
W. Indian. *Ochroma Lagopus*
Corn, Amel. *Triticum amyleum*
Chinese. *Setaria italica*
Gero. *Penicillaria spicata*
Goose. *Bromus mollis*
Kaffir. *Sorghum saccharatum*
Marsh. *Potentilla anserina*
Pharaoh's. *Triticum compositum*
Pop. A small-grained kind of Maize, used for parching
St. Peter's. *Triticum monococum*
Corn-bells. *Nidularia campanulata*
Corn-berries. *Vaccinium Oxycoccos*
Corn-bind. *Convolvulus arvensis* and *Polygonum Convolvulus*
Corn-bottle. *Centaurea Cyanus*
Corn-Flag, African. *Antholyza* (*Gladiolus*) *æthiopica*
European. *Gladiolus segetum*
Corn-flower, Blue. *Centaurea Cyanus*
Golden or Yellow. *Chrysanthemum segetum*
Red. *Lychnis Githago* and *Papaver Rhœas*
Corn-leaves. *Cotyledon Umbilicus*
Corn-Lily. *Convolvulus arvensis* and *C. sepium*
Corn-Marigold. *Chrysanthemum segetum*
Corn-Mint. *Mentha arvensis*
Corn-Mustard. *Sinapis arvensis*
Corn-Pink. *Lychnis Githago*
Corn-Poppy. *Papaver Rhœas*

Corn-Rose. *Papaver Rhœas*
Corn-salad. *Valerianella olitoria*
Italian. *Valerianella eriocarpa*
Keeled. *Valerianella carinata*
Narrow-fruited. *Valerianella dentata*
Sharp-fruited. *Valerianella Auricula*
Corn-Thistle. *Carduus arvensis*
Corn-Violet. *Campanula hybrida*
Corn-wood. The wood of *Pterocarpus erinaceus* (*P. echinatus*)
Cornel. *Cornus sanguinea*
Dwarf. *Cornus suecica*
Cornish Heath. *Erica vagans*
Cornish Moneywort. *Sibthorpia europæa*
Coromandel - wood. See Calamander-wood.
Coronation. *Dianthus Caryophyllus*
Corpse-plant. *Monotropa uniflora*
Cosmos, Feathery. *Cosmos bipinnatus*
Cost, or Costmary. *Tanacetum Balsamita*
Costus, of the Ancients. The roots of *Aplotaxis auriculata*
Cotoneaster, Allied. *Cotoneaster affinis*
Black-fruited. *Cotoneaster vulgaris melanocarpa*
Box-leaved. *Cotoneaster buxifolia*
Common. *Cotoneaster vulgaris*
Loose-flowered. *Cotoneaster laxiflora*
Money-wort-leaved. *Cotoneaster Nummularia*
Mountain. *Cotoneaster frigida*
Pointed-leaved. *Cotoneaster acuminata*
Red-fruited. *Cotoneaster vulgaris erythrocarpa*
Round-leaved. *Cotoneaster rotundifolia*
Small-leaved. *Cotoneaster microphylla*
Woolly. *Cotoneaster tomentosa*
Cotton, French. See French Cotton
Cotton-plant, American. *Gossypium herbaceum*
Australian Wild. *Gomphocarpus fruticosus*
Bahia. A variety of *Gossypium barbadense*
Barbadoes. *Gossypium barbadense*
Cape. *Gomphocarpus fruticosus*
Indian. *Gossypium indicum*
Kidney. *Gossypium peruvianum*
Nankin. *Gossypium religiosum*
Natal Wild. *Ipomœa Gerrardi*
New Zealand. *Plagianthus betulina* (*P. urticina*) and the genus *Celmisia*
Otago. Various species of *Astelia*
Peruvian. *Gossypium peruvianum* (a variety of *G. barbadense*)
Silk. The genera *Bombax*, *Calotropis*, *Eriodendron*, and some other cottony plants
Tree. *Gossypium arboreum*
Wild. The genus *Eriophorum*
Cottoner, The. *Viburnum Lantana*
Cotton-bush, Australian. *Kochia villosa*
Cotton-grass, or Cotton-rush. The genus *Eriophorum*
Hare's-tail. *Eriophorum vaginatum*
Tassel. *Eriophorum polystachyon*
Cotton-nettle, Chinese. *Urtica* (*Bœhmeria*) *nivea*
Cotton-rose. The genus *Filago*
Pigmy. *Filago pygmæa*
Cotton-sedge, Common. *Eriophorum polystachyum*

Cotton - sedge, Sheathed. *Eriophorum vaginatum*
Cotton-thistle. *Onopordon Acanthium*
Cotton - tree. *Populus heterophylla* and *Viburnum Lantana*
Cotton-weed. *Diotis maritima* and the genus *Gnaphalium*
Purple Mountain. *Antennaria dioica*
Cotton-wood, American. *Populus monilifera*
British Columbia. Various species of *Populus* and *Salix*
Californian. *Populus Fremontii*
Missouri. *Populus angulata*
Couch, Couch-grass, or Couch-wheat. *Triticum repens*
Couch Onion. *Avena elatior*
Cough-wort. *Tussilago Farfara*
Countryman's Treacle. *Ruta graveolens*
Country Pepper. *Sedum acre*
Courtship - and - Matrimony. *Spiræa Ulmaria*
Coven Tree. *Viburnum Lantana*
Coventry-bells. *Campanula Medium* and *C. Trachelium*; also *Anemone Pulsatilla*
Coventry Rapes. *Campanula Medium*
Cover Keys, or Covey Keys. See Culverkeys
Cowage. See Cow-itch
Cow-bane. *Cicuta virosa*
American. *Archemora rigida*
Spotted. *Cicuta maculata*
Cow-basil, or Cow-fat. *Saponaria Vaccaria*
Cow-bell. *Silene inflata*
Cow-berry. *Vaccinium Vitis-Idæa* and *Comarum palustre*
Cow-fat. See Cow-basil
Cowhage. See Cow-itch
Cow-herb. *Saponaria Vaccaria*
Cow-itch, Cowage, or Cowhage. *Mucuna pruriens*, *M. urens*, and *Acidoton urens*
New Zealand. *Bidens pilosa*
Twining. *Tragia volubilis*
Cow-leaf of New Zealand. The genus *Melicytus*
Cow-parsley. *Anthriscus sylvestris*
Cow-parsnip. The genus *Heracleum*
American. *Heracleum lanatum*
Blunt-lobed. *Heracleum eminens*
Broad-leaved. *Heracleum latifolium*
Common. *Heracleum Sphondylium*
Downy. *Heracleum pubescens*
Fig-leaved. *Heracleum Panaces*
Giant. *Heracleum giganteum*
Persian. *Heracleum persicum*
Rough-leaved. *Heracleum elegans*
Wilhelms'. *Heracleum Wilhelmsii*
Yellowish. *Heracleum flavescens* (*H. austriacum*)
Cow-plant, Ceylon. *Gymnema lactiferum*
Cow-poison, Californian. *Delphinium troliifolium*
Cowslip, Bedlam. *Pulmonaria officinalis*
Also the l'aigle, or Larger Cowslip
Blue. *Pulmonaria angustifolia*
Californian. *Primula suffrutescens*
Cape. The genus *Lachenalia*
Common. *Primula veris*
Common American. *Dodecatheon Meadia*
Deep - rose-coloured American. *Dodecatheon Meadia var. splendens*

Cowslip, Elegant-flowered American. *Dodecatheon elegans*
Entire-leaved American. *Dodecatheon integrifolium*
French. *Primula Auricula*
Giant American. *Dodecatheon Jeffreyanum*
Great. *Primula elatior*
Jerusalem. *Pulmonaria officinalis*
Mountain. *Primula Auricula*
Pyrenean. *Primula intricata*
Sikkim. *Primula Sikkimensis*
Virginian. *Mertensia (Pulmonaria) virginica*
Cow-tree, or Palo de Vaca. *Drosimum Galactodendron*
Pará. *Mimusops elata*
Rio Negro. A species of *Callophora*
Cow Vetch. *Vicia Cracca*
Cow-weed. *Anthriscus sylvestris*
Cow-wheat. The genus *Melampyrum*
Common. *Melampyrum pratense*
Crested. *Melampyrum cristatum*
Eye-bright. See Eye-bright-Cow-wheat
Purple. *Melampyrum arvense*
Small-flowered. *Melampyrum sylvaticum*
Crab, Crab-apple, or Crab-tree, Common. *Pyrus Malus* var. *acerba*
Chinese. *Pyrus spectabilis*
Garland. *Pyrus coronaria*
Minshull. *Mespilus germanica*
Oregon. *Pyrus rivularis*
Profuse-flowering Chinese. *Pyrus Malus* var. *floribunda*
Scarlet-flowered. *Pyrus baccata*
Siberian. *Pyrus baccata* and *P. prunifolia*
Sweet-scented. *Pyrus coronaria*
Crab-Cherry. *Prunus Avium*
Crab-grass, or Crab-weed. *Polygonum aviculare*
Crab-oil-tree. *Carapa guianensis*
Crab-weed. See Crab-grass
Crab-wood, Guiana. *Carapa guianensis*
W. Indian. *Schaefferia frutescens*
Crab's-claw. *Polygonum Persicaria* and *Stratiotes aloides*
Crab's-eyes. The seeds of *Abrus precatorius*
Crab's-eye Lichen. See Lichen
Crack Willow. *Salix fragilis*
Crake-berry. See Crow-berry
Crake-feet. *Orchis mascula* and *Scilla nutans*
Cram-berries. See Cran-berry
Crambling-rocket. *Reseda lutea* and *Sisymbrium officinale*
Cran-berry, or Crane-berry, American, or Large. *Vaccinium macrocarpon*
Australian. *Astroloma humifusum,* and *Lissanthe sapida*
Common. *Vaccinium Oxycoccos.* Applied in Scotland to *V. Vitis-Idaea*
High. *Viburnum Oxycoccos*
Cran-berry Tree. *Viburnum Opulus*
Crane's-bill, Anemone-leaved. *Geranium anemonaefolium*
Backhouse's. *Geranium Backhousianum (G. armenum)*
Broad-petalled. *Geranium platypetalum*

Crane's-bill, Carolina. *Geranium Carolinianum*
Crested. *Geranium cristatum*
Dwarf. *Geranium subcaulescens (G. asphodeloides)*
Endres's. *Geranium Endresi*
Fremont's. *Geranium Fremontii*
Gray. *Geranium cinereum*
Hill. *Geranium collinum*
Iberian. *Geranium ibericum*
Lambert's. *Geranium Lamberti*
Large Rosy-purple. *Geranium armenum*
Long-rooted. *Geranium macrorrhizum*
Pinkish. *Geranium sanguineum* var. *lancastriense*
Silvery. *Geranium argenteum*
Striped. *Geranium striatum*
Tuberous-rooted. *Geranium tuberosum*
Walney. *Geranium sanguineum* var. *lancastriense*
White-flowered. *Geranium sylvaticum album*
Wild American. *Geranium maculatum*
Cranberry. *Empetrum nigrum* and *Vaccinium Oxycoccos*
Crawley. *Corallorrhiza odontorrhiza*
Cray-fish. *Doronicum Pardalianches*
Crazy. Another name for the Buttercup
Cream-cups. *Platystemon californicus*
Cream-fruit of Sierra Leone. The fruit of *Roupellia grata*
Cream-of-Tartar-tree. *Adansonia Gregorii*
Creashak. *Arctostaphylos Uva-ursi*
Creasote-plant. *Larrea mexicana*
Creeper, Cayenne Red. *Tetracera Tigarea*
China. *Quamoclit vulgaris*
Tasmanian Blue. *Comesperma volubile*
Virginian. *Ampelopsis quinquefolia*
Virginian, Cut-leaved. *Ampelopsis dissecta*
Virginian, Two-winged. *Ampelopsis bipinnata*
West Coast. *Pergularia odoratissima*
Creeping-bur. *Lycopodium clavatum*
Creeping Jack. *Sedum acre*
Creeping Jenny. *Lysimachia Nummularia*
Golden. *Lysimachia Nummularia aurea*
Creeping Sailor. *Saxifraga sarmentosa*
Cress. The genus *Lepidium*
American. *Barbarea praecox*
Lake. *Nasturtium lacustre*
Mountain Water. *Cardamine rotundifolia*
Spring. *Cardamine rhomboidea*
Bitter. *Cardamine amara* and other species
Broad-leaved. *Lepidium latifolium*
Creeping Water. *Nasturtium sylvestre*
Garden, or Town. *Lepidium sativum*
Great Water. *Nasturtium amphibium*
Hairy Bitter. *Cardamine hirsuta*
Hoary. *Lepidium Draba*
Lamb's. *Cardamine hirsuta*
Land. *Barbarea praecox*
Marsh Water. *Nasturtium palustre*
Meadow, or Meadow Bitter. *Cardamine pratensis*
Narrow-leaved. *Lepidium ruderale*
New Zealand. *Lepidium oleraceum*
Normandy. *Barbarea praecox*
Pará. *Spilanthes oleracea*

Cress, Penny. *Thlaspi arvense*
Pepper. *Teesdalia nudicaulis*
Peter's. *Crithmum maritimum*
Rock. The genus *Arabis*. (See also Rock-cress)
Sciatica. *Iberis amara*
Showy Bastard. *Thlaspi latifolium*
Spanish. *Lepidium Cardamines* and *Vella annua*
Sun. *Heliophila pectinata*
Swine's, or Wart. *Senebiera Coronopus*
Thale. *Arabis Thaliana*
Town. *Lepidium sativum*
Tower. *Arabis Turrita*
Violet-flowered. *Ionopsidion acaule*
Wart. See Cress, Swine's
Water. *Nasturtium officinale*
Wild. *Thlaspi arvense*
Winter. *Barbarea vulgaris*
Yellow. The genus *Barbarea*; also *Nasturtium amphibium* and *N. palustre*
Cress-Rocket. *Vella annua* and *V. Pseudocytisus*
Crestmarine. *Crithmum maritimum*
Creyat, or Kariyat. *Andrographis (Justicia) paniculata*
Crimson-berry-plant. *Phytolacca decandra*
Cristaldre. An old name for *Erythræa Centaurium*
Crocus, Adriatic. *Crocus Hadriaticus*
Aucher's. *Crocus Aucheri (C. Olivieri)*
Autumn, Fog, Meadow, Michaelmas, or Purple. *Colchicum autumnale*
Bottle-flowered. *Crocus lagenæflorus*
Byzantine. *Crocus Byzantinus*
Cape. The genus *Gethyllis*
Cartwright's. *Crocus Cartwrightianus*
Chilian. *Tecophylæa cyanocrocus*
Cloth of Gold. *Crocus reticulatus (C. susianus)*
Clusius's. *Crocus Clusii*
Common Yellow. *Crocus luteus*
Cream-coloured. *Crocus lacteus*
Cross-barred. *Crocus cancellatus*
Dalmatian. *Crocus dalmaticus*
Damascus. *Crocus Damascenus*
Dark-flowered. *Crocus obscurus*
Dwarf. *Crocus pusillus*
Elwes's. *Crocus Elwesi*
Fleischer's. *Crocus Fleischeri*
Fog. See Crocus, Autumn
Imperati's. *Crocus Imperati*
Indian. The genus *Pleione*
Intermediate. *Crocus medius*
Kotschy's. *Crocus Kotschyanus*
Large Autumn. *Crocus speciosus*
Late-blooming. *Crocus serotinus (C. autumnalis)*
Long-flowered. *Crocus longiflorus*
Meadow. See Crocus, Autumn
Michaelmas. See Crocus, Autumn
Morean. *Crocus peloponnesiacus*
Mount Athos. *Crocus pulchellus*
Naked-flowered. *Crocus nudiflorus (C. multifidus)*
Orphanides'. *Crocus Orphanidesi*
Parti-coloured. *Crocus versicolor*
Pigmy. *Crocus minimus*

Crocus, Purple. See Crocus, Autumn
Saffron. *Crocus sativus*
Saltzmann's. *Crocus Saltzmannianus*
Scented. *Crocus odorus*
Scotch. *Crocus biflorus*
Sieber's. *Crocus Sieberi (C. nivalis)*
Spring. *Crocus vernus*
Star-flowered. *Crocus stellaris*
Straw-coloured. *Crocus ochroleucus*
Sulphur-flowered. *Crocus sulphureus*
Tuscan. *Crocus etruscus*
Two-flowered. Varieties of *Crocus biflorus*
Various-coloured. *Crocus versicolor*
Welden's. *Crocus Weldeni*
White Autumn. *Crocus Boryanus*
White Spring. *Crocus vernus var. niveus*
Crone-berry. *Vaccinium Oxycoccos*
Crop-weed. *Centaurea nigra*
Cross-flower. *Polygala vulgaris*
Cross of Jerusalem. *Lychnis chalcedonica*
Cross-spine. *Stauracanthus aphyllus*
Cross-wood. *Jacquinia ruscifolia*
Cross-wort. Any cruciferous plant; also applied to *Eupatorium perfoliatum*, *Galium cruciatum*, and the genus *Cruciaanella*
Common. *Galium cruciatum*
Large-styled. *Crucianella stylosa*
Sea. *Crucianella maritima*
Crotal. See Crottle
Croton-oil-plant. *Croton Tiglium*
Crottle, Crotal, or Crottles. *Parmelia Omphalodes* and various other Lichens
Crottles, Hazel. *Sticta pulmonacea*
Crow-bells. *Scilla nutans*
Yellow. *Narcissus Pseudo-Narcissus*
Crow-berry, or Crake-berry. *Empetrum nigrum*
Broom. *Corema (Empetrum) Conradii*
Portugal. *Corema lusitanicum*
Crow-corn. *Aletris farinosa*
Crow-flower. *Lychnis Flos-cuculi*; also another name for Buttercup
Crow-foot. The genus *Ranunculus*
Alpine. *Ranunculus alpestris*
American. *Geranium maculatum*
Anemone-flowered. *Ranunculus anemonoides*
Blistered-leaved. *Ranunculus bullatus*
Bulbous-rooted. *Ranunculus bulbosus*
Celery-leaved. *Ranunculus sceleratus*
Corn-field. *Ranunculus arvensis*
Cortusa-leaved. *Ranunculus cortusæfolius*
Crane's-bill. *Geranium pratense*
Creeping. *Ranunculus repens*
Glacier. *Ranunculus glacialis*
Hairy. *Ranunculus hirsutus*
Ivy-leaved. *Ranunculus Cymbalaria*
Kidney-leaved. *Ranunculus Thora*
Large Double-flowered. *Ranunculus speciosus (R. grandiflorus fl.-pl., R. bullatus fl.-pl.)*
Madeira. *Ranunculus megaphyllus (R. grandifolius)*
New Zealand. *Ranunculus Lyalli*
One-flowered. *Ranunculus uniflorus*
Parnassia-leaved. *Ranunculus parnassifolius*

D

Crow-foot, Plane-tree-leaved. *Ranunculus platanifolius*
Portugal. *Ranunculus bullatus*
Profuse-flowering. *Ranunculus floribundus*
Pyrenean. *Ranunculus pyrenæus*
Red, or Tripoli. *Ranunculus asiaticus var. sanguineus*
Rue-leaved. *Ranunculus rutæfolius* (*Callianthemum rutæfolium*)
Sea-side. *Ranunculus Cymbalaria*
Small-flowered. *Ranunculus parviflorus*
Snake-tongue. *Ranunculus ophioglossifolius*
Snowy-flowered. *Ranunculus amplexicaulis*
Steven's. *Ranunculus Stevenii*
Thousand-leaved. *Ranunculus millefoliatus*
Tuberous. *Adoxa Moschatellina*
Turkey. An old name for *Ranunculus asiaticus*
Upright. *Ranunculus acris*
Water. *Ranunculus aquaticus*
Wind-flower. *Ranunculus anemonoides*
Wood. *Ranunculus auricomus*
Yellow - tinted Alpine. *Ranunculus alpinus*
Yellow Water. *Ranunculus multifidus*
Crow-garlic. *Allium vineale*
Crow-leek. An old name for *Scilla nutans*
Crow-silk. The *Confervæ* and other delicate green-spored *Algæ*
Crow-soap. *Saponaria officinalis*
Crow-toe. Another name for Crow-foot
Crown-beard. The genus *Verbesina*
Crown Imperial. *Fritillaria imperialis*
Crown-of-thorns. *Medicago Echinus*
Crown-Vetch. The genus *Coronilla*
Crow's-foot. *Lotus corniculatus*
Crow's-foot Grass. See Grass
Cubebs. *Piper Cubeba*
African. *Piper Cubeba Clusii*
Cuckle, or **Cuckold.** The fruit of *Arctium Lappa*
Cuckold-tree. *Acacia cornigera*
Cuckoo-bread, or **Cuckoo-sorrel.** *Oxalis Acetosella*
Cuckoo-bud. *Cardamine pratensis;* also an old name for either the Cowslip or the Buttercup
Cuckoo-flower. *Cardamine pratensis,* Orchis *masculo, Scilla nutans,* and some other early spring flowers
Asarum-leaved. *Cardamine asarifolia*
Broad-leaved. *Cardamine latifolia*
Double. *Cardamine pratensis fl.-pl.*
Large-leaved. *Cardamine macrophylla*
Round-leaved. *Cardamine rotundifolia*
Three-leaved. *Cardamine trifolia*
Cuckoo-grass. *Luzula campestris*
Cuckoo-pint. *Arum maculatum*
Cuckoo-sorrel. See Cuckoo-bread
Cuckoo-spit. *Cardamine pratensis*
Cucumber. The genus *Cucumis*
Apple. *Cucumis Dudaim*
Climbing. *Cyclanthera pedata*
Common. *Cucumis sativus*
One-seeded Star. *Sicyos angulatus*
Globe. *Cucumis prophetarum*
Hairy. *Cucumis Chate*
Prickly-fruited Gherkin. *Cucumis Anguria*

Cucumber, Snake. *Cucumis flexuosus*
Spanish Wild. *Clematis cirrhosa*
Squirting. *Ecbalium agreste* (*Momordica Elaterium*)
Cucumber-root, Indian. *Medeola virginica*
Cucumber-tree. *Magnolia acuminata*
Long-leaved. *Magnolia Fraseri*
Yellow. *Magnolia cordata*
Cudbear. *Lecanora tartarea*
Cud-weed. The genus *Gnaphalium*
American. *Antennaria margaritacea*
Common. *Gnaphalium germanicum*
Dwarf. *Gnaphalium supinum*
Field. *Gnaphalium arvense*
Golden. *Pterocaulon virgatum*
Jersey. *Gnaphalium luteo-album*
Lion's-paw. *Gnaphalium Leontopodium*
Marsh. *Gnaphalium uliginosum*
Mountain. *Antennaria dioica*
Narrow-leaved. *Gnaphalium gallicum*
Pearl. *Antennaria margaritacea*
Purplish. *Gnaphalium purpureum*
Sea-side. *Diotis maritima*
Silvery. *Antennaria tomentosa*
Wood. *Gnaphalium sylvaticum*
Cuji-pods. The pods of *Acacia macrantha*
Culcit Fern. See Fern
Cull-me-to-you. *Viola tricolor*
Cullay-tree, or Quillai-tree, of Chili. *Quillaia Saponaria*
Culrage, or Curage. *Polygonum Hydropiper*
Culverkeys. *Scilla nutans,* the Oxlip (*Primula variabilis*), and the fruit of *Fraxinus excelsior*
Culver's-root. *Veronica virginica*
Culver-wort. An old name for Columbine
Cumin, or Cummin. *Cuminum Cyminum*
Black. The seeds of *Nigella sativa*
Sweet. *Pimpinella Anisum*
Wild. *Lagœcia cuminoides*
Cupidone, Blue. *Catananche cœrulea*
Cupid's-flower. *Quamoclit vulgaris*
Cup-Fern, Tasmanian. *Cyathea arborea*
Cup-flower, Chili. *Scyphanthus elegans*
Trailing White. *Nierembergia rivularis*
Cup-Goldilocks. *Trichomanes radicans*
Cup-Lichen. *Scyphophorus pyxidatus*
Cup-Mint. *Calamintha officinalis*
Cup-Moss. *Lecanora tartarea* and *Scyphophorus pyxidatus*
Cup-Mushroom. The genus *Peziza*
Cup-plant. *Silphium perforatum*
Cup-seed, Lyon's. *Calycocarpum Lyoni*
Cups-and-saucers. *Cobœa scandens*
Curage. See Culrage
Curana-wood. The wood of *Icica altissima*
Curra Tow. The fibre of *Bromelia* (*Anannassa*) *Karatas*
Curragong-bark, Moreton Bay. *Plagianthus sidioides*
Currant, Alpine Red. *Ribes alpinum*
American Black. *Ribes floridum*
Australian. *Leptomeria acerba* and *Leucopogon Richei*
Buffalo, or Missouri. *Ribes aureum*
Californian Black. *Ribes bracteosum*
Carpathian Red. *Ribes carpaticum*
Common Black. *Ribes nigrum*

Currant, Common Red. *Ribes rubrum*
Common White. *Ribes rubrum var. album*
Crimson-flowered. *Ribes sanguineum var. atro-rubens*
Dark-flowered, Black. *Ribes triste*
Dark-purple-flowered Black. *Ribes atropurpureum*
Dotted-leaved Red. *Ribes punctatum*
Double-flowered Red. *Ribes sanguineum plenum*
Fetid. *Ribes prostratum*
Glanded-calyxed Red. *Ribes glandulosum*
Golden-flowered. *Ribes aureum*
Guelder-rose-leaved Black. *Ribes opulifolium*
Hudson's Bay Black. *Ribes Hudsonianum*
Indian. *Symphoricarpus vulgaris*
Intoxicating Black. *Ribes inebrians*
Late-flowering Golden. *Ribes aureum serotinum*
Long-flowered Black. *Ribes longiflorum*
Many-flowered Red. *Ribes multiflorum*
Missouri. See Currant, Buffalo
Mountain, or Wild. *Ribes alpinum*
New Zealand. *Aristotelia fruticosa*
Native, Tasmanian. Some species of *Coprosma*
Native, Victoria. *Leptomeria Billardieri*
Nepaul Black. *Ribes glaciale*
Pennsylvanian Black. *Ribes floridum*
Procumbent Red. *Ribes procumbens*
Red-flowered. *Ribes sanguineum*
Resinous Red. *Ribes resinosum*
Rock Red. *Ribes petræum*
Slender-flowered Black. *Ribes tenuiflorum*
Spiked-flowered Red. *Ribes spicatum*
Stiff-racemed Red. *Ribes rigens*
Tasteless Mountain. *Ribes alpinum*
Very Clammy. *Ribes viscosissimum*
Waxy-leaved Black. *Ribes cereum*
W. Indian. *Jacquinia armillaris, Beureria havanensis,* and *B. succulenta*
White-flowered. *Ribes sanguineum var. album*
White-fruited. *Ribes rubrum var. album*
White-ribbed-leaved Red. *Ribes albinervium*
Yellow-flowered. *Ribes aureum*
Yellow-flowered Black. *Ribes flavum*
Yellow-flowered Red. *Ribes tenellum*
Currant-bush, Indian, of Tropical America. The genera *Miconia* and *Clidemia*
Currant-galls. Galls formed in the male flowers of the Oak to be by an insect named *Spathegaster baccarum*
Currants (grocers'). The fruit of the Black Corinth, or Zante, Grape-vine (a variety of *Vitis vinifera*)
Curry-leaf-tree. *Bergera Königi*
Cusco-bark. A kind of Cinchona-bark
Cus-cus, or Khus-khus, Grass. *Andropogon muricatus*
Cushion, Lady's. *Armeria maritima*
Cushion-pink. *Armeria maritima*
Cushion-plant, or Lion's-ear, of the Andes. Various species of *Culcitium* and *Espeletia*
Cusparia Bark. See Angostura Bark-plant
Cusso, Abyssinian. *Hagenia abyssinica* (*Brayera anthelmintica*)

Custard-apple. The genus *Anona*, especially *A. squamosa*
Long-leaved. *Anona hexapetala*
Netted. *Anona reticulata*
Peruvian. *Anona Cherimolia*
Prickly (Sour-sop). *Anona muricata*
Scaly (Sweet-sop). *Anona squamosa*
Shining-leaved. *Anona palustris*
Custard-cups. *Epilobium hirsutum*
Cutberdole, or Cutbertill. An old name for *Acanthus*
Cutch. See Catechu, Black
Cut-finger. *Valeriana pyrenaica* and *Vinca major*
Cut-grass. *Leersia oryzoides*
Cut-heal. *Valeriana officinalis*
Cuvy. *Laminaria digitata*
Cyananth, Hoary-leaved. *Cyananthus incanus*
Lobed-leaved. *Cyananthus lobatus*
Cyclamen, Atkins's. *Cyclamen Atkinsii*
European. *Cyclamen europæum*
Iberian. *Cyclamen ibericum*
Ivy-leaved. *Cyclamen hederæfolium*
Large-leaved. *Cyclamen africanum* (*C. algeriense macrophyllum*)
Neapolitan. *Cyclamen neapolitanum*
Persian. *Cyclamen persicum*
Round-leaved. *Cyclamen Coum*
Spring. *Cyclamen vernum* (*C. repandum*)
Variegated Round-leaved. *Cyclamen Coum vernum* (*C. Coum zonale*)
White-flowered Round-leaved. *Cyclamen Coum album*
Cyphel. *Cherleria sedoides*
Cypress, African. The genus *Widdringtonia*
Bald, Black, or Deciduous. *Taxodium distichum*
Bhotan. *Cupressus torulosa*
Black. See Cypress, Bald
Blunt-leaved Japan. *Retinospora obtusa*
Broom, or Summer. *Kochia* (*Chenopodium*) *scoparia*
Chinese Deciduous. *Taxodium sinense*
Common Pyramidal. *Cupressus sempervirens*
Compact Japan. *Retinospora obtusa var. compacta*
Deciduous. See Cypress, Bald
Embossed. The genus *Glyptostrobus*
Evergreen. See Cypress, Italian
Fern-like Japan. *Retinospora filicoides*
Field. *Ajuga Chamæpitys*
Fragrant. *Cupressus fragrans*
Funereal. *Cupressus funebris*
Garden. *Artemisia maritima*
Golden Plume - like Japan. *Retinospora plumosa var. aurea*
Golden-variegated Japan. *Retinospora obtusa var. aurea*
Gowen's Californian. *Cupressus Goveniana*
Ground. *Santolina Chamæcyparissus*
Himalayan. *Cupressus torulosa*
Horizontal. *Cupressus sempervirens var. horizontalis*
Incense-bearing Mexican. *Cupressus thurifera*
Italian, or Evergreen. *Cupressus sempervirens*

Cypress, Japan. The genus *Retinospora*, especially *R. obtusa*
Lawson's. *Cupressus (Chamæcyparis) Lawsoniana*
Monterey. *Cupressus macrocarpa*
Montezuma. *Taxodium distichum var. mexicanum (T. mucronatum)*
Nootka Sound. *Cupressus nutkaensis*
Oregon. *Cupressus (Chamæcyparis) Lawsoniana*
Pea-fruited Japan. *Retinospora pisifera*
Pigmy Japan. *Retinospora obtusa var. pygmæa*
Plume-like Japan. *Retinospora plumosa*
Portugal. *Cupressus pendula*
Silver-variegated Japan. *Retinospora obtusa var. argentea*
Silvery Plume-like Japan. *Retinospora plumosa var. argentea*
Standing. *Ipomopsis elegans*
Strawberry-fruited Tasmanian. *Microcachrys tetragona*
Summer. See Cypress, Broom
Swamp. The genus *Chamæcyparis*
Swan River. *Actinostrobus acuminatus* and *A. pyramidalis*
Tall Guatemala. *Cupressus excelsa*
Thread-leaved Japan. *Retinospora filifera*
Upright. *Cupressus sempervirens var. stricta*
Upright Indian. *Cupressus Whitleyana*
Weeping. *Cupressus funebris*
Weeping Deciduous. *Taxodium distichum var. pendulum*
Yellow. *Thujopsis borealis (Cupressus nutkaensis)* and *Thuja Lobbii (T. Menziesii, T. gigantea)*
Cypress-Broom. See Broom
Cypress "knees." Tumours produced by disease on the roots of *Taxodium distichum*
Cypress-Moss. *Lycopodium alpinum*
Cypress-root, or Sweet Cypress. *Cyperus longus*
Cypress-Spurge. *Euphorbia Cyparissias*
Cypress-Vine. *Quamoclit vulgaris*
Carolina. *Ipomæa caroliniana*
Ivy-leaved. *Ipomæa hederacea coccinea*

Daffadowndilly. *Narcissus Pseudo-narcissus*
Daffodil, Ajax. *Narcissus Ajax*
Bazelman major. *Narcissus Trewianus*
Bazelman minor. *Narcissus crenulatus*
Chequered. *Fritillaria Meleagris*
Cliff. *Narcissus rupicola*
Clusius's. *Narcissus Clusii*
Common. *Narcissus Pseudo-narcissus*
Cyclamen-flowered. *Narcissus calathinus pulchellus, N. triandrus*, and other species which have the perianth-segments reversed
Dwarf. *Narcissus minor*
Dwarfest. *Narcissus minimus (N. pumilus)*
Eastern. *Narcissus orientalis*
French. *Narcissus Tazetta*
Golden. *Narcissus maximus*
Graells's. *Narcissus Graellsi*
Green-flowered. *Narcissus viridiflorus*

Daffodil, Hoop-petticoat. *Narcissus Bulbocodium*
Horsfield's. *Narcissus Horsfieldi*
Incomparable. *Narcissus incomparabilis*
Indian Autumn. *Pancratium indicum*
Jonquil. *Narcissus Jonquilla*
Jonquil, Queen Anne's. *Narcissus pusillus plenus*
Jonquil, Small. *Narcissus pusillus*
Larger Curled-cup. *Narcissus interjectus*
Mountain. *Narcissus montanus*
Musk-scented. *Narcissus moschatus*
Narrow-leaved. *Narcissus angustifolius*
One-leaved. *Narcissus monophyllus*
Pale-flowered. *Narcissus pallidulus*
Paper-white. *Narcissus papyraceus*
Peerless. *Narcissus incomparabilis*
Peruvian. *Ismene Amancaes*
Pigmy. *Narcissus minimus (N. pumilus)*
Polyanthus. *Narcissus Tazetta*
Rush-leaved. *Narcissus juncifolius* and *N. triandrus*
Sea-shore. *Narcissus calathinus;* also applied to *Pancratium maritimum*
Sibthorp's. *Narcissus obrallaris*
Slender. *Narcissus tenuior*
Smaller Curled-cup. *Narcissus heminalis*
Telamon. *Narcissus Telamonius*
Tenby Six-lobed. *Narcissus lobularis*
Thick-bulbed. *Narcissus pachybulbos*
Three-stamened. *Narcissus triandrus*
Twisted. *Narcissus tortuosus*
Two-coloured. *Narcissus bicolor*
Two-flowered. *Narcissus biflorus*
White Musk-scented. *Narcissus patulus*
Winter. *Sternbergia lutea* and *S. lutea angustifolia*
Yellow Rush-leaved. *Narcissus gracilis*
Dagger flower. *Machæranthera tanacetifolia (Aster tanacetifolius)*
Dagger-plant. The genus *Yucca*
Dahlia, Black. *Dahlia Zimapani (Cosmos atropurpureus)*
Cactus. *Dahlia Juarezi*
Common. Varieties of *Dahlia (Georgina) variabilis*
Florists'. Varieties of *Dahlia superflua*
Giant. *Dahlia imperialis*
Imperial. *Dahlia imperialis*
Long-flower-stemmed. *Dahlia scapigera*
Scarlet. *Dahlia coccinea*
Slender. *Dahlia gracilis*
Tree. *Dahlia imperialis*
Yuarez's. *Dahlia Yuarezi*
Daisy, African. *Athanasia annua*
American False. *Eclipta brachypoda*
Annual. *Bellis annua*
Arctic Ox-eye. *Leucanthemum (Chrysanthemum) arcticum*
Aucuba-leaved. *Bellis aucubæfolia*
Big. *Chrysanthemum Leucanthemum*
Blue. *Aster Tripolium* and various species of *Globularia*
Blue Alpine. *Aster alpinus*
Blue Round-leaved. *Bellis rotundifolia var. cærulea*
Christmas. *Aster grandiflorus*
Chusan. The small, or Pompone, variety of *Chrysanthemum sinense*

Daisy, Common. *Bellis perennis*
Common Small. *Bellium bellidioides*
Crown. *Chrysanthemum coronarium* and vars.
Dog, Horse, Moon, or Ox-eye. *Chrysanthemum Leucanthemum*
Double Red. *Bellis perennis rubra plena*
Double White. *Bellis perennis alba plena*
Dwarf Small. *Bellium minutum*
False. *Bellium bellidioides*
Globe. The genus *Globularia*
Hen-and-Chickens. *Bellis perennis var. prolifera*
Horse. See Daisy, Dog
Italian. *Bellis hybrida*
Marsh. *Armeria maritima*
Marsh Ox-eye. *Chrysanthemum lacustre*
Mexican. *Erigeron (Leptostelma) maximum*
Michaelmas. *Aster Tripolium* and *A. Tradescanti*
Moon. See Daisy, Dog
New Holland. *Vittadenia australis (V. triloba)*
New Zealand. The genus *Lagenophora*
Ox-eye. See Daisy, Dog
Ox-eye, Tricoloured. *Chrysanthemum Burridgeanum*
Paris. *Chrysanthemum frutescens*
Portugal Wood. *Bellis sylvestris*
Small. The genus *Bellium*
Swan River. *Brachycome iberidifolia*
Texan. *Bellis integrifolia*
Thick-leaved Small. *Bellium crassifolium*
Turfing. *Pyrethrum Tchihatchewi*
Western. *Bellis integrifolia*
Daisy Bush, New Zealand. *Olearia Haastii* and other species
Daisy Star. The genus *Bellidiastrum*
Daisy Tree, New Zealand. Various species of *Olearia*
Tasmanian. *Eurybia lyrata*
Dalmatian Cap. An old name for the genus *Tulipa*
Damasse. Another name for the Damson
Dame's Violet. *Hesperis matronalis*
Dammar Tree, Black. *Canarium strictum*
White. *Vateria indica*
Damsel. Another name for the Damson
Damson, or Damson-Plum. *Prunus communis var. damascena*
Bitter, or Mountain. *Simaruba amara*
W. Indian. *Chrysophyllum oliviferum*
Wild. *Prunus insititia*
Danchi, or Dhunchi-plant. *Sesbania aculeata*
Dancing Girls. See Opera Girls
Dandelion. *Leontodon Taraxacum*
Blue. *Lactuca sonchifolia*
Dwarf American. *Krigia virginica*
False. *Pyrrhoppus carolinianus*
Dane-ball, or Dane-wort. *Sambucus Ebulus*
Danes'-blood. *Sambucus Ebulus, Anemone Pulsatilla,* and *Campanula glomerata*
Danes'-flower. *Anemone Pulsatilla*
Danes'-weed. *Eryngium campestre* and *Sambucus Ebulus*
Dane-wort. See Dane-ball
Dangle-berry, or Blue Tangle. *Gaylussaccia frondosa*
Daphne, Alpine. *Daphne alpina*

Daphne, Altaian. *Daphne altaica*
Australian. *Wickstrœmia indica*
Chinese. *Daphne chinensis*
Downy. *Daphne pubescens*
Flax-leaved. *Daphne Gnidium*
Fortune's. *Daphne Fortunei*
Green-flowered. *Daphne viridiflora*
Hill. *Daphne collina*
Indian. *Daphne indica*
Japan. *Daphne japonica*
Lady Auckland's. *Daphne Aucklandiœ*
Neapolitan. *Daphne collina*
Olive-leaved. *Daphne oleoides*
Paper. *Daphne papyracea*
Pontic. *Daphne pontica*
Red-flowered Indian. *Daphne indica var. rubra*
Rock. *Daphne rupestris*
Silky. *Daphne sericea*
Silvery-leaved. *Daphne Tartonraira*
Smooth-leaved. *Daphne Thymelœa*
Streaked-barked. *Daphne striata*
Sweet-scented. *Daphne odora*
Tinus-leaved. *Daphne tinifolia*
Variegated. *Daphne odora var. variegata*
Willow-leaved. *Daphne salicifolia*
Darnel. *Lolium temulentum*
Red. *Lolium perenne*
Darning-needle, Devil's. *Scandix Pecten*
Darsham Fern. *Nephrodium cristatum*
Dart-Grass. *Holcus mollis* and *H. lanatus*
Date, Chinese. The fruit of a species of *Zizyphus*
Date-Coffee. See Coffee
Date-Palm. *Phœnix dactylifera* and other species
Date-Plum, Virginian. *Diospyros virginiana*
Virginian Downy - leaved. *Diospyros virginiana var. pubescens*
Dattock. *Detarium senegalense*
David's-harp. *Polygonatum multiflorum*
David's-root. *Chiococca racemosa* and *Celastrus scandens*
Dawa. The fruit of *Nephelium pinnatum*
Day-berry. The wild Gooseberry
Day Nettle. See Dead Nettle
Day-flower. The genus *Commelina*
Day-lily. The genus *Hemerocallis*
Dark Yellow. *Hemerocallis lutea*
Dumortier's. *Hemerocallis Dumortieri*
Grass-leaved. *Hemerocallis graminea*
Tawny. *Hemerocallis fulva*
Two-rowed. *Hemerocallis disticha*
Variegated-leaved. *Hemerocallis Kwanso variegata*
Yellow. *Hemerocallis flava*
Dea Nettle, or Dee Nettle. See Dead Nettle
Dead-Man's-Toe. *Laminaria digitata*
Dead-Men's-Fingers. *Orchis mascula, O. maculata,* and other species
Dead Nettle, or Deaf Nettle. A general name for various species of *Lamium*
Balm-leaved Red. *Lamium Orvala*
Gargano. *Lamium garganicum*
Golden. *Lamium aureum*
Long-flowered. *Lamium longiflorum*
Pyrenean. *Horminum pyrenaicum*
Red. *Lamium purpureum*
Spotted. *Lamium maculatum*

Dead Nettle, White. *Lamium album*
 Yellow. *Lamium Galeobdolon*
Dead-tongue. *Œnanthe crocata*
Dead-wort. *Sambucus Ebulus*
Deaf Nettle. See Dead Nettle
Deal Trees. Various species of *Pinus* and *Abies*
Deberries. Another name for Gooseberries
Deer Balls. *Elaphomyces*
Deer-berry. *Mitchella repens*, and *Vaccinium stamineum*
Deer-food. *Brasenia peltata*
Deer-grass. *Rhexia virginica*
Deer's-hair. *Scirpus (Eleocharis) cæspitosus*
Deer's-foot Grass. *Agrostis setacea*
Deer's-weed. See Dyer's-weed
Deil's-Meal. *Anthriscus sylvestris*
Deil's-Oatmeal. *Bunium flexuosum*
Deil's-Spoons. *Potamogeton natans* and *Alisma Plantago*
Delt-Orache. *Atriplex patula*
Demigod's-food. An old name for *Ambrina ambrosioides*
Deptford Pink. *Dianthus Armeria*
Deodar. *Cedrus Deodara*
 Heavy-branched. *Cedrus Deodara robusta*
 Silvery-leaved. *Cedrus Deodara nirea*
Desert-Rod. The genus *Eremostachys*
Devil-in-a-bush. *Nigella damascena*
Devil-tree. *Alstonia scholaris*
Devil-wood. *Osmanthus americanus (Olea americana)*
Devil's-Apron. *Laminaria saccharina*
Devil's-Bean. *Capparis cynophallophora*
Devil's-Bit, or Devil's-Bit Scabious. *Scabiosa succisa*
 American. *Helonias dioica (Chamælirion luteum)* and *Liatris squarrosa*
 Swamp. *Ptelea trifoliata*
Devil's-Candlestick. *Nepeta Glechoma*
Devil's-Claws. *Ranunculus arvensis*
Devil's-Coach-wheel. *Ranunculus arvensis*
Devil's-Curry-comb. *Ranunculus arvensis*
Devil's-Darning-needle. *Scandix Pecten*
Devil's-Flower. *Lychnis diurna*
Devil's-Garter. *Convolvulus sepium*
Devil's-Herb, W. Indian. *Plumbago scandens*
Devil's-Horn. *Phallus impudicus*
Devil's-Leaf. *Urtica urentissima*
Devil's-Milk. *Euphorbia Helioscopia* and other species
Devil's - Snuff-box. Various species of *Lycoperdon*
Devil's-Tree. *Alstonia scholaris*
Devil's-Trumpet. *Datura Stramonium*
Dew-berry. *Rubus cæsius*
 American. *Rubus canadensis*
Dew-Grass. *Dactylis glomerata*
Dew-plant. *Mesembryanthemum glabrum*
Dewtry. *Datura Stramonium*
Deye-Nettle. *Galeopsis Tetrahit* and *Stachys sylvatica*
Dhak-tree. *Butea frondosa*
Dhal, or Dhol. The seeds of *Cajanus indicus*
Dhourra, Dhurra, Doura, or Durra. *Sorghum vulgare*

Dhunchi. See Danchi
Dhurra. See Dhourra
Dika, or Udika, Bread-plant of W. Africa. *Irvingia Barteri*
Dikamali Resin-plant. *Gardenia lucida*
Dill, or Dill-seed. *Anethum graveolens*
Dillisk. *Rhodymenia palmata*
Dilnote. An old name for *Cyclamen*
Dirt-weed, or Dirty Dick. *Chenopodium album*
Distaff Cane. *Arundo Donax*
Diss. *Arundo (Ampelodesmos) tenax (A. festucoides)*
 Arab's. *Festuca altissima*
Ditch-Bur. *Xanthium strumarium*
Ditch-Fern. *Osmunda regalis*
Ditch-Grass, American. *Ruppia maritima*
Ditch-Reed. *Phragmites communis*
Dithering (Shivering) **Grass.** *Briza media*
Dittander. *Lepidium latifolium*
Dittany. *Dictamnus Fraxinella*
 American. *Cunila Mariana*
 Of Amorgos. *Origanum Tournefortii*
 Of Crete. *Origanum Dictamnus*
Divi-Divi-tree. *Cæsalpinia coriaria*
Dock, or Docken. Various species of *Rumex*
 Bon. *Œnanthe crocata*
 Bladder. *Rumex vesicarius*
 Broad-leaved. *Rumex obtusifolius*
 Can. *Nymphæa alba* and *Nuphar lutea*
 Clustered. *Rumex conglomeratus*
 Curled. *Rumex crispus*
 Fiddle. *Rumex pulcher*
 Flatter. *Nymphæa alba* and other aquatic plants with floating leaves
 Gentle. *Polygonum Bistorta*
 Golden. *Rumex maritimus*
 Grainless. *Rumex aquaticus*
 Grove. *Rumex Nemolapathum*
 New Zealand. *Rumex flexuosus*
 Pale, of N. America. *Rumex Britannica*
 Patient, or Patience. *Polygonum Bistorta*
 Prairie. *Silphium terebinthinaceum*
 Red-veined. *Rumex sanguineus*
 Round. *Malva sylvestris*
 Sea. An old name for *Acanthus*
 Sharp. See Dock, Sour
 Sorrel. *Rumex Acetosa*
 Sour, or Sharp. *Rumex Acetosa*
 Swamp, of N. America. *Rumex verticillatus*
 Tuberous-rooted. *Rumex tuberosus*
 Velvet. *Verbascum Thapsus*
 Water. *Rumex Hydrolapathum*
 Water, American. *Rumex orbiculatus*
 White, or Willow-leaved, Californian. *Rumex salicifolius*
Docken, Eldin. *Rumex aquaticus*
 Flowery. *Chenopodium Bonus-Henricus*
 Water. *Petasites vulgaris*
Dockmackie. *Viburnum acerifolium*
Doctor's-gum-tree. *Rhus Metopium*
Dod. *Typha latifolia*
Dodder. The genus *Cuscuta*
 Bengal. *Cuscuta capitata*
 Flax. *Cuscuta Epilinum*
 Greater. *Cuscuta europæa*
 Small. *Cuscuta Epithynum*
Dodder-cake-plant. *Camelina sativa*
Dodder-grass. *Briza media*

Dog-berry, or Dog-cherry. The fruit of *Cornus sanguinea*
Dog-Daisy. See Daisy
Dog-Eller. *Viburnum Opulus*
Dog-Fennel. *Anthemis Cotula*
Dog-Gowan. *Matricaria inodora*
Dog-Grass. *Triticum repens*
Dog-Nettle. See Dead Nettle
Dog-Oak. *Acer campestre*
Dog-Rose. *Rosa canina*
Dog-Rowan-tree. *Viburnum Opulus*
Dog-Thistle. *Carduus arvensis*
Dog-Tree. *Cornus sanguinea*
Dog-Violet. *Viola canina* and *V. sylvatica*
Dog-wood. *Cornus sanguinea*; also applied to *Euonymus europæus*, *Rhamnus Frangula*, and *Viburnum Opulus*
 Alternate-leaved. *Cornus alternifolia*
 Black. *Piscidia carthaginensis*
 Blue-berried. *Cornus sericea*
 Flowering. *Cornus florida*
 Headed-flowered. *Cornus capitata*
 Illawarra. *Emmenosperma alphitonioides*
 Jamaica. *Piscidia Erythrina*
 Large-leaved. *Cornus macrophylla*
 Male. *Cornus mas*
 N. S. Wales. *Jacksonia scoparia*
 Oblong-leaved. *Cornus oblonga*
 Panicle-flowered. *Cornus paniculata*
 Pond. See Pond-Dogwood
 Red Osier. *Cornus stolonifera*
 Rough-leaved. *Cornus asperifolia*
 Round-leaved. *Cornus circinata*
 Siberian Scarlet. *Cornus sibirica*
 Silky, or Kinnikinnik. *Cornus sericea*
 Stiff. *Cornus stricta*
 Striped. *Acer pennsylvanicum*
 Tall Mexican. *Cornus grandis*
 Upright-branched. *Cornus stricta*
 Tasmanian. *Bedfordia salicina*
 White-berried. *Cornus alba*
 White Jamaica. *Piscidia Erythrina*
Dog's-bane. The genus *Apocynum*
 Climbing. *Periploca græca*
 Spreading, or Tutsan-leaved. *Apocynum androsæmifolium*
 Venetian. *Apocynum Venetum*
Dog's-Chamomile. *Matricaria Chamomilla*
Dog's-chop. *Mesembryanthemum caninum*
Dog's Grass. See Dog's-tail Grass
Dog's Leek. See Leek
Dog's Mercury. *Mercurialis perennis*
Dog's Parsley. *Æthusa Cynapium*
Dog's-rib. See Rib-grass
Dog's-tail Grass. *Cynosurus cristatus*
Dog's Tansy. *Potentilla anserina*
Dog's-tongue. *Cynoglossum officinale*
Dog's-tooth Grass. *Triticum caninum* and *Cynodon Dactylon*
Dollee-wood. *Myristica surinamensis*
Dolphin-flower. *Delphinium Consolida*
Donkey's-eye. The seed of *Mucuna pruriens*
Donninethell. An old name for *Galeopsis Ladanum*
Doon-tree. *Doona zeylanica*
Doorda, or Doorwa, Grass. *Cynodon Dactylon*

Door-weed. *Polygonum aviculare*
Dorn-boom, of S. Africa. *Acacia horrida*
Double-leaf. *Listera ovata*
Double-tongue. *Ruscus Hypoglossum*
Doura. See Dhourra
Dove-Dock. *Tussilago Farfara*
Dove-flower, Waxen. *Peristeria cerina*
Dove-Orchid. See Orchid
Dove-plant. *Peristeria elata*
Dove-wood. *Alchornea latifolia*
Dove's-foot. *Geranium molle* and *G. maculatum*
Down-thistle. *Onopordon Acanthium*
Down-tree. *Ochroma Lagopus*
Down-weed. *Filago germanica*
Downy Ling. *Eriophorum polystachyon*
Drake, Drawk, Dravick, or Droke. Applied to Darnel, Cockle, or weeds in general
Dragon, Green. *Arisæma Dracontium*
Dragon-Arum. *Arum Dracunculus*
Dragon-bushes. *Linaria vulgaris*
Dragon-claw. *Pterospora Andromeda* and *Corallorrhiza odontorrhiza*
Dragon-plant. The genus *Dracæna*
 Hardy. *Dracæna (Cordyline) indivisa*
 New Zealand. *Dracæna (Cordyline) australis*
 Oval-leaved. *Dracæna borealis*
 Zebra-striped. *Dracæna Goldieana*
Dragon-tree, Canary Islands. *Dracæna Draco*
Dragon-root. *Arisæma Dracontium*
Dragon-wort. *Polygonum Bistorta*
Dragon's Female. An old name for *Arum Dracunculus*
Dragon's-blood-plant. *Calamus (Dæmonorops) Draco*; also applied to *Geranium Robertianum*
Dragon's-blood-tree. *Dracæna Draco*
Dragon's-eye. *Nephelium Longanum*
Dragon's-head. The genus *Dracocephalum*
 Argunsk. *Dracocephalum argunense*
 Austrian. *Dracocephalum austriacum*
 Betony-leaved. *Dracocephalum grandiflorum*
 False. The genus *Physostegia*
 Hyssop-leaved. *Dracocephalum Ruyschianum*
 Imbricated False. *Physostegia imbricata*
 Large-flowered. *Dracocephalum grandiflorum*
 Prickly-leaved. *Dracocephalum peregrinum*
 Tasmanian. *Diuris sulphurea*
 Toothed False. *Physostegia denticulata*
 Twin-flowered. *Dracocephalum peregrinum*
Dragon's-mouth. *Antirrhinum majus*, *Arum crinitum*, and *Epidendrum macrochilum*
Dragons' Water. *Calla palustris*
Draper's Teasel. *Dipsacus fullonum*
Drias-plant. *Thapsia garganica*
Drooping Avens. *Geum rivale*
Drooping Tulip. *Fritillaria Meleagris*
Drop-seed Grass. See Grass
Drop-wort. *Spiræa Filipendula*
 False Water. *Tiedemannia teretifolia*
 Double-flowered. *Spiræa Filipendula plena*
 Hemlock. *Œnanthe crocata*

Dropwort, Water. *Œnanthe fistulosa*
Western. *Gillenia trifoliata*
Drum-stick-tree. *Cathartocarpus conspicua*
Drunken-wort. An old name for Tobacco
Dryad, White-flowered. *Dryas octopetala*
Yellow-flowered. *Dryas Drummondi*
Dry-rot Fungus. *Merulius lachrymans*
Duck's-foot. *Alchemilla vulgaris* and *Podophyllum peltatum*
Duck-weed, or Duck-meat. The genus *Lemna*
Common. *Lemna minor*
Coral-berried, or Fruiting. *Nertera depressa*
Greater. *Lemna polyrrhiza*
Ivy-leaved. *Lemna trisulca*
Lesser. *Lemna minor*
Thick-leaved. *Lemna gibba*
Tropical. *Pistia Stratiotes*
Duck-meat. See Duck-weed
Duck-mud. *Conferva rivularis* and other delicate green-spored *Algæ*
Duffle. *Verbascum Thapsus*
Dulse. *Rhodymenia palmata* and *Iridæa edulis*
Pepper. *Laurencia pinnatifida*
Dumb Cane. *Caladium seguinum*
Dumpling, Greater. *Mesembryanthemum obcordellum*
Small. *Mesembryanthemum minimum*
Dunga-runga-tree. *Notelæa ovata*
Dunny Nettle. Another name for Dead Nettle
Durgan Wheat. Bearded Wheat
Durian, or Duryon, Tree. *Durio zibethinus*
Durra. See Dhourra
Dusty Miller. *Primula Auricula* and *Senecio Cineraria*
Dutch Agrimony. *Eupatorium cannabinum*
Dutch-Clover. *Trifolium repens*
Dutch-Mice. *Lathyrus tuberosus*
Dutch-Myrtle. *Myrica Gale*
Dutch-Rush. *Equisetum hyemale*
Dutchman's-breeches. *Dicentra cucullaria*
Dutchman's-butter. *Cassia glandulosa*
Dutchman's-pipe. *Aristolochia Sipho*
Dutch-Pink-plant. *Reseda Luteola*
Dwale. *Atropa Belladonna*
Deadly. *Acnistus arborescens*
Dwarf Bay. See Bay
Dwarf Cornel. See Cornel
Dwarf Elder. See Elder
Dway-berries. *Atropa Belladonna*
Dyer's Broom. See Broom
Dyer's-grapes. *Phytolacca decandra*
Dyer's Green-weed. *Genista tinctoria*
Dyer's Rocket. *Reseda Luteola*
Dyer's-weed. *Reseda Luteola*
Double. *Genista tinctoria fl.-pl.*
Dyer's Yellow-weed. *Reseda Luteola*

Eagle-wood. *Aquilaria Agallochum* and *A. ovata*
Ear-cockle, or Purples. A disease in Wheat, in which the grain is attacked by multitudes of very minute worms of the genus *Vibrio*

Ear-drops, Lady's. The garden *Fuchsia*
Ear-wort. *Rhavicallis rupestris*
Ear-ring Flower. The garden *Fuchsia*
Earning-grass. *Pinguicula vulgaris*
Earth-balls (Truffles). *Tuber cibarium*
Earth-gall. *Erythræa Centaurium* and other plants of the *Gentian* tribe
Chinese. *Picria Fel-terræ*
Of the Malays. *Ophiorrhiza Mungos*
Earth-jelly. *Tremella Auricula*
Earth-moss. The genus *Phascum*
Earth-nut, or Earth-chestnut. *Bunium flexuosum*
Large. *Bunium Bulbocastanum*
Earth-quakes. *Briza media*
Earth-smoke. *Fumaria officinalis*
Earth-star. The genus *Geaster*
Easter-flower, Mexican. *Poinsettia pulcherrima*
Easter Giant, Easter Ledges, Easter Magiants, or Easter Mangiants. *Polygonum Bistorta*
Eboe-tree. *Dipterix eboënsis*
Ebony-tree, Ceylon. *Diospyros Ebenum*
Coromandel. *Diospyros Melanoxylon*
East Indian. *Dalbergia latifolia, Diospyros Ebenaster,* and *D. Melanoxylon*
False. *Cytisus Laburnum*
Green. *Excœcaria glandulosa, Bignonia leucoxylon,* and *Jacarandu ovalifolia*
Jamaica. *Brya Ebenus*
Mauritius. *Diospyros reticulata*
Mountain. *Bauhinia variegata*
St. Helena. *Melhania Melanoxylon*
Senegal. *Dalbergia Melanoxylon*
W. Indian. *Brya Ebenus*
Edder-wort. An old name for *Arum Dracontium*
Edelweiss. *Gnaphalium Leontopodium*
New Zealand. *Gnaphalium (Helichrysum) Colensoi* and *G. grandiceps*
Edge-weed. *Œnanthe Phellandrium*
Eel-grass. See Grass
Eel-ware, or Eel-beds. *Ranunculus fluitans* and various pond-weeds
Egg-berry. *Prunus Padus*
Egg-plant, or Bringall. *Solanum Melongena (S. origerum)*
Red-fruited. *Solanum Melongena fructu-rubro*
Violet-fruited. *Solanum Melongena fructu-violaceo*
White-fruited. *Solanum Melongena fructu-albo*
Yellow-fruited. *Solanum Melongena fructu-luteo*
Eglantine. *Rosa Eglanteria, Rosa rubiginosa,* and *Rubus Eglanteria*
Eggs-and-Bacon. *Linaria vulgaris* and *Lotus corniculatus*
Eggs-and-Butter. *Linaria vulgaris* and the Buttercup
Eileber. An old name for *Erysimum Alliaria*
Egyptian Rose. *Scabiosa arvensis* and *S. atropurpurea*
Egyptian Thorn. *Cratægus Pyracantha*
Elder, or Elder-berry. *Sambucus nigra*
American. *Sambucus Canadensis*

Elder, Autumn-flowering. *Sambucus Canadensis*
Bishop's. See Bishop's Elder
Californian. *Sambucus glauca*
Chinese. *Sambucus chinensis*
Cut-leaved. *Sambucus nigra var. laciniata*
Dwarf. *Sambucus humilis* and *Pilea grandis*
Ground. *Sambucus Ebulus*
Hart's. *Sambucus racemosa*
Herbaceous. *Sambucus Ebulus*
Horse. See Horse Elder
Marsh, or Water. *Viburnum Opulus*
N. American. The genus *Iva*
Parsley-leaved. *Sambucus nigra var. laciniata*
Poison. *Rhus venenatum*
Red-berried. *Sambucus pubens* and *S. racemosa*
Rose. See Rose-Elder
Rosy-flowered. *Sambucus nigra var. roseæflora*
Spanish. *Artanthe adunca*
Variegated Cut-leaved. *Sambucus laciniata variegata*
Water. See Elder, Marsh
Wild, of N. America. *Aralia hispida*
Elecampane. *Inula Helenium*
Elemi-Resin-plant, Brazilian. Several species of *Icica*
Mauritius. *Colophonia Mauritiana*
Mexican. *Amyris elemifera*
Elephant's-Apple. *Feronia elephantum*
Elephant's-ear. The genus *Begonia* and *Siphonanthus hastatus*
Elephant's-foot. *Testudinaria elephantipes* and the genus *Elephantopus*
Carolina. *Elephantopus carolinensis*
Rough. *Elephantopus scaber*
Elephant's-grass. See Grass
Elephant's-trunk-plant. *Martynia proboscidea* and *Adenium namaquanum*
Elephant's Vine. *Cissus latifolia*
Eleven-o'Clock-Lady. *Ornithogalum umbellatum*
Elf-Dock. *Inula Helenium*
Elk-bark. *Magnolia glauca*
Elk-tree, Sour-leaved. *Andromeda arborea*
Elk's-horn Fern. See Fern
Elm. The genus *Ulmus*
Acute-leaved. *Ulmus campestris var. acutifolia*
American, or White. *Ulmus americana*
American Cork. *Ulmus racemosa*
American False. *Celtis occidentalis*
American Rock. *Ulmus racemosa*
American Small-leaved. *Ulmus alata*
Birch-leaved. *Ulmus campestris var. betulæfolia*
Black Irish. *Ulmus montana var. nigra*
Broad-leaved Common. *Ulmus campestris var. latifolia*
Camperdown Weeping. *Ulmus montana var. pendula*
Canterbury Seedling. *Ulmus glabra var. major*
Cevennes. *Ulmus montana var. cebenensis*
Chichester. *Ulmus glabra var. vegeta*

Elm, Chinese. *Ulmus campestris var. chinensis*
Clammy. *Ulmus campestris var. viscosa*
Common. *Ulmus campestris*
Concave-leaved. *Ulmus campestris var. concavæfolia*
Cork-barked. *Ulmus suberosa*
Corky White. *Ulmus racemosa*
Cornish. *Ulmus campestris var. cornubiensis*
Curled-leaved. *Ulmus montana var. crispa*
Declining-branched. *Ulmus major*
Downton. *Ulmus glabra var. pendula*
Dutch, or Sand. *Ulmus suberosa*
Dwarf. *Ulmus campestris var. nana*
Dwarf Siberian. *Ulmus pumila*
E. Indian. *Ulmus (Holoptelea) integrifolia*
English. *Ulmus campestris*
Exeter. *Ulmus montana var. fastigiata*
Feathered. *Ulmus glabra*
Fleetbeck. *Ulmus glabra var. ramulosa*
Glanded-leaved. *Ulmus glabra var. glandulosa*
Golden. *Ulmus Dampieri aurea*
Golden-variegated. *Ulmus campestris foliis aureis*
Greater Wych. *Ulmus montana var. major*
Hertfordshire. *Ulmus suberosa vars. latifolia* and *angustifolia*
Himalayan. *Ulmus Wallichiana*
Hooded-leaved. *Ulmus campestris var. cucullata*
Hornbeam-leaved. *Ulmus carpinifolia*
Huntingdon. *Ulmus glabra var. vegeta*
Japanese. *Ulmus Kaki*
Jersey. *Ulmus campestris var. sarniensis*
Kidbrook. *Ulmus campestris var. virens*
Moose. *Ulmus fulva*
Monumental. *Ulmus monumentalis*
Purple-leaved. *Ulmus campestris var. purpurea*
Red, or Slippery. *Ulmus fulva*
Red English. *Ulmus campestris var. stricta*
Rough-leaved Wych. *Ulmus montana var. rugosa*
Sand. See Elm, Dutch
Scampston. A variety of *Ulmus glabra*
Scotch. *Ulmus montana*
Slippery. See Elm, Red
Slippery, of California. *Fremontia californica*
Smaller Wych. *Ulmus montana var. minor*
Small-leaved Common. *Ulmus campestris var. parvifolia*
Smooth-leaved. *Ulmus glabra*
Smooth-leaved Wych. *Ulmus montana var. glabra*
Spanish. *Cordia Geraschanthus (Geraschanthus vulgaris)*
Spreading-flowered. *Ulmus effusa*
Twiggy. *Ulmus campestris var. viminalis*
Twisted. *Ulmus campestris var. tortuosa*
Upright-branched. *Ulmus suberosa var. erecta*
Variegated. *Ulmus campestris var. viminalis variegata*
Weeping. *Ulmus glabra var. pendula*
Weeping Wych. *Ulmus montana var. pendula*

Elm, White. See Elm, American White-barked. *Ulmus suberosa var. alba*
White-variegated. *Ulmus campestris foliis variegatis*
Winged. *Ulmus alata*
Witch or Wych. *Ulmus montana*
Yoke. *Carpinus Betulus*
Else-Dock. An old name for *Elecampane*
Eltrot. *Heracleum Sphondylium*
Emetic-weed. *Lobelia inflata*
Emony. A corruption of "Anemone"
Enchanter's Nightshade. *Circæa lutetiana*
Endive. *Cichorium Endivia*
Enemy. A corruption of "Anemone"
English Maidenhair. *Asplenium Trichomanes*
English Mercury. *Chenopodium Bonus-Henricus*
English Sea Grape. *Salicornia herbacea*
Ergot of Rye. See Rye
Ensilage. Forage plants and grasses (chiefly Maize, Clover, Millet, and Alfalfa) preserved in a green state through the winter by storing them in water-tight and air-tight pits (silos). Extensively used in the United States
Eringo, or Eryngo. The genus *Eryngium*
Ermine-chop. *Mesembryanthemum erminium*
Er-nut. See Earth-nut
Ers. *Ervum Ervilia*
Ervalenta (Revalenta) plant. *Ervum Lens*
Eryngo, or Sea-Holly. The genus *Eryngium*
Alpine. *Eryngium alpinum*
Amethyst. *Eryngium amethystinum*
Dwarf. *Eryngium pusillum*
Field. *Eryngium campestre*
Flat-leaved. *Eryngium planum*
Giant. *Eryngium giganteum*
Ivory. *Eryngium eburneum*
Palm-leaved. *Eryngium pandanifolium*
Pine-apple-leaved. *Eryngium bromeliæfolium*
Sea. *Eryngium maritimum*
White-spined. *Eryngium spinâ-albâ* and *Eryngium Bourgati*
Esparto-grass. See Grass
Espibawn. *Chrysanthemum Leucanthemum*
Euphorbium Gum-plant. *Euphorbia resinifera*
Euphrasy, or Eye-bright. *Euphrasia officinalis*
Evening-Flower. The genus *Hesperantha*
Grass-leaved. *Hesperantha graminifolia*
Evening-Primrose. The genus *Œnothera*
Common. *Œnothera biennis*
Dandelion-leaved. *Œnothera taraxacifolia*
Dwarf. *Œnothera pumila*
Glaucous. *Œnothera glauca*
Fraser's. *Œnothera Fraseri*
Hairy-calyxed. *Œnothera trichocalyx*
James's. *Œnothera Jamesii*
Large-flowered. *Œnothera Lamarckiana*
Large Rose-tinted. *Œnothera marginata*
Missouri. *Œnothera missouriensis* (*Œ. macrocarpa*)
Mountain. *Œnothera montana*
Narrow-leaved. *Œnothera linearis*

Evening - Primrose, Orange - flowered. *Œnothera bistorta Veitchi*
Scalloped-leaved. *Œnothera sinuata*
Small-toothed-leaved. *Œnothera serrulata*
Swamp. *Œnothera riparia*
Tall White. *Œnothera speciosa*
Tufted. *Œnothera cœspitosa*
Young's. *Œnothera Youngi*
Ever Fern. *Polypodium vulgare*
Evergreen Cliver. *Rubia peregrina*
Everlasting, or Everlasting-Flower. Various species of *Helichrysum, Gnaphalium,* and *Antennaria*
Annual. *Xeranthemum annuum*
Australian. *Helichrysum lucidum* (*H. bracteatum*) and *Helipterum Manglesii*
Bracted. *Helichrysum bracteatum*
Bridal. *Gnaphalium Leontopodium*
Carpathian. *Antennaria carpatica*
Common American. *Gnaphalium polycephalum* and *G. decurrens*
Common Shrubby. *Helichrysum Stœchas*
Crimson. *Gnaphalium sanguineum*
Crimson Cape. *Helichrysum* (*Astelma*) *eximium*
Dwarf. *Helichrysum bracteatum nanum*
Jamaica. *Gnaphalium americanum*
Large-flowered. *Helichrysum macranthum*
Moor, or Mountain. *Antennaria dioica*
Mouse-ear. *Antennaria plantaginifolia*
Pearly. *Antennaria margaritacea*
Pink-rosette. *Rhodanthe Manglesii*
Plantain - leaved. *Antennaria plantaginifolia*
Red Globe. *Gomphrena globosa*
Rosy. *Rhodanthe Manglesii maculata*
Rosy-flowered Mountain. *Antennaria dioica minor*
Swan River. *Rhodanthe Manglesii*
Sweet-scented. *Gnaphalium polycephalum*
Tasmanian. *Helichrysum apiculatum*
Tree. *Gnaphalium arboreum*
Winged. *Ammobium alatum*
Yellow. *Helichrysum orientale* and *H.* (*Gnaphalium*) *arenarium*
Everlasting Pea. *Lathyrus latifolius* and other species
Eve's-cushion. *Saxifraga hypnoides*
Ewe-daisy. *Potentilla Tormentilla*
Ewe-gowan. *Bellis perennis*
Extinguisher-Moss. See Moss
Eye (The Pink). *Dianthus*
Eye-bright. *Euphrasia officinalis*
Blue. An old name for *Myosotis repens*
Red. *Bartsia Odontites*
W. Indian. *Euphorbia maculata*
Eye-bright-Cow-wheat. The genus *Bartsia*
Alpine. *Bartsia alpina*
Red. *Bartsia Odontites*
Viscid. *Bartsia viscosa*
Eye-seed. The seed of *Salvia Verbenacea*
Fae-berry. See Fay-berry
Fair Grass, or Fair Days. *Potentilla anserina*
Fair Maids of February. *Galanthus nivalis*

Fair Maids of France. *Ranunculus aconitifolius, Saxifraga granulata,* and *Achillea Ptarmica*
Fair Maids of Kent. *Ranunculus aconitifolius*
Fairies' Butter, or Fairy Butter. *Tremella albida* and *T. arborea*
Fairy Fingers. *Digitalis purpurea*
Fairy Flax. *Linum catharticum*
Fairy-ring Champignon. See Champignon
Faitour's Grass. Probably *Euphorbia Esula*
Fallen Stars. *Nostoc commune*
Fall-poison, American. *Amianthium muscætoxicum*
False-Asphodel, Downy. *Tofieldia pubens*
Marsh. *Tofieldia palustris*
Small-calyxed. *Tofieldia calyculata*
False-Indigo, Downy. *Amorpha herbacea* (*A. pubescens*)
Dwarf. *Amorpha nana*
Fragrant. *Amorpha fragrans*
Shrubby. *Amorpha fruticosa*
Smooth. *Amorpha glabra*
Yellow-woolled. *Amorpha croceo-lanata*
False-Goat's-beard, Japanese. *Astilbe japonica*
Red-flowered. *Astilbe rubra*
River-side. *Astilbe rivularis*
False-Mermaid. *Floerkia proserpinacoides*
False-Oat. See Oat
False-Parsley. *Æthusa Cynapium*
False-Rhubarb. *Thalictrum flavum*
Fame-flower. *Talinum teretifolium*
Fancy (Pansy). *Viola tricolor*
Fan-flower, Tasmanian. *Scævola cuneiformis*
Fare-nut, or Vare-nut. *Bunium flexuosum*
Fare-well-Summer. *Saponaria officinalis* and some species of *Aster*
Farkle-berry. *Vaccinium arboreum*
Faselles. An old name for Kidney-beans.
Fat Hen. *Chenopodium album* and other species of *Chenopodium* and *Atriplex*
Fat Pork. *Clusia flava*
Faverell. An old name for *Veronica Anagallis*
Faverole. An old name for *Arum Dracunculus*
Fay-berry, Fae-berry, or Fea-berry. Another name for the Goose-berry
Fea-berry. See Fay-berry
Feather-beds. *Chara vulgaris*
Feather-crown. *Ptilostephium trifidum*
Feather-few. Another name for Fever-few
Feather-foil. *Hottonia palustris*
Feather-grass. *Stipa pennata*
Feather-Moss. See Moss
Feathered Columbine. *Thalictrum aquilegifolium*
Feather-top Grass. *Calamagrostis Epigejos*
Feld-wood. *Verbascum Thapsus* (?), or a species of *Gentian* (?)
Fellon-grass. *Imperatoria Ostruthium*
Fellon-herb. *Artemisia vulgaris*
Fellon-weed. *Senecio Jacobæa*
Fellon-wood, or Fellon-wort. *Solanum Dulcamara*

Felt. *Triticum repens*
Felt-wort. *Verbascum Thapsus*
Fel-wort. *Gentiana Amarella* and the genus *Swertia*
Marsh. *Swertia perennis*
Fen-berry. *Vaccinium Oxycoccos*
Fen-Orchis. See Orchis
Fen-Rue. *Thalictrum flavum*
Fennel, or Fenkelle. *Fœniculum vulgare*
Asparagus-leaved Giant. *Ferula asparagifolia*
Broad-leaved Giant. *Ferula Ferulago*
Common Giant. *Ferula communis*
Dog's. *Athemis Cotula*
Furrowed Giant. *Ferula sulcata*
Giant. The genus *Ferula*
Glaucous Giant. *Ferula glauca*
Hog's. *Peucedanum officinale*
Horse. *Seseli Hippomarathrum*
Knotted Giant. *Ferula nodiflora*
Sea. *Crithmum maritimum*
Sweet. *Fœniculum officinale*
Tangier Giant. *Ferula tingitana*
Water. *Callitriche verna*
Fennel-flower. The genus *Nigella*
Common. *Nigella damascena*
Small. *Nigella sativa*
Spanish. *Nigella hispanica*
Wild. *Nigella arvensis*
Fenugreek, Common. *Trigonella Fœnumgræcum*
Fern. A general name for plants of the *Filix* Family
Adder's. *Polypodium vulgare*
Adder's-tongue. *Ophioglossum vulgatum*
Alpine Bladder. *Cystopteris alpina*
American Grape. *Botrychium lunarioides*
American Maiden-hair. *Adiantum pedatum*
Ash-leaf. *Marattia fraxinea*
Australian Tree. *Dicksonia antarctica*
Basket. *Nephrodium Filix-mas*
Bladder. The genus *Cystopteris*
Beech. *Polypodium Phegopteris*
Bird's-foot Rock-Brake. *Pellæa ornithopus*
Bird's-nest. *Asplenium Nidus*
Black Oak. *Adiantum nigrum*
Boss. Various species of *Nephrodium*
Brake, or Bracken. *Pteris aquilina*
Branching Maiden-hair. *Adiantum formosum*
Broad Prickly-toothed. *Lastrea dilatata*
Bristle. The genus *Trichomanes*
Brittle Bladder. *Cystopteris fragilis*
Buckler. The genus *Lastrea*
Burke's Tree. *Cyathea Burkei*
Californian Chain. *Woodwardia radicans*
Californian Lace. *Cheilanthes gracillima*
Chain. The genus *Woodwardia*
Chignon. *Cibotium regale*
Christmas Shield. *Aspidium acrostichoides*
Cinnamon. *Osmunda cinnamomea*
Clayton's Flowering. *Osmunda Claytoniana*
Cliff Brake. The genus *Pellæa*
Climbing Polypody. *Niphobolus heteractis*
Climbing Shield. *Aspidium capense*
Climbing Snake's-tongue. *Lygodium scandens*

Fern, Cloak. The genus *Nothochlæna*
Comb, or Rush. The genus *Schizæa*
"Conjuror of Chalgrave's." *Puccinia Anemones*
Crested Shield. *Lastrea cristata*
Crisped Hart's-tongue. *Scolopendrium crispum*
Culcit. See Fern, Cushion
Curled Rock-Brake. *Allosorus crispus*
Cushion, or Culcit. *Dicksonia (Balantium) Culcita*
Darsham. *Lastrea cristata*
Deer. *Lomaria Spicant*
Dwarf Brazilian Tree. *Blechnum brasiliense*
Eagle. *Pteris aquilina*
Edible. *Pteris esculenta*
Elk's-horn. *Platycerium alcicorne*
Elk's-horn, E. Indian. *Platycerium biforme*
Elk's-horn, Guinea. *Platycerium Stemmaria*
Elk's-horn, Queensland. *Platycerium grande*
Elk's-horn, Wallich's. *Platycerium Wallichii*
Elk's-horn, Willinck's. *Platycerium Willincki*
English Maiden-hair, of New Zealand. *Adiantum æthiopicum*
Ever. *Polypodium vulgare*
Fijian Hare's-foot. *Davallia fijiensis*
Filmy. A name applied to those kinds which have pellucid or transparent fronds, as various species of *Hymenophyllum, Todea,* and *Trichomanes*
Finger. *Asplenium Ceterach*
Florida Ribbon. *Vittaria lineata*
Flowering. *Osmunda regalis*
Grape. The genus *Botrychium*
Gray Tree. *Cyathea medullaris*
Great Maiden-hair. *Adiantum Farleyense*
Ground. *Nephrodium (Lastrea) Thelypteris*
Grove. The genus *Alsophila*
Hairy Lip. *Cheilanthes vestita*
Hard. *Blechnum boreale (B. Spicant)*
Hare's-foot. The genus *Davallia*
Hartford. *Lygodium palmatum*
Hart's-tongue. The genus *Scolopendrium*
Hay-scented. *Cheilanthes odora* and *Dicksonia pilosiuscula*
Heath. *Lastrea (Aspidium) Oreopteris*
Herring-bone. *Blechnum boreale*
Holly. *Aspidium (Polystichum) Lonchitis*
Jamaica Tree. *Dicksonia dissecta*
Japan Climbing. *Lygodium scandens*
Japan Hare's-foot. *Davallia Mariesi*
Japan Ostrich. *Struthiopteris japonica*
Killarney. *Trichomanes radicans*
King. *Osmunda regalis*
Lace. *Cheilanthes gracillima*
Ladder. *Nephrolepis cordifolia*
Lady. *Athyrium Filix-fœmina*
Lesser Adder's-tongue. *Ophioglossum lusitanicum*
Limestone. *Polypodium calcareum*
Lip. The genus *Cheilanthes*
Lobed Prickly Shield. *Polystichum lobatum*
Maiden-hair. *Adiantum Capillus-Veneris*
Maiden-hair, American. *Adiantum pedatum*

Fern, Male. *Lastrea Filix-mas*
Marsh, or Marsh Shield. *Lastrea (Aspidium) Thelypteris*
Miniature Basket. *Adiantum dolabriforme*
Moss. *Polypodium Dryopteris*
Mountain Bladder. *Cystopteris montana*
Mountain Shield. *Lastrea (Aspidium) Oreopteris*
Moon. *Botrychium Lunaria*
Moon-wort, of New Zealand. *Botrychium ternatum*
Mule. *Hemionitis palmata* and *Scolopendrium Hemionitis*
Narrow-fronded Bladder. *Cystopteris angustata*
Narrow-fronded Chain. *Woodwardia angustifolia*
Narrow Prickly-toothed. *Lastrea spinulosa*
Netted Chain. *Woodwardia angustifolia*
Nevada Wood. *Aspidium nevadense*
New Zealand Tree. *Dicksonia antarctica, D. lanata,* and *D. squarrosa*
Norfolk Island Tree. *Alsophila excelsa*
Oak. *Polypodium Dryopteris.* Formerly applied to *P. vulgare*
Oregon Cliff-Brake. *Pellæa densa*
Oregon Rock-Brake. *Allosorus acrostichoides*
Ostrich. *Struthiopteris pennsylvanica (S. germanica)*
Para, of New Zealand. *Marattia salicina*
Parsley. *Allosorus crispus*
Pod. *Ceratopteris thalictroides* and *Ellobocarpus oleraceus*
Polypody. See Polypody
Prickly Shield. *Aspidium (Polystichum) aculeatum*
Queensland Elk's-horn. *Platycerium grande*
Rattle-snake. *Botrychium virginicum*
Recurved Prickly-toothed. *Lastrea æmula (L. fœniseeii)*
Rigid Shield. *Lastrea rigida*
Rock-Brake. *Allosorus crispus*
Roseate Maiden-hair. *Adiantum rubellum*
Royal, or Royal Osmund. *Osmunda regalis*
Royal Calabar. *Litobrochia Currori*
Rue-leaved. The genus *Gymnogramma*
Rush. See Fern, Comb
St. Helena Tree. *Dicksonia arborescens*
Scale, or Scaly. *Asplenium Ceterach*
Scented. *Lastrea Oreopteris*
Sensitive. *Onoclea sensibilis*
Shield. The genus *Aspidium (Lastrea)*
Sierra Shield. *Aspidium nevadense*
Silvery Tree. *Cyathea dealbata*
Snake. *Blechnum boreale (B. Spicant)*
Soft Prickly Shield. *Polystichum angulare*
Spider. *Pteris serrulata*
Spleen-wort. See Spleen-wort
Stag's-horn. *Platycerium grande* and other species
Stone. *Ceterach officinarum*
Sweet. *Myrrhis odorata*
Sword. The genus *Niphopteris*
Tall Hart's-tongue. *Scolopendrium erectum*
Tara, of Tasmania. *Pteris esculenta*
Tree. Various species of *Dicksonia, Alsophila,* and *Cyathea*

and Foreign Plants, Trees, and Shrubs. 45

Fern, Tree, Australian. *Dicksonia antarctica*
Tree, Burke's. *Cyathea Burkei*
Tree, Dwarf Brazilian. *Blechnum brasiliense*
Tree, Jamaica. *Dicksonia dissecta*
Tree, New Zealand. *Dicksonia antarctica, D. lanata,* and *D. squarrosa*
Tree, New Zealand Gray. *Cyathea medullaris*
Tree, Norfolk Island. *Alsophila excelsa*
Tree, St. Helena. *Dicksonia arborescens*
Tree, Silvery. *Cyathea dealbata*
Tree, Tasmanian. *Dicksonia antarctica*
Toothed Bladder. *Cystopteris dentata*
Tunbridge, or Common Filmy. *Hymenophyllum Tunbridgense*
Twining String. *Lygodium articulatum*
Virginian Rattle-snake. *Botrychium virginicum*
Walking-leaf. *Camptosorus rhizophyllus*
Walking American. *Lycopodium alopecuroides*
Wall. *Polypodium vulgare*
Water. *Osmunda regalis*
Whip-stick. *Alsophila Leichardtiana*
White-Maiden-hair. *Asplenium Ruta muraria*
Wilson's Filmy. *Hymenophyllum Wilsoni*
Wood. *Polypodium vulgare*
Wood, American. The genus *Aspidium (Lastrea)*
Woolly Cloak. *Nothochlæna distans*
Zig-zag Cliff-Brake. *Pellæa flexuosa*
Fern-bush, Sweet. *Comptonia asplenifolia*
Fernambuc-wood. The wood of *Cæsalpinia echinata*
Fescue-grass. See Grass
Festoon Pine. *Lycopodium rupestre*
Fetch (Vetch). *Vicia sativa*
Fetches, Wild. *Vicia cracca*
Fever-bush. *Lindera Benzoin*
Feverfew, Alpine. *Pyrethrum alpinum*
Bastard. *Parthenium Hysterophorus*
Common. *Pyrethrum Parthenium*
Late-flowering. *Pyrethrum serotinum*
Marsh. *Pyrethrum (Chrysanthemum) lacustre*
Narrow-leaved. *Pyrethrum achilleæfolium*
Rosy-flowered. *Pyrethrum carneum (P. roseum)*
Willemot's. *Pyrethrum Willemoti*
Fever Gum-tree. See Fever-tree
Fever-plant. *Brugmansia (Datura) arborea* and *Pæderia fœtida*
Sierra Leone. *Ocymum viride*
Fever-root. *Pterospora Andromeda* and *Triosteum perfoliatum*
Fever-tree, or Fever-Gum-tree. *Eucalyptus globulus*
Georgia. *Pinckneya pubens*
Fever-twig. *Celastrus scandens*
Fever-weed. *Gerardia pedicularia*
Fever-wort. *Erythræa Centaurium*
American. The genus *Triosteum*
Field Ash. *Pyrus Aucuparia*
Field Cypress. *Ajuga Chamæpitys*
Field Kale. *Sinapis arvensis*
Field Madder. *Sherardia arvensis*
Fig-Marigold. The genus *Mesembryanthemum*

Fig-Marigold, Annual. *Mesembryanthemum tricolor*
Bright-red. *Mesembryanthemum conspicuum*
Broad White. *Mesembryanthemum testiculare*
Golden. *Mesembryanthemum aureum*
Heart-leaved. *Mesembryanthemum cordifolium*
Orange. *Mesembryanthemum aurantiacum*
Scimetar-leaved. *Mesembryanthemum aciniforme*
Small Truncated. *Mesembryanthemum truncatellum*
Smallest. *Mesembryanthemum minimum*
Three-coloured. *Mesembryanthemum tricolor*
Wolf's-chop. *Mesembryanthemum lupinum*
Fig-tree. The genus *Ficus*
Adam's. *Musa paradisiaca*
Balsam, of the W. Indies. *Clusia alba, C. flava, C. rosea,* and other species
Barbary. *Opuntia vulgaris*
Bastard. The genus *Opuntia*
Black. *Ficus laurifolia*
Common. *Ficus Carica*
Creeping. *Ficus repens*
Devil's, or Infernal. *Argemone mexicana*
Dwarf Creeping. *Ficus repens var. minima*
Dyer's. *Ficus tinctoria*
Hottentot's. *Mesembryanthemum edule*
Indian. The genus *Opuntia,* especially *O. vulgaris*
Infernal. See Fig-tree, Devil's
Keg, of Japan. *Diospyros Kaki*
Laurel-leaved S. American. *Ficus pertusa*
Laurel-leaved, W. Indian. *Ficus laurifolia*
Mangrove. *Rhizophora Mangle*
Moreton Bay. *Ficus macrophylla*
Pharaoh's. *Sycomorus antiquorum*
Poplar-leaved. *Ficus religiosa*
Red. *Ficus pedunculata*
Sacred. *Ficus religiosa*
White. *Ficus ochroleuca*
Wild, W. Indian. *Clusia flava*
Willow-leaved. *Ficus pedunculata*
Figs. *Callithamnion floridulum*
Fig-wort. The genus *Scrophularia;* also *Ranunculus Ficaria*
Balm-leaved. *Scrophularia Scorodonia*
Barbary. *Scrophularia mellifera*
Cape. *Phygelius capensis*
Knotted-rooted. *Scrophularia nodosa*
Variegated. *Scrophularia nodosa variegata*
Yellow-flowered. *Scrophularia vernalis*
Water. *Scrophularia aquatica*
Fiji, or Fuji, of Japan. *Wistaria sinensis*
Filbert, Common. *Corylus Avellana*
Cut-leaved. *Corylus Avellana var. laciniata*
Frizzled. *Corylus Avellana var. crispa*
Purple-leaved. *Corylus Avellana var. purpurea*
Red. *Corylus Avellana var. tubulosa*
White. *Corylus Avellana var. tubulosa alba*
Filberts, W. Indian. The seeds of *Entada scandens*
Fimble. The male plant of *Cannabis sativa*
Fin. *Ononis arvensis*
Finger-flower. *Digitalis purpurea*
Finger-grass. See Grass

Fingers-and-Thumbs, or Fingers-and-Toes. *Lotus corniculatus*
Finocchio, or Finicho. *Fœniculum dulce*
Asses'. *Fœniculum piperitum*
Fiorin, or Fiorin-Grass (Butter-Grass). *Agrostis stolonifera*
Fir. A general name for various species of *Abies*, *Picea*, and *Pinus*
Algerian Silver. *Picea numidica*
American Black, or Double Spruce. *Abies nigra*
American White, or Single Spruce. *Abies alba*
Balm of Gilead, or Balsam. *Abies balsamea*
Black Mountain. *Picea cephalonica*
Black Spruce. *Abies nigra*
Californian Silver. *Picea concolor*
Common Silver. *Picea pectinata*
Crimean Silver. *Picea Nordmanniana*
Double, or Double Balsam. See Fir, Fraser's
Double Spruce. *Abies nigra*
Douglas Spruce. *Abies (Pseudo-tsuga) Douglasii*
Fraser's, Double, or Southern, Balsam. *Abies Fraseri*
Great Californian Silver. *Picea (Pinus) grandis*
Hemlock Spruce. *Abies Canadensis*
Intermediate. *Abies Fortunei*
Japan Silver. *Picea firma*
Leafy-bracted Silver. *Picea bracteata*
Manchurian Silver. *Picea holophylla*
Mexican Silver. *Picea religiosa* and var. *glaucescens*
Mount Enos. *Picea cephalonica*
Noble Silver. *Picea nobilis*
Nootka. *Abies Douglasii*
Norway Spruce. *Abies excelsa*
Oyamel. *Pinus religiosa*
Parasol. The genus *Sciadopitys*
Patton's Californian. *Abies Pattoniana*
Pigmy Scotch. *Pinus sylvestris var. nana*
Pitch, or Siberian Silver. *Picea Pichta*
Plum. *Podocarpus andina* and *Prumnopitys elegans*
Prickly. *Abies Morinda (A. Smithiana)*
Prussian. *Abies excelsa*
Red. *Abies excelsa var. nigra*
Sacred Silver. *Picea religiosa*
Sapindus. *Pinus orientalis*
Scotch. *Pinus sylvestris*
Siberian Silver. See Fir, Pitch
Silver. *Picea pectinata*
Single Spruce. *Abies alba*
Southern Balsam. See Fir, Fraser's
Spanish Silver. *Picea Pinsapo*
Spruce. *Abies excelsa*
Umbrella. The genus *Sciadopitys*
Upright Indian Silver. *Picea Pindrow*
Western Red. *Abies magnifica*
Western White. *Abies (Picea) concolor*
White. *Abies excelsa var. communis*
White Spruce. *Abies alba*
Woolly-coned Silver. *Picea amabilis*
Fir-apples. Another name for Fir-cones
Fir-moss. *Lycopodium Selago*
Fire-bush. *Crataegus Pyracantha*
Fire-cracker, Vegetable. *Brodiæa coccinea* and *Cuphea platycentra*
Fire-flower, Jamaica. *Euphorbia punicea*

Fire-grass. *Alchemilla arvensis*
Fire-leaves, or Fire-weed. *Plantago media*
Fire-tree, New Zealand. *Metrosideros tomentosa*
Fire-weed. See Fire-leaves
American. *Erechthites hieracifolia (Senecio hieracifolius)* and *Lactuca elongata*
First-of-May. *Saxifraga granulata*
Fish-bone-Thistle. See Thistle
Fish-Mint. *Mentha aquatica*
Fish-poison-plant, Brazilian. *Serjania lethalis* and *Paullinia pinnata*
Ceylon. *Hydnocarpus venenatus*
E. Indian. *Walsura Piscidia*
S. Sea Islands. *Lepidium Piscidium*
Timboc. *Serjania lethalis*
W. Indian. *Tephrosia (Galega) toxicaria*
Fist-balls. *Lycoperdon Bovista*
Fitch, or Fitches. *Vicia sativa*
Fit-plant. *Monotropa uniflora*
Fitt-weed. *Eryngium fœtidum*
Five-finger-grass, or Five-leaf. *Potentilla reptans*
Five-fingers, Canadian. *Potentilla Canadensis*
Marsh. *Potentilla palustris*
W. Indian. *Syngonium auritum*
Five-leaf. See Five-finger-grass
Five-leaves. *Ampelopsis quinquefolia*
Flag. A general name for the genus *Iris*
American Blue. *Iris versicolor*
Corn. The genus *Gladiolus*
Crimson. *Schizostylis coccinea*
Pyrenean. *Iris xiphioides*
Sweet. *Acorus Calamus*
Variegated Sweet. *Acorus gramineus variegatus*
Yellow, or Water. *Iris Pseud-acorus*
Slender Blue American. *Iris virginica*
Flag-flower, Virginian. *Vexillaria virginica*
Flame-flower. The genus *Tritoma*
Flame-tree. *Brachychiton acerifolium*
Or Fire-tree, of S. W. Australia. *Nuytsia floribunda*
Flamingo-plant. The genus *Anthurium*
André's. *Anthurium Andreanum*
Baker's. *Anthurium Bakeri*
Lindig's. *Anthurium Lindigi*
Palmer's. *Anthurium Palmeri*
Saunders's. *Anthurium Saundersi*
Scherzer's. *Anthurium Scherzerianum*
Splendid. *Anthurium magnificum*
Veitch's. *Anthurium Veitchi*
Warocque's. *Anthurium Warocqueanum*
White-flowered. *Anthurium candidum*
Flamy. *Viola tricolor*
Flannel, or Flannel-Plant. *Verbascum Thapsus*
Flap-Dock. *Digitalis purpurea*
Flaps, or Flats. Large broad Mushrooms (probably *Agaricus arvensis*); also *Peziza cochleata*
Flat-Pea, Large-flowered. *Platylobium formosum*
Flatter-Dock. See Dock
Flat-tops. *Vernonia noveboracensis*
Flats. See Flaps
Flaver. *Avena fatua*
Flaw-flower. *Anemone Pulsatilla*

and Foreign Plants, Trees, and Shrubs. 47

Flax. The genus *Linum*
Alpine. *Linum alpinum*
American Bog. *Linum striatum*
American False. *Camelina sativa*
Berlandier's. *Linum Berlandieri*
Bright Blue Perennial. *Linum provinciale*
Common. *Linum usitatissimum*
Crimean. *Linum tauricum*
Crimson-flowered. *Linum grandiflorum*
Dwarf, Fairy, Mountain, or Purging. *Linum catharticum*
E. Indian. *Linum trigynum*
Evergreen. *Linum arboreum*
Fairy. See Flax, Dwarf
Heath-like. *Linum salsoloides*
Larger Blue, of N. America. *Iris versicolor*
Large Yellow-flowered American. *Linum sulcatum*
Mountain. See Flax, Dwarf
Narbonne. *Linum narbonnense*
Native, of New Zealand. *Linum monogynum*
New Zealand. *Phormium tenax* and *P. Colensoi*
New Zealand, Variegated. *Phormium tenax variegatum*
Orange - flowered. *Linum Macraei* (*L. Chamissonis*)
Pale-flowered. *Linum angustifolium*
Perennial. *Linum perenne*
Purging. See Flax, Dwarf
Sicilian Red-flowered. *Linum rubrum*
Toad. The genus *Linaria*; also applied to *Spergula arvensis*
Virginian Yellow-flowered. *Linum virginianum*
Viscid. *Linum viscosum*
White-flowered. *Linum monogynum*
White-flowered Dwarf. *Linum alpinum album*
Wild. *Linaria vulgaris* and *Cuscuta Epilinum*
Winter-flowering. *Linum trigynum*
Yellow Herbaceous. *Linum flavum* (*L. campanulatum*)
Yellow Toad. *Linaria vulgaris*
Flax-lily. *Phormium tenax*
Australian. The genus *Dianella*
Smaller. *Phormium Colensoi*
Variegated. *Phormium tenax variegatum*
Flax-seed. *Radiola Millegrana*
Flax-tail. *Typha latifolia*
Flax-star. *Lysimachia Linum - stellatum* (*Asterolinum stellatum*)
Flax-weed. *Linaria vulgaris*
Flea-bane, African. The genus *Tarchonanthus*
American. *Erigeron Philadelphicus*
Australian. *Erigeron mucronatus*
Blue-flowered. *Erigeron acris*
Canadian. *Erigeron Canadensis*
Common. *Inula dysenterica* and *I. Pulicaria*
Daisy. *Erigeron annuus* and *E. strigosus*
Georgian. *Inula glandulosa*
Glaucous-leaved. *Erigeron glaucus*
Glutinous. *Jasonia glutinosa*
Large-flowered. *Erigeron grandiflorus*
Many-rayed. *Erigeron multiradiatus*

Flea-bane, Marsh. The genus *Pluchea*
Mountain. *Erigeron alpinus* (*E. uniflorus*)
Pyrenean. *Inula montana*
Royle's. *Erigeron Roylei*
St. Helena. *Conyza rugosa*
Salt-marsh. *Pluchea camphorata*
Showy. *Erigeron speciosus* (*Stenactis speciosa*)
Shrubby African. *Tarchonanthus camphoratus*
Smoothish. *Erigeron glabellus*
W. Indian. *Vernonia arborescens*
White. *Inula candida*
Flea - bane - Mullet. An old name for *Inula Pulicaria*
Flea-grass. *Carex pulicaris*
Flea-seed. The seed of *Plantago Psyllium*
Flea-wort. *Inula Conyza*, *Pulicaria vulgaris*, and *Plantago Psyllium*
Flint-wood, N. S. Wales. *Eucalyptus pilularis*
Flix - weed, or Flux - weed. *Sisymbrium Sophia*
Float-grass. *Glyceria fluitans*
Floating-Heart. *Limnanthemum lacunosum* (*Villarsia cordata*)
Floramor, or Florimer. *Amarantus caudatus* and *A. tricolor*
Flower de Luce. A corruption of Fleur de Lis, a French name for the *Iris*
Yellow. *Iris Pseud-acorus*
Flower-Gentle. *Amarantus tricolor*
Flower-fence. *Cæsalpinia* (*Poinciana*) *pulcherrima*
Flower-of-a-day. *Tradescantia virginica*
Flower - of-an - hour. *Hibiscus* (*Ketmia*) *Trionum*
Flower-of-Bristow. *Lychnis chalcedonica*
Flower-of-Constantinople. *Lychnis chalcedonica*
Flower-of-the-Axe. *Lobelia urens*
Flower-of-the-Dead. *Oncidium tigrinum*
Flower-of-Tigris. *Tigridia Pavonia*
Flowering Box. See Box
Flowering Fern. See Fern
Flowering Grass. *Anomatheca cruenta*
Flowering Rush. *Butomus umbellatus*
Flowering Shot. The genus *Canna*
Flowering Spurge. See Spurge
Flowery Docken. See Docken
Flowk-wort. *Hydrocotyle vulgaris*
Fluellen. *Veronica officinalis*
Female. *Veronica Chamædrys*
Male. *Linaria spuria*
Fluff-weed. *Verbascum Thapsus*
Flux-weed. See Flix-weed
Fly-bane. *Agaricus muscarius*
Fly-Honeysuckle. *Lonicera Xylosteum*
Fly-poison, American. *Amianthium muscætoxicum*
Fly-trap, American, or Border. *Apocynum androsæmifolium*
Venus's. *Dionæa muscipula*
Fly-Orchis. See Orchis
Fly-wort. The genus *Myanthus* (*Catasetum*)
Foal-foot, or Fole-foot. *Tussilago Farfara* and *Asarum europæum*
Sea. *Convolvulus Soldanella*

Fog. A general term in the north of England for Moss; also applied to the second crop of Grass, or aftermath Yorkshire. *Holcus lanatus*
Fog-fruit. *Lippia (Zapania) lanceolata*
Fold Meadow-grass. See Fowl-grass
Fool's-Cicely. *Æthusa Cynapium*
Fool's-Parsley, Common. *Æthusa Cynapium*
Fine-leaved. *Æthusa fatua*
Fool's-Watercress. *Helosciadium nodiflorum*
Forbidden Fruit. Small-sized Shaddocks (*Citrus decumana*)
Forcible Plant. An old name for Hare's-ear
Fore-bit, or **Fore-bitten More.** *Scabiosa succisa*
Forget-me-not. *Myosotis palustris* and, generally, the genus *Myosotis*. Also an old name for *Ajuga Chamæpitys*
Alpine. *Myosotis alpestris* (*M. rupicola*)
American. *Myosotis verna*
Azorean. *Myosotis azorica*
Cape. *Anchusa capensis*
Chatham Island. *Myosotidium nobile*
Colour-changing. *Myosotis versicolor*
Creeping. *Omphalodes verna* and *Myosotis repens*
Early. *Myosotis dissitiflora* (*M. montana*)
Early Hill. *Myosotis collina*
Field. *Myosotis arvensis*
Indian. *Quamoclit vulgaris*
Long-flowering. *Myosotis semperflorens*
Mountain. *Myosotis rupicola*
Rock. *Omphalodes Luciliæ*
Tongue-leaved. *Myosotis lingulata*
White-flowered Early. *Myosotis dissitiflora var. alba*
Wood. *Myosotis sylvatica*
Fork-Moss. See Moss
Fountain-plant. *Amarantus salicifolius*
Fountain-tree. *Cedrus Deodara*
Four-leaved-grass. *Paris quadrifolia*
Four-o'clock-flower. *Mirabilis dichotoma*
Californian. *Mirabilis californica, M. multiflora*, and *M. Greenei*
Fowl-grass, or **Fold Meadow-grass.** *Poa trivialis*
Fox-bane. *Aconitum Vulparia*
Fox-chop. *Mesembryanthemum vulpinum*
Fox-Geranium, or **Fox-Grass.** *Geranium Robertianum*
Fox-glove, Blue. *Campanula Trachelium*
Common. *Digitalis purpurea*
Cream-coloured. *Digitalis ochroleuca*
Downy False. *Gerardia flava*
False. The genera *Gerardia* and *Dasystoma*
Fern-leaved False. *Gerardia pedicularia*
Great Yellow. *Digitalis ambigua*
Ladies'. *Verbascum Thapsus*
Large-flowered. *Digitalis grandiflora*
Mullein. *Digitalis Thapsi* and *Seymeria macrophylla*
Sierra Morena. *Digitalis Mariana*
Smooth False. *Gerardia quercifolia*
W. Indian. The genus *Phytolacca*
Willow-leaved. *Digitalis obscura*

Fox-glove, Woolly. *Digitalis lanata*
Yellow. *Digitalis lutea*
Fox-grass. See Fox-Geranium
Fox Rose. *Rosa spinosissima*
Fox's Brush. *Centranthus ruber*
Fox's-brush Saxifrage. See Saxifrage
Fox-tail. *Lycopodium clavatum*
Fox-tail-grass. *Alopecurus pratensis*
Fox-tailed Asparagus. *Equisetum maximum*
Fraghan, Frocken, or Frughan. *Vaccinium Myrtillus*
Frail Rush. *Scirpus lacustris*
Framboise, or Framboys. *Rubus Idæus*
Franck. An old name for Milk-wort
Frangipani-shrub. *Plumieria alba* and *P. rubra*
Frankincense, Rosemary. *Cachrys Libanotis*
Frankincense, or Olibanum, Tree. *Boswellia Carteri* and various other species
Frankincense-tree, Common, or Gum-Thus-tree. Various species of Pine-trees
Fraxinell. *Polygonatum multiflorum*
Fraxinella. *Dictamnus Fraxinella*
Caucasian. *Dictamnus caucasicus*
White-flowered. *Dictamnus Fraxinella var. albus*
Free-stone. A term applied to some varieties of Peaches and Nectarines, the flesh of which parts freely from the stone. (See Cling-stone.)
Freiser. An old name for the Strawberry-plant
French-Asparagus. *Ornithogalum pyrenaicum*
French-Beans. *Phaseolus vulgaris*
French-Berries. The fruit of *Rhamnus infectorius* and *R. catharticus*
French-Cotton. *Calotropis procera*
French-Cowslip. *Primula Auricula*
French-Furze. *Ulex europæus*
French-Grass. *Onobrychis sativa* and *Phalaris arundinacea variegata*
French-Honeysuckle. *Hedysarum coronarium*
French-Lavender. *Lavandula Stœchas*
French-Lungwort. *Hieracium murorum*
French-Nut. Another name for the Walnut
French-Peas. An old name for garden Peas
French-Sorrel. *Oxalis Acetosella*
French-Sparrowgrass. *Ornithogalum pyrenaicum*
French-Wheat. *Polygonum Fagopyrum*
French-weed. *Commelina cayennensis*
French-Willow. *Epilobium angustifolium*
Fresh-water-Soldier. *Stratiotes aloides*
Friar's-Caps. *Aconitum Napellus*
Friar's-Cowl. *Arum Arisarum* and *A. maculatum*
Friar's-Crown. *Carduus eriophorus*
Frijol-bean. *Phaseolus Hernandezii*
Fringe, Water. *Limnanthemum* (*Villarsia*) *nymphæoides*
Fringe-flower, Graham's. *Schizanthus Grahami*

and Foreign Plants, Trees, and Shrubs.

Fringe-flower, Notched. *Schizanthus retusus*
Pinnate. *Schizanthus pinnatus*
Fringe-Moss. See Moss
Fringe-pod. *Thysanocarpus laciniatus var. crenatus*
Fringe-tree. *Chionanthus virginica*
Fritillary, Broad-leaved. *Fritillaria latifolia*
Double-flowered. *Fritillaria Meleagris plena*
Pyrenean. *Fritillaria pyrenaica*
Scarlet. *Fritillaria recurva*
Snake's-head. *Fritillaria Meleagris*
Tulip-leaved. *Fritillaria tulipifolia*
White. *Fritillaria Meleagris alba*
Frocken. See Fraghan
Frog-bit. *Hydrocharis Morsus-ranæ*
American. *Limnobium Spongia*
Frog-cheese. A name applied to the large Puff-balls (*Lycoperdon Borista*) while they are young and soft; also to *Boleti* growing on decayed wood
Frog-flower. The genus *Ranunculus*
Frog-grass. *Salicornia herbacea*
Frog-stools. Another name for Toad-stools
Frog-wort. *Ranunculus hederaceus*
Frog's-Lettuce. *Potamogeton densus*
Frost-blite. *Chenopodium album*
Frost-weed. *Euerigeron philadelphicus* and *Helianthemum canadense*
Frughan. See Fraghan
Fuchsia, Australian. The genus *Correa*, especially *C. speciosa*
Basket. *Fuchsia procumbens*
Box-thorn-leaved. *Fuchsia lycioides*
Brilliant. *Fuchsia fulgens*
Californian. *Zauschneria californica*
Conical-tubed. *Fuchsia conica*
Corymbose. *Fuchsia corymbiflora*
Dwarf Trailing. *Fuchsia minima*
Edible-fruited. *Fuchsia racemosa*
Fuegian. *Fuchsia magellanica*
Globe-flowered. *Fuchsia globosa*
New Zealand. *Fuchsia Colensoi*, *F. excorticata*, and *F. procumbens*
Riccarton's. *Fuchsia Riccartoni*
Saw-leaved. *Fuchsia serratifolia*
Scarlet. *Fuchsia coccinea*
Slender. *Fuchsia gracilis*
Small-leaved. *Fuchsia microphylla*
Thompson's. *Fuchsia Thompsoni*
Thyme-leaved. *Fuchsia thymifolia*
Tom Thumb. *Fuchsia pumila*
Trailing. *Fuchsia procumbens*
Fuchsias, Hardy. *Fuchsia globosa*, *F. gracilis*, *F. minima*, *F. procumbens*, *F. Riccartoni*, and *F. Thompsoni*
Fuet. A Scotch name for House-leek (*Sempervivum tectorum*)
Fuller's-Herb. *Saponaria officinalis*
Fuller's-Teasel. *Dipsacus Fullonum*
Fume-wort, Golden. *Corydalis aurea*
Great-flowered. *Corydalis nobilis*
Hollow-rooted. *Corydalis tuberosa* (*C. cava*)
Ledebour's. *Corydalis Ledebouriana*
Marschall's. *Corydalis Marschalliana*

Fume-wort, Pale. *Corydalis glauca*
Solid-rooted. *Corydalis solida* (*Fumaria bulbosa*)
White-flowered. *Corydalis cava var. albiflora*
Fumitory, or Fumiterre. *Fumaria officinalis*
Bladdered. *Fumaria vesicaria*
Climbing. *Corydalis* (*Fumaria*) *claviculata*
Climbing, of N America. *Adlumia cirrhosa*
Dense-flowered. *Fumaria densiflora*
Yellow-flowered. *Corydalis* (*Fumaria*) *lutea*
Funeral-flower, Swan River. The genus *Anigosanthus*
Fungus, Barberry. *Æcidium Berberidis*
Beef-steak. *Fistulina Hepatica*
Black-knot. *Sphæria mortosa*
Bunt. *Ustilago fœtida*
Cellar. *Lentinus lepideus*
Club-root. *Plasmodiophora brassicæ*
Coffee Blight. *Hemileia vastatrix*
Dry-rot, of Conifers. *Merulius lachrymans*
Dry-rot, of the Elm. *Polyporus ulmarius*
Dry-rot, of the Oak. *Polyporus hybridus*
Earth-star. The genus *Geaster*
Edible Tree, of Tierra del Fuego. A species of *Cyttaria*
Ergot. *Claviceps purpurea*
Jew's-Ear. *Hirneola* (*Exidia*) *Auricula-Judæ*
Man. The genus *Geaster*
"Melitensis." *Cynomorium coccineum*
"Oak-leather." *Xylostroma giganteum*
Pepper-brand. *Ustilago fœtida*
Potato-disease. *Peronospora infestans*
"Rust." *Trichobasis Rubigo vera*
Sap-ball. *Polyporus squamosus*, *P. betulinus*, and other species which are found on trees.
Smut. *Ustilago segetum*
Stinking-Pole-cat. *Phallus impudicus*
Wine-cellar. *Zasmidium cellare*
Witch's-butter. *Exidia glandulosa*
Yeast-plant. A species of *Penicillium* or *Mucor*
Fungus-tinder. *Polyporus* (*Boletus*) *igniarius* and *P. fomentarius*
Furbelows. *Laminaria bulbosa*
Furze, Autumn-flowering. *Ulex nanus*
Common. *Ulex europæus*
Double-blossomed. *Ulex europæus plenus*
Dwarf. *Ulex nanus*
Ground. *Ononis arvensis*
Irish. *Ulex strictus* (*U. hibernicus*)
Jam. *Ulex nanus*
Needle. *Genista anglica*
Provence. *Ulex provincialis*
Spanish. *Genista hispanica*
Fuss-balls, or Fuzz-balls. The genus *Lycoperdon*, especially *L. Bovista*
Fustic-wood. The wood of *Maclura tinctoria*
"Young." The wood of *Rhus Cotinus*

E

Fuzz-balls. See Fuss-balls
Fyams (Seaweed). The genus *Laminaria*
Gadrise, Gaiter, Gatteridge, Gatten, or Gatter-tree. *Cornus sanguinea*, *Euonymus europæus*, and *Viburnum Opulus*
Gaiter-berries. The fruit of *Cornus sanguinea* and *Euonymus europæus*
Gaiter-tree. See Gadrise
Galangal, or Galingale. *Cyperus longus*, and the rhizome of *Alpinia officinarum* (*A. Galanga*)
Galapee-tree. *Sciadophyllum Brownei*
Galbanum-resin-plant. *Ferula galbaniflua*
Gale, or Sweet Gale. *Myrica Gale*
Fern-leaved. *Comptonia asplenifolia*
Galingale See Galangal
E. Indian. *Kæmpferia Galanga*
Galingale-Rush. See Rush
Galimeta-wood. The timber of *Bumelia salicifolia*
Gall, Artichoke. A gall formed in the male flower-buds of the Oak by an insect named *Andricus pilosus*
Bedeguar. A gall produced on Rose-trees by *Rhodites rosæ*
Button. Formed on Oak leaves by *Spathegaster vesicatrix*
Currant. Formed in the male flowers of the Oak by *Spathegaster buccarum*
Marble. Formed on Oak twigs by *Cynips Kollari*
Oak-apple. Formed on Oak twigs by *Biorhiza aptera*
Oak-root. Formed on the roots of the Oak by *Andricus radulli*
Oak-spangle. Formed on Oak-leaves by *Neuroterus lenticularis*
Gall-of-the-earth. *Nabalus Fraseri*
Gall-wort. *Linaria vulgaris*
Galls, Aleppo, of commerce. From *Quercus lusitanica var. infectoria*
Gallant. An old name for Anemone
Gallegaskins, or Gaskins. *Primula veris flore et calyce crispo*
Gallow-grass. *Cannabis sativa*
Gambier, or Gambier Catechu-plant *Uncaria Gambier* and *U. acida*
Gamboge-tree. *Garcinia Morella var. pedicellata*
American. *Vismia guianensis*
Ceylon. *Garcinia Morella*
Gamote, of New Mexico. The tuberous roots of *Cymopterus montanus*
Gandergosses. An old name for *Orchis Morio*
Gang Flower. *Polygala vulgaris*
Garavance (the Chick-pea). *Cicer arietinum*
Garden-gate. *Viola tricolor*
Garden-Ginger. An old name for *Capsicum*
Garden Speedwell. *Veronica agrestis*
Gardener's-Delight, or Gardener's-Eye. *Lychnis coronaria*
Gardener's-Garters. *Phalaris arundinacea variegata*
Gardenia, Lemon-scented. *Gardenia citriodora*

Gard-robe. *Osyris alba* (*Cassia poetica*)
Garget. *Phytolacca decandra*
Garland-Crab. See Crab-tree
Garland-flower, *Daphne Cneorum*
Australian. *Calocephalus Brownii*
Fragrant. *Hedychium coronarium*
Garlic, or Garlick. The genus *Allium*
Alpine. *Allium alpinum*
Azure-flowered. *Allium azureum*
Bear's. *Allium ursinum*
Black. *Allium nigrum*
Blue-flowered. *Allium cœruleum*
Carolina. *Allium inodorum*
Common. *Allium sativum*
Daffodil. *Allium neapolitanum*
Drooping-flowered. *Allium cernuum*
Field. *Allium oleraceum*
Fragrant-flowered. *Allium odorum*
Fringed. *Allium ciliatum*
Golden-flowered. *Allium Moly*
Great-headed. *Allium Ampeloprasum*
Hedge. *Alliaria officinalis*
Honey. The genus *Nectaroscordum*
Keeled. *Allium carinatum*
Large-flowered. *Allium grandiflorum*
Levant. *Allium Ampeloprasum*
Long-rooted. *Allium Victoriale*
One-flowered. *Allium uniflorum*
Piedmont. *Allium pedemontanum*
Quaint. *Allium paradoxum*
Rosy-flowered. *Allium roseum*
Round-headed. *Allium sphærocephalum*
Sickle-leaved. *Allium falcifolium*
Sorcerer's. *Allium Moly*
Spanish. *Allium Scorodoprasum*
Stag's. *Allium vineale*
Sweet-scented. *Allium* (*Nothoscordon*) *fragrans*
Triangular-stalked. *Allium triquetrum*
Wild. *Allium vineale* and *A. ursinum*
Wild American. *Allium canadense*
Yellow-flowered. *Allium flavum*
Garlic-Germander. *Teucrium Scordium*
Garlic-Pear. *Cratæva gynandra*
Garlic-Sage. *Teucrium Scorodonia*
Garlic-Shrub. *Bignonia alliacea* and *Petiveria alliacea*
Garlic-wort. *Alliaria officinalis*
Garnet-berry. The fruit of *Ribes rubrum*
Garten-berries. The fruit of *Rubus fruticosus*
Gaskins. See Gallegaskins
Gas-plant. *Dictamnus Fraxinella*
Gatten-tree, or Gatter-tree. See Gadrise
Gatteridge-tree. See Gadrise
Gay-bine. *Convolvulus* (*Ipomæa*) *Nil*
Gay-feather, Kansas. *Liatris pycnostachya*
Gauze-tree. *Lagetta lintearia*
Gazels, or Gazles. The fruit of *Ribes nigrum*
Gean-tree. *Prunus Avium*
Geckdor. An old name for *Galium Aparine*
Gelsemin. *Gelseminum sempervirens*
Genipap-tree, or Genip-tree. *Genipa americana*
W. Indian. *Melicocca bijuga* and *M. paniculata*

Genisaro-tree. *Pithecolobium Saman*
Gentian. The genus *Gentiana*
 Autumn. *Gentiana Amarella*
 Ascending. *Gentiana adscendens*
 Barrel-flowered. *Gentiana Saponaria*
 Bastard. *Hypericum Sarothra (Sarothra gentianoides)*
 Bavarian. *Gentiana bavarica*
 Burser's. *Gentiana Burseri*
 Caucasian. *Gentiana caucasica*
 Climbing. *Crawfurdia japonica*
 Closed-flowered. *Gentiana Andrewsii*
 Cream-coloured. *Gentiana gelida*
 Crested. *Gentiana septemfida*
 Cross-wort. *Gentiana Cruciata*
 Dotted-flowered. *Gentiana punctata*
 Field. *Gentiana campestris*
 Five-flowered. *Gentiana quinqueflora*
 Fringed. *Gentiana crinita*
 Horse. The genus *Triosteum*
 Marsh. *Gentiana Pneumonanthe*
 Milkweed. *Gentiana asclepiadea*
 New Zealand Mountain. *Gentiana saxosa*
 New Zealand Sea-shore. *Gentiana cerina*
 New Zealand Yellow. *Gentiana pleurogynoides*
 Purple-flowered. *Gentiana purpurea*
 Pyrenean. *Gentiana pyrenaica*
 Round-petalled. *Gentiana pannonica*
 Small-leaved. *Gentiana brachyphylla*
 Small Mountain. *Gentiana nivalis*
 Smaller Fringed. *Gentiana detonsa*
 Soapwort. *Gentiana Saponaria* and *Saponaria officinalis*
 Spring. *Gentiana verna*
 Spurred. The genus *Halenia*
 Vernal. *Gentiana verna*
 Whitish. *Gentiana alba*
 Willow. *Gentiana asclepiadea*
 Yellow-flowered. *Gentiana lutea*
 Yellowish-white. *Gentiana ochroleuca*
Gentian-root. The root of *Gentiana lutea*
Gentianella. *Gentiana acaulis*
Gentle Dock. See Dock
George's (St.) Herb. *Valeriana officinalis*
George's (St.) Mushroom. *Agaricus gambosus*
Georgia-bark-tree. *Pinckneya pubens*
Geranium (See also Crane's-bill.) Australian. *Geranium dissectum*
 Bassinet. *Geranium sylvaticum*
 Blood-red-flowered. *Geranium sanguineum*
 Cut-leaved. *Geranium dissectum*
 Dove's-foot. *Geranium molle*
 Dusky-flowered. *Geranium phæum*
 Feather. *Chenopodium Botrys*
 Indian. *Andropogon Nardus*
 Long-stalked. *Geranium columbinum*
 Meadow. *Geranium pratense*
 Mountain. *Geranium pyrenaicum*
 Nettle. *Coleus fruticosus*
 Round-leaved. *Geranium rotundifolium*
 Shining-leaved. *Geranium lucidum*
 Small-flowered. *Geranium pusillum*
 Strawberry. *Saxifraga sarmentosa*
 Wild. *Geranium pratense* and *G. Robertianum*
 Wood. *Geranium sylvaticum*
Geranium-Oil-plant. See Oil-plant
Gerard. See Herb-Gerard
German Lilac (Valerian). *Centranthus ruber?*
German Mad-wort. *Asperugo procumbens*
German Tinder. See Amadou
Germander. *Teucrium Chamædrys*
 American. *Teucrium canadense*
 Bastard, or Sea-side. *Stemodia maritima*
 Garlic. *Teucrium Scordium*
 Golden. *Teucrium aureum*
 Madeira. *Teucrium betonicum*
 Poly. *Teucrium Polium*
 Pyrenean. *Teucrium pyrenaicum*
 Sea-side. See Germander, Bastard
 Shining. *Teucrium lucidum*
 Small-flowered. *Teucrium campanulatum*
 Sweet-scented. *Teucrium Massiliense*
 Tree. *Teucrium fruticans*
 Wall, or Wild. *Teucrium Chamædrys*
 Water. *Teucrium Scordium*
 Wood. *Teucrium Scorodonia*
Germander-Chick-weed. *Veronica agrestis*
Germander-Speedwell. *Veronica Chamædrys*
Geslins, or Goslings. See Goslings
Gesse. An old name for Jasmine
Gethsemane. *Orchis mascula*
Gherkin. A small-fruited variety of *Cucumis sativa*
 W. Indian. The unripe fruit of *Cucumis Anguria*
Gil-cup, or Gilty-cup. Another name for Buttercup
Gill, or Gill-go-by-Ground. *Nepeta Glechoma*
Gill, or Gule, Gowan. *Chrysanthemum segetum*
Gilliflower, or Gillyflower. *Dianthus Caryophyllus;* also the genera *Matthiola* and *Cheiranthus*
 Clove. *Dianthus Caryophyllus*
 Marsh. *Lychnis Flos-cuculi*
 Queen's. *Hesperis matronalis*
 Rogue's. *Hesperis matronalis*
 Sea. *Armeria vulgaris (A. maritima)*
 Stock. *Matthiola annua, M. incana,* and *M. græca*
 Wall. *Cheiranthus Cheiri*
 Water. *Hottonia palustris*
 Winter. *Hesperis matronalis*
 Yellow. *Cheiranthus Cheiri*
Gilliflower-Grass. See Grass
Gimlet-wood. *Eucalyptus salubris*
Ginkgo-tree. *Salisburia adiantifolia*
Gingelly, or Gingili, Oil-plant. *Sesamum orientale*
Ginger, Amada, or Mango. *Curcuma Amada*
 Broad-leaved. *Zingiber Zerumbet*
 Common. *Zingiber officinale*
 Egyptian. *Colocasia esculenta*
 Garden. See Garden-Ginger
 Indian. *Asarum canadense*
 Mango. *Curcuma Amada*
 Red, or E. Indian. *Zingiber officinale*
 Wild American. *Asarum canadense*
 Wild W. Indian. The genera *Costus* and *Renealmia*
 Wood. *Anemone ranunculoides*

Ginger-bread-tree, Egyptian. *Hyphæne thebaica*
Sierra Leone. *Parinarium macrophyllum*
Ginger-Grass-Oil-plant. *Andropogon Schœnanthus*
Gingili-Oil-plant. See Gingelly-Oil-plant
Ginseng-plant. *Panax quinquefolium* (*Aralia quinquefolia*)
Blue. *Caulophyllum thalictroides*
Dwarf. *Panax* (*Aralia*) *trifolium*
Gipsies'-Rose. *Scabiosa arvensis*
Gipsy-flower. *Cynoglossum officinale*
Gipsy-wort. *Lycopus europæus*
Girdle, Sea. *Laminaria digitata*
Girls' Mercury. The male plant of *Mercurialis annua*
Gith. An old name for Nigella
Gladden, Gladdon, Gladin, Glader, or Gladwyn. *Iris fœtidissima*
Gladiole, Water. *Butomus umbellatus*
Gladiole, or Sword-lily, Blood-red. *Gladiolus cruentus*
Branching. *Gladiolus ramosus*
Brenchley. *Gladiolus Brenchleyensis*
Butterfly. *Gladiolus Papilio*
Cardinal-flowered. *Gladiolus Cardinalis*
Colville's. *Gladiolus Colvillei*
Cooper's. *Gladiolus Cooperi*
Corn-field. *Gladiolus segetum*
Dragon's-head. *Gladiolus dracocephalus*
Early. *Gladiolus præcox*
Fox-glove. *Gladiolus communis*
Ghent Hybrid. Varieties of *Gladiolus Gandavensis*
Hart-wort. *Gladiolus Libanotis*
Narrow-leaved. *Gladiolus angustus*
Purple-and-gold. *Gladiolus purpureus-auratus*
Sad-coloured. *Gladiolus tristis*
Saunders's. *Gladiolus Saundersi*
Sulphur-flowered. *Gladiolus ochroleucus*
Gladwyn, Stinking. *Iris fœtidissima*
Variegated. *Iris fœtidissima variegata*
Gland-pod. The genus *Adenocarpus*
Glass-wort. *Salicornia herbacea*
Prickly. *Salsola Kali*
White. *Suæda maritima*
Glass-wrack. *Zostera* (*Alga*) *marina*
Glastonbury Thorn. *Cratægus Oxyacantha præcox*
Globe-Daisy, Common. *Globularia vulgaris*
Hair-flowered. *Globularia trichosantha*
Heart-leaved. *Globularia cordifolia*
Naked-stalked. *Globularia nudicaulis*
Thyme-leaved. *Globularia nana*
Globe-flower. The genus *Trollius*
American. *Trollius laxus*
Asiatic. *Trollius asiaticus*
Bush. *Cephalanthus occidentalis*
Common. *Trollius europæus*
Double Japanese. *Trollius japonicus fl.-pl.*
Fortune's. *Trollius Fortunei*
Giant. *Trollius Loddigesi*
Napellus-leaved. *Trollius napellifolius*
Spreading. *Trollius laxus*
Swamp. *Cephalanthus occidentalis*
Globe-Thistle. The genus *Echinops*
Hungarian. *Echinops bannaticus*

Globe-Thistle, Russian. *Echinops ruthenicus*
Small. *Echinops Ritro*
Tall. *Echinops exaltatus*
Glond. An old name for *Saponaria Vaccaria*
Glory-bush, Peruvian. *Pleroma* (*Lasiundra*) *sarmentosa*
Glory-flower, Chilian. *Eccremocarpus scaber*
Glory-of-the-Snow. *Chionodoxa Luciliæ*
Glory-Pea, Dampier's. *Clianthus Dampieri*
New Zealand. *Clianthus puniceus*
Norfolk Island. *Clianthus carneus*
Glory-tree. *Clerodendron fragrans* and other species
Gluttony, Plant of. *Cornus suecica*
Gnat-wort. *Triumfetta Lappula*
Go-to-bed-at-noon. *Tragopogon pratensis*
Goat-bush, W. Indian. *Castelea Nicholsoni*
Goat-root. *Ononis Natrix*
Goat-weed. The genus *Capraria*; also *Ægopodium Podagraria* and *Stemodia durantæfolia*
Goat-willow. *Salix Caprea*
Goat's-bane. *Aconitum Tragoctonum*
Goat's-beard. *Tragopogon pratensis*
False. See False-Goat's-beard
Goat's-foot. *Oxalis caprina*
Goat's-horn. *Astragalus Ægiceras*
Goat's-Rue. *Galega officinalis*
Oriental. *Galega orientalis*
Virginian. *Tephrosia Virginiana*
Goat's-thorn. *Astragalus Trayacantha*
and *A. Poterium*
Goat's-wheat. The genus *Tragopyrum*
Box-leaved. *Tragopyrum buxifolium*
Lance-leaved. *Tragopyrum lanceolatum*
Polygamous. *Tragopyrum polygamum*
Gobbo. *Hibiscus* (*Abelmoschus*) *esculentus*
Gobo, of Japan. *Lappa edulis*
God-tree. See Ceiba-tree
God's-eye. *Veronica Chamædrys*
God's-flower, or Gold-flower. *Helichrysum Stœchas*
Gold-balls, or Gold-cups. Another name for Buttercups
Gold-basket. *Alyssum saxatile*
Gold-dust. *Alyssum saxatile* and *Sedum acre*
Gold-flower. See God's-flower
Gold Heath. The genus *Sphagnum*
Gold-Leaf Plant. *Aucuba japonica*
Gold-of-Pleasure. *Camelina sativa*
Perennial. *Myagrum perenne*
Gold-seal. *Frasera verticillata* and *Hydrastis canadensis*
Gold-shrub. *Palicourea speciosa*
Gold-thread, Three-leaved. *Coptis trifoliata*
Golden-ball. *Trollius europæus*
Golden-ball Tree. *Forsythia suspensa* and *F. viridissima*
Golden-bush. *Cassinia fulgida*
Golden-chain. *Cytisus Laburnum*
Golden-club. *Orontium aquaticum*
Golden-crown. The genus *Chrysostemma*
Golden-drop. *Onosma tauricum*

Golden-Flower of Peru. *Helianthus annuus*
Golden-locks. *Polypodium vulgare*
Golden Moss. *Sedum acre*
Golden Moth-wort. See Moth-wort
Golden Mouse-ear. *Hieracium aurantiacum*
Golden Osier. *Salix vitellina* and *Myrica Gale*
Golden-pert. *Gratiola aurea*
Golden-rain. *Cytisus Laburnum*
Golden-rod. *Solidago Virgaurea*
Canadian. *Solidago canadensis*
Canary Islands. *Bosea Yervamora*
Drooping. *Solidago nutans*
Dwarf. *Solidago Virgaurea var. cambrica*
False. *Brachychæta cordata*
Large-flowered. *Solidago grandiflora*
Many-flowered. *Solidago multiflora*
Many-rayed. *Solidago multiradiata*
Rayless. *Bigelovia nudata*
Upright. *Solidago stricta*
W. Indian. *Neurolæna lobata*
Golden-rod Tree. *Bosea Yervamora*
Golden-seal. See Gold-seal
Golden Stœchas. *Helichrysum Stœchas plenum*
Golden-tuft. *Alyssum saxatile* and *Helichrysum Stœchas*
Golden Saxifrage. The genus *Chrysosplenium*
Golden Samphire. *Inula crithmoides*
Golden-spoon. *Byrsonima cinerea*
Golden-star, Maryland. *Chrysopsis Mariana*
Golden-wand, Hooker's. *Chrysobactron Hookeri*
Goldilocks. *Ranunculus auricomus, Chrysocoma Linosyris,* and *Polytrichum commune*
Broad-leaved. *Chrysocoma latifolia*
Tunbridge. *Hymenophyllum tunbridgense*
Golds, or Goldins *Calendula officinalis, Caltha palustris,* and *Chrysanthemum segetum*
Gombo. See Okro
Good Henry, or Good King Henry. *Chenopodium Bonus-Henricus*
Good-night-at-noon. *Hibiscus Trionum*
Googul. See Bdellium
Goompany-tree. *Odina Wodier*
Goora-nut. See Nut
Goose-and-Goslings. The catkins of Willows, especially of *Salix Caprea*
Gooseberry. *Ribes Grossularia*
American, or W. Indian. *Pereskia aculeata* and *Heterotrichum patens*
Barbadoes. *Pereskia aculeata*
Bristly. *Ribes setosum*
Cape. *Physalis peruviana* and *P. pubescens*
Caucasian. *Ribes caucasicum*
Coromandel, or Country. *Averrhoa Carambola*
Dog-bramble. *Ribes Cynosbati*
Fragrant-flowered. *Ribes aureum (R. fragrans)*
Fuchsia-flowered. *Ribes speciosum*
Hairy-branched. *Ribes hirtellum*
Gooseberry, Hawthorn-leaved. *Ribes oxyacanthoides*
Hedge-hog. *Ribes lacustre var. echinatum*
Hill, of India. *Rhodomyrtus tomentosa*
Large-spined. *Ribes macracanthum*
Lobb's. *Ribes Lobbii*
Malabar. *Melastoma malabathrica*
Menzies'. *Ribes Menziesii*
Needle-spined. *Ribes aciculare*
Otaheite. *Cicca disticha*
Rock. *Ribes saxatile*
Round-leaved. *Ribes rotundifolium*
Slender-branched. *Ribes gracile*
Small-leaved. *Ribes microphyllum*
Spreading-branched. *Ribes divaricatum*
Swamp. *Ribes lacustre*
Syrian. *Ribes orientale*
Three-flowered. *Ribes triflorum*
Two-spined. *Ribes diacantha*
Vancouver's Island. *Ribes subvestitum (R. Lobbii)*
White-flowered. *Ribes niveum*
Yellow-flowered. *Ribes leptanthum*
Gooseberry-Pie. *Epilobium hirsutum*
Goose-bill. *Galium Aparine*
Goose-corn. *Juncus squarrosus*
Goose-foot. The genus *Chenopodium*
Bearded. *Chenopodium aristatum*
Broom. *Kochia Scoparia*
Cut-leaved. *Chenopodium Botrys*
Glaucous. *Chenopodium glaucum*
Guinea. *Chenopodium quinensee*
Many-seeded. *Chenopodium polyspermum*
Maple-leaved. *Chenopodium hybridum*
Mexican. *Chenopodium ambrosioides*
Nettle-leaved. *Chenopodium murale*
Oak-leaved. *Chenopodium glaucum*
Perennial. *Chenopodium Bonus-Henricus*
Purple. *Chenopodium Atriplicis*
Red. *Chenopodium rubrum*
Sea-side. *Suæda maritima*
Stinking. *Chenopodium Vulvaria*
Upright. *Chenopodium urbicum*
White. *Chenopodium album*
Goose-grass. *Galium Aparine* and *Potentilla anserina*
Goose-nest. *Neottia Nidus-avis*
Goose-Tansy. *Potentilla anserina*
Goose-tongue. *Galium Aparine* and *Achillea Ptarmica*
Goose-tree, or Barnacle-tree. A name applied by Gerard to "certaine trees in the north parts of Scotland," on which barnacles were said to grow, from which it was believed in his time that "Barnacle Geese" were produced
Gorgon-plant. *Euryale ferox*
Gorse, or Goss. *Ulex europæus*
Hen. *Ononis arvensis* and *Bartsia Odontites*
Gory Dew. *Palmella cruenta*
Goss. See Gorse
Gourd. The genus *Cucurbita*
Bitter. *Citrullus Colocynthis*
Bottle. *Lagenaria vulgaris*
Elector's-cap. A variety of *Cucurbita Melopepo*
Gooseberry. *Cucurbita grossularioides*

English Names of Cultivated, Native,

Gourd, Jerusalem-Artichoke. A variety of *Cucurbita Melopepo*
Orange. *Cucurbita aurantia*
Perennial. *Cucurbita perennis*
Scarlet-fruited. *Coccinea indica*
Snake. *Trichosanthes anguina*
Sour. *Adansonia digitata*
Sour, of Australia. *Adansonia Gregorii*
Succade. *Cucurbita ovifera var. Succada*
Towel, or Washing. *Luffa œgyptiaca*
Trumpet. *Lagenaria vulgaris clavata*
Viper. *Trichosanthes colubrina*
Washing. See Gourd, Towel
Wax, or White. *Benincasa (Cucurbita) cerifera*
Gout Ivy. *Ajuga Chamæpitys*
Gout-weed, or Gout-wort. *Ægopodium Podagraria*
Variegated. *Ægopodium Podagraria variegatum*
Gouty-stemmed Tree, of Australia. *Adansonia Gregorii* and *Sterculia (Delabechea) rupestris*
Gowan, or Gowlan. *Bellis perennis*; the name is also applied to Buttercups and various other yellow flowers.
Horse. *Chrysanthemum Leucanthemum, Matricaria Chamomilla*, and *Leontodon Taraxacum*
Lapper, Lopper, or Lockin. *Trollius europæus*
Large White. *Chrysanthemum Leucanthemum*
Lockin or Luckin. *Trollius europæus*
May. *Bellis perennis*
Meadow. *Caltha palustris*
Milk. *Leontodon Taraxacum*
Water, or Open. *Caltha palustris*
Yellow. Buttercups and various other yellow flowers
Gowk-meat. *Oxalis Acetosella*
Gowlan, or Goulan. See Gowan.
Grace-of-God. An old name for St. John's-wort
Grains of Paradise. *Amomum Grana-Paradisi*
Gram. *Cicer arietinum*
Black. *Phaseolus Mungo melanospermus*
Green. *Phaseolus Mungo chlorospermus* (*P. Max*)
Horse. *Dolichos biflorus* and *D. uniflorus*
Red. *Dolichos Catjang*
Turkish. *Phaseolus aconitifolius*
White. *Soja hispida*
Granadilla. *Passiflora quadrangularis* and *P. edulis*
Grape, Bear's. *Arctostaphylos Uva-ursi*
Bland's. *Vitis Labrusca*
Bull. *Vitis rotundifolia*
Bullet. *Vitis vulpina*
Chicken. *Vitis cordifolia*
Delaware. A variety of *Vitis riparia*
False. *Ampelopsis quinquefolia*
Isabella. *Vitis Labrusca*
Kangaroo. *Vitis (Cissus) antarctica*
Muscadine. *Vitis vulpina*
Northern Fox. *Vitis Labrusca*
Oregon. *Berberis Aquifolium*
Plum. *Vitis Labrusca*

Grape, Port Jackson Black. *Vitis (Cissus) antarctica*
Scuppernong. A variety of *Vitis vulpina*
Sea. *Salicornia herbacea, Ephedra distachya*, and *Sargassum bacciferum*
Southern Fox. *Vitis vulpina*
Summer. *Vitis æstivalis*
Tail. The genus *Artabotrys*
Taylor Bullet. A variety of *Vitis riparia*
Wild, of the Peruvians. *Chondrodendrum convolvulaceum*
Winter, or Frost. *Vitis cordifolia*
Grape-flower. *Muscari racemosum*
Grape-Hyacinth, Armenian. *Muscari armeniacum*
Common. *Muscari racemosum*
Dark-purple. *Muscari commutatum*
Feathery. *Muscari comosum monstruosum* (*Hyacinthus monstruosus*)
Greek. *Muscari Heldreichii*
Musk. *Muscari moschatum*
Sky-blue. *Muscari botryoides*
Yellow. *Muscari luteum*
Grape-pear. *Amelanchier Botryapium*
Grape-plant, Sea-side. The genus *Coccoloba*
Macquarie Harbour. *Coccoloba (Polygonum) adpressa*
Grape-vine. *Vitis vinifera* and vars. (See Vine)
Grape-wort. *Actæa spicata* and *Bryonia dioica*
Grapes, Dyer's. *Phytolacca decandra*
Fen. *Vaccinium Oxycoccos*
Grapple-plant, Cape. *Uncaria procumbens*
Grass. A general name for plants of the order *Gramineæ*, but often used in the sense of "herb"
Adder's. *Orchis mascula* and *O. maculata*
Aleppo Millet. *Sorghum halepense*
Alfa, Alpha, or Halfa. *Macrochloa tenacissima*
American. *Eleusine indica*
Amoor River Giant Silvery. *Imperata saccharifora*
Ant-hill. *Festuca sylvatica*
Artificial. Rye-grass and the Clovers, which are sown in rotation
Australian Meadow. *Poa cæspitosa (P. australis)*
Australian Ornamental. *Poa australis*
Awned Wheat. *Triticum caninum*
Bahama. *Cynodon Dactylon*
Ballock. *Orchis mascula*
Bastard Knot. *Corrigiola littoralis*
Beard. *Polypogon monspeliensis* and the genus *Andropogon*
Bengal. *Setaria italica*
Bent. *Agrostis vulgaris, Psamma arenaria*, and some other wiry grasses
Bermuda. *Cynodon Dactylon*
Bird. *Poa trivalis*
Black. *Alopecurus agrestis, A. geniculatus, Bromus sterilis*, and *Medicago lupulina*
Black-head. *Luzula campestris*
Black Oat. *Stipa avenacea*
Black Quitch. *Agrostis vulgaris*

Grass, Blubber. The genus *Bromus*
Blue. Various species of *Carex*
Kentucky. *Poa pratensis*
Blue-eyed. *Sisyrinchium Bermudianum*
Blue Joint. *Calmagrostis canadensis*
Bottle. *Setaria viridis*
Bottle-brush. *Gymnostichum Hystrix*
Bowel-hive. *Alchemilla arvensis*
Brazilian. Strips of the leaves of *Thrinax argentea*
Brome. The genus *Bromus*
Broom. *Andropogon scoparius*
Brush. *Andropogon Gryllus*
Buck. *Lycopodium clavatum*
Buffalo, American. *Buchloë dactyloides*; also applied to *Medicago sativa* and *M. lupulina*
Buffalo, Australian, *Stenotaphrum americanum* (*S. glabrum*)
Bull. *Bromus mollis* and other species
Bunch. *Elymus condensatus* and *Festuca scabrella*
Bunkuss. *Spodiopogon angustifolius*
Bur. See Grass, Hedge-hog
Bur, or Ginger, W. Indian. *Panicum glutinosum*
Bush. *Calamagrostis Epigejos*
Butter. *Agrostis stolonifera*
Button. *Avena elatior*
Californian Prairie. *Ceratochloa unioloides*
Canary. *Phalaris canariensis*
Cape Thatch. *Restio chondropetalus*
Capon's-tail. *Valeriana pyrenaica*
Carpenter. *Prunella vulgaris*
Carnation. *Carex glauca* and *C. panicea*
Cat's-tail. *Phleum pratense*
Causeway. *Poa annua*
Cheesecake. *Lotus corniculatus*
China or Chinese. *Boehmeria* (*Urtica*) *nivea*
Claver or Clever. *Galium Aparine*
Cloud. The genus *Agrostis*
Clover. *Trifolium pratense* and *T. repens*. Sometimes applied to *Medicago lupulina*
Club. The genera *Scirpus* and *Corynephorus*
Cock. *Bromus mollis*, *B. secalinus*, and *Plantago lanceolata*
Cock's-comb. *Cynosurus echinatus*
Cock's-foot. *Dactylis glomerata*
Cock's-foot, Golden-edged. *Dactylis glomerata aurea*
Cock-shin. *Panicum Crus-galli*
Comb Finger. The genus *Dactyloctenium*
Cord. *Lygeum Spartum* and *Spartina stricta*
Cord, Fresh-water. *Spartina cynosuroides*
Cord, Twin-spiked. *Spartina stricta*
Corn. *Apera Spica-venti*
Cotton. The genus *Eriophorum*
Couch. *Triticum repens*
Cow. *Trifolium medium* and *T. pratense*
Crab. *Salicornia herbacea*, and *Polygonum aviculare*
Crow's-foot. *Echinochloa Crus-corvi*
Cuba. *Andropogon halepensis*
Cuckoo. *Luzula campestris*
Cut. *Leersia oryzoides*

Grass, Cutting. *Scleria flagellum*
Dadder, or Dodder. *Briza media*
Dart. *Holcus mollis* and *H. lanatus*
Deccan. *Panicum frumentaceum*
Deer's-foot. *Agrostis setacea*
Desert. The genus *Eremochloë*
Dew. *Dactylis glomerata* and *Setaria germanica*
Dithering, or Dothering. *Briza media*
Doddle. *Briza media*
Dog. *Triticum repens*
Dog's-tail. *Cynosurus cristatus*
Dog's-tooth. *Triticum caninum* and *Cynodon Dactylon*
Doob, or Doorva. *Cynodon Dactylon*
Dothering. See Grass, Dithering
Dover. *Festuca elatior*
Drop-seed. The genera *Muhlenbergia* and *Sporobolus*
Dudder. *Adiantum Capillus-Veneris*
Duffel. *Holcus mollis* and *H. lanatus*
Dutch. *Panicum molle*
Earning. *Pinguicula vulgaris*
East Indian Matting. *Cyperus corymbosa*
Eaver. See Grass, Ever
Eccle. *Pinguicula vulgaris*
Ecl. See Grass, Tape
Egyptian. *Dactyloctenium ægyptiacum*
Elbowit. *Alopecurus geniculatus*
Elephant's. *Typha elephantum*
Esparto. *Stipa* (*Macrochloa*) *tenacissima*
Ever, or Eaver. *Lolium perenne*
Fair. *Potentilla anserina*
"Faitour's." Probably *Euphorbia Esula*
False Brome. The genus *Brachypodium*
Feather. *Stipa pennata* and *Eragrostis elegans*
Feather-top. *Calamagrostis Epigejos*
Fescue. The genus *Festuca*
Fescue, Blue. *Festuca glauca* and *F. amethystina*
Fescue, Fine-leaved. *Festuca tenuifolia*
Fescue, Hard. *Festuca duriuscula*
Fescue, Meadow. *Festuca pratensis*
Fescue, Red or Creeping. *Festuca rubra*
Fescue, Sheep's. *Festuca ovina*
Fiddle. *Epilobium hirsutum*
Fine-top. *Agrostis alba*
Finger. The genus *Digitaria*
Finger, Red. *Digitaria sanguinalis*
Fiorin. *Agrostis stolonifera*
Fire. *Alchemilla arvensis*
Five-finger. *Potentilla reptans*
Flea. *Carex pulicaris*
Flote. *Glyceria fluitans*
Flowering. *Anomatheca cruenta*
Fly-away. *Agrostis scabra*
Fly-catch. *Leersia lenticularis*
Fodder. *Onobrychis sativa* and the genus *Chilochloa*
Four-leaved. *Lotus tetraphyllus*
Fowl-Meadow. *Glyceria nervata* and *Poa serotina*
Fox. *Geranium Robertianum*
Fox-tail. *Alopecurus pratensis*
Fox-tail, Gold-striped. *Alopecurus pratensis variegatus*
Fox-tail, W. Indian. *Anatherum bicorne* and *A. macrurum*

Grass, French. *Onobrychis sativa* and *Phalaris arundinacea variegata*
Frog. *Salicornia herbacea* and *Juncus bufonius*
Gallow. *Cannabis sativa*
Gama, or Sesame. *Tripsacum dactyloides*
Giant Woolly-beard. *Erianthus Ravennæ*
Gilliflower. *Carex glauca* and *C. panicea*
Ginger. *Andropogon Schœnanthus*
Ginger, W. Indian. See Grass, Bur
Globe Cotton. *Eriophorum capitatum* (*E. Scheuchzeri*)
Glow-worm, of Australia. *Luzula campestris*
Golden-edged Cock's-foot. *Dactylis glomerata aurea*
Gold-striped Fox-tail. *Alopecurus pratensis variegatus*
Goat. The genus *Ægilops*
Goose. See Goose-grass
Grama. The genus *Bouteloua*
Grip. *Galium Aparine*
Guinea. *Panicum jumentorum* (*P. maximum*)
Gull. *Galium Aparine*
Hair. The genus *Aira*
Hair, American. *Agrostis scabra* and *Muhlenbergia capillaris*
Halfa. See Grass, Alfa
Hard. *Dactylis glomerata*
Hard, or Goat. The genus *Ægilops*
Hare's-tail. *Lagurus ovatus*
Hassock. *Aira cœspitosa*, *Carex cœspitosa*, and *C. paniculata*
Haver. *Avena elatior*, *Bromus sterilis*, and *B. mollis*
Heath. *Triodia decumbens*
Hedge-hog. *Panicum stagninum* and *Carex flava*
Hedge-hog, or Bur, of N. America. *Cenchrus tribuloides*
Hen-penny. *Rhinanthus Crista-galli*
Herd's. *Agrostis vulgaris*
Hog. *Senebiera Coronopus*
Holy. *Hierochloë borealis* and other species
Honey-suckle. *Trifolium repens*
Hooded. *Bromus mollis*
Horn. The genus *Ceratochloa*
Horn-of-Plenty. *Cornucopiæ cucullatum*
Horse-well. *Veronica Beccabunga*
Hose. *Holcus lanatus*
Hundred-leaved. *Achillea Millefolium*
Hunger. *Alopecurus agrestis*
Indian. *Sorghum nutans* and *Molinia cœrulea*
Irby-dale. *Euphorbia Helioscopia*
Iron. *Polygonum aviculare*
Italian Rye. *Lolium italicum* (*L. perenne* var. *multiflorum*)
Jockey. *Briza media*
Joint. Various species of *Equisetum*
Kangaroo. *Anthistiria ciliata* (*A. australis*)
Kentucky Blue. *Poa pratensis*
Knife. *Scleria latifolia*
Knot. *Polygonum aviculare*
Ladies'-laces, or Lady. *Phalaris arundinacea variegata*
Lagoon. *Triticum repens*

Grass. Lalong. *Imperata arundinacea*
Lamb's. Various Spring grasses
Land. *Alopecurus agrestis*
Lavender. *Molinia cœrulea*
Lemon. *Andropogon Schœnanthus*
Lily. *Arum maculatum* and *Butomus umbellatus*
Lob, or Lop. *Bromus mollis*
Long. The genus *Macrochloa*
Love. *Eragrostis elegans*
Low Spear. *Poa annua*
Lyme. *Elymus arenarius*
Maiden-hair. *Briza media*
Manatu. See Grass, Turtle
Manna. *Glyceria fluitans*
Marl. *Trifolium pratense*, or *T. medium*
Marrem. *Psamma arenaria*
Marsh Arrow. *Triglochin palustre*
Marsh Saw. *Cladium Mariscus*
Mat. *Psamma arenaria* and *Nardus stricta*
May. *Stellaria Holostea*
Meadow. The genus *Poa*
Meadow, Rough-stalked. *Poa trivialis*
Meadow, Smooth-stalked. *Poa pratensis*
Meadow-Fescue. *Festuca pratensis*
Meadow Soft. *Holcus lanatus* and *H. mollis*
Melick. The genus *Melica*
Merlin's. *Isoëtes lacustris*
Mesquite. See Grass, Mosquito
Midge. *Holcus lanatus*
Millet. *Milium effusum*
Mitchell, of Australia. *Astrebla* (*Danthonia*) *triticoides*
Monkey, or Pari. The fibre of *Attalea funifera*
Monkey's. *Agrostis vulgaris*
Moor. *Sesleria cœrulea*
Mosquito, or Mesquite. *Panicum obtusum*
Mountain, Jamaica. *Andropogon bicornis*
Mouse. *Aira caryophylla*
Mouse, Australian. *Dichelachne crinita*
Mouse-tail. *Festuca Myurus* and *Alopecurus agrestis*
Mulga. *Neurachne Mitchelliana*
Murrain. *Scrophularia nodosa*
Muskit. The genus *Bouteloua*
Myrtle. *Acorus Calamus*
Natural. *Poa trivialis* and *P. pratensis*; also applied to Grasses generally (except Rye-grass), because they are found in natural pastures
New Zealand Cutting. Various species of *Gahnia*
New Zealand Plume. *Arundo conspicua*
New Zealand Spear. *Aciphylla squarrosa*
"Nimble Will." *Muhlenbergia diffusa*
Nit. *Gastridium lendigerum*
Nut. *Cyperus rotundus* var. *Hydra*
Oat. *Avena pratensis* and *Bromus mollis*
Oat, Yellow. *Avena flavescens*
Of Parnassus. *Parnassia palustris*
Of Parnassus, Asarum-leaved. *Parnassia asarifolia*
Of Parnassus, Fringed. *Parnassia fimbriata*
Of Parnassus, Himalayan. *Parnassia nubicola*

Grass, of Parnassus, Large. *Parnassia Caroliniana*
Old-witch. *Panicum capillare*
Onion. *Avena elatior*
Orange. *Hypericum Sarothra*
Orchard. *Dactylis glomerata*
Orcheston. *Agrostis stolonifera*
Ornamental Cloud. *Agrostis nebulosa* and *A. pulchella*
Otago Tupak. *Carex appressa*
Painted. *Phalaris arundinacea variegata*
Pampas. *Gynerium argenteum*
Panick. The genus *Panicum*
Panick, Bulbous, *Panicum bulbosum*
Panick, Tall. *Panicum altissimum*
Panick, Twiggy. *Panicum virgatum*
Parà. See Grass, Monkey
Pearl. *Briza maxima* and *Avena elatior*
Penny. *Rhinanthus Crista-galli*
Pen-reed. *Saccharum Sara*
Pepper. *Pilularia globulifera*
Pepper, American. *Lepidium virginicum*
Pigeon's. *Verbena officinalis*
Pin. See Pin-grass
Pink. Another name for Gilliflower-grass
Porcupine. *Stipa spartea*
Porcupine, Australian. *Festuca (Triodia) irritans*
Poverty. *Aristida dichotoma*
Prairie. *Bromus ciliatus* and *Spartina cynosuroides*
Prairie, or Rescue. *Bromus Schraderi* and *Ceratochloa unioloides*
Pudding. *Mentha Pulegium*
Purple Moor. *Molinia cærulea*
Quack. *Triticum repens*
Quake, or Quaking. *Briza media*
Quitch. *Triticum repens*
Ramee, Ramie, or Rheea. *Bœhmeria (Urtica) nivea*
Rattle. *Rhinanthus Crista-galli*
Rattle-snake. *Glyceria canadensis*
Ravenna. *Erianthus Ravennæ*
Ray, or Rye. *Lolium perenne*
Ray, or Rye, Italian. *Lolium perenne var. multiflorum (L. italicum)*
Razor. *Scleria scindens*
Red Rattle. *Pedicularis palustris*
Red-top. *Agrostis vulgaris*
Red-top, Tall. *Tricuspis sesleroides*
Reed. The genus *Arundo*
Reed, W. Indian. *Arundo occidentalis*
Reed, Meadow. *Glyceria aquatica*
Rheea. See Grass, Prairie
Rescue. See Grass, Prairie
Rib. *Plantago lanceolata*
Ribbon. *Phalaris arundinacea variegata*
Rice Cut. *Leersia oryzoides*
Ridging. *Anatherum bicorne*
Rie, or Rye. See Grass, Ray
Ripple. *Plantago lanceolata*
Roosa, or Roussa. *Andropogon Ivarancusa*, and *A. Martini*
Rope. The genus *Restio*
Rot. *Pinguicula vulgaris, Holcus lanatus,* and *H. mollis*
Rye. See Grass, Ray
Rush. The genus *Vilfa*
St. John's. Various species of *Hypericum*

Grass, Salt-water Reed. *Spartina polystachya*
Sand. *Tricuspis purpurea*
Scented, New Zealand. *Hierochloë redolens*
Scorpion. The genus *Myosotis*
"Scotch," of Jamaica, *Panicum molle*
Scrub. *Equisetum hyemale* and other species
Scurvy. *Cochlearia officinalis*
Scutch. *Triticum repens*
Sea. *Armeria maritima* and *Ruppia maritima*
Sea Arrow. *Triglochin maritimum*
Sea Hard. *Lepturus incurvatus* and the genus *Ophiurus*
Sea-shore Bent. *Psamma arenaria (Ammophila arundinacea)*
Sea Spear. *Glyceria maritima*
Serpent. *Polygonum riviparum*
Shadow. *Luzula sylvatica*
Shaking. *Briza media*
Shamalo. *Panicum frumentaceum*
Shave. *Equisetum hyemale*
Shear, or Shere. *Cladium Mariscus;* also some species of *Carex*
Sheath-flowering. *Coleanthus subtilis*
Sheep's-Fescue. *Festuca ovina*
Shelly. *Triticum repens*
Shere. See Grass, Shear
Shore. *Littorella lacustris*
Shrubby. The genus *Thamnochortus*
Silk. The fibre of *Bromelia (Nidularium) Karatas* and of *Agave vivipara* and *A. yuccæfolia*
Silk, Carolina. *Yucca filamentosa*
Silk, of N. America. *Eriocoma cuspidata*
Silt. *Paspalum distichum*
Silver. *Phalaris arundinacea variegata*
Skally. *Triticum repens*
Snake. *Myosotis palustris*
Sour. *Panicum leucophæum* and *Rumex Acetosa*
Sour, W. Indian. *Paspalum conjugatum*
Southern Holy. *Hierochloë australis*
Sparrow. *Asparagus officinalis*
Sparrow, French. *Ornithogalum pyrenaicum*
Spart. *Spartina stricta*
Spear. Various species of *Agrostis*
Spear, of New Zealand. *Aciphylla squarrosa*, and *A. Colensoi*
Spike. *Brizopyrum spicatum*, and the genus *Uniola*
Spire. A species of *Carex*
Spring. See Grass, Vernal
Spurt. *Scirpus lacustris* and *S. maritimus*
Squirrel-tail. *Hordeum jubatum, H. maritimum,* and *H. murinum*
Squitch. *Triticum repens*
Star. The genus *Callitriche*
Steep. *Pinguicula vulgaris*
Sticky. *Galium Mollugo* and *Dactylis glomerata*
Suffolk. *Poa annua*
Sweet. *Asperula odorata* and the genus *Glyceria*
Swine's. *Polygonum aviculare*
Sword. The genus *Gladiolus;* also *Arenaria segetalis, Melilotus segetalis,* and *Phalaris arundinacea*

English Names of Cultivated, Native,

Grass, Tape, or Eel. *Vallisneria spiralis*
Tassel. *Ruppia maritima*
Tassel Cotton. *Eriophorum polystachyon*
Teosinte. *Euchlaena (Reeana) luxurians*
Thin. *Agrostis elata*, and *A. perennans*
Timothy *Phleum pratense*
Toad. *Juncus bufonius*
Tongue. *Lepidium sativum* and *Stellaria media*
Tooth-ache. *Ctenium americanum*
Tranecn. *Cynosurus cristatus*
Tupak. *Carex appressa*
Turkey. *Galium Aparine*
Turk's-head. *Lagurus ovatus*
Turtle. *Zostera marina*
Turtle, or Manatu. *Thalassia testudinum*
Twig. The genus *Rhabdochloa*
Tussock. *Aira cæspitosa*
Tussock, of Australia. *Xerotes longifolia*
Twopenny. *Lysimachia Nummularia*
Umbrella. *Fuirena squarrosa* and *Panicum decompositum*
Vanilla. *Hierochloë fragrans* and other species
Vanilla, Large-leaved. *Hierochloë macrophylla*
Variegated Japanese. *Eulalia japonica variegata*
Velvet. *Holcus lanatus* and *H. mollis*
Vernal, or Sweet Vernal. *Anthoxanthum odoratum*
Vetch. *Lathyrus Nissolia*
Viper's. *Echium vulgare* and the genus *Scorzonera*
Wall Penny. *Cotyledon Umbilicus*
Wallaby. *Danthonia penicillata*
Wart. *Euphorbia Helioscopia*
Warted, of Australia. *Chloris ventricosa*
Water. *Nasturtium officinale*
Water Hair. *Catabrosa (Aira) aquatica*
Water Star. *Leptanthus gramineus*
Water Whorl. *Catabrosa (Aira) aquatica*
White. *Leersia virginica*
Whitlow. The genus *Draba*; also *Saxifraga tridactylites*
Wild Oat. Various species of *Avena*
Wild Oat, American. The genus *Danthonia*
Willow. *Polygonum amphibium*
Wind. *Apera Spica-venti*
Wind-mill. *Chloris truncata*
Winter-green. The genus *Trichodium*
Wire. *Polygonum aviculare*
Wire, American. *Eleusine indica* and *Poa compressa*
Wire, E. Indian. *Cynosurus indicus*
Wire Bent. *Nardus stricta*
Wood. *Luzula sylvatica* and *Sorghum nutans*
Wood Meadow. *Poa nemoralis*
Wood Reed. *Cinna arundinacea*
Wool. *Scirpus Eriophorum*
Woolly Beard. The genus *Erianthus*
Worm. *Sedum album* and the genus *Spigelia*
Wrack. *Zostera marina*
Yellow. *Narthecium ossifragum*
Yellow-eyed. The genus *Xyris*

Grass-cloth-plant. *Bœhmeria (Urtica) nivea*
Queensland. *Pipturus argenteus (P. propinquus)*
Silver-leaved. *Bœhmeria argentea*
Grasses, Artificial, See Grass, Artificial
Grass-Ivy. See Ivy
Grass-Nettle, Wild. *Stachys sylvatica*
Grass-Pink. *Calopogon pulchellus*
Grass-Poly. *Lythrum hyssopifolium*
Grass-root. *Eupatorium purpureum*
Grass-Thistle. See Thistle
Grass-tree, Australian. The genus *Xanthorrhæa*; also applied to *Richea dracophylla*, and *Kingia australis*
Grass-wrack, or Grass-weed, *Zostera marina*
Dwarf. *Zostera nana*
Gravel-root. *Eupatorium purpureum*
Grease-wood. The genera *Sarcobatus* and *Grayia*
Californian. *Sarcobatus vermiculatus*
Great Celandine. *Chelidonium majus*
Great Pilewort. *Scrophularia nodosa*
Great Sanicle. *Alchemilla vulgaris*
Gree-Gree-tree, or Sassy-tree, of Sierra Leone. *Erythrophlæum guineense*
Of Trinidad. *Astrocaryum aculeatum* and *Acrocomia sclerocarpa*
Greeds. *Lemna minor*
Greek Valerian. *Polemonium cæruleum*
Green Dragons. *Arum Dracontium*
Green-heart, Bastard, W. Indian. *Calyptranthes Chytraculia*
Guinea. *Nectandra Rodiæi*
W. Indian. *Colubrina ferruginea*
Green-heart-Bark, or Bibiru-Bark-tree. *Nectandra Rodiæi*
Green-man Orchis. *Aceras anthropophora*
Green-sauce. *Rumex Acetosa*
Green-weed, or Greening-weed. *Genista tinctoria*
Hairy. *Genista pilosa*
Hare's-foot. *Genista sagittalis*
Green-withe. *Vanilla clariculata*
Greens. *Lemna minor*
Grim-the-Collier. *Hieracium aurantiacum*
Grip-grass. See Grass
Grit-berry. The genus *Comarostaphylis*
Gromwell, Grummel, or Graymile. *Lithospermum officinale*
Corn-field. *Lithospermum arvense*
Creeping. *Lithospermum purpureo-cæruleum*
Deep-yellow-flowered. *Lithospermum (Batschia) canescens*
False. The genus *Onosmodium*
Gaston's. *Lithospermum Gastoni*
Gentian, or Purple. *Lithospermum prostratum*
Rock. *Lithospermum petræum*
Rosemary-leaved. *Lithospermum rosmarinifolium*
Shrubby. *Lithospermum fruticosum*
Gromwell-Reed. An old name for *Coix Lachryma*
Ground-Ash. See Ash

Ground-Enell. An old name for *Scandix Pecten-Veneris*
Ground-hele. *Veronica officinalis*
Ground-Honey-suckle. *Lotus corniculatus*
Ground-Ivy. *Nepeta Glechoma*
Ground-needle. An old name for *Erodium moschatum*
Ground-Nut, American. *Arachis hypogæa, Apios tuberosa,* and *Aralia* (*Panax*) *trifolia*
Bambarra. *Voandzeia subterranea*
Or Earth-nut, Oil-plant. *Arachis hypogæa*
Ground-Pine. *Ajuga Chamæpitys*
Groundsel. The genus *Senecio*
Adonis-leaved. *Senecio admifolius*
Bolander's. *Senecio Bolanderi*
Broad-leaved. *Senecio saracenicus*
Clammy. *Senecio viscosus*
Climbing. *Senecio mikanoides*
Common. *Senecio vulgaris*
Fen. *Senecio paludosus*
Field. *Senecio campestris*
Hoary. *Senecio incanus*
Ivy-leaved. *Senecio macroglossus*
Japanese. *Senecio japonicus*
Large Crimson-flowered. *Senecio pulcher*
Large-flowered. *Senecio Doronicum*
Leopard's-bane. *Senecio Doronicum*
Marsh. *Senecio palustris*
Narrow-leaved. *Senecio erucæfolius*
One-flowered. *Senecio uniflorus*
Orange-flowered. *Senecio abrotanifolius*
Oxford. *Senecio squalidus*
Plume-leaved. *Senecio Petasites*
Purple-flowered. *Senecio cruentus*
Showy. *Senecio speciosus*
Silvery. *Senecio argenteus*
Spoon-leaved. *Senecio spatulæfolius*
Tree. *Baccharis halimifolia*
Tyerman's. *Senecio pulcher*
Wood. *Senecio sylvaticus*
Woolly. *Senecio campestris*
Wormwood-leaved. *Senecio artemisiæfolius*
Groundsel-tree. *Baccharis halimifolia*
Ground-swell, or Ground-will. An old name for Groundsel.
Grove-tree, Indian. *Ficus indica*
"Guaco - plant." *Aristolochia Guaco* and *Mikania Guaco*
Guarana-plant. *Paullinia sorbilis*
Guava-berry. *Eugenia lineata*
Guava-Real. *Inga spectabilis*
Guava-tree. *Psidium Guaiava, P. pomiferum,* and *P. pyriferum*
Black. *Guettarda argentea*
E. Indian. *Psidium indicum*
Hill. *Rhodomyrtus tomentosus*
Mountain. *Psidium montanum*
Purple. *Psidium Cattleyanum*
Spice. *Psidium cordatum*
Guelder-Rose. *Viburnum Opulus*
Dahurian. *Viburnum dahuricum*
Dwarf. *Viburnum Opulus nanum*
Eastern. *Viburnum orientale*
Virginian. *Spiræa opulifolia*
Guernsey Lily. *Nerine sarniensis*
Guimauve. *Althæa officinalis*

Guinea-corn. *Sorghum vulgare*
Guinea-fowl-plant. *Petiveria alliacea*
Guinea-fowl-wood. *Badula* (*Anguillaria*) *Barthesia*
Guinea-hen-flower. *Fritillaria Meleagris*
Guinea-hen-weed. *Petiveria alliacea*
Dwarf. *Petiveria octandra*
Guitar-wood, or Fiddle-wood, Tree. The genus *Citharexylon*
New Zealand, *Myoporum lætum*
Gulf-weed. *Sargassum bacciferum*
Gull-grass. See Grass
Gum-Box, Chilian. *Escallonia macrantha*
Gum-Myrtle. See Myrtle
Gum-plant, Acaroid, or Yellow. *Xanthorrhœa arborea* and *X. hostilis*
Alk. *Pistacia Terebinthus*
Ammoniac. *Peucedanum ammoniacum*
Angico. *Piptadenia rigida*
Arabic. *Acacia arabica, A. Verek,* and several other species of *Acacia*
Barbary, or Morocco. *Acacia gummifera*
Butea. *Butea frondosa* and *B. superba*
Californian. The genus *Grindelia*
Cape. *Acacia horrida*
Carana. *Icica Carana*
Cashew. *Anacardium occidentale*
Cedar, of the Cape. *Widdringtonia juniperoides*
Cistus. *Cistus ladaniferus*
Conina. *Icica heptaphylla*
Copal. *Trachylobium Hornemannianum, T. Coubaril,* and *T. verrucosum;* also *Guibourtia copalifera* and *Vateria indica*
Copal, Brazilian. *Hymenæa martiana*
"Doctor's." *Rhus Metopium*
Dragon. *Pterocarpus Draco;* also a commercial name for Gum Tragacanth
Elemi. *Amyris elemifera* (*A. Plumieri*) and *Icica Abilo* (?)
Euphorbium. See Euphorbium
Guaiacum. *Guaiacum officinale*
Gutta (American). *Vismea guianensis.*
Hog. *Moronobea coccinea* (*Symhond a globulifera*)
Ivy. *Hedera Helix*
Juniper, or Sandarach. *Callitris quadrivalvis*
Kino. *Pterocarpus erinaceus* and *P. Marsupium*
Kos. *Artocarpus integrifolia*
Kuteera. *Acacia leucophlæa, Cochlospermum Gossypium,* and *Sterculia urens*
Lac. *Croton lacciferum, Erythrina monosperma,* and *Schleichera trijuga*
Ladanum. *Cistus creticus, C. ladaniferus, C. villosus, C. salviæfolius,* and other species
Ledon. *Cistus Ledon*
Morocco. See Gum, Barbary
Myrrh. *Balsamodendron Myrrha*
Opocalpasum. *Acacia gummifera*
Orenberg. *Abies Larix*
Red, or Yellow, of W. Africa. *Gibourtia copallina*
Sandarach. See Gum-plant, Juniper
Sarcocol. *Penca Sarcocolla*

Gum-plant, Sassa. *Inga Sassa*
Senegal. A variety of Gum-Arabic
Seraphic. See Sagapenum
Soudan. A variety of Gum-Arabic
Succory. *Chondrilla juncea*
Sweet. *Liquidambar styraciflua*
Thur. A variety of Gum-Arabic
Thus. See Frankincense, Common
Tragacanth. *Sterculia Tragacantha*
Wadalee. *Acacia Catechu*
Wattle. *Acacia mollissima*
Yellow. See Gum-plant, Acaroid
Gum-shrub, of St. Helena. *Commidendron rugosum*
Gum-thistle, Poisonous. *Euphorbia officinarum*
"Gum-top." *Eucalyptus Sieberiana (E. virgata)*
Gum-tree, American. *Bursera gummifera*
Apple-scented. *Eucalyptus Stuartiana*
Black, or Sour. *Nyssa multiflora (N. sylvatica)*
Blue. *Eucalyptus globulus, E. botryoides, E. diversicolor, E. hæmastoma, E. megacarpa, E. tereticornis,* and *E. riminalis*
Brown. *Eucalyptus robusta*
Cape. *Acacia horrida*
Cluster-leaved, of St. Helena. *Commidendron spurium*
Cotton. *Nyssa uniflora*
Dominica. *Dacryodes hexandra*
Drooping. *Eucalyptus Risdoni* and *E. riminalis*
Fever. *Eucalyptus globulus*
Flooded. *Eucalyptus coriacea, E. decipiens, E. rostrata,* and *E. rudis*
Fluted. *Eucalyptus salubris*
Giant. *Eucalyptus amygdalina*
Gray. *Eucalyptus resinifera,* and *E. saligna*
Green, Olive-green, Lead, or White. *Eucalyptus stellulata*
Lead. See Gum-tree, Green
Lemon-scented. *Eucalyptus maculata var. citriodora*
Manna. *Eucalyptus viminalis*
Olive-green. See Gum-tree, Green
Red. *Eucalyptus amygdalina, E. calophylla, E. melliodora, E. odorata, E. resinifera, E. rostrata, E. Stuartiana,* and *E. tereticornis*
Risdon. *Eucalpytus Risdoni*
River. *Eucalyptus dealbata* and *E. rostrata*
Rusty. *Eucalyptus eximia*
Salmon-barked. *Eucalyptus salmoniphloia*
Scarlet-flowered. *Eucalyptus ficifolia*
Sour. *Nyssa multiflora*
Spotted. *Eucalyptus goniocolyx, E. hæmastoma,* and *E. maculata*
Sugar. *Eucalyptus corynocalyx*
Swamp. *Eucalyptus coriacea* and *E. rudis*
Sweet. *Liquidambar styraciflua*
Turpentine. *Eucalpytus Stuartiana longifolia*
Water. *Tristania neriifolia*
Weeping. *Eucalyptus coriacea,* and *E. riminalis*
W. Indian. *Sapium laurifolium*

Gum-tree, White. *Eucalyptus albens, E. coriacea, E. goniocalyx, E. leucoxylon minor, E. paniculata fasciculosa, E. rostrata, E. saligna, E. stellulata,* and *E. Stuartiana*
York. *Eucalyptus loxophleba*
Gum-wood, Little Bastard, of St. Helena. *Commidendron spurium*
Gum Wattle-tree. *Acacia decurrens* and *A. mollissima*
Gunny-bag Plant. *Corchorus capsularis*
Gunyang, of S. E. Australia. *Solanum vescum*
Gurjun-tree. *Dipterocarpus turbinatus*
Gutta-Percha Tree, E. Indian. *Dichopsis (Isonandra) Gutta*
Of Guiana. *Mimusops globosa*
Gymnogram, Small. *Gymnogramma leptophylla*

Hack-berry, or Hag-berry. The fruit of *Prunus Padus*
American. *Celtis occidentalis*
Hackmatack. *Larix americana*
Hag-berry. See Hack-berry
Hag-taper, Hedge-taper, Hig-taper, or High-taper. *Verbascum Thapsus*
Hair, African. The fibre of *Chamærops humilis*
Hair-bell. See Hare-bell
Austrian. *Campanula pulla*
Double-fringed. *Campanula soldanellæflora*
Hair-branch Tree. *Trichocladus crinitus*
Hair-cup-flower. *Calythrix tetragona*
Haireve. See Harif
Hair-grass. See Grass
Hair-Moss. See Moss
Halbert-weed. The genus *Neurolæna*
W. Indian. *Neurolæna (Calea) lobata*
Hale-nut. *Corylus Avellana*
Hallelujah. *Oxalis Acetosella*
Hammer-wort. An old name for *Parietaria officinalis*
Hand-flower-tree, or Hand-plant. *Cheirostemon platanoides (Cheiranthodendron pentadactylon)*
Hard-beam. An old name for Horn-beam
Harbinger-of-spring. *Erigenia bulbosa*
Hard Fern. See Fern
Hard Grass. See Grass
Hard-hack. *Spiræa tomentosa*
Hard-hay. *Hypericum quadrangulare*
Hard-heads. *Centaurea nigra*
Hard-peer. *Olinia cymosa*
Hard-wood-tree. *Ixora ferrea*
Hare-bell, or Hair-bell. *Campanula rotundifolia* and *Scilla nutans*
African. *Roella ciliata* and other species
Australian. *Wahlenbergia gracilis*
Ivy-leaved. *Wahlenbergia (Campanula) hederacea*
Loefling's. *Campanula Loeflingi*
Lorey's. *Campanula Loreyi*
New Zealand. Various species of *Wahlenbergia,* especially *W. saxicola*
Silvery Dwarf. *Edraianthus Pumilio*
Thrift-leaved Dwarf. *Edraianthus Pumiliorum*

Hare-nut. *Bunium flexuosum*
Hare's-bane. *Aconitum Lycoctonum*
Hare's-beard. *Verbascum Thapsus*
Hare's-Cole-wort. Another name for Hare's-Lettuce
Hare's-ear. *Bupleurum rotundifolium*
 Bastard. *Phyllis Nobla*
 Shrubby. *Bupleurum fruticosum*
Hare's-eye. An old name for *Lychnis diurna*
Hare's-foot Fern. See Fern
Hare's-Lettuce, or Hare's-palace. *Sonchus oleraceus*
Hare's-tail Grass. See Grass
Hare's-tail Rush. *Eriophorum vaginatum*
Harif, Heiriff, Haireve, or Haritch. *Galium Aparine*
Haritch. See Harif
Harlequin-flower, African. The genus *Sparaxis*
Harlock. *Arctium Lappa*
Harstrong, or Horestrang. *Peucedanum officinale*
Hart-berries. *Vaccinium Myrtillus*
Hart's-balls. See Deer-balls
Hart's-horn Plantain. *Plantago Coronopus*
Hart's-thorn. *Rhamnus catharticus*
Hart's-tongue Fern. See Fern
Hart-wort, Common. *Tordylium officinale*
 Great. *Tordylium maximum*
 Mountain. *Athamanta (Peucedanum) Cervaria*
 Small. *Tordylium apulum*
Harvest-bells. *Gentiana Pneumonanthe*
Hask-wort. *Campanula latifolia*
Hassagay-tree. See Assagay
Hassock-grass. See Grass
Hat-plant, Singapore. *Æschynomene aspera*
Hautbois. *Fragaria elatior*
Haver, or Havver-grass. *Avena sativa*
Haw, or Haws. The fruit of *Crataegus Oxyacantha*
 Apple. See Haw, May
 Black. *Viburnum prunifolium*
 May, or Apple. *Crataegus æstivalis*
 Summer. *Crataegus flava*
Hawk-bit. The genus *Apargia*
Hawk-nut. See Hog-nut
Hawk-weed. The genus *Hieracium*
 Canada. *Hieracium canadense*
 Hairy. *Hieracium Gronovii*
 Honey-wort. *Hieracium cerinthoides*
 Long-bearded. *Hieracium longipilum*
 Mountain. *Hieracium alpinum*
 Mouse-ear. *Hieracium Pilosella*
 Orange-flowered. *Hieracium aurantiacum*
 Panicled. *Hieracium paniculatum*
 Pink-flowered. *Hieracium incarnatum*
 Red. *Boerkhausia rubra*
 Rough. *Hieracium scabrum*
 Shrubby. *Hieracium sabaudum*
 Umbellate. *Hieracium umbellatum*
 Wall. *Hieracium murorum*
 Wall-Lettuce. *Hieracium prenanthoides*
 Yellow Garden. *Tolpis barbata*
Hawk's-beard. The genus *Crepis*
 Beaked. *Crepis taraxacifolia*
 Fetid. *Crepis fœtida*
 Golden. *Crepis aurea*
 Hawk-weed. *Crepis hieracioides*

Hawk's-beard, Marsh. *Crepis paludosa*
 Red-flowered. *Crepis rubra*
 Rough. *Crepis biennis*
 Smooth. *Crepis virens*
Hawk's-eye. A name suggested by Mr. Ruskin for the genus *Hieracium*
Hawthorn, *Crataegus Oxyacantha*
 Chinese. *Photinia serrulata (Crataegus glabra)* and other species
 Double Pink. *Crataegus Oxyacantha rosea plena*
 Double Scarlet. *Crataegus Oxyacantha coccinea*
 Double White. *Crataegus Oxyacantha plena*
 Douglas's. *Crataegus Douglasii*
 E. Indian. *Raphiolepis (Crataegus) indica*
 Gooseberry-leaved. *Crataegus parvifolia*
 New Zealand. See Thorn
 Parsley-leaved. *Crataegus Azarolus*
 Scarlet. *Crataegus Oxyacantha coccinea*
 Scalloped-leaved. *Crataegus crenulata*
 Tansy-leaved. *Crataegus tanacetifolia*
 Weeping. *Crataegus Oxyacantha var. pendula*
Hay-plant, Tibet or Prangos. *Prangos pabularia*
Hazel. *Corylus Avellana*
 American Beaked. *Corylus rostrata*
 American Wild. *Corylus americana*
 American Witch. *Hamamelis virginica*
 Constantinople. *Corylus Colurna*
 Cuckold. *Corylus rostrata (C. cornuta)*
 Evergreen. *Guevina Avellana (Quadria heterophylla)*
 Japan. *Corylus heterophylla*
 N. S. Wales. *Pomaderris lanigera*
 Victorian. *Pomaderris apetala*
 Witch, or Wych. *Ulmus montana*
Hazel-Crottles, Hazel-Rag, or Hazel-Raw. *Sticta pulmonacea*
Hazel-nut, Snapping. *Hamamelis virginica*
Hazel-rag, or Hazel-Raw. See Hazel-Crottles
Hazel-wort. *Asarum europæum*
Head-ache. *Papaver Rhœas*
Head-ache-tree. *Premna integrifolia*
Head-ache-weed. *Hedyosmum nutans*
Heal-all. *Collinsonia canadensis* and *Rhodiola rosea*
Heal-dog. An old name for Mad-wort
Heart-clover. See Heart-trefoil
Heart of-the-Earth. *Prunella vulgaris*
Heart-seed. The genus *Cardiospermum*
Heart-trefoil, or Heart-clover. *Medicago maculata*
Heart's-ease. *Viola tricolor*
Heath, or Heather. A general name for the genera *Erica* and *Calluna*
 American False. *Hudsonia ericoides*
 Bell-flowered. *E. codonodes,* and *E. Tetralix*
 Berried. *Erica baccans*
 Black-berried. *Empetrum nigrum*
 Ciliated. *Erica ciliaris*
 Connemara. *Erica carnea var. hibernica*
 Cornish. *Erica vagans*
 Cross-leaved. *Erica Tetralix*
 False. *Fabiana imbricata*
 Irish, or St. Dabeoc's. *Dabœcia (Menziesia-polifolia*

Heath, Ling. *Calluna vulgaris*
Large-flowered Australian. *Epacris grandiflora*
Many-flowered. *Erica multiflora*
Mediterranean. *Erica mediterranea* (*E. carnea*)
Otago. *Leucopogon Fraseri*
Palm. *Richea pandanifolia*
Polytrichum-leaved. *Erica polytrichifolia*
Prickly. *Pernettya angustifolia*
St. Dabeoc's. See Heath, Irish
Sea. *Frankenia lævis*
Sicilian. *Erica sicula*
Spanish. *Erica australis*
Tasmanian. *Epacris exserta*
Tree. *Erica arborea*
Upright. *Erica stricta*
White-flowered Irish. *Dabœcia* (*Menziesia*) *alba*
White-flowered Spring. *Erica carnea alba*
Winter. *Erica carnea*
Heath-cup, Fringed, of Australia. *Artanema fimbriata*
Heath-Cypress. *Lycopodium alpinum*
Heath-pea. *Lathyrus macrorrhizus*
Heather, He. *Calluna vulgaris*
Himalayan. *Andromeda fastigiata*
Monox, or Monnaghs. *Empetrum nigrum*
Scotch. *Erica cinerea*
She. *Erica cinerea*
Heavy-wood, Red. *Baroxylon rufum*
Hedge-bells. *Convolvulus sepium*
Hedge-berry. *Prunus Padus* and *Cerasus avium*
Hedge-hog. *Ranunculus arvensis*
Hedge-hog Grass. See Grass
Hedge-hog Parsley. *Caucalis daucoides*
Hedge-hog-plant. *Anthyllis erinacea* and *Echinaria capitata*
Hedge-Hyssop. *Scutellaria minor*
Hedge-maids. *Glechoma hederacea*
Hedge-Mustard. *Sisymbrium officinale*
Sweet-scented. *Erysimum odoratum*
Hedge-nettle. *Stachys sylvatica*
Hedge-parsley. *Torilis Anthriscus*
Hedge-Pink. *Saponaria officinalis*
Hedge-Taper, or Hag-Taper. *Verbascum Thapsus*
Hedge-Vine. *Clematis Vitalba*
Hedge-Violet. *Viola sylvatica*
Heiriff. See Hariff
Helen-flower. The genus *Helenium*
Dark-purple. *Helenium atro-purpureum*
Hoopes's. *Helenium Hoopesii*
Smooth. *Helenium autumnale*
Heliotrope, Common. *Heliotropium europæum*
Indian. *Heliotropium indicum*
Peruvian. *Heliotropium peruvianum*
Summer. *Tournefortia heliotropioides*
Winter. *Nardosma* (*Tussilago*) *fragrans*
Hellebore, Abchasian. *Helleborus abchasicus*
Bastard. *Helleborus viridis*
Bear's-foot. *Helleborus fœtidus*
Black. *Helleborus niger*
" Black," of the Ancients. *Helleborus officinalis*
Dark Purple. *Helleborus atrorubens*

Hellebore, Fetid. *Helleborus fœtidus*
Green. *Helleborus viridis*
Holly-leaved. *Helleborus argutifolius*
Olympian. *Helleborus olympicus*
Oriental. *Helleborus orientalis*
Purple-flowered. *Helleborus purpurascens*
Swamp. *Veratrum viride*
Sweet-scented. *Helleborus odorus*
White. See White Hellebore
Winter. *Eranthis hyemalis*
Helleborine. The genera *Epipactis* and *Cephalanthera*
Marsh. *Epipactis palustris*
Red. *Cephalanthera rubra*
White. *Cephalanthera grandiflora*
Hell-weed. *Cuscuta europœa* and *C. Epithymum*
Helmet-flower. *Aconitum Napellus* and the genus *Scutellaria*
Brazilian. *Coryanthes speciosa*
Yellow. *Aconitum Anthora*
Hemlock. *Conium maculatum*
Ground. *Taxus baccata var. canadensis*
Lesser. *Æthusa Cynapium*
Mountain. *Levisticum officinale*
Water. *Cicuta virosa* and *C. maculata*: also *Phellandrium aquaticum* and *Œnanthe crocata*
Hemlock-Dropwort. *Œnanthe fistulosa*
Hemp. The genus *Cannabis*. Applied also to various other plants
African. *Sparmannia africana*
American False. *Datisca hirta*
Angola. *Sanseviera angolensis*
Banded Bowstring. *Sanseviera fasciata*
Bengal. *Crotalaria juncea*
Black-fellows', of Australia. *Commersonia Fraseri*
Bombay. *Crotalaria juncea*
Bow-string, of Africa. *Sanseviera guineensis*
Bow-string, of India. *Sanseviera Zeylanica* and *Calotropis gigantea*
Brown Indian. *Hibiscus cannabinus*
Canada, or Indian. *Apocynum cannabinum*
Common. *Cannabis sativa*
Cretan. *Datisca cannabina*
E. Indian. *Cannabis sativa* and *Hibiscus cannabinus*
Holy. See Holy Hemp
Indian, or Canada. *Apocynum cannabinum*
Jubbulpore. *Crotalaria tenuifolia*
Kentucky. *Urtica* (*Laportea*) *canadensis* and *U. cannabina*
Koffo. The fibre of *Musa textilis*
Madras. *Crotalaria juncea*
Manilla. The fibre of *Musa textilis*
Peruvian. *Buonapartea juncea*
Sisal. *Agave Sisalana*
Sunn. *Crotalaria juncea*
Water. *Eupatorium cannabinum*, *Acnida cannabina*, and *Bidens tripartita*
Willow. *Acnida cannabina*
Hemp-Agrimony. *Eupatorium cannabinum*
Aromatic. *Eupatorium aromaticum*
Nettle-leaved. *Eupatorium ageratoides*
Purple. *Eupatorium purpureum*

and Foreign Plants, Trees, and Shrubs. 63

Hemp-bush, Victorian. *Plagianthus pulchellus* (*Sida pulchella*)
Hemp-Nettle, Common. *Galeopsis Tetrahit*
Downy. *Galeopsis ochroleuca*
Red-flowered. *Galeopsis Ladanum*
Hemp-tree. An old name for *Vitex Agnus-castus*
Hemp-weed. *Eupatorium cannabinum*
Climbing. *Mikania scandens*
Hen-and-Chickens. The proliferous variety of *Bellis perennis*. Applied locally to various other flowers
Hen-bane, Canary Island. *Hyoscyamus Canariensis*
Common. *Hyoscyamus niger*
Dwarf. *Hyoscyamus pusillus*
Egyptian. *Hyoscyamus reticulatus*
Nightshade-leaved. *Hyoscyamus Scopolia*
Purple-flowered. *Hyoscyamus physaloides*
White. *Hyoscyamus albus*
Hen-bell. An old name for Hen-bane
Hen-bit, Greater *Lamium amplexicaule*
Small. *Veronica hederæfolia*
Henna-plant. *Lawsonia alba* and *L. inermis*
Henne. Another name for Shear-grass
Hen-penny-grass. See Grass
Hen-plant. *Plantago lanceolata*
Great. *Plantago lanceolata major*
Hen's-bill. An old name for the genus *Onobrychis*
Hen's-foot. *Caucalis daucoides*
Hen-ware. *Alaria esculenta*
Hep, or **Hip.** The fruit of *Rosa canina* and other species
Hep-Briar, Hep-Rose, or Hep-Tree. *Rosa canina* and other species
Hepatica. *Anemone Hepatica*
Large. *Anemone Hepatica var. angulosa*
Herb, Poor-Man's. *Gratiola officinalis*
Willow. The genus *Epilobium*
Herb-Bennet. *Geum urbanum*, *Conium maculatum*, and *Valeriana officinalis*
Herb-Carpenter. *Prunella vulgaris*
Herb-Christopher. *Actæa spicata* and *Osmunda regalis*
Herb-Eve, or Ivy. *Ajuga Iva* and *Plantago Coronopus*
Herb-Frankincense. *Laserpitium latifolium*
Herb-Gerard. *Ægopodium Podagraria*
Herb-Impious. *Filago germanica*
Herb-Ivy. See Herb-Eve
Herb-Lily. See Lily
Herb-Louisa. *Aloysia citriodora*
Herb-Margaret. *Bellis perennis*
Herb-Mastick. *Thymus Mastichina*
Herb-of-Friendship. *Sedum Anacampseros*
Herb-of-Grace, or Herb-of-Repentance. *Ruta graveolens*
Herb-Paris. *Paris quadrifolia*
Herb-Peter. *Primula veris*
Herb-Robert. *Geranium Robertianum*
Herb-Terrible. *Daphne Tartonraira*
Herb-Trinity. *Viola tricolor* and *Hepatica trilobe*
Herb-Twopence. *Lysimachia Nummularia*

Herb-William. *Ammi majus*
Hercules' All-heal. See All-heal
Hercules'-club. *Xanthoxylon Clava-Herculis* and *Aralia spinosa*
Hernant-seeds. The seeds of *Hernandia ovigera*
Heron's-bill. The genus *Erodium*
Alpine. *Erodium alpinum*
Black-eyed. *Erodium macradenum*
Caraway-leaved. *Erodium caruifolium*
Cheilanthes-leaved. *Erodium cheilanthifolium*
Common. *Erodium cicutarium*
Fairy. *Erodium Reichardi*
Fern-leaved. *Erodium trichomanefolium*
Long-beaked. *Erodium Ciconium*
Musk. *Erodium moschatum*
Pelargonium. *Erodium hymenodes*
Rock. *Erodium petræum*
Roman. *Erodium romanum*
Sea-side. *Erodium maritimum*
Showy. *Erodium Manescavi*
Three-leaved. *Erodium hymenodes*
Herring-bone Fern. See Fern
Thistle. See Thistle
Hert-wort. An old name for the Ash-tree
Hickory. The genus *Carya*
Broom, or Brown. *Carya porcina*
Compressed-fruited. *Carya compressa*
Entire-leafleted. *Carya integrifolia*
Nutmeg. *Carya myristicæformis*
Queensland Marsh. *Cupania xylocarpa*
Shell-bark, or Shag-bark. *Carya alba*
Small-fruited. *Carya microcarpa*
Thick Shell-bark. *Carya sulcata*
Water. *Carya aquatica*
Western Shell-bark. *Carya sulcata*
White-heart. *Carya tomentosa*
Hickory-Eucalypt, of N. S. Wales. *Eucalyptus punctata*, *E. resinifera*, and *E. Stuartiana*
High-Mallow. See Mallow
Hig-taper, or High-taper. See Hag-taper
High-water Shrub. The genus *Iva*
Hill Margosa. *Melia Azedarach*
Hill-wort. An old name for *Thymus Serpyllum*
Hinau, or Hino-tree. *Elæocarpus Hinau*
Hind-berry, or Hine-berry. *Rubus Idæus*
Hind-heal. *Chenopodium Botrys* and *Teucrium Scorodonia*
Hine-berry. See Hind-berry
Hip. See Hep
Hip-rose. See Hip-briar
Hip-wort. *Cotyledon Umbilicus*
Hobble-bush. *Viburnum lantanoides*
Hockes. An old name for Holly-hock
Hood-wort. *Scutellaria lateriflora* and other species
Hog-cherry. *Prunus Padus*
Hog-Gum-tree. *Moronobea coccinea*
Hog-meat, Jamaica. *Boerhaavia decumbens*
Hog-nut, or Hawk-nut. *Bunium flexuosum*
Hog-plum. The genus *Spondias*
Hog-weed. *Heracleum Sphondylium* and the genus *Boerhaavia*
American. *Ambrosia artemisiæfolia*, *Boerhaavia erecta*, and other species

Hog-weed, Climbing. *Boerhaavia scandens* Poisonous. *Aristolochia grandiflora*
Hog's-beans. An old name for *Globularia*
Hog's-Garlic. *Allium ursinum*
Hole-wort, Hollow-wort, or Hollow-root. *Corydalis tuberosa* and *Adoxa Moschatellina*
Hollow-root. See Hole-wort
Hollow-wort. See Hole-wort
Holly, or Holm. *Ilex Aquifolium*
American. *Ilex opaca*
American Mountain. *Nemopantes canadensis*
Black-berried. *Ilex Aquifolium fructû nigro*
Box. *Ruscus aculeatus*
Broad-leaved. *Ilex latifolia*
Broad-spined. *Ilex latispina*
Canary Island. *Ilex canariensis*
Chinese. *Ilex chinensis*
Dahoon. *Ilex Dahoon*
Dwarf Golden. *Ilex crenata variegata*
Dwarf Japanese, or Dwarf Rock. *Ilex crenata*
Emetic. *Ilex vomitoria*
Fortune's Japanese. *Ilex Fortunei*
Gold-Striped. *Ilex Aquifolium aureo-variegatum*
Ground. *Chimaphila umbellata*
Hedge-hog. *Ilex Aquifolium ferox*
High-Clere. *Ilex Aquifolium Altaclarensis*
Knee. *Ruscus aculeatus ferox*
Laurel-leaved Dahoon. *Ilex Dahoon var. laurifolia*
Loose-flowered. *Ilex laxiflora*
Milk-maid. *Ilex Aquifolium var. ferox albo-pictum*
Minorca. *Ilex balearica*
Myrtle-leaved. *Ilex Aquifolium myrtifolia*
Narrow-leaved. *Ilex angustifolia*
New Zealand. *Olearia ilicifolia*
Native, of Australia. *Lomatia ilicifolia* and other species
Sea. The genus *Eryngium*. (See also Eryngo)
Silver-striped. *Ilex Aquifolium argenteo-variegata*
S. American. *Ilex paraguayensis*
Striped Hedge-hog. *Ilex Aquifolium ferox argenteo-variegata*
Two-seeded. *Ilex dipyrena*
Variegated Weeping. *Ilex Aquifolium pendula variegata*
Various-leaved. *Ilex Aquifolium heterophylla*
White-berried. *Ilex Aquifolium fructû albo*
Whorled. *Ilex Aquifolium verticillata*
Yellow-berried. *Ilex Aquifolium fructû luteo*
Holly Fern. See Fern
Hollyhock, or Holy Hoke. *Althæa rosea*
Antwerp. *Althæa ficifolia*
Hollyhock-tree, Queensland. *Hibiscus splendens*
Holly-Oak, or Holm Oak. *Quercus Ilex*
Holly-rose. *Turnera ulmifolia*; also an old name for the genus *Cistus*

Holme, Knee. See Holly, Knee
Holm-Oak. See Holly-Oak
Holy Ghost. *Angelica sylvestris*
Holy-Ghost-flower. *Peristeria elata*
Holy Grass. See Grass
Holy Hay. *Medicago sativa*
Holy Hemp. An old name for *Galeopsis Ladanum*
Holy Herb. *Verbena officinalis*
Holy Hoke. See Hollyhock
Holy Rose, Marsh. *Andromeda polifolia*
Holy Rope. *Eupatorium cannabinum*
Holy Seed. An old name for Worm-seed
Holy Tree. *Melia Azedarach*
Home-wort. *Sempervivum tectorum*
Hominy. Meal of Indian Corn (*Zea Mays*)
Homlock. An old name for Hemlock
Honesty, Common. *Lunaria biennis*
Maiden's. *Clematis Vitalba*
Perennial. *Lunaria rediviva*
White-flowered. *Lunaria biennis albiflora*
Hone-wort. *Trinia vulgaris*
American. *Cryptotænia canadensis*
Corn. *Petroselinum segetum*
Hedge. *Sison Amomum*
Honey-Balm. *Melittis Melissophyllum*
Honey-berry, of Greece. *Celtis australis*
W. Indian. *Melicocca bijuga* and *M. paniculata*
Honey-flower, Cape. *Protea mellifera*
Great. *Melianthus major*
Honey-Locust-tree, Broad-podded. *Gleditschia latisiliqua*
Caspian. *Gleditschia caspica*
Chinese. *Gleditschia sinensis*
Common. *Gleditschia triacanthos*
Curved-spined. *Gleditschia brachycarpa*
E. Indian. *Gleditschia indica*
Flat-spined. *Gleditschia ferox*
Long-spined. *Gleditschia macrospina*
Small-spined. *Gleditschia microspina*
Smooth. *Gleditschia lævis*
South Western. *Prosopis juliflora*
Honey-plant. The genus *Hoya*
Honey-suckle. *Lonicera Periclymenum*; also applied to the flowers of *Trifolium pratense* and some other plants
African Fly. *Halleria lucida*
Alpine. *Lonicera alpigena*
Australian, or Heath. *Banksia serrata*
Black-berried. *Lonicera nigra*
Blue-berried. *Lonicera cærulea*
Bush. *Diervilla canadensis* (*D. trifida*); also *Weigela rosea* and vars.
Canadian. *Lonicera canadensis*
Cape, of the W. Indies. *Tecoma capensis*
Castellamare. *Lonicera Stabiana*
Chinese. *Lonicera flexuosa*
Crimson-flowered. *Lonicera punicea*
Douglas's. *Lonicera Douglasii*
Downy. *Lonicera tormentilla*
Downy American. *Lonicera pubescens*
Dutch. *Lonicera Periclymenum belgica*
Dutch, Red. *Lonicera Periclymenum rubra*
Dwarf. *Cornus succica*
Early-flowering. *Lonicera fragrantissima* and *L. Standishi*

and Foreign Plants, Trees, and Shrubs.

Honey-suckle, Eastern. *Lonicera orientalis*
Evergreen. *Lonicera grata*
False. The genus *Azalea*
Fly. *Lonicera Xylosteum*
Fly, American. *Lonicera ciliata*
Four-leaved. *Lonicera quadrifolia*
French. *Hedysarum coronarium*
Gold-netted. *Lonicera brachypoda aureo-reticulata*
Ground. *Lotus corniculatus*
Hairy. *Lonicera hirsuta*
Heath. See Honey-suckle, Australian
Himalayan. *Leycesteria formosa*
Iberian. *Lonicera iberica*
Jamaica. *Passiflora laurifolia*
Japan. *Lonicera japonica*
Ledebour's. *Lonicera Ledebourii*
Minorca. *Lonicera implexa*
Narrow-leaved. *Lonicera angustifolia*
Perfoliate. *Lonicera Caprifolium*
Privet-leaved. *Lonicera ligustrina*
Pyrenean. *Lonicera pyrenaica*
Red Italian. *Lonicera etrusca rubra*
Shaggy. *Lonicera villosa*
Short-stalked. *Lonicera brachypoda*
Small-flowered. *Lonicera parviflora*
Small-leaved. *Lonicera microphylla*
Swamp Fly. *Lonicera oblongifolia*
Sweetest. *Lonicera odoratissima*
Tartarian. *Lonicera tatarica* and vars.
Tasmanian. *Banksia australis*
Three-flowered. *Lonicera triflora*
Trumpet. *Lonicera sempervirens*
Trumpet, Small. *Lonicera sempervirens minor*
Tuscan. *Lonicera etrusca*
Two-coloured. *Lonicera discolor*
Various-leaved. *Lonicera diversifolia*
Virgin Mary's. *Pulmonaria officinalis*
Western. *Lonicera occidentalis*
W. Indian. *Tecoma capensis* and various species of *Desmodium*
White. *Azalea viscosa*
White Italian. *Lonicera Caprifolium alba*
White Swamp. *Azalea viscosa*
Winter-flowering. *Lonicera fragrantissima*
Yellow. *Lonicera flava*
Yellow, Italian. *Lonicera Caprifolium flava*
Yellow, Upright. *Lonicera Diervilla*
Honey-suckle Clover. *Trifolium repens*
Honey-suckle-tree. *Banksia australis*, *B. Cunninghamii*, and *B. ericæfolia*
Honey-ware. *Alarea esculenta* and *Laminaria saccharina*
Honey-wort. *Galium cruciatum* and the genus *Cerinthe*
Hedge. *Sison Amomum*
Rough. *Cerinthe aspera*
Honghel-bush. *Adenium Honghel*
Hood-wort. *Scutellaria lateriflora* and other species
Hooded Grass. See Grass
Hooded Water-Milfoil. *Utricularia vulgaris*
Hoofs. *Tussilago Farfara*
Hook-heal. An old name for *Prunella vulgaris*
Hoop-Ash. *Celtis crassifolia*

"Hoop-koop"-plant. *Lespedeza striata*
Hoop-tree. *Melia sempervirens*
Hoop-withe, or Hoop-withy. *Colubrina asiatica*
Jamaica. *Rivina octandra*
W. Indian. The genus *Rivina*
Hop, Bog. *Menyanthes trifoliata*
Common. *Humulus Lupulus*
Little. *Origanum sipyleum*
Native, of Australia. The seed-vessels of *Dodonæa*; also various species of *Daviesia*
Native, of Victoria. *Daviesia latifolia*
Wild. *Bryonia dioica*
Hop-Clover. *Trifolium procumbens* and *Medicago lupulina*
Hop-Horn-beam. *Ostrya virginica*
Hop-tree. *Ptelea trifoliata*
Yellow-leaved. *Ptelea trifoliata aurea*
Hopes. *Matthiola incana*
Horehound, Black. *Ballota nigra*
Common. *Marrubium vulgare*
Stinking. *Ballota nigra*
Water. *Lycopus europæus* and other species
White. *Marrubium vulgare*
Wild. *Eupatorium teucrifolium*
Horestrang. See Harstrong
Horn, Devil's. *Phallus impudicus*
Horn-beam. *Carpinus Betulus*
American. *Carpinus americana*
Cut-leaved. *Carpinus Betulus incisa*
Hop. *Carpinus Ostrya*
Horn-of-Plenty. *Fedia Cornucopiæ*
Horn-plant. *Ecklonia buccinalis*
Horn-wort, or Horn-weed. *Ceratophyllum demersum*
Horned-Poppy, Common Yellow-flowered. *Glaucium luteum*
Orange-flowered. *Glaucium fulvum*
Red-flowered. *Glaucium phæniceum*
Three-coloured. *Glaucium tricolor*
Various-coloured. *Glaucium corniculatum*
Violet-flowered. *Ræmeria hybrida* (*Glaucium violaceum*)
Horned-Rampion. The genus *Phyteuma*
Linear-leaved. *Phyteuma hemisphæricum*
Round-headed. *Phyteuma orbiculare*
Spiked. *Phyteuma spicatum*
Tufted. *Phyteuma comosum*
Horse-bane. *Œnanthe Phellandrium*
Horse-bean. *Faba vulgaris var. equina*
Jamaica. *Canavalia ensiformis*
Horse, Hirst, or Hurst, Beech. *Carpinus Betulus*
Horse-chestnut. *Æsculus Hippocastanum*
Horse-Daisy. See Daisy
Horse-Elder. An old name for *Elecampane*
Horse-flower. *Melampyrum sylvaticum*
Horse-fly-weed. *Baptisia tinctoria*
Horse-Gowan. See Gowan
Horse-Gram. *Dolichos biflorus* and *D. uniflorus*
Horse-hoof. *Tussilago Farfara*
Horse-Mint. *Mentha sylvestris* and other species
Bradbury's. *Monarda Bradburyana*
Canadian. *Collinsonia canadensis*

F

Horse-Mint, Panicled. *Monarda paniculata*
Horse-Mushroom. *Agaricus arvensis*
Horse-Parsley. *Smyrnium Olusatrum*
Horse-Radish. *Cochlearia Armoracia*
  Kerguelen. *Pringlea antiscorbutica*
Horse-Radish-tree. *Moringa pterygosperma*
Horse-shoe Vetch. *Hippocrepis comosa*
Horse-Sorrel. *Rumex Hydrolapathum*
Horse-sugar. *Symplocos (Hopea) tinctoria*
Horse-tail. The genus *Equisetum*
  American. *Equisetum robustum* and *E. lævigatum*
  Blunt-topped. *Equisetum umbrosum*
  Dwarf. *Equisetum scirpoides*
  Field. *Equisetum arvense*
  Great. *Equisetum Telmateia*
  Great Shrubby. *Ephedra distachya*
  Long-branched. *Equisetum ramosum*
  Mackay's. *Equisetum Mackayi*
  Marsh. *Equisetum palustre*
  Moore's. *Equisetum Moorei*
  Rough. *Equisetum hyemale*
  Small Shrubby. *Ephedra monostachya*
  Smooth. *Equisetum limosum*
  Tree. *Casuarina equisetifolia*
  Variegated. *Equisetum variegatum*
  Wilson's. *Equisetum Wilsoni*
  Wood. *Equisetum sylvaticum*
Horse-Thyme. *Calamintha Clinopodium*
Horse-tongue. *Scolopendrium vulgare* and *Ruscus Hypoglossum*
Horse-weed. *Collinsonia canadensis* and *Erigeron canadensis*
Horse-well Grass. See Grass
Horse-wood. *Calliandra comosa*
  W. Indian. Various species of *Calliandra*
Horst-Beech. See Horse-Beech
Horts. *Vaccinium Myrtillus*
Hose-Grass. See Grass
Hose-in-hose. A variety of Polyanthus
Hottentot-Bread. *Testudinaria elephantipes*
Hottentot's-head. *Stangeria paradoxa*
Houmiri-tree. *Humirium balsamiferum*
Hound, or Hound's, berry. *Solanum nigrum*
Hound-berry-tree. *Cornus sanguinea*
Hound's-tongue, Alpine. *Cynoglossum alpinum*
  Common. *Cynoglossum officinale*
  Green. *Cynoglossum montanum* (*C. sylvaticum*)
  Stock-leaved. *Cynoglossum cheirifolium*
House-leek. The genus *Sempervivum*
  Anomalous. *Sempervivum anomalum*
  Bearded. *Sempervivum barbulatum*
  Boutigni's. *Sempervivum Boutignianum*
  Cob-web. *Sempervivum arachnoideum*
  Common. *Sempervivum tectorum*
  Fringed. *Sempervivum ciliatum* and *S. fimbriatum*
  Funck's. *Sempervivum Funckii*
  Gouty-stalked. *Sempervivum tortuosum*
  Hair-tipped. *Sempervivum heterotrichum*
  Hairy. *Sempervivum hirtum*
  Hairy-tufted. *Sempervivum piliferum*
  Hen-and-Chickens. *Sempervivum globiferum* (*S. soboliferum*)
  Heuffel's. *Sempervivum Heuffeli*
House-leek, Lagger's. *Sempervivum Laggeri*
  Long-runnered. *Sempervivum flagelliforme*
  Metten's. *Sempervivum Mettenianum*
  Mountain. *Sempervivum montanum*
  Pitton's. *Sempervivum Pittoni*
  Purple-tipped. *Sempervivum calcareum* (*S. californicum*)
  Red-leaved. *Sempervivum triste*
  Russian. *Sempervivum ruthenicum*
  Sand. *Sempervivum arenarium*
  Sea. An old name for *Aloe*
  Spiny. *Sempervivum spinosum*
  Table-shaped. *Sempervivum tabulæforme*
  Teneriffe. *Sempervivum ciliatum*
  Tree. *Sempervivum* (*Æonium*) *arboreum*
  Woolly. *Sempervivum tomentosum*
  Wulfen's. *Sempervivum Wulfeni*
Huck-berry. See Hack-berry
Huckle-berry. *Vaccinium Myrtillus*
  Black. *Gaylussacia resinosa*
  Box. *Gaylussacia brachycera*
  Dwarf. *Gaylussacia dumosa*
  Squaw. *Vaccinium stamineum*
Hulver. An old name for Holly
Hul-wort. An old name for *Teucrium Polium*
Humble Plant. *Mimosa pudica*
Humming-bird, bush. *Æschynomene monteridensis*
Hundred-fold. *Galium rerum*
Hundred-leaved Grass. See Grass
Hunger-grass. See Grass
Hunger-weed. *Ranunculus arvensis*
Huntsman's-Cup. *Sarracenia purpurea*
Huntsman's-Horn, Yellow-flowered. *Sarracenia flava*
Hursingar-tree. *Nyctanthes Arbor-tristis*
Hurst-Beech. *Carpinus Betulus*
Hurtle-berry. *Vaccinium Myrtillus*
Hurts. See Horts
Hurt-sickle. *Centaurea Cyanus*
Hutu, or Futu, tree of Tahiti. *Barringtonia speciosa*
Hyacinth. The genus *Hyacinthus*. Applied also to some species of *Scilla* and other Liliaceous plants
  Amethyst, or Spanish. *Hyacinthus amethystinus*
  Californian, or Missouri. The genus *Brodiæa*
  Californian, Allium-like. *Brodiæa congesta*
  Californian, Crimson-flowered. *Brodiæa coccinea*
  Californian, Ixia-like. *Brodiæa ixioides* (*Calliprora lutea*)
  Californian, Large-flowered. *Brodiæa grandiflora*
  Californian, Twining. *Brodiæa volubilis*
  Cape. *Scilla brachyphylla* and *S. corymbosa*
  Common Garden. Varieties of *Hyacinthus orientalis*
  Fair-haired. *Muscari comosum*
  Grape. See Grape-Hyacinth
  Late-flowering. *Hyacinthus serotinus*
  Lily. *Scilla Lilio-Hyacinthus*
  Missouri. The genus *Brodiæa*
  Roman. *Hyacinthus romanus* (*Belleralia romana*)

Hyacinth, Spanish. *Hyacinthus amethystinus*
Star. *Scilla amœna*
Starch. *Muscari racemosum*
Tasmanian. *Thelymitra nuda*
Tassel. *Muscari comosum*
White Cape. *Hyacinthus (Galtonia) candicans*
Wild. *Scilla nutans*
Wild, American. *Camassia (Scilla) Fraseri*
Winter. *Scilla autumnalis*
Hyacinth-Bean. The genus *Dolichos*
Purple. *Dolichos Lab-lab*
White. *Dolichos albus*
Hyæna-poison, Cape. *Hyænanche globosa*
Hyawa-tree. *Icica heptaphylla*
Hydrangea, Wild American. *Hydrangea arborescens*
Blue-flowered. *Hydrangea hortensis* var. *cærulea* and *H. japonica* var. *cærulea*
Climbing. *Schizophragma hydrangeoides*
Common. *Hydrangea hortensis*
Golden-edged. *Hydrangea japonica aurea superba*
Hardy Flesh-coloured. *Hydrangea Otaksa*
Hardy White-flowered. *Hydrangea grandiflora*
Heart-leaved. *Hydrangea cordata*
Japan. *Hydrangea japonica*
Oak-leaved. *Hydrangea quercifolia*
Rosy-flowered. *Hydrangea japonica rosea*
Silver-edged. *Hydrangea japonica variegata*
Starry-flowered. *Hydrangea stellata*
Two-coloured-leaved. *Hydrangea arborescens* var. *discolor*
White-leaved. *Hydrangea nivea*
Woolly-leaved. *Hydrangea heteromalla*
Hyssop, Anise. *Lophanthus anisatus*
Bastard. *Teucrium Pseudo-hyssopus*
Common. *Hyssopus officinalis*
Giant. The genus *Lophanthus*
Hedge. The genus *Gratiola*
Nettle-leaved Giant. *Lophanthus urticæfolius*
Square-stalked. *Hyssopus nepetoides*
Water. *Herpestis Monnieria*
Wild. *Verbena hastata*

Iceland Moss. *Cetraria islandica*
Ice-plant. *Mesembryanthemum crystallinum*
American. *Monotropa uniflora*
New Zealand. *Tetragonia expansa* and *Mesembryanthemum australe*
Small. *Mesembryanthemum sessiliflorum album*
Tasmanian. *Tetragonia implexicoma*
Ilang-Ilang-tree. *Cananga odorata*
Imbreke. An old name for House-leek
Immortelle-flower. The genus *Helichrysum,* especially *H. orientale.* Also the genera *Helipterum* and *Xeranthemum*
Incense-tree. *Boswellia thurifera* and *B. serrata*
Incense-wood, Guiana. *Icica heptaphylla*
India, Pride of. See Pride-of-India
India-rubber Plant or Tree. *Ficus elastica.* (See also Caoutchouc)
Guiana. *Hevea guianensis* (*Siphonia elastica*)
Indian Balm. *Trillium pendulum*

Indian Corn. *Zea Mays*
Indian Cress. The genus *Tropæolum*
Five-leaved. *Tropæolum pentaphyllum*
Flame-flowered. *Tropæolum speciosum*
Tuberous-rooted. *Tropæolum tuberosum*
Yellow Rock. *Tropæolum polyphyllum*
Indian Crocus. The genus *Pleione*
Bottle-gourd. *Pleione Lagenaria*
Dwarf. *Pleione humilis*
Early-flowering. *Pleione præcox*
Hooker's. *Pleione Hookeriana*
Spotted. *Pleione maculata*
Three-coloured. *Pleione tricolor*
Wallich's. *Pleione Wallichii*
Indian Cups. The genus *Sarracenia*
Indian Eye. *Dianthus plumarius*
Indian Fig. *Cactus Opuntia*
Downy. *Opuntia pubescens*
Thick-lobed. *Opuntia crassa*
Indian Forget-me-not. See Forget-me-not
Indian Grass. See Grass
Indian Grass-Oil-plant. Various species of *Andropogon*
Indian Hawthorn. See Hawthorn
Indian Heart. *Cardiospermum Corindum*
Indian Kale, Large-leaved. *Caladium grandiflorum*
Indian Leaf. *Cinnamomum malabathricum*
Indian-matting Plant. *Papyrus corymbosus*
Indian Melissa-Oil-plant. *Andropogon citratus*
Indian Moss. *Saxifraga hypnoides*
Indian Mourner, or Sad-tree. *Nyctanthes Arbor-tristis*
Indian Paper-tree. *Daphne cannabina* and *Edgeworthia Gardneri*
Indian Physic. *Gillenia trifoliata* and *Magnolia Fraseri*
Indian Pink. *Dianthus chinensis* and *Quamoclit vulgaris*
Indian Pipe. *Monotropa uniflora*
Indian Posy. *Gnaphalium polycephalum*
Indian Shot. The genus *Canna*
Indian Turnip. *Arum dracontium*
Indigo-berry. *Randia latifolia*
Indigo-plant, American Wild. *Baptisia tinctoria*
Australian. *Indigofera australis*
Blue False. *Baptisia australis*
Chinese. *Polygonum tinctorium*
Chinese, "Green." *Rhamnus utilis* and *R. chlorophorus*
Common Dyer's. *Indigofera tinctoria*
Egyptian. *Tephrosia Apollinea*
False. The genus *Baptisia*
Japanese. *Polygonum tinctorium*
Of the Niger. *Tephrosia toxicaria*
Pala. *Wrightia tinctoria*
Pegu. *Marsdenia tinctoria*
Purple-flowered. *Indigofera floribunda*
Shrubby False. *Amorpha fruticosa*
W. Indian. *Randia aculeata*
Indigo-weed. *Baptisia tinctoria*
Ink-berry. *Prinos glaber*
Queensland Black. *Kibara macrophylla*
W. Indian. *Randia aculeata*
Ink-plant, New Zealand. *Coriaria ruscifolia* and *C. thymifolia*

English Names of Cultivated, Native,

"**Insane Root.**" Canon Ellacombe, in his "Plant-lore of Shakespeare," says that this term most probably refers to the Hemlock
Inverted-flower. The genus *Parastranthus*
Iodine-plants. Various kinds of *Algæ* or Sea-weeds
Ipecacuanha, American. *Gillenia stipulacea*
 Bastard, or Wild. *Asclepias curassavica*
 Black Peruvian, or Striated. *Psychotria emetica*
 Country, or E. Indian. *Tylophora (Asclepias) asthmatica*
 E. Indian. See Ipecacuanha, Country
 False Brazilian, or White. *Ionidium Ipecacuanha*
 Guiana. *Boerhaavia decumbens*
 Peruvian. See Ipecacuanha, Black
 Striated. See Ipecacuanha, Black
 Undulated, or White. *Richardsonia scabra*
 White. See Ipecacuanha, False Brazilian, and Undulated
 Wild. *Asclepias curassavica*
 Venezuelan. *Sarcostemma glaucum*
Irby-dale Grass. See Grass
Iris, Algerian. *Iris stylosa*
 Bermuda. *Marica irioides*
 Black. The genus *Ferraria*
 Blue-eyed Peacock. *Vieusseuxia glaucopis (Iris Pavonia)*
 Bobart's Orange. *Bobartia aurantiaca*
 Boston. *Iris virginica*
 Brown-flowered. *Iris squalens*
 Bristle-pointed. *Iris setosa*
 Butterfly. *Moræa papiliomacea*
 Caucasian. *Iris caucasica*
 Christmas-flowering. *Iris alata*
 Clouded. *Iris Xiphium*
 Common Garden. Varieties of *Iris germanica*
 Cream-coloured. *Iris stenogyna*
 De Bergh's. *Iris De Berghi*
 Delicately-tinted. *Iris amœna*
 Dwarf. *Iris pumila*
 Dwarf American. *Iris verna*
 Dwarf Lake. *Iris lacustris*
 Early Bulbous. *Iris reticulata*
 Elder-scented. *Iris sambucina*
 "English." *Iris xiphioides (Xiphion latifolium)*
 Ever-blooming. *Iris ruthenica*
 Florentine. *Iris florentina*
 German. *Iris germanica*
 Golden. *Iris Monnieri*
 Grass-leaved. *Iris graminea*
 Great Bulbous. *Iris Xiphioides*
 Great Spotted. *Iris susiana*
 Greek. *Iris attica*
 Iberian. *Iris iberica (Onocyclus ibericus)*
 Italian. *Iris italica*
 Kæmpfer's. *Iris Kæmpferi*
 Labrador. *Iris tridentata*
 Long-petalled. *Iris longipetala*
 Missouri. *Iris missouriensis*
 Monnier's. *Iris Monnieri*
 Naked-stemmed. *Iris nudicaulis*
 Onion. *Iris tuberosa*

Iris, Pale Blue. *Iris pallida*
 Peacock. *Iris (Vieusseuxia) Pavonia*
 Pretty-flowered. *Iris pulchella*
 Red-stemmed. *Iris rubricaulis*
 Robinson's. *Iris (Moræa) Robinsoniana*
 Rush-leaved. *Iris juncea (I. lusitanica)*
 Russian. *Iris ruthenica*
 Saar's. *Iris Saari*
 Sad-flowered. *Iris susiana*
 Sardinian. *Iris Olbiensis*
 Scorpion. *Iris alata*
 Siberian. *Iris sibirica*
 Small Bulbous. *Iris Xiphium*
 Smooth. *Iris lævigata*
 Snake's-head. *Iris tuberosa (Hermodactylus tuberosus)*
 Spanish-nut. *Iris (Moræa) Sisyrinchium*
 Spurious. *Iris spuria*
 Sweet-scented. *Iris spatulata (I. desertorum)*
 Swert's. *Iris Swertii*
 Sword-leaved. *Iris ensata*
 Tangier. *Iris Tingitana*
 Thunderbolt. *Iris Xiphium*
 Telford's. *Iris Telfordi*
 Tiger. The genus *Tigridia*
 Tough-leaved. *Iris tenax*
 Twice-flowering. *Iris scorpioides (Xiphium planifolium)*
 Van Houtte's. *Iris Van Houttei*
 Variegated. *Iris variegata*
 Vernal. *Iris verna*
 Wall. *Iris tectorum*
 Winter-blooming. *Iris scorpioides (Xiphium planifolium)*
 Yellow-banded. *Iris ochroleuca*
 Yellowish. *Iris flavescens*
Irish Heath. See Heath
 Daisy. *Leontodon Taraxacum*
 Moss. *Chondrus crispus*
Iron-bark-tree. Various species of *Eucalyptus*, especially *E. resinifera* and *E. Sideroxylon*
 Red. *Eucalyptus Sideroxylon*
 She. *Eucalyptus paniculata*
 Silver-leaved. *Eucalyptus pulverulenta*
 Queensland. *Eucalyptus Boucherii*
Iron Grass. See Grass
Iron-hard. An old name for *Centaurea nigra*
Iron-heads. See Iron-weed
Iron-shrub. *Sauragesia erecta*
Iron-tree, Norfolk Island. *Notelæa longifolia*
 True. *Metrosideros vera*
 W. Indian. *Siderodendron triflorum*
Iron-weed, or Iron-heads. *Centaurea nigra*
 American. The genus *Vernonia*
Iron-wood, American. *Bromelia lycioides, Carpinus americana,* and *Ostrya virginica*
 Bastard. *Xanthoxylon Pterota*
 Bastard, of the W. Indies. *Fagara lentiscifolia* and *Trichilia hirta*
 Bourbon. *Stadtmannia (Cupania) Sideroxylon*
 Burmah. *Xylia dolabriformis*
 Dutch E. Indies. *Cassia florida, Dodonæa Waitziana, Eusideroxylon Zwageri, Intsia amboinensis, Memecylon ferreum, Namia vera, Stadtmannia Sideroxylon,* and *Sloëtia Sideroxylon*

and Foreign Plants, Trees, and Shrubs. 69

Iron-wood, E. Indian. *Xylia dolabriformis* (*Inga Xylocarpa*) and *Mesua ferrea*
E. Tropical African. *Copaifera Mopane*
Jamaica. *Erythroxylon areolatum*
Morocco. *Argania Sideroxylon*
N. S. Wales. *Notelæa ligustrina*
Norfolk Island. *Notelæa longifolia* and *Olea apetala*
Persian. *Parrotia persica*
W. Indian. *Sloanea jamaicensis* and *Fagara Pterota*
Small-leaved. *Mouriria myrtilloides*
S. African. *Olea capensis, O. undulata,* and *Sideroxylon capense*
S. Sea Islands. *Casuarina equisetifolia*
Tasmanian. *Notelæa ligustrina*
White. *Vepris lanceolata*
Iron-wort. The genus *Sideritis*
Canary Island. *Sideritis canariensis*
Hyssop-leaved. *Sideritis hyssopifolia*
Syrian. *Sideritis syriaca*
Yellow. *Galeopsis villosa*
Isabella-wood. *Persea carolinensis*
Isle-of-Wight-Vine. *Bryonia dioica* and *Tamus communis*
Itaka-wood. The timber of *Machærium Schomburgkii*
Italian Rye-Grass or Ray-Grass. See Grass
Itch-tree, Queensland. *Davidsonia pruriens*
Itch-weed. *Veratrum viride*
Itch-wood-tree. *Onocarpus vitiensis*
Ithuriel's Spear. *Tritelcia laxa*
Ivory. A corruption of "Ivy"
Ivory-plant, Vegetable. *Phytelephas macrocarpa*
Ivory-tree, E. Indian. Various species of *Wrightia*
Ivray. An old name for *Lolium temulentum*
Ivy. *Hedera Helix*
American. *Ampelopsis hederacea* (*A. quinquefolia*)
American Poison. *Rhus Toxicodendron*
Cape. *Senecio macroglossus*
Chinese. *Rhynchospermum jasminoides* (*Parechites Thunbergii*)
Clustered. *Hedera conglomerata*
Coliseum. See Ivy, Kenilworth
Crimean. *Hedera Helix taurica*
German. *Senecio mikanoides*
German, Yellow. *Senecio scandens*
Giant. *Hedera Helix Rægneriana* (*H. H. colchica*)
Gold-blotched. *Hedera Helix aureo-maculata*
Gold-edged Tree. *Hedera Helix arborea aureo-marginata*
Grass, of Australia. *Pericampylos incanus* (*Cocculus Moorei*)
Ground. See Ground-Ivy
Irish. *Hedera Helix canariensis* (*H. H. hibernica*)
Indian. *Scindapsus pertusus* (*Monstera deliciosa*) and other species
Kenilworth, or Coliseum. *Linaria Cymbalaria*
Marbled-leaved. *Hedera Helix latifolia maculata*

Ivy, Mexican. *Cobæa scandens*
Native, of Australia. *Muhlenbeckia adpressa*
New Zealand. *Panax Colensoi*
Palmate-leaved. *Hedera Helix palmata*
Parlour. *Mikania scandens*
Poet's. *Hedera Helix poetica*
Prickly. *Smilax aspera*
Queensland. *Hedera (Irvingia) australiana*
Rægner's. *Hedera Helix Rægneriana*
Silver-edged Tree. *Hedera Helix argenteo-marginata*
Three-coloured. *Hedera Helix tricolor*
Variegated. *Hedera Helix variegata*
Variegated Japan. *Hedera Helix rhombea variegata*
Water. *Ranunculus hederaceus*
W. Indian. *Marcgraavia umbellata*
Yellow-berried Roman. *Hedera Helix chrysocarpa*
Yellow German. *Senecio scandens*
Ivy-Bindweed. *Polygonum Convolvulus*
Ivy-Grape-vine. See Vine
Ivy-leaved Chick-weed. *Veronica hederæfolia*
Ivy-tree, Otago. *Panax Colensoi*
Ivy-wort. *Linaria Cymbalaria*
Ixia-lily. The genus *Ixiolirion*
Ledebour's. *Ixiolirion Ledebouri*
Mountain. *Ixiolirion montanum*
Pallas's. *Ixiolirion Pallasi*
Tartarian. *Ixiolirion tataricum*

Jaborandi-plant. *Pilocarpus pinnatus*
Jaca, or Jack-tree. *Artocarpus integrifolia*
Jacinth. Another name for Hyacinth
Jack-by-the-hedge. *Alliaria officinalis*
Jack, Creeping. *Sedum acre*
Jack-go-to-bed-at-noon. *Ornithogalum umbellatum*
Jack-in-a-box. *Hernandia sonora*
Jack-in-prison. *Nigella damascena*
Jack-in-the-green. A variety of *Primula vulgaris*
Jack-in-the-pulpit. *Arisæma triphyllum*
Jack-of-the-buttery. *Sedum acre*
Jack-straws. *Plantago lanceolata*
Jack-tree. *Artocarpus integrifolia*
Jackal's-Kost. *Hydnora africana*
Jacobæa, Purple. *Senecio elegans*
Jacob's-ladder. *Polemonium cæruleum*
Dense-clustered-flowered. *Polemonium confertum*
Dwarf. *Polemonium humile*
Richardson's. *Polemonium cæruleum var. Richardsoni*
Jacob's-rod. *Asphodelus luteus*
Jacob's-staff. *Verbascum Thapsus*
Jacob's-sword. *Iris Pseud-acorus*
Jacoby. *Senecio elegans*
Jalap-plant, Garden. *Mirabilis Jalapa*
E. Indian. *Ipomæa Turpethum*
False. *Mirabilis Jalapa*
Male, or Jalap-tops. *Ipomæa Orizabensis* (*I. batatoides*)
Mechoacan. *Batatas (Ipomæa) Jalapa*
True. *Exogonium Purga*
Wild. *Convolvulus panduratus*
Jalap-tops. See Jalap, Male

"Jamabuki"-shrub. *Rhodotypos kerrioides*
Jamaica Pepper. Another name for Allspice (*Eugenia Pimenta*)
Jambolan-tree. *Calyptranthes Jambolana*
James's (St.)-Flower. *Lotus Jacobæus*
James's (St.)-wort, or James's-weed. *Senecio Jacobæa*
Jamestown-weed. *Datura Stramonium*
Janca-tree. *Amyris toxifera*
Japan-Allspice. *Chimonanthus fragrans*
Large-flowered. *Chimonanthus grandiflora*
Japan-Lacquer Tree. *Rhus vernicifera*
Japan-Varnish Tree. *Ailantus glandulosa*
Japanese Grass, Variegated. See Grass
Japura Tree, of Brazil. *Erisma Japura*
Jarool-wood. The timber of *Lagerstræmia reginæ*
Jarrah-wood. The timber of *Eucalyptus marginata* and *E. rostrata*
Jasmine, or Jessamine. The genus *Jasminum*
Arabian. *Jasminum Sambac*
Bastard W. Indian. The genus *Cestrum*
Bay-leaved. *Jasminum laurifolium*
Cape. *Gardenia florida*
Carolina. *Gelsemium nitidum* (*Bignonia sempervirens*)
Catalonian, or Spanish. *Jasminum grandiflorum*
Chili. *Mandevilla suaveolens*
Chinese. *Rhynchospermum jasminoides*
Churchill Island. *Tecoma australis*
Common White-flowered. *Jasminum officinale*
Double-flowered. *Jasminum officinale fl.-pl.*
French. *Calotropis procera*
Golden-leaved. *Jasminum aureum*
Ground. *Passerina Stelleri*
Hardy White. *Jasminum grandiflorum*
Italian Yellow. *Jasminum humile*
Jonquil-scented. *Jasminum odoratissimum*
Large White-flowered. *Jasminum affine* (*J. ochroleucum*)
Madagascar. *Stephanotis floribunda*
Moreton Bay Trumpet. *Tecoma jasminoides*
Narrow-leaved. *Jasminum angustifolium*
Nepaul. *Jasminum revolutum*
Nepaul, Downy. *Jasminum pubigerum*
Night. *Nyctanthes Arbor-tristis*
Rock. *Androsace Chamæjasme*
Rosy-flowered. *Mascarenhasia Curnowiana*
Shang-hae. *Rhynchospermum jasminoides*
Silver-leaved. *Jasminum officinale foliis argenteis*
Spanish. See Jasmine, Catalonian
White Azorean. *Jasminum azoricum*
White-flowered Indian. *Jasminum Sambac*
Wild, of the W. Indies. *Faramea odoratissima* and the genus *Ixora*
Winter-flowering. *Jasminum nudiflorum*
Yellow-flowered. *Jasminum fruticans*
Yellow Azorean. *Jasminum odoratissimum*
Yellow Carolina. See Jasmine, Carolina
Jasmine-Box. The genus *Phillyrea*
Jasmine-Mango. *Plumieria rubra*

Jasmine-scented-wood Tree. *Erithalis fruticosa*
Jasmine-wood. *Ochna Mauritiana*
Jaundice-berry, or Jaundice-tree. *Berberis vulgaris*
Javance-plant. See Ajowan
Jean Cherry. See Cherry, Gean
"Jellico," of St. Helena. *Sium helenianum*
Jelly-plant, Australian. *Eucheuma speciosum*
Jersey Live-long. *Gnaphalium luteo-album*
Jerusalem Artichoke. *Helianthus tuberosus*
Jerusalem Cowslip. See Cowslip
Jerusalem Cross. *Lychnis chalcedonica*
Jerusalem Oak. *Teucrium Botrys*
Jerusalem Sage. The genus *Phlomis*
Jerusalem Star. *Cerastium tomentosum* and *Tragopogon porrifolium*
Jew-bush. *Pedilanthus tithymaloides*
Jew's-Apple. See Apple
Jew's-Ears. *Exidia Auricula-Judæ.* Also applied to some species of *Peziza*
Jew's Mallow. See Mallow
Jew's-Myrtle. *Ruscus aculeatus*
Jew's-Thorn. Another name for Christ's-Thorn
Jewel-weed, Spotted. *Impatiens fulva*
Jim-Crow's-nose. *Phyllocoryne jamaicensis*
Joan, or John, Silver-pin. An old name for *Papaver Rhœas fl.-pl.*
Job's-Tears, or Job's-Drops. *Coix Lachryma*
Joe-Pye-weed. *Eupatorium purpureum*
Jockey Grass. See Grass
John-Crow's-nose. See Jim-Crow's-nose
John's (St.)-bread. *Ceratonia Siliqua*
John's (St.)-wort. The genus *Hypericum* more especially *H. perforatum*
Aspalathus-like. *Hypericum fasciculatum*
Attenuated. *Hypericum attenuatum*
Axillary-flowered. *Hypericum axillare*
Burser's. *Hypericum Burseri*
Canadian. *Hypericum canadense*
Canary Island. *Hypericum canariense*
Chinese. *Hypericum chinense*
Close-panicled. *Hypericum elodioides*
Cochin-China. *Hypericum cochin-chinense*
Common. *Hypericum hircinum*
Creeping. *Hypericum repens*
Cross-leaved. *Hypericum decussatum*
Curled-leaved. *Hypericum crispum*
Dense-foliaged. *Hypericum frondosum*
Dotted-flowered. *Hypericum punctatum*
Downy. *Hypericum lanuginosum*
Egyptian. *Hypericum ægyptiacum*
Elegant. *Hypericum elegans*
Empetrum-leaved. *Hypericum empetrifolium*
Fringed. *Hypericum fimbriatum*
Galium-leaved. *Hypericum galioides*
Glandular. *Hypericum glandulosum*
Glaucous-leaved. *Hypericum glaucum*
Glossy-flowered. *Hypericum Hookerianum*
Goat-scented. *Hypericum hircinum*
Great American. *Hypericum pyramidatum*

John's (St.)-wort, Hair-fringed-flowered. *Hypericum ciliatum*
Hairy. *Hypericum hirsutum*
Hatchet-like. *Hypericum dolabriforme*
Heart-leaved. *Hypericum cordifolium*
Heath-leaved. *Hypericum Coris*
Honey-suckle-leaved. *Hypericum Caprifolium*
Imperforate. *Hypericum dubium*
Involute-flowered. *Hypericum involutum*
Japanese. *Hypericum japonicum*
Kalm's. *Hypericum Kalmianum*
Large-capsuled. *Hypericum ascyroides*
Large-flowered. *Hypericum calycinum*
Majorca. *Hypericum balearicum*
Many-flowered. *Hypericum floribundum*
Marsh. *Hypericum Elodes*
Mexican. *Hypericum mexicanum*
Money-wort. *Hypericum Nummularium*
Mt. Lebanon. *Hypericum cuneatum*
Mt. Olympus. *Hypericum olympicum*
Mountain. *Hypericum montanum*
Naked-flowered. *Hypericum nudiflorum*
Nepaul. *Hypericum nepalense*
Oblong-leaved. *Hypericum oblongifolium*
"Park-leaves." *Hypericum Androsæmum*
Perfoliate. *Hypericum perfoliatum*
Pink-flowered. *Hypericum virginicum*
Pretty. *Hypericum amœnum*
Proliferous. *Hypericum prolificum*
Pyramidal. *Hypericum pyramidatum*
Red-leaved. *Hypericum Ascyron*
Reflexed-leaved. *Hypericum reflexum*
Small. *Hypericum pusillum*
Small-flowered. *Hypericum quinquenervium*
Shining, Leafy. *Hypericum foliosum*
Shrubby. *Hypericum prolificum*
Siberian. *Hypericum Ascyron*
Simple-stalked. *Hypericum simplex*
Slender. *Hypericum pulchrum*
Smooth. *Hypericum lævigatum*
Spotted-flowered. *Hypericum maculatum*
Spreading. *Hypericum patulum*
Square-stalked. *Hypericum quadrangulum*
Stiff-haired. *Hypericum pilosum*
Tall. *Hypericum elatum*
Three-flowered. *Hypericum triflorum*
Three-nerved. *Hypericum triplinerve*
Toothed-flowered. *Hypericum angulosum*
Trailing. *Hypericum humifusum*
Twiggy. *Hypericum virgatum*
Urala. *Hypericum Uralum*
Various-leaved. *Hypericum heterophyllum*
Verona. *Hypericum veronense*
Thyme-leaved. *Hypericum serpyllifolium*
Whorled-leaved. *Hypericum verticillatum*
Woolly. *Hypericum tomentosum*
Joint Grass. See Grass
Joint-weed. *Polygonum articulatum*
Jointed Charlock. *Raphanus Raphanistrum*
Jolly, Brown, of the W. Indies. *Solanum Melongena*
Jonquil, Common. *Narcissus Jonquilla*
Large. *Narcissus odorus*
Queen Anne's. *Narcissus pusillus plenus*
Small. *Narcissus pusillus*

Joseph's-coat. *Amarantus tricolor*
Joseph's-Flower. *Tragopogon pratensis*
Jove's-beard or Jupiter's-beard. *Anthyllis Barba-Jovis, Hydnum Barba-Jovis*, and *Sempervivum tectorum*
Jove's-Nuts (Acorns). The fruit of *Quercus Robur*
Jove's-Fruit. *Lindera melissæfolia*
Joy-weed. The genus *Alternanthera*
Juba's-brush. *Iresine celosioides*
Judas's-Ear. See Jew's-Ears
Judas-tree. *Cercis Siliquastrum*
American. *Cercis canadensis*
Californian. *Cercis occidentalis*
Jujube-tree. *Zizyphus Jujuba*
Incurved-spined. *Zizyphus incurva*
Zig-zag. *Zizyphus flexuosa*
July-flower. Another name for Gilliflower
Jamaica. *Prosopis Juliflora*
Jump-up-and-kiss-me. *Viola tricolor*
June-berry. *Amelanchier canadensis*
Jungle-bendy-tree. A species of *Tetrameles*
Jungle-nail. *Acacia tomentosa*
Juniper. The genus *Juniperus*
Abyssinian. *Juniperus procera*
Azores. *Juniperus Cedrus var. brevifolia*
Brown-berried. *Juniperus Oxycedrus*
Californian. *Juniperus californica* and *J. occidentalis*
Canary Island. *Juniperus Cedrus*
Carpet. *Juniperus prostrata*
Chinese. *Juniperus chinensis*
Common. *Juniperus communis*
Drooping Indian. *Juniperus recurva*
Dwarf. *Juniperus nana*
Golden-variegated Chinese. *Juniperus chinensis var. aurea*
Golden-variegated Japan. *Juniperus japonica var. aurea*
Grey-carpet. *Juniperus sabinoides*
Green-carpet. *Juniperus Sabina var. nana*
Hedge-hog. *Juniperus echiniformis (J. hemisphærica)*
Incense. *Juniperus religiosa (J. excelsa)*
Irish. *Juniperus communis var. hibernica (J. stricta)*
Japan. *Juniperus japonica*
Large Purple-fruited. *Juniperus macrocarpa*
Mexican. *Juniperus tetragona*
Plum-fruited. *Juniperus drupacea*
Spanish. *Juniperus thurifera (J. hispanica)*
Swedish. *Juniperus communis var. succica*
Sweet-fruited. *Juniperus pachyphlœa*
Tall Upright. *Juniperus excelsa var. stricta*
Tamarisk-leaved. *Juniperus tamariscifolia (J. sabinoides)*
Virginian. *Juniperus virginiana*
White-variegated Japan. *Juniperus japonica var. alba*
Juno's-Rose. *Lilium candidum*
Juno's-Tears. *Verbena officinalis*
Jupiter's-Beard. See Jove's-beard
Jupiter's-Distaff. *Salvia glutinosa*
Jupiter's-Eye. *Sempervivum tectorum*
Jupiter's-Flower. *Lychnis Flos-Jovis*

Juray. An old name for Darnel
Jur-nut. *Bunium flexuosum*
Jute-plant (True). *Corchorus capsularis*

Kaffir-bread, or Caffre-bread. The genus *Encephalartos*
Kaffir's-tree. *Erythrina Caffra*
Kaladana-plant. *Pharbitis (Convolvulus) Nil*
Kale, or Borecole. See Cabbage
Indian. *Arum divaricatum, Caladium nymphæifolium,* and *C. sagittæfolium*
Indian, Large-leaved. *Caladium grandiflorum*
Large-leaved Ornamental. *Crambe palmatifida*
Scotch. *Brassica oleracea var. sabellica*
Sea. *Crambe maritima*
W. Indian. *Xanthosoma atrovirens* and the genus *Colocasia*
Wild. *Sinapis arvensis*
Kamala, or Kamela, Tree. *Rottlera tinctoria* (*Mallotus philippinensis*)
Kambala-tree. *Sonneratia apetala*
Kandlegosses. An old name for *Galium Aparine*
Kangaroo's-foot Plant. *Anigozanthus Manglesii*
Kangaroo-Grass. *Anthistiria australis*
Kangaroo, or Kanguru, Vine. *Cissus antarctica*
Karaka, or Kopi, tree. *Corynocarpus lævigata*
Kariyat. See Creyat
Karri-tree. *Eucalyptus diversicolor*
Karse. An old name for Cress
Kat, or Kâth, tea-shrub. *Catha edulis*
Katharine's (St.) Flower. *Nigella damascena*
Kauri, or Kawrie, Pine. See Pine
Kava-plant. See Ava-plant
Kawaka-tree. *Libocedrus Doniana*
Kawrie-Pine. See Kauri
Kaw-Tabua-tree, of Fiji. *Podocarpus cupressina*
"Keatlegs." A Kentish name for Orchises
Kecks, Kex, or Keks. The dry hollow stalks of various large umbelliferous plants; also applied to the plants themselves
Keddle, or Kettle, Dock. *Senecio Jacobæa* and *Rumex obtusifolius*
Kedlock, Kerlock, or Ketlock. A name applied, like Charlock, in a general way, to *Sinapis arvensis* and some other cruciferous plants
Keeslip. *Galium verum*
Keeso-flowers. The flowers of *Butea frondosa*
Kei-apple. See Apple
"Keklam," of Bengal. The fruit of *Limonia acidissima* (?)
Keks. See Kecks
Kelp-ware, or Kelp-wrack. *Fucus vesiculosus* and *F. nodosus*
Kemps. *Plantago media*
Kendal Green. *Genista tinctoria*
Kentish Balsam. *Mercurialis perennis*
Kerlock. See Kedlock
Kernel-wort. *Scrophularia nodosa*

Ketlock. See Kedlock
Ketmia, African. *Hibiscus africanus*
Ketmia, Bladder. See Bladder-Ketmia
Kettle-Dock. See Keddle-Dock
Kex. See Kecks
Keys. The fruit of the Ash, Maple, and Sycamore
Khair-tree. *Acacia Catechu*
Khus-Khus, or Cus-cus, Plant. *Andropogon muricatus*
Kidney-bean. See Bean
Egyptian. *Dolichos Lab-lab*
Malacca. The seed of *Semecarpus Anacardium*
Kidney-bean Tree, American. *Wistaria* (*Glycine*) *frutescens*
Chinese. *Wistaria (Glycine) sinensis*
Kidney-Vetch, Common. *Anthyllis Vulneraria*
Mountain. *Anthyllis montana*
Pink-flowered. *Anthyllis Vulneraria rubra*
Rushy. *Anthyllis cinacea*
Kidney-wort. *Cotyledon Umbilicus* and *Saxifraga stellaris*
"Kindly Savin." See Savin
King-cups, or King-cobs. Another name for Buttercups
King Fern. *Osmunda regalis*
King-plant. *Anœctochilus setaceus*
King's-Clover. *Melilotus officinalis*
King's-Feather. *Saxifraga umbrosa*
King's-Flower. *Eucomis regia*
King's-spear. *Asphodelus luteus* and *A. ramosus*
Kinnikinnick. *Arctostaphylos Uva-ursi* (See also Dogwood, Silky)
Kino Gum-tree. *Pterocarpus Marsupium* and *P. indicus*
African, or Gambia. *Pterocarpus erinaceus*
Australian. *Eucalyptus amygdalina*
Bengal. *Butea frondosa* and *B. superba*
Botany Bay. *Eucalyptus resinifera*
Burmese. *Pterocarpus dalbergioides*
E. Indian, or Amboyna. *Pterocarpus Marsupium*
Jamaica. *Coccoloba uvifera*
Palas, or Pulas. *Butea frondosa*
Kiss-me, Kiss-me-ere-I-rise, or Kiss-me-at-the-garden-gate. *Viola tricolor*
Kite-flower. *Hyoscyamus (Atropa) physaloides*
Kite-keys. An old name for the "Keys" or seed-vessels of the Ash
Knap-weed, or Knob-weed. *Centaurea nigra*
Russian. *Centaurea ruthenica*
Silvery-leaved. *Centaurea gymnocarpa*
Knawel. *Scleranthus annuus*
Knee-Holly. See Holly, Knee
Knee-Holme. See Holme, Knee
Knight's-spur. An old name for Larkspur
Knit-back. *Symphytum officinale*
Knob-Sedge. *Sparganium ramosum*
Knob-tang. *Fucus nodosus*
Knob-weed. See Knap-weed; also applied to *Centaurea Cyanus* and *C. Scabiosa*
Knot-berry. *Rubus Chamæmorus*
Knot-grass. The genus *Illecebrum* (See also Grass)
Slender. *Polygonum minus*

Knot-weed. *Centaurea nigra, C. Cyanus,* and *C. Scabiosa*
Alpine. *Polygonum alpinum*
Amphibious. *Polygonum amphibium*
Copse. *Polygonum dumetorum*
E. Indian. *Polygonum Brunonis*
E. Indian, Annual. *Polygonum orientale*
Giant. *Polygonum cuspidatum (P. Sieboldi)* and *P. sachalinense*
Pale-flowered. *Polygonum lapathifolium*
Rock. *Polygonum vaccinifolium*
Sachalin. *Polygonum sachalinense*
Sea-side. *Polygonum maritimum*
Virginian. *Polygonum virginianum*
Viviparous. *Polygonum viviparum*
Water-pepper. *Polygonum Hydropiper*
Whortle-berry-leaved. *Polygonum vaccinifolium*
Koa-tree, of the Sandwich Islands. *Acacia Koa (A. heterophylla?)*
Kohe, or Wahahé-tree. *Hartighsea spectabilis*
Kohl-Rabi. *Brassica oleracea var. gongylodes*
Kokoon-tree. *Kokoona zeylanica*
Kokra-wood. *Alnus integrifolia*
Kokum-butter-tree. *Garcinia indica*
"Kolkas," of the Arabians and Egyptians. *Colocasia antiquorum*
Kolla-nut. See Nut
Kopi-tree. See Karaka
Koso, or Cusso, Flowers. The flowers of *Hagenia abyssinica*
Kotukutuki-tree, of New Zealand. *Fuchsia excorticata*
Kumahou-tree. *Pomaderris elliptica*
Kum-quat. A variety of *Citrus japonica*
Kureel-tree. *Capparis aphylla*
Kurrajong. A native Australian name for various fibrous plants.
Black, of Illawarra. *Rulingia ramosa*
Brown. *Commersonia platyphylla*
Green. *Hibiscus heterophyllus*
Tasmanian. *Plagianthus sidoides*

Labaria - plant. *Dracontium polyphyllum*
Lab-lab. *Dolichos (Lablab) vulgaris* and *D. (L.) cultratus*
White China. *Dolichos (Lablab) perennans*
Laburnum-tree, Adam's. *Cytisus Adami*
Austrian. *Cytisus austriacus*
Black-rooted. *Cytisus nigricans*
Cluster-flowered. *Cytisus capitatus*
Common. *Cytisus Laburnum*
Dwarf. *Cytisus nanus*
Evergreen. *Cytisus racemosus*
Flesh-coloured. *Cytisus incarnatus*
Fringed-podded. *Cytisus ciliatus*
Hairy. *Cytisus hirsutus*
Indian. *Cassia Fistula*
Large-calyxed. *Cytisus calycinus*
Large-flowered. *Cytisus grandiflorus*
Late-flowering. *Cytisus serotinus*
Lilliputian. *Cytisus Ardoinii*
Long-branched. *Cytisus elongatus*
Many-flowered. *Cytisus multiflorus*
Nepaul. *Piptanthus nepalensis*

Laburnum-tree, New Zealand. *Edwardsia (Sophora) grandiflora* and *E. microphylla (Sophora tetraptera var. microphylla)*
Oriental. *Cytisus orientalis*
Pigmy. *Cytisus Ardoinii*
Portugal. *Cytisus albus*
Prickly. *Cytisus spinosus*
Proliferous. *Cytisus proliferus*
Purple-flowered. *Cytisus purpureus*
Queensland. *Cassia Brewsteri*
Scotch. *Cytisus alpinus*
Sickle-podded. *Cytisus falcatus*
Silver-leaved. *Cytisus argenteus*
Soft-leaved. *Cytisus mollis*
Spreading. *Cytisus patens*
Stalkless-flowered. *Cytisus sessiliflorus*
Sweet-scented. *Cytisus fragrans*
Teneriffe. *Cytisus nubigenus*
Three-flowered. *Cytisus triflorus*
Trailing. *Cytisus supinus*
Two-flowered. *Cytisus biflorus*
Weeping. *Cytisus Laburnum pendulus*
White-flowered. *Cytisus leucanthus*
Whitish-flowered. *Cytisus albidus*
Wild. *Melilotus officinalis*
Woolly. *Cytisus lanigerus*
Lace-bark-tree, Jamaica. *Lagetta lintearia*
New Zealand. *Plagianthus betulinus*
Lace-leaf, or Lattice-leaf, Plant, White-flowered. *Ouvirandra fenestralis*
Pink-flowered. *Ouvirandra Berneriana*
Lace-pod. *Thysanocarpus curvipes*
Ladder-Fern. See Fern
Ladder-to-Heaven. *Polygonatum multiflorum* and *Polemonium cœruleum*
Ladle-wood-tree. *Cassine Colpoon*
Lad's-love. *Artemisia Abrotanum*
Lady-Birch. *Betula alba*
Lady-Fern. See Fern
Lady-Grass. See Grass
Lady - in - the - Bower. *Nigella damascena*
Lady - nut. The seed of *Entada Purseetha*
Lady-of-the-Meadow. *Spirœa Ulmaria*
Lady-Poplar. *Populus fastigiata*
Lady's-Bedstraw. *Galium verum*
Lady's-Bower. *Clematis Vitalba*
Lady's-Calamus. *Iris cristata*
Lady's-Cushion. *Arabis albida* and *Armeria maritima*
Lady's-Ear-drops. The flowers of the common *Fuchsia*
Lady's-Fingers. *Anthyllis Vulneraria.* Also locally applied to various other flowers
Lady's-Garters. See Lady's-Laces
Lady's-Gloves. *Inula Conyza*
Lady's-Hair. *Briza media*
Lady's-Laces, or Lady's-Garters. *Phalaris arundinacea variegata*
Lady's-Looking-glass. *Campanula Speculum*
Lady's-Mantle. *Alchemilla vulgaris*
Silky. *Alchemilla sericea*
Lady's-Pincushion. *Armeria maritima*
Lady's-Seal, or Lady's-Signet. *Polygonatum multiflorum* and *Tamus communis*

Lady's-Signet. See Lady's-Seal
Lady's-Slipper. The genus *Cypripedium*, especially *C. Calceolus*
Bearded. *Cypripedium barbatum*
Japanese. *Cypripedium japonicum*
Large-flowered. *Cypripedium macranthum*
Large Yellow. *Cypripedium pubescens*
Long-tailed. *Cypripedium caudatum*
Mexican. *Cypripedium Irapeanum*
Ram's-head. *Cypripedium arietinum*
Showy. *Cypripedium spectabile*
Siberian. *Cypripedium macranthum*
Small White. *Cypripedium candidum*
Small Yellow. *Cypripedium parviflorum*
Snow White. *Cypripedium niveum*
S. American. The genus *Selenipedium*
Spotted. *Cypripedium guttatum*
Stemless. *Cypripedium acaule*
Twin-flowered. *Cypripedium insigne*
White. *Cypripedium candidum*
Lady's-Smock. *Cardamine pratensis*
Lady's-Thimble. *Campanula rotundifolia*
Lady's (Our) Thistle. *Carduus Marianus* (*Silybum Marianum*)
Lady's-Thumb. *Polygonum Persicaria*
Lady's-Tresses, or Lady's-Traces. The genus *Spiranthes*
Autumn-flowering. *Spiranthes autumnalis*
Drooping. *Spiranthes cernua*
Irish. *Spiranthes gemmipara*
Summer. *Spiranthes æstivalis*
Lake-weed. *Polygonum Hydropiper*
Lamb-kill. *Andromeda Mariana* and *Kalmia angustifolia*
Lamb-in-a-pulpit. *Arum maculatum*
Lamb's-chop. *Mesembryanthemum agninum*
Lamb's-Cress. *Cardamine hirsuta*
Lamb's-ear. *Stachys germanica*
Lamb's-Grass. Various Spring Grasses
Lamb's-Lettuce. *Valerianella olitoria*
Lamb's-tails. The Catkins of *Salix Caprea* and *Corylus Avellana*
Lamb's-toe. *Lotus corniculatus* and *Anthyllis Vulneraria*
Lamb's-quarters. *Chenopodium album* and *Trillium erectum*
Lamb's-tongue. *Plantago media*; locally applied to a few other plants
Lamp-flower. The genus *Lychnis*
Lamp-wick. *Phlomis Lychnites*
Lance-pod, Queensland. *Lonchocarpus* (*Milletia*) *Blackii*
Lance-wood, Australian. *Backhousia australis*
Caffre's. *Guatteria Caffra*
Cuba. *Duguetia quitarensis*
Guiana. *Duguetia quitarensis, Guatteria virgata, Rollinia multiflora*, and *R. longifolia*
Jamaica. *Guatteria virgata* (*Uvaria lanceolata*)
New Zealand. *Panax crassifolium* (*Aralia crassifolia*)
Land-Grass. See Grass
"Landra." *Raphanus Landra*
Land-Whin. *Ononis arvensis*

Langdebeefe. *Helminthia echioides*
"Langsat." *Lansium domesticum*
Lang-wort, or Lyng-wort. An old name for *Veratrum album*
Lanseh-tree. *Lansium domesticum*
Lantern-flower. The genus *Abutilon*
Larch. The genus *Larix*
Altaian. *Larix Ledebourii*
American Black. *Larix americana*
American Red. *Larix microcarpa*
Chinese, or False. The genus *Pseudo-Larix*
Common. *Larix europæa*
Drooping. *Larix pendula*
False. See Larch, Chinese
Golden. *Abies* (*Larix*) *Kœmpferi*
Himalayan. *Larix Griffithii*
Oregon. *Larix occidentalis*
Siberian. *Larix sibirica*
Sikkim. *Larix Griffithii*
Lark's-claw, or Lark's-heel. Another name for Larkspur
Larkspur. A general name for the garden species of *Delphinium*
Alpine. *Delphinium alpinum*
American. *Delphinium exaltatum*
Azure. *Delphinium azureum*
Beatson's. *Delphinium Beatsoni*
Belladonna. *Delphinium Belladonna*
Blue-and-white-flowered. *Delphinium bicolor*
Branching. *Delphinium Consolida*
Cashmere. *Delphinium cashmerianum*
Common Bee. *Delphinium elatum*
Double Siberian. *Delphinium grandiflorum fl.-pl.*
Dwarf Red. *Delphinium nudicaule*
Elegant. *Delphinium elegans*
Hyacinth-flowered. *Delphinium hyacinthiflorum*
Intermediate. *Delphinium intermedium*
Keteleer's. *Delphinium Keteleeri*
Large-flowered. *Delphinium grandiflorum*
Musk-scented. *Delphinium moschatum*
Red-flowered. *Delphinium grandiflorum var. rubrum*
Rocket. *Delphinium Ajacis* and vars.
Scarlet-flowered. *Delphinium puniceum*
Showy. *Delphinium formosum*
Siberian. *Delphinium grandiflorum*
Tall. *Delphinium exaltatum*
Three-horned. *Delphinium tricorne*
White-flowered. *Delphinium albiflorum*
Wild. *Delphinium Ajacis* (*D. Consolida*)
Yellow. Various species of *Tropæolum*
Laser-wort. The genus *Laserpitium*; also applied to *Thapsus Laserpitii*
Lattice-leaf-plant. See Lace-Leaf
Lattice-Moss. See Moss
Laurel, or Cherry-Laurel. *Cerasus Lauro-cerasus*
Alexandrian. *Calophyllum Inophyllum* and *Ruscus racemosus*
American. The genus *Kalmia*
American Great. *Rhododendron maximum*
American Mountain. *Kalmia latifolia*
American Swamp. *Kalmia glauca*
Azores. *Laurus azorica*

and Foreign Plants, Trees, and Shrubs. 75

Laurel, Blotched-leaved. *Aucuba japonica*
Bunch-flowered. *Laurus thyrsiflora*
Californian. *Oreodaphne (Umbellularia) californica*
Canary Island. *Laurus canariensis*
Cape. *Laurus bullata*
Carolina. *Laurus (Persea) carolinensis*
Catesby's. *Laurus Catesbiana*
Catkin-bearing. The genus *Nageia*
Chili. *Laurelia aromatica*
Chinese. *Stilago Bunias*
Common, or Cherry. *Cerasus Lauro-cerasus*
Copse. *Daphne Laureola*
Crowded-flowered. *Laurus aggregata*
Diamond-leaved, Queensland. *Pittosporum rhombifolium*
Ground. *Epigæa repens*
Himalayan. *Aucuba himalaica*
Jamaica. *Laurus Chloroxylon*
Japan. *Aucuba japonica* and *Nageia japonica*
Jointed. *Laurus geniculata*
Jove's-fruit. *Laurus Diospyros*
Leather-leaved. *Laurus coriacea*
Lofty. *Laurus exaltata*
Madeira. *Laurus indica* and *L. fœtens*
Malabar. *Melastoma malabaricum*
Many-flowered. *Laurus floribunda*
Moreton Bay. *Cryptocarya (Laurus) australis*
Mountain. *Oreodaphne bullata*
N. S. Wales. *Cryptocarya glaucescens*
New Zealand. *Corynocarpus lævigata, Laurelia Novæ-Zelandiæ,* and *L. Kohekohe*
New Zealand Hedge. *Pittosporum eugenioides* and other species
Panama. *Cordia Geraschanthus*
Portugal. *Cerasus lusitanica*
Portugal, Miniature. *Cerasus lusitanica var. myrtifolia*
Poet's, or Roman. *Laurus nobilis*
Red. *Persea carolinensis*
Royal. *Laurus regalis*
Sassafras. *Oreodaphne californica*
Sea-side. *Xylophylla latifolia*
Sheep's-poison. *Kalmia angustifolia*
Shining-leaved. *Laurus splendens*
Snow-white. *Laurus nivea*
South Sea. The genus *Codiæum (Croton)*
Spotted. *Aucuba japonica*
Spreading. *Laurus patens*
Spurge. *Daphne Laureola*
Strong-scented. *Laurus fœtens*
Summer. *Laurus æstivalis*
Tasmanian. *Anopterus glandulosa*
Thick-leaved. *Laurus crassifolia*
Til. *Laurus fœtens*
Variegated. *Aucuba japonica*
Variegated Indian. *Croton variegatus*
Versailles. *Cerasus Lauro-cerasus var. latifolia*
Victorian. *Pittosporum undulatum*
Victor's. *Laurus nobilis*
Weeping. *Laurus pendula*
White. *Magnolia glauca*
Whitish-leaved. *Laurus albida*
Willow-leaved. *Laurus salicifolia*
Yellow-flowered Brisbane. *Pittosporum revolutum*

Laurustinus, Common. *Viburnum Tinus*
Shining-leaved. *Viburnum Tinus var. lucidum*
Upright-branched. *Viburnum Tinus var. strictum*
Lavender, Common. *Lavandula Spica*
French. *Lavandula Stœchas*
Sea. See Sea-Lavender
Sweet-scented. *Lavandula dentata*
True. *Lavandula vera*
Lavender-Cotton. The genus *Santolina*
Alpine. *Santolina alpina*
Chamomile-leaved. *Santolina anthemoides*
Clammy. *Santolina viscosa*
Comb-leaved. *Santolina pectinata*
Common. *Santolina Chamæcyparissus*
Green. *Santolina viridis*
Heath-like. *Santolina ericoides*
Hoary. *Santolina canescens*
Pinnate-leaved. *Santolina pinnata*
Rosemary-leaved. *Santolina rosmarinifolia*
Samphire-leaved. *Santolina crithmoides*
Scallop-leaved. *Santolina pectinata*
Spreading. *Santolina squarrosa*
Woolly. *Santolina incana*
Lavender-Grass. See Grass
Laver, Green. *Ulva latissima*
Red. *Porphyra laciniata*
Turkey-feather. *Padina pavonia*
Lawyer, Bush. *Rubus australis*
Lawyers. Old thorny stems of Briars and Brambles
Penang. *Licuala acutifida*
Laylock. A corruption of "Lilac"
Lead-plant. *Amorpha canescens*
Lead-tree, W. Indian. *Leucæna (Acacia) glauca*
Lead-wort, European. *Plumbago europæa*
Cape. *Plumbago capensis*
Hardy Blue. *Plumbago Larpentæ*
Rosy-flowered. *Plumbago rosea*
Scarlet-flowered. *Plumbago coccinea*
Leaf-bellows Plant. *Cochlospermum Gossypium*
Leaf-cup. The genus *Polymnia*
Leaf-of-St. Patrick. *Saxifraga umbrosa*
Leather-coat-leaf-tree. *Coccoloba pubescens*
Leather-flower. *Clematis Viorna*
Leather-jacket. *Eucalyptus punctata* and *E. resinifera*
Leather-leaf. *Cassandra (Andromeda) calyculata*
Leather-plant, New Zealand. Various species of *Celmisia*
Leather-wood, American. *Dirca palustris*
Australian. The genus *Ceratopetalum*
Leba-tree. *Eugenia neurocalyx*
Lee-chee. See Li-tchi
Leed. *Glyceria aquatica*
Leek, American Mountain. *Allium triflorum*
Blue. *Allium Ampeloprasum*
Canker. *Pyrola rotundifolia*
Common Garden. *Allium Porrum*
Crow. See Crow-leek
Dog's. An old name for *Scilla nutans*
Hollow. An old name for *Corydalis cava*
Lily. *Allium Moly*
Sand. *Allium Scorodoprasum*

Leek, Sour. *Rumex Acetosa*
Stone. *Allium fistulosum*
Wild. *Allium Ampeloprasum* and *A. ursinum*
Wild American. *Allium tricoccum*
Lemon-Bergamot Tree. *Citrus Bergamia*
Lemon-grass. *Andropogon Schœnanthus*
Lemon-tree. *Citrus Limonum.* Applied also to *Lippia (Aloysia) citriodora*
Desert of N. S. Wales and Queensland. *Atalantia glauca*
Java. *Citrus javanica*
Median (Citron). *Citrus Medica*
Pear. *Citrus Lumia*
Pearl. *Citrus Margarita*
Sweet. *Citrus Lumia*
Water. *Passiflora laurifolia* and *P. maliformis*
Wild. *Podophyllum peltatum*
Wild Water. *Passiflora fœtida*
Lens, Water. See Lentils, Water
Lent Lily, or Lent Rose. *Narcissus Pseudonarcissus*
Lentil, Common. *Ervum Lens*
Sea. *Sargassum bacciferum*
Single-flowered. *Ervum monanthos*
Lentils, or Lens, Water. *Lemna minor*
Leopard-flower, Chinese. *Pardanthus chinensis*
Leopard's-bane. The genus *Doronicum*
American. *Doronicum nudicaule*
Austrian. *Doronicum austriacum*
Caucasian. *Doronicum caucasicum*
Clusius's. *Doronicum (Aronicum, Arnica) Clusii*
Columna's. *Doronicum Columnæ*
Common. *Doronicum Pardalianches*
Cray-fish. *Doronicum Pardalianches*
Medicinal. *Arnica montana*
Plantain-leaved. *Doronicum plantagineum*
Letter-leaf, or Letter-plant. The genus *Grammatophyllum*
Letter-wood Tree. *Brosimum Aubletii*
Lettuce. The genus *Lactuca*
American "White." *Nabalus albus*
American Wild. *Lactuca canadensis (L. elongata)*
Blue, or False. The genus *Mulgedium*
Cabbage. *Lactuca capitata*
Californian. *Lactuca pulchella*
Common Garden. Varieties of *Lactuca sativa*
Cos. The erect-growing varieties of *Lactuca sativa*
Drum-head. Varieties of *Lactuca capitata*
Endive-leaved, or Lombard. *Lactuca intybacea*
False. See Lettuce, Blue
Frog's. *Potamogeton densus*
Hare's. *Sonchus oleraceus*
Indian. *Frasera verticillata*
Lamb's. See Lamb's-Lettuce
Least. *Lactuca saligna*
Lombard. See Lettuce, Endive-leaved
Oak-leaved. *Lactuca quercina*
Perennial. *Lactuca perennis*
Prickly. *Lactuca Scariola* and *L. virosa*

Lettuce, Sea. *Ulva Lactuca* and *Fucus vesiculosus*
Strong-scented. *Lactuca virosa*
Wall. *Lactuca (Prenanthes) muralis*
Water. *Pistia Stratiotes*
Wild. *Lactuca virosa*
Willow. *Lactuca saligna*
Lettuce-tree. *Pisonia morindifolia*
Lever-wood. *Ostrya virginica*
Licca-tree. *Sapindus spinosus (Xanthoxylum sapindoides)*
Lichen, Crab's-eye. *Lecanora pallescens*
Horse-hair. *Cornicularia jubata*
Letter. The genus *Opegrapha*
Lung. *Sticta pulmonacea*
Map. *Lecidea geographica*
Red Snow. *Protococcus nivalis*
Rock-Tripe. Various species of *Gyrophora* and *Umbilicaria*
Tree-hair. *Cornicularia jubata* and *Usnea jubata*
"Tripe de Roche." See Lichen, Rock-Tripe
Lichwale, or Lychwale. *Lithospermum officinale*
Lich-wort. *Parietaria officinalis*
Lid-flower. The genus *Calyptranthes*
Life-everlasting. *Gnaphalium americanum*
Life-plant. *Bryophyllum calycinum (B. proliferum)*
Life-root. *Senecio aureus*
Lign-Aloes. The wood of *Aloexylon Agallochum*
Light-wood, Australian. *Acacia Melanoxylon* and *Ceratopetalum apetalum*
Lignum Rhodium. The wood of *Amyris balsamifera* and *Rhodorrhiza scoparia*
Lignum-vitæ-tree. *Guaiacum officinale*
Bastard. *Badiera diversifolia*
N. S. Wales. *Acacia falcata* and *Eucalyptus polyanthemos*
New Zealand. *Metrosideros scandens*
Pegu. *Melanorrhæa usitatissima*
Queensland. *Vitex Lignum-vitæ*
Lilac. The genus *Syringa*
African. *Melia Azedarach*
Australian. *Hardenbergia monophylla, Prostanthera violacea,* and *P. lasianthos*
Californian. *Ceanothus integerrimus* and *C. thyrsiflorus*
Charles X. *Syringa vulgaris var. grandiflora*
Chinese. *Syringa chinensis*
Common Blue. *Syringa vulgaris var. cærulea*
Common Double Red. *Syringa vulgaris var. rubra-plena*
Common Double White. *Syringa vulgaris var. alba-plena*
Common Purple, or Scotch. *Syringa vulgaris var. violacea*
Common Red-flowered. *Syringa vulgaris var. rubra*
Common Violet-flowered. *Syringa vulgaris var. violacea*
Common White. *Syringa vulgaris var. alba*

and Foreign Plants, Trees, and Shrubs. 77

Lilac, Dwarf. *Syringa vulgaris var. nana*
Guinea. *Melia guineensis*
Himalayan. *Syringa Emodi*
Indian. *Melia Azedarach* and the genus *Lagerstræmia*
Lady Josika's. *Syringa Josikæa*
Large-red-flowered. *Syringa vulgaris var. rubra major*
Large-white-flowered. *Syringa vulgaris var. alba major*
Persian, Blue. *Syringa persica*
Persian, Cut-leaved. *Syringa persica var. laciniata*
Persian, Sage-leaved. *Syringa persica var. salviæfolia*
Persian, White. *Syringa persica var. alba*
Rouen (Hybrid). *Syringa rothomagensis*
Scotch. See Lilac, Common Purple
Lily. A general name for plants of the genus *Lilium*; applied also to various other plants
African, Blue. *Agapanthus umbellatus*
African Corn. The genus *Ixia*
African, Small. *Agapanthus minor*
African, White. *Agapanthus umbellatus albus*
Amazon. *Eucharis amazonica*
Amboyna. *Eurycles amboinensis*
Arum. *Calla (Richardia) æthiopica*
Atamasco. *Zephyranthes (Amaryllis) Atamasco*
Australian Giant. *Doryanthes excelsa*
Australian Purple. The genus *Patersonia*
Barbadoes. *Hippeastrum equestre*
Belladonna. *Amaryllis Belladonna*
Bengal. *Crinum longifolium*
Bermuda. *Lilium Harrisi* (*L. japonicum floribundum?*)
Black. *Lilium camtschatcense* (*Fritillaria camtschatcensis*)
Black Martagon. *Lilium dalmaticum*
Blackberry. *Pardanthus (Ixia) chinensis*
Blue African. *Agapanthus umbellatus*
Blue Grass. *Cæsia vittata*
Bourbon. *Lilium candidum*
Brisbane. *Eurycles australasica* (*E. Cunninghamii*)
Brodie's. The genus *Brodiæa*
Brownish-red. *Lilium Parthencion*
Brown's. *Lilium Brownii*
Buff-coloured. *Lilium testaceum*
Bugle. The genus *Watsonia*
Bulb-bearing. *Lilium bulbiferum*
Busch's. *Lilium Buschianum*
Caffre. *Schizostylis coccinea* and the genus *Cliria*
Californian. *Lilium californicum*
Cape. *Crinum capense*
Cape Coast. *Crinum spectabile*
Carniola. *Lilium carniolicum*
Carolina. *Lilium carolinianum*
Caucasian. *Lilium monadelphum*
Chequered. *Fritillaria Meleagris*
Chinese Blackberry. *Pardanthus chinensis*
Club. The genus *Tritoma*
Common Orange. *Lilium croceum*
Corfu. *Funkia subcordata*
Crimson-anthered. *Lilium Szovitzianum* (*L. colchicum*)

Lily, Cuban. *Scilla peruviana*
Daffodil. The genus *Amaryllis*
Egyptian. *Calla (Richardia) æthiopica*
Flame, or Fire. The genus *Pyrolirion*
Flax. See Flax-Lily
Fire. See Lily, Flame
Fringed. The genus *Thysanotus*
Giant. *Lilium giganteum*
Giant Heart-leaved. *Lilium cordifolium giganteum*
Giant Mexican. *Fourcroya gigantea*
Giant St. Bruno's. *Anthericum Liliastrum majus*
Golden. *Amaryllis (Lycoris) aurea*
Gold-striped. *Lilium auratum*
Ground. *Trillium latifolium*
Guernsey. *Nerine Sarniensis*
Guernsey, Scarlet. *Nerine Fothergilli*
Heart-leaved. *Lilium cordifolium*
Hedge. *Convolvulus sepium*
Herb. The genus *Alstræmeria*
Himalayan White. *Lilium polyphyllum*
Hulch. *Lilium Martagon*
Humboldt's. *Lilium Humboldtii*
Jacobea. *Sprekelia (Amaryllis) formosissima*
Japan. *Lilium lancifolium*
Knight's-star. The genus *Hippeastrum*
Kramer's. *Lilium Krameri*
Large African. *Crinum giganteum*
Lent. See Lent-Lily
Leopard. The genus *Lachenalia*
Madonna. *Lilium candidum*
Malabar Glory. *Gloriosa superba*
Mariposa. See Mariposa-Lily
Martagon. *Lilium Martagon*
Max Leichtlin's. *Lilium Leichtlinii*
May. *Convallaria majalis*
Medeola. *Lilium medeoloides*
Mediterranean. *Pancratium maritimum*
Mexican, or Queen's. *Amaryllis Reginæ*
Midnight. The genus *Calonyction*
Moreton Bay. *Eurycles Cunninghamii*
Mount Etna. *Sternbergia ætnensis*
Mozambique. *Gloriosa virescens*
Murray. *Crinum australe* (*C. pedunculatum*)
Nankeen. *Lilium testaceum*
Narrow-leaved Queensland. *Crinum angustifolium*
Natal. *Imantophyllum miniatum*
Neilgherry. *Lilium neilgherrense*
Nepaul. *Gloriosa nepalensis*
Nevada. *Lilium Washingtonianum*
Oat-bulbed. *Lilium avenaceum*
"Of-the-Field." *Sternbergia lutea* (?)
Of-the-Incas. *Alstræmeria peregrina*
Of-the-Mountain. *Polygonatum multiflorum*
Of-the-Nile. *Calla (Richardia) æthiopica*
Of-the-Nile, Spotted-leaved. *Richardia maculata*
Of-the-Nile, Yellow-flowered. *Richardia hastata*
Of-the-Valley. *Convallaria majalis*
Of-the-Valley, Japanese. *Convallaria japonica*
Of-the-Valley, Star-flowered. *Smilacina stellata*
Of-the-Valley-Tree. *Andromeda floribunda*

Lily, of the Valley, Two-leaved. *Maianthemum bifolium*
Orange. *Lilium bulbiferum var. aurantium*
Orange, Common. *Lilium croceum*
Orange, Late. *Lilium venustum*
Orange-Red, American. *Lilium philadelphicum*
Oregon. *Lilium columbianum*
Palm. The genus *Cordyline*
Panther. *Lilium pardalinum*
Paroo. *Dianella cærulea*
Parry's. *Lilium Parryi*
Persian. *Fritillaria persica*
Peruvian Swamp. *Zephyranthes candida*
Plantain. See Plantain-Lily
Prairie. *Mentzelia ornata*
Pyrenean. *Lilium pyrenaicum*
Queen. The genus *Phædranassa*
Queen's. See Lily, Mexican
Queensland Spear. *Doryanthes Palmeri*
Red Star. *Lilium concolor*
Robinson's. *Lilium Robinsoni*
Rock. *Selaginella convoluta*
Rock, N. S. Wales. *Dendrobium speciosum*
Rocky Mountain. *Leucocrinum montanum*
Rock-wood. *Ranunculus Lyalli*
Royal Brunswick. *Brunsvigia* (*Amaryllis*) *Josephinæ*
Rush. See Rush-Lily
Rush-leaved, Black. *Melanthium junceum*
Russian. *Lilium pulchellum*
St. Bernard's. *Anthericum* (*Czackia, Paradisia*) *Liliago*
St. Bruno's. *Anthericum Liliastrum*
St. James's Cross. *Amaryllis formosissima*
Sander's. *Eucharis Sanderi*
Scarborough. *Vallota purpurea*
Scarborough, Scarlet. *Vallota purpurea var. eximia*
Scarlet Martagon. *Lilium chalcedonicum*
Scarlet-striped. *Lilium auratum var. rubro-vittatum*
Senegal. *Gloriosa simplex*
Siberian Orange. *Lilium davuricum*
Siberian Scarlet. *Lilium pulchellum*
Small. *Lilium parvum*
Small African. *Agapanthus minor*
Southern Red. *Lilium Catesbæi*
Spear-leaved. *Lilium lancifolium*
Spire. *Hyacinthus* (*Galtonia*) *candicans*
Spotted. *Lilium speciosum* (*L. lancifolium*)
Swamp. *Lilium superbum*. Applied also to *Saururus cernuus*
Sword. The genus *Gladiolus*
Tall Sulphur-flowered. *Lilium colchicum* (*L. Szovitzianum*)
Thong. The genus *Imantophyllum*
Thunberg's. *Lilium venustum* (*L. Thunbergianum*)
Tiger. *Lilium tigrinum*
Tiger, Fortune's. *Lilium tigrinum Fortunei*
Toad. See Toad-Lily
Toad-cup. *Marica cærulea*
Tom Thumb. *Lilium tenuifolium*

Lily, Torch. See Torch-Lily
Triplet. The genus *Tritelcia*
Trumpet. *Lilium longiflorum*; also applied to *Richardia* (*Calla*) *æthiopica*
Trumpet, Transparent. *Lilium eximium*
Trumpet, Wallich's. *Lilium Wallichianum*
Turban. *Lilium Pomponium*
Turk's-cap. Varieties of *Lilium Martagon*
Turk's-cap, Great American. *Lilium superbum*
Umbel-flowered. *Lilium umbellatum*
Warty Red Japanese. *Lilium callosum*
Washington. *Lilium Washingtonianum*
Washington, Purple. *Lilium Washingtonianum var. purpureum* (*L. rubescens*)
Water. See Water-Lily
White. *Lilium candidum*
White, of Tasmania. *Diplarrhena Moræa*
Whitsun. *Narcissus poeticus*
Whorled-leaved American. *Lilium philadelphicum*
Wilson's. *Lilium Wilsoni*
Wood. *Pyrola minor*. (See also Wood-Lily)
Wreath. *Myrsiphyllum asparagoides*
Yellow American. *Lilium canadense*
Yellow Pond. *Nuphar advena*
Yellow Star. *Lilium Coridion*
Lily-Conval. *Convallaria majalis*
Lily-grass. See Grass
Lily-Leek. *Allium Moly*
Lily-oak. A corruption of Lilac
"Lily Pillies," of Victoria. *Eugenia Smithii*
Lily-pilly-tree, of Australia. The genus *Acmena*
Lily-pink, of Montpelier. *Aphyllanthes monspeliensis*
Lily-riall. An old name for Penny-royal
Lily-thorn. The genus *Catesbæa*
Of the Bahamas. *Catesbæa spinosa*
Lima-wood. The timber of *Cæsalpinia echinata*
Lime-fruit-tree. *Citrus acida*
Coromandel, or Indian Wild. *Limonia* (*Atalantia*) *monophylla*
Native, of Queensland. *Citrus australasica*
Sweet. *Citrus Limetta*
Lime-berries. The fruit of *Triphasia trifoliata*
Lime-stone Fern. See Fern
Lime-tree, Lime, or Linden. The genus *Tilia*
Broad-leaved. *Tilia europæa var. platyphylla*
Common. *Tilia europæa var. intermedia*
Cut-leaved. *Tilia europæa var. laciniata*
Downy-leaved American. *Tilia americana var. pubescens*
Downy Narrow-leaved American. *Tilia americana var. pubescens leptophylla*
Fringed-leaved. *Tilia platyphylla laciniata*
Golden-twigged. *Tilia europæa var. aurea*
Golden-twigged Broad-leaved. *Tilia europæa var. platyphylla aurea*
Hairy-styled. *Tilia europæa var. dasystyla*
Loose-flowered American. *Tilia americana var. laxiflora*

Lime-tree, Ogechee. *Nyssa capitata (N-candicans)*
Red-twigged. *Tilia europœa* var. *rubra*, or *corallina*
Silver-leaved. *Tilia argentea*
Small-leaved. *Tilia parvifolia*
White-leaved. *Tilia europœa* var. *alba*
White American. *Tilia americana* var. *heterophylla*
Lime-wort. An old name for *Lychnis Viscaria*
Linaloa, or Linaloe, wood. See Lign-Aloes
Lin, Line, or Linde. *Tilia europœa*
Linde. See Lin
Linden, or Lime-tree. *Tilia europœa*
Line, or Lint. *Linum usitatissimum*
Ling. *Calluna vulgaris*; sometimes applied to Heather of any sort
Chinese. *Trapa natans*
Downy. *Eriophorum polystachyon*
Ling-berry. The fruit of *Empetrum nigrum*, *Vaccinium Vitis-Idœa*, and *Calluna vulgaris*
"Lingo"-tree. *Pterocarpus indicus*
Lin-seed. The seed of Flax (*Linum usitatissimum*)
Lint. See Line
Fairy. *Linum catharticum*
Lints. *Vicia sativa*
Lion's-ear. The genus *Leonotis*
Of the Andes. See Cushion-plant
Lion's-foot. *Leontopodium alpinum* and *Alchemilla vulgaris*
American. *Nabalus albus* and *N. Fraseri*
Lion's-leaf. *Leontice Leontopetalum*
Lion's-mouth. *Antirrhinum majus*. Applied locally to *Linaria vulgaris* and one or two other plants
Lion's-paw. *Alchemilla vulgaris*
Lion's-snap. *Lamium amplexicaule*
Lion's-tail. *Leonotis Leonurus*
Lion's-tooth. The genus *Leontodon*
Lion's-Turnip. *Leontice Leontopetalum*
Liquorice-bush, of the Cape. *Vascoa amplexicaulis*
Liquorice-plant (cultivated). *Glycyrrhiza glabra*
Wild. *Ononis arvensis*
Wild, American. *Galium circœzans*, *G. lanceolatum*, and *Glycyrrhiza lepidota*
Wild, of Australia. *Scoparia australis* (*Teucrium corymbosum*)
Wild, of India. *Abrus precatorius*
Wild, Oregon. *Polypodium falcatum*
Liquorice-Vetch. *Astragalus glycyphyllus*
Liriconfancy. *Convallaria majalis*
Litchi, or Lee-chee. The fruit of *Nephelium Litchi*
Lithy-tree. *Viburnum Lantana*
Of Chili. *Rhus caustica*
Litmus-plant. *Roccella tinctoria* and *Croton tinctorum*
Little-good, or Little Goody. *Euphorbia Helioscopia*
Little Snow-balls. *Cephalanthus occidentalis*
Little-Wale. An old name for Gromwell
Live-long. *Sedum Telephium*
Jersey. *Gnaphalium luteo-album*

Liver-grass. *Marchantia polymorpha*
Liver-leaf, Round-lobed. *Hepatica triloba*
Sharp-lobed. *Hepatica acutiloba*
Liver-wort. *Agrimonia Eupatoria* and the genus *Marchantia*, especially *M. polymorpha*
Brook, or Common. *Marchantia polymorpha*
Conical. *Marchantia conica*
Ground. *Peltidea canina*
Hemispherical. *Marchantia hemisphœrica*
Noble. *Hepatica triloba*
Live-in-idleness. An old name for *Viola tricolor*
Lizard Orchis. *Orchis hircina*
Lizard's-herb. *Goniophlebium trilobum*
Lizard's-tail. The genus *Saururus*
Common. *Saururus cernuus*
W. Indian. *Piper peltatum*
Lizard's-tongue. The genus *Sauroglossum*
Lob-Grass. See Grass
Lobelia, Acrid. *Lobelia urens*
Blue-and-white. *Lobelia ramosa* and *L. heterophylla major*
Branching. *Lobelia ramosa*
Brilliant. *Lobelia fulgens*
Cardinal-flower. *Lobelia cardinalis*
Dwarf. *Lobelia pumila*
Dwarf Blue. *Lobelia azurea nana*
Dwarf Blue, Close-growing. *Lobelia Erinus compacta*
Dwarf Blue-and-white. *Lobelia Paxtoniana*
Great. *Lobelia syphilitica*
Holly-leaved. *Lobelia ilicifolia*
Italian. *Lobelia Laurentia*
Mullein-leaved. *Lobelia Tupa*
Scarlet. *Lobelia cardinalis*
Shore. *Lobelia littoralis*
Showy. *Lobelia speciosa*
Slender. *Lobelia gracilis*
Small Spreading. *Lobelia Erinus*
Stag's-horn-leaved. *Lobelia coronopifolia*
Tall Blue. *Lobelia syphilitica*
Trailing. *Lobelia erinoides*
Tree. *Lobelia assurgens*
Water. *Lobelia Dortmanna*
Yellow-flowered. *Lobelia lutea*
Loblolly-Bay, Common. *Gordonia lasianthra*
Downy. *Gordonia pubescens*
Loblolly - Sweet - wood. *Sciadophyllum Jacquinii*
Loblolly-wood, Jamaica. *Cupania glabra* and other species. Applied also to *Pisonia cordata*
Lobster-flower. *Poinsettia pulcherrima*
Lockin Gowan. See Gowan
Locks - and - keys. *Dielytra spectabilis* and the fruit of the Ash and Sycamore
"Loco." *Astragalus Hornii*
Locust-berry. *Malpighia coriacea*
Locust-tree, African. *Parkia africana*
Bastard. *Clethra tinifolia*
Bastard, of Jamaica. *Ratonia apetala*
Bristly. *Robinia hispida*
Decaisne's. *Robinia Pseud-acacia Decaisneana*
Doubtful. *Robinia dubia*

Locust-tree, European. *Ceratonia Siliqua*
Fragrant White-flowered. *Robinia Pseudacacia*
Honey. See Honey-Locust
S. American. *Hymenæa Courbaril*
Swamp, or Water. *Gleditschia monosperma*
W. Indian. *Hymenæa Courbaril;* also *Byrsonima cinerea* and *B. coriacea*
Loddon Lilies. *Leucojum æstivum*
Lode-wort. An old name for *Ranunculus aquatilis*
Logger-heads. *Centaurea nigra* and *C. Cyanus*
Log-wood. *Hæmatoxylon Campechianum*
Texan. *Condalia obovata*
Lokao-Dye-plant. *Rhamnus utilis*
Lombardy Poplar. *Populus fastigiata*
London Pride. *Saxifraga umbrosa*
London Rocket. *Sisymbrium Irio*
London Tufts. *Dianthus barbatus*
Longan. The fruit of *Nephelium Longanum*
"Long Purples." *Orchis mascula*
Long-wort. An old name for *Anacyclus Pyrethrum*
Looking-glass Plant or Tree, Large-leaved. *Heritiera littoralis*
Small-leaved. *Heritiera minor*
Loose-strife. The genus *Lysimachia*
False. The genus *Ludwigia*
Carmine-flowered. *Lysimachia Leschenaultii*
Catch-fly. *Cuphea silenoides*
Clethra-leaved. *Lysimachia clethroides*
Common. *Lysimachia vulgaris*
Dotted. *Lysimachia punctata*
Four-leaved. *Lysimachia quadrifolia*
Fringed. *Lysimachia ciliata*
Heavy-spiked. *Lysimachia barystachya*
Hyssop-leaved. *Lythrum hyssopifolium*
Large-flowered. *Lysimachia grandiflora*
Money-wort. *Lysimachia Nummularia*
Narrow-leaved. *Lysimachia angustifolia*
Purple, or Red. *Lythrum Salicaria*
Rosy-flowered. *Lythrum Salicaria* var. *roseum*
Short-spiked. *Lysimachia brachystachys*
Siberian. *Lysimachia dahurica*
Swamp. *Nesæa (Decodon) verticillata*
Tufted. *Lysimachia thyrsiflora*
Twiggy Purple. *Lythrum virgatum*
W. Indian. *Jussiæa suffruticosa*
Willow-leaved. *Lysimachia Ephemerum*
Winged. *Lythrum alatum*
Wood. *Lysimachia nemorum*
Lopez Root. The root of *Toddalia aculeata*
Lop-grass. See Lob-grass
Lop-seed. *Phryma leptostachya*
Loquat, or Japan Medlar. *Eriobotrya (Mespilus) japonica*
Lords-and-Ladies. *Arum maculatum*
Lord-wood. *Liquidambar orientale*
Lot-tree. *Pyrus Aria*
Lote-tree. *Celtis australis*
Lotus, E. Indian. *Nymphæa pubescens*
Egyptian. *Nymphæa Lotus*
Hungarian. *Nymphæa thermalis*

Lotus, or Lotos, Tree, African. *Zizyphus Lotus*
European. *Diospyros Lotus*
Of the Ancients. Probably *Nitraria tridentata*
Louisa, Herb. *Aloysia citriodora*
Louse-berry-tree. *Euonymus europæus*
Louse-bur. *Xanthium strumarium*
Louse-wort. The genus *Pedicularis;* also an old name for *Delphinium Staphisagria*
American. *Gerardia pedicularia*
Common. *Pedicularis sylvatica*
Marsh. *Pedicularis palustris*
Spiked. *Pedicularis comosa*
Lovage, or Lovache, Common. *Ligusticum scoticum*
Italian. *Levisticum officinale (Ligusticum Levisticum)*
Parsley-leaved. *Ligusticum peregrinum*
Love-apple. *Lycopersicum esculentum*
Love-grove. The genus *Nemophila*
Love-in-a-mist, or Love-in-a-puzzle. *Nigella damascena*
W. Indian. *Passiflora fœtida*
Love-in-idleness. *Viola tricolor*
Love-grass, Whorled. *Poa amabilis.* (See also Grass)
Love-lies-bleeding. *Amarantus caudatus*
Love-man. *Galium Aparine*
Love-Pea. See Pea
Love-plant. The genus *Anacampseros*
Tasmanian. *Comesperma volubile*
Love-tree. *Cercis Siliquastrum*
Lucerne. *Medicago sativa*
Luckie's-mutch. *Aconitum Napellus*
Lucy-of-Teesdale. A name suggested by Mr. Ruskin for *Gentiana verna*
Lunary. *Botrychium Lunaria*
Lung-flower. *Gentiana Pneumonanthe*
Lung-wort. The genus *Pulmonaria*
Alpine. *Mertensia alpina*
Azure-flowered. *Pulmonaria azurea*
Blunt-leaved. *Mertensia oblongifolia*
Bullock's, Cow's, or Clown's. *Verbascum Thapsus*
Clown's. See Lung-wort, Bullock's
Common. *Pulmonaria officinalis*
Cow's. See Lung-wort, Bullock's
French. See Lung-wort, Golden
Golden, or French. *Hieracium murorum*
Himalayan. *Lindelophia spectabilis*
Narrow-leaved. *Pulmonaria angustifolia*
Oblong-leaved. *Mertensia oblongifolia*
Siberian. *Pulmonaria (Mertensia) sibirica* or *Dahurica*
Smooth. The genus *Mertensia*
Tree. *Sticta pulmonacea*
White-flowered. *Pulmonaria saccharata alba*
Lupin, or Lupine. The genus *Lupinus*
Bastard. *Trifolium Lupinaster*
Blue-flowered. *Lupinus insignis, L. polyphyllus* and *L. nanus*
Blue-and-white. *Lupinus Moritzianus*
Changeable-flowered. *Lupinus mutabilis*
Dwarf Perennial. *Lupinus Nutkaensis*
Hartweg's. *Lupinus Hartwegi*
Flax-leaved. *Lupinus linifolius*
Long-leaved. *Lupinus macrophyllus*
Perennial. *Lupinus polyphyllus*

and Foreign Plants, Trees, and Shrubs. 81

Lupin, Rosy-flowered. *Lupinus venustus*
Small. *Psoralea Lupinella*
Tree. *Lupinus arboreus*
White-and-Rose-flowered. *Lupinus subcarnosus*
Lusmore. *Digitalis purpurea*
Lust-wort. The genus *Drosera*
Common. *Drosera rotundifolia*
Lychnis, Alpine. *Lychnis alpina*
Brilliant. *Lychnis fulgens*
Clammy. *Lychnis Viscaria*
Large-flowered. *Lychnis grandiflora (L. coronata)*
Meadow. *Lychnis Flos-cuculi*
Presl's. *Lychnis Preslii*
Pyrenean. *Lychnis pyrenaica*
Red-flowered. *Lychnis diurna*
Rock. *Lychnis Lagascæ*
Scarlet. *Lychnis chalcedonica*
Shaggy. *Lychnis Haageana*
Siebold's. *Lychnis Sieboldii*
Umbelled. *Lychnis Flos-Jovis*
White-flowered. *Lychnis vespertina*
Lychwale. See Lichwale
Lycoperdon-nut. See Nut
Lyday-wood. *Kageneckia oblonga*
Lyme Grass. See Grass
Lyng-wort. See Lang-wort
Lyre-tree. *Liriodendron tulipiferum*

Maalok-tree, of Australia. *Eucalyptus platypus*
Mabolo-tree. *Diospyros Mabola (D. discolor)*
Macarius(St.)Flower of. *Senecio vulgaris*
Macary-Bitter. *Picramnia Antidesma*
Macaw-bush. *Solanum mammosum*
"**Macaw-fat.**" *Elæis guineensis*
Macaw-tree, Great. *Cocos (Acrocomia) fusiformis*
Prickly. *Cocos (Acrocomia) aculeata*
Mace. Common. The aril or fleshy envelope of the Nutmeg (*Myristica moschata*)
"Red." The aril of *Pyrrhosa tingens*
Reed. *Typha latifolia*
"White." The aril of *Myristica Otoba*
Mackarel Mint. *Mentha viridis*
Madder. The genus *Rubia*
Bengal. *Rubia cordifolia*
Common Dyer's. *Rubia tinctorum*
E. Indian. *Hedyotis (Oldenlandia) umbellata*
Field. *Sherardia arvensis*
Indian. The genus *Hedyotis*
Petty. The genus *Crucianella*
Wild. *Rubia peregrina* and *Galium Mollugo*
Madders. *Anthemis Cotula*
Madder-wort. See Mad-wort
Madeira-wood. *Hypelate paniculata*
Madnep. *Heracleum Sphondylium*
Madonna Lily. See Lily
Madrona, or Madrono, Tree. *Arbutus Menziesii*
Mad-dog Weed. *Alisma Plantago*
Mad-wort, or Madder-wort. The genus *Alyssum*; also *Asperugo procumbens*
Alpine. *Alyssum alpestre*
Bladder-podded. *Alyssum utriculatum*

Mad-wort, Dense-flowered. *Alyssum saxatile var. compactum*
Dwarf. *Alyssum olympicum*
Eastern. *Alyssum orientale*
Fragile. *Alyssum olympicum*
Hoary German. *Alyssum gemonense*
Mountain. *Alyssum montanum*
Purple. *Alyssum deltoideum*
Rock. *Alyssum saxatile*
Russian. *Schiverckia podolica*
Spring. *Alyssum spinosum*
Sweet-scented. *Alyssum maritimum*
Toothed-leaved. *Alyssum denticulatum*
Variegated-leaved. *Alyssum saxatile variegatum*
Wiersbeck's. *Alyssum Wiersbeckii*
Yellow Rock. *Alyssum saxatile*
Magic-tree, Peruvian. *Cantua buxifolia*
Magnolia, "Blue." *Magnolia acuminata*
Brown-stalked. *Magnolia fuscata*
Campbell's. *Magnolia Campbelli*
"Exmouth." *Magnolia grandiflora var. lanceolata*
Great Laurel-leaved. *Magnolia grandiflora*
Halle's. *Magnolia Halleana*
Lemon-scented. *Magnolia citriodora*
Lenné's. *Magnolia Lenné*
Lily-flowered. *Magnolia conspicua*
Purple-flowered. *Magnolia purpurea*
Slender. *Magnolia gracilis*
Small-flowered. *Magnolia parviflora*
Small Laurel-leaved. *Magnolia glauca*
Yulan. *Magnolia conspicua*
Maguey-fibre-plant. Various species of *Agave*
Magydare. An old name for Laser-wort
Maharanga-dye-plant. *Onosma Emodi*
Mahoe-tree. *Paritium tiliaceum* and *Sterculia caribæa*
Common Blue, or Gray. *Paritium elatum*
Congo. *Hibiscus clypeatus*
Mountain. *Paritium elatum*
New Zealand. *Melicytus ramiflorus*
Sea-side. *Thespesia populnea*
Mahogany-tree. *Swietenia Mahagoni*
African. *Khaya senegalensis*
American Mountain. *Betula lenta*
Australian. Various species of *Eucalyptus* and *Angophora*
Bastard. *Ratonia apetala*
E. Indian. *Soymida febrifuga*
Forest, of N. S. Wales. A variety of *Eucalyptus resinifera*
Mountain, of California. The genus *Cercocarpus*
Red, of N. S. Wales. *Eucalyptus resinifera*
S. W. Australian. *Eucalyptus marginata*
Spanish. *Swietenia Mahagoni*
Swamp, of N. S. Wales. *Eucalyptus botryoides* and *E. robusta*
Victoria. *Eucalyptus botryoides*
Wasatch Mountain. *Cercocarpus ledifolius*
W. Australian. *Eucalyptus marginata (E. Mahagoni)*
White Australian. *Eucalyptus triantha (E. acmenioides), E. pilularis acmenioides,* and *E. robusta*

G

Mahogany-tree, Wild, or White, of the W. Indies. *Stenostomum bifurcatum*
Mahwah-tree. *Bassia latifolia*
Maid-of-the-Meadow. *Spiræa Ulmaria*
Maid-weed. See May-weed
Maiden-hair. *Adiantum Capillus-Veneris;* also applied locally to *Narthecium ossifragum* and some other plants
Golden. *Polytrichum commune* and *Polypodium vulgare*
Maiden-hair Grass. See Grass
Maiden-hair Tree. *Salisburia adiantifolia*
Maiden Oak. *Quercus sessiliflora*
Maiden Pink. *Dianthus deltoides*
Maiden's-Honesty. *Clematis Vitalba*
Maid's-hair. *Galium rerum*
Maid's-love. *Artemisia Abrotanum*
Maise. See Mathes
Makebate. *Jasminum fruticans*
Makinboy, or Makin-bwee. *Euphorbia hiberna*
Makooloo-tree. *Hydnocarpus venenata* (*H. inebrians*)
Maize, Common. *Zea Mays*
Japanese. *Zea japonica*
Saw-leaved. *Zea Curagua*
Mountain. The genus *Ombrophytum*
Water. *Victoria regia*
Majo-bitter Tree. *Picramnia Antidesma*
Malabar-Leaf. *Cinnamomum malabathrum*
Malabar-Nightshade. The genus *Basella*
Red. *Basella rubra*
White. *Basella alba*
Malabar-Nut. *Adhatoda vasica*
Male Fern. See Fern
Malla-tree. *Olax zeylanica*
Mallee-tree. *Eucalyptus dumosa*
Victorin. *Eucalyptus oleosa*
Mallet-flower. The genus *Tupistra*
Mallow. The genus *Malva*
Bell-flowered. *Malva campanulata*
Birch-leaved. *Malva scoparia*
Blue-flowered. *Malva limensis*
Buff-coloured. *Malva lateritia*
Curled-leaved. *Malva crispa*
Dwarf. *Malva rotundifolia*
E. Indian Musk. *Hibiscus moschatus*
False. The genus *Malvastrum*
Glade. *Napæa dioica*
Globe. The genus *Sphæralcea*
Hemp. *Hibiscus cannabinus*
High. *Malva sylvestris*
Hollyhock. *Malva Alcea*
Indian. The genera *Abutilon, Sida,* and *Urena*
Jews'. *Corchorus olitorius*
Marsh. *Malva sylvestris;* applied locally to *Althæa officinalis*
Moren's. *Malva Morenii*
Musk. *Malva moschata*
Poppy. See Poppy-Mallow
Rose. See Rose-Mallow
Scarlet. *Pavonia coccinea*
Thorny. *Hibiscus Sabdariffa*
Tree, or Sea. See Tree-Mallow
Venice. *Hibiscus Trionum*
Vervain. *Malva Alcea*
White. *Althæa officinalis*
White Musk. *Malva moschata* var. *alba*

Mallow-wort, Barbary. *Malope malacoides*
Maloo Creeper. *Bauhinia racemosa*
Mammee-Apple-tree. *Mammea americana*
Wild, of Jamaica. *Rheedia lateriflora*
Mammoth Tree, of California. *Sequoia* (*Wellingtonia*) *gigantea*
Manchineel-tree. *Hippomane Mancinella*
False. *Cameraria latifolia*
Mountain. *Rhus Metopium*
Mandioc-plant. *Manihot* (*Jatropha*) *utilissima*
Mandrake. *Mandragora officinalis, Bryonia dioica, Tamus communis,* and *Arum maculatum*
American. *Podophyllum peltatum*
Autumn-flowering. *Mandragora autumnalis*
Medicinal. *Mandragora officinalis*
Man-Fungus. See Fungus
Mangaba, or Mangava, Tree. *Hancornia speciosa*
Mangel, or Mangold, Wurtzel. *Beta vulgaris macrorrhiza*
Mango-tree. *Mangifera indica*
Mountain, or Wild. *Clusia flava* and various species of *Irringia*
Mangold. See Mangel
Mangosteen-tree. *Garcinia Mangostana*
Wild. *Embryopteris glutinifera*
Mangrove-tree. *Rhizophora Mangle*
New Zealand. *Avicennia officinalis*
Red. *Rhizophora racemosa*
White. *Avicennia tomentosa, A. nitida,* and *Laguncularia racemosa*
Zaragoza. *Conocarpus erecta*
Manioc-plant. *Jatropha Manihot*
Manjack. *Cordia macrophylla* and *C. elliptica*
Manna-tree. *Fraxinus Ornus* var. *rotundifolia*
Alhagi. *Alhagi Camelorum*
Australian. *Eucalyptus mannifera*
Briançon. *Pinus Larix*
Hebrew, or Persian. *Alhagi Maurorum*
Mt. Sinai. *Tamarix mannifera*
Oak. *Quercus Vallonea* and *Q. persica*
Persian. See Manna, Hebrew
Poland. *Glyceria fluitans*
Tamarisk. *Tamarix gallica* var. *mannifera*
Manna-croup. *Glyceria fluitans*
Manna-Grass. See Grass
Man-in-the-Ground, or Man-in-the-Earth. *Ipomœa* (*Convolvulus*) *pandurata*
Man-of-the-earth. *Ipomœa* (*Convolvulus*) *pandurata*
Man Orchis. *Aceras anthropophora*
Man-root, Colorado. *Ipomœa leptophylla*
Manuka-scrub. *Leptospermum scoparium*
Many-root. *Ruellia tuberosa*
Manzanita-shrub. *Arctostaphylos glauca* and *A. pungens*
Map-Lichen. See Lichen
Maple. The genus *Acer*
Acute-leaved. *Acer acuminatum*
Ash-leaved. *Acer Negundo* (*Negundo aceroides*)
Athenian. *Acer atheniense*
Austrian. *Acer campestre* var. *austriacum*

Maple, Bearded-calyxed. *Acer barbatum*
" Bird's-eye." A variety of *Acer saccharinum*
Black Sugar. *Acer saccharinum var. nigrum*
Bohemian. *Acer Loudoni*
Carolina. *Acer barbatum*
Common. *Acer campestre*
Cretan. *Acer creticum*
Crimson-leaved. *Acer Ginnala*
Crimson-leaved Japanese. *Acer polymorphum atro-purpureum*
" Curled." A variety of *Acer saccharinum*
Cut-leaved. *Acer dissectum*
Cut-leaved Norway. See Maple, Eagle's-claw
Downy Norway. *Acer platanoides var. pubescens*
Downy-fruited Field. *Acer campestre var hebecarpum*
Eagle's-claw, or Hawk's-foot. *Acer platanoides var. laciniatum*
Evergreen. *Acer heterophyllum (A. sempervirens)*
French, or Guelder-rose-leaved. *Acer opulifolium*
Glaucous-leaved Himalayan. *Acer glaucum*
Guelder-rose-leaved. See Maple, French
Hawk's-foot. See Maple, Eagle's-claw
Hungarian. *Acer obtusatum*
Hyrcanian. *Acer hyrcanum*
Iberian. *Acer ibericum*
Italian. *Acer Opalus*
Japanese. *Acer palmatum*
Japanese, Purple. *Acer palmatum var. dissectum atro-purpureum*
Lobel's Platanus-leaved. *Acer platanoides var. Lobelii*
Loudon's. *Acer Loudoni*
Montpelier. *Acer monspessulanum*
Mountain. *Acer spicatum*
Neapolitan. *Acer neapolitanum*
Norway, or Platanus-leaved. *Acer platanoides*
Oblong-leaved. *Acer oblongum*
Palmate-leaved. *Acer palmatum*
Platanus-leaved. See Maple, Norway
Red, or Swamp. *Acer rubrum*
Rock. *Acer saccharinum*
Round-leaved. *Acer circinatum*
Siberian. *Acer lobatum*
Sikkim. *Acer Sikkimense*
Silver, White, or Soft. *Acer dasycarpum*
Silver-variegated Platanus-leaved. *Acer platanoides var. variegatum*
Sir Charles Wager's. *Acer eriocarpum (A. dasycarpum)*
Soft. See Maple, Silver
Spanish. *Acer granatense*
Spiked-flowered. *Acer spicatum*
Striped-barked. *Acer pennsylvanicum (A. striatum)*
Sugar. *Acer saccharinum*
Swamp. See Maple, Red
Tartarian. *Acer tataricum*
Thick-leaved. *Acer coriaceum*
Trifid-leaved. *Acer trifidum*
Truncate-leaved. *Acer truncatum*

Maple, Variegated Field. *Acer campestre foliis variegatis*
Vine. *Acer circinatum*
White, or Silver. *Acer dasycarpum*
Wier's Cut-leaved. *Acer Wieri laciniatum*
Woolly-leaved Nepaul. *Acer villosum*
Maple Service. *Pyrus torminalis*
Maram, or Marram. *Psamma arenaria*
Marble-gall. See Gall
Marble-wood. The wood of *Diospyros Kurzii*
" **Marcella.**" *Grangea (Cotula) maderaspatana*
" **Marche.**" An old name for *Apium graveolens*
March, or Marish, Beetle. *Typha latifolia*
March, or Marish, Berries. *Vaccinium Oxycoccos*
March Violet. *Viola odorata*
Mare's-tail, Common. *Hippuris vulgaris*
Sea-shore. *Hippuris maritima*
Margosa-Bark-plant. See Nim-Bark
Margosa-tree. *Melia Azadirachta*
Marguerite. *Bellis perennis* and *Chrysanthemum frutescens*
Blue. *Agathæa cœlestis*
Mariettes. An old name for *Campanula Medium*
Marigold. The genera *Calendula* and *Tagetes*
African. *Tagetes erecta*
American Water. *Bidens Beckii*
Bur, American. *Bidens frondosa*
Bur, Common. *Bidens tripartita*
Bur, Larger. *Bidens chrysanthemoides*
Corn, Field, or Wild. *Chrysanthemum segetum*
Dwarf. *Tagetes tenuifolia*
Dwarf, Double French. *Tagetes patula pumila*
Dwarf, Striped. *Tagetes signata pumila*
Fetid. *Dysodia chrysanthemoides*
Field. See Marigold, Corn
Fig. See Fig-Marigold
French. *Tagetes patula*
Large Cape. *Calendula (Dimorphotheca) hybrida*
Marsh. See Marsh-Marigold
Mountain. *Senecio Lyalli*
Of Peru. An old name for the Sun-flower
Parry's. *Tagetes Parryi*
Pot. *Calendula officinalis*
Small Cape. *Calendula (Dimorphotheca) pluvialis*
Sweet-scented Mexican. *Tagetes lucida*
W. Indian. *Pectis punctata* and *Wedelia carnosa*
Wild. See Marigold, Corn
Mariposa-lily. The genus *Calochortus*
Blue. *Calochortus cœruleus*
Brilliant. *Calochortus splendens*
Elegant. *Calochortus elegans*
Gunnison's. *Calochortus Gunnisoni*
Large-eyed. *Calochortus luteus var. oculatus*
Lemon-coloured. *Calochortus citrinus*
Lilac. *Calochortus lilacinus*
Pretty. *Calochortus pulchellus (Cyclobothra pulchella)*

Mariposa-lily, White. *Calochortus venustus*
Yellow. *Calochortus luteus*
Marjoram, Common, English, Grove, or Wild. *Origanum vulgare*
Goat's. *Thymus Tragoriganum*
Knotted. *Origanum Marjorana*
Mt. Sipylus. *Origanum sipyleum*
Pot. *Origanum Onites*
Sweet. *Origanum Marjorana*
Wild. See Marjoram, Common
Winter. *Origanum heracleoticum*
Marking-nut Tree. *Semecarpus Anacardium*
Marl-Grass. See Grass
Marmalade-Box-tree. *Genipa americana*
Marmalade, Natural. The fruit of *Lucuma mammosa*
Marmalade-tree. *Lucuma mammosa*
"**Marmeladinha**." *Alibertia edulis*
Marool. See Moorva
Marram. See Maram
Marrem-grass. See Grass
Marsh-Beetle, or Marsh-Pestle. *Typha latifolia*
Marsh Cistus. *Ledum palustre*
Marsh Daisy. See Daisy
Marsh, or Marish, Elder. *Viburnum Opulus*
Marsh-flower. The genus *Limnanthemum*
Marsh-Gilliflower. See Gilliflower
Marsh-Marigold. *Caltha palustris*
Double. *Caltha palustris flore-pleno*
Large-flowered. *Caltha grandiflora*
Purplish. *Caltha purpurascens*
White-flowered. *Caltha leptosepala*
Marsh-Parsley. *Apium graveolens*
Marsh-Penny-wort. *Hydrocotyle vulgaris*
Marsh-Pestle. See Marsh-Beetle
Marsh-Pile-wort. *Ranunculus Ficaria*
Marsh-Samphire. *Salicornia herbacea*
Marsh-Trefoil. *Menyanthes trifoliata*
Marsh-wort. *Helosciadium (Sium) nodiflorum*
Martin's (St.)-flower. *Alstræmeria Flos-Martini*
Martin's (St.)-herb. *Sauragesia erecta*
Marvel-of-Peru. *Mirabilis Jalapa*
Many-flowered. *Mirabilis multiflora*
Sweet-scented. *Mirabilis longiflora*
Marvellous-Apples. An old name for *Momordica Balsamina*
Mary-bud. *Calendula officinalis*
Maryland Golden-star. *Chrysopsis Mariana*
Mary's (St.)-seed. An old name for Sow-thistle-seed
Mary's (St.)-wood. *Calophyllum Inophyllum*
Master-tree. *Acer campestre*
Mask-flower. The genus *Alonsoa*
Cut-leaved. *Alonsoa incisifolia*
Mast. The fruit of *Fagus sylvatica* and *Quercus Robur*
Mast-tree, E. Indian. *Guatteria longifolia*
Mast-wood, Yellow. *Robinia (Acacia) coriacea*
Master-wort. *Peucedanum (Imperatoria) Ostruthium*
American. *Heracleum lanatum*

Master-wort, Dwarf. *Dondia (Hacquetia) Epipactis (Astrantia Epipactis)*
Great Black. *Astrantia major*
Small Black. *Astrantia minor*
Mastich-tree, Algerian, or Barbary. *Pistacia Terebinthus (P. atlantica)*
Common. *Pistacia Lentiscus*
E. Indian, or Bombay. *Pistacia Khinjuk* and *P. Cabulica*
Peruvian. *Schinus molle*
Scian. *Pistacia Lentiscus*
W. Indian. *Bursera gummifera*
Match-wood. An old name for Amadou or German Tinder. (See Amadou)
Mate. *Ilex Paraguariensis*
Matfellon. An old name for Knap-weed (*Centaurea nigra*)
Mat-Grass. See Grass
Mather. See Mathes
Mathes, Maithes, Mather, Mauthern, Mavin, Maythig, Mawthen, Mawther, Maise, Meliden, Mazes, or Moithern. *Anthemis Cotula*
Matico-plant. *Piper angustifolium (Artanthe elongata)*
Peruvian. *Eupatorium glutinosum*
Mat-reed. *Typha latifolia*
Matrimony-vine. *Lycium barbarum*
Mat-rush. *Scirpus lacustris*
Mat-weed. *Psamma arenaria*, *Nardus stricta*, and *Spartina stricta*
Hooded. *Lygeum Spartum*
Sea. *Ammophila arenaria*
Small. *Nardus stricta*
Mat-wood-tree. *Imbricaria borbonica*
Matting-grass, E. Indian. See Grass
Maudlin, Sweet. *Achillea Ageratum*
Maul, Mauls, or Maws. *Malva sylvestris*
Mauthern. See Mathes
Mauritius-weed. *Rocella fuciformis*
Mavin. See Mathes
Mawroll. An old name for *Marrubium vulgare*
Maw-seed. The seeds of *Papaver somniferum*
Mawthen, or Mawther. See Mathes
May, or May-bush. *Cratægus Oxyacantha*
American. A variety of *Spiræa hypericifolia*
Blackthorn. *Prunus spinosa*
Californian. *Photinia arbutifolia*
Italian. *Spiræa Filipendula*
N. W. Wales. *Spiræa corymbosa*
May-apple. *Podophyllum peltatum*
Himalayan. *Podophyllum Emodi*
Oregon. *Achlys triphylla*
May-blobs. *Caltha palustris*
May-bush. See May
May-flower. A name applied locally to *Cardamine pratensis* and various other plants which bloom in May
American. *Azalea nudiflora* and *Epigæa repens*
New England. *Epigæa repens*
W. Indian. *Dalbergia (Ecastaphyllum) Brownei*
May-Lily. *Convallaria majalis*
May-pole, of Jamaica. *Spathelia simplex*
Maythig. See Mathes

and Foreign Plants, Trees, and Shrubs. 85

May-weed. *Anthemis Cotula, Pyrethrum Parthenium,* and *Matricaria (Anthemis) inodora*
Double. *Matricaria (Anthemis) inodora fl.-pl.*
Stinking. *Anthemis Cotula*
May-wort. *Galium Cruciata*
Mazard, or Mazzards. *Prunus Avium*
Maze-berry. *Sapota Achras*
Mazes. See Mathes
Meaden. See Mathes
Meadow-Beauty. *Rhexia virginica*
Meadow-Bright, or Meadow-bout. *Caltha palustris*
Meadow-Cress. *Cardamine pratensis*
Meadow-Crocus. See Meadow-Saffron
Meadow-Distaff. *Cirsium oleraceum*
Meadow-Fescue-grass. See Grass
Meadow-Grass. See Grass
Meadow-Nut. *Comarum palustre*
Meadow-Parsnip. *Heracleum Sphondylium*
Meadow-Pink. *Lychnis Flos-cuculi*
Meadow-Rocket. *Orchis latifolia*
Meadow-Rue. The genus *Thalictrum*
Alpine. *Thalictrum alpinum*
Canadian. *Thalictrum Cornuti*
Columbine. *Thalictrum aquilegifolium*
Common Yellow-flowered. *Thalictrum flavum*
Dark-flowered. *Thalictrum atropurpureum*
Early. *Thalictrum dioicum*
Fetid. *Thalictrum foetidum*
Maiden-hair. *Thalictrum minus*
Purplish. *Thalictrum purpurascens*
Tall. *Thalictrum elatum* and *T. Cornuti*
Tuberous-rooted. *Thalictrum tuberosum*
Wind-flower. *Thalictrum anemonoides*
Meadow-Saffron, or Meadow-Crocus. *Colchicum autumnale*
Alpine. *Colchicum alpinum*
Bivona's. *Colchicum Bivonæ*
Byzantine. *Colchicum byzantinum*
Chequered. *Colchicum tessellatum*
Chian. *Colchicum chionense*
Giant. *Colchicum speciosum*
Parkinson's Chequered. *Colchicum Parkinsoni*
Pyrenean. *Merendera Bulbocodium (Colchicum montanum)*
Slender-leaved. *Bulbocodium tenuifolium*
Spring. *Bulbocodium vernum*
Spring, Caucasian. *Bulbocodium trigynum*
Variegated. *Colchicum variegatum (C. Agrippinæ)*
Meadow-Saxifrage. *Silaus pratensis*
Meadow-Soft-Grass. See Grass
Meadow-Sweet. *Spiræa Ulmaria*
Alleghany. *Spiræa corymbosa*
American. *Spiræa salicifolia*
Long-sprayed. *Spiræa flagelliformis*
Pigmy. *Spiræa crispifolia*
Plumy. *Spiræa pachystachya*
Shrubby. *Spiræa aricæfolia*
Variegated. *Spiræa Ulmaria variegata*
"Mealies." *Zea Mays*
Mealy-tree. *Viburnum Lantana*
"Mechameck," of the N. American Indians. *Ipomæa pandurata*

Medick. *Medicago sativa* and other species
Black. *Medicago lupulina*
Bur. *Medicago minima*
Denticulate. *Medicago denticulata*
Hedge-hog. *Medicago intertexta*
Purple. *Medicago sativa*
Sickle-podded. *Medicago falcata*
Spotted. *Medicago maculata*
Tree. *Medicago arborea*
Medlar, Common, or Dutch. *Mespilus germanica*
Dwarf. *Pyrus (Mespilus) Chamæmespilus*
Japan. *Eriobotrya japonica*
Lobed-leaved. *Mespilus lobata*
Neapolitan. *Cratægus Azarolus*
Quince-leaved. The genus *Cotoneaster*
Smith's. *Mespilus Smithii*
Snowy. *Mespilus grandiflora*
Spreading. *Mespilus germanica* var. *diffusa*
Surinam. *Achras mammosa*
Upright. *Mespilus germanica* var. *stricta*
W. Indian. *Mimusops Elengi*
Wild. *Mespilus germanica* var. *sylvestris*
Medlar-bush. *Amelanchier ovalis*
Medlar-wood. *Myrtus mespiloides*
Medusa's-head. *Euphorbia Caput-Medusæ*
Melancholy-gentleman. *Hesperis tristis*
Melick Grass. See Grass
Melilot. *Melilotus officinalis*
Field. *Melilotus arvensis*
White-flowered. *Melilotus alba*
Trefoil. *Medicago lupulina*
Melon, Common. *Cucumis Melo*
Sweet-scented. *Cucumis Dudaim*
Water. *Cucumis Citrullus*
Melon-cactus. The genus *Melocactus*
Melon-thistle, Turk's-cap. *Melocactus communis*
Melon-tree. *Carica Papaya*
Menow-weed. *Ruellia tuberosa*
Mercury, Baron's. See Baron's Mercury
Boy's, or Girl's. *Mercurialis annua*
Dog's. *Mercurialis perennis*
English, False, or Wild. *Chenopodium Bonus-Henricus*
French. *Mercurialis annua*
Golden. *Mercurialis perennis* rar. *aurea*
Three-seeded. The genus *Acalypha*
Vegetable. *Franciscea uniflora*
Wild. See Mercury, English
Mercury's Violets. An old name for *Campanula Medium*
Merlin's Grass. See Grass
Mermaid, False. *Flœrkia proserpinacoides*
Mermaid-weed. *Proserpinaca palustris*
Merman's-Shaving-brush. Various species of *Chamædoris* and *Penicillus*
Merry-tree. *Prunus Avium*
Meskit, Mesquit, or Mezquit, Honey. *Prosopis juliflora*
Screw-pod. *Prosopis pubescens*
Mespilus, Snowy. *Amelanchier Botryapium*
Mess-mate-tree. *Eucalyptus obliqua*
Meu. *Meum Athamanticum*
Mezereon. *Daphne Mezereum*
Mezquit. See Meskit

Mice, Dutch. *Lathyrus tuberosus*
Michaelmas Daisy. *Aster Tripolium*
"**Mid-day-flower,**" of Australia. The genus *Mesembryanthemum*
Midnapore Creeper. *Rivea Bona-nox*
Midshipman's-butter. The fruit of *Persea gratissima*
Midsummer Daisy. *Chrysanthemum Leucanthemum*
Midsummer-Men. *Sedum Telephium*
Mignonette. *Reseda odorata*
  Jamaica. *Lawsonia alba*
  Of the French. *Dianthus chinensis* and *Saxifraga umbrosa*
  Pyramidal. *Reseda odorata var. pyramidalis*
  Red-flowered. *Reseda odorata var. rosea*
  Tree. *Reseda odorata*, grown under a special mode of treatment
  White. *Reseda alba*
  Wild. *Reseda Luteola*
Mignonette-vine. *Madaria elegans* and *M. corymbosa*
Mildew. Various species of minute *Fungi*
  Vine. *Oidium Tuckeri*
  Wheat. *Puccinia graminis*
Milfoil, Alpine. *Achillea alpina*
  Black. *Achillea atrata*
  Chamomile-leaved. *Achillea atrata*
  Common. *Achillea Millefolium*
  Downy. *Achillea pubescens*
  Fever-few-leaved. *Achillea macrophylla*
  Great. *Achillea magna*
  Hooded. The genus *Utricularia*
  Musk. *Achillea moschata*
  Rough-headed. *Achillea squarrosa*
  Showy. *Achillea nobilis*
  Sweet. *Achillea Ageratum*
  Water. The genus *Myriophyllum*
  Woolly. *Achillea tomentosa*
Milk-bush, African. The genus *Synadenium*
Milk-Gowan. *Leontodon Taraxacum*
Milk-Grass. *Valerianella olitoria*
Milk-Parsley, or Milk-weed. *Peucedanum palustre*
  Caraway-leaved. *Selinum carvifolium*
Milk-Pea. The genus *Galactia*
Milk-Thistle. *Silybum Marianum*
  Ivory. *Silybum eburneum*
Milk-tree, Jamaica. *Brosimum spurium*
  Madagascar. *Tanghinia (Cerbera) lactaria*
  S. American. *Brosimum Galactodendron*
Milk-Trefoil, or Shrub-Trefoil. An old name for the genus *Cytisus*
Milk-Vetch. The genus *Astragalus*
  Alpine. *Astragalus alpinus*
  Clover. *Astragalus dasyglottis*
  Egyptian. *Astragalus trimestris*
  Goat's-Rue. *Astragalus galegæformis*
  Goat's-Thorn. *Astragalus Tragacantha*
  Montpelier. *Astragalus monspessulanus*
  Mountain. *Astragalus alpinus*
  Pontic. *Astragalus ponticus*
  Purple-flowered. *Astragalus hypoglottis*
  Red-fruited. *Astragalus caryocarpus*
  Saint-foin. *Astragalus Onobrychis*
  Shaggy. *Astragalus pannosus*
  Sheathed. *Astragalus vaginatus*

Milk-Vetch, Silvery. *Astragalus vimineus*
  Starry. *Astragalus sesameus*
  Sweet-leaved. *Astragalus glycyphyllos*
  Triangular-podded. *Astragalus bæticus*
Milk-vine. *Periploca græca*
Milk-weed. *Sonchus oleraceus, Peucedanum palustre,* and *Euphorbia corollata*
  American. The genus *Asclepias*
  Bitter. *Polygala amara*
  Common American. *Asclepias Cornuti*
  Douglas's. *Asclepias Douglasii*
  Green. *Asclepiadora decumbens* and the genus *Acerates*
  Four-leaved. *Asclepias quadrifolia*
  Poke. *Asclepias phytolaccoides*
  Purple. *Asclepias purpurascens*
  Sea-side. *Glaux maritima*
  Swamp. *Asclepias incarnata*
  Whorled. *Asclepias verticillata*
  Variegated. *Asclepias variegata*
Milk-wood, Horny-leaved. *Rauwolfia canescens*
  Jamaica. *Sapium laurifolium*
Milk-wort. The genus *Polygala*
  Box-leaved. *Polygala Chamæbuxus*
  Chalk. *Polygala calcarea*
  Common. *Polygala vulgaris*
  Fringed. *Polygala paucifolia*
  Ground-flowering. *Polygala polygama*
  Sea-side. *Glaux maritima*
Miller's-Star. *Stellaria Holostea*
Millet. *Panicum miliaceum*; also applied to various species of the genera *Milium, Paspalum,* and *Sorghum*
  Australian. *Panicum decompositum*
  "Bajree." *Panicum cæruleum*
  E. Indian. *Sorghum vulgare*
  German. *Setaria germanica*
  Italian. *Setaria italica*
  Sugar. *Sorghum saccharatum*
  Texas. *Sorghum cernuum*
Millet Grass. See Grass
Millet Khoda, of India. The grains of *Paspalum scrobiculatum*
Mill-mountain. *Linum catharticum*
Miltwaste. *Asplenium Ceterach*
"**Minarta.**" *Geum urbanum*
Minshull Crab. See Crab
Mint. The genus *Mentha*
  Apple. *Mentha rotundifolia*
  Bergamot. *Mentha citrata*
  Black. *Mentha piperita vulgaris*
  Brandy. See Mint, Pepper
  Brook, or Water. *Mentha sylvestris*
  Cat. *Nepeta Cataria* and *Calamintha officinalis*
  Corn. *Mentha arvensis*
  Curled, Crisped, or Cross. *Mentha crispa*
  Field. *Mentha arvensis*
  Flea. *Mentha Pulegium*
  Garden, or Spear. *Mentha viridis*
  Horse. *Mentha sylvestris*
  Horse, American. *Monarda punctata*
  Hyssop-leaved. *Mentha cervina*
  Mackarel. *Mentha viridis*
  Mountain. *Monarda didyma* and the genus *Pycnanthemum*
  Pepper, or Brandy. *Mentha piperita*
  Spear. *Mentha viridis*

and Foreign Plants, Trees, and Shrubs. 87

Mint, Squaw. *Hedeoma pulegioides*
Stag. *Mentha cervina*
Variegated Round-leaved. *Mentha rotundifolia variegata*
Water. See Mint, Brook
White. *Mentha piperita officinalis*
Whorled Water. *Mentha sativa*
Wild, American. *Mentha canadensis*
Mint-bush, or Mint-tree, Australian. The genus *Prostanthera*
Victorian. *Prostanthera lasianthos*
Mio-Mio-shrub. *Baccharis cordifolia*
Mirabel. A French name for candied or preserved Plums
Miraculous-berry, of W. Africa. The fruit of *Sideroxylon dulcificum*
"Miro"-tree. *Podocarpus ferruginea*
Mistletoe, American. *Phoradendron flavescens*
Australian. Various species of *Loranthus*
Californian. Various species of *Phoradendron*
E. Indian. Various species of *Loranthus*
European. *Viscum album* and *Loranthus europæus*
New Zealand. Various species of *Loranthus*
W. Indian. Various species of *Loranthus* and *Phoradendron*
Mishmi-Bitter. *Coptis Teeta*
Missel-tree, Guiana. *Blakea quinquenervia*
Mist-flower. *Conoclinium cœlestinum*
Mist-tree, Brush-land, of Australia. *Litsæa dealbata*
Mithridate Mustard, or Pepper-wort. *Lepidium campestre*
Mitre-flower. The genus *Cyclamen*
Mitre-pod. The genus *Mitraria*
Scarlet. *Mitraria coccinea*
Mitre-wort, or Bishop's-cap, American. The genus *Mitella*
False. *Tiarella cordifolia*
Mocan-shrub. *Vismia Mocanera*
Moccason, or Mocassin, Flower. The genus *Cypripedium*
Moccasin-flower, Yellow. *Cypripedium pubescens*
Mock-apple, Canadian. *Echinocystis lobata*
Mocker-nut. *Carya tomentosa*
Mock-Orange. The genus *Philadelphus*
Broad-leaved. *Philadelphus latifolius*
Carolina. *Prunus caroliniana*
Common. *Philadelphus coronarius*
Dwarf. *Philadelphus coronarius nanus*
Gordon's. *Philadelphus Gordonianus*
Hairy-leaved. *Philadelphus hirsutus*
Large-flowered. *Philadelphus grandiflorus*
Lewis's. *Philadelphus Lewisii*
Loose-branched. *Philadelphus laxus*
Many-flowered. *Philadelphus floribundus*
Mexican. *Philadelphus mexicanus*
Scentless. *Philadelphus inodorus*
Showy. *Philadelphus speciosus*
Warted. *Philadelphus verrucosus*
Woolly-leaved. *Philadelphus tomentosus*
Zeyher's. *Philadelphus Zeyheri*
Mock-Plane. *Acer Pseudo-platanus*

Model-wood, E. Indian. The wood of *Nauclea cordifolia*
Modesty, of N. America. *Bupleurum rotundifolium*
Mohoe. *Hibiscus arboreus*
Mohomoho-plant. *Piper angustifolium*
Moithern. See Mathes
Moly. *Allium Moly*
Dwarf. *Allium Chamæmoly*
Of Dioscorides. *Allium subhirsutum*
Molompi-wood. The wood of *Pterocarpus erinaceus*
Money-flower. *Lunaria biennis*
Money-wort. *Lysimachia Nummularia*; also applied to *Anagallis tenella*, *Thymus Nummularius*, *Tavernicra Nummularia*, and *Dioscorea Nummularia*
Cornish. *Sibthorpia europæa*
Cornish, Variegated. *Sibthorpia europæa variegata*
Golden. *Lysimachia Nummularia aurea*
Monkey-apple. *Anisophyllum laurinum*
Monkey-cup. The genus *Nepenthes*
Monkey-grass. The fibre of *Attalea funifera*
Monkey-flower. Various species of *Mimulus*
Blotched. *Mimulus maculosus*
Cardinal, or Scarlet. *Mimulus cardinalis*
Copper-coloured. *Mimulus cupreus*
Creeping. *Mimulus repens*
James's. *Mimulus Jamesii*
Orange. *Mimulus glutinosus*
Primrose. *Mimulus primuloides*
Scarlet. See Monkey-flower, Cardinal
Spotted. *Mimulus guttatus*
Tiling's. *Mimulus Tilingi*
Yellow. *Mimulus luteus*
Monkey-pot Tree. *Lecythis Ollaria*
Monkey-puzzle. *Araucaria imbricata*
Monkey-wort. *Evolvulus Nummularius*
Monkey's-bread. *Adansonia digitata*
Monkey's-dinner-bell. The fruit of *Hura crepitans*
Monkey's-face. The genus *Mimusops*
Monk's-hood, or Monk's-cowl. The genus *Aconitum*, especially *A. Napellus*
American Wild. *Aconitum uncinatum*
Autumn-flowering. *Aconitum autumnale*
Chinese. *Aconitum chinense*
Common. *Aconitum Napellus*
Jacquin's Yellow-flowered. *Aconitum Anthora*
Japanese. *Aconitum japonicum*
Northern. *Aconitum septentrionale*
Panicled. *Aconitum paniculatum*
Pyrenean. *Aconitum pyrenaicum*
Two-coloured. *Aconitum bicolor*
Variegated-flowered. *Aconitum variegatum*
Monk's-Rhubarb. *Rumex Patientia*
Monox, or Moonog, Heather. *Empetrum nigrum* and *Vaccinium Oxycoccos*
Moon-creeper. *Menispermum canadense* and *Ipomæa Bona-nox*
Moon-Daisy, or Moon-Flower. *Chrysanthemum Leucanthemum*
Moon-Fern, or Moon-wort *Botrychium Lunaria*

Moon-Flower. See Moon-Daisy
Moon-fruit Pine. *Lycopodium lucidulum*
Moonog. See Monox
Moon-penny. *Chrysanthemum Leucanthemum*
Moon-seed, Canadian. *Menispermum canadense*
Daurian. *Menispermum dauricum*
Smilax-like. *Menispermum smilacinum*
Moon-Trefoil. *Medicago arborea*
Moon-wort. See Moon-Fern
Blue. *Soldanella alpina*
Moor-balls. *Conferva ægagropila*
Moor-berries, or Moss-berries. *Vaccinium Oxycoccos*
Moor-Grass. See Grass
Moor-Heath. *Erica vagans*
Moor-Myrtle. *Myrica Gale*
Moor-pawns. The flowers of *Eriophorum* and various species of *Carex*
Moor-silk. *Polytrichum commune*
Moorva, or Marool, plant. *Sanseviera Roxburghii*
Moor-Whin, or Moss-Whin. *Genista anglica*
Moor-wort. *Andromeda polifolia*
Moose-wood. *Dirca palustris* and *Acer pennsylvanicum*
Mootchie-wood. *Erythrina indica*
Morass-weed. *Ceratophyllum demersum*
Mora-wood. *Mora excelsa*
Morel. *Morchella esculenta* and *Peziza coccinea*
Great. *Atropa Belladonna*
Petty. *Solanum nigrum*
Morello. A variety of Cherry
Morning-flower, Australian. *Orthrosanthus multiflorus*
Morning-glory. The genus *Ipomæa*
Common. *Ipomæa purpurea*
Smaller. *Ipomæa Nil*
Morocco, Red. *Adonis autumnalis*
Mortal. *Solanum Dulcamara*
Moschatel. *Adoxa Moschatellina*
Moss. A general name for plants of the *Musci* family, but popularly applied to various other plants
Apple. The genus *Bartramia*
Apple, Common. *Bartramia pomiformis*
Beard. See Moss, Tree
Black. *Tillandsia usneoides*
Bladder. The genus *Gymnostomum*
Bog. The genus *Sphagnum*
Bog, Blunt-leaved. *Sphagnum cymbifolium*
Bog, Long-leaved Floating. *Sphagnum cuspidatum*
Bog, Slender. *Sphagnum acutifolium*
Bog, Spreading-leaved. *Sphagnum squarrosum*
Bon-grace. *Splachnum rubrum*
Branched Beardless. The genus *Anæctangium*
Bristle. The genus *Orthotrichum*
Bristle, Club-fruited. *Orthotrichum Ludwigii*
Bristle, Elegant. *Orthotrichum pulchellum*
Bristle, River. *Orthotrichum rivulare*
Bristle, Rock. *Orthotrichum rupicola*
Bristle, Showy. *Orthotrichum speciosum*

Moss, Bristle, Straw-like. *Orthotrichum stramineum*
Bristle, White-tipped. *Orthotrichum diaphanum*
Canary. *Parmelia perlata*
Carrageen. See Moss, Irish
Cavern, or Feather Cavern. *Schistostega pennata*
Ceylon, or Jaffna. *Sphærococcus (Gracilaria) lichenoides (Plocaria candida)*
Club, or Stag's-horn. *Lycopodium clavatum*
Club, Creeping. *Selaginella apus*
Club, Golden. *Selaginella Kraussiana aurea*
Club-stalked. The genus *Ædipodium*
Collar. The genus *Splachnum*
Cord. The genus *Funaria*
Cord, Irish. *Funaria hibernica*
Corsican. *Gracilaria Helminthocorton* and *Laurencia obtusa*
Cup. *Scyphophorus pyxidatus* and *Lecanora tartarea*
"Dovedale." *Saxifraga hypnoides*
Dwarf-Apple. The genus *Bartramidula*
Earth. The genus *Phascum*
Earth, Common Dwarf. *Phascum muticum*
Earth, Tall. *Phascum bryoides*.
Extinguisher. The genus *Encalypta*
Feather. The genus *Hypnum*
Feather, Beaded. *Hypnum moniliforme*
Feather, Clustered. *Hypnum confertum*
Feather, Creeping. *Hypnum repens*
Feather, Curled. *Hypnum commutatum*
Feather, Cypress-leaved. *Hypnum cupressiforme*
Feather, Dwarf. *Hypnum pumilum*
Feather, Dark-green. *Hypnum atro-virens*
Feather, Elegant. *Hypnum pulchellum*
Feather, Fern-like. *Hypnum trichomanoides*
Feather, Flat. *Hypnum complanatum*
Feather, Floating. *Hypnum fluitans*
Feather, Fox-tail. *Hypnum alopecurum*
Feather, Glittering. *Hypnum splendens*
Feather, Green-patch. *Hypnum cæspitosum*
Feather, Lesser Golden Fern. *Hypnum filicinum*
Feather, Long-headed. *Hypnum medium*
Feather, Marsh. *Hypnum palustre*
Feather, Matted. *Hypnum populeum*
Feather, Mouse-tail. *Hypnum myosuroides*
Feather, Neat Meadow. *Hypnum purum*
Feather, Ostrich-plume. *Hypnum Cristacastrensis*
Feather, Plumy-crested. *Hypnum molluscum*
Feather, Red Mountain. *Hypnum rufescens*
Feather, Rusty. *Hypnum plumosum*
Feather, Scorpion. *Hypnum scorpioides*
Feather, Shaded. *Hypnum umbratum*
Feather, Sharp Fern-like. *Hypnum denticulatum*
Feather, Shining. *Hypnum nitens*
Feather, Showy. *Hypnum speciosum*
Feather, Silky. *Hypnum sericeum*
Feather, Soft Water. *Hypnum molle*

Moss, Feather, Sparkling. *Hypnum micans*
Feather, Spruce-tree. *Hypnum abietinum*
Feather, Tree-like. *Hypnum dendroides*
Feather, Velvet. *Hypnum velutinum*
Feather, Wall. *Hypnum murale*
Feather, Wavy. *Hypnum undulatum*
Feather, Whitish. *Hypnum albicans*
Feather, Yellow. *Hypnum lutescens* and *H. salebrosum*
Feather, Yellow Starry. *Hypnum stellatum*
Fork. The genus *Dicranum*
Fork, Broom. *Dicranum scoparium*
Fork, Maiden-hair. *Dicranum adiantoides*
Fork, Pellucid. *Dicranum pellucidum*
Fork, Red-necked. *Dicranum cerviculatum*
Fork, Silky-leaved. *Dicranum heteromallum*
Fork, Tawny. *Dicranum fulvellum*
Fork, White. *Dicranum glaucum*
Fork, Yellowish. *Dicranum flavescens*
Fringe. The genus *Trichostomum*
Fringe, Cord-like. *Trichostomum funale*
Fringe, Dark Mountain. *Trichostomum aciculare*
Fringe, Hoary. *Trichostomum canescens*
Fringe, Spreading. *Trichostomum patens*
Fringe, Woolly. *Trichostomum lanuginosum*
Golden. *Sedum acre*
Ground. *Polytrichum juniperinum*
Hair-cap. *Polytrichum juniperinum*
Hair. The genus *Polytrichum*
Hair, Alpine. *Polytrichum alpinum*
Hair, Common. *Polytrichum commune*
Hair, Dwarf Long-headed. *Polytrichum aloides*
Hair, Dwarf Round-headed. *Polytrichum nanum*
Hair, Juniper-leaved. *Polytrichum juniperinum*
Hair, Slender. *Polytrichum gracile*
Hair, Urn-bearing. *Polytrichum urnigerum*
Hygrometric. *Funalia hygrometrica*
Iceland. *Cetraria islandica*
Indian. *Saxifraga hypnoides*
Iridescent. *Schistostega pennata*
Irish or Carrageen. *Chondrus crispus*
Irish Cord. *Funaria hibernica*
Jaffna. See Moss, Ceylon
Lattice. The genus *Cinclidotus*
Lattice, Fountain. *Cinclidotus fontinaloides*
Long, or Black. *Tillandsia usneoides*
Mungo Park's. *Dicranum bryoides*
Necklace. *Usnea barbata*
New Orleans. *Tillandsia usneoides*
Pearl. *Chondrus crispus*
Pepper. *Pilularia globulifera*
Reindeer. *Cenomyce rangiferina*
Rock. *Roccella tinctoria*
Rock-hair. *Alectoria jubata*
Scale. The genus *Jungermannia*
Sea. *Ulva latissima*
Screw. The genus *Tortula*
Screw, Aloe-like. *Tortula rigida*
Screw, Bird's-claw. *Tortula unguiculata*
Screw, Frizzled Mountain. *Tortula tortuosa*

Moss, Screw, Great Hairy. *Tortula ruralis*
Screw, Hoary. *Tortula canescens*
Screw, Wall. *Tortula muralis*
Serpent. *Selaginella serpens*
Silvery. *Bryum argenteum*
Spanish. *Tillandsia usneoides*
Split. The genus *Andreæa*
Stag's-horn. See Moss, Club
Thatch. *Bryum rurale*
Thread. The genus *Bryum*
Thread, Golden. *Bryum pyriforme*
Thread, Greater Matted. *Bryum capillare*
Thread, Lesser Matted. *Bryum cæspititium*
Thread, Long-stalked. *Bryum triquetrum*
Thread, Many-stalked. *Bryum affine*
Thread, Marsh. *Bryum palustre*
Thread, Narrow-leaved. *Bryum androgynum*
Thread, Pale-leaved. *Bryum albicans* and *B. dealbatum*
Thread, Pink-fruited. *Bryum carneum*
Thread, Red Alpine. *Bryum alpinum*
Thread, Rosette. *Bryum roseum*
Thread, Silvery. *Bryum argenteum*
Thread, Slender. *Bryum gracile*
Thread, Slender-branched. *Bryum julaceum*
Thread, Swan's-neck. *Bryum hornum*
Thread, Swelling Bog. *Bryum ventricosum*
Tree, or Beard. A name applied to various Lichens of the genera *Usnea*, *Ramalina*, *Cornicularia*, &c. Also to *Lycopodium Selago*
Urn. *Polytrichum urnigerum* and various Mosses of the family *Bryaceæ*
Velvet. *Gyrophora murina*
Wall. *Sedum acre*
Water. The genus *Fontinalis*
Water, Alpine. *Fontinalis squamosa*
Water, Greater. *Fontinalis antipyretica*
Wing. The genus *Pterigonium*
Wood. *Bryum cuspidatum* and various species of *Hypnum*
Yoke. The genus *Zygodon*
Yoke, Lesser. *Zygodon conoideus*
Moss-berry, or Moor-berry. *Vaccinium Oxycoccos*
Moss-Campion. *Silene acaulis*
Moss-crops. *Eriophorum vaginatum* and other species
Moss-Fern. See Fern
Moss-Pink. *Phlox subulata*
Moss-Rush. *Juncus squarrosus*
Moss-Whin. See Moor-Whin
Mossy-Red-shanks. *Tillæa muscosa*
Moth-Mullein. *Verbascum Blattaria*
Moth-plant. The genus *Phalænopsis*
Mother-of-Thousands. *Linaria Cymbalaria* and *Saxifraga sarmentosa*
Mother-of-Thyme. *Thymus Serpyllum*
Mother-of-Vinegar. See Vinegar-plant
Mother-of-Wheat. *Veronica hederæfolia*
Mother-wort. *Leonurus Cardiaca* and *Artemisia vulgaris*
Golden. Various species of *Helichrysum*
Siberian. *Leonurus sibiricus*
Mother's-heart. *Capsella Bursa-pastoris*

Mould. Various minute *Fungi*
Blue, of Cheese. *Aspergillus glaucus*
Bread. *Ascophora elegans*
Mount-Caper. *Orchis latifolia*
Mountain-Ash. *Pyrus Aucuparia*
Western. *Pyrus sambucifolia*
Mountain-Avens. *Dryas octopetala*
Mountain-Cowslip. *Primula Auricula*
Mountain-Currant. See Currant
Mountain-Elm. *Ulmus montana*
Mountain-Fern. *Nephrodium Oreopteris*
Mountain-Flax. See Flax
Mountain-fringe. *Adlumia cirrhosa*
Mountain-green. See Mountain-pride
Mountain-lover, Canby's. *Pachystigma Canbyi*
Mountain-Marigold. See Marigold
Mountain - pride, or Mountain - green. *Spathelia simplex*
Mountain-Rocket. See Rocket
Mountain-Sage. *Teucrium Scorodonia*
Mountain-snow. *Arabis albida*
Mountain-Sorrel. *Oxyria reniformis*
Mountain-Spikenard. See Spikenard
Mountain-sweet. *Ceanothus americanus*
Mountain - Wine - berry. *Aristotelia Colensoi*
Mournful-Widow. *Scabiosa atropurpurea*
Mourning-Widow. *Geranium phæum*
Mouse-bane. *Aconitum myoctonum*
Mouse-Barley. See Barley, Mouse
Mouse-chop. *Mesembryanthemum murinum*
Mouse-ear. *Hieracium Pilosella* and *Cerastium triviale*
Bastard. *Hieracium Pseudo-Pilosella*
Golden. *Hieracium aurantiacum*
Mouse - ear - Chick - weed. The genus *Cerastium*
Alpine. *Cerastium alpinum*
Bieberstein's. *Cerastium Biebersteinii*
Boissier's. *Cerastium Boissieri*
Broad-leaved. *Cerastium latifolium*
Common Woolly. *Cerastium tomentosum*
Glacier. *Cerastium glaciale*
Large-flowered. *Cerastium grandiflorum*
Mouse-ear-Scorpion-grass. *Myosotis palustris*
Mouse-Grass. See Grass
Mouse-Pea. See Pea
Mouse-tail. *Myosurus minimus*
Mouse-tail Grass. See Grass
Mouse-thorn. *Centaurea myacantha*
Mousseron, Autumn. *Agaricus prunulus*
Moutan (the Tree-Peony). *Pæonia Moutan*
Mouth-root. *Coptis trifoliata*
Moving-plant. *Desmodium gyrans*
Moxa. *Polyporus fomentarius* and *Artemisia chinensis*
Chinese, or Japanese. *Artemisia Moxa* and *A. chinensis*
Muccaady-tree. *Schrebera swietenioides*
"Much-good." An old name for *Athamantha Cervaria*
Muck-weed. *Chenopodium album* and other species
Mudar-plant. *Calotropis gigantea* and *C. procera*
Mud-wort, or Mud-weed. *Limosella aquatica* and *Heloseiadium inundatum*

Mug-weed. *Galium Cruciata*
Mugwet, or Mugget. *Asperula odorata* and *Convallaria majalis*
Mug-wort, Common. *Artemisia vulgaris*
E. Indian. *Artemisia hirsuta*
Silvery. *Artemisia argentea*
Western. *Artemisia Ludoviciana*
W. Indian. *Parthenium Hysterophorus*
Mulberry-tree. The genus *Morus*
Australian. *Hedycarya Pseudo-morus*
Black-fruited. *Morus nigra*
Cock-spur. *Morus Calear-galli*
Columba. *Morus alba var. Columbassa*
Common. *Morus nigra*
Constantinople. *Morus Constantinopolitana*
Dandolo's. *Morus alba var. Morettiana*
Dwarf. *Morus alba var. pumila*
Dyer's. *Morinda tinctoria*
E. Indian. *Morus indica* and *Morinda citrifolia*
French. *Callicarpa americana*
Italian. *Morus alba var. italica*
Large-leaved. *Morus alba var. macrophylla*
Many-stemmed. *Morus alba var. multicaulis*
Mauritian. *Morus mauritiana*
Membranous. *Morus alba var. membranacea*
Native, of Victoria. *Hedycarya angustifolia*
New Zealand. *Entelea arborescens*
Paper. *Broussonetia papyrifera*
Red-fruited. *Morus rubra*
Roman. *Morus alba var. romana*
Rosy-leaved. *Morus alba var. rosea*
Rough-leaved. *Morus scabra* (*M. canadensis*)
Tartarian. *Morus tatarica*
White-fruited. *Morus alba*
White-fruited, Chinese. *Morus alba var. sinensis*
Mullein. The genus *Verbascum*
Annual. *Verbascum Boerhaavii*
Black-rooted. *Verbascum nigrum*
Borage-leaved. *Verbascum Myconi*
Cretan. *Celsia cretica*
Great-flowered. *Verbascum grandiflorum*
Hoary. *Verbascum pulverulentum*
Long-flowered. *Verbascum macranthum*
Moth. *Verbascum Blattaria*
Nettle-leaved. *Verbascum Chaixii*
Olympian. *Verbascum olympicum*
Purple-flowered. *Verbascum phœniceum*
Rosette. *Ramondia pyrenaica*
Shepherd's-club. *Verbascum Thapsus*
Snow-white. *Verbascum nivcum*
Tall. *Verbascum vernale*
Twiggy. *Verbascum virgatum*
White-flowered. *Verbascum Lychnitis*
Woolly. *Verbascum phlomoides*
Mullein-Dock. *Verbascum Thapsus*
Mummy-case-wood, Egyptian. Supposed to be the timber of *Cordia Myxa*
Mummy-Wheat. See Wheat
Mundi-root. *Chlorocodon Whitei*
Mungeet, or Bengal Madder. *Rubia cordifolia*

Munshock. *Vaccinium Vitis-Idæa*
Murrain-berries. *Tamus communis* and *Bryonia dioica*
Murrain Grass. See Grass
Murram. See Marram.
Mushroom. Various species of *Fungi*, especially the genus *Agaricus*
  Bishop's-Mitre. *Helvella Mitra*
  Brown Warty. *Agaricus rubescens*
  Chantarelle. *Cantharellus cibarius*
  Clouded. *Agaricus nebularis*
  Common. *Agaricus campestris*
  Fairy. See Toad-stool
  Fairy-ring. *Marasmius oreades* and *M. urens*
  Hedge. A large-sized variety of *Agaricus arvensis*
  Hedge-hog. See Mushroom, Spiny
  Horse. *Agaricus arvensis*
  Imperial. *Agaricus Cæsareus*
  Luminous. *Agaricus olearius*
  Maned. *Coprinus comatus*
  Mitre. *Helvella crispa*
  Orange-milk. *Lactarius deliciosus*
  Ox. A large-sized variety of *Agaricus campestris*
  Oyster. *Agaricus ostreatus*
  Parasol. *Agaricus procerus*
  Plum. *Agaricus prunulus* and *A. Orcella*
  Polish. *Boletus edulis*
  Red-fleshed. *Agaricus rubescens*
  St. George's. *Agaricus gambosus*
  Scaly. *Agaricus procerus*
  Scarlet, of Malta. *Cynomorium coccineum*
  Snow-ball. *Agaricus arvensis*
  Spiny, or Hedge-hog. *Hydnum repandum*
  Viscid White. *Hygrophorus virgineus*
Musk, or Musk-plant. *Mimulus moschatus* and *Erodium moschatum*
  Large-flowered. *Mimulus moschatus var. Harrisoni*
  Scarlet-flowered. *Mimulus cardinalis*
  Wild. *Erodium cicutarium*
Musk-deer Plant. *Limonia acidissima*
Musk Mallow. See Mallow
Musk Okro. *Hibiscus Abelmoschus*
Musk Orchis. *Herminium Monorchis*
Musk-root. *Adoxa Moschatellina*
Musk Thistle. *Carduus nutans*
Musk-tree, Silver-leaved. *Eurybia argophylla* (*Aster argophyllus*)
Musk-wood, Australian. *Eurybia argophylla* (*Aster argophyllus*)
  Jamaica. *Guarea Swartzii* and *Trichilia moschata*
  Tasmanian. *Eurybia argophylla*
Musquash-root. *Cicuta maculata*
Mustard. Various species of *Sinapis*
  Black, Brown, or Grocer's. *Sinapis nigra*
  Boor's. *Thlaspi arvense*
  Buckler. The genus *Biscutella*; also applied to *Clypeola Jonthlaspi*
  Candy. An old name for *Æthionema saxatile*
  Churl's. *Lepidium campestre*
  Clown's. *Iberis amara*
  Corn, or Wild. *Sinapis arvensis*
  Cultivated. *Sinapis alba*
  Dish. *Thlaspi arvense*
  Garlic. *Alliaria officinalis*
  Hedge. *Sisymbrium officinale*
  Hedge, of the W. Indies. *Ambrina ambrosioides*
  Hill. *Bunias orientalis*
  Mithridate. See Mithridate Mustard
  Tansy. *Sisymbrium canescens*
  Treacle. *Lepidium campestre* and *Erysimum cheiranthoides*
  White, or Salad. *Sinapis alba*
  Wild. *Sinapis arvensis*
Mustard-shrub. *Capparis ferruginea*
"**Mustard-tree**," of Scripture. Supposed to be the common Mustard-plant (*Sinapis alba* or *nigra*), which in Palestine is said to attain the height of 10 to 15 feet. The late Dr. Royle endeavoured to prove that *Salvadora persica* was meant, but this tree does not grow in Galilee
Mutton-chops, or Mutton-tops. The young shoots of *Galium Aparine*
Myall-wood. The wood of *Acacia homalophylla, A. pendula, A. acuminata,* and *A. glaucescens*
Myrobalan. The fruit of various species of *Terminalia*
Myrobella Plum. See Plum
Myrrh, British. *Myrrhis odorata*
Myrrh-seed. *Myrospermum pubescens*
Myrrh-tree. *Balsamodendron Myrrha*
  Abyssinian. *Acacia Sassa*
  False. *Amyris commiphora* (*Balsamodendron Roxburghii*)
Myrtle. The genus *Myrtus*
  African. *Myrtus africana*
  Australian. *Acmena floribunda*
  Australian, Slender-leaved. *Myrtus tenuifolia*
  Australian, Three-nerved. *Myrtus trinervis*
  Backhouse's. *Backhousia myrtifolia*
  Beaufort's. *Beaufortia decussata*
  Bird's-nest, or Cock's-comb. A variety of *Myrtus communis*
  Bog, Moor, Devonshire, or Dutch. *Myrica Gale*
  Box-leaved. *Myrtus communis var. Tarentina*
  Broad-leaved, or Dutch. *Myrtus communis var. belgica.* (See also Myrtle, Roman)
  Burren. *Arctostaphylos Uva-ursi*
  Candleberry. *Myrica Gale*
  Candleberry, Azorean. *Myrica Faya*
  Candleberry, Sharp - toothed - leaved. *Myrica arguta*
  Cock's-comb. See Myrtle, Bird's-nest
  Common. *Myrtus communis*
  Cranberry. *Myrtus Nummularia*
  Crape. The genus *Lagerstrœmia*, especially *L. indica*
  Devonshire. See Myrtle, Bog
  Double-flowered. *Myrtus communis var. belgica fl.-pl.*
  Dutch. See Myrtle, Bog, and Myrtle, Broad-leaved
  Edible-fruited. *Myrtus* (*Myrcianthes*) *edulis*

Myrtle, Fringe. The genus *Chamælaucium*
Fruiting. *Eugenia Ugni*
Golden, of Queensland. *Xanthostemon (Metrosideros) chrysantha*
Gum. The genus *Angophora*
Jew's. See Jew's-Myrtle
Juniper. The genus *Verticordia*
Moor. See Myrtle, Bog
Moreton Bay. *Myrtus melastomoides*
Nutmeg. *Myrtus communis var. lusitanica acuta*
Orange-leaved. *Myrtus communis var. bætica*
Otaheite. *Securinega nitida (Lithoxylon nitidum)*
Peach. The genus *Hypocalymma*
Portugal. *Myrtus communis var. lusitanica*
Roman, or Broad-leaved. *Myrtus communis var. romana*
Rosemary-leaved. *Myrtus communis var. mucronata*
Sand. *Leiophyllum (Ledum) buxifolium*
Scarlet-berried Scrub. *Nelitris ingens*
Shepherd's. *Ruscus aculeatus*
Small-leaved. *Myrtus Tarentina* and *M. communis var. mucronata*
Slender-leaved. *Myrtus tenuifolia*
South Sea. The genus *Leptospermum*
Spotted-leaved. *Myrtus communis var. maculata*
Swan River. *Hypocalymma robusta*
Tasmanian. *Fagus Cunninghamii*
Tasmanian Mountain. *Phebalium montanum*
Upright Italian. *Myrtus communis var. italica*
Variegated. *Myrtus communis variegata*
Wax. *Myrica cerifera*
Wax, Californian. *Myrica Californica*
W. Indian. The genus *Eugenia*
White-berried. *Myrtus communis var. leucocarpa*
Wild. *Ruscus aculeatus*
Woolly. *Myrtus tomentosa*
Myrtle-Flag. *Acorus Calamus*
Myrtle-Grass. See Grass
Myrtle-Sedge. *Acorus Calamus*
Myrtle-Spurge. *Euphorbia Lathyris*
Mysore Thorn. See Thorn
"Mysterious Plant." *Daphne Mezereum*

" Nagkushur," or **"Nagkesur."** The flowers of *Mesua ferrea*
Naiad. The genus *Naias*
Slender. *Naias flexilis*
Nail-wort. The genus *Paronychia*, *Draba verna*, and *Saxifraga tridactylis*
Thyme-leaved. *Paronychia serpyllifolia*
Naked-Ladies. *Colchicum autumnale*
Nancy-Pretty. *Saxifraga umbrosa*
Nap-at-noon. *Tragopogon pratensis*
Narawael-shrub. *Nararelia zeylanica*
Nard. Various aromatic plants, chiefly of the Valerian tribe
Common. *Nardus stricta*
Wild. *Asarum europæum*
Nardoo-plant. *Marsilea macropus (M. hirsuta, M. salvatrix)*
Narcissus. (See also Daffodil)

Narcissus, Mock. The genus *Queltia*
Of Japan. *Nerine sarniensis*
Poet's. *Narcissus poeticus*
Polyanthus. *Narcissus Tazetta*
Nase-berry, Nees-berry, or Nis-berry, Tree. *Achras Sapota*
Nase-berry-bully Tree. *Achras Sideroxylon*
Broad-leaved. *Lucuma multiflora*
Nasuta, or Tong-pang-chong-shrub. *Justicia Nasuta*
Nasturtium, Canary-bird. *Tropæolum peregrinum (T. aduncum)*
Common Garden. *Tropæolum majus, T. minus*, and *T. atro-sanguineum*
Dwarf. *Tropæolum minus*
Flame-flowered. *Tropæolum speciosum*
Peruvian. *Tropæolum tuberosum*
Tall. *Tropæolum majus*
Tuberous-rooted. *Tropæolum tuberosum*
Yellow Rock. *Tropæolum polyphyllum*
Natchnee-plant. *Eleusine coracana*
Natural Grasses. See Grasses
Naughty-man's-Cherry. *Atropa Belladonna*
Navel-wort. The genera *Omphalodes* and *Cotyledon*
Common. *Cotyledon Umbilicus*
Rock. *Omphalodes Luciliæ*
Venus's. *Omphalodes linifolia*
Navew. *Brassica campestris*
Necklace-tree. *Ormosia dasycarpa*
Neck-weed. *Veronica peregrina*; also an old name for Hemp
Nectarine-tree. *Amygdalus persica var. lævis*
Needle, Adam's. See Adam's-needle
Shepherd's, Adam's, Beggar's, Clock, Crake, Crow, Poke, Puck, Pink, Tailor's, Venus's, or Witches'. *Scandix Pecten-Veneris*
Needle-and-Thread, Adam's. *Yucca filamentosa*
Needle-Chervil. *Scandix Pecten-Veneris*
Needle-Furze. *Genista anglica*
Nees-berry. See Nase-berry
Negro's-head. *Phytelephas macrocarpa*
Negro's-slippers. *Euphorbia myrtifolia*
"Nei-Nei"-plant. *Dracophyllum Traversi*
Neem-tree. *Melia Azadirachta*
Neese-wort. See Sneeze-wort
Nelumbo. The genus *Nelumbium*
"Neminies." A corruption of *Anemones*
Nep, or Neps. *Nepeta Cataria*
Neroli-Oil-plant. *Citrus Aurantium var. amara*
Nerve-root. *Cypripedium pubescens*
Net-bush, Australian. The genus *Calothamnus*
Nettle. The genus *Urtica*; the name is also applied to some other plants, as the *Lamiums*, &c.
American. *Urtica gracilis* and *U. chamædryoides*
False. *Bœhmeria cylindrica*
Blind. See Blind-Nettle
Chili. Various species of *Loasa*
Chinese Cotton. *Urtica (Bœhmeria) nivea*
Common, or Stinging. *Urtica dioica*
Dead. See Dead-Nettle

Nettle, Deaf. See Deaf-Nettle
Deye. See Deye-Nettle
Dog. See Dog-Nettle
Dumb. *Lamium album*
Flame. The genus *Coleus*
Giant, of N. S. Wales. *Urtica (Laportea) Gigas*
Hedge. *Stachys sylvatica*
Hedge, Great Spanish. *Prasium majus*
Hedge, Small Spanish. *Prasium minus*
Hemp. See Hemp-Nettle
Hemp-leaved. *Urtica cannabina*
Horse. *Solanum carolinense*
Roman. *Urtica pilulifera*
Small. *Urtica urens*
Stingless. The genus *Pilea*
Variegated. *Lamium maculatum*
White. See Dead-Nettle, White
Wood, of N. America. *Laportea canadensis*
Nettle-tree. The genus *Celtis*
American. *Celtis occidentalis*
Australian. *Urtica Gigas, U. photiniæphylla*, and *U. moroides*
Caucasian. *Celtis caucasica*
Chinese. *Celtis sinensis*
Dwarf. *Celtis pumila*
Eastern. *Celtis orientalis*
European. *Celtis australis*
File-leaved. *Celtis Lima*
Jamaica. *Celtis micrantha*
Prickly. *Celtis aculeata*
Small-flowered. *Celtis parviflora*
Smooth. *Celtis lævigata*
Thick-leaved. *Celtis crassifolia*
Tournefort's. *Celtis Tourneforti*
New-Chapel-flower. *Orobanche major*
"New-Year-flower," of the Chinese. Various species of *Narcissus*
Nicaragua-wood. *Cæsalpinia echinata*
Nicker-tree, Common. *Guilandina Bonducella*
Sumatra. *Guilandina Bonduc*
Niger-seed. The seed of *Guizotia oleifera*
Nigger's-cord. The genus *Antidesma*
Nightshade. The genus *Solanum*
Beet-leaved. *Solanum betaceum*
Bell-flowered. *Solanum campanulatum*
Bittersweet, or Woody. *Solanum Dulcamara*
Black-spined. *Solanum sodomeum*
Broad-leaved. *Solanum stramonifolium*
Canary. *Solanum Vespertilio*
Climbing. *Solanum scandens* and *S. radicans*
Deadly. *Atropa Belladonna*
Enchanter's. *Circæa Lutetiana*
Fringed. *Solanum ciliatum*
Garden. *Solanum nigrum*
Hairy. *Solanum hirsutum*
Hybrid Guinea. *Solanum hybridum*
Jasmine. *Solanum jasminoides*
Large Black-berried. *Solanum guineense*
Malabar. See Malabar-Nightshade
Purple-spined. *Solanum campechiense*
Nipple. *Solanum mammosum*
Red. *Physalis Alkekengi*
Red-spined. *Solanum igneum*
Scarlet-berried. *Solanum coccineum*
Star-Capsicum. *Solanum Capsicastrum*

Nightshade, Texan. *Solanum Texanum*
Three-leaved. The genus *Trillium*
Tree. *Solanum arboreum*
Two-coloured. *Solanum discolor*
White-eyed. *Solanum marginatum*
Woody. See Nightshade, Bittersweet
Yellow-berried. *Solanum flavum, S. villosum*, and *S. xanthocarpum*
Nightingales. *Geranium Robertianum* and *Arum maculatum*
Nikau Palm-tree. *Kentia sapida*
Nim-Bark, or Margosa-Bark-plant. *Melia indica*
Nimble Will. *Muhlenbergia diffusa*
Ninety-knot. *Polygonum aviculare*
Niopo-tree. *Piptadenia peregrina*
Nipa-tree. *Nipa fruticans*
Nipple-wort. *Lapsana communis*
Dwarf. *Arnoseris pusilla*
Nis-berry. See Nase-berry
Nit. Scotch for *Nut*
Nit-grass. See Grass
Nitre-bush. *Nitraria Billardieri, N. tridentata*, and *N. Schoberi*
Nitta-tree. *Parkia africana*
Noah's-Ark. *Cypripedium pubescens*
Nonda-tree. *Parinarium Nonda*
Nondo. *Ligusticum actæifolium*
None-so-pretty. *Saxifraga umbrosa*
None-such. *Lychnis chalcedonica*
Black. *Medicago lupulina*
White. *Lolium perenne*
Noon-tide, or Noon-flower. *Tragopogon pratensis*
Nopal-plant. *Nopalea (Opuntia) coccinellifera*
Nose-bleed. *Achillea Millefolium*
Nose-burn-tree. *Daphnopsis tenuifolia*
Nosegay-tree, Red. *Plumieria rubra*
White. *Plumieria alba*
Notch-weed. *Chenopodium Vulvaria*
Nubk-tree, of Palestine. *Zizyphus Spina-Christi*
Nunnari-Root. See Sarsaparilla, E. Indian
Nuns. *Orchis Morio* and *Impatiens glandulifera*
Nut, or Nut-tree, Acajou. See Nut, Cashew
Anthony. See Nut, St. Anthony's
Antilles. *Omphalea triandra*
Avellano, of Chili. *Guevina Avellana*
Bag. See Nut, Bladder
Bambarra Ground. *Voandzeia subterranea*
Ban. See Ban-nut
Barbadoes. *Curcas purgans*
Barcelona. See Nut, Spanish
Beazor. See Nut, Bonduc
Bedda. *Terminalia Bellerica*
Beetle. See Beetle-nut-plant
Ben. *Moringa pterygosperma*
Betel. *Areca Catechu*
Bladder, or Bag. See Bladder-nut
Bonduc, or Beazor. *Guilandina Bonduc*
Bog. See Bog-nut
Bomah. *Pycnocoma macrophylla*
Brazil. *Bertholletia excelsa*
Buffalo, or Elk. *Pyrularia (Hamiltonia) oleifera*
Cahoun. *Attalea Cohune*

Nut, Candle. *Aleurites moluccensis* and *A. triloba*
Cashew, or Acajou. *Anacardium occidentale*
Castanha, or Brazil. *Bertholletia excelsa*
Cat. See Cat-nut
Chop. *Physostigma venenosum*
Cluster. *Corylus Avellana var. glomerata*
Cob. *Corylus Avellana var. grandis*
Cob, Jamaica. *Omphalea triandra*
Cocoa. *Cocos nucifera*
Cocoa, Double, or Sea. *Lodoicea Seychellarum*
Cola, Kolla, or Goora. *Cola acuminata*
Coquilla. *Attalea funifera*
Cosford, or Thin-shelled. *Corylus Avellana var. tenuis*
Earth. See Earth-nut
Eboe. *Dipterix oleifera*
Edible Rush. *Cyperus esculentus*
Elk. See Nut, Buffalo
Fare. See Fare-nut
Filbert. See Filbert
"Fisticke." An old name for the Pistachio-nut
French. See French Nut
Goora. *Cola acuminata*
Ground. See Earth-nut
Guiana Snake. *Ophiocaryon paradoxum*
Hale. See Hale-nut
Hara. *Terminalia citrina*
Hare. See Hare-nut
Hazel. *Corylus Avellana*
Hazel, American. *Corylus americana*
Hazel, American Beaked. *Corylus rostrata*
Hog. See Hog-nut
Hog-pea. *Amphicarpaea monoica*
Illinois. *Carya oliviformis*
Ivory. *Phytelephas macrocarpa*
Jesuit's. *Trapa natans*
Jove's. See Jove's Nuts
Karaka. *Corynocarpus laevigata*
Keena. *Calophyllum Calaba*
Kisky Thomas. *Carya alba*
Kola, or Kolla. See Nut, Cola
Kundoo, or Mote. *Carapa Touloucouna*
Lady. See Lady-nut
Lambert's, Large Bond, Spanish, or Toker. *Corylus Avellana var. Lamberti*
Large Bond. See Nut, Lambert's
Levant. *Anamirta Cocculus*
Little Coker. The fruit of *Jubaea spectabilis*
Lumbang. *Aleurites triloba*
Lycoperdon. *Elaphomyces*
Malabar. *Adhatoda vasica*
Manilla. *Arachis hypogaea*
Marany. See Nut, Marking
Marking, or Marany. The Seed of *Semecarpus Anacardium*
Marsh. The Seed of *Semecarpus Anacardium*
Meadow. See Meadow-Nut
Mocker. See Mocker-nut
Monkey. *Arachis hypogaea* and the seeds of *Anacardium*
Mote. See Nut, Kundoo
Mt. Atlas. *Corylus algeriensis*
Oak. See Oak-nuts

Nut, Oil. *Pyrularia (Hamiltonia) oleifera* and *Ricinus communis*
Pará, or Brazil. *Bertholletia excelsa*
Pecan, or Illinois. *Carya oliviformis*
Physic. *Jatropha Curcas (Curcas purgans)*
Pig. See Pig-nut
Pistacia, or Pistachio. *Pistacia vera*
Quandang. *Fusanus acuminatus*
Queensland. *Macadamia (Helicia) ternifolia*
Ravensara. *Agathophyllum aromaticum*
Rush. *Cyperus esculentus*
Rush, of N. America. *Cyperus rotundus var. Hydra*
St. Anthony's. *Staphylaea pinnata* and *Bunium flexuosum*
Sapucaia. *Lecythis Zabucajo*
Sardian. *Castanea vesca*
Sassafras. *Nectandra Puchury*
Singhara. *Trapa bicornis* and *T. bispinosa*
Snake. *Ophiocaryon paradoxum*
Soap. *Mimosa abstergens*
Souari. See Nut, Suwarrow
Spanish, or Barcelona. *Corylus Avellana barcelonensis*
Springfield. *Carya sulcata*
Suwarrow, or Souari. *Caryocar nuciferum* and *C. butyrosum*
Taqua. *Phytelephas macrocarpa*
Toker. See Nut, Lambert's
Vare. See Fare-nut
Vegetable-Ivory. *Phytelephas macrocarpa*
Vomit. *Strychnos Nux-vomica*
Water. Various species of *Trapa*
Water-filter. *Strychnos potatorum*
Welsh. *Juglans regia*
W. Australian. *Santalum cygnorum*
Wood. *Corylus Avellana*
Nut-Bush. *Corylus Avellana*
Nut-gall-tree, Chinese or Japanese. *Rhus semialata*
Nut-galls. Galls produced by insects on the bark of *Quercus infectoria*
Nut-grass. See Grass
Nut-Palms. Catkins of *Corylus Avellana*
Nut-rush. See Rush
Nutmeg-tree. *Myristica moschata*
Ackawai. *Acrodiclidium camara*
American, Calabash, or Jamaica. *Monodora Myristica*
Australian. Various species of *Cryptocarya*
Brazilian. *Cryptocarya moschata*
Calabash. See Nutmeg, American
Californian. *Torreya Myristica*
Clove, or Madagascar. *Agathophyllum aromaticum*
Fijian. *Myristica castanaefolia*
Flowering. *Leycesteria formosa*
Jamaica Calabash. *Monodora Myristica*
Long. *Myristica fatua*
Madagascar. See Nutmeg, Clove
Male. *Myristica tomentosa*
Mexican. *Monodora Myristica*
Peruvian. *Laurelia sempervirens*
Plume. *Atherosperma moschata*
Queensland. *Myristica insipida*
Santa Fé. *Myristica Otoba*
Tallow. *Myristica sebifera*

and Foreign Plants, Trees, and Shrubs. 95

Nutmeg, Wild. *Myristica fatua* and *M. tomentosa*
Nutmeg-wood. The wood of *Borassus flabelliformis*
Nux-vomica-tree. *Strychnos Nux-vomica*
Oadal, or Oo'dhall, tree. *Sterculia villosa*
Oak. The genus *Quercus*
Abram's, of Mamre. *Quercus pseudo-coccifera*
African. *Oldfieldia africana* and *Laurus bullata*
African Evergreen. *Cliffortia ilicifolia*
Agnostus-leaved. *Quercus agnostifolia*
Aleppo Gall. *Quercus lusitanica var. infectoria*
American Black. *Quercus tinctoria*
American Live. *Quercus virens*
American Scrub. *Quercus Catesbæi*
American "Spanish." *Quercus falcata*
American Swamp Spanish. *Quercus palustris*
Anderson's. *Quercus Andersoni*
Barbary. *Quercus Ballota*
Barren. *Quercus nigra*
Barren Scrub. *Quercus Catesbæi*
Bartram's. *Quercus coccinea var. tinctoria*
Bay. *Quercus sessiliflora*
Bear, or Black Scrub. *Quercus ilicifolia* (*Q. Banisteri*)
Belote. *Quercus Ballota*
Black. *Quercus Robur*
Black Jack. *Quercus nigra*
Black Jack, Fork-leaved. *Quercus Catesbæi*
Black Scrub. See Oak, Bear
Blue Jack. *Quercus cinerea*
Botany Bay Forest. *Casuarina torulosa*
Buerger's. *Quercus Buergeri*
Burr. *Quercus macrocarpa*
Californian Blue. *Quercus Douglasii*
Californian "Desert." *Quercus Wislizeni var. frutescens*
Californian Evergreen White. *Quercus oblongifolia*
Californian Live. *Quercus chrysolepis* and *Q. oblongifolia*
Californian Lobed-leaved. *Quercus lobata*
Californian Mountain White. *Quercus Douglasii*
Cappadocian. *Ambrosia maritima* (*Ambrina ambrosioides*)
"Chapparal." *Quercus Breweri*
Chestnut. *Quercus sessiliflora*
Chestnut, American. *Quercus Prinus*
Chestnut, American Dwarf. *Quercus Prinus var. humilis*
Chestnut, American Rock. *Quercus Prinus var. monticola*
Chestnut, American Yellow. *Quercus Prinus var. acuminata*
Chinquapin. *Quercus Prinus var. humilis*
Common (Long-flower-stalked). *Quercus Robur pedunculata*
Common (Stalkless-flowered). *Quercus Robur sessiliflora*
Cork. *Quercus Suber*
Cypress. *Quercus pedunculata fastigiata*
"Desert," of California. *Quercus Wislizeni var. frutescens*
Devonshire, Exeter, or Lucombe. *Quercus Cerris var. Lucombeana* (*Q. exoniensis*)

Oak, Dominica. *Ilex sideroxyloides*
Dog. *Acer campestre*
Durmast. *Quercus pubescens*
Dyer's. *Quercus tinctoria*
E. Indian, or Teak. *Tectona grandis*
Enceno. *Quercus agrifolia*
Evergreen, or Holm. *Quercus Ilex*
Exeter. See Oak, Devonshire
Female. *Quercus pedunculata*
"French." *Catalpa longissima* and *Bucida Buceras*
Fulham. *Quercus Cerris var. Fulhamensis*
Golden. *Quercus pedunculata Concordia*
Golden, of Cyprus. *Quercus alnifolia*
Golden-cup. *Quercus chrysolepis*
Gray. *Quercus coccinea var. ambigua*
He. *Casuarina stricta*
Himalayan. See Oak, Woolly
Holm, or Evergreen. *Quercus Ilex*
Holly-leaved. *Quercus Gramuntia*
Hungarian. *Quercus conferta*
Iron. *Quercus Cerris* and *Q. obtusiloba*
Italian. *Quercus esculus*
Japanese. *Quercus glabra*
Japanese Silkworm. *Quercus serrata*
"Jerusalem." *Chenopodium Botrys*
Kermes. *Quercus coccifera*
Laugh-lady. A variety of *Quercus pedunculata*
Laurel. *Quercus laurifolia* and *Q. imbricaria*
Lea's. *Quercus Leana*
Lucombe. See Oak, Devonshire
Lucombe, New. *Quercus Cerris var. Lucombeana crispa*
Maiden. See Oak, White
Male. *Quercus sessiliflora*
Mirbeck's. *Quercus Mirbecki*
Mongolian. *Quercus mongolica* and *Q. dentata*
Mossy-cup. *Quercus Cerris*
Mossy-cup, White. *Quercus macrocarpa*
Mount Lebanon. *Quercus Libani*
Nepaul. *Quercus lanuginosa*
New Zealand. *Alectryon excelsum* and *Knightia excelsa*
Nutgall. *Quercus lusitanica var. infectoria*
"Of Cappadocia." *Ambrosia maritima*
"Of Jerusalem," or "of Paradise." *Chenopodium Botrys*
Over-cup. *Quercus lyrata* and *Q. macrocarpa*
Pin. *Quercus palustris*
Poison. *Rhus Toxicodendron*
Post. *Quercus obtusiloba* (*Q. stellata*)
"Quebec." *Quercus alba*
Quercitron. *Quercus tinctoria*
Ragnal. *Quercus Cerris var. Ragnal*
Red. *Quercus rubra*
River, of N. S. Wales. *Casuarina leptoclada*
Rocky Mountain Scrub. *Quercus undulata*
Rough, or Box, White. *Quercus obtusiloba* (*Q. stellata*)
Running. *Quercus sericea*
St. Domingo. *Catalpa longissima*
Scarlet. *Quercus coccinea*
"Sea." *Fucus vesiculosus*

Oak, She. The genus *Casuarina*
She, Coast. *Casuarina quadrivalvis*
She, Desert. *Casuarina glauca*
She, Erect. *Casuarina tuberosa*
She, Tasmanian. *Casuarina quadrivalvis*
Shingle. *Quercus imbricaria*
Sierra Leone Scrubby. *Lophira alata*
Silky, or Silk-bark. *Grevillea robusta*
Stone. *Lithocarpus javensis*
Striped. *Quercus striata*
Swamp, Australian. *Casuarina paludosa*, *C. suberosa*, and *C. equisetifolia*
Swamp Post, of America. *Quercus lyrata*
Sweet Acorn. *Quercus Ballota*
Truffle. *Quercus pubescens*, *Q. Robur*, and *Q. lanuginosa*
Turkey. *Quercus Cerris*
Turkey, American. *Quercus Catesbæi*
Upland Willow. *Quercus cinerea*
Vallonea, or Velani. *Quercus Ægilops*
Victorian Swamp. *Viminaria denudata*
Wainscot. *Quercus Cerris*
Water. *Quercus aquatica*
White, or Maiden. *Quercus sessiliflora*
White American. *Quercus alba*
White American Water. *Quercus lyrata*
White, of Norfolk Island. *Lagunaria Patersoni*
White, Sacramento. *Quercus lobata*
White Swamp. *Quercus bicolor*
Willow. *Quercus Phellos* and *Q. salicina*
Woolly, or Himalayan. *Quercus lanata*
Yellow-barked. *Quercus tinctoria*
Oak-apple-gall. See Gall
Oak-currant. See Currant-Gall
Oak-Fern. See Fern
Oak-nuts (Acorns). Fruit of *Quercus Robur*
Oak-root-gall. See Gall
Oak-spangles. Galls formed on Oak-leaves by an insect named *Neuroterus lenticularis*
Oak-wood, Green. Oak-timber impregnated with the spawn of *Peziza æruginosa*
Oar-weed, or Ore-weed. See Ore
Oat. The genus *Avena*
Animal. *Avena sterilis*
Bristle-pointed. *Danthonia strigosa*
Common Cultivated. Varieties of *Avena sativa*
False. *Arrhenatherum avenaceum*
Fly, or Hygrometric. *Avena sterilis*
Hygrometric. See Oat, Fly
Naked. *Avena nuda*
Perennial. *Avena pratensis*
Sea-side. The genus *Uniola*
Short. *Avena brevis*
Tartarian, or Tartary. *Avena orientalis*
Water. *Zizania aquatica*
W. Indian. *Pharus latifolius*
Wild. *Avena fatua* and *Bromus secalinus*
Wild, of N. America. *Avena striata* and *A. Smithii*
Oat-grass. See Grass
Oats, "Sea." See "Sea-oats"
Oblionker-tree. *Æsculus Hippocastanum*
"Oca-plant." *Oxalis crenata* and *O. tuberosa*

Ochro, or Ochra, Plant. *Hibiscus esculentus*
Ogechee-Lime. *Nyssa candicans* (*N. capitata*)
"Ohelo," of the Sandwich Islands. *Vaccinium penduliflorum*
Oil-nut, American. *Pyrularia* (*Hamiltonia*) *oleifera*
Oil-Palm of Guinea. *Elæis guineensis*
Oil-plant, Adul, or Odal. *Sarcostigma Kleinii*
Allspice, or Pimento. *Eugenia Pimenta*
Almond. *Amygdalus communis*
Andiroba. See Oil-plant, Carap
Anise. *Pimpinella Anisum*
Asafœtida. *Narthex Asafœtida*
Aspic. See Oil-plant, Spike
Bacaba. *Œnocarpus Bacaba*
Balm *Melissa officinalis*
Bancoul. See Oil-plant, Lumbang
Bay. *Laurus nobilis*
Beech-nut. *Fagus sylvatica*
Ben. *Moringa pterygosperma*
Benné. *Sesamum indicum*
Bergamot. *Citrus Bergamia*
Birch-bark. *Betula alba*
Cade. *Juniperus Oxycedrus*
Cajeput, or Cajuput. *Melaleuca minor*
Camphor. *Laurus Camphora* and *Dryobalanops aromatica*
Carap, Crab, or Andiroba. *Carapa guianensis*
Caraway *Carum Carui*
Cardamom. *Elettaria Cardamomum*
Cashew-apple, or Cashew-nut. *Anacardium occidentale*
Cassia. *Cinnamomum Cassia*
Cassié. *Acacia Farnesiana*
Castanha. *Bertholletia excelsa*
Castor. *Ricinus communis*
Cebadilla. *Asagræa officinalis*
Cedar. *Abies Cedrus* and *Juniperus virginiana*
Cedrat. See Oil-plant, Citron
Chamomile. *Anthemis nobilis*
Chaulmoogra. *Gynocardia odorata*
Cherojee, or Cheeroonjee. *Buchanania latifolia*
Cherry. *Cerasus serotina*
Cherry-Laurel. *Cerasus Lauro-cerasus*
Cinnamon. *Cinnamomum Zeylanicum*
Citron, or Cedrat. *Citrus medica*
Citronelle. See Oil-plant, Lemon-grass
Clove. *Caryophyllus aromaticus*
Cocoa-nut. *Cocos nucifera*
Cocum, or Kokum. *Garcinia purpurea*
Cohune. *Attalea Cohune*
Colza. *Brassica campestris Napus*
Coondi. See Oil-plant, Kundah
Copaiba. Various species of *Copaifera*
Corooko. *Argemone mexicana*
Cotton-seed. Various species of *Gossypium*
Crab. See Oil-plant, Carap
Croton. *Croton Tiglium*
Cubeb. *Piper Cubeba* (*Cubeba officinalis*)
Cumaru. See Oil-plant, Tonquin
Cumin. *Cuminum Cyminum*
Dill. *Anethum graveolens*
Dilo. See Oil plant, Tamana

and Foreign Plants, Trees, and Shrubs.

Oil-plant, Domba. See Oil-plant, Poon-seed
Epie, or Mahowa-seed. *Bassia latifolia*
Ergot. *Claviceps purpurea*
Euphorbia. *Euphorbia Lathyris*
Exile. *Theretia nereifolia*
Fennel. *Fœniculum dulce* and *F. vulgare*
Florence. *Olea europæa*
Garlic. *Allium sativum*
Gentian. *Gentiana lutea*
Geranium. *Geranium odoratissimum* and *Pelargonium roseum*. See also Oil-plant, Grass
Gingelly, or Gingilie. *Sesamum indicum*
Grass, or Geranium. *Andropogon Schœnanthus*
Ground-nut. *Arachis hypogæa*
Hemp-seed. *Cannabis sativa*
Hop. *Humulus Lupulus*
Huts-yellow. *Guizotia oleifera*
Ilpa, Illipoo, or Illupie. *Bassia longifolia*
Jasmine, or Mogree. Various species of *Jasminum*
Jatropha. *Curcas purgans* and *C. multifidus*
Juniper. *Juniperus communis*
Kanari. *Canarium commune*
Katjang. *Arachis hypogæa*
Keena. A species of *Calophyllum*
Kekune. See Oil-plant, Lumbang
Keora. *Pandanus odoratissimus*
Khatzum. *Vernonia anthelmintica*
Khus-Khus, Cus-cus, or Vetti-ver. *Andropogon muricatus*
Kikuel. *Salvadora persica*
Kinka. *Vernonia anthelmintica*
Kokum. See Oil-plant, Cocum
Kossumba, or Safflower. *Carthamus tinctorius*
Kukin. See Oil-plant, Lumbang
Kundah, Coondi, or Tallin-Coonah. *Carapa guineensis* (*C. Touloucouna*)
Kurring, Kurunj, or Poonga. *Pongamia glabra*
Laurel. *Laurus nobilis*
Lavender. *Lavandula vera*
Lemon. *Citrus Limonum*
Lemon-grass, or Siri. *Andropogon citratus*
Lily. *Lilium candidum*
Limbolee. *Bergera* (*Murraya*) *Kœnigii*
Linseed. *Linum usitatissimum*
Lumbang, Kekune, Bancoul, or Kukin. *Aleurites triloba*
Mace. *Myristica moschata*
Macuja. *Acrocomia sclerocarpa*
Madia. *Madia sativa*
Mahoua, Mahowa-seed, or Yallah. *Bassia latifolia*
Male Fern. *Lastrea Filix-mas*
Mallee. *Eucalyptus oleosa*
Margosa, or Neem. *Melia Azadirachta*
Marjoram, or Origanum. *Origanum vulgare*
Marking-nut. *Semecarpus Anacardium*
Marmottes. *Prunus Brigantiaca*
Mezereon. *Daphne Mezereum*
Mogree. See Oil-plant, Jasmine
Mustard. *Sinapis nigra* and other species
Myrrh. *Balsamodendron Myrrha*

Oil-plant, Nahor. *Mesua ferrea*
Namur, or Nemaur. *Andropogon Schœnanthus*
Napala. *Curcas purgans*
Narcissus. *Narcissus odorus*
Narpaulah. A species of Croton
Neem. See Oil-plant, Margosa
Nemaur. See Oil-plant, Namur
Neroli. *Citrus Bigaradia* and *C. Aurantium*
Nut. *Corylus Avellana* and *Juglans regia*; also *Arachis hypogæa*
Nutmeg. *Myristica moschata*
Odal. See Oil-plant, Adul
Olive, or Sweet. *Olea europæa*
Onion. *Allium Cepa*
Oondee. See Oil-plant, Poon-seed
Orange. *Citrus Aurantiam* and *C. Bigaradia*
Origanum. See Oil-plant, Marjoram
Ouabe. *Omphalea diandra*
Palm. *Elæis guineensis* and *E. melanococca*
Palmarosa. *Andropogon Schœnanthus*
Pand. *Michelia Champaca*
Pandang. *Pandanus odoratissimus*
Patawa, or Patava. *(Enocarpus Batava*
Patchouli. *Pogostemon Patchouli.*
Penny-royal. *Mentha Pulegium*
Phoolwa. *Bassia butyracea*
Physic-nut. *Curcas purgans*
Pimento. See Oil-plant, Allspice
Pinhoën. *Curcas multifidus*
Pinnacottay. See Oil-plant, Poon-seed
Piquia. *Caryocar brasiliense*
Poonay. See Oil-plant, Poon-seed
Poonga. See Oil-plant, Kurring
Poongum. *Sapindus emarginatus*
Poon-seed, Poonay, Pinnacottay, or Oondee. *Calophyllum Inophyllum*
Pootungee. *Calophyllum spurium*
Poppy. *Papaver somniferum*
Portia-nut. *Thespesia populnea*
Potato. *Solanum tuberosum*
Provence. *Olea europæa*
Ram-til, or Valisaloo. *Guizotia oleifera*
Rape-seed. *Brassica Napus*
Rhodium. *Rhodorrhiza scoparia* and *R. florida*
Rose. *Rosa damascena, R. centifolia*, and other species
Rosemary. *Rosmarinus officinalis*
Rosin. Various species of Pine-tree (*Pinus*)
Rue. *Ruta graveolens*
Safflower. See Oil-plant, Kossumba
Sandal-wood, or Sander's-wood. *Santalum album*
Sapucaia. *Lecythis Zabucajo*
Sassafras. *Laurus Sassafras* and *Nectandra cymbarum*
Savin. *Juniperus Sabina*
Seed. Till, Poppy, and other Indian plants
Senna. *Cassia Senna*
Seringa. *Siphonia elastica*
Serpolet. *Thymus Serpyllum*
Sesamum. A variety of *Sesamum orientale*
Shanghae. *Brassica chinensis*
Siri. See Oil-plant, Lemon-grass
Soap-nut. *Sapindus emarginatus*

H

Oil-plant, Spear-mint. *Mentha viridis*
Spike, or Aspic. *Lavandula Spica* and *L. Stœchas*
Spikenard. *Andropogon Schœnanthus*
Spurrey. *Spergula sativa*
Star-Anise. *Illicium anisatum*
Sun-flower. *Helianthus annuus*
Sweet. See Oil-plant, Olive
Sweet Marjoram. *Origanum Marjorana*
Tallincoonah. See Oil-plant, Kundah
Tamana, or Dilo. *Calophyllum Inophyllum*
Tar. Various species of Pine-trees
Tea. *Camellia Sasanqua*
Teuss. *Arachis hypogœa*
Thistle. *Argemone mexicana*
Thyme. *Thymus vulgaris*
Tobacco. *Nicotiana Tabacum*
Tonquin. *Dipterix odorata*
Tuberose. *Polianthes tuberosa*
Tumika. *Diospyros Embryopteris*
Tung. *Aleurites cordata*
Turpentine. Various species of Pinus
Uggur. *Aquilaria Agallocha*
Valisaloo. See Oil-plant, Ram-til
Verbena. *Aloysia citriodora* and *Andropogon citratus*
Vetti-ver. See Oil-plant, Khus-Khus
Violet. *Viola odorata*
Walnut. *Juglans regia*
Winter-green. *Gaultheria procumbens*
Wood. *Dipterocarpus turbinatus* and *Chloroxylon Swietenia*
Worm-seed. *Ambrina anthelmintica*
Yallah. See Oil-plant, Mahoua
Yamadou. *Myristica sebifera*
Zachun. *Balanites œgyptiaca*
Zakkoum. *Elœagnus hortensis angustifolia*
Oil-seed. *Guizotia oleifera*, *Ricinus communis*, and *Camelina sativa*
"Oily-grain-plant." *Sesamum orientale*
Okro, or Gombo, Plant. *Hibiscus (Abelmoschus) esculentus*
Old-maid. *Vinca rosea*
Old-man. *Artemisia Abrotanum*, *Clematis Vitalba*, and *Rosmarinus officinalis*
Old-man's-Beard. *Clematis Vitalba*, *Saxifraga sarmentosa*, *Tillandsia usneoides*, and various species of *Equisetum*
Old-man's-Eye-brow. *Drosera binata*
Old-man's-Head. *Pilocereus senilis*
Old-woman. *Artemisia Absinthium* and *A. argentea*
Old-woman's-Bitter. *Picramnia Antidesma* and *Citharexylon cinereum*
Oleander. The genus *Nerium*
Common. *Nerium Oleander*
Dyer's. *Nerium tinctorium*
Sweet-scented. *Nerium odorum*
Oleaster. The genus *Elœagnus*
Bohemian. *Elœagnus angustifolia*
Silvery-leaved. *Elœagnus argentea*
Small-leaved. *Elœagnus parvifolia*
Variegated. *Elœagnus reflexa variegata*
Olibanum. See Frankincense-tree
Olive-bark Tree. *Bucida Buceras*
Queensland. *Olea paniculata*
Olive-tree. The genus Olea
Barbadoes, Wild. *Bontia daphnoides*

Olive-tree, Bastard, of Victoria. *Notelœa ligustrina*
Black, or Wild, of the W. Indies. *Bucida Buceras* and *B. capitata.* Also *Ximenia americana*
Botany Bay. *Olea apetala*
Californian. *Oreodaphne (Umbellularia) californica*
Chinese. *Canarium commune*
Cultivated. *Olea sativa*
False, of the Antilles. *Bontia daphnoides*
Holly-leaved. *Olea ilicifolia (Osmanthus ilicifolius)*
Negro's. *Terminalia Chebula*
Spurge. *Daphne Mezereum* and *Cneorum tricoccum*
Sweet-scented. *Olea (Osmanthus) fragrans*
Wild. *Olea Oleaster* and *Elœagnus angustifolius.* Also applied to *Daphne Thymelœa* and *Rhus Cotinus*
Wild, of India. *Olea dioica* and *Putranjiva Roxburghii*
Wild, of the W. Indies. See Olive, Black
Olive-wood, Australian. *Elœodendron australe* and *E. integrifolium*
East Indian. *Elœodendron orientale*
Omander-wood. *Diospyros Ebenaster*
Ombu-tree. *Pircunia dioica*
Ome-tree. A corruption of "Elm-tree"
Omime-root. *Plectranthus ternatus*
One-berry. *Paris quadrifolia*
One-blade, or One-leaf. *Smilacina bifolia*
Onion. *Allium Cepa*
Barbadoes. *Ornithogalum scilloides*
Bog. *Osmunda regalis*
Burn, or Potato. *Allium Cepa var. aggregatum*
Canada. See Onion, Tree
Catawissa. A variety of the Tree-Onion
Ciboul. See Onion, Welsh
Common. Varieties of *Allium Cepa*
Crow. *Allium vineale*
Gipsy. *Allium ursinum*
Himalayan. *Allium leptophyllum*
Potato. *Allium Cepa var. aggregatum*
Pearl. *Allium Ampeloprasum*
Purple-flowered. *Allium acuminatum*
Sea. *Urginea (Scilla) maritima* ; also applied to *Scilla verna*
Tree, or Canada. *Allium Cepa proliferum (Cepa bulbifera)*
Underground. *Allium Cepa var. aggregatum*
Welsh, or Ciboul. *Allium fistulosum*
Wild, American. *Allium cernuum*
Onion-grass, Onion-Couch, or Onion-Twitch. See Grass
Oo'dhall. See Ondal
Opera-girls, or Dancing-girls. *Mantisia saltatoria*
Ople-tree. *Viburnum Opulus*
Opopanax-plant. *Opopanax Chironium (Pastinaca Opopanax)*
Orache. *Atriplex hortensis*
Common, or Delt. *Atriplex patula*
Dog's. *Chenopodium Vulvaria*
Frosted. *Atriplex rosea*
Grass-leaved. *Atriplex littoralis*
Lesser Shrubby. *Atriplex portulacoides*

Orache, Red-leaved. *Atriplex hortensis rubra*
Stalked. *Atriplex pedunculata*
White. *Atriplex albicans*
Wild. Various species of *Atriplex* and *Chenopodium*
Orange-ball Tree. *Buddlea globosa*
Orange-flower-tree. *Philadelphus coronarius*
Mexican. *Choisya ternata*
Orange-grass. *Hypericum Sarothra*
Orange-leaf, Otago. *Coprosma lucida*
Orange Lily. See Lily
Orange-root. *Hydrastis canadensis*
Orange-thorn. The genus *Citriobatus*
Orange-tree. *Citrus Aurantium*
Australian. The genus *Citriobatus*
Bergamot. *Citrus Bergamia*
Bigarade, or Bitter. See Orange, Seville
Blood, or Maltese. *Citrus Aurantium var. melitensis*
Box-leaved. *Citrus buxifolia*
Jamaica. *Glycosmis citrifolia*
Maltese. See Orange, Blood
Mandarin. *Citrus nobilis var. major*
Mock. *Philadelphus coronarius*
Osage. *Maclura aurantiaca*
Queensland. *Citrus australis (C. Planchoni)*
Seville, Bigarade, or Bitter. *Citrus Aurantium var. amara*
Sumatra. *Murraya sumatrana*
Sweet. *Citrus Aurantium*
Tangerine. *Citrus nobilis var. minor* or *Tangeriana*
Wild, of the W. Indies. *Drypetes glauca*
Orchal, Orchel, Orchil, or Archall. *Roccella tinctoria*
Orchanet. A corruption of "Alkanet"
Orchard-grass. See Grass
Orchard-weed. *Anthriscus sylvestris*
Orchella-weed. Various species of *Roccella*
Orcheston Grass. See Grass
Orchid, Almond-scented. *Odontoglossum madrense*
Armadillo's-tail. *Oncidium Sprucei*
Bartram's Tree. *Epidendrum conopseum*
Boat-lip. The genus *Sophyglottis*
Butterfly. *Oncidium Papilio*
Citron-scented. *Odontoglossum citrosmum*
Coral-root. The genus *Corallorrhiza*
Cow-horn. *Schomburghia tibicinis*
Cowslip-scented. *Vanda furva*
Dove. The genus *Peristeria*
Dragon's-mouth. *Epidendrum macrochilum*
Earthy-scented. The genus *Geodorum*
Florida Tree. *Epidendrum conopseum* and *E. venosum*
Hair. The genus *Trichosma*
Helmet. The genus *Coryanthes*
"King of the Woods." *Anœctochilus setaceus*
Long-spurred. *Angræcum sesquipedale*
Medusa's-head. *Cirrhopetalum Caput-Medusæ*
Monk's-Cowl. The genus *Pterygodium*
Mouse-tail. *Dendrobium Myosurus*
Purple-lip. *Vanilla claviculata*

Orchid, Rattle-snake. The genus *Pholidota*
Scarlet-flowered. *Sophronitis grandiflora* and *Lælia harpophylla*
Spectral-flowered. *Masdevallia Chimæra*
Spread-eagle. *Oncidium carthaginense*
Table Mountain. *Disa grandiflora*
Violet-scented. *Odontoglossum Warneri*
Woman's-cap. The genus *Thelymitra*
Orchids, Queen of the. *Grammatophyllum speciosum*
Orchis, Adder's-mouth. The genus *Microstylis*
Ape. *Orchis Simia*
Aromatic. *Gymnadenia Conopsea*
Ash-coloured. *Orchis tephrosanthos*
Bee. *Ophrys apifera*
Bell. The genus *Codonorchis*
Bird's-head. *Ornithocephalus gladiatus*
Bird's-nest. *Neottia Nidus-avis*
Black Spider. *Ophrys arachnites*
Bog. *Malaxis paludosa*
Breeze-fly. *Ophrys tabanifera*
Brown. *Ophrys fusca*
Bug. *Orchis coriophora*
Bumble-bee. *Ophrys bombylifera*
Butterfly. *Habenaria chlorantha* and *H. bifolia*
Calypso. *Calypso borealis*
Crane-fly. *Tipularia discolor*
Cuckoo. *Orchis mascula*
Dark-flowered. *Orchis nigra (Nigritella angustifolia)*
Dog. The genus *Cynorchis*
Drone. *Ophrys fucifera*
Dwarf. *Orchis ustulata*
Elder-scented. *Orchis sambucina*
False. The genus *Platanthera*
Fen. *Liparis Læselii*
Fly. *Ophrys muscifera*
Fool. *Orchis Morio*
Fragrant. *Gymnadenia Conopsea* and *G. odoratissima*
Frog. *Habenaria viridis*
Grass-Pink. *Calopogon pulchellus*
Great Fringed. *Habenaria (Platanthera) fimbriata*
Green-Man. See Orchis, Man
Green-winged. *Orchis Morio*
Guernsey. *Orchis laxiflora*
Hand. *Orchis maculata*
Heart-flowered. *Serapias cordigera*
Horned. *Ophrys cornuta*
Horse-shoe. *Ophrys Ferrum-equinum*
Lady. *Orchis purpurea*
Leafy. *Orchis foliosa*
Lizard. *Orchis hircina*
Lœsel's. *Ophrys Læselii*
Long-bracted. *Orchis Robertiana* and *O. longibracteata*
Long-spurred Algerian. *Orchis longicalcarata*
Looking-glass. *Ophrys Speculum*
Loose-flowered. *Orchis laxiflora*
Man, or Green-Man. *Aceras anthropophora*
Madeira. *Orchis foliosa*
Marsh. *Orchis latifolia*
May. *Orchis majalis*
Military. *Orchis militaris*

Orchis, Monkey. *Orchis tephrosanthos*
Musk. *Herminium Monorchis*
Pale-flowered. *Orchis pallens*
Purple. *Orchis mascula*
Purple Butterfly. *Orchis papilionacea*
Purple Fringed. *Habenaria fimbriata*
Purple-spotted Broad-leaved. *Orchis latifolia var. Lagotis*
Pyramidal. *Orchis pyramidalis*
Ragged Fringed. *Habenaria lacera*
Rein. The genus *Habenaria*
Salep. *Orchis Morio*
Saw-fly. *Ophrys tenthredinifera*
Saw-fly, Small. *Ophrys tenthredinifera var. minor*
Scarlet Dark-flowered. *Nigritella coccinea*
Showy American. *Orchis spectabilis*
Snake's-mouth. *Pogonia ophioglossoides*
Spider. *Ophrys aranifera* and *O. arachnites*
Spider, Bordered. *Ophrys aranifera var. limbata*
Spider, Common or Early. *Ophrys aranifera*
Spider, Late. *Ophrys arachnites*
Spider-like. *Ophrys arachnoides*
Spotted. *Orchis maculata*
Surrey. *Habenaria nivea*
Tawny. *Orchis fusca*
Three-birds. *Pogonia (Triphora) pendula*
Tongue-flowered. *Serapias Lingua*
Wasp. *Ophrys vespifera*
White Butterfly. *Habenaria nivea*
White Fringed. *Habenaria blephariglottis*
White Prairie. *Habenaria leucophæa*
Woodcock. *Ophrys Scolopax*
Yellow. *Ophrys lutea*
Yellow Fringed. *Habenaria ciliaris*
Yellow Fringed, Small. *Habenaria cristata*
Ordeal - tree, Hottentot's. *Toxicophlæa (Acocanthera) Thunbergii*
Of Madagascar. *Tanghinia (Cerbera) venenifera*
Of W. Africa. *Erythrophlæum guineense*
Ore, Ore-weed, or Oar-weed. *Fucus vesiculosus, F. serratus,* and *Laminaria digitata*
Organ, Organs, or Organy. *Origanum vulgare* and *Mentha Pulegium*
Orphan-John. See Orpine
Orpine, or Orphan-John. *Sedum Telephium*
Bastard, or False. *Andrachne telephioides*
Evergreen. *Sedum Anacampseros*
Stone. *Sedum reflexum*
Tree. *Telephium Imperati*
Orris-root. The root of *Iris germanica, I. pallida,* and *I. Florentina*
Osier, or Ozier, Common. *Salix viminalis*
Fine Basket. *Salix purpurea var. Forbyana*
Golden. *Salix vitellina* and *Myrica Gale*
Green. *Cornus florida*
Purple. *Salix purpurea*
Red. *Salix rubra*
Velvet. *Salix viminalis*
Osmund the Water-man. *Osmunda regalis*
Oso-berry-tree. *Nuttallia cerasiformis*
Ouler, or Owler. A corruption of "Alder"
Our-Lady-of-New-Chapel's Flower. See New-Chapel-flower

Ova-ova. *Monotropa uniflora*
"Overlook," of Jamaica. *Canavalia gladiata*
Owm. A corruption of "Elm"
Owala-tree. *Pentaclethra macrophylla*
Owl's Crown. *Filago germanica*
Ox-balm. See Balm
Ox-berry. *Tamus communis*
Ox-eye. The genus *Buphthalmum*; also *Adonis vernalis*
American. *Heliopsis lævis*
Creeping. *Wedelia carnosa*
Great. *Pyrethrum uliginosum*
Heart-leaved. *Buphthalmum speciosum*
Miniature Sun-flower. *Heliopsis lævis*
Sea-side. *Borrichia frutescens*
Sweet-scented. *Buphthalmum aquaticum*
Willow-leaved. *Buphthalmum salicifolium*
Ox-eye-Daisy. See Daisy
Yellow. *Buphthalmum salicifolium*
Oxford Weed. *Linaria Cymbalaria*
Ox-heel. *Helleborus fœtidus*
Ox-horn. *Bucida Buceras*
Ox-lip. *Primula elatior*; also applied to *P. variabilis* and *P. vulgaris caulescens*
Bardfield, or True. *Primula elatior*
Blue. *Primula elatior var. cærulea*
Ox-tongue. *Helminthia echioides*; also applied to several plants with rough leaves, such as *Borago* and *Anchusa*
Oxytrope, Brilliant-flowered. *Oxytropis splendens*
Fetid. *Oxytropis fœtida*
Purple-flowered. *Oxytropis uralensis*
Pyrenean. *Oxytropis pyrenaica*
Yellow-flowered. *Oxytropis campestris*
Oyster, Vegetable. *Tragopogon porrifolius*
Oyster-green. *Ulva Lactuca*
Broad-leaved. *Ulva latissima*
"Oysterloit." An old name for *Polygonum Bistorta*
Oyster-plant. *Mertensia maritima*
Spanish. *Scolymus hispanicus*
Ozier. See Osier

Paardepis-tree. *Hippobromus alatus*
Pacay-tree. *Inga Feuillei*
Paddle-wood. *Aspidosperma excelsum*
Paddock-pipes. The genus *Equisetum*
Paddock-stools. See Toad-stool
Paddy. Another name for unhusked Rice
Pæony. The genus *Pæonia*
Anemone - flowered. *Pæonia anemonæflora*
Chinese Tree. *Pæonia Moutan*
Common Garden. Varieties of *Pæonia officinalis*
Dwarf. *Pæonia humilis*
Edible-rooted. *Pæonia edulis*
"English." *Pæonia corallina*
Fennel-leaved. *Pæonia tenuifolia*
Large-flowered Scented. *Pæonia odorata grandiflora*
Lobed-leaved. *Pæonia lobata*
Ram's-horn-fruited. *Pæonia arietina*
Siberian. *Pæonia sibirica*
Slender-leaved. *Pæonia tenuifolia*
Soft-leaved. *Pæonia mollis*
Steep Holmes, or Wild. *Pæonia corallina*

and Foreign Plants, Trees, and Shrubs. 101

Pæony, Tree. *Pæonia Moutan*
Tree, Banks's. *Pæonia Moutan Banksii*
Tree, Poppy-flowered. *Pæonia Moutan papaveracea*
Tree, Rosy-flowered. *Pæonia Moutan rosea*
White-flowered. *Pæonia albiflora*
Pagoda-tree, Chinese or Japanese. *Sophora japonica*
E. Indian. *Ficus indica*
W. Indian. *Plumieria alba*
Pai-cha-wood. *Euonymus Sieboldianus*
Pagle. An old name for *Stellaria*
Paigle. *Primula veris*
Paint-root. *Lachnanthes tinctoria*
Painted-Cup, Scarlet. *Castilleia coccinea*
Painted-grass. See Grass
Pak-choi. See Cabbage, Chinese
Palay, or Ivory-tree, of the E. Indies. Various species of *Wrightia*
Palissander, or Palixander-wood. Another name for Rose-wood and Violet-wood
Palillos-tree. *Compomanesia linearifolia*
Palm. A general name for trees of the *Palmæ* family; applied also to *Salix Caprea* and *Taxus baccata*
Acuyuru. *Astrocaryum aculeatum*
Alexandra. *Ptychosperma Alexandræ*
Areng. *Saguerus (Gomutus) saccharifera*
Aricuri, or Aracuri. *Cocos schizophylla*
Assai. *Euterpe edulis*
Australian Feather. The genus *Ptychosperma*
Bacaba. *Œnocarpus Patava*
Bamboo. *Raphia vinifera*
Bangalow. *Seaforthia (Ptychosperma) elegans*
Betel-nut. *Areca Catechu*
Blowing-cane. *Iriartea setigera*
Bourbon. The genus *Latania*
Brazilian Coast. *Diplothemium maritimum*
Broom. *Attalea funifera* and *Thrinax argentea*
Buriti. *Mauritia vinifera*
Bussu. *Manicaria saccifera*
Cabbage. *Oreodoxa (Areca) oleracea* and *Chamærops Palmetto*
Cabbage, Australian. *Livistona (Corypha) australis*
Cabbage, N. S. Wales. *Seaforthia (Ptychosperma) elegans*
Cabbage, New Zealand. *Cordyline (Dracæna) australis*
Caiané. *Elæis melanococca*
Carana. *Mauritia Carana*
Carnaüba. *Copernicia cerifera*
Catechu. *Areca Catechu*
Central and W. Australian. *Livistona Mariæ*
Chiqui-Chiqui. *Mauritia flexuosa*
Chusan. *Chamærops Fortunei*
Club. The genus *Cordyline*
Cocoa-nut. *Cocos nucifera*
Cocoa-nut, Double, or Sea. *Lodoicea Seychellarum*
Cohune. *Attalea Cohune*
Coquito. *Jubæa spectabilis*
Coyoli. *Acrocomia mexicana*
Crown. *Maximiliana caribæa*

Palm, Cumari. *Astrocaryum vulgare*
Curly. *Kentia Belmoreana*
Cusi, or Cusich. *Orbignya phalerata*
Date. *Phœnix dactylifera*
Date, Slender. *Phœnix tenuis*
Date, Wild. *Phœnix sylvestris*
Deleb. *Borassus (?) Æthiopum*
Dragon's-blood. *Calamus Draco*
European. *Chamærops humilis*
Fan. A name applied to any Palm which has fan-shaped leaves
Fan, Dwarf. *Chamærops humilis*
Fan, Great. *Borassus flabelliformis* and *Corypha umbraculifera*
Fan, Jamaica. *Sabal Blackburniana* and *S. umbraculifera*
Fan, Jamaica, Small-flowered. *Thrinax parviflora*
Fan, Nepaul. *Chamærops excelsa*
Fan, N. S. Wales. *Livistona australis*
Fan, Shang-hae. *Chamærops Fortunei*
Fern. *Cycas revoluta* and other species
Gebang. *Corypha Gebanga*
Gipp's-land. *Livistona australis*
Gomuti. *Saguerus (Gomutus) saccharifera*
Grigi. *Aiphanes corallina*
Gru-gru. *Astrocaryum vulgare* and *Acrocomia sclerocarpa*
Hemp. *Chamærops excelsa*
Illawarra. *Ptychosperma Cunninghamii*
Inija. *Maximiliana regia*
Iraiba. *Cocos oleracea*
Ita. *Mauritia flexuosa*
Iú. *Astrocaryum acaule*
Ivory-nut. *Phytelephas macrocarpa*
Jacitara. *Desmoncus macracanthos*
Jaggery. *Caryota urens*
Jagua. *Maximiliana regia*
Jará. *Leopoldinia pulchra*
Jará-assu. *Leopoldinia major*
Jupati. *Raphia tædigera*
Khujjoor, or Khurjurah. *Phœnix sylvestris*
Lord Bentinck's. *Bentinckia coddapanna*
Macaw. *Acrocomia fusiformis (A. sclerocarpa)*
Marajah. *Bactris Maraja*
Miriti, or Morici. *Mauritia flexuosa*
Mucuja. *Acrocomia lasiospatha*
Murumuru. *Astrocaryum Murumuru*
New Zealand. *Areca sapida*
Nibung. *Oncosperma filamentosa*
Nikau. *Areca (Kentia) sapida*
Norfolk Island. *Kentia Baueri*
"Northamptonshire." *Salix caprea*
Oil. *Elæis guineensis* and *Cocos butyracea*
Palmetto. *Sabal (Chamærops) Palmetto*
Palmetto, Blue. *Chamærops (Rapidophyllum) Hystrix*
Palmetto, Cabbage. *Chamærops (Sabal) Palmetto*
Palmetto, Humble. *Carludovica insignis*
Palmetto, Royal. *Thrinax parviflora* and *Sabal umbraculifera*
Palmetto, Saw. *Chamærops serrulata*
Palmetto, Silver-leaved. *Thrinax argentea*
Palmetto, Small. The genus *Carludovica*
Palmyra. *Borassus flabelliformis*
Panama-hat. *Carludovica palmata*

Palm, Parlour. *Aspidistra lurida*
Pashiuba. *Iriartea exorrhiza*
Patawa. *Œnocarpus Patava*
Peach. *Guilielma speciosa*
Piassaba. *Attalea funifera* and *Leopoldinia Piassaba*
Pinang. *Areca Catechu*
Pindova. *Attalea compta*
Princess Alexandra's. *Ptychosperma Alexandræ*
Pumos. *Copernicia Pumos*
Raffia, or Roffia. *Raphia Ruffia* and *R. tædigera*
Rasp. *Iriartea exorrhiza*
Rope. The genus *Dæmonorops*
Sago. *Sagus lævis*, *S. Rumphii*, and *Cycas circinalis*
Sago, Bastard. *Caryota urens*
Sago, Japanese. *Cycas revoluta*
Sago, Moluccas. *Sagus farinifera*
Sago, Prickly. *Sagus Rumphii*
Tal, or Tala. *Borassus flabelliformis*
Taliera, or Tara. *Corypha Taliera*
Talipot. *Corypha umbraculifera*
Thatch. *Sabal Blackburniana*
Tiger-grass. *Chamærops Ritchieana*
Toddy. *Caryota urens*
Toko-pat. *Livistona Jenkinsiana*
Troolie. *Manicaria saccifera*
Tucumu. *Astrocaryum Tucuma*
Umbrella. *Kentia Canterburyana*
Urucuri. *Attalea excelsa*
Walking-stick. *Areca (Kentia) monostachya*
Wax, Brazil. *Copernicia cerifera*
Wax, New Granada. *Ceroxylon (Iriartea) andicola*
Whip-stick. *Kentia (Areca) monostachya*
Wine, Brazilian. *Mauritia vinifera*
Wine, E. Indian. *Caryota urens*, *Phœnix sylvestris*, and *Borassus flabelliformis*
Wine, Guiana. *Œnocarpus Patava* and *Manicaria saccifera*
Wine, Moluccas. *Gomutus saccharifera*
Wine, New Granada. *Cocos butyracea*
Wine, Royal. *Oreodoxa regia*
Wine, W. African. *Raphia vinifera*
Zanora. *Iriartea exorrhiza*
Palm-bark-tree. *Melaleuca Wilsoni*
Palm-butter. Another name for Palm-oil
Palm-honey-tree. *Jubæa spectabilis*
Palma-Christi. *Ricinus communis*
Palmetto. See Palm, Palmetto
Palmite-Rush. *Prionium Palmita*
Palmyra-tree. *Borassus flabelliformis*
Palo-de-Vaca, or Cow-tree. *Brosimum Galactodendron*
Palo-Santo-tree. *Swartzia tomentosa*
Palsy-wort. An old name for *Primula veris*
Pameroon-bark Tree. *Moschoxylon Swartzii*
Pampas Grass. *Gynerium argenteum*
Rosy-spiked. *Gynerium roseum*
Panama-Hat-tree. *Carludovica palmata*
Panay. *Prunella vulgaris*
Panick-grass. See Grass
Pansy. *Viola tricolor* and vars.
Australian. *Erpetion reniforme*

Papaw-tree. *Carica Papaya*
Virginian. *Asimina (Anona) triloba*
Paper-bark-tree. Various species of *Callistemon*
Paper-Mulberry-tree. *Broussonetia papyrifera*
Paper-reed, or Paper-rush, of the Nile, or of the Ancients. *Papyrus antiquorum (Cyperus Papyrus)*
Fragrant. *Papyrus odoratus*
Sicilian, or Syrian. *Papyrus syriacus*
Paper-tree, of Siam. *Trophis aspera*
Pappoose-root. *Caulophyllum thalictroides*
Paradise, Grains of. The seeds of *Amomum Melegueta*
Parasol, Chinese. *Sterculia platanifolia*
Parasol Fir-tree. See Fir-tree
Parchment-bark. *Pittosporum crassifolium*
Pareira Brava-plant, White. *Abuta rufescens*
False. *Cissampelos Pareira*
Parilla, Yellow. *Menispermum canadense*
Paris. See Herb-Paris
Park-leaves. *Hypericum Androsæmum*
Parnassus, Grass of. See Grass
Paroquet-Bur. The genus *Triumfetta*
Parrot-beak-plant. The genus *Clianthus*
Parrot-weed. *Bocconia frutescens*
Parrot's-bill, of New Zealand. *Clianthus puniceus*
Parsley. The genus *Petroselinum*. Applied also to various other plants
Asses'. See Parsley, Fool's
Bastard, or Bur. The genus *Caucalis*, especially *C. daucoides*
Beaked. The genus *Anthriscus*
Black. *Melanoselinum decipiens*
Bur. See Parsley, Bastard
Corn. *Petroselinum segetum*
Cow. *Chærophyllum temulum* and *C. sylvestre*
Dog's. See Parsley, Fool's
Fool's, Asses', or Dog's. *Æthusa Cynapium*
Garden. *Petroselinum sativum*
Hamburgh. A variety of *Petroselinum sativum*
Hedge. *Torilis Anthriscus*
Hemlock. The genus *Conioselinum*
Horse. *Smyrnium Olusatrum*
Macedonian. *Athamanta macedonica*
Marsh. *Œnanthe Lachenalii* and the genus *Eleoselinum*
Meadow. *Œnanthe pimpinelloides*
Milk. *Peucedanum palustre*
Mountain. *Peucedanum Oreoselinum*
Square. *Ptychotis heterophylla*
Stone. *Sison Amomum*
Thorough-bored. An old name for *Smyrnium apiifolium*
Parsley-Fern. See Fern
Parsley-Piert. *Alchemilla arvensis*
Parsnip, Common. *Pastinaca sativa*
Giant. The genus *Heracleum*
Cow. *Heracleum Sphondylium*
Maori. *Ligusticum Lyalli*
Meadow. The genus *Thaspium*
Rough. *Pastinaca Opopanax*

Parsnip, Sea. The genus *Echinophora*, especially *E. spinosa*
Victorian. *Trachymene australis*
Water. The genus *Sium*
Water, Common. *Sium latifolium*
Wild. *Peucedanum sativum*
Partridge-berry. *Gaultheria procumbens* and *Mitchella repens*
"**Partridge-wood.**" Supposed to be the wood of *Andira inermis*
Pasque-flower. *Anemone Pulsatilla*
American. *Anemone patens* var. *Nuttalliana*
Haller's. *Anemone Halleri*
Mountain. *Anemone montana*
Passe-flower. Another name for Pasque-flower
Passion-flower. The genus *Passiflora*
Australian. *Disemma (Passiflora) Herbertiana* and *D. coccinea (Passiflora) Bauksii*
Bat-winged. *Passiflora Vespertilio*
Blood-red. *Tacsonia sanguinea*
Buchanan's. *Tacsonia Buchanani*
Common Blue. *Passiflora cærulea*
Crimson. *Passiflora hermesina*
Edible-fruited. *Passiflora edulis*
Hardy. *Passiflora cærulea, P. glaucophylla*, and *P. incarnata*
Lime-tree-leaved. *Passiflora tiliæfolia*
New Zealand. *Passiflora tetrandra*
Norfolk Island. *Passiflora (Disemma) aurantia*
Pumpkin, or Large-fruited. *Passiflora macrocarpa*
Racemed. *Passiflora racemosa (P. princeps)*
Square-stalked. *Passiflora quadrangularis*
Van Volxem's. *Tacsonia Van Volxemi*
Vine-leaved. *Passiflora vitifolia*
Patchouli Plant. *Coleus aromaticus*
Patience-Dock. *Rumex Patientia*
Pecan-nut-tree. See Nut-tree
Peg-wood. *Cornus sanguinea* and *Euonymus europæus*
Pea. A name applied to various plants of the *Papilionaceæ* family
Angola. *Cajanus indicus*
Black-eyed. *Dolichos sphærospermus*
Black-rooted. *Orobus niger*
Butterfly. *Clitoria Mariana*
Congo. *Cajanus indicus bicolor*
Darling River. *Swainsona Greyana*
Earth. *Lathyrus amphicarpus*
Earth-nut. *Lathyrus tuberosus*
Egyptian. *Cicer arietinum*
Everlasting. *Lathyrus sylvestris* var. *latifolius*
Everlasting, Californian. *Lathyrus californicus*
Everlasting, Large-flowered. *Lathyrus grandiflorus*
Everlasting, Mountain. *Lathyrus montanus*
Everlasting, Rosy-flowered. *Lathyrus roseus*
Everlasting, Round-leaved. *Lathyrus rotundifolius*
Everlasting, Siberian. *Lathyrus pisiformis*
Pea, Everlasting, Sibthorpe's. *Lathyrus Sibthorpei*
Everlasting, Tuberous-rooted. *Lathyrus tuberosus*
Field, or Gray. *Pisum sativum* var. *arvense*
Flat. The genus *Platylobium*
French. See French Peas
Garden. Varieties of *Pisum sativum*
Grass. *Lathyrus Nissolia*
Gray. See Pea, Field
Ground-squirrel. *Jeffersonia diphylla*
Heart. *Cardiospermum Halicacabum*
Heath. *Orobus tuberosus*
Hoary. The genus *Tephrosia*
Lord Anson's. *Lathyrus magellanicus*
Love. *Abrus precatorius*
Marrow-fat. A variety of *Pisum sativum*
Marsh. *Lathyrus palustris*
Meadow. *Lathyrus pratensis*
Milk. The genus *Galactia*
Mouse. *Lathyrus macrorrhizus*
"No-eye." *Cajanus indicus flavus*
Orange. Small unripe fruit of the Curaçao Orange
Partridge. *Heisteria coccinea* and *Cassia Chamæcrista*
Pigeon. *Ervum Ervilia*
Pigeon, of the W. Indies. *Cajanus indicus*
Poison, of Australia. The genus *Swainsona*
Red Pottage. *Ervum Lens*
Rosary. The seeds of *Abrus precatorius*
Rough. *Lathyrus hirsutus*
Salisbury, of Australia. *Goodia latifolia*
Scurfy. The genus *Psoralea*
Sea-side. *Pisum maritimum (Lathyrus maritimus)*
Sensitive. *Cassia nictitans*
Shamrock. *Parochetus communis*
Spear. *Dorycnium fruticosum*
Spurred Butterfly. The genus *Centrosema*
Sturt's Desert. *Clianthus Dampieri*
Sugar. A name given to some varieties of *Pisum sativum* which have tender edible pods
Swainson. The genus *Swainsona*
Sweet. *Lathyrus odoratus*
Sweet, Everlasting. *Lathyrus latifolius* and vars.
Tangier. *Lathyrus tingitanus*
Tuberous-rooted. *Orobus tuberosus*
Wild Turkey. *Corydalis formosa*
Winged. The genus *Tetragonolobus*
Winged, Purple-flowered. *Tetragonolobus purpureus*
Wood. *Lathyrus sylvestris* and *Orobus sylvaticus*
Yellow-flowered. *Pisum Ochrus* and *Lathyrus Aphaca*
Pea-bush, Burton's. *Burtonia scabra*
Pea-flower, Vilmorin's Purple. *Vilmorinia multiflora*
Pea-nut, American. *Arachis hypogæa*
Hog. *Amphicarpæa monoica*
Madagascar, or Bambarra. *Voandzeia subterranea*
Pea-tree. The genus *Sesbania*
Chinese. *Caragana Chamlagu*
Coral. *Adenanthera pavonina*
Siberian. The genus *Caragana*

English Names of Cultivated, Native,

Pea-tree, Siberian, Common. *Caragana arborescens*
Siberian, Flat-podded. *Caragana Altagana*
Siberian, Goat's-horn-like. *Caragana tragacanthoides*
Siberian, Large-flowered. *Caragana grandiflora*
Siberian, Maned. *Caragana jubata*
Siberian, Pigmy. *Caragana pygmaea*
Siberian, Redlowski's. *Caragana Redlowskii*
Siberian, Sand. *Caragana arenaria*
Siberian, Shrubby. *Caragana frutescens*
Siberian, Small-leaved. *Caragana microphylla*
Siberian, Soft-leaved. *Caragana mollis*
Siberian, Thorny. *Caragana spinosa*
West Indian. *Æschynomene (Agati) grandiflora*
W. Indian Swamp. *Sesbania occidentalis*
"**Pea-vine,**" Californian. The genus *Vicia*
Peach-tree. *Amygdalus persica* and *vars*.
African. *Sarcocephalus esculentus*
Almond. *Amygdalus communis var. persicoides*
Australian. *Santalum acuminatum*
Camellia-flowered. *Amygdalus persica sinensis camelliaeflora*
Carnation-flowered. *Amygdalus persica sinensis caryophylliflora*
Common. *Amygdalus persica vulgaris*
Double-blossomed. *Amygdalus persica vulgaris flore-pleno*
Double Crimson Chinese. *Amygdalus persica sinensis fl.-pl. sanguineo*
Double White Chinese. *Amygdalus persica sinensis fl.-pl. albo*
Double-fruited. *Amygdalus persica vulgaris fructu-pleno*
Flat-fruited. *Amygdalus persica vulgaris var. compressa*
Guinea, Negro, or Sierra Leone. *Sarcocephalus esculentus*
Negro. See Peach, Guinea
Purple-leaved. *Amygdalus persica foliis purpureis*
Sierra Leone. See Peach, Guinea
Spanish. *Amygdalus persica vulgaris var. hispanica*
Variegated-leaved. *Amygdalus persica vulgaris foliis variegatis*
Weeping. *Amygdalus persica vulgaris var. pendula*
White-flowered. *Amygdalus persica vulgaris var. alba*
Peach-Myrtle. *Hypocalymma robustum*
Peach-wood, or Nicaragua-wood. *Caesalpinia echinata*
Peach-wort. *Polygonum Persicaria*
Peacock-flower. *Caesalpinia pulcherrima*
Royal. *Poinciana regia*
Peacock Flower-fence. *Adenanthera pavonina*
Peacock Treasure-flower. *Gazania pavonia*
Peacock's-tail Sea-weed. *Padina pavonia*
Pear-tree. The genus *Pyrus*
Alligator. *Persea carolinensis*
Alpine. *Pyrus nivalis*
Anchovy. *Grias cauliflora*

Pear-tree, Aurelian, or Sage-leaved. *Pyrus salviaefolia*
Avocado. *Persea gratissima*
Birch-leaved. *Pyrus betulaefolia*
Bollwyller. *Pyrus Bollwylleriana*
Chinese. *Pyrus sinensis*
Common. Varieties of *Pyrus communis sativa*
Garlic. *Cratæva gynandra* and *C. Tapia*
Grape. *Amelanchier Botryapium*
Mount Sinai. *Pyrus sinaica*
Notched-leaved. *Pyrus crenata*
Oleaster-leaved. *Pyrus elaeagnifolia*
Prickly. *Cactus Opuntia*
Prickly, Hardy. *Opuntia Rafinesquiana, O. humilis,* and *O. missouriensis*
Sage-leaved. See Pear, Aurelian
Snow. *Pyrus nivalis*
Straw-berry. *Cereus triangularis*
Three-lobed-leaved. *Pyrus trilobata*
Variable-leaved. *Pyrus variolosa*
Weeping. *Pyrus spuria pendula*
White-leaved. *Pyrus nivalis*
Wild, of the W. Indies. *Clethra tinifolia*
Willow-leaved. *Pyrus salicifolia*
Wooden, of Australia. *Xylomelum pyriforme*
Pear-main. A variety of Apple
Pearl-berry. *Margyricarpus setosus*
Pearl-bush. *Exochorda (Spiraea) grandiflora*
Pearl-grass. See Grass
Pearl-plant. *Lithospermum officinale*
Pearl-wort. The genus *Sagina*
Lawn. *Sagina glabra var. corsica (Spergula pilifera)*
"**Pearls-of-Spain.**" *Muscari botryoides var. album*
Pee-put, or Piput, tree. *Ficus religiosa*
Pelican-flower. *Aristolochia grandiflora*
Mexican. *Cypripedium Irapeanum*
"**Pellamountain.**" An old name for *Thymus Serpyllum*
Pellitory. *Pyrethrum Parthenium*
American. *Parietaria pennsylvanica*
Bastard. *Achillea Ptarmica*
New Zealand. *Parietaria debilis*
Of-Spain. *Anacyclus Pyrethrum*
Of-Spain, False. *Imperatoria Ostruthium*
Of-the-Wall. *Parietaria officinalis*
Pelu-tree. *Sophora tetraptera*
Penang Lawyers. *Licuala acutifida*
Pencil-flower. The genus *Stylosanthes*
Carolina. *Stylosanthes elatior*
Pencil-tree. *Baccharis halimifolia*
Penguin, or Pinguin, Plant. *Bromelia Pinguin*
Penny-cress. *Thlaspi arvense*
"**Penny-pies.**" The leaves of *Cotyledon Umbilicus*
Penny-rot. *Hydrocotyle vulgaris*
Penny-royal. *Mentha Pulegium*
American. *Hedeoma pulegioides*
Bastard, or False. *Trichostemma dichotomum* and *Isanthus coeruleus*
Requien's. *Mentha Requieni (Thymus corsicus)*
Penny-royal-tree. *Satureia riminea*

Penny-wort, or Penny-leaf. *Linaria Cymbalaria*
House-leek. *Umbilicus chrysanthus*
Indian. *Hydrocotyle asiatica*
Marsh. *Hydrocotyle vulgaris*
Spiny. *Umbilicus spinosus*
Wall. *Cotyledon Umbilicus*
Pentstemon, Bearded. *Pentstemon barbatus*
Bell-flowered. *Pentstemon campanulatus*
Centranthus-leaved. *Pentstemon centranthifolius*
Cobæa-flowered. *Pentstemon Cobœa*
Common. *Pentstemon gentianoides (P. Hartwegi)*
Dwarf. *Pentstemon humilis* and *P. glaber*
Flame-coloured. *Pentstemon lœtus*
Foxglove. *Pentstemon Digitalis*
Gentian-blue. *Pentstemon Jaffrayanus*
Glaucous. *Pentstemon glaucus*
Graceful. *Pentstemon argutus*
Lewis's. *Pentstemon Lewisii*
Murray's. *Pentstemon Murrayanus*
Oval-leaved. *Pentstemon ovatus*
Palmer's. *Pentstemon Palmeri*
Pointed-leaved. *Pentstemon acuminatus*
Scarlet-flowered. *Pentstemon puniceus*
Scouler's. *Pentstemon Scouleri*
Showy. *Pentstemon speciosus*
Side-flowered. *Pentstemon secundiflorus*
Spreading. *Pentstemon diffusus*
Thick-leaved. *Pentstemon crassifolius*
Torrey's. *Pentstemon Torreyi*
Whorled. *Pentstemon confertus (P. procerus)*
Wright's. *Pentstemon Wrightii*
Peony. Another name for Pæony
Pepper. The genus *Piper*. Applied also to various other plants
African. *Habzelia æthiopica (Xylopia aromatica)*
Anise. *Xanthoxylon mandschuricum*
Australian. *Tasmannia (Drimys) aromatica* and *Schinus molle*
Bell. *Capsicum grossum*
Betel. *Chavica Betel*
Bird. *Capsicum baccatum*
Bird's-eye. The genus *Capsicum*
Bitter, or Star. *Xanthoxylon Daniellii*
Black, of W. Africa. *Piper (Cubeba) Clusii*
Bonnet. *Capsicum tetragonum*
Boulon. *Habzelia æthiopica*
Californian. *Schinus molle*
Cayenne, Chilli, Guinea, or Red. *Capsicum annuum* and *C. fastigiatum*
Cherry. *Capsicum cerasiforme*
Chilli. See Pepper, Cayenne
Chinese, or Japanese. *Xanthoxylon piperitum*
Cubeb, or Java. *Cubeba officinalis*
Ethiopian. *Habzelia æthiopica*
False. *Schinus molle*
Goat. *Capsicum frutescens*
Guinea. *Capsicum annuum.* (See also Pepper, Negro)
Japanese. See Pepper, Chinese
Java. See Pepper, Cubeb

Pepper, Long. *Chavica officinarum* and *C. Roxburghii*
Malaguetta, or Melegueta. *Amomum Melegueta*
Monkey. *Habzelia æthiopica*
Monks'. *Vitex Agnus-castus*
Mountain. *Capparis sinaica*
Negro, or Guinea. *Habzelia æthiopica* and *Xanthoxylon guineense*
N. S. Wales. *Drimys dipetala*
New Zealand. *Drimys axillaris*
New Zealand Native. *Macropiper excelsum*
Red. See Pepper, Cayenne
Spur. *Capsicum frutescens*
Star. *Xanthoxylon Daniellii*
Tasmanian. *Drimys aromatica*
Wall. *Sedum acre*
Water. *Polygonum Hydropiper* and *Elatine Hydropiper*
Wild, E. Indian. *Vitex trifolia*
Pepper-brand Fungus. See Fungus
Pepper-bush. *Andromeda racemosa*
Sweet. *Clethra tinifolia*
Pepper-dulse. *Laurencia pinnatifida*
Pepper Elder. The genus *Piperoma*
Pepper-grass. See Grass
Pepperidge. *Nyssa multiflora (N. sylvatica)*
Pepper-mint, Australian. *Mentha australis*
Chinese. *Mentha arvensis glabrata*
Common. *Mentha piperita*
Japanese. *Mentha arvensis* var. *piperascens*
Small. *Thymus Piperella*
Peppermint-tree, Australian. *Eucalyptus amygdalina, E. capitellata, E. coriacea, E. odorata,* and *E. piperita*
Tasmanian. *Eucalyptus piperita* and *E. amygdalina*
Pepper-rod. *Croton humilis*
Pepper-root, American. The genus *Dentaria*
Pepper-Saxifrage. *Silaus pratensis*
Pepper-wood. *Licania guianensis* and *Dicypellium caryophyllatum*
Pepper-wort. The genus *Lepidium*, especially *L. latifolium*
Water. *Polygonum Hydropiper*
Water, Small. *Elatine Hydropiper*
Peri-root, of New Zealand. *Gastrodia Cunninghamii*
Periwinkle. The genus *Vinca*
Dark-purple-flowered. *Vinca minor* var. *atropurpurea*
Double Purple. *Vinca minor fl.-pl. purpureo*
E. Indian, or Rosy-flowered. *Vinca rosea*
Herbaceous. *Vinca herbacea*
Italian. *Vinca acutiloba*
Large. *Vinca major*
Madagascar. *Vinca rosea*
Small. *Vinca minor*
Variegated-leaved. *Vinca major* var. *elegantissima* and other varieties
White-flowered. *Vinca minor* var. *alba*
White-flowered Indian. *Vinca rosea* var. *alba*

"Persicaria," or Persicary. *Polygonum Persicaria*
Persimmon-tree. *Diospyros virginiana*
Mexican. *Diospyros Texana*
Peru, Balsam-of. *Myroxylon* (*Myrospermum*) *Peruiferum*
Peruvian-bark Tree. Various species of *Cinchona*, the best of which are *C. officinalis*, *C. Calisaya*, and *C. succirubra*
Peruvian-Glory-bush. See Glory-bush
Peter's (St.)-wort. The genus *Symphoricarpus, Hypericum quadrangulum, Ascyrum stans,* and *Primula veris*
Mountain. *Symphoricarpus montana*
Peter's-sperm. A corruption of *Pittosporum*
Pe-tsai. See Cabbage, Chinese
Pettigree, or Pettigrue. *Ruscus aculeatus*
Petty-Whin. See Whin
Petunia, Common white. *Petunia nyctaginiflora*
Pet-wood. *Berrya mollis*
Pewter-wort. *Equisetum hyemale*
Pharaoh's-Corn. See Corn
Pheasant-wood. Another name for Partridge-wood
Pheasant's-eye. *Adonis autumnalis*
Summer. *Adonis æstivalis*
Phlox, Carolina. *Phlox Carolina*
Cleft-petalled. *Phlox bifida*
Creeping. *Phlox reptans* (*P. verna, P. stolonifera*)
Cross-leaved. *Phlox decussata*
Gray-leaved. *Phlox canescens*
Hairy. *Phlox pilosa*
Leafy. *Phlox frondosa*
Long-flowered. *Phlox longiflora*
Mossy. *Phlox subulata, P. muscoides,* and *P. bryoides*
Nelson's. *Phlox Nelsoni*
Ovate-leaved. *Phlox ovata*
Panicled. *Phlox paniculata*
Procumbent. *Phlox procumbens*
Straggling. *Phlox divaricata* (*P. canadensis*)
Tall Garden. Varieties of *Phlox paniculata*
Tufted. *Phlox cæspitosa*
Very smooth. *Phlox glaberrima*
White-flowered Dwarf. *Phlox nivalis*
Physic-nut-tree. *Curcas* (*Jatropha*) *purgans*
Pi-plant, of the Sandwich Islands. *Tacca pinnatifida*
Pickerel-weed. The genus *Pontederia*
Bladder-stalked. *Pontederia crassipes*
Heart-leaved. *Pontederia cordata*
Sky-blue. *Pontederia azurea*
Pick-purse. *Capsella Bursa-pastoris*
Pick-tooth. *Daucus Visnaga*
Picotee. A variety of *Dianthus Caryophyllus*, which has the flowers coloured only on the edges of the petals
Picræna-wood. See Quassia
Pie-plant, Californian. *Rumex hymenosepalus*
Pig-root. The genus *Sisyrinchium*
Pigeon-berry. *Phytolacca decandra*
Pigeon-pea. See Pea

Pigeon-wood. *Guettarda speciosa, Diphilos salicifolia,* and *Diospyros tetrasperma*
Long-leaved. *Coccoloba diversifolia*
Small-leaved. *Coccoloba punctata* and *C. leoganensis*
Pigeon's-foot. *Geranium columbinum*
Pigeon's-grass. See Grass
Pig-nut. *Bunium flexuosum* and *B.* (*Carum*) *Bulbocastanum*
American. *Carya porcina*
Pig-weed. *Chenopodium album*
Winged. *Cycloloma platyphyllum* (*Salsola platyphylla*)
Pig-wood. *Hedwigia balsamifera*
Pig's-face. *Mesembryanthemum æquilaterale* and *M. australe*
Pile-wort. *Ranunculus Ficaria*
Great. *Ficaria grandiflora* (*Ranunculus calthæfolius*). Also an old name for *Scrophularia nodosa*
Pill-corn, or Pil-corn. *Avena nuda*
Pill-wort. *Pilularia globulifera*
Pilot-weed. *Silphium laciniatum*
Pimento-bush. *Eugenia Pimenta* and *E. acris*
Pimpernel. *Pimpinella Saxifraga*. Formerly applied to *Poterium Sanguisorba* and *Prunella vulgaris*
Bastard, or False. *Centunculus minimus* and *Ilysanthis gratioloides*
Bog. *Anagallis tenella*
Indian. *Anagallis indica*
Italian. *Anagallis Monelli*
Large-flowered. *Anagallis grandiflora* (*A. collina, A. fruticosa*)
Red. *Anagallis arvensis*
Sea-side. *Honkenya peploides*
Water. *Samolus Valerandi.* Also applied to *Veronica Beccabunga* and *V. Anagallis*
Water, of Tasmania. *Samolus littoralis*
Yellow. *Lysimachia nemorum*
Pincushion-flower. The genera *Scabiosa* and *Asterocephalus*
Starry. *Asterocephalus stellatus*
Pin-grass, or Pin-clover, of California. *Erodium cicutarium*
Pine-apple. *Ananassa sativa*
Wild. *Bromelia Pinguin*
Pine-apple - flower. *Eucomis punctata* and other species
Pine-barren Beauty. *Pyxidanthera barbulata*
Pine-drops. *Pterospora Andromedea*
Pine-knots. An American name for the cones of Pine-trees
Pine-sap. *Monotropa Hypopitys*
Sweet. *Schweinitzia odorata*
Pine-tree. The genus *Pinus*
Adventure Bay. *Phyllocladus rhomboidalis*
Aleppo, or Jerusalem. *Pinus Halepensis*
Amboyna. *Dammara orientalis*
Austrian. *Pinus austriaca*
Awned-coned. *Pinus aristata*
Bhotan. *Pinus excelsa*
Bishop's. *Pinus muricata*
Black. *Pinus austriaca*
Black, New Zealand. *Podocarpus ferruginea* and *P. spicata*

Pine-tree, Brazilian. *Araucaria brasiliensis*
Broom, or Brown. *Pinus australis*
"Bull." *Pinus ponderosa*
Bunya-Bunya. *Araucaria Bidwillii*
Calabrian. *Pinus Laricio*
Calabrian Cluster. *Pinus Brutia*
Californian Giant. *Pinus Lambertiana*
Candle-wood. *Pinus Teocote*
Celery-leaved. The genus *Phyllocladus*
Celery-leaved, of New Zealand. *Phyllocladus trichomanoides*
Celery-leaved, of Tasmania. *Phyllocladus rhomboidalis*
Cheer, or Emodi. *Pinus longifolia*
Chilian. *Araucaria imbricata*
Chinese. *Pinus sinensis*
Chinese Lace-bark. *Pinus Bungeana*
Cluster, or Star. *Pinus Pinaster*
Corean. *Pinus Koraiensis*
Corsican. *Pinus Laricio*
Corsican Dwarf. *Pinus Laricio var. pygmæa*
Cortcan. *Pinus Pinaster var. minor*
Crimean. *Pinus Pallasiana*
Cypress. *Frenela verrucosa*
Dammar. *Dammara orientalis*
Digger. *Pinus Sabiniana*
Dye, or King. *Pinus Webbiana*
Emodi. See Pine, Cheer
Festoon. *Lycopodium rupestre*
Fox-tail. *Pinus Balfouriana* and *Pinus serotina*
Frankincense. *Pinus Tæda*
Georgia. *Pinus australis*
Ginger. *Cupressus (Chamæcyparis) Lawsoniana*
Golden. *Pinus Kæmpferi*
Grey or Northern Scrub. *Pinus Banksiana*
Ground. *Lycopodium dendroideum* and *Ajuga Chamæpitys*
Ground, Bastard. *Teucrium Pseudo-Chamæpitys*
Hard. *Pinus australis*
Heavy-wooded. *Pinus ponderosa*
Hickory. *Pinus Balfouriana*
Highland. *Pinus sylvestris var. horizontalis*
Hudson's Bay. *Pinus Banksiana*
Huon. *Dacrydium Franklini*
Illawarra. *Podocarpus spinulosa*
Imou. *Dacrydium cupressinum*
Italian Stone. *Pinus Pinea*
Jerusalem. See Pine, Aleppo
Kauri, or Kowrie. *Dammara (Agathis) australis*
Kauri, New Caledonian. *Dammara ovata*
Kauri, Queensland. *Dammara robusta*
King. *Abies Webbiana*
Knee. *Pinus Mugho var. nana* and *P. Pumilio nana*
Labrador, or Hudson's Bay. *Pinus Banksiana*
Lace-bark. *Pinus Bungeana*
Larch. *Pinus Laricio*
Loblolly. *Pinus Tæda*
Lofty Bhotan. *Pinus excelsa*
Long-leaved. *Pinus australis*
Lord Aberdeen's. *Pinus Pinaster var. Hamiltoni*

Pine-tree, Mahogany. *Podocarpus (Nageia) Totara*
Monterey. *Pinus insignis*
"Moon-fruit." *Lycopodium lucidulum*
Moreton Bay. *Araucaria Cunninghamii*
Mountain. *Pinus monticola* and *P. Pumilio*
Murray. *Frenela robusta*
Neoza. *Pinus Gerardiana*
Nepaul. *Pinus Gerardiana*
New Caledonia. *Araucaria Cookii* and *A. Rulei*
New Jersey Scrub. *Pinus inops*
New Zealand. *Dacrydium cupressinum*
Norfolk Island. *Araucaria excelsa*
Norway. *Pinus (Abies) excelsa*
"Norway," of N. America. *Pinus resinosa*
Nut. *Pinus edulis* and *P. monophylla (P. Fremontiana)*
Nut, Californian. *Pinus Fremontiana*
Old Field. *Pinus Tæda*
Oyster Bay. *Callitris australis*
Parasol. See Fir, Parasol
Pinon. *Pinus edulis*
Pitch, or Sap. *Pinus rigida*
Pitch, Bahamas. *Pinus bahamensis*
Pitch, Georgia. *Pinus australis*
Pond. *Pinus serotina*
"Pumpkin," of Canada. *Pinus Strobus*
Red. *Pinus resinosa* and *Abies rubra*
Red, New Zealand. *Dacrydium cupressinum*
Rosemary. *Pinus Tæda*
Sap, or Pitch. *Pinus rigida*
Screw. The genus *Pandanus*
Screw, E. Australian. *Pandanus pedunculatus*
Screw, Fragrant. *Pandanus odoratissimus*
Sea-side. *Pinus maritima*
Shake. *Pinus Lambertiana*
Siberian. *Pinus Cembra var. sibirica*
Silver. *Pinus Picea*. (See also Fir, Silver)
Slash. *Pinus Tæda*
Snow, or White Weymouth. *Pinus Strobus var. nivea*
S. African. *Leucadendron argenteum*
Southern. *Pinus australis*
Spey-side. *Pinus sylvestris var. horizontalis*
Spruce. See Fir, Spruce
Spruce, American. *Pinus mitis* and *P. glabra*
Star. See Pine, Cluster
Stone. *Pinus Pinea*
Stone, Swiss. *Pinus Cembra*
Sugar. *Pinus Lambertiana*
Swamp. *Pinus Tæda*
Swiss. *Pinus Cembra*
Table Mountain. *Pinus pungens*
"Tamarack." *Pinus contorta var. Murrayana*
Tarentina. *Pinus Pinea var. fragilis*
Tartarian. *Pinus Pallasiana*
Thread-leaved. *Pinus filifolia*
Timor. *Pinus insularis*
Totara. *Podocarpus Totara*
"Trucker." *Pinus ponderosa*
"Tuck-tuck." *Abies (Pinus) nobilis*
Twisted-branched. *Pinus contorta*
Twisted Mexican. *Pinus Teocote*
Umbrella. The genus *Sciadopitys*
Virginian. *Pinus australis (P. palustris)*

Pine-tree, Water. *Glyptostrobus heterophylla*
Wax. The genus *Dammara*
W. Indian. *Pinus occidentalis*
Weymouth. *Pinus Strobus*
Weymouth, White. See Pine, Snow
White. *Pinus Strobus* and *P. flexilis*
White, N. S. Wales. *Podocarpus (Nageia) spinulosus*
White, New Zealand. *Podocarpus (Nageia) dacrydioides*
Yellow. *Pinus australis*, *P. mitis*, and *P. ponderosa*
Yellow Arizona. *Pinus Arizonica*
Puget Sound. *Pinus (Abies) Douglasii*
Pine-weed. *Hypericum Sarothra*
Piney-tree. *Calophyllum angustifolium*
Pink. The genus *Dianthus*
Alpine. *Dianthus alpinus*
American, Wild. *Silene pennsylvanica*
Amoor. *Dianthus dentosus*
Bush. *Dianthus ramosissimus*
Carolina. *Spigelia marilandica*
Caucasian. *Dianthus caucasicus*
Cheddar. *Dianthus cæsius*
Chinese. *Dianthus chinensis*
China Tree. *Dianthus arbuscula*
Clove. *Dianthus Caryophyllus*
Clove, Shrubby. *Dianthus Caryophyllus var. fruticosus*
Colour-changing. *Dianthus versicolor*
Common Garden. *Dianthus plumarius*
Corsican. *Dianthus corsicus*
Cushion. *Silene acaulis*
Deptford. *Dianthus Armeria*
Feathered. *Dianthus plumarius*
Fire. *Silene virginica*
Fischer's. *Dianthus Fischeri*
Fragrant White. *Dianthus pungens*
Free-flowering. *Dianthus floribundus*
Fringed. *Dianthus superbus*
German. *Dianthus Carthusianorum*
Glacier. *Dianthus glacialis*
Grass. *Calopogon pulchellus*
Grass, Rose. *Dianthus neglectus*
Hedge. *Saponaria officinalis*
Hungarian. *Dianthus collinus*
Indian. *Dianthus chinensis* and vars.
Indian, of North America. *Quamoclit vulgaris*
Italian, Dark-red. *Dianthus atro-rubens*
Japan, Cut-flowered. *Dianthus Heddewigi var. laciniatus*
Japan, Double-flowered. *Dianthus Heddewigi var. diadematus*
Japan, Single-flowered. *Dianthus Heddewigi*
Lily. See Lily-Pink
Maroon-coloured. *Dianthus Dunnettii superba*
Meadow. *Dianthus deltoides*
Moss. *Phlox subulata*
Mountain. *Dianthus cæsius*
Mt. Tymphrestus. *Dianthus Tymphresteus*
"Mule." Hybrid varieties of *Dianthus*
Pheasant's-eye. *Dianthus plumarius var. annulatus*
Pine-leaved. *Dianthus pinifolius*
Proliferous. *Dianthus prolifer*

Pink, Ringed-flowered. *Dianthus plumarius var. annulatus*
Rock. *Dianthus petræus*
Russian. *Dianthus ruthenicus*
Sand. *Dianthus arenarius*
Sea. *Armeria maritima*
Sheathed. *Dianthus vaginatus*
Spanish. *Dianthus hispanicus*
Sweet-scented. *Dianthus fragrans*, *D. pungens*, and *D. suavis*
Toothed. *Dianthus dentosus*
Viscid. *Dianthus viscidus*
Wood. *Dianthus sylvestris*
Pink-of-my-John. *Viola tricolor*
Pink-root, Demerara. *Spigelia Anthelmia*
Maryland. *Spigelia marilandica*
Pink-weed. *Polygonum aviculare*
Pink-wood-tree. *Persea caryophyllata*
Brazilian. *Physocalymma floridum*
Pin-pillow. *Opuntia curassavica*
Pin-weed. The genus *Lechea*
Pinxter-flower. *Azalea nudiflora*
Piony. Another name for Peony
Pipe-stem-wood, American. *Andromeda acuminata*
Pipe-tree. *Syringa vulgaris*
Pipe-Vine. *Aristolochia Sipho*
Pipe-wort. *Eriocaulon septangulare*
Hairy. *Lachnocaulon Michauxii*
Pipi-pod-tree. *Cæsalpinia Pipai*
Pipperidge, or Piprage. *Berberis vulgaris*
"Pipsissewa." *Chimaphila umbellata*
Piquillin-bush. *Condalia microphylla*
Piri-jiri-shrub. *Haloragis citriodora*
"Pisang." *Musa paradisaica*
Pishamin. Another name for Persimmon
Sour. *Carpodiscus acidus*
Sweet. *Carpodiscus dulcis*
"Pismire-tree." An old name for *Erythrina monosperma*
Pistachio-nut-tree. *Pistacia vera*
Pita-fibre, or Pita-thread-plant. *Agave americana* and some other species
Pitch-tree, Amboyna. *Dammara (Agathis) loranthifolia*
Burgundy. *Pinus Abies (Abies excelsa)*
Pitcher-plant, American. The genus *Sarracenia*; also *Darlingtonia californica*
Australian. *Cephalotus follicularis*
Blood-red. *Nepenthes sanguinea*
Bottle-like. *Nepenthes ampullacea*
Brilliant-red. *Nepenthes superba*
Californian. *Darlingtonia californica*
Chinese. *Nepenthes distillatoria*
E. Indian. The genus *Nepenthes*
Guiana. *Heliamphora nutans*
Hairy. *Nepenthes villosa*
Hooker's. *Nepenthes Hookeriana*
Intermediate. *Nepenthes intermedia*
Java Smooth. *Nepenthes lævis*
Lawrence's. *Nepenthes Lawrenciana*
Lindley's. *Nepenthes Lindleyana*
Loddige's. *Nepenthes Loddigesii*
Miss Morgan's. *Nepenthes Morganiana*
Rajah. *Nepenthes Rajah*
Singapore, Yellow. *Nepenthes Rafflesiana*
Singapore, White-margined. *Nepenthes albo-marginata*
South American. *Heliamphora nutans*

and Foreign Plants, Trees, and Shrubs. 109

Pitcher-plant, Two-spurred. *Nepenthes bicalcarata*
Ventricose. *Nepenthes Phyllamphora*
"Pitchery-Bidgery." *Duboisia Hopwoodii*
Pith-hat-plant. *Æschynomene aspera*
Pith-tree, of the Nile. *Herminiera Elaphroxylon (Ædemone mirabilis)*
Pitury-shrub. *Duboisia Hopwoodii*
Plane-tree. The genus *Platanus*
American. *Platanus occidentalis*
Californian. *Platanus racemosa*
"Canopy." *Platanus acerifolia umbellata*
Caucasian. *Platanus digitata*
Corstorphine. A yellow-leaved variety of *Acer Pseudo-platanus*
Eastern. *Platanus orientalis*
Maple-leaved. *Platanus acerifolia*
Mexican. *Platanus mexicana*
Oriental. *Platanus orientalis*
Scotch. The genus *Acer*
Spanish. *Platanus acerifolia var. hispanica*
Streaked-barked. *Platanus striata*
Various-leaved. *Platanus heterophylla*
Wedge-leaved. *Platanus cuneata*
Western. *Platanus occidentalis*
Planer-tree. *Planera aquatica*
Gmelin's. *Planera Gmelini*
Hornbeam-leaved. *Planera carpinifolia*
Small-leaved. *Planera parvifolia*
Plank-plant, Australian. *Bossiæa Scolopendrium*
Plant-of-Gluttony. *Cornus succica*
Plantage. An old name for Plantain (*Plantago*)
Plantain. The genus *Plantago*; applied also to some other plants
Broad-leaved. *Plantago maxima*
Buck's-horn. *Plantago Coronopus*
False. *Heliconia Bihai*
Greater. *Plantago major*
Hart's-horn. See Plantain, Buck's-horn
Hoary. *Plantago media*
Indian, of N. America. *Cacalia reniformis*
Indian, Tuberous. *Cacalia tuberosa*
Mud. The genus *Heteranthera*
Not-leaf. *Goodyera pubescens*
Parrot's. *Heliconia psittacorum*
Purple-leaved. *Plantago media var. atropurpurea*
Rattle-snake. See Rattlesnake-Plantain
Rib-wort. *Plantago lanceolata*
Robin's. *Erigeron bellidifolius*
Rose, or Rose-bracted. *Plantago major var. rosea*
Sea-side. *Plantago maritima*
Shrubby. *Plantago Cynops*
Water. *Alisma Plantago*
Water, Floating. *Alisma natans*
Water, Jamaica. *Pontederia azurea*
Water, Small. *Alisma ranunculoides*
Plantain-Lily. The genus *Funkia*
Blue. *Funkia cærulea (F. ovata)*
Fortune's. *Funkia Fortunei*
Golden-variegated. *Funkia undulata foliis aureo-variegatis*
Large-flowered. *Funkia grandiflora*
Siebold's. *Funkia Sieboldi*
Silver-variegated. *Funkia undulata foliis argenteo-variegatis*

Plantain-Lily, Spear-leaved. *Funkia lanceolata*
Sweet-scented. *Funkia japonica (F. grandiflora)*
White-flowered. *Funkia sub-cordata*
White-margined. *Funkia albo-marginata*
Plantain-Shore-weed. *Littorella lacustris*
Plantain-tree. *Musa paradisiaca*
Plants, "American." See "American Plants"
Pleurisy-root. *Asclepias tuberosa*
Ploughman's-Spikenard. *Inula Conyza*
American. *Baccharis angustifolia*
Plum-tree. The genus *Prunus*
American Wild. *Prunus americana*
"Assyrian." *Cordia Myxa* and *C. latifolia*
Australian. *Cargillia arborea* and *C. australis*
Batoko. *Flacourtia Ramontchi*
Beach. *Prunus maritima*
Black, of Illawarra. *Cargillia australis*
Blood. *Hæmatostaphis Barteri*
Californian Wild. *Prunus subcordata*
Canada. *Prunus americana*
Carolina. *Prunus caroliniana*
Ceylon. *Flacourtia sapida*
Cherry. *Prunus Myrobalana*
Chicasaw. *Prunus Chicasa*
Cocoa. *Chrysobalanus Icaco*
Cocomilla. *Prunus Cocomilla*
Common Cultivated. Varieties of *Prunus domestica (P. insititia)*
Damson. *Prunus domestica var. damascena*
Date, American. *Diospyros virginiana*
Date, Chinese. *Diospyros Kaki*
Date, European. *Diospyros Lotus*
Double-blossomed. *Prunus domestica flore-pleno*
Double-flowered Chinese. *Prunus sinensis fl.-pl.*
Downy. *Prunus pubescens*
E. Indian. *Flacourtia cataphracta* and *F. Ramontchi*
Gingerbread. *Parinarium macrophyllum*
Gray, Australian. *Cargillia arborea*
Gray, Sierra Leone. *Parinarium excelsum*
Green gage. *Prunus Claudiana*
Ground. *Astragalus caryocarpus*
Hog. Various species of *Spondias*; also *Ximenia americana*
Hog, Purple. *Spondias purpurea*
"Islay," of California. *Prunus ilicifolia*
Jamaica. *Spondias lutea*
Japanese. *Prunus japonica (P. sinensis)*
Japanese "Mume." *Prunus Mume*
Java. *Calyptranthes Jambolana*
Late-flowering. *Prunus serotina*
Maiden. The genus *Comocladia*
Malabar. *Eugenia Jambos*
Mobola, or Mola. *Parinarium Mobola*
Mountain, or Hog. *Ximenia americana*
"Mume." *Prunus Mume*
Myrobella. *Prunus Myrobalana*
Natal. *Arduina grandiflora*
Native, of N. S. Wales. *Podocarpus spinulosa*
Orleans. A variety of *Prunus domestica*
Pigeon. *Coccoluba Floridana*

Plum-tree, Pigeon, Sierra Leone. *Chrysobalanus ellipticus* and *C. luteus*
Port Arthur, or Tasmanian. *Cenarrhenes nitida*
Premier Swiss. *Prunus domestica var. Turonensis*
Queensland, Sour. *Owenia venosa*
Queensland, Sweet. *Owenia cerasifera*
Rough-skinned. *Parinarium excelsum*
Sand, or Beach. *Prunus maritima*
Sapodilla, or Sapotilla. *Achras Sapota*
Sea-side, of the W. Indies. *Ximenia americana*
Sebestens. *Cordia Myxa* and *C. latifolia*
"Spanish," of the W. Indies. *Spondias Mombin* and *Mammea humilis*
Spreading. *Prunus divaricata*
Sugar. *Malpighia saccharina*
Tamarind. *Dialium indicum*
Tasmanian. *Cenarrhenes nitida*
Three-lobed-leaved. *Prunus triloba*
Variegated-leaved. *Prunus domestica variegata*
Various-leaved. *Prunus domestica var. heterophylla*
Weeping. *Prunus cerasifera*
Whitish-leaved. *Prunus candicans*
Wild. *Prunus communis*
Wild, American. *Prunus americana*
Wild, British Columbia. *Prunus subcordata*
Wild, N. S. Wales. *Achras (Sapota) australis*
Wild, of the Cape. *Pappea capensis*
Wild-Goose. An improved variety of *Prunus Chicasa*
Plum-bush, Australian. *Astrotricha pterocarpa*
Plume-grass. See Grass
Plume-Nutmeg. *Atherosperma moschata*
Plume-thistle. See Thistle
Podophyllin-plant. *Podophyllum peltatum*
Poe-plant, of the Sandwich Islands. *Colocasia esculenta*
Pohutu-kawa-tree, of New Zealand. *Metrosideros tomentosa*
Poison-berry. W. Indian. *Beurreria esculenta* and various species of *Cestrum*
Poison-bulb, Asiatic. *Crinum asiaticum*
Cape. *Buphane (Hæmanthus) toxicaria*
Poison-dart, of the Philippine Islands. *Aglaonema commutata*
Poison-Oak, Californian. *Rhus diversiloba*
Poison-plant, Horse, of Australia. *Swainsona Greyana*
Sheep, of Australia. *Gastrolobium bilobum, G. Callistachys*, and other species; also *Lotus australis*
Tame. *Vincetoxicum officinale*
York Road, of W. Australia. *Gastrolobium calycinum*
Poison-tree, Jamaica. *Rhus arborea*
Queensland. *Croton Verreauxii*
Poison-wood, W. Indian. *Sebastiania lucida*
Poke, or Poke-weed. The genus *Phytolacca*
E. Indian. *Phytolacca acinosa*
Electrical. *Phytolacca electrica*

Poke, Hydrangea-leaved. *Phytolacca icosandra*
Indian. *Veratrum viride*
Purple-flowered. *Phytolacca purpurea*
Red-stemmed. *Phytolacca icosandra*
Tree. *Phytolacca dioica*
Virginian. *Phytolacca decandra*
Polar-Plant. *Silphium laciniatum*
Pole-cat-weed. *Symplocarpus fœtidus*
Wild. *Convolvulus panduratus*
Poly, Golden. *Teucrium aureum*
Mountain. *Bartsia alpina*
Yellow. *Teucrium flavescens*
Polyanthus. A garden variety of the Ox-lip (*Primula elatior*)
Polypody. The genus *Polypodium*
Alpine. *Polypodium alpestre*
American. *Polypodium incanum*
Climbing. *Niphobolus Heteractis*
Female. *Athyrium Filix-fœmina*
Golden. *Polypodium (Phlebodium) aureum*
Hoary. *Polypodium incanum*
Irish. *Polypodium hibernicum*
Mule. *Lastrea Filix-mas*
Rigid Three-branched. *Polypodium calcareum*
Scented. *Polypodium pustulatum*
Twining. *Polypodium serpens*
Welsh. *Polypodium Cambricum*
Pomegranate-tree. *Punica Granatum*
Double-flowered. *Punica Granatum fl.-pl.*
Dwarf. *Punica Granatum var. nana*
"**Pomme de Prairie**," or "Pomme Blanche." *Psoralea esculenta*
Pomeloes (Large-sized Shaddocks). *Citrus decumana*
Pond-Dog-wood, Louisiana White. *Cephalanthus occidentalis*
Pond-spice. *Tetranthera geniculata*
Pond-weed. The genus *Potamogeton*
Broad-leaved. *Potamogeton natans*
Cape. *Aponogeton distachyon*
Choke. *Anacharis Alsinastrum*
Close-leaved. *Potamogeton densus*
Curly. *Potamogeton crispus*
Fennel-leaved. *Potamogeton pectinatus*
Flat-stalked. *Potamogeton compressus*
Grass-leaved. *Potamogeton gramineus*
Hawthorn-scented. *Aponogeton distachyon*
Horned. *Zannichellia palustris*
Long. *Potamogeton prælongus*
Perfoliate. *Potamogeton perfoliatus*
Reddish. *Potamogeton rufescens*
Shining. *Potamogeton lucens*
Slender. *Potamogeton pusillus*
Tassel. *Ruppia maritima*
Various-leaved. *Potamogeton heterophyllus*
Wading. *Stratiotes aloides*
Pond-wort, Knight's. An old name for *Stratiotes aloides*
Pony. *Tecoma serratifolia*
Pool-rush. Another name for Bul-rush
Poon-wood-tree. A species of *Calophyllum*
Poor-man's-Parmacetie. An old name for *Capsella Bursa-pastoris*
Poor-man's-Rhubarb. *Thalictrum flavum*

Poor-man's-Weather-glass. *Anagallis arvensis*
Pop-corn. A variety of *Zea Mays*, used for parching
Popering, or Poperin. A variety of Pear mentioned by Shakespeare (" Romeo and Juliet," Act ii., scene 1)
Pope's-head. *Melocactus communis*
Poplar. The genus *Populus*
 Athenian. *Populus græca*
 Berry-bearing. *Populus monilifera*
 Black. *Populus nigra*
 Black Italian. *Populus acladesca*
 Californian. *Populus trichocarpa* and *P. Fremontii*
 Carolina. *Populus monilifera*
 Downy. *Populus heterophylla*
 Fragrant. *Populus suaveolens*
 Gray. *Populus alba var. canescens*
 Laurel-leaved. *Populus laurifolia*
 Lombardy. *Populus fastigiata*
 Necklace. *Populus monilifera*
 Ontario. *Populus balsamifera var. candicans* (*P. macrophylla*)
 Queensland. *Carumbium populifolium* (*Omalanthus populifolius*)
 Rocky Mountain. *Populus angustifolia*
 Soft, or Paper. *Populus grandidentata*
 "Tacamahac." *Populus balsamifera*
 Weeping. *Populus pendula*
 Weeping Gray. *Populus canescens var. pendula*
 "Western." *Liriodendron tulipiferum*
 White. *Populus alba*
 Willow-leaved. *Populus nigra var. salicifolia*
 "Yellow." *Liriodendron tulipiferum*
Poppy. The genus *Papaver*
 Alpine. *Papaver alpinum*
 Blue Himalayan. *Meconopsis aculeata*
 Blue, Wallich's. *Meconopsis Wallichii*
 Californian. *Platystemon californicus* and the genus *Eschscholtzia*
 Californian Orange. *Eschscholtzia californica crocea*
 Californian Pink. *Eschscholtzia californica crocea rosea*
 Californian White. *Eschscholtzia californica alba*
 Carnation. *Papaver somniferum*
 Cathcart's. *Cathcartia villosa*
 Caucasian Scarlet. *Papaver umbrosum*
 Celandine. *Stylophorum diphyllum*
 Corn. *Papaver Rhœas*
 Corn, Smooth-fruited. *Papaver dubium*
 Danebrog. A variety of *Papaver somniferum* (?)
 Dark-spotted. *Papaver umbrosum*
 "Frothy." *Silene inflata*
 Golden. *Papaver croceum*
 Great Scarlet. *Papaver bracteatum*
 Hairy-stemmed. *Papaver pilosum*
 Horned. See Horned-Poppy
 Iceland. A variety of *Papaver nudicaule*
 Joan, or John, Silver-pin. *Papaver Rhœas fl.-pl.*
 Long-headed. *Papaver dubium*
 Mexican, or Prickly. *Argemone mexicana*
 Mexican Large-flowered. *Argemone grandiflora*
 Mexican White. *Argemone grandiflora*
 Mexican Yellow. *Argemone speciosa*
 Nepaul. *Meconopsis nepalensis*
 Opium. *Papaver somniferum*
 Opium, Black. *Papaver somniferum var. nigrum*
 Opium, Double. *Papaver somniferum fl.-pl.*
 Orange. *Papaver lateritium*
 Oriental. *Papaver orientale*
 Pæony. *Papaver somniferum*
 Pale. *Papaver Argemone*
 Plume. The genus *Bocconia*
 Prickly. See Poppy, Mexican
 Prickly, Himalayan. *Meconopsis aculeata*
 Rough. *Papaver hybridum*
 Sea-side. *Glaucium luteum*
 Sikkim. *Meconopsis simplicifolia*
 "Spatling." *Silene inflata*
 Spiked. *Papaver spicatum*
 Thompson's. *Papaver umbrosum*
 Tree. *Dendromecon rigidum*
 Welsh. *Meconopsis cambrica*
 White. *Papaver somniferum*
 Yellow Arctic. *Papaver nudicaule*
Poppy-Mallow. The genus *Callirrhoë*
 Crimson-flowered. *Callirrhoë involucrata*
 Finger-leaved. *Callirrhoë digitata*
 Long-stalked. *Callirrhoë pedata*
 White-flowered. *Callirrhoë macrorrhiza*
Poppy-tree, Californian. *Dendromecon rigidum*
Poppy-wort, Satin. *Meconopsis Wallichiana*
Porcelain. An old name for Purslane
Porcupine-wood. The hard outer part of the trunk of *Cocos nucifera*
Portland-Starch-root. *Arum maculatum*
Potato, Canada. *Helianthus tuberosus*
 Common. Varieties of *Solanum tuberosum*
 Creeping-stemmed. *Solanum stoloniferum*
 Hog's. *Zygadenus venenosus*
 Madagascar. *Solanum Anguivi*
 Mic-mac. *Apios tuberosa*
 Native, of N. S. Wales. *Marsdenia viridiflora*
 Spanish, or Sweet. *Batatas edulis*
 Sea-side, of India. *Ipomœa Pes-capræ*
 Sweet. See Potato, Spanish
 Sweet, Giant. *Batatas paniculata*
 Tasmanian Native. *Gastrodia sesamoides*
 Telinga. *Amorphophallus campanulatus*
 Wild. *Convolvulus panduratus*
 Wild, of the W. Indies. *Ipomœa fastigiata*
Potato-Fungus. See Fungus
Potato-tree. *Solanum crispum*
Potato-Vine, Wild. *Ipomœa pandurata*
Potentil, Marsh. *Potentilla Comarum* (*Comarum palustre*)
Pot-herb, Black. *Smyrnium Olusatrum*
 White. *Valerianella olitoria*
Pot-plant. *Lecythis Ollaria*
Pottery-tree, French Guiana. *Caraipa angustifolia*
 Pará. *Moquilea utilis*
Pounce-tree. *Callitris quadrivalvis*
Pow-itch. The fruit of *Pyrus rivularis*
Prangos Hay-plant. *Prangos pabularia*
"Prattling Parnell." An old name for *Saxifraga Geum var. serrata*

Preacher-in-the-pulpit. *Orchis spectabilis*
Pretty-grass. The genus *Calochortus*
  Yellow. *Calochortus luteus*
Pricket. An old name for *Sedum acre*
Prickly Pear. *Cactus Opuntia*
Prickly Pettigree. *Ruscus aculeatus*
Prickly-pole. *Bactris Plumieriana*
Prickly Poppy. See Poppy
Prickly-withe. *Cereus triangularis*
Prick-wood. *Cornus sanguinea* and *Euonymus europæus*
Pride-of-Barbadoes. *Cæsalpinia pulcherrima*
Pride-of-China. *Melia Azedarach*
Pride-of-Columbia. *Phlox speciosa*
Pride-of-India. *Melia Azedarach*
Pride-weed. *Erigeron canadensis*
Priest's-crown. *Taraxacum officinale*
Prim-print, or **Prim.** *Ligustrum vulgare*
Primrose. The genus *Primula;* also an old name for Privet
  Abyssinian. *Primula verticillata*
  Allioni's. *Primula Allioni*
  Altaian. *Primula altaica*
  Bird's-eye. *Primula farinosa*
  Bird's-eye, Brilliant. *Primula farinosa var. superba*
  Bird's-eye, Scotch. *Primula scotica*
  Californian. *Primula suffrutescens*
  Cape. The genus *Streptocarpus*
  Carniolic. *Primula carniolica*
  Cashmere. *Primula cashmeriana*
  Caucasian. *Primula amœna*
  Changeable. *Primula commutata*
  Chinese. *Primula sinensis*
  Chinese, Fringed. *Primula sinensis fimbriata*
  Clammy. *Primula viscosa*
  Common. *Primula acaulis*
  Cortusa-leaved. *Primula cortusoides*
  Creamy-flowered. *Primula involucrata*
  De Candolle's. *Primula Candolleana*
  Ear-leaved. *Primula auriculata*
  Entire-leaved. *Primula integrifolia*
  Evening. See Evening-Primrose
  Fairy. *Primula minima*
  Floerk's. *Primula Floerkiana*
  Fringed. *Primula ciliata*
  Fortune's. *Primula Fortunei (P. erosa)*
  Glaucous-leaved. *Primula glaucescens*
  Glutinous. *Primula glutinosa*
  Hairy-leaved. *Primula hirsuta*
  Heavy-scented. *Primula graveolens*
  Henry's. *Primula Henryi*
  Himalayan. *Primula Sikkimensis*
  Japanese. *Primula japonica*
  Java. *Primula imperialis*
  Kitaibel's. *Primula Kitaibelii*
  Large-calyxed. *Primula calycina*
  Munro's. *Primula Munroi*
  Neapolitan. *Primula Palinuri*
  Parry's. *Primula Parryi*
  Piedmont. *Primula Pedemontana*
  Round-headed Himalayan. *Primula capitata*
  Round-headed Purple. *Primula purpurea*
  Round-headed Yellow-eyed Purple. *Primula pulcherrima*
  Rosy-flowered. *Primula rosea*
  Scotch. *Primula scotica*
  Shaggy-leaved. *Primula villosa*
  Showy. *Primula spectabilis*
  Siebold's. *Primula Sieboldi*
  Silver-edged. *Primula marginata*
  Snow-white. *Primula nivalis* and *P. nivea*
  Soft-leaved. *Primula mollis*
  Stein's. *Primula Steini*
  Strong-scented. *Primula graveolens*
  Stuart's. *Primula Stuartii*
  Toothed-leaved. *Primula denticulata*
  Tree. The genus *Œnothera*
  Tyrolese. *Primula Tyrolensis*
  Veitch's. *Primula cortusoides amœna*
  Whorled. *Primula verticillata*
  Wulfen's. *Primula Wulfeniana*
  Yellowish. *Primula luteola*
Primrose-Peerless. *Narcissus biflorus*
Prince's-Feather. *Amarantus hypochondriacus*
Prince's-Pine. *Chimaphila umbellata*
Prince-wood. The wood of *Cordia gerascanthoides* and *Hamelia ventricosa*
Privet. The genus *Ligustrum*
  Barren. *Rhamnus Alaternus*
  Box-leaved. *Ligustrum vulgare var. buxifolium*
  Chinese. *Ligustrum sinense*
  Chinese Dwarf. *Ligustrum sinense nanum*
  Chinese Wax. *Ligustrum lucidum*
  Common. *Ligustrum vulgare*
  Egyptian. *Lawsonia alba*
  Evergreen. *Ligustrum vulgare sempervirens (L. v. italicum)*
  Fortune's. *Ligustrum Fortunei*
  Gold-blotched. *Ligustrum japonicum variegatum*
  Hairy. *Ligustrum villosum*
  Japanese. *Ligustrum japonicum*
  Mock. The genus *Phillyrea*
  Oval-leaved. *Ligustrum ovalifolium*
  Pipe. *Syringa vulgaris*
  Shining-leaved. *Ligustrum japonicum lucidum*
  Spike-flowered. *Ligustrum spicatum*
  Syrian. *Fontanesia phillyræoides*
  Variegated-leaved. *Ligustrum vulgare variegatum*
  White-berried. *Ligustrum vulgare leucocarpum*
  Yellow-berried. *Ligustrum vulgare xanthocarpum*
Procession-flower. *Polygala vulgaris*
Prophet-flower. *Arnebia echioides*
Prune-tree. *Prunus domestica*
  W. Indian. *Prunus occidentalis*
  Wild, of the Cape. *Sapindus Pappea*
Puchero-plant. *Talinum patens*
Puccoon, Hairy. *Lithospermum hirtum*
  Hoary. *Lithospermum canescens*
  Red. *Sanguinaria canadensis*
  Yellow. *Hydrastis canadensis*
Pucha-pat. *Pogostemon Patchouly*
Pudding-berry. *Cornus canadensis*
Pudding-grass. An old name for Pennyroyal
Pudding-pipe-tree. *Cassia Fistula*
Puff-ball. The genus *Lycoperdon*
  Common. *Lycoperdon Bovista*

and Foreign Plants, Trees, and Shrubs.

Puff-ball, False. *Scleroderma Cepa*
False, Warty. *Scleroderma verrucosum*
Giant. *Lycoperdon giganteum*
Pukatea-tree. *Laurelia Novæ-Zelandiæ* and *Griselinia lucida*
Pulas-tree. *Butea frondosa* and *B. superba*
Pulque-plant. *Agave americana* and other species
Pumpkin, Squash, or Gourd. Various species of *Cucurbita*
Common. *Cucurbita Pepo*
Elephant. *Cucurbita maxima*
Melon. *Cucurbita Melopepo*
Punk. See Amadou
Purification-flower. *Galanthus nivalis*
Puriri-tree. *Vitex littoralis*
Purple-fringe. *Rhus Cotinus*
Purple-heart Tree, Guiana. *Copaifera bracteata* and *C. pubiflora*
Trinidad. *Peltogyne paniculata*
W. Indian. *Copaifera officinalis*
Purple Jacobæa. *Senecio elegans*
Purple-tassels. *Muscari comosum*
Purples. See Ear-cockle
Long. See Long-Purples
Purple-wort. *Comarum palustre*
Purslane. The genus *Portulaca*
Common Garden. *Portulaca oleracea*
Crimson-flowered. *Portulaca Thellussoni*
Cuban Winter. *Claytonia perfoliata*
Milk. *Euphorbia maculata*
Red-flowered. *Portulaca splendens*
Rock. *Calandrinia umbellata*
Rock, Shining. *Calandrinia nitida*
Sea. *Arenaria peploides* and *Atriplex portulacoides*
Sea, American. *Sesuvium Portulacastrum*
Siberian. *Claytonia sibirica*
Water. *Peplis Portula* and *Isnardia palustris*
Water, American. *Ludwigia palustris*
Winter. *Claytonia perfoliata*
Yellow-flowered. *Portulaca aurea*
Purslane-tree. *Portulacaria afra*
Broad-leaved, Sea. *Atriplex Halimus*
Pussy's-foot. *Antennaria plantaginifolia*
"Putchuck." *Aristolochia recurvilabra*
Putty-root. *Aplectrum hyemale*
Puya-fibre-plant. *Bæhmeria Puya*
"Pyracantha." *Cratægus Pyracantha*
White-berried. *Cratægus Pyracantha fructū-albo*
Pyramid-tree, Golden. *Guilfoylia (Cadellia) monostylis*

Quacksalver's Turbith, or Quacksalver's Spurge. *Euphorbia Esula*
Quamash, or Camash. *Camassia esculenta*
Californian. *Camassia Leichtlini*
"Death." *Zygadenus venenosus*
Eastern. *Scilla Fraseri*
White-flowered. *Camassia esculenta var. alba* and *C. Leichtlini*
Quandang-nut. See Nut
Quandong-tree. *Santalum Preissianum (S. acuminatum)*
Quassia-tree, or Picræna-wood, Jamaica. *Picræna (Quassia) excelsa*
Surinam. *Quassia amara*

Quebracho-tree, Red. *Loxopterygium Lorentzii*
White. *Aspidosperma Quebracho*
Queen-Lily. See Lily
Queen-Mother-Herb. An old name for Tobacco
Queen-of-the-Meadow. *Spiræa salicifolia*
Queen-of-the-Orchids. See Orchids
Queen-of-the-Prairie. *Spiræa lobata*
Queen's-cushion. *Saxifraga hypnoides*
Queen's-delight. See Queen's-root
Queen's-flower. *Lagerstræmia Reginæ*
Queen's-root, or Queen's-delight. *Styllingia sylvatica*
Quick, or Quick-set, Thorn. See Thorn
Quick-beam, Quicken, or Wicken-tree. *Pyrus Aucuparia*
Quillai-tree. See Cullary
Quill-wort. *Isoëtes lacustris*
Mountain. *Isoëtes alpinus*
Quinancy-wort, Quinsy-wort, or Squinancywort. *Asperula Cynanchica*
Quince-tree. *Pyrus Cydonia*
Bastard. *Pyrus Chamæmespilus*
Bengal. *Ægle Marmelos*
Common. *Cydonia vulgaris*
Maule's Japanese. *Pyrus Maulei*
Portugal. *Cydonia vulgaris var. lusitanica*
Quinine-plants. Chiefly *Cinchona lancifolia, C. Palton, C. Pitayensis,* and *C. cordifolia*
Quinoa, or White Quinoa. *Chenopodium Quinoa*
Quinsy-berry. The fruit of *Ribes nigrum*
Quinsy-wort. *Asperula Cynanchica*
Quirinca-pods. The pods of *Acacia Cavenia*
Quiver-tree. *Aloë dichotoma*

Rabbit-berry. *Shepherdia argentea*
Rabbit-foot. *Trifolium arvense*
Rabbit-root. *Aralia nudicaulis*
Rabone. An old name for Radish
Raccoon-berry. *Podophyllum peltatum*
Races. A commercial name for Ginger
Radish, Cultivated. *Raphanus sativus*
Italian. *Raphanus Landra*
Rat-tail. *Raphanus caudatus*
Seaside. *Raphanus maritimus*
Water. *Nasturtium amphibium*
Water, Annual. *Nasturtium palustre*
Wild. *Raphanus Raphanistrum*
Raffia, or Roffia-plant. *Raphia Ruffia* and *R. tædigera*
Rag, or Rags, Hazel. *Sticta pulmonacea*
Ragee-plant. *Eleusine coracana*
Ragged-Robin. *Lychnis Flos-cuculi*
Rag-weed, or Rag-wort, African. The genus *Othonna*
American Golden. *Senecio aureus*
Barbary. *Othonna cheirifolia*
Common. *Senecio Jacobæa*
Great American. *Ambrosia trifida*
Oak-leaved. *Senecio aquaticus*
Oxford. *Senecio squalidus*
Sea-side. *Cineraria maritima*
Water. *Senecio squalidus*
Woolly American. *Senecio tomentosus*

Rain-berry, or Rhine-berry-thorn. *Rhamnus catharticus*
Rainbow-flower. The genus *Iris*
Rain-tree. *Pithecolobium Saman*
Raisin-tree, Japanese. *Hovenia dulcis* and *Ribes rubrum*
Ram. An old name for the genus *Rhamnus*
Ram-goat. *Fagara microphylla*
Ram-of-Libya. *Paliurus aculeatus*
Rambutan, or Rampostan. The fruit of *Nephelium lappaceum*
Ramee, or Ramie-grass. See Grass
"Ramleh," of Rangoon. The fruit of *Picrardia sapida*
Ramoon-tree. The genus *Trophis*
Rampion, Garden. *Campanula Rapunculus*
Horned. See Horned-Rampion
Large. *Œnothera biennis*
Rampostan. See Rambutan
Ram's-foot. An old name for *Ranunculus aquatilis*
Ram's-head. *Cypripedium arietinum*. Also applied to the seeds of *Cicer arietinum*
Ram's-horns. *Orchis mascula*
Ramson's. *Allium ursinum*
"Ramsted," of N. America. *Linaria vulgaris*
Ram-til-plant. *Guizotia oleifera*
Rangoon-Creeper. See Creeper
Ransted. See Ramsted
Rantry. *Pyrus Aucuparia*
Ranunculus. (See also Crow-foot)
American Early. *Ranunculus fascicularis*
Bristly. *Ranunculus pennsylvanicus*
Bulbous-rooted. *Ranunculus bulbosus*
Celery-leaved. *Ranunculus sceleratus*
Common Garden. Varieties of *Ranunculus asiaticus*
Corn-field. *Ranunculus arvensis*
Creeping. *Ranunculus repens*
Double Persian. *Ranunculus asiaticus fl.-pl.* in var.
Fig-wort. *Ranunculus Ficaria*
Globe. The genus *Trollius*
Great. *Ranunculus Lingua*
Hairy. *Ranunculus hirsutus*
Hooked. *Ranunculus recurvatus*
Ivy-leaved. *Ranunculus hederaceus*
Lenormand's. *Ranunculus Lenormandi*
Meadow. *Ranunculus acris*
Sea-side. *Ranunculus Cymbalaria*
Small-flowered. *Ranunculus parviflorus*
Snake-tongue. *Ranunculus ophioglossifolius*
Spear. *Ranunculus Flammula*
Water. *Ranunculus aquatilis*
Water, Yellow. *Ranunculus multifidus*
Wood. *Ranunculus auricomus*
Rape. *Brassica Napus*
Summer. *Brassica campestris*
Rasp-berry, American Wild Red. *Rubus strigosus*
Black American. *Rubus occidentalis*
Common. *Rubus Idæus*
Dwarf American. *Rubus triflorus*
Ground. *Hydrastis canadensis*
Himalayan. *Rubus rugosus*

Rasp-berry, Nootka-Sound. *Rubus Nutkanus*
Purple-flowered American. *Rubus odoratus*
Showy-flowered. *Rubus spectabilis*
Victorian. *Rubus rosæfolius*
White-skinned. *Rubus leucodermis*
Wild Red American. *Rubus strigosus*
Rasp-berry-jam Tree. *Acacia acuminata*
Rasp-pod, Queensland. *Flindersia australis*
Raspis. An old name for Rasp-berry
Rata-tree, of New Zealand. *Metrosideros florida* and *M. robusta*
Rat-poison-plant, of Sierra Leone. *Chailletia toxicaria*
Of the W. Indies. *Hamelia patens*
Rattan-Cane. *Calamus Draco* (*C. Rotang*)
Rattle, Red. *Pedicularis sylvatica*
Yellow, or Rattle-box. *Rhinanthus Crista-galli*
Rattle-box. *Crotalaria sagittalis*. (See also Rattle, Yellow)
Rattle-root. *Cimicifuga racemosa*
Rattle-snake-leaf. *Goodyera pubescens*
Rattle-snake-plant, Cape. *Crotalaria arborescens*
Rattle-snake-Plantain, Creeping. *Goodyera repens*
Silvery. *Goodyera pubescens*
Rattle-snake-root. *Trillium latifolium*, *Nabalus albus*, and *N. virgatus*
Rattle-snake-weed. *Hieracium renosum*
Rattle-snake's-Master. *Agave virginica*, *Eryngium yuccæfolium*, and *Liatris pilosa*
Rattle-weed. *Astragalus lentiginosus*
Californian. The genus *Astragalus*
Ray-flower, Australian. The genus *Anthocercis*
Ray-grass, or Rye-grass. See Grass
Ray-pod. *Actinocarpus Damasonium*
Red-berry, or Sea-berry, Australian. The genus *Rhagodia*
Red-bud. *Cercis canadensis*
Californian. *Cercis occidentalis*
Red-head. *Asclepias curassavica*
Red-hot-poker-plant. *Tritoma Uvaria* and other species
Red-ink-plant. *Phytolacca decandra*
Red-knees. *Polygonum Hydropiper*
Red-legs. *Polygonum Bistorta*
Red Morocco. *Adonis autumnalis*
Red-ray. An old name for Darnel
Red-root. *Ceanothus americanus*
Blue-flowered. *Ceanothus azureus*
Box-leaved. *Ceanothus buxifolius*
Carolina. *Lachnanthes tinctoria*
Intermediate. *Ceanothus intermedius*
Ovate-leaved. *Ceanothus ovatus*
Red-twigged. *Ceanothus sanguineus*
Small-leaved. *Ceanothus microphyllus*
Red-rot. *Drosera rotundifolia*
Red-shanks. *Polygonum Persicaria* and *Geranium Robertianum*
Mossy. *Tillæa muscosa*
Red-weed. *Papaver Rhœas*
Red-water-tree, of Sierra Leone. *Erythrophlœum leonense*

Red - wood - tree, Andaman. *Pterocarpus dalbergioides*
Bahama. *Ceanothus colubrinus* (*Colubrina ferruginosa*)
Californian. *Ceanothus spinosus*
Californian Evergreen. *Sequoia* (*Taxodium*) *sempervirens*
Coromandel. *Soymida febrifuga*
E. Indian. *Pterocarpus santalinus*
Jamaica. *Gordonia Hæmatoxylon*
Laurel-leaved. *Erythroxylon laurifolium*
St. Helena. *Melhania Erythroxylon*
Santa Barbara. *Ceanothus spinosus*
Reed. The genus *Arundo*
Canary. *Phalaris* (*Digraphis*) *arundinacea*
Common. *Arundo Phragmites*
Great. *Arundo Donax*
Gromwell. See Gromwell-reed
Indian, or Flowering Indian. *Canna indica*
New Zealand. *Arundo conspicua*
Paper. See Paper-reed
Sea-sand. *Calamagrostis arenaria*
Silvery. *Arundo conspicua*
Sweet. *Sorghum saccharatum*
"Thorny," of Peru. *Cactus peruvianus*
Trumpet. *Arundo occidentalis*
Wood. *Calamagrostis lanceolata*
Wood, Silvery. *Calamagrostis argentea*
Writing. *Calamagrostis Epigeios*
Reed-mace. *Typha latifolia*
Reem. *Calamus Flagellum*
Reree-plant, of India. *Typha angustifolia*
Resin-bush, of S. Africa. *Euryops speciosissimus*
"Resino," of Juan Fernandez. *Robinsonia thurifera*
Resin-plant, Carana. *Bursera acuminata*
Chibou, or Cachibou. *Bursera gummifera*
Copal, Brazilian. *Hymenæa Courbaril*
Copal, Madagascar. *Trachylobium* (*Hymenæa*) *verrucosum*
Copal, New Zealand. *Dammara australis*
Copal, N. American. *Rhus copallina*
Copal, W. African. *Guibourtia copalifera*
Copal, Zanzibar. *Trachylobium* (*Hymenæa*) *Hornemannianum*
Coumia. *Icica Tacamahaca*
Elemi. *Amyris Plumieri*
Guaiac. *Guaiacum officinale*
Hotai. Supposed to be *Balsamodendron Playfairii*
Manawa. *Avicennia tomentosa*
Mastich. *Pistacia Lentiscus*
Maynas. *Calophyllum Calaba*
Rest-harrow. The genus *Ononis*
Apulian. *Ononis Apula*
Aragon. *Ononis arragonensis*
Bristly. *Ononis hispida*
Cape. *Ononis capensis*
Clammy. *Ononis viscosa*
Common. *Ononis arvensis*
Denhardt's. *Ononis Denhardtii*
Drooping. *Ononis pendula*
Few-leaved. *Ononis oligophylla*
Long-flower-stalked. *Ononis peduncularis*
Long-leaved. *Ononis longifolia*
Mt. Cenis. *Ononis cenisia*

Rest-harrow, Notch-leaved. *Ononis emarginata*
Painted-flowered. *Ononis picta*
Pigmy. *Ononis minutissima*
Pointed-leaved. *Ononis cuspidata*
Ram. *Ononis Natrix*
Rooting-branched. *Ononis procurrens*
Round-headed. *Ononis capitata*
Round-leaved. *Ononis rotundifolia*
Sand. *Ononis arenaria*
Short-flowered. *Ononis breviflora*
Short-podded. *Ononis brachycarpa*
Shrubby. *Ononis fruticosa*
Sickle-podded. *Ononis falcata*
Small. *Ononis reclinata*
Smooth. *Ononis glabra*
Spanish. *Ononis hispanica*
Spreading. *Ononis diffusa*
Strong-scented. *Ononis fœtida*
Tall. *Ononis antiquorum*
Three-toothed. *Ononis tridentata*
Tree. *Ononis arborescens*
Twin-flowered. *Ononis geminiflora*
Two-flowered. *Ononis biflora*
Very branching. *Ononis ramosissima*
Very narrow-leaved. *Ononis angustissima*
White-flowered. *Ononis alba*
Yellow-flowered. *Ononis Natrix*
Resurrection - plant. *Anastatica Hierochuntina, Mesembryanthemum Tripolium,* and *Selaginella lepidophylla*
Rewa - Rewa - tree, of New Zealand. *Knightia excelsa*
Rhatany-root. The root of *Krameria triandra*
Savanilla. The root of *Krameria Ixina*
Rheea-grass. See Grass
Rheumatism-root. *Jeffersonia diphylla*
Rhine-berry. See Rain-berry
Rhodes-wood. The wood of *Amyris balsamifera*
Rhododendron, Azalea-like. *Rhododendron azaleoides*
Bearded-flowered. *Rhododendron anthopogon*
Bearded-leaf-stalked. *Rhododendron barbatum*
Bell-flowered. *Rhododendron campanulatum*
Blandfordia-flowered. *Rhododendron blandfordiæflorum*
Blunt-leaved. *Rhododendron retusum*
Bristly. *Rhododendron setosum*
Californian. *Rhododendron californicum*
Canadian. *Rhodora canadensis*
Catawba. *Rhododendron Catawbiense*
Catesby's. *Rhododendron Catesbæi*
Caucasian. *Rhododendron caucasicum*
Citron-flowered. *Rhododendron citrinum*
Curved-podded. *Rhododendron campylocarpum*
Daurian. *Rhododendron dauricum*
Dotted-leaved. *Rhododendron punctatum*
Early-blooming. *Rhododendron præcox*
Epiphytal. *Rhododendron Dalhousiæ*
Fringed. *Rhododendron ciliatum*
Gibson's. *Rhododendron Gibsonii*
Glaucous-leaved. *Rhododendron glaucum*

Rhododendron, Golden-flowered. *Rhododendron chrysanthum*
Hairy-leaved. *Rhododendron hirsutum*
Herbert's Hybrid. *Rhododendron hybridum*
Javanese. *Rhododendron javanicum*
Jessamine - flowered. *Rhododendron jasminiflorum*
Kamtschatka. *Rhododendron camtchaticum*
Lady Dalhousie's. *Rhododendron Dalhousiæ*
Lapland. *Rhododendron lapponicum*
Largest. *Rhododendron maximum*
Long-flowered. *Rhododendron longiflorum*
Major Madden's. *Rhododendron Maddeni*
Metternich's. *Rhododendron Metternichii*
Mrs. Champion's. *Rhododendron Championæ*
Mrs. Farrer's. *Rhododendron Farreræ*
Myrtle-leaved. *Rhododendron myrtifolium*
Neilgherry. *Rhododendron Nilagiricum*
Pontic. *Rhododendron ponticum*
Purple-flowered. *Rhododendron purpureum* and *R. ponticum*
Pursh's. *Rhododendron Purshii*
Rajah Brooke's. *Rhododendron Brookeanum*
Rusty-leaved. *Rhododendron ferrugineum*
Scaly. *Rhododendron lepidotum*
Silvery. *Rhododendron argenteum*
Slender. *Rhododendron gracile*
Slender-branched. *Rhododendron virgatum*
Snowy-leaved. *Rhododendron niveum*
Sweet-scented. *Rhododendron ponticum var. odoratum*
Thyme-leaved. *Rhododendron Chamæcistus*
Tree. *Rhododendron arboreum*
Vermilion-flowered. *Rhododendron cinnabarinum*
White-flowered. *Rhododendron albiflorum*
Whorl-leaved. *Rhododendron verticillatum*
Yellow-flowered. *Rhododendron chrysanthum*

Rhubarb. The genus *Rheum*
Bog. *Petasites vulgaris*
Bucharian. *Rheum undulatum*
Curled-leaved. *Rheum crispum*
Currant-fruit. *Rheum Ribes*
Garden, or Tart. Varieties of *Rheum Rhaponticum*
Monk's. *Rumex alpinus* and *R. Patientia*
Nepaul. *Rheum australe*
Palmate-leaved. *Rheum palmatum*
Poor-man's. *Thalictrum flavum*
Red-veined. *Rheum Emodi*
Sikkim. *Rheum nobile*
Tart. See Rhubarb, Garden
Thick-leaved. *Rheum compactum*
Turkey. *Rheum palmatum*
Warted-leaved. *Rheum Ribes*
Wavy-leaved. *Rheum undulatum*
Rib-grass. See Grass
Ribbon-flower, Cape. *Spatalanthus speciosus*
Ribbon-grass. See Grass
Ribbon-tree. *Plagianthus betulinus*
Ribbon-weed. *Laminaria saccharina*
Ribbon-wood, New Zealand. *Plagianthus Lyalli*
Otago. *Hoheria populnea*

Rice, American Mountain. The genus *Oryzopsis*
Canada. See Rice, Indian
Cultivated. *Oryza sativa*
Hungary. *Paspalum exile*
Indian, or Canada. *Zizania aquatica* (*Hydropyrum esculentum*)
Millet. *Panicum colonum*
Mountain. *Oryza mutica*
Petty, of Peru. The seeds of *Chenopodium Quinoa*
Rice Cut-grass. See Grass
Rice-flower. The genus *Pimelea*
Rice-paper-plant, Chinese. *Aralia papyrifera*
Malay. *Scævola Taccada*
Rich-weed. *Pilea pumila* and *Collinsonia canadensis*
Rimu-tree. *Dacrydium cupressinum*
Ring-worm-root. *Rhinacanthus communis* (*Justicia nasuta*)
Ring-worm-shrub. *Cassia alata*
Rivulet-tree, Australian. *Glochidion australe* (*Bradleia australis*)
Roan-tree. See Rowan-tree
Roast-beef-plant. *Iris fœtidissima*
Robin-run-in-the-hedge. *Nepeta Glechoma*
Robin's-pincushion. The Bedeguar of the Rose
Robin's-Rye. *Polytrichum juniperinum*
"Roble"-tree, of Mexico. *Quercus lobata*
Rocambole. *Allium Ophioscorodon*
Wild. *Allium Scorodoprasum*
Rock-Beauty. *Petrocallis pyrenaica*
Rock-Cress. The genus *Arabis*
Blue. *Arabis deltoidea* and varieties
Bristol. *Arabis stricta*
Early-flowering White. *Arabis albida*
Fringed. *Arabis ciliata*
Glabrous. *Arabis perfoliata*
Hairy. *Arabis hirsuta*
Northern. *Arabis petræa*
Pink-tinted. *Arabis aubrictioides*
Purple. *Aubrietia purpurea*
Rosette. *Arabis Androsace*
Rosy-flowered. *Arabis rosea* and *A. blepharophylla*
Shining-leaved. *Arabis lucida*
Spreading. *Arabis procurrens*
Thale. *Arabis Thaliana*
Tower. *Arabis Turrita*
Variegated. *Arabis lucida variegata*
Rocket, Base. *Reseda lutea*
Bastard. *Brassica Erucastrum*
Blue. *Aconitum pyramidalis* and other species
Crambling. See Crambling-Rocket
Cress. *Vella Pseudo-Cytisus* and *V. annua*
Dame's. See Rocket, Garden
Double Purple. *Hesperis matronalis purpurea plena*
Double White. *Hesperis matronalis alba plena*
Dyer's. *Reseda Luteola*
Edible. *Eruca sativa*
Evening-scented. *Hesperis fragrans*
Garden, Dame's, or White. *Hesperis matronalis*

Rocket, Italian. *Reseda lutea*
London. *Sisymbrium Irio*
Mountain. *Saxifraga granulata fl.-pl.*
Night-scented. *Hesperis tristis*
Salad. *Eruca sativa*
Sea-side. *Cakile maritima*
Sea-side, American. *Cakile americana*
Violet-flowered. *Hesperis violacea*
Wall. *Diplotaxis tenuifolia*
Water. *Nasturtium sylvestre*
White. See Rocket, Garden
Wild. *Brassica muralis*
Winter, or Yellow. *Barbarea vulgaris*
Yellow. See Rocket, Winter
Rock-foil. A name suggested by Mr. Ruskin for the genus *Saxifraga*
Rock-plant, of St. Helena. *Petrobium arboreum*
Rock-rose. The genus *Cistus*
Blunt-leaved. *Cistus obtusifolius*
Broad-leaved. *Cistus latifolius*
Canary Island. *Cistus candidissimus*
Clusius's. *Cistus Clusii*
Common. *Cistus vulgaris*
Corbières. *Cistus corbariensis*
Cretan. *Cistus creticus*
Curled-leaved. *Cistus crispus*
Cyme-flowered. *Cistus cymosus*
Dwarf. *Cistus lusitanicus*
Florentine. *Cistus florentinus*
Hairy. *Cistus hirsutus*
Heart-leaved. *Cistus Cupanianus*
Hoary. *Cistus incanus*
Laurel-leaved. *Cistus laurifolius*
Long-leaved. *Cistus longifolius*
Loose-flowered. *Cistus laxus*
Many-flowered. *Cistus Ledon*
Montpelier. *Cistus monspeliensis*
Oblong-leaved. *Cistus oblongifolius*
Poplar-leaved. *Cistus populifolius*
Purple-flowered. *Cistus purpureus*
Rough-leaved. *Cistus asperifolius*
Sage-leaved. *Cistus salviæfolius*
Sheathed-stalked. *Cistus vaginatus*
Shrubby. *Cistus villosus* and *C. frutescens*
Silky-leaved. *Cistus sericeus*
Smooth-calyxed. *Cistus psilosepalus*
Various-leaved. *Cistus heterophyllus*
Waved-leaved. *Cistus undulatus*
Whitish-leaved *Cistus albidus*
Rock-tripe. *Gyrophora vellea*
Rods-gold. An old name for Marigold
Rod-wood, Black. *Eugenia pallens*
Jamaica. *Lætia Guidonia*
Red. *Eugenia axillaris*
White. *Calyptranthes Chytraculia*
Roe-buck-berry. The fruit of *Rubus saxatilis*
Roffla. See Raffia
Rogation-flower. *Polygala vulgaris*
Rohun-Bark-plant. *Soymida (Swietenia) febrifuga*
"Roi," of New Zealand. The root of *Pteris esculenta*
Roka-tree. *Trichilia emetica*
Romerillo - dye - plant. *Heterothalamus brunioides*
Roof-foil. A name suggested by Mr. Ruskin for the House-leek

Room, or Roum-plant. *Ruellia tinctoria*
Root-blossom, Scarlet. *Agalmyla longistyla*
Root-of-Scarcity. Another name for the Mangel-Wurzel
Rope-grass, Tasmanian. *Restio australis*
Rosary-plant. *Abrus precatorius*
Mexican. *Rhynchosia precatoria*
Rose. The genus *Rosa*
Alpine. *Rhododendron ferrugineum*, *R. hirsutum*, and *R. striatum*
Anemone-flowered. *Rosa anemonæflora*
Apple-bearing, or Apple. *Rosa villosa var. pomifera*
Ash-leaved. *Rosa fraxinifolia*
Australian Native. *Boronia serrulata*
Austrian. *Rosa lutea var. punicea*
Austrian, Dwarf. *Rosa pumila*
Ayrshire. *Rosa arvensis var. scandens*
Barberry-leaved. *Rosa berberidifolia*
Barrow. *Rosa spinosissima*
Bengal. *Rosa Bengalensis*
Boursault. *Rosa Boursaulti*
Bramble. *Rosa polyantha*
Bramble-leaved. *Rosa rubifolia*
Brown's. *Rosa Brunonii*
Burnet, Pink-shaded. *Rosa spinosissima var. carnea*
Burnet, White. *Rosa spinosissima*
Burnet, Pyrenean White. *Rosa pimpinellifolia var. pyrenaica*
Cabbage. *Rosa centifolia*
Canker. *Papaver Rhœas*
Cayenne. *Licuria guianensis*, or *Dicypellium caryophyllatum*
Celestial. A variety of *Rosa alba*
Cherokee. *Rosa lævigata*
China, or Monthly. *Rosa indica*
China, Red. *Rosa semperflorens*
Cinnamon. *Rosa cinnamomea*
Cinnamon, Double. *Rosa cinnamomea var. fœcundissima*
Cinnamon, Dwarf. *Rosa majalis*
Cliff. See Cliff-rose
Clove-scented. *Rosa caryophyllacea*
Cretan. *Rosa glutinosa*
Crimean. *Rosa taurica*
Damask. *Rosa damascena*
Diverse-leaved. *Rosa diversifolia*
Dwarf Wild American. *Rosa lucida*
Early Wild American. *Rosa blanda*
Evergreen. *Rosa sempervirens*
Fairy. *Rosa Lawrenceana*
Field. *Rosa arvensis*
Frankfort. *Rosa turbinata var. Francofurtana*
French. *Rosa gallica*
Green-flowered. *Rosa viridiflora*
Guelder. *Viburnum Opulus*
Guiland. *Rosa Pissarti*
Holly. The genus *Helianthemum*
Hudson's Bay. *Rosa blanda*
Hundred-leaved. *Rosa centifolia*
Hyacinth-scented. *Rosa hyacinthina*
Irish. *Rosa hibernica*
Jamaica. The genus *Meriania*. Also applied to *Blakea trinervis*
Japanese. *Rosa Yrara* and the genus *Camellia*

Rose, "Juno's." *Lilium candidum*
Lady Banks's. *Rosa Banksiæ*
Lenten. The species of *Helleborus* which bloom in Lent
Macartney. *Rosa bracteata*
Malabar. *Hibiscus Rosa-malabarica*
Many-flowered. *Rosa polyantha*
Mexican. *Rosa Montezumæ*
Miss Lawrence's. *Rosa Lawranceana*
Monthly. See Rose, China
Moss. *Rosa centifolia var. muscosa*
Mossy-crested. *Rosa centifolia var. muscosa-cristata*
Mountain, of the W. Indies. *Antigonon leptopus*
Musk-scented. *Rosa moschata*
Noisette. *Rosa indica var. Noisettiana*
"Of Heaven." *Lychnis Cœli-rosa*
"Of Jericho." *Anastatica Hierochuntina*
"Of Sharon." *Hibiscus syriacus*
Pea-fruited. *Rosa pisocarpa*
Prairie, or Climbing. *Rosa setigera*
Provins. *Rosa centifolia var. pomponia*
Pyrenean Dwarf. *Rosa alpina var. pyrenaica*
Ramanas. *Rosa rugosa var. alba*
Red-leaved. *Rosa rubrifolia*
Redouté's. *Rosa rubrifolia var. Redoutéa*
Regel's. *Rosa Regeliana*
River, of Tasmania. *Bauera rubioides*
Sabine's. *Rosa Sabini*
Sage. See Sage-Rose
Scotch. *Rosa spinosissima*
Seven-sisters. *Rosa Grevillei*
Shaggy-fruited. *Rosa villosa*
Shining-leaved. *Rosa lucida*
Silky-leaved. *Rosa sericea*
"South Sea," of Jamaica. *Nerium Oleander*
Sulphur-flowered. *Rosa sulphurea*
Sweet-briar. *Rosa rubiginosa*
Tea, or Tea-scented. A variety of *Rosa indica*
Three-leaved. *Rosa sinica*
Turkestan Dwarf. *Rosa rugosa*
Turpentine. *Rosa terebinthinacea*
Vilmorin. *Rosa gallica var. inaperta*
"Vinegar," of Germany. *Pæonia officinalis*
Water. See Water-Rose
West Indian Mountain. *Brownea Rosa*
Willow-leaved. *Rosa longifolia*
Wind. *Papaver Rhœas* and *Rœmeria hybrida*
Yellow, Double. *Rosa sulphurea*
Yellow, Single. *Rosa lutea*
York-and-Lancaster. *Rosa versicolor* (a variety of *R. Damascena*) and *Rosa mundi* (or *Gloria-mundi*), a variety of *R. gallica*
Rose-Acacia. *Robinia hispida*
Rose-apple. *Eugenia Jambos*
Malay. *Eugenia malaccensis*
Rose-a-ruby. *Adonis autumnalis*
Rose-Bay. *Nerium Oleander* and *Epilobium angustifolium*
American. *Rhododendron maximum*
Lapland. *Rhododendron Lapponicum*
Rose-Box. The genus *Cotoneaster*

Rose-Campion, Common. *Lychnis coronaria*
Double-blossomed. *Lychnis coronaria fl.-pl.*
Italian. *Lychnis Nicæensis*
Smooth-leaved. *Lychnis Cœli-rosa*
Rose-Elder. *Viburnum Opulus*
"Roselle"-plant. *Hibiscus Sabdariffa*
Rose-Mallow, American Scarlet. *Hibiscus coccineus*
Chinese. *Hibiscus Rosa-sinensis*
Fringed-flowered. *Rosa-sinensis schizopetalus*
Halberd-leaved. *Hibiscus militaris*
Marsh. *Hibiscus palustris*
Rosy-flowered. *Hibiscus roseus*
Swamp. *Hibiscus Moscheutos*
Rose-Pink. *Sabbatia angularis*
Rosemary, Common. *Rosmarinus officinalis*
Marsh. *Ledum palustre* and *Andromeda polifolia*. Also applied, in America, to the genus *Statice*
Poet's. *Osyris alba* (*Cassia poetica*)
Sand-hill. *Ceratiola ericoides*
Sea-side. *Schoberia fruticosa*
Victorian. *Westringia rosmariniformis*
Wild, American. *Andromeda polifolia*
Wild, Jamaica. *Croton Cascarilla*
Rosemary-Frankincense. *Cachrys Libanotis*
Rose-root. *Rhodiola rosea*
Rose-wood-tree, African. *Pterocarpus erinaceus*
Australian. *Synoum glandulosum* (*Trichilia glandulosa*)
Brazil. *Dalbergia nigra*
Burmese. *Pterocarpus indicus*
Canary. *Rhodorrhiza scoparia*
Dominica. *Cordia Geraschanthus*
E. Indian. *Dalbergia latifolia* and *D. sissoides*
Jamaica. *Linociera ligustrina* and *Amyris balsamifera*
Moulmein. A species of *Millettia*
Tasmanian. A species of *Acacia*
Rosin-weed. *Silphium laciniatum*
Rouge-berry, or Rouge-plant. *Rivina humilis* and other species
Roum. See Room
"Rounsivals." An old name for Garden Peas
Rowan-tree, or Roan-tree. *Pyrus Aucuparia*
"Rozelle"-plant. *Hibiscus Sabdariffa*
Rubber-plant, E. Indian. *Ficus elastica*
Rubber-tree, African. The genus *Landolphia*
India. See "India-rubber" and "Caoutchouc"
Ruby-wood. Another name for Red Sanders-wood
"Ruddes." An old name for Marigolds
Rue. The genus *Ruta*
Black, of New Zealand. *Podocarpus spicata*
Common. *Ruta graveolens*
Fen. See Rue, Meadow
Goat's. *Galega officinalis*

and Foreign Plants, Trees, and Shrubs. 119

Rue, Goat's, of the W. Indies. *Tephrosia cinerea*
Meadow, or Fen. *Thalictrum flavum*
Mountain. *Ruta montana*
Otago Black. *Podocarpus spicata*
Padua. *Ruta patavina*
Sicilian. *Ruta bracteosa*
Simple-leaved. The genus *Aplophyllum*
Syrian. *Peganum Harmala*
Wall. *Asplenium Ruta-muraria*
White-flowered. *Ruta albiflora*
Rue-Anemone. *Thalictrum anemonoides*
Runch. *Raphanus Raphanistrum*
Rupture-wort. *Herniaria glabra* and *Alternanthera polygonoides*
Rusa-Oil-plant. *Andropogon Schœnanthus*
Rush. The genus *Juncus*
Baltic. *Juncus balticus*
Beak. The genus *Rhynchospora*
Black-spiked. *Juncus castaneus*
Blunt-flowered. *Juncus obtusiflorus*
Bul, Club, Mat, or Pool. *Scirpus lacustris*
Bog. *Schœnus nigricans*
Candle, or Pin. *Juncus effusus*
Club. See Rush, Bul, and Club-rush
Club, American Hedgehog. *Mariscus cylindricus*
Cluster-headed. *Juncus capitatus*
Common Soft. *Juncus communis*
Cord, of Victoria. *Schœnus brevifolius*
Cork-screw. *Juncus conglomeratus spiralis*
Dutch. See Rush, Scouring
Flowering. *Butomus umbellatus*
Galingale. *Cyperus tegetum*
Goose-corn. *Juncus squarrosus*
Hard. *Juncus glaucus*
Heath, or Moss. *Juncus squarrosus*
Highland. *Juncus trifidus*
Horned. *Rhynchospora corniculata*
Jersey. *Juncus capitatus*
Jointed. *Juncus articulatus*
Mat. See Rush, Bul
Mat, Chinese. *Lepironia mucronata*
Moss. See Rush, Heath
Nut. *Cyperus esculentus*
Nut, American. *Cyperus rotundus var. Hydra* and the genus *Scleria*
Palmite. *Prionium Palmita*
Paper. *Papyrus antiquorum* and other species
Petsi. *Heleocharis tuberosa*
Pin. See Rush, Candle
Pool. See Rush, Bul
Round-fruited. *Juncus compressus*
Scouring, or Dutch. *Equisetum hyemale*
Sea-side. *Juncus maritimus*
Sea-side, Great Sharp. *Juncus acutus*
Spike. The genus *Eleocharis*
Sweet. *Andropogon Schœnanthus* and *A. laniger*
Thread. *Juncus filiformis*
Three-flowered. *Juncus triglumis*
Toad. *Juncus bufonius*
Twig. *Cladium mariscoides*
Two-flowered. *Juncus biglumis*
Wood. *Luzula sylvatica*. (See also Wood-rush)
Zebra-striped. *Eulalia japonica zebrina*

Rush-Broom, Australian. *Viminaria denudata*
Rush-Fern. See Fern
Rush-grass. See Grass
Rush-Lily. The genus *Sisyrinchium*
Purple. *Sisyrinchium grandiflorum*
White. *Sisyrinchium grandiflorum var. album*
Rush-plant, Knife-leaved. *Cyperus elegans*
Rush-nut, Edible. *Cyperus esculentus*
American. *Cyperus rotundus var. Hydra*
"Rust"-Fungus. See Fungus
Rusty-back. *Blechnum Spicant* and *Ceterach officinarum*
Rutton-root. *Onosma (Maharanga) Emodi*
Rye, Common. *Secale cereale*
Ergot of. The *sclerotium* (or compact *mycelium*) of *Claviceps purpurea*
Perennial. A variety of *Secale fragile*
Spurred. *Secale cornutum* (Rye-grain attacked by the Ergot-Fungus)
Wild, of N. America. The genus *Elymus*

Sabadilla. Another name for Cevadilla
Sabicu, Savacú, or Savicú-wood, of Cuba. The timber of *Lysiloma Sabicú*
Sabino-tree. *Taxodium distichum*
Sack-tree. *Antiaris (Lepurandra) saccidora (A. toxicaria)*
Sad-tree, or Indian Mourner. *Nyctanthes Arbor-tristis*
Saddle-tree. *Liriodendron tulipiferum*
Safflower. *Carthamus tinctorius*
Saffron-plant, African. *Lyperia crocea*
Bastard, or False. *Carthamus tinctorius*
Common. *Crocus sativus*
Indian. Various species of *Curcuma*
Sicilian. *Crocus odorus*
Saffron-wood. *Elæodendron croceum*
Sagapenum, or Seraphic-Gum-plant. Supposed to be *Ferula persica* or *F. Szovitziana*
Sage. The genus *Salvia*
Apple-bearing. *Salvia pomifera*
Ash-leaved. *Salvia interrupta*
Beetle. *Cuphea Jorullensis*
Bengal. *Meriandra bengalensis*
Black. *Cordia cylindrostachya*
Black, of California. *Trichostema lanatum*
Blue-flowered. *Salvia azurea, S. campestris, S. officinalis, S. patens* and *S. verticillata*
Camerton's. *Salvia Camertoni*
Common Garden. *Salvia officinalis*
Crimson-flowered Dwarf. *Salvia Hæmeriana*
Egyptian. *Salvia ægyptiaca*
Gaping Cashmere. *Salvia hians*
Garlic. *Teucrium Scorodonia*
Gesnera-leaved. *Salvia gesneræfolia*
Graham's. *Salvia Grahami*
Green-topped. *Salvia viridis*
Jamaica Mountain. *Lantana aculeata*
"Jerusalem." *Phlomis fruticosa* and other species
Long-branched. *Salvia virgata*
Lupine. *Salvia pratensis var. lupinoides*
Lyre-leaved. *Salvia lyrata*
Meadow. *Salvia pratensis*

Sage, Nettle-leaved. *Salvia urticæfolia*
"Of Bethlehem." *Pulmonaria officinalis*
Orange-flowered. *Salvia colorata*
Paper. *Salvia canariensis*
Pine-apple-scented. *Salvia rutilans*
Rape-leaved. *Salvia napifolia*
Red-and-white-flowered. *Salvia bicolor*
Red-flowered. *Salvia sanguinea*
Red-flowered, Large. *Salvia sanguinea grandiflora*
Red-topped. *Salvia Horminum*
Scarlet-flowered. *Salvia coccinea* and *S. splendens*
Scarlet Mexican. *Salvia fulgens*
"Sea-side." *Croton balsamiferum*
Silvery-leaved. *Salvia argentea* (*S. patula*)
Southern Blue. *Salvia azurea*
Spreading. *Salvia patens*
Texan. *Salvia Texana*
Tree. *Salvia arborea*
Variegated. *Salvia officinalis var. tricolor*
Violet-flowered. *Salvia violacea*
White, of California. *Audibertia polystachya*
White-flowered. *Salvia argentea, S. chionantha, S. officinalis alba, S. patens alba,* and *S. patula*
Wild. *Salvia Verbenaca*
Wild, of the Cape. *Tarchonanthus camphoratus*
Wood. *Teucrium Scorodonia* and *Salvia sylvestris*
Yellow-flowered Greenhouse. *Salvia nubicola*
Yellow-flowered Hardy. *Salvia glutinosa*
Sage-bush, or Sage-brush, of N. America. *Artemisia tridentata, A. trifida,* and *A. arbuscula*
Sage-Rose. An old name for the genus *Cistus* Of the W. Indies. *Turnera ulmifolia*
Sage-tree, Brush-land, of Australia. *Psychotria daphnoides*
Sago-Palm. See Palm
Sago-plant, Portland. *Arum maculatum*
Sago-tree, Jamaica. *Zamia integrifolia* and *Z. furfuracea*
Saintfoin. *Onobrychis sativa*
Mountain. *Onobrychis montana*
Sal, or Saul-tree. *Shorea robusta*
Salad, Shawanese. The leaves of *Hydrophyllum virginicum*
Salad-Burnet. See Burnet
Salad-Rocket. See Rocket
Salal, or Shallon-shrub. *Gaultheria Shallon*
Salep. The dried tubers of various species of *Orchis*
E. Indian. *Eulophia campestris* and other species
Otaheite. *Tacca pinnatifida*
"**Saligot.**" An old name for *Trapa natans*
Sallow. *Salix Caprea, S. cinerea,* and allied species
Gray. *Salix cinerea*
Great. *Salix Caprea*
Water. *Salix aquatica*
Salmon-berry. *Rubus spectabilis* and *R. Nutkanus*

Saloop-bush, of Australia. *Rhagodia hastata*
Salsify, or Vegetable Oyster. *Tragopogon porrifolius*
Salt-bush, Australian. *Atriplex halimoides* and other species
Salt-Lake-tree, Queensland. *Ægiceras majus*
Salt-rheum-weed. *Chelone glabra*
Salt-tree, Greenish. *Halimodendron subvirescens*
Silvery-leaved. *Halimodendron argenteum*
Three-flowered. *Halimodendron triflorum*
Salt-wort, Black. *Glaux maritima*
Prickly. *Salsola Kali*
Shrubby. *Salsola fruticosa*
W. Indian. *Batis maritima*
Saman, or Zamang-tree. *Pithecolobium Saman*
Samara-wood. *Icica altissima*
Samphire, Common. *Crithmum maritimum*
Golden *Inula crithmoides*
Jamaica. *Batis maritima* and *Borrichia frutescens*
Longwood, of St. Helena. *Pharnaceum acidum*
Marsh. *Salicornia herbacea*
Prickly. *Echinophora spinosa*
W. Indian. *Sesuvium Portulacastrum*
Sand-box-tree, Rattling. *Hura crepitans*
Sand-flower, Winged-stalked. *Ammobium alatum*
Sand-fly-bush, of Australia. *Zieria Smithii*
Sand-Myrtle, Box-leaved. *Leiophyllum buxifolium* (*Ammyrsine buxifolia*)
Sand-paper-tree. *Curatella americana, Dillenia scabrella,* and *D. sarmentosa*
"**Sand-stay,**" of Australia. *Leptospermum lævigatum*
Sand-wood-shrub. *Bremontiera Ammoxylon*
Sand-wort. The genus *Arenaria*
Balearic. *Arenaria balearica*
Bog. *Arenaria uliginosa*
Fringed. *Arenaria ciliata*
Grass-leaved. *Arenaria graminifolia*
Larch-leaved. *Arenaria laricifolia*
Large-flowered. *Arenaria grandiflora*
Many-stemmed. *Arenaria multicaulis*
Mountain. *Arenaria montana* and *A. Groenlandica*
Pine-barren. *Arenaria squarrosa*
Purplish-flowered. *Arenaria purpurascens*
Red. *Arenaria rubra* and *Spergularia rubra*
Square-stemmed. *Arenaria tetraquetra*
Tufted. *Arenaria cæspitosa*
Vernal. *Arenaria verna*
Sandal-tree. The genus *Sandoricum*
Sandal-wood-Oil-plant. *Santalum album*
Sandal-wood-tree. The genus *Santalum*
Citron, or Yellow. *Santalum Freycinetianum*
False. *Ximenia americana* and *Myoporum tenuifolium*
False, of Crete. *Quercus abelicea*
Indian. See Sandal-wood, White

Sandal-wood-tree, Queensland. *Eremophila Mitchelli*
Red. *Adenanthera pavonina* and *Pterocarpus santalinus*
Sandwich Islands. *Santalum Freycinetianum* and *S. paniculatum*
W. Australian. *Fusanus spicatus* (*Santalum spicatum*)
White, or Indian. *Santalum album*
Yellow. See Sandal-wood, Citron
Sandarac-gum-tree. *Callitris quadrivalvis* (*Thuja articulata*)
Sanders-wood, Red, or Ruby-wood. *Pterocarpus santalinus*
Yellow. *Bucida capitata*
Sanicle, American. *Heuchera americana*
Alpine. The genus *Cortusa*
Bear's-ear. *Cortusa Matthioli*
Common. *Sanicula europœa*
Downy. *Cortusa pubens*
Yorkshire. *Pinguicula vulgaris*
Santa-Maria-tree. *Calophyllum Calaba*
Saouari, or Souari-wood. The wood of *Caryocar nuciferum* and *C. tomentosum*
Sap-ball-Fungus. See Fungus
Sapodilla, or Sapotilla, Plum. *Achras Sapota*
Sapota, White. *Casimiroa edulis*
Sapote, or Chupa-Chupa-tree. *Mutisia cordata*
Sappan-fruit. The fruit of *Inga dulcis*
Sappan-wood. *Cæsalpinia Sappan*
Sapucaia-nut. See Nut
Sarcocolla-tree. *Penœa Sarcocolla*
Sarsaparilla-plant, Australian. *Smilax glycyphylla* and *Hardenbergia monophylla*
Bristly. *Aralia hispida*
E. Indian, or Nunnari-Root. *Hemidesmus indicus* (*Periploca indica*)
Jamaica. *Smilax officinalis*
New Zealand. *Rhipogonum scandens*
Of Commerce. Various species of *Smilax*
Wild. *Aralia nudicaulis*
Sassafras-tree. *Laurus Sassafras*
Australian. *Doryphora Sassafras* and *Atherosperma moschata*
Brazilian, or Orinoco. *Nectandra cymbarum*
Californian. *Oreodaphne californica*
Cayenne. *Licania guianensis*
Chilian. *Laurelia sempervirens* and *Boldoa fragans*
New Zealand. *Laurelia Novæ-Zealandiæ*
Oriental. *Sassafras Parthenoxylon*
Orinoco. See Sassafras, Brazilian
Swamp. *Magnolia glauca*
Tasmanian. *Atherosperma moschata*
White. *Laurus albida*
Sassy-tree. *Erythrophlœum guineense*
Satin-bush, African. *Podalyria sericea*
Satin-flower. *Lunaria biennis*
Bermuda. *Sisyrinchium Bermudianum*
Broad-leaved. *Sisyrinchium latifolium*
Crimson. *Brodiæa coccinea*
Grass-leaved. *Sisyrinchium anceps*
Purple. See Satin-flower, Spring
Red. *Hedysarum coronarium*
Spring, or Purple. *Sisyrinchium grandiflorum*

Satin-flower, Yellow. *Sisyrinchium californicum* and *S. striatum*
Satin-leaf. *Heuchera Richardsoni* (*H. americana, H. ribifolia*)
"**Satin-leaves**." The dried seed-vessels of *Lunaria biennis*
Satin-wood-tree. *Xanthoxylum caribæum*
Bahamas. *Maba guineensis* (?)
E. Indian. *Maba buxifolia*
E. Indian, Yellow. *Swietenia Chloroxylon*
Sauce-alone. *Erysimum Alliaria*
Saul, or Sâl-tree. *Shorea robusta*
Saunders-wood. See Sanders-wood
Savacu. See Sabicù
Savannah-flower. *Echites suberecta* and other species
Savannah-wood. *Hosta cœrulea*
Savico. See Sabicù
Savin-tree, Common. *Juniperus Sabina*
Dwarf. *Juniperus Sabina var. nana*
"Kindly." *Juniperus Sabina var. cupressifolia*
Siberian. *Juniperus Pseudo-Sabina*
W. Indian. *Cæsalpinia bijuga* and *Fagara lentiscifolia*
Savonette-tree. *Pithecolobium micradenium*
Savory, Canadian. *Satureia Thymbra*
Dyer's. *Serratula tinctoria*
Summer. *Satureia hortensis*
Winter. *Satureia montana*
Savoy-Cabbage. See Cabbage
Saw-grass, Marsh. See Grass
Saw-wort. The genus *Serratula*
Alpine. *Saussurea alpina*
Common. *Serratula tinctoria*
Crowned. *Serratula coronata*
Large-leaved. *Saussurea macrophylla*
Saxifrage. The genus *Saxifraga*
"Aaron's-beard." *Saxifraga sarmentosa*
Aizoon. *Saxifraga Aizoon*
Ambiguous. *Saxifraga ambigua*
Andrews's. *Saxifraga Andrewsii*
Aromatic. *Saxifraga aromatica*
Arctic. *Saxifraga arctioides*
Brook. *Saxifraga rivularis*
Bryum-leaved. *Saxifraga bryoides*
Buckland's. *Saxifraga Bucklandii*
Bugle-leaved. *Saxifraga ajugœfolia*
Burnet. *Pimpinella Saxifraga*
Burser's. *Saxifraga Burseriana*
Carinthian. *Saxifraga carinthiaca*
Channelled-leaved. *Saxifraga canaliculata*
Churchill's. *Saxifraga Churchillii*
Creeping. *Saxifraga sarmentosa*
Crusted-leaved. *Saxifraga crustata*
Diapensia. *Saxifraga diapensioides*
Downy. *Saxifraga pubescens*
Drooping. *Saxifraga cernua*
Early American. *Saxifraga virginiensis*
Early Silver. *Saxifraga coriophylla*
Early White-flowered. *Saxifraga Burseriana*
Elegant. *Saxifraga elegans*
Five-fingered. *Saxifraga pentadactylis*
Fox's-brush. *Saxifraga lantoscana*
Furrowed. *Saxifraga exarata*
Gibraltar. *Saxifraga gibraltarica*
Golden. The genus *Chrysosplenium*

Saxifrage, Gray. *Saxifraga cæsia*
Great Strap-leaved. *Saxifraga ligulata*
Great Californian. *Saxifraga peltata*
Greenland. *Saxifraga groenlandica*
Guthrie's. *Saxifraga Guthrieana*
Hairy. *Saxifraga hirsuta*
Hausmann's *Saxifraga Hausmanniana*
Heart-leaved. *Saxifraga (Megasea) cordifolia*
Horn-leaved. *Saxifraga ceratophylla*
Incurved-leaved. *Saxifraga incurvifolia*
Ivy. *Saxifraga Cymbalaria*
Juniper. *Saxifraga juniperina*
Kerry. *Saxifraga affinis* (a variety of *S. hypnoides*)
Kidney-leaved. *Saxifraga Geum*
Lance-leaved. *Saxifraga lanceolata*
Large Purple-flowered. *Saxifraga biflora*
Large-flowered Pyrenean. *Saxifraga oppositifolia pyrenaica var. maxima*
Lettuce. *Saxifraga erosa*
London-pride. *Saxifraga umbrosa*
Long-leaved. *Saxifraga longifolia*
Long-leaved, Tall. *Saxifraga longifolia var. elatior*
Marsh. *Saxifraga Hirculus*
Mawe's. *Saxifraga Mawreana*
Mayl's. *Saxifraga Maylii*
Meadow. *Saxifraga granulata, Silaus pratensis,* and the genus *Seseli*
Mossy. *Saxifraga hypnoides*
Mountain. *Saxifraga nivalis*
Mountain, Large. *Saxifraga aizoides var. autumnalis*
Mountain, Large Rosy-flowered. *Saxifraga oppositifolia var. pyrenaica maxima*
Mountain, Small. *Saxifraga aizoides*
Nepaul. *Saxifraga nepalensis*
Palmate-leaved. *Saxifraga palmata*
Pepper. *Silaus pratensis*
Piedmont. *Saxifraga pedemontana*
Polished. *Saxifraga polita*
Pretty. *Saxifraga pulchella*
Purple-flowered. *Saxifraga oppositifolia* and *S. retusa*
Purple Himalayan. *Saxifraga purpurascens*
Pyramidal. *Saxifraga Cotyledon* and *S. pyramidalis*
Pyramidal, House-leek-leaved. *Saxifraga mutata*
Pyramidal, Margined. *Saxifraga Aizoon*
Pyrenean. *Saxifraga longifolia*
Rochel's. *Saxifraga Rocheliana*
Rosetted. *Saxifraga rosularis*
Rosy-calyxed. *Saxifraga calyciflora*
Rough. *Saxifraga aspera*
Round-leaved. *Saxifraga rotundifolia*
Rue-leaved. *Saxifraga tridactylites*
Saffron-flowered. *Saxifraga mutata*
Scallop-leaved. *Saxifraga pectinata*
Schrader's. *Saxifraga Schraderi*
Shield-leaved. *Saxifraga peltata*
Silver Moss. *Saxifraga cæsia*
Slender. *Saxifraga tenella*
Spoon-leaved. *Saxifraga spatulata*
Spreading. *Saxifraga repanda*
Stag's-horn. *Saxifraga ceratophylla*
Starry. *Saxifraga stellaris*

Saxifrage, Sternberg's. *Saxifraga Sternbergii*
Sturm's. *Saxifraga Sturmiana*
Swamp. *Saxifraga pennsylvanica*
Thick-leaved. *Saxifraga (Megasea) crassifolia*
Three-spined. *Saxifraga tricuspidata*
Tombe's. *Saxifraga Tombeana*
Tufted. *Saxifraga cæspitosa*
Vandel's. *Saxifraga Vandelii*
Various-leaved. *Saxifraga diversifolia*
Vaudois. *Saxifraga valdensis*
Wallace's. *Saxifraga Wallacei*
Water. *Saxifraga rivularis*
Wedge-leaved. *Saxifraga cuneifolia*
Whip-cord. *Saxifraga flagellaris*
Wilkomm's. *Saxifraga Wilkommiana*
Yellow-flowered Dwarf. *Saxifraga sancta*
Yellowish-leaved. *Saxifraga tenella*
Yellow Mountain. *Saxifraga aizoides*
Scabious. The genus *Scabiosa*
Blue. *Scabiosa succisa*
Catch-fly-leaved. *Scabiosa silenifolia*
Caucasian. *Scabiosa caucasica*
Devil's-bit. *Scabiosa succisa*
Field. *Scabiosa arvensis*
Fischer's. *Scabiosa Fischeri*
Grass-leaved. *Scabiosa graminifolia*
Mt. Parnassus. *Pterocephalus Parnassi*
Pale Yellow. *Scabiosa ochroleuca*
Sheep's-bit. *Jasione montana*
Silvery. *Centaurea argentea*
Small. *Scabiosa Columbaria*
Starry. The genus *Crupina*
Sweet. *Scabiosa atropurpurea* and varieties
Sweet, of N. America. *Erigeron annuus*
Webb's. *Scabiosa Webbiana*
Scab-wort. An old name for *Elecampane*
Scale-fern. See Fern
Scallion. *Allium ascalonicum var. majus.* The name is also generally applied to all Onions that do not bulb, but form long necks like Leeks
Scammony - plant. *Convolvulus Scammonia*
Montpelier. *Cynanchum monspeliacum* and *C. acutum*
Scariole. An old name for Endive
Scarlet-Runner. See Bean
Scarlet - seed. *Ternströmia ovoalis* and *Lætia Thamnia*
Scent-wood, Tasmanian. *Alyxia buxifolia*
Sceptre-flower. The genus *Sceptranthus*
Scimitar - pods. The pods of *Entada scandens*
Scoke, or Poke-weed. *Phytolacca decandra*
Scorpion-grass. See Grass
Scorpion - plant. *Renanthera arachnitis* and *Genista scorpius*
Scorpion-Senna. *Coronilla Emerus*
Scorzonera, French. *Scorzonera picroides* (*Picridium vulgare*)
Garden. *Scorzonera hispanica*
Scotch-Asphodel. See Asphodel
Scotch-Attorney, of the W. Indies. The genus *Clusia*
Scotch-bonnets. *Agaricus (Marasmius) Oreades*
Scotch Fir. See Fir

Scotch Primrose. See Primrose
"Scotino." *Rhus Cotinus*
Scouring-Rush. See Rush
Screw-Moss. See Moss
Screw-Pine. The genus *Pandanus*
Screw-tree. The genus *Helicteres*
Jamaica. *Helicteres jamaicensis*
Large-fruited. *Helicteres Isora*
Scripture-wort. See Chink-wort
Scrofula-leaf, or Scrofula-weed. *Goodyera pubescens*
Scrub-Oak. See Oak
Scrub-wood, St. Helena. *Commidendron rugosum*
Scurvy-grass. *Cochlearia officinalis*
Rock. *Kernera (Cochlearia) saxatilis*
"Scotch." *Conrolvulus Soldanella*
Scythian Lamb. The woolly rhizome of *Cibotium Barometz*
Sea-Beet. *Beta maritima*
Sea-bells. *Conrolvulus Soldanella*
Sea-belt. *Laminaria saccharina*
Sea-berry, of Australia. The genera *Haloragis* and *Rhagodia*
Sea-Bindweed. *Conrolvulus Soldanella*
Sea-Buckthorn. *Hippophaë rhamnoides*
Sea - Bugloss. *Pulmonaria (Mertensia) maritima*
Sea-coal. *Conrolvulus Soldanella*
Sea-colander. *Agarum Turneri*
Sea-Dock. See Dock
Sea-egg. *Medicago Echinus*
Sea-Foal-foot. See Foal-foot
Sea-Furbelows. *Laminaria bulbosa*
Sea-girdles. *Laminaria digitata*
Sea-Goose-foot. See Goose-foot
Sea-grape. *Salicornia herbacea*
Sea-grass. See Grass
Sea-hangers. *Laminaria bulbosa*
Sea-hay. *Zostera (Alga) marina*
Sea-Heath. *Frankenia pulverulenta*
Sea-Holly. *Eryngium maritimum* and other species
Sea-kale. *Crambe maritima*
Sea-laces. *Chorda Filum*
Sea-Lavender. Various species of *Statice*, especially *S. Limonium*
Bonduelle's. *Statice Bonduellei*
Caspian. *Statice caspia*
Common. *Statice Limonium*
Crimean. *Statice taurica*
Daisy-leaved. *Statice bellidifolia*
Downy. *Statice puberula*
Dwarf. *Statice nana*
Globularia - leaved. *Statice globulariæfolia*
Gmelin's. *Statice Gmelini*
Great. *Statice latifolia*
Hoary. *Statice incana (Goniolimon callicomum)*
Holford's. *Statice Holfordii*
Kaufmann's. *Statice Kaufmanniana*
Matted. *Statice reticulata*
Narrow-leaved. *Statice angustifolia*
Olive-leaved. *Statice oleæfolia (S. virgata)*
Pigmy. *Statice minuta*
Profuse. *Statice profusa*
Rock. *Statice auriculæfolia*

Sea - Lavender, Rosy-flowered. *Statice eximia*
Showy. *Statice speciosa*
Small. *Statice minuta*
Spiked. *Statice spicata*
Tall. *Statice elata (Goniolimon elatum)*
Tartarian. *Statice tatarica*
Sea-Leaf. *Bryophyllum calycinum*
"Sea-oats." *Uniola paniculata*
Sea - otter's - Cabbage. *Nereocystis Lutkeana*
Sea-Pink. *Statice Armeria*
Sea-Poppy. See Poppy
Sea-Purslane. See Purslane
Sea-Purslane-tree, Broad-leaved. *Atriplex Halimus*
Sea-Reed. *Psamma arenaria*
Sea-Rocket. See Rocket
Sea-Starwort. *Aster Tripolium*
Sea-thongs. *Himanthalia lorea*
Sea-wand. *Laminaria digitata*
Sea-weed. A general name for the marine *Algæ*
Cape Trumpet. *Ecklonia buccinalis*
Corsican Anthelmintic. *Gigartina helminthocorthon*
Cultivated, Japan. *Porphyra vulgaris*
Glaziers'. *Zostera mediterranea*
Gulf. *Sargassum bacciferum*
Swallow's-nest. *Plocaria tenax*
Whip-cord. *Chordaria flagelliformis*
Seal-flower. *Dielytra (Dicentra) spectabilis*
Sebestens. The fruit of *Cordia latifolia* and *C. Myxa*
Sedge. The genus *Carex*
Alpine. *Carex alpina*
Axillary. *Carex axillaris*
Beak. The genus *Rhynchospora*
Black. *Carex atrata*
Bladder. *Carex vesicaria*
Bottle. *Carex ampullacea*
Bracteated Marsh. *Carex divisa*
Bracteated Sea. *Carex extensa*
Capillary. *Carex capillaris*
Carpeting. *Carex divulsa*
Cotton. See Cotton-sedge
Crimson-fruited. *Carex baccans*
Curved. *Carex incurva*
Cyperus-like. *Carex Pseudo-cyperus*
Dioecious. *Carex dioica*
Dotted. *Carex punctata*
Downy. *Carex tomentosa*
Drooping Bog. *Carex limosa*
Dwarf. *Carex humilis*
Elongated. *Carex elongata*
Few-flowered. *Carex pauciflorus*
Fingered. *Carex digitata*
Flea. *Carex pulicaris*
Glaucous Heath. *Carex glauca*
Great Pendulous. *Carex pendula*
Great Spiked. *Carex vulpina*
Greater Bank. *Carex riparia*
Greater Tufted. *Carex stricta*
Green-ribbed. *Carex binervis*
Hairy. *Carex hirta*
Hammer. *Carex hirta*
Hare's-foot. *Carex lagopina*
Highland. *Carex saxatilis*

Sedge, Hoary. *Carex Buxbaumii*
Lesser Bank. *Carex paludosa*
Lesser Prickly. *Carex stellulata*
Lesser Tufted. *Carex vulgaris*
Loose-flowered. *Carex distans* and *C. strigosa*
Mountain. *Carex montana*
Mountain, Stiff. *Carex rigida*
Oval-spiked. *Carex leporina* (*C. ovalis*)
Pale. *Carex pallescens*
Panicled. *Carex paniculata*
Paradoxical. *Carex paradoxa*
Pendulous. *Carex pendula*
Pink-leaved. *Carex panicea*
Prickly. *Carex muricata*
Remote-spikeleted. *Carex remota*
Rock. *Carex rupestris*
Round-headed. *Carex pilulifera*
Sea-side. *Carex arenaria*
Slender-leaved. *Carex filiformis*
Slender-spiked. *Carex acuta*
Smooth. *Carex lævigata*
Spring. *Carex præcox*
Starveling. *Carex depauperata*
Sword, of Australia. *Lepidosperma gladiatum*
Tawny. *Carex fulva*
Tufted. *Carex cœspitosa*
Tussock. *Carex stricta*
White. *Carex canescens* (*C. curta*)
Wood. *Carex sylvatica*
Yellow. *Carex flava*
Seed-box. *Ludwigia* (*Isnardia*) *alternifolia* and *L. hirtella*
Sego. *Calochortus Nuttalli*
Segra-seed. *Feuillea cordifolia*
Selaginella-tree. *Selaginella cæsia arborea*
Self-heal. *Prunella vulgaris*
Hyssop-leaved. *Prunella hyssopifolia*
Large-flowered. *Prunella grandiflora*
Large-flowered, Cut-leaved. *Prunella grandiflora var. laciniata*
Pyrenean. *Prunella pyrenaica*
Seneca Snake-root. *Polygala Senega*
Senegal-root. The root of *Cocculus Bakis*
Sengreen. *Sempervivum tectorum* and *Saxifraga nivalis*
Water. *Stratiotes aloides*
Senna-plant, Aleppo. *Cassia obovata*
Alexandrian. *Cassia lanceolata*
Bladder. *Colutea arborescens*
Chili. *Myoschilos oblongus*
Italian. *Cassia obovata*
Jamaica. *Cassia emarginata*
Of Commerce. *Cassia acutifolia* and *C. angustifolia*
Scorpion. *Coronilla Emerus*
Tinnevelly. *Cassia angustifolia*
Winged. *Cassia alata*
Sensitive Briar. See Briar
Sensitive Plant (true). *Mimosa sensitiva*
"Wild," of N. America. *Cassia nictitans*
Septfoil. *Potentilla Tormentilla*
Serapias Turbith. An old name for *Aster Tripolium*
Serpent-Moss. See Moss
Serpent-withe. *Aristolochia odoratissima*
Serpent-wood. *Ophioxylon serpentinum*
Serpent's-bane. *Cerbera Ahouai*

Serpent's-beard. *Ophiopogon japonicus*
Serpent's-tongue. *Erythronium americanum*
Serpentary-Root. *Aristolochia Serpentaria*
Serradella, or Serratella. *Ornithopus sativus*
Service-berry. *Amelanchier canadensis*
Service-tree, Auricled. *Pyrus Sorbus auriculata*
Bastard. *Pyrus pinnatifida*
Common. *Pyrus Sorbus* (*P. domestica*)
Maple. *Pyrus torminalis*
Swedish. *Pyrus scandica* (*P. intermedia*)
Wild. *Pyrus torminalis*
Sesame, or Oily-grain-plant. *Sesamum orientale* and *S. indicum*
Sesban. *Sesbania ægyptiaca*
Seseli, Gum. *Seseli gummiferum*
Setter-wort. *Helleborus fœtidus*
Set-wall. An old name for *Valerian*
Shad-bush, or Shad-flower. *Amelanchier canadensis*
Shaddock-tree. *Citrus decumana*
Shade-leaf, Norfolk Island. *Botryodendron latifolium*
Shadow-grass. See Grass
Shaking-grass. See Grass
Shallon-bush, or Salal-bush. *Gaultheria Shallon*
Shallot. *Allium ascalonicum*
Shamrock, Blue-flowered. *Parochetus communis*
Four-leaved. *Trifolium repens var. purpureum*
Indian. *Trillium latifolium*
Of Ireland. *Trifolium repens,* for which *Medicago lupulina* is sometimes substituted
Share-wort. *Aster Tripolium*
Shave-grass. See Grass
Shawanese Salad. See Salad
Shear-grass. See Grass
Shea-tree. *Bassia Parkii*
Sheep-berry. *Viburnum Lentago*
Sheep-fodder-plant, S. African. *Pentzia virgata*
Sheep-pest, of Australia and New Zealand. *Acæna ovina*
"**Sheep-poison,**" Californian. *Lupinus densiflorus*
Sheep's-bane. *Hydrocotyle vulgaris*
Sheep's-beard. The genus *Arnopogon*
Sheep's-bit-Scabious. *Jasione montana*
Dwarf. *Jasione humilis*
Perennial. *Jasione perennis*
Sheep's-Fescue-grass. See Grass
Sheep's-Sorrel. *Rumex Acetosella*
Shell-flower. *Moluccella lævis* and the genus *Chelone*
Brush. *Alpinia* (*Hellenia*) *cærulea*
Indian. *Alpinia nutans*
Lyon's. *Chelone Lyoni*
Twisted. *Chelone obliqua*
White. *Chelone glabra*
Shelter-tree, Organ Mountain. *Adenostephanes organensis*
Shepherd's-club. *Verbascum Thapsus*
Shepherd's-cress. *Teesdalia Iberis*
Shepherd's-joy, Australian. The genus *Geitonoplesium*

Shepherd's-knot. *Tormentilla officinalis*
Shepherd's-purse. *Capsella Bursa-pastoris*
Shepherd's-rod, or Shepherd's-staff. *Dipsacus pilosus* and *D. sylvestris*
Shepherd's-scrip. Another name for Shepherd's-purse
Shepherd's-staff. See Shepherd's-rod
Shield-flower. The genus *Aspidistra*
Shingle-wood. *Nectandra leucantha*
Shin-leaf, American. *Pyrola elliptica*
"Shireesh," of India. *Acacia Serissa*
Shittim-wood, of Scripture. Supposed to be the wood of *Acacia nilotica*
Shoe-black-plant, or Shoe-flower. *Hibiscus Rosa-sinensis*
Shoe - maker's - bark - tree. *Byrsonima spicata*
Shola, or Solah-plant. *Æschynomene aspera*
Shooting-star. *Dodecatheon Meadia*
Shore-grass. See Grass
Shot, Indian. See Indian-Shot
Shumee-tree. *Acacia Suma*
Shuttle-cock-plant. *Periptera punicea* (*Sida Periptera*)
Sickle-pod. *Arabis canadensis*
S. American. *Drepanocarpus lunatus*
Sickle-wort. *Prunella vulgaris*
Side-saddle-flower. The genus *Sarracenia*
Dark-red. *Sarracenia atrosanguinea*
Drummond's. *Sarracenia Drummondi*
Handsome. *Sarracenia formosa*
Hook-leaved. *Sarracenia variolaris*
Purple. *Sarracenia purpurea*
Red. *Sarracenia rubra* (*S. psittacina*)
Small. *Sarracenia minor*
Yellow. *Sarracenia flava*
Silk-cotton-tree. The genera *Bombax* and *Eriodendron*
Brazilian. *Pachira macrantha*
Malabar. *Bombax malabaricum*
New Granada. *Pachira (Carolinea) alba*
Of the Amazon. *Eriodendron Samauma*
Seven-leaved. *Bombax malabaricum*
W. Indian. *Eriodendron anfractuosum*
White, East Indian. *Eriodendron orientale*
Silk-flower. *Calliandra trinervia*
Silk-fruit, Maryland. *Seriocarpus conyzoides*
Silk-grass. See Grass
Silk-plant, Vegetable, Syrian. *Periploca græca*
Vegetable, Brazil. *Chorissa speciosa*
Silk-tree, of Constantinople. *Acacia Julibrissin*
Silk-vine. *Periploca græca*
Silk-weed, American. The genus *Asclepias*
Silk-wood Tree. *Muntingia Calabura*
Silken Cissy. An old name for the genus *Asclepias*
"Silphium," of the Ancients. Supposed to be a species of *Ferula* (*Thapsia*), probably *F. glauca*, or *F. garganica*
Silver-berry, Missouri. *Elæagnus argentea*
Silver-bracts. *Pachyphytum bracteosum*
Silver-bush. *Anthyllis Barba-Jovis*
Silver Fir. See Fir
Silver-head. *Paronychia argyrocoma*

Silver-leaf, Japanese. *Ligularia Kæmpferi var. argentea*
Silver-rod. *Asphodelus ramosus*
Silver-tree. The genus *Elæagnus*
Cape. *Leucadendron argenteum*
Missouri. *Elæagnus argentea*
Queensland. *Tarrietia Argyrodendron* (*Argyrodendron trifoliatum*)
Silver-weed. *Potentilla anserina* and the genus *Argyreia*
Shining. *Argyreia acuta* (*Convolvulus splendens*)
Silver-wood. *Guettarda argentea*, *Quelania lætioides*, and the genus *Mouriria*
Simaruba-bark Tree. *Quassia* (*Picræna*) *excelsa*
Simool-tree. *Salmalia malabarica*
Simpler's-joy. *Verbena officinalis*
Of N. America. *Verbena hastata*
Simson. *Senecio vulgaris*
Singapore Hat - plant. *Æschynomene aspera*
"Sink-field." An old name for *Potentilla reptans*
Sipiri-tree. *Nectandra Rodiæi*
Siris-Acacia. *Albizzia Lebbek*
Sissoo-tree. *Dalbergia Sissoo*
Skewer - wood. *Cornus sanguinea* and *Euonymus europæus*
Skirret. *Sium Sisarum*
Skir-wort. An old name for Skirret
Skoke. *Phytolacca decandra*
Skull-cap. The genus *Scutellaria*
Alpine. *Scutellaria alpina*
Caucasian. *Scutellaria caucasica*
Common. *Scutellaria galericulata*
Entire-leaved. *Scutellaria integrifolia*
Japanese. *Scutellaria japonica*
Large-flowered. *Scutellaria macrantha*
"Mad-dog." *Scutellaria lateriflora*
Purplish - flowered. *Scutellaria purpurascens*
Scarlet-flowered. *Scutellaria Mocciniana*
Scordium-leaved. *Scutellaria scordifolia*
Siberian. *Scutellaria macrantha*
Small. *Scutellaria minor*
Snap-dragon. *Scutellaria antirrhinoides*
Wright's. *Scutellaria Wrightii*
Yellow-flowered. *Scutellaria orientalis*
Skunk Cabbage, or Skunk-weed. *Symplocarpus fœtidus*
Sky-flower. The genus *Duranta*
Sleep - wort. An old name for *Lactuca virosa*
Sleet-bush, African. *Diosma* (*Coleonema*) *alba*
Slipper-flower, or Slipper-wort. The genus *Calceolaria*
Celandine. *Calceolaria chelidonioides*
Hardy. *Calceolaria Kellyana*
Wing-leaved. *Calceolaria pinnata*
Slipper-Spurge. The genus *Pedilanthus*
Sloe-tree. *Prunus spinosa*
Slog-wood. *Hufelandia pendula*
Sloke. *Ulva Lactuca* and some other edible species of *Ulva* and *Porphyra*
Slow-match Tree. *Careya arborea*
Smallage, or Smalledge. *Apium graveolens*

Smart-weed. *Polygonum Hydropiper*
American. *Polygonum punctatum*
Water. *Polygonum acre*
Smilax, Boston. *Myrsiphyllum asparagoides*
Smoke-tree. *Rhus Cotinus*
Smoke-wood. *Clematis Vitalba*
Smooth - Flower. *Leianthus longifolius* and other species
Smut-Fungus. *Uredo segetum*
Snail-Clover, or Snails. *Medicago scutellata*
Snail-flower. *Phaseolus Caracalla*
Snake-charm. *Bauhinia anguina*
Snake-nut. See Nut
Snake-plant. *Arum Dracunculus*
Snake-poison-Antidote, of S. America. *Mikania Guaco*
Snake-root, Black. *Actæa racemosa* and *Sanicula canadensis*.
Brazilian. *Chiococca angustifolia* and *Cascaria ulmifolia*
Button. Various species of *Liatris*; also *Eryngium yuccæfolium*
Canadian. *Asarum canadense*
Ceylon. The tubers of *Arisæma papillosum*
E. Indian. *Ophiorrhiza Mungos*
Heart. *Asarum virginicum*
Red River, or Texan. *Aristolochia reticulata*
Seneca. *Polygala Senega*
South American *Dorstenia Contrayerba*
Tailed. *Asarum caudatum*
Virginian. *Aristolochia Serpentaria*
White. *Eupatorium ageratoides*
Snake - seed. The seed of *Ophiocaryon paradoxum*
Snake-weed. *Polygonum Bistorta*
Little. *Anacharis Alsinastrum* (*Elodea canadensis*)
Snake-wood, Bahamas. *Colubrina ferruginosa*
E. Indian. *Strychnos Colubrina*
Guiana. *Brosimum Aubletii*
Jamaica. *Brosimum Alicastrum* and *Cecropia peltata*
Snake's-beard. The genus *Ophiopogon*
Golden-variegated. *Ophiopogon Jaburan aureo-variegatus*
Japanese. *Ophiopogon japonicus*
Silver - variegated. *Ophiopogon spicatus argenteo-marginatus*
Spiked. *Ophiopogon spicatus*
Snake's-head Fritillary. See Fritillary
Snake's-mouth. *Pogonia ophioglossoides*
Snake's-tail. *Lepturus incurvatus*
Snake's-tongue. The genus *Lygodium*
Snap-dragon. The genus *Antirrhinum*
Common Garden. Varieties of *Antirrhinum majus*
Heart-leaved. *Antirrhinum Asarina*
Jamaica. *Ruellia tuberosa* (*Cryphiacanthus barbadensis*)
Rock. *Antirrhinum rupestre* (*Linaria rupestris*)
Small. *Antirrhinum Orontium*
Snap-tree. *Justicia hyssopifolia*
Snap-weed, American. Various species of *Impatiens*

Sneeze-weed, American. *Helenium autumnale* and other species
Sneeze-wood. *Pteroxylon utile*
Sneeze-wort. *Achillea Ptarmica*
Sniddel. *Carex paludosa* and other species
Snow, Mountain. *Arabis albida*
Snow-ball, Wild. *Ceanothus americanus*
Snow-ball-tree. *Viburnum Opulus*
Large-flowered. *Viburnum macrocephalum*
Nine-bark. *Spiræa opulifolia*
Snow-berry, Creeping. *Chiogenes hispidula*
Snow-berry - tree. *Symphoricarpus racemosus*
Small - leaved. *Symphoricarpus microphyllus*
Snow-bush, Californian. *Ceanothus cordulatus*
Victorian. *Olearia stellulata* (*Eurybia Gunniana*)
Snow-cups, Water. *Ranunculus aquatilis*
Snow-drop. The genus *Galanthus*
Broad-leaved. *Galanthus latifolius*
Common. *Galanthus nivalis*
Cup-flowered. *Galanthus poculiformis*
Crimean. *Galanthus plicatus*
Early-flowering. *Galanthus præcox*
Elwes's. *Galanthus Elwesii*
Greenish-flowered. *Galanthus virescens*
Imperati's. *Galanthus Imperati*
Late-flowering. *Galanthus serotinus*
Queen Olga's. *Galanthus Reginæ Olgæ*
Redouté's. *Galanthus Redoutéi*
Reflexed. *Galanthus reflexus*
Shaylock's. *Galanthus Shaylocki*
Summer. *Leucojum æstivum*
Yellowish. *Galanthus lutescens*
Snow-drop-tree. *Halesia tetraptera*
African. *Royena lucida*
Small-flowered. *Halesia parviflora*
Two-winged-fruited. *Halesia diptera*
W. Indian. *Hemanthus* (?) *incrassatus*
Snow-flake. The genus *Leucojum*
Autumn. *Leucojum autumnale*
Carpathian. *Leucojum carpaticum*
Many-flowered. *Leucojum trichophyllum* (*Acis trichophylla*)
Rosy-flowered. *Leucojum roseum*
Spring. *Leucojum vernum*
Summer. *Leucojum æstivum*
Summer, Small. *Leucojum Hernandezii* (*L. pulchellum*)
Winter. *Leucojum hyemale*
Snow-flower. *Chionanthus virginica*
Japanese. *Deutzia gracilis*
Snow-glory. *Chionodoxa Luciliæ*
Dwarf. *Chionodoxa nana*
Snow-in-Summer. *Cerastium tomentosum*
Snow-mould. *Lanosa nivalis*
Snow-plant. *Cerastium tomentosum*
Californian. *Sarcodes sanguinea*
Snow-Rosette. *Primula minima*
Soap-bark-tree, Chili. *Quillaia Saponaria*
Venezuela. *Pithecolobium bigeminum*
Soap-berry-tree. The genus *Sapindus*
American. *Sapindus marginatus*
W. Indian. *Sapindus Saponaria*
Soap-nut. The fruit of *Sapindus Saponaria* and *Acacia concinna* (*Mimosa abstergens*)

Soap-plant, Indian. *Chlorogalum pomeridianum*
Mexican. *Agave saponaria*
Soap-pods, Chinese. The pods of various species of *Cæsalpinia*
E. Indian. The pods of *Acacia concinna*
Soap-root, Californian. *Leucocrinum montanum*
Spanish, or Egyptian. *Gypsophila Struthium*
Soap-wood. *Clethra tinifolia*
Soap-wort. The genus *Saponaria*; also applied to *Vaccaria vulgaris*
Brilliant-flowered. *Saponaria ocymoides var. splendens*
Common. *Saponaria officinalis*
Cow-herb. *Saponaria Vaccaria*
Daisy-leaved. *Saponaria bellidifolia*
Double - flowered. *Saponaria caucasica fl.-pl.*
Rock. *Saponaria ocymoides*
Tufted. *Saponaria cæspitosa*
Soft-wood, Black. *Myrsine læta*
"Solah"-plant. *Æschynomene aspera*
Soldier-bush. *Inga purpurea*
Soldier-wood. *Calliandra purpurea*
Soldier's-herb. *Piper angustifolium (Artanthe elongata)*
Solomon's-seal. The genus *Polygonatum*
Common. *Polygonatum officinale*
False. The genus *Smilacina*
Giant. *Polygonatum giganteum*
Many-flowered. *Polygonatum multiflorum*
Rosy-flowered. *Polygonatum roseum (Convallaria rosea)*
Whorled. *Polygonatum verticillatum*
Sooly-Qua Gourd. See Gourd
Sophee-shrub, of Sylhet. *Myrica integrifolia*
Sorrel. The genus *Oxalis*, some species of *Rumex*, and a few other plants
Buckler-shaped. See Sorrel, French
Climbing. *Begonia scandens*
French, or Buckler - shaped. *Rumex scutatus*
Indian, or Red. *Hibiscus Sabdariffa*
Maiden. *Rumex montanus*
Mountain. *Oxyria reniformis*
Red. See Sorrel, Indian
Switch. *Dodonæa viscosa*
Sheep's. *Rumex Acetosella*
Tree. *Rumex Lunaria*
Wood. See Wood-Sorrel
Sorrel-tree. *Andromeda arborea*
Queensland. *Hibiscus heterophyllus*
Sorrowful Tree. *Nyctanthes Arbor-tristis*
Souari, or Suwarrow-nut-tree. *Caryocar nuciferum*
Souari-wood. See Saouari-wood
Sour-Gourd-tree. *Adansonia digitata*
Sour Sop. *Anona muricata*
Southern-wood. *Artemisia Abrotanum*
Tartarian. *Artemisia Santonica*
"Sowa"-plant, Bengal. *Anethum Sowa (Peucedanum graveolens)*
Sow-bane. *Chenopodium rubrum*
Sow-bread. *Cyclamen europæum*
Sowd-wort. *Salsola Kali*; also an old name for Columbine

Sow-thistle, Blue - flowered. *Mulgedium alpinum, M. macrorrhizum,* and the genus *Agathyrsus*
Common. *Sonchus oleraceus*
Corn-field. *Sonchus arvensis*
Cut-leaved. *Sonchus laciniatus* and *S. elegantissimus*
Marsh. *Sonchus palustris*
Mountain. *Sonchus alpinus*
Purple-flowered. *Mulgedium Plumieri*
Shrubby. *Sonchus fruticosus*
Tree. *Atalanthus arboreus*
Soy-plant, or Soy-bean. *Dolichos Soja (Soja hispida)*
"Spadic"-bush. *Erythroxylon Coca*
Spætlum, or Spatlum. *Lewisia rediviva*
"Spaniards," of New Zealand. *Aciphylla squarrosa* and *A. Colensoi*
Spanish-Arbour-Vine. *Ipomæa tuberosa*
Spanish - Bayonet, or Spanish-Dagger. *Yucca aloifolia* and other species
Spanish - juice - plant. The Liquorice-plant (*Glycyrrhiza glabra*)
Spanish-Needles. *Bidens bipinnata*
Spanish-Pick-tooth. *Daucus Visnaga*
Spanish-Tuft. *Thalictrum aquilegifolium*
Sparrow-grass. A corruption of "Asparagus"
Sparrow-tongue. *Polygonum aviculare*
Sparrow-wort. *Passerina nivalis*
Spart-grass. See Grass
Spatlum. See Spætlum.
Spear-flower. The genus *Ardisia*
Spear-grass. See Grass
Spear-mint. See Mint
Spear-wood, Australian. *Acacia doratoxylon* and *Eucalyptus doratoxylon*
Spear-wort, Adder's-tongue. *Ranunculus ophioglossifolius*
Great. *Ranunculus Lingua*
Small. *Ranunculus Flammula*
Water-Plantain. *Ranunculus alismæfolius*
"Speck-boom," of S. Africa. *Portulacaria afra*
Spelt. *Triticum Spelta*
Speedwell. The genus *Veronica*
Alpine. *Veronica alpina*
Amethyst. *Veronica amethystina (V. paniculata, V. spuria)*
Anderson's. *Veronica Andersoni*
Austrian. *Veronica austriaca*
Bastard. *Veronica spuria*
Brook-lime. *Veronica Beccabunga*
Buchanan's. *Veronica Buchanani*
Buxbaum's. *Veronica Buxbaumii*
Caucasian. *Veronica caucasica*
Caucasian, White. *Veronica peduncularis*
Clustered. *Veronica spicata corymbosa*
Common. *Veronica officinalis*
Corn-field. *Veronica arvensis*
Creeping. *Veronica repens*
Crimean. *Veronica taurica*
Cut-leaved. *Veronica incisa*
Daisy-leaved. *Veronica bellidioides*
Digger's, of Australia. *Veronica perfoliata*
Falkland Islands. *Veronica decussata*
Fern-leaved. *Veronica laciniata*
Flesh-coloured. *Veronica incarnata*
Gentian. *Veronica gentianoides*

Speedwell, Gray-leaved. *Veronica neglecta*
Great Virginian. *Veronica (Leptandra) virginica*
Guthrie's. *Veronica Guthrieana*
Hungarian. *Veronica Teucrium*
Ivy-leaved. *Veronica hederaefolia*
Japanese. *Veronica longifolia subsessilis*
Leafy. *Veronica foliosa*
Many-spiked. *Veronica corymbosa*
Marsh. *Veronica scutellata*
Money-wort. *Veronica Nummularia*
Mountain. *Veronica montana*
Naked-stalked. *Veronica aphylla*
Narrow-leaved. *Veronica multifida*
Pink - flowered. *Veronica officinalis var. rosea*
Procumbent. *Veronica agrestis*
Prostrate. *Veronica prostrata*
Purslane. *Veronica peregrina*
Rock. *Veronica saxatilis* and *V. rupestris*
Rock, Rosy-flowered. *Veronica saxatilis var. Grierei*
Savory-leaved. *Veronica satureiaefolia*
Saw-leaved. *Veronica Teucrium*
Scallop-leaved. *Veronica pectinata*
Sea-side. *Veronica maritima*
Shrubby-stalked. *Veronica fruticulosa*
Siberian. *Veronica sibirica*
Silvery. *Veronica candida*
Thick-leaved. *Veronica pinguifolia*
Thyme-leaved. *Veronica serpyllifolia*
Variable-leaved. *Veronica triphyllos*
Vernal. *Veronica verna*
Vervain. *Veronica verbenacea*
Virginian. *Veronica (Leptandra) virginica*
Water. *Veronica Anagallis*
Wedge-leaved. *Veronica cuneifolia*
Welsh. *Veronica hybrida*
White - flowered. *Veronica lactea (V. repens)*
Willow-leaved. *Veronica salicifolia*
Sperage. An old name for Asparagus
Spice-bush. *Laurus (Lindera) Benzoin (Benzoin odoriferum)*
Spice-tree, Californian. *Oreodaphne (Umbellularia) californica*
Spider-flower. The genus *Cleome*
Brazilian. The genus *Lasiandra*
Spider-Orchis. See Orchis
Spider-wort, Blue. *Commelina coelestis*
Branched. *Anthericum (Czackia) Liliago*
Crowded-flowered. *Tradescantia congesta*
Dwarf. *Tradescantia pilosa*
Gentian-leaved. *Tradescantia Zanonia*
Grass-leaved. *Tradescantia malabarica*
Great Savoy. *Anthericum (Czackia) Liliastrum*
Mountain. *Lloydia serotina*
Purple-leaved. *Tradescantia discolor*
Virginian. *Tradescantia virginica*
White. *Tradescantia virginica var. alba*
Spignel, Common. *Meum Athamanticum*
Broad-leaved. *Athamanta Cervaria*
Fine-leaved. *Athamanta cretensis*
Flix-weed-leaved. *Athamanta sicula*
Mountain. *Athamanta Libanotis*
Spike-flower, Australian Blue. *Calectasia cyanea*

Spike-grass. See Grass
Spike-oil-plant. *Lavandula Spica*
Spike-rush. See Rush
Spikenard, American. *Aralia racemosa*
Cretan. *Valeriana Phu*
False. *Smilacina racemosa*
Mountain. *Valeriana tuberosa*
Ploughman's. *Conyza squarrosa* and the genus *Baccharis*
W. Indian. *Hyptis suaveolens*
Spinach, African. *Rumex vesicarius roseus*
Australian. *Tetragonia implexicoma (T. trigyna)*, *Chenopodium erosum*, and *C. auricomum*
Beet, or Perpetual. *Beta maritima*
Common Garden. *Spinacia oleracea*
Cuban. *Claytonia cubensis*
E. Indian. *Basella alba* and *B. rubra*
Mountain. *Atriplex hortensis*
New Zealand. *Tetragonia expansa* and *T. implexicoma (T. trigyna)*
Prickly-seeded. *Spinacia oleracea var. spinosa*
Round - seeded. *Spinacia oleracea var. glabra*
Straw-berry. *Blitum capitatum*
Victorian Bower. *Tetragonia implexicoma*
Wild. *Chenopodium Bonus-Henricus*
Spindle-tree. The genus *Euonymus*
Broad-leaved. *Euonymus latifolius*
Dwarf. *Euonymus nanus*
Fringed. *Euonymus fimbriatus*
Garcinia-leaved. *Euonymus garciniaefolius*
Golden-leaved. *Euonymus japonicus aureovariegatus*
Hamilton's. *Euonymus Hamiltonianus*
Japanese. *Euonymus japonicus*
Large-flowered. *Euonymus grandiflorus*
Large-fruited. *Euonymus latifolius*
Narrow-leaved. *Euonymus angustifolius*
Obovate-leaved. *Euonymus obovatus*
Silver-edged-leaved. *Euonymus japonicus argenteus*
Three - coloured - leaved. *Euonymus japonicus tricolor*
Trailing-stemmed. *Euonymus sarmentosus*
Variegated Dwarf. *Euonymus japonicus radicans rariegatus*
Warty-barked. *Euonymus verrucosus*
White-fruited. *Euonymus europaeus fructu albo*
Spindle-wort. *Atractylis acaulis*
Spire-lily. See Lily
Spire-reed. *Arundo Phragmites*
Spirit-leaf, or Spirit-weed. *Ruellia tuberosa (Cryphiacanthus barbadensis)*
Spleen-wort. The genus *Asplenium*
Alternate-leaved. *Asplenium alternifolium*
Black. *Asplenium Adiantum-nigrum*
Common, or Maiden - hair. *Asplenium Trichomanes*
Ebony. *Asplenium ebeneum*
Ebony, Smaller. *Asplenium parvulum*
Forked. *Asplenium septentrionale*
Green. *Asplenium viride*
Lance-fronded. *Asplenium lanceolatum*
Maiden-hair. See Spleen-wort, Common
New Zealand Common. *Asplenium bulbiferum*

Spleen-wort, New Zealand Hanging-tree. *Asplenium flaccidum*
New Zealand Shore. *Asplenium obtusatum*
Scaly. *Asplenium Ceterach*
Sea-side. *Asplenium marinum*
Smooth Rock. *Asplenium fontanum*
Wall-Rue. *Asplenium Ruta-muraria*
Split Moss. See Moss
Spogel - Seeds. The seeds of *Plantago decumbens*
Sponge-tree. *Acacia Farnesiana*
Sponge-wood, Bengal. *Æschynomene aspera*
Isle of France. *Gastonia cutispongia*
Spoon-flower, Arrow-leaved. *Xanthosoma sagittæfolia*
Spoon-wood-tree. *Kalmia latifolia*
Spoon-wort. The genus *Cochlearia*
Common. *Cochlearia officinalis*
Sportsman's Climber. *Cissus venatorum*
"Spotted-tree," of Queensland. *Flindersia maculata*
Spread-Eagle. *Oncidium Carthaginense*
Spring-Beauty. *Claytonia virginica*
Spring-Bell. *Sisyrinchium grandiflorum*
Sprit. *Juncus articulatus*
Sprout-leaf. *Bryophyllum calycinum*
Spruce, or Spruce Fir. The genus *Abies*
Alcock. *Abies Alcockiana*
Black, or Double. *Abies nigra*
Black, of British Columbia. *Abies Menziesii*
Blue, or Tideland. *Pinus sitchensis (P. Douglasii)*
Common, or Norway. *Abies excelsa*
Dinsdale's Silver. *Abies alba var. glauca*
Double. See Spruce, Black
Dwarf. *Abies excelsa var. pygmæa*
Eastern. *Abies orientalis*
Engelmann's. *Abies commutata*
Finedon Hall. *Abies excelsa var. Finedonensis*
Hemlock. *Abies canadensis*
Hemlock, Californian. *Tsuga (Abies) Mertensiana* and *T. Pattoniana*
Hemlock, Indian. *Abies Brunoniana*
Hemlock, Japan. *Abies Tsuga*
Indian, or Himalayan. *Abies Morinda (A. Smithiana)*
Lord Clanbrasil's Dwarf. *Abies excelsa var. Clanbrasiliana*
Menzies'. *Abies Menziesii*
Norway. *Abies excelsa*
Norway, Weeping. *Abies excelsa var. pendula*
New Zealand. *Dacrydium cupressinum*
Rocky Mountain. *Picea pungens*
Siberian. *Picea sibirica* and *P. oborata*
Single. *Abies alba*
Tideland. See Spruce, Blue
Tiger's-tail. *Abies polita*
White. *Abies alba*
White, Dwarf. *Abies alba var. nana*
White Hedge-hog. *Abies alba var. minima*
Spunk. See Amadou
Spurge. The genus *Euphorbia*
Alleghany Mountain. *Pachysandra procumbens*

Spurge, Balsam. *Euphorbia balsamifera*
Branched. *Ernodea littoralis*
Broad-leaved. *Euphorbia platyphyllos*
Caper. *Euphorbia Lathyris*
Cucumber. *Euphorbia cucumerina*
Cypress. *Euphorbia Cyparissias*
Dwarf. *Euphorbia exigua*
Flowering. *Euphorbia corollata*
Gromwell-leaved. *Euphorbia Esula*
Hairy. *Euphorbia pilosa*
Hyssop. *Euphorbia Peplis*
Indian Tree. *Euphorbia Tirucalli*
Irish. *Euphorbia hibernica*
Leafy. *Euphorbia Esula*
Marsh. *Euphorbia palustris*
Petty. *Euphorbia Peplis*
Porcupine. *Euphorbia Hystrix*
Portland. *Euphorbia segetalis (E. Portlandica)*
Purple. *Euphorbia Peplis*
Scarlet-flowered. *Euphorbia punicea*
Sea-side. *Euphorbia Paralias*
Slipper. The genus *Pedilanthus*
Sun. *Euphorbia Helioscopia*
Sweet. *Euphorbia dulcis*
Trailing Red. *Euphorbia prostrata*
White-flowered. *Euphorbia corollata*
Wood. *Euphorbia amygdaloides*
Sprue. A market name for the smallest sprouts of Asparagus
Spurge-Flax. An old name for *Daphne Mezereum* and other species
Spurge-Laurel. See Laurel
Spurge - Nettle. *Jatropha urens var. stimulosa*
Spurge-Olive. See Olive
Spur-tree. *Petitia domingensis*
Spur-wort. *Sherardia arvensis*
Spurrey. The genus *Spergula*
Corn-field. *Spergula arvensis*
Knotted. *Sagina nodosa*
Lawn. *Spergula pilifera*
Sand, or Sand-wort. The genera *Spergularia* and *Lepigonum*
Spurt-grass. See Grass
Squash, Long. *Cucurbita verrucosa*
Summer. *Cucurbita Pepo*
Winter. *Cucurbita maxima*
Squaw-Mint. *Hedeoma pulegioides*
Squaw-root. *Cimicifuga racemosa*
Squaw-weed. *Enerigeron Philadelphicus* and *Senecio aureus*
Squill. The genus *Scilla*
Amethyst. *Scilla amethystina*
Autumn-flowering. *Scilla autumnalis*
Californian. *Scilla (Camassia) Fraseri*
Chinese. The genus *Barnardia*
Crimean. *Scilla taurica*
Dwarf Australian. *Chamæscilla (Cæsia) corymbosa*
Early-flowering. *Scilla bifolia*
Medicinal. *Urginea (Scilla) maritima*
Pyramidal-flowered. *Scilla peruviana*
Roman. The genus *Bellevalia*
Sea-side. *Urginea (Scilla) maritima*
Siberian. *Scilla sibirica*
Spanish. *Scilla campanulata*
Spring-flowering. *Scilla verna*
Star-flowered. *Scilla amæna*

K

Squill, Striped-flowered. *Puschkinia scilloides*
Umbel-flowered. *Scilla umbellata*
Squinancy-wort. *Asperula cynanchica*
"Squinant." An old name for *Andropogon Schœnanthus*
Squirrel-corn. *Dicentra canadensis*
Squirrel-tail-grass. See Grass
Squitch-grass. See Grass
Staff-tree. The genus *Celastrus*
Scarlet-fruited. *Celastrus bullatus*
Staff-Vine. *Celastrus scandens*
Stagger-bush. *Andromeda Mariana*
Stagger-wort, or Staver-wort. *Senecio Jacobæa*
Star, Sea. *Aster Tripolium*
Stone-Mountain. See Stone-Mountain-Star
Star-Anise. See Anise
Star-apple. *Chrysophyllum cæruleum*
Star-bush, African. *Grewia occidentalis*
Star-fish-flower. *Stapelia Asterias* and other species
Star-flower. *Trientalis europæa*
American. *Trientalis americana*
Bouquet. *Aster ptarmicoides*
Green. *Ornithogalum prasinum*
Lilac. *Tritelcia lilacina*
Murray's. *Tritelcia Murrayana*
Spring. *Tritelcia uniflora*
Turkish. *Sternbergia Clusiana*
Yellow. *Sternbergia lutea*
Star-Glory. *Ipomœa coccinea*
Star-grass. *Hypoxis erecta* and the genus *Aletris*
Common. *Aletris farinosa*
Golden-tipped. *Aletris aurea*
Star-head. The genus *Asterocephalus*
Star-jelly, or Fairies'-butter. *Nostoc commune* and *N. edule*
Star-of-Bethlehem, Common. *Ornithogalum umbellatum*
Bouche's. *Ornithogalum Boucheanum*
Bright-yellow. *Gagea fistulosa*
Broad-leaved. *Ornithogalum latifolium*
Drooping. *Ornithogalum nutans*
Golden-flowered. *Ornithogalum aureum*
Narbonne. *Ornithogalum narbonnense*
Pyrenean. *Ornithogalum pyrenaicum*
Short-spiked. *Ornithogalum comosum*
Tall. *Ornithogalum pyramidale*
Star-of-Hungary. *Ornithogalum pannonicum*
Star-of-Jerusalem. *Tragopogon porrifolius*
Star-of-night. *Clusia rosea*
Star-of-the-earth. *Plantago Coronopus*
Star-thistle. See Thistle
Star-wort. The genera *Aster* and *Stellaria*
Australian. *Aster (Eurybia) argophylla*
Bog. *Stellaria uliginosa*
Branching. *Aster ramosus*
Brittle. *Aster fragilis*
Bushy. *Aster dumosus*
Christmas-flowering. *Aster Datschyi*
Crimson-flowered. *Aster coccineus*
E. Indian. *Aster cabulicus*
Elegant. *Aster elegans*
Fortune's. *Aster Fortunei*

Star-wort, Greater. *Stellaria Holostea*
Heart-leaved. *Aster cordifolius*
Heath-leaved. *Aster ericoides*
Hoary. *Aster canescens*
Horizontal-branched. *Aster horizontalis*
Hyssop-leaved. *Aster hyssopifolius*
Italian. *Aster Amellus*
Large-flowered Altaian. *Aster altaicus*
Late-flowering. *Aster tardiflorus*
Long-leaved. *Aster longifolius*
Loose-branched. *Aster laxus*
Lesser. *Stellaria graminea*
Many-flowered. *Aster multiflorus*
Mauve-flowered. *Aster turbinellus*
New England. *Aster Novæ-Angliæ*
N. S. Wales. *Olearia dentata*
New York. *Aster Novi-Belgii*
Northern. *Stellaria borealis*
Oblique. *Aster obliquus*
Pendulous. *Aster pendulus*
Purple-stemmed. *Aster puniceus*
Pyrenean. *Aster pyrenæus*
Red-stemmed. *Aster rubricaulis*
Reeves's. *Aster Reevesi*
Scorzonera-leaved. *Aster scorzoneræfolius*
Sea-side. *Aster Tripolium*
Short's. *Aster Shortii*
Showy. *Aster spectabilis*
Shrubby. *Aster albescens* and *Eurybia ramulosa*
Silky. *Aster sericeus*
Slender. *Felicia tenella*
Slender-leaved. *Aster tenuifolius*
Smooth. *Aster lævis var. lævigatus*
Spreading. *Aster patens*
Tarragon-like. *Aster (Galatella) dracunculoides*
Townshend's. *Aster Townshendi*
Various-coloured. *Aster versicolor*
Water. *Stellaria aquatica* and the genus *Callitriche*
Wood. *Stellaria nemorum*
Yarrow-leaved. *Aster ptarmicoides*
Stave-wood. *Simaruba amara*
Staver-wort. See Stagger-wort
Staves-acre. *Delphinium Staphisagria*
Stay-plough. *Ononis arvensis*
Steeple-bush. *Spiræa tomentosa*
Sterile-wood, Otago. *Coprosma fœtidissima*
Sticadoue, or Sticados. *Lavandula Stœchas*
Stick-seed. The genus *Echinospermum*
Stick-tight. See Beggar-ticks
Stickle-wort. *Agrimonia Eupatoria*
Stiff-stalk, Mexican. *Rigidella flammea*
Stinging-bush. *Jatropha stimulans*
Stink-horn, or Stinking-Polecat. *Phallus impudicus* and *P. fœtidus*
Lattice. *Clathrus cancellatus*
Stinking-wood. *Anagyris fœtida* and *Cassia occidentalis*
Stink-weed. *Diplotaxis muralis*
Stink-wood. The wood of *Oreodaphne bullata, Fœtidia mauritiana,* and *Zieria macrophylla*
Tasmanian. *Zieria lanceolata*
Stitch-wort, or Stitch-grass. The genus *Stellaria*
Golden. *Stellaria graminea aurea*

Stock, or Stock-Gilliflower. Various species of *Matthiola*
Brompton. *Matthiola incana var. coccinea*
Cape. The genus *Heliophila*
Cluster-leaved. *Matthiola fenestralis*
Dark-flowered. *Matthiola tristis*
Mediterranean. *Hesperis maritima*
Persian. *Matthiola odoratissima*
Queen's. *Matthiola incana*
Sea-side. *Matthiola sinuata*
Ten-weeks. *Matthiola annua*
Ten-weeks, Double. *Matthiola annua fl.-pl.*
Virgin, or Virginia. *Malcolmia maritima*
Wall-flower-leaved, or Smooth-leaved. *Matthiola græca*
Stone-Basil. See Basil
Stone-crop. The genus *Sedum*
Asiatic. *Sedum asiaticum*
Azure. *Sedum cyaneum*
Bird's-foot. *Sedum pulchellum*
Blood-red. *Sedum cruentum*
Blunt-leaved. *Sedum obtusatum*
Brilliant. *Sedum spectabile (S. Fabarium)*
Brown's. *Sedum Brownii*
Cock's-comb. *Sedum Crista-galli*
Common. *Sedum acre*
Corsican. *Sedum corsicum*
Crimson. *Sedum spurium*
Dark-purple. *Sedum atropurpureum*
Deep-green-flowered. *Sedum virens*
Ditch, of N. America. *Penthorum sedoides*
Elongated. *Sedum elongatum*
English. *Sedum anglicum*
Ewers's. *Sedum Ewersii*
Forster's. *Sedum Forsterianum*
Fringed. *Sedum ciliare (S. oppositifolium)*
Glaucous. *Sedum glaucum*
Greenish-flowered. *Sedum virescens*
Hairy. *Sedum villosum*
Iberian. *Sedum ibericum*
Large-fringed. *Sedum spurium*
Lydian. *Sedum Lydium*
Mealy. *Sedum farinosum*
Meehan's. *Sedum Meehani*
Nævius's. *Sedum Nævii*
Narrow-petalled. *Sedum stenopetalum*
Noble. *Sedum spectabile*
Orange. *Sedum camtschaticum*
Orpine-like. *Sedum telephioides*
Poplar-leaved. *Sedum populifolium*
Purple. *Sedum purpureum*
Purple, American. *Sedum pulchellum*
Purplish. *Sedum purpurascens*
Recurved-leaved. *Sedum recurvatum*
Rock. *Sedum rupestre*
Rosy-flowered. *Sedum roseum* and *S. rhodanthum*
Rose-coloured, Pale. *Sedum pallidum roseum*
Round-leaved. *Sedum rotundifolium*
St. Vincent's Rocks. *Sedum rupestre*
Scarlet. *Sedum sempervivoides*
Short-leaved. *Sedum brevifolium*
Siebold's. *Sedum Sieboldi*
Six-angled. *Sedum sexangulare*
Spiral. *Sedum spirale*
Spoon-leaved. *Sedum spatulæfolium*

Stone-crop, Thick-leaved. *Sedum dasyphyllum* and *S. turgidum*
Variegated Japanese. *Sedum japonicum variegatum*
Welsh. *Sedum Forsterianum*
White-flowered. *Sedum album*
Wightmann's. *Sedum Wightmannianum*
Stone-hore, or Stone-Orpine. *Sedum reflexum*
Stone-Mountain-Star. *Gymnolomia Porteri*
Stone-root. *Collinsonia canadensis*
Stone-wort. The genus *Chara*
Stonnord. An old name for Stone-crop
Storax-plant. The genus *Styrax*
Large-leaved. *Styrax grandifolia*
Liquid. *Liquidambar orientalis*
Officinal. *Styrax officinalis*
Powdery. *Styrax pulverulenta*
Smooth-leaved. *Styrax lævigata*
Stork's-bill. The genus *Pelargonium*
Stover. An old name for dried Grass or Fodder
Strangle-Tare. *Vicia sylvestris, V. lathyroides,* and *Cuscuta europæa*
Strap-wort. *Corrigiola littoralis*
Straw, Leghorn-bonnet. The straw of *Triticum Spelta*
Straw-berry. The genus *Fragaria*
Alpine. *Fragaria collina*
Barren. *Waldsteinia (Comaropsis) fragarioides*
Chili. *Fragaria chilensis*
Common. *Fragaria vesca* and vars.
Dalmatian. *Arbutus Unedo*
Green Pine. *Fragaria collina*
Hautbois. *Fragaria elatior*
Indian. *Fragaria indica*
Old Scarlet, or Scarlet Virginia. *Fragaria virginiana*
Rock. *Fragaria indica*
Virginian Scarlet. *Fragaria virginiana*
Straw-berry-Blite. *Blitum capitatum*
Straw-berry-bush, American. *Euonymus americanus*
"Straw-berry-Geranium." *Saxifraga sarmentosa*
Straw-berry-Saxifrage. See Saxifrage
Straw-berry-Spinach. *Blitum capitatum*
Straw-berry-Tomato. See Tomato
Straw-berry-tree. *Arbutus Unedo*
Alpine. *Arbutus alpina*
Canadian. *Arbutus alpina*
Menzies's. *Arbutus Menziesii*
Oriental. *Arbutus Andrachne*
Red-flowered. *Arbutus Unedo rubra*
Scarlet-barked. *Arbutus Unedo var. Croomi*
String-wood, of St. Helena. *Acalypha rubra*
Stringy-bark-tree. *Eucalyptus obliqua, E. capitellata, E. macrorrhynca, E. pilularis,* and *E. piperita*
Strong-man's-weed. *Petiveria alliacea*
Strychnine-plant. *Strychnos Nux-vomica*
Stub-wort. An old name for *Oxalis Acetosella*
Stud-flower. *Helonias bullata (H. latifolia)*

Succory. *Cichorium Intybus*
Blue. *Catananche cœrulea*
Gum. The genus *Chondrilla*
Lamb's, or Swine's. *Arnoseris pusilla*
Yellow. *Catananche lutea*
Suckles. An old name for Honeysuckles
Sugar-berry. *Celtis occidentalis*
Sugar-bush, of the Cape. *Protea mellifera*
Sugar-cane, Bengal. *Saccharum Bengalense*
Common. *Saccharum officinarum*
Sugar-Pea. See Pea
Sugar-tree. *Myoporum platycarpum*
Sulphur-wort, or Sulphur-weed. *Peucedanum officinale*
Sultan, Sweet. See Sweet-Sultan
Sultan's-Parasol. *Sterculia platanifolia*
Sumach. The genus *Rhus*
Carolina. *Rhus elegans*
Coral. *Rhus Metopium*
Cut-leaved. *Rhus glabra rar. laciniata*
Dwarf. *Rhus pumila* and *R. copallina*
Elm-leaved. *Rhus Coriaria*
False Venetian. *Rhus cotinoides*
Fern-leaved. *Rhus glabra var. laciniata*
Five-leaved. *Rhus pentaphylla*
Fragrant. *Rhus aromatica*
Green-flowered. *Rhus viridiflora*
Jamaica. *Rhus Metopium*
Myrtle-leaved. *Coriaria myrtifolia*
Osbeck's. *Rhus Osbecki*
Poison. *Rhus venenata*
Purple-fringed. *Rhus Cotinus*
Red Lac. *Rhus succedanea*
Rooting-branched. *Rhus radicans*
Scarlet. *Rhus glabra*
Stag's-horn. *Rhus typhina*
Swamp. *Rhus venenata*
Sweet-scented. *Rhus suaveolens*
Tanner's. *Rhus Coriaria*
Venetian, or Venus's. *Rhus Cotinus*
Venetian, False. *Rhus cotinoides*
Virginian. *Rhus typhina*
Walnut - leaved. *Rhus juglandifolia* (*R. vernicifera*)
W. Indian. *Brunellia comocladifolia*
Zizyphus-like. *Rhus zizyphina*
Sumbul-root. The root of *Ferula* (*Euryangium*) *Sumbul*. Also applied to the root of *Nardostachys Jatamansi*
Sun-dew. The genus *Drosera*
Common, or Round-leaved. *Drosera rotundifolia*
Double-leaved. *Drosera dichotoma*
English. *Drosera anglica*
Long-leaved. *Drosera longifolia*
Portuguese. *Drosophyllum lusitanicum*
Slender American. *Drosera linearis*
Thread-leaved. *Drosera filiformis*
Sun-drops. *Œnothera fruticosa*
Sun-flower. The genus *Helianthus*
Carrot-rooted. *Helianthus strumosus*
Common. *Helianthus annuus*
Cucumber - leaved. *Helianthus cucumerifolius*
Dark-red. *Helianthus atrorubens*
Dwarf Annual. *Helianthus indicus*
Dwarf Tufted. *Actinella scaposa*
False. See Sun-flower, Swamp

Sun-flower, Giant. *Helianthus giganteus*
Graceful. *Helianthus orgyalis*
Jungle, of the Cape. *Osteospermum moniliferum*
Many-flowered. *Helianthus multiflorus*
Maximilian's. *Helianthus Maximiliani*
Miniature. *Heliopsis lœvis*
Nuttall's. *Helianthus Nuttallii*
Perennial. *Helianthus multiflorus* and vars.
Pigmy. *Actinella grandiflora*
Prairie. *Helianthus rigidus* (*Harpalium rigidum*)
Primrose. A variety of *Helianthus annuus*
Rough-leaved. *Helianthus rigidus*
Silvery-leaved. *Helianthus argophyllus*
Swamp, or False. The genus *Helenium*
Ten-petalled. *Helianthus decapetalus*
Tick-seed. *Coreopsis trichosperma*
Zig-zag. *Helianthus flexuosus*
Sun-plant. *Portulaca grandiflora* and other species
Sun-rose. The genus *Helianthemum*
Alpine. *Helianthemum alpestre*
Alyssum-like. *Helianthemum alyssoides*
Apennine. *Helianthemum apenninum*
Basil-like. *Helianthemum ocymoides*
Beautiful. *Helianthemum formosum*
Bristly. *Helianthemum hispidum*
Bryony-scented. *Helianthemum fœtidum*
Cairo. *Helianthemum Kahiricum*
Cluster - flowered. *Helianthemum glomeratum*
Cluster-leaved. *Helianthemum lœvipes*
Colour-changing. *Helianthemum mutabile*
Common. *Helianthemum vulgare* (*Cistus Helianthemum*)
Copper - colour - flowered. *Helianthemum cupreum*
Downy. *Helianthemum Pilosella*
Glaucous-leaved. *Helianthemum glaucum*
Hairy. *Helianthemum hirsutum*
Hairy-flowered. *Helianthemum lasianthum*
Heath-like. *Helianthemum Fumana*
Hoary - leaved. *Helianthemum canum* (*Cistus marifolius*)
Italian. *Helianthemum italicum*
Juniper-leaved. *Helianthemum juniperinum*
Lagasca's. *Helianthemum hirtum rar. Lagascæ*
Large - flowered. *Helianthemum grandiflorum* and *H. macranthum*
Lavender-leaved. *Helianthemum larandulæfolium*
Marjoram-leaved. *Helianthemum origanifolium* and *H. marjoranifolium*
Marum-leaved. *Helianthemum marifolium*
Mealy-leaved. *Helianthemum farinosum*
Money-wort-leaved. *Helianthemum Nummularium*
Naked-stemmed. *Helianthemum nudicaule*
Narrow-leaved. *Helianthemum angustifolium*
Orache-leaved. *Helianthemum atriplicifolium*
Ovate-leaved. *Helianthemum ovatum*
Parti - coloured - flowered. *Helianthemum versicolor*

Sun-rose, Pencilled. *Helianthemum penicillatum*
Plantain-leaved. *Helianthemum Tuberaria*
Polium-leaved. *Helianthemum poliifolium*
Powdered-leaved. *Helianthemum pulverulentum*
Procumbent. *Helianthemum procumbens*
Red-flowered. *Helianthemum rhodanthum*
Rosemary-leaved. *Helianthemum rosmarinifolium*
Rosy-flowered. *Helianthemum roseum*
Rough. *Helianthemum scabrosum*
Saffron-flowered. *Helianthemum croceum*
Sea-Purslane-leaved. *Helianthemum halimifolium*
Serpyllum-leaved. *Helianthemum serpyllifolium*
Shining-leaved. *Helianthemum lucidum*
Showy. *Helianthemum venustum*
Slender Trailing. *Helianthemum vineale*
Small-leaved. *Helianthemum microphyllum*
Smooth. *Helianthemum læve*
Soft-leaved. *Helianthemum molle*
Spear-leaved. *Helianthemum lanceolatum*
Spotted. *Helianthemum guttatum*
Stiff-haired. *Helianthemum pilosum*
Straw-coloured - flowered. *Helianthemum stramineum*
Striped - flowered. *Helianthemum variegatum*
Sulphur - flowered. *Helianthemum sulphureum*
Surrey. *Helianthemum surrejanum*
Thick-leaved. *Helianthemum crassifolium*
Thyme-leaved. *Helianthemum thymifolium*
Truffle. *Helianthemum Tuberaria*
Twiggy. *Helianthemum virgatum*
Umbel - flowered. *Helianthemum umbellatum*
Upright-branched. *Helianthemum strictum*
Various - leaved. *Helianthemum diversifolium*
Violet-calyxed. *Helianthemum violaceum*
Wall-flower-like. *Helianthemum cheiranthoides*
White-leaved. *Helianthemum candidum* and *H. polifolium*
Whitish-leaved. *Helianthemum canescens*
Woody-stemmed. *Helianthemum lignosum*
Woolly-leaved. *Helianthemum stœchadifolium* and *H. tomentosum*
Sunn-Hemp. See Hemp
Sunshine-plant, Australian. *Acacia discolor*
Sunt-wood. The wood of *Acacia arabica*
Supple-Jack, American. *Berchemia volubilis*
Australian. *Clematis aristata*
New Zealand. *Rubus australis*; also applied to the genera *Parsonsia* and *Lygodium*
W. Indian. *Paullinia barbadensis*, *P. curassavica*, and *P. polyphylla*; also applied to *Berchemia volubilis* and *Cardiospermum grandiflorum*
Surinam-poison. *Tephrosia toxicaria*
Swallow-wort. The genera *Chelidonium* and *Asclepias*
Tuberous-rooted. *Asclepias tuberosa*

Swallow's - nest Seaweed. *Plocaria tenax*
Swamp-weed, Victorian. *Selliera radicans*
Swan-neck. See Swan-wort
Swan-flower, of Surinam. *Cycnoches Loddigesii*
Swan-wort, or Swan-neck. The genus *Cycnoches*
Sweet-briar. *Rosa rubiginosa* and *R. Eglanteria*
Sweet-briar Sponge. See Bedeguar
Sweet-Cicely. *Myrrhis odorata*
Californian. The genus *Osmorrhiza*
Sweet-Flag, or Sweet-Sedge. *Acorus Calamus*
Grass-leaved. *Acorus gramineus*
Sweet-Gale, or Sweet-Willow. *Myrica Gale*
Sweet-John. *Dianthus barbatus var. angustifolius*
Sweet-leaf. *Symplocos (Hopea) tinctoria*
Sweet-Maudlin. *Achillea Ageratum*
Sweet-Nancy. *Narcissus biflorus fl.-pl.*
Sweet-Pea. *Lathyrus odoratus*
Sweet-Sedge. See Sweet Flag
Sweet-Sop. The fruit of *Anona squamosa* and *A. sericea*
Sweet-Sultan, Purple. *Amberboa (Centaurea) moschata*
Red-flowered. *Amberboa (Centaurea) moschata var. rubra*
Yellow. *Amberboa (Centaurea) suaveolens*
Sweet-weed. *Capraria biflora*
Sweet-William. *Dianthus barbatus*
Barbadoes. *Ipomœa Quamoclit*
Double-red. *Dianthus barbatus var. magnificus*
Wild, of N. America. *Phlox maculata*
Sweet-Willow. See Sweet-Gale
Sweet-wood, Black. *Strychnodaphne floribunda*
Jamaica. *Oreodaphne (Nectandra) exaltata*
Loblolly, or Rio Grande. *Oreodaphne Leucoxylum*
Lowland, Pepper, White, or Yellow. *Nectandra sanguinea*
Mountain. *Acrodiclidium jamaicense*
Pepper. See Sweet-wood, Lowland
Rio Grande. *Oreodaphne Leucoxylum*
Shrubby. The genus *Amyris*
Timber. *Acrodiclidium jamaicense*, *Nectandra exaltata*, and *N. leucantha*
W. Indian. *Croton Eluteria*
White. *Nectandra leucantha* and *N. sanguinea*
Yellow. See Sweet-wood, Lowland
Swimming-Herb. An old name for Duckweed
Swine's-bane. *Chenopodium rubrum*
Swine's-cress. *Senebiera Coronopus*
Swine's-grass. See Grass
Swine's-snout. *Taraxacum Dens-leonis*
Swine's-Succory. See Succory
Swiss Genipi. *Achillea moschata*
Sword-Lily. The genus *Gladiolus*. (See also Gladiole)
Australian Yellow. *Anigozanthus flavidus*
Sycamine-tree, of Scripture. Supposed to be either the Mulberry-tree or the wild Fig

English Names of Cultivated, Native,

Sycamore-tree. *Acer Pseudo-platanus*
American. *Platanus occidentalis*
Australian. *Brachychiton lucidum*
Californian. *Platanus racemosa*
False. *Melia Azedarach*
Purple-leaved. *Acer Pseudo-platanus purpureum*
White-variegated. *Acer Pseudo-platanus albo-variegatum*
Yellow-variegated. *Acer Pseudo-platanus flavo-variegatum*
Sycomore-tree. *Ficus Sycomorus*
Syringa, Common. *Philadelphus coronarius*
Large-flowered. *Philadelphus grandiflorus*
Showy. *Philadelphus speciosus*

Taccada-plant. *Scævola Lobelia (S. Taccada)*
"**Tagasaste.**" *Cyperus proliferus*
Tail-flower. The genus *Anthurium*
Tale-wort. An old name for *Borago officinalis*
Tallow-shrub. *Myrica cerifera*
Tallow-tree, of China. *Stillingia sebifera*
Sierra Leone. *Pentadesma butyracea*
Tamarack. *Larix americana*
Tamarind-tree, Australian. *Cupania australis*
Bastard. *Acacia Julibrissin*
Bastard, Jamaica. *Acacia trichophylloides*
Black, Brown, Velvet, or Wild, of Sierra Leone. *Codarium acutifolium*
Brown. See Tamarind-tree, Black
Manilla. *Pithecolobium dulce*
N. S. Wales. *Cupania australis*
Sweet. *Inga edulis*
Velvet. *Codarium acutifolium*
Wild. The genus *Codarium*; also *Pithecolobium filicifolium*
Wild, Jamaica. *Acacia arborea*
Wild, Trinidad. *Pentaclethra filamentosa*
Yellow. *Acacia villosa*
Tamarisk-tree, Common. *Tamarix gallica*
E. Indian. *Tamarix indica*
Feathery. *Tamarix plumosa*
French. *Tamarix gallica*
German. *Myricaria germanica*
Showy. *Tamarix spectabilis*
Small-flowered. *Tamarix parviflora*
Tamarisk-Salt-tree. *Tamarix orientalis*
Tanekaha-tree, of New Zealand. *Phyllocladus trichomanoides*
Tang. *Fucus nodosus*
Tangle, Sweet. *Laminaria saccharina*
"**Tanke.**" An old name for *Peucedanum sativum*
Tanner's-tree. *Coriaria myrtifolia* and other species
Tansy, Common. *Tanacetum vulgare*
Goose, or Wild. *Potentilla Anserina*
Shrubby. *Tanacetum suffruticosum*
Wild. See Tansy, Goose
"**Tanya,**" of S. Carolina. *Caladium esculentum*
Tapa-cloth-tree, of Polynesia. *Broussonetia papyrifera*
Tape-grass. See Grass

Tape-worm-plant. *Brayera anthelmintica*
Tapioca-plant. *Janipha (Manihot) utilissima*
Tar-bush, Californian. *Eriodictyon californicum*
Tar-weed, Californian. The genera *Madia* and *Hemizonia*
Taraire-tree. *Nesodaphne Tarairi*
Tare, Common. *Vicia hirsuta*
One-flowered. *Ervum monanthos*
Rough-podded. *Ervum hirsutum*
Tine. *Lathyrus tuberosus*
Taro, or Tara-plant, Sandwich Islands. *Caladium esculentum*
Of Tahiti. *Colocasia macrorrhiza*
Tarragon-plant. *Artemisia Dracunculus*
Tartar-Bread-plant. *Crambe tatarica*
Tartarian-Lamb. The woolly rhizome of *Cibotium (Polypodium) Barometz*
Tassel-flower, Orange. *Cacalia aurantiaca*
Scarlet. *Cacalia coccinea*
Tassel-grass. See Grass
"**Tata**"-shrub, of Brazil. *Eugenia supra-axillaris*
Tawa-tree, of New Zealand. *Nesodaphne Tawa*
Tawhai-tree, of New Zealand. *Fagus fusca*
Tea-berry. *Gaultheria procumbens*
Tea-plant, or Tea-tree, Abyssinian, or Arabian. *Catha (Celastrus) edulis*
African. *Lycium afrum*
American Mountain. *Gaultheria procumbens*
Appalachian. *Viburnum cassinoides* and *Prinos glaber*
Arabian. See Tea-plant, Abyssinian
Assam. *Thea Assamensis*
Australian. Various species of *Leptospermum* and *Melaleuca*
Bee-balm. *Monarda didyma*
Bencoolen, or Malay. *Glaphyria nitida*
Bohea. *Thea Bohea*
Blue Mountain, or Golden-rod. *Solidago odora*
Bottle-green, of Victoria. *Kunzea corifolia*
Bourbon, or Faham. *Angræcum fragrans*
Botany Bay, or Sweet. *Smilax glycyphylla*
Brazilian. *Stachytarpheta jamaicensis*
Broussa. *Vaccinium Arctostaphylos*
Bush, of the Cape. *Cyclopia genistoides*
Canary. *Sida canariensis*
Cape Colony. *Helichrysum serpyllifolium*
Captain Cook's. *Leptospermum scoparium*
Carolina, or South Sea. *Ilex vomitoria*
Ceylon. *Elæodendron glaucum*
Cochin-China. *Teucrium Thea*
Duke of Argyll's. *Lycium barbarum*
Faham. *Angræcum fragrans*
Green. *Thea viridis*
Golden-rod. See Tea, Blue Mountain
Gout, of the W. Indies. *Cordia globosa*
Isle of Bourbon. *Angræcum fragrans*
Jesuit's, of Chili. *Psoralea glandulosa*
Labrador. *Ledum latifolium*
Malay. *Eugenia variabilis*; see also Tea, Bencoolen
Martinique. *Capraria biflora*

Tea-plant, Mexican. *Psoralea glandulosa* and *Chenopodium (Ambrina)ambrosioides*
Mountain. *Gaultheria procumbens*
New Jersey. *Ceanothus americanus*
N. S. Wales. *Melaleuca uncinata, Callistemon pallidum,* and *C. salignum*
New Zealand. *Leptospermum flavescens* and *L. scoparium*
New Zealand Sweet-scented. *Philadelphus aromaticus*
Oily. *Thea oleosa*
Oswego. *Monarda didyma*
Paraguay. *Ilex Paraguariensis*
St. Helena. *Beatsonia portulacæfolia*
South Sea. See Tea, Carolina
Surinam. Various species of *Lantana*
Swamp, of Australia. *Melaleuca squarrosa*
Sweet. See Tea, Botany Bay
Tasmanian. *Melaleuca squarrosa*
Theezan. *Rhamnus Theezans*
W. Indian. *Capraria biflora*
White, of Australia. *Melaleuca genistifolia*
Wild, of N. America. *Amorpha canescens*
Winter-berry. *Prinos glaber*
Teak-tree, African. *Oldfieldia africana*
Ben. *Lagerstræmia microcarpa*
E. Indian. *Tectona grandis*
N. S. Wales. *Endiandra glauca*
New Zealand. *Vitex littoralis*
Tear-thumb, Arrow-leaved. *Polygonum sagittatum*
Halberd-leaved. *Polygonum arifolium*
Tear's, Job's. See Job's-Tears
Juno's. See Juno's-Tears
Of St. Peter. *Anthacanthus microphyllus*
Teasel, or Thistle, Fullers'. *Dipsacus Fullonum*
Wild. *Dipsacus sylvestris*
Wild, Small. *Dipsacus pilosus*
Teel, or Til, Oil-plant. See Oil-plant, Sesame
"Teff," of Abyssinia. *Poa abyssinica*
Teil, Teyl, or Til-tree. *Tilia europæa*
Telegraph-plant. *Desmodium gyrans*
Tench-weed. *Potamogeton nutans* and other species
Tent-tree, of Lord Howe's Island. *Pandanus Forsteri*
Tent-wort. *Asplenium Ruta-muraria*
Teosinte-grass. See Grass
Terebinth-tree. *Pistacia Terebinthus*
Terra-japonica-plant. *Uncaria Gambier* and *U. acida*
Merita (a kind of Turmeric). *Curcuma longa*
Tetter-berry. *Bryonia dioica*
Tetter-wort. *Chelidonium majus*
Tettigass, or Tettigaha tree. *Trichadenia zeylanica*
Texan Pride. *Phlox Drummondi*
Teyl-tree. See Teil-tree
"Thatch," of the W. Indies. *Calyptronoma Swartzii* and *Copernicia tectorum*
Palmetto. *Thrinax parviflora*
Silver. *Thrinax argentea*
Thatch-grass, Cape. See Grass
Thick-leaf. The genus *Crassula*
Thick-stamen, American. *Pachysandra procumbens*

Thimble-berry. *Rubus occidentalis*
Thistle. The genus *Carduus*; applied also to some other plants
Barnaby's, or St. Barnaby's. *Centaurea solstitialis*
Blessed, or Holy. *Carduus benedictus* and *Silybum Marianum*
Blume's. *Blumea longifolia*
Brilliant. *Cirsium Douglasii*
Canada. *Cirsium arvense*
Carline. *Carlina vulgaris* and other species
Cornfield or Way. *Serratula arvensis*
Cotton. *Onopordon Acanthium*
Cotton, Stemless. *Onopordon acaule*
Creeping. *Carduus (Cnicus) arvensis*
"Cruel." *Cirsium ferox*
"Cursed," of N. America. *Cirsium arvense*
Distaff. *Carthamus lanatus*
Dwarf. *Carduus (Cnicus) acaulis*
Elephant, or Ivory. *Silybum eburneum*
Fish-bone. *Chamæpeuce Casabonæ*
Fuller's. *Dipsacus Fullonum*
"Gentle." *Carduus anglicus*
Globe. *Echinops Ritro*
Globe, E. Indian. *Sphæranthus hirtus*
Globe, Great. *Echinops sphærocephalus*
Golden. *Scolymus hispanicus*
Golden-spotted. *Scolymus maculatus*
Grass. *Agathæa spathulata*
Gum. Various species of *Euphorbia*
Hare's. *Sonchus oleraceus*
"Hedgehog." *Cactus Echinocactus*
Herring-bone. *Chamæpeuce Casabonæ*
Holy. See Thistle, Blessed
Horse. The genus *Cirsium*
Ivory. See Thistle, Elephant
Jersey. *Centaurea Isnardi*
Marsh. *Carduus (Cnicus) palustris*
Meadow. *Carduus (Cnicus) pratensis*
Melancholy. *Carduus heterophyllus*
Melon. The genus *Melocactus*
Mexican. *Erythrolæna conspicua*
Milk, or Our Lady's. *Carduus Marianus*
Musk. *Carduus nutans*
Our Lady's. See Thistle, Milk
Plume. The genera *Cirsium* and *Cnicus;* also *Carduus lanceolatus*
Plume, Yellow. *Cnicus Acarna*
Saffron. *Carthamus tinctorius*
St. Barnaby's. *Centaurea solstitialis*
Scarlet Mexican. *Erythrolæna conspicua*
Scotch (of artists). *Carduus nutans*
Scotch (of gardeners). *Onopordon Acanthium*
Slender. *Carduus pycnocephalus (C. tenuiflorus)*
Sow. *Sonchus oleraceus.* (See also Sow-thistle)
Spear. *Carduus (Cnicus) lanceolatus*
Star. *Centaurea Calcitrapa*
Syrian. *Notobasis syriaca*
Torch. The genus *Cereus*
Tuberous-rooted. *Carduus (Cnicus) tuberosus*
Waste. *Carduus arvensis*
Way. See Thistle, Cornfield
Welted. *Carduus acanthoides*
Woolly-headed. *Carduus eriophorus*
Yellow. *Argemone mexicana*

"**Thistle - upon - Thistle.**" *Onopordon Acanthium*
Thomas's (St.)-Tree. *Bauhinia variegata* and *B. tomentosa*
Thorn, African Cockspur. *Plectronia ventosa*
American Black, or Pear. *Cratægus tomentosa*
American Dwarf. *Cratægus parvifolia*
Aronia. *Cratægus Aronia*
Azarole. *Cratægus Azarolus*
Black. *Prunus spinosa*
Black-fruited. *Cratægus nigra*
Buck, or Way. The genus *Rhamnus*
Buffalo. *Acacia latronum*
Camel's. *Hedysarum Alhagi*
Christ's. *Paliurus aculeatus* and *Zizyphus Spina-Christi*
Dotted-fruited. *Cratægus punctata*
Douglas's. *Cratægus Douglasii*
Eastern. *Cratægus orientalis*
Egyptian. *Acacia vera (Mimosa nilotica)*
Elephant. *Acacia tomentosa*
Evergreen. *Cratægus Pyracantha*
Evergreen Glaucous. *Cratægus Pyracantha glauca*
Evergreen,White-berried. *Cratægus Pyracantha fructu-albo*
Fontainebleau. *Cratægus latifolia*
Garland. *Paliurus aculeatus*
Glastonbury. *Cratægus Oxyacantha* var. *præcox*
Goat's. *Astragalus Tragacantha*
Haw, or White. *Cratægus Oxyacantha*. (See also Haw-thorn)
Haw, New Zealand. *Discaria Toumatou*
Hook. *Uncaria procumbens*
Jerusalem. *Parkinsonia aculeata*
Kangaroo. *Acacia armata*
Lily. *Catesbæa spinosa*
Lobed-leaved. *Cratægus lobata*
Long-spined. *Cratægus macrantha*
Maple-leaved. *Cratægus cordata*
Mexican. *Cratægus mexicana*
Morocco. *Cratægus maroccana*
Mysore. *Cæsalpinia sepiaria*
Oval-leaved. *Cratægus ovalifolia*
Parsley-leaved. *Cratægus apiifolia*
Pear. See Thorn, American Black
Pear-leaved. *Cratægus pyrifolia*
Plum-leaved. *Cratægus prunifolia*
Purple-bunched. *Cratægus purpurea*
Quick, or Quick-set. *Cratægus Oxyacantha*
Rain-berry. See Rain-berry-thorn
St. Joseph of Arimathea's. *Cratægus Oxyacantha* var. *præcox*
Scarlet-fruited. *Cratægus coccinea*
Sea Buck, or Willow. *Hippophaë rhamnoides*
Tasmanian Native. *Bursaria spinosa*
Thirsty. *Acacia Seyal*
Three-lobed-leaved. *Cratægus trilobata*
Various-leaved. *Cratægus heterophylla*
Virginian. *Cratægus virginica*
"Wait-a-bit." *Uncaria procumbens*
Washington. *Cratægus cordata*
Way. See Thorn, Buck
W. Indian. *Macromerium jamaicense*

Thorn, White. See Thorn, Haw
Willow, or Sallow. *Hippophaë rhamnoides*
Yellow-fruited. *Cratægus flava*
Thorn-apple, Common. *Datura Stramonium*
Blue-flowered. *Datura Tatula*
Downy. *Datura Metel*
E. Indian. *Datura alba*
Horn-stalked. *Datura ceratocaula*
Long-spined. *Datura ferox*
Purple-flowered. *Datura fastuosa*
Tree. *Datura arborea*
Thorn-Broom. *Ulex europæus*
Thorough-wax, or Thorow-wax. *Bupleurum rotundifolium*
Tree. *Bupleurum fruticosum*
Thorough - wort. *Eupatorium hyssopifolium*
Thousand-Seal. *Achillea Millefolium*
Thread-flower,Crimson. *Poinciana (Cæsalpinia) Gillicsii*
Thread-foot. *Podostemon ceratophyllus*
Thread-Moss. See Moss
Three-faces-under-a-hood. *Viola tricolor*
Thrift, Alpine. *Armeria alpina*
Alpine Rosy-flowered. *Armeria alpina rosea*
Broom. *Armeria Scoparia*
Common. *Armeria vulgaris*
Garlic. *Armeria alliacea*
Great. *Armeria Cephalotes*
Large-flowered. *Armeria grandiflora*
Plantain-leaved. *Armeria plantaginea*
Prickly. *Acantholimon glumaceum*
White - flowered. *Armeria maritima alba*
Throat-root. *Geum virginianum*
Throat-wort. The genus *Trachelium*; also applied to *Campanula Cervicaria* and *Digitalis purpurea*
Blue. *Trachelium cæruleum*
Great. *Campanula Trachelium*
Thrum-wort. The genus *Actinocarpus*; also applied to *Amarantus caudatus*
Thunder-dirt-Fungus of the New Zealanders. *Ileodictyon oibarium*
Thyme, Alpine. *Thymus alpinus*
Azorean. *Thymus azoricus*
Azure. *Thymus azureus*
Balm-leaved. *Thymus melissoides*
Basil. *Calamintha (Melissa) Acinos*
Cat. *Teucrium Marum* and *T. Polium*
Common Garden. *Thymus vulgaris* and vars.
Corsican. *Thymus corsicus*
Downy. *Thymus lanuginosus*
Incense. *Thymus thuriferus*
"Laced." An old name for Thyme-Dodder
Lemon-scented. *Thymus citriodorus*
Lemon-scented, Variegated. *Thymus citriodorus aureo-variegatus*
Marjoram-leaved. *Thymus (Acinos) patarinus*
Peppermint. *Thymus Piperella*
Pot-herb. *Thymus vulgaris*
Round-leaved. *Thymus rotundifolius*

Thyme, Virginian. *Pycnanthemum linifolium*
Virginian, Narrow-leaved. *Pycnanthemum virginicum*
Water. *Anacharis Alsinastrum (Elodea canadensis)*
Wild. *Thymus Serpyllum*
Wild Otago. *Samolus littoralis*
Wild White-flowered. *Thymus Serpyllum var. album*
Thyme-Dodder. *Cuscuta Epithymum*
Thyrse-flower. The genus *Thyrsacanthus*
Ti-plant, N. Zealand. *Cordyline australis* and *C. indivisa*
Tickle-my-fancy. *Viola tricolor*
Tick-seed. The genus *Coreopsis*
Dyers'. *Coreopsis tinctoria*
Ear-leaved. *Coreopsis auriculata*
Early-flowering. *Coreopsis præcox*
Lance-leaved. *Coreopsis lanceolata*
Large-flowered. *Coreopsis grandiflora*
Philadelphian. *Coreopsis philadelphica*
Six-leafleted. *Coreopsis senifolia*
Slender-leaved. *Coreopsis tenuifolia*
Whorled-leaved. *Coreopsis verticillata*
Tick-Trefoil, Canadian. *Desmodium canadense*
Dillon's. *Desmodium Dilleni*
Japanese. *Desmodium japonicum*
Tick-weed. *Hedeoma pulegioides*
"Tidy-tips," of San Francisco. *Layia platyglossa*
Tigarea-tree. *Tetracera Tigarea*
Tiger-chop. *Mesembryanthemum tigrinum*
Tiger-flower, Peacock. *Tigridia Pavonia*
White. *Tigridia speciosa alba*
Tiger-grass Palm. See Palm
Tiger-wood. The wood of *Machærium Schomburgkii*
Tigris, Flower of. An old name for *Tigridia Pavonia*
Til-tree. *Laurus fœtens*
Tile-root. The genus *Geissorrhiza*
Rusby. *Geissorrhiza juncea*
Tile-seed, of Australia. The genus *Geissois*
Till. Another name for Lentil
Timboe Fish-poison-plant, of Brazil. *Serjania lethalis*
Tinder, German. See Amadou
Tine. *Vicia sylvestris*
Tipsy-wood. *Galega frutescens*
Brazil. *Phyllanthus Conami*
Tithymale. An old name for Spurge
Toad-flax. The genus *Linaria*
Alpine. *Linaria alpina*
American. *Linaria canadensis*
Bastard. *Thesium linophyllum*
Bastard, American. The genus *Comandra*
Broom-leaved. *Linaria genistæfolia*
Cloven-flowered. *Linaria bipartita*
Common. *Linaria vulgaris*
Creeping. *Linaria repens*
Gold-and-purple. *Linaria reticulata aurea purpurea*
Hairy. *Linaria pilosa*
Hepatica-leaved. *Linaria hepaticæfolia*
Ivy-leaved. *Linaria Cymbalaria*
Large Yellow. *Linaria dalmatica*

Toad-flax, Marjoram-leaved. *Linaria origanifolia*
Netted-flowered. *Linaria reticulata*
Pelisser's. *Linaria Pelisseriana*
Perez's. *Linaria Perezii*
Pointed. *Linaria Elatine*
Purple. *Linaria purpurea*
Pyrenean. *Linaria pyrenaica*
Round-leaved. *Linaria spuria*
Shaggy. *Linaria villosa*
Small. *Linaria minor*
Tendrilled. *Linaria cirrhosa*
Thick-leaved. *Linaria crassifolia*
Three-birds. *Linaria triornithophora*
Three-leaved. *Linaria triphylla*
Trailing. *Linaria supina*
Toad-flower, African. *Stapelia bufonia*
Toad-lily. *Fritillaria nigra*
Japanese. *Tricyrtis hirta*
Toad-root. *Actæa alba*
Toad-stool, Paddock-stool, or Fairy-Mushroom. Any of the poisonous *Fungi* or Mushrooms
Tobacco-plant. The genus *Nicotiana*
Acute-leaved. *Nicotiana acutifolia*
"English." An old name for *Hyoscyamus*
Giant. *Nicotiana macrophylla gigantea*
Glaucous-leaved perennial. *Nicotiana glauca*
Havannah. *Nicotiana repanda*
Indian. *Lobelia inflata*
Latakia. *Nicotiana rustica*
Long-flowered. *Nicotiana longiflora*
Mountain. *Arnica montana*
Native Australian. *Nicotiana suaveolens*
Night-flowering. *Nicotiana noctiflora*
Persian or Shiraz. *Nicotiana persica*
Shiraz. See Tobacco-plant, Persian
Shrubby. *Nicotiana fruticosa*
Syrian, or Wild. *Nicotiana rustica*
Tube-flowered. *Nicotiana tubiflora*
Tuberose-flowered. *Nicotiana affinis*
Vinca-flowered (perennial). *Nicotiana vincæflora*
Virginian. *Nicotiana Tabacum*
Wild. *Nicotiana rustica*
Toddy-Palm. See Palm
Tollon, or Toyon, of California. *Heteromeles arbutifolia*
Tolosa-wood. *Pittosporum bicolor*
Tolu-Balsam-plant. *Myroxylon (Myrospermum) toluiferum*
Tomato-plant. *Solanum Lycopersicum*
Cannibal's. *Solanum anthropophagorum*
Cherry. *Solanum Lycopersicum var. cerasiforme*
Strawberry. *Physalis Alkekengi* and *P. pubescens*
Tong-pang-chong. See Nasuta
Tonga-bean-wood. *Alyxia buxifolia*
"Tonga"-plant. *Epipremnum mirabile*
Tongue-blade. *Ruscus Hypoglossum*
Tongue-flower. *Glossula tentaculata*
Australian. The genus *Glossodia*
Tongue-grass. *Lepidium sativum*
Tonquin-bean-tree. *Dipterix odorata*
Tontel-tree. *Tontelea (Salacia) pyriformis*

"Tooart"-tree. *Eucalyptus gomphocephala*
"Toot," or **"Tutu"-plant,** New Zealand. *Coriaria ruscifolia*
Tooth-ache-tree. The genus *Xanthoxylum*
Thornless. *Xanthoxylum mite*
Three-fruited. *Xanthoxylum tricarpum*
Tooth-brush-tree. *Salvadora persica*
Tooth-cress, or Tooth-Violet. *Dentaria bulbifera*
Tooth-Cup. *Ammania humilis*
Tooth-flower, Australian. *Dentella repens*
Tooth-pick. *Ammi Visnaga*
Tooth-Violet. See Tooth-cress
Tooth-wort. *Lathræa squamaria* and the genus *Dentaria*; also an old name for *Capsella Bursa-pastoris*
Bulb-bearing. *Dentaria bulbifera*
Five-leaved. *Dentaria pentaphylla*
Hidden. *Lathræa clandestina*
Many-leaved. *Dentaria polyphylla*
Nine-leaved. *Dentaria enneaphylla*
Pinnate-leaved. *Dentaria pinnata*
Seven-leaved. *Dentaria pinnata*
Showy. *Dentaria digitata*
Three-leaved. *Dentaria triphylla*
Two-leaved. *Dentaria diphylla*
W. Indian. *Plumbago scandens*
Tornilla-plant. *Prosopis pubescens*
"Torches." An old name for *Verbascum Thapsus*
Torch-lily. The genus *Tritoma*
Burchell's. *Tritoma Burchelli*
Common. *Tritoma Uvaria*
Dwarf. *Tritoma pumila*
Dwarf Glaucous-leaved. *Tritoma Uvaria glaucescens*
Intermediate. *Tritoma media*
Large-flowered. *Tritoma Uvaria grandiflora*
Rooper's. *Tritoma Rooperi*
Tall. *Tritoma grandis*
Torch-wood-tree. *Amyris sylvatica*
Mountain. *Amyris balsamifera*
Tormentil-root. *Potentilla Tormentilla*
Tortoise-plant. *Testudinaria elephantipes*
Tortoise-wood. A variety of Zebra-wood
Touch-me-not, Common. *Impatiens Noli-me-tangere*
Pale. *Impatiens pallida*
Touch-wood. *Polyporus (Boletus) fomentarius* and *P. igniarius*
Tow-tree, Tahiti. *Cordia subcordata*
Towai-tree, New Zealand. *Fagus Menziesii*
Tower-cress. *Arabis Turrita*
Town-cress. *Lepidium sativum*
Toyon. See Tollon
Toy-wort. An old name for *Capsella Bursa-pastoris*
Tragacanth-Gum-plant. *Astragalus Tragacantha*
Traveller's-Joy. *Clematis Vitalba*
Spanish. *Clematis cirrhosa*
Traveller's-tree, of Madagascar. *Ravenala madagascariensis (Urania speciosa)*
Treacle, Countryman's. *Ruta graveolens*
Poor-Man's. *Allium sativum*

Treacle-Clover. See Clover
Treacle-Mustard. *Erysimum cheiranthoides* and *Lepidium campestre*
Treacle-Wormseed. *Erysimum cheiranthoides*
"Tread-softly," Virginian. *Jatropha urens var. stimulosa (Cnidoscolus stimulosus)*
Treasure-flower. The genus *Gazania*
Pea-cock. *Gazania Pavonia*
Tree-Celandine, Shrubby. *Bocconia frutescens*
Tree-Fern. See Fern
Tree-hair Lichen. See Lichen
Tree-Mallow, or Sea-Mallow. *Lavatera arborea*
African. *Lavatera africana*
Common Annual. *Lavatera trimestris*
Garden. *Lavatera Olbia*
Variegated. *Lavatera arborea variegata*
Wild. *Lavatera arborea*
Tree-of-Chastity. *Vitex Agnus-castus*
Tree-of-Heaven. *Ailantus glandulosa*
Tree-of-Life. The genus *Arbor-vitæ*
Tree-of-Long-Life. *Glaphira nitida*
Tree-of-Sadness, Night-scented. *Nyctanthes Arbor-tristis*
Tree-of-the-Gods. *Ailantus glandulosa*
Tree-of-the-Sun. *Retinospora obtusa*
Trefoil. The genus *Trifolium*. (See also Clover)
Alpine. *Trifolium alpinum*
Bean. *Cytisus Laburnum* and the genus *Anagyris*
Bird's-foot. *Lotus corniculatus*
Bird's-foot, Double. *Lotus corniculatus fl.-pl.*
Bird's-foot, Edible. *Lotus edulis*
Bird's-foot, Greater. *Lotus major*
Bird's-foot, Red-flowered. *Lotus arabicus* and *L. australis*
Bird's-foot, Shrubby. *Lotus Dorycnium*
Bird's-foot, Silvery. *Lotus argenteus*
Bird's-foot, Silvery Evergreen. *Lotus creticus*
Bitumen. *Psoralea bituminosa*
Bladder-podded. *Trifolium spumosum*
Golden. An old name for *Hepatica triloba*
Hare's-foot. *Trifolium arvense*
Milk. See Milk-Trefoil
Moon. *Medicago arborea*
One-flowered. *Trifolium uniflorum*
Red. *Trifolium rubens*
Sand. *Trifolium suffocatum*
Scented, of Australia. *Melilotus parviflora*
Shrubby. *Ptelea trifoliata* and *Jasminum fruticans*
Starry. *Trifolium stellatum*
Striped-flowered. *Trifolium involucratum*
Sweet. *Melilotus cœrulea*
Tick. The genus *Desmodium*
Tree. The genus *Cytisus*
Venomous. The genus *Dorycnium*
Water. *Menyanthes trifoliata*
Yellow. *Trifolium minus*
"Tricolor." *Amarantus tricolor*
Trigger-plant. *Stylidium graminifolium* and other species

Trincomalee-wood. The wood of *Berrya Ammonilla*
Trip-madam. *Sedum reflexum*
Tripoly. *Aster Tripolium*
Troll-flower. An old name for the genus *Trollius*
"Trompet Marin." *Narcissus albicans*
Trottles. *Symphytum asperrimum*
True-love. *Paris quadrifolia*
Truffle. *Tuber cibarium*
 African. The genus *Terfezia*
 English. *Tuber æstivum*
 French. *Tuber melanosporum*
 Piedmontese. *Tuber magnatum*
 Red. *Tuber rufum* and *Melanogaster variegatus*
 Tuckahoe. *Pachyma cocos*
 White, or False. *Tuber album* (*Choiromyces meandriformis*)
Truffles, Hart's. *Elaphomyces*
Trumpet-Creeper. *Tecoma radicans* and other species
Trumpet-flower. A name applied to various large trumpet-shaped flowers, as those of *Tecoma*, *Bignonia*, &c.
 Evergreen. *Bignonia sempervirens*
 Fern-leaved. *Bignonia procera* (*Jacaranda Copaia*)
 Four-leaved. *Bignonia Unguis*
 Hardy. *Bignonia radicans*
 Hardy Four-leaved. *Bignonia capreolata*
 Hardy Large. *Tecoma* (*Bignonia*) *grandiflora*
 Nepaul. *Beaumontia grandiflora*
 Peach-coloured. *Solandra grandiflora*
 Peruvian. *Brugmansia suaveolens*
 Rooting-branched. *Tecoma radicans*
 Shrubby. *Tecoma stans*
 Tendrilled. *Bignonia capreolata*
 Virginian. *Tecoma radicans*
Trumpet - Honeysuckle. See Honeysuckle
Trumpet-leaf. The genus *Sarracenia*
Trumpet-tree. *Cecropia peltata*
Trumpet-weed. *Eupatorium purpureum*
Trumpets. *Sarracenia flava*
Tuberose. *Polianthes tuberosa*
Tube-flower. *Clerodendron Siphonanthus*
Tube-tongue. *Salpiglossis sinuata* and other species
Tuber-root. *Asclepias tuberosa*
"Tule"-plant, of California. *Scirpus lacustris var. occidentalis*
Tulip. The genus *Tulipa*
 African. The genus *Hæmanthus*
 Bithynian. *Tulipa bithynica*
 Brilliant. *Tulipa fulgens*
 Butterfly. *Calochortus uniflorus*
 Cape. *Melanthium uniflorum* (*Tulipa Breyiana*)
 Cape, Red. *Hæmanthus coccineus*
 Cels's. *Tulipa Celsiana*
 Clusius's. *Tulipa Clusiana*
 Common Garden. Varieties of *Tulipa Gesneriana*
 Cowslip-scented. *Tulipa iliensis*
 Dog's-tooth-Violet. *Tulipa erythronioides*
 Drooping. *Fritillaria Meleagris*
 Dwarf Rosy-purple. *Tulipa pulchella*
Tulip, Dwarf Yellow. *Tulipa Celsiana*
 Elegant-flowered. *Tulipa elegans*
 Four-leaved. *Tulipa tetraphylla*
 Golden Star. *Cyclobothra pulchella*
 Greig's. *Tulipa Greigi*
 Hager's. *Tulipa Hageri*
 Kolpakowsky's. *Tulipa Kolpakowskyana*
 Korolkow's. *Tulipa Korolkowi*
 Orphanides's. *Tulipa Orphanidesi*
 Parrot. *Tulipa Gesneriana var. laciniata*
 Parrot, Florentine. *Tulipa turcica*
 Persian. *Tulipa persica*
 Star-flowered. *Tulipa stellata*
 Sun's-eye. *Tulipa Oculus-solis*
 Sun's-eye, Large. *Tulipa præcox*
 Sweet-scented. *Tulipa fragrans*
 Three-leaved. *Tulipa triphylla*
 Turkestan. *Tulipa turkestanica* and *T. Greigi*
 Two-flowered. *Tulipa biflora*
 Van Thol. *Tulipa suaveolens*
 Van Thol, Double. *Tulipa suaveolens fl.-pl.*
 Wild. *Tulipa sylvestris*
 Wild, of California. The genus *Calochortus*
Tulip-tree. *Liriodendron tulipiferum*
 Chinese. *Magnolia fuscata*
 Laurel-leaved. The genus *Magnolia*
 Queensland. *Stenocarpus sinuatus* (*S. Cunninghamii*)
Tulip-wood-tree. *Physocalymma floribundum*
 Queensland. *Harpalia Hillii* and *H. pendula*
"Tulp," of the Cape. Various species of *Morea* and *Vieusseuxia*, especially *Morea polyanthos* and *Vieusseuxia tripetaloides*
Tunhoof. *Nepeta Glechoma*
Tupelo-tree. The genus *Nyssa*
 Common. *Nyssa multiflora* (*N. sylvatica*)
 Large. *Nyssa uniflora*
 Mountain. *Nyssa biflora*
 Water. *Nyssa aquatica*
Turanira-wood. The wood of *Bumelia retusa*
Turbith, Quacksalver's. *Euphorbia Esula*
 Serapias. *Aster Tripolium*
Turkey-berry. *Solanum mammosum* and *S. torvum*
Turkey-berry-tree. *Cordia Collococca*
Turkey-blossom. *Tribulus cistoides*
Turkey-corn. *Corydalis formosa*
Turkey-pea, Wild. *Corydalis formosa*
Turkey's-beard. *Xerophyllum asphodeloides*
Turk's-herb. An old name for *Herniaria glabra*
Turmeric-plant. *Curcuma longa*
 Wild. *Curcuma aromatica*
Turmeric-root. *Hydrastis canadensis*
Turmeric-tree, Australian. A species of *Zieria*
Turnip, Common. *Brassica Rapa var. depressa*
 Devil's. *Bryonia dioica*
 Hungarian. *Brassica oleracea gongylodes*
 Indian. *Arisæma triphyllum* and *Arum dracontium*

Turnip, Lion's. The roots of *Leontice Leontopetalum*
Prairie. *Psoralea esculenta*
St. Anthony's. *Ranunculus bulbosus*
Swedish. *Brassica campestris var. Rutabaga*
Teltow. A variety of *Brassica Napus*
Turnip-Radish. A variety of *Raphanus Raphanistrum*
Turnsole. *Croton (Chrozophora) tinctorium* and the genus *Heliotropium*
Peruvian. *Heliotropium peruvianum*
Turpentine-tree, American. *Pinus australis* and *P. Tæda*
Australian. *Tristania conferta, T. albicans,* and *Syncarpia laurifolia*
Brea. *Pinus Teocote*
Chian or Cyprian. *Pistacia Terebinthus* (*P. atlantica*)
European. *Pinus sylvestris, P. Laricio,* and *P. Pinaster*
Mt. Atlas. *Pistacia atlantica*
N. S. Wales. *Syncarpia latifolia*
Queensland. *Syncarpia laurifolia*
Strasburg. *Abies (Picea) pectinata*
Venice. *Larix europæa*
Turtle-head. *Chelone glabra*
Turtle-bloom. *Chelone glabra*
"Tusca"-tree. *Acacia moniliformis*
Tussock-grass. See Grass
Tutsan. The genus *Hypericum.* (See St. John's-wort)
Glossy-flowered. *Hypericum Hookerianum*
Tutu-poison-plant, of New Zealand. *Coriaria ruscifolia* or *C. thymifolia*
Tway-blade, American. *Listera convallarioides*
British. *Listera ovata*
Green-flowered. *Liparis Loeselii*
Purple-flowered. *Liparis liliifolia*
Small. *Listera cordata*
Twelve-o'clock-Flower. *Abutilon americanum*
Twig-rush. See Rush
Twin-flower. *Linnæa borealis*
Scarlet. *Bravoa geminiflora*
Twin-leaf. *Jeffersonia diphylla*
Twisted-stalk. The genus *Streptopus*
Twisted Stick, Twisted Horn, or Twisty, of S. India. The fruit of *Helicteres Isora*
Twitch-grass. See Grass
Two-pence. See Herb Two-pence
Tyle-berry. *Jatropha multifida*

Udika-bread. *Irvingia Barteri*
Ugni-shrub. *Eugenia Ugni*
Umbel. *Cypripedium pubescens*
Umbra-tree. *Pircunia (Phytolacca) dioica*
Umbrella Grass. See Grass
Umbrella-leaf. *Diphylleia cymosa*
Umbrella-Pine. The genus *Sciadopitys*
Whorled-leaved. *Sciadopitys verticillata*
Umbrella-plant. *Saxifraga peltata*
Umbrella-tree. *Magnolia tripetala* (*M. Umbrella*) and *Thespesia populnea*
Ear-leaved. *Magnolia Fraseri*
Guinea. *Paritium guineense*
Queensland. *Brassaia actinophylla*

Umbrella-wort. The genus *Oxybaphus*
Umire-tree. *Humirium floribundum*
Unicorn-plant. *Martynia lutea* and *M. proboscidea*
Unicorn-root. *Aletris farinosa*
False. *Helonias dioica*
Unicorn's-horn. *Helonias dioica*
Upas Tree, of Java. *Antiaris toxicaria*
Urari. See Wourali
Urn-flower, Drooping. *Urceolina pendula*
Golden. *Urceolina aurea*
Urn-fruit Tree. *Callicarpa incana*
Urn-Moss. See Moss
Ushoka-tree, of Bengal. *Jonesia Asoca*

Valerian. The genus *Valeriana*
American. *Cypripedium pubescens*
Cat's. *Valeriana officinalis*
Elder-leaved. *Valeriana sambucifolia*
False. *Senecio aureus*
Garden. *Valeriana Phu*
Globularia-leaved. *Valeriana globulariæfolia*
Long-spurred. *Centranthus macrosiphon*
Marsh. *Valeriana dioica*
Medicinal. *Valeriana officinalis*
Mountain. *Valeriana montana*
Pyrenean. *Valeriana pyrenaica*
Red, or Spur. *Centranthus ruber*
Three-winged. *Valeriana tripteris*
Valonia. A commercial name for the acorn-cups of *Quercus Ægilops,* which are used for tanning, dyeing, and making ink
Vanilla-plant. *Epidendrum Vanilla*
N. America. *Liatris odoratissima*
Varnish-tree, Black, Burmah, or Martaban. *Melanorrhæa usitatissima*
False. *Ailantus glandulosa*
Japan. *Rhus vernicifera*
Martaban. See Varnish-tree, Black
Moreton Bay. *Pentaceras australis* (*Ailantus punctatus*)
New Granada. *Elæagia utilis*
Sylhet. *Semecarpus Anacardium*
Vegetable Antimony. *Eupatorium perfoliatum*
Vegetable Brimstone. *Lycopodium* powder
Vegetable Butter, or Phoolwa Oil, plant. *Bassia butyracea*
Vegetable Egg. *Lucuma (Sapota) mammosa*
Vegetable Fire-cracker. *Brodiæa coccinea*
Vegetable-Glue-plant. *Combretum Guayca*
Vegetable Hair. *Tillandsia usneoides*
Vegetable Horse-hair. The fibre of *Chamærops humilis*
Vegetable Ivory. The nuts of *Phytelephas macrocarpa* and of *Sagus amicarum*
Vegetable Leather. *Euphorbia punicea*
Vegetable Marrow. *Cucurbita ovifera* (a variety of *C. maxima*); also the fruit of *Persea gratissima*
Vegetable Oyster. *Tragopogon porrifolius*
Vegetable Sheep, of New Zealand. *Raoulia eximia*
Vegetable Silk. Obtained from the seed-pods of *Chorisia speciosa*

and Foreign Plants, Trees, and Shrubs. 141

Vegetable Sweet-bread. *Agaricus Orcella*
Vegetable - Tallow - plant. *Stillingia sebifera* and *Vateria indica*
Vegetable Tinder. See Amadou
Vegetable Wax. See Wax-plant and Wax-tree
Velvet-bur. *Priva echinata*
Velvet-flower. *Amarantus caudatus*
Velvet-leaf. *Lavatera arborea*
E. Indian. *Tourneforia argentea*
N. American. *Abutilon Avicennæ*
S. American. *Cissampelos Pareira*
Velvet-seed. *Guettarda elliptica*
Venatica, or Vinatico, wood. The wood of *Persea indica*
Venus's-Basin. An old name for *Dipsacus sylvestris*
Venus's-Comb. *Scandix Pecten-Veneris*
Venus's Fly-trap. *Dionæa muscipula*
Venus's Golden Apple. *Atalantia (Limonia) monophylla*
Venus's Hair. *Adiantum Capillus-Veneris*
Venus's Looking - glass. *Specularia Speculum*
N. America. *Specularia perfoliata*
Venus's Navel-wort. The genus *Omphalodes*
Venus's, or Venice, Sumach. *Rhus Cotinus*
Verbena, Lemon-scented. *Aloysia citriodora*
Sand. The genus *Abronia*
Sand, Purple. *Abronia umbellata*
Verbena-Oil-plant. *Andropogon citratus*
Vernal-grass. See Grass
Vervain. The genus *Verbena*
Bastard. The genus *Stachytarpheta*
Blue. *Verbena hastata*
Blue, American. *Verbena hastata*
Common. *Verbena officinalis*
Creeping. *Zapania (Lippia) nodiflora*
Cut-leaved. *Verbena Aubletia*
Hardy, or Large-veined. *Verbena venosa*
Rocky Mountain. *Verbena montana (V. Aubletia)*
Rose. *Verbena Aubletia*
White, or Nettle-leaved. *Verbena urticæfolia*
Vetch. The genus *Vicia*
Bastard. The genus *Phaca*
Bastard, Spanish. *Phaca Bætica*
Bastard, Trailing. *Phaca australis*
Bithynian. *Vicia bithynica*
Bitter. The genus *Orobus*
Bitter, Black. *Ervum Ervilia*
Bitter, Blue. *Orobus (Platystylis) cyaneus*
Bitter, Crimean. *Orobus tauricus*
Bitter, Dark Purple. *Orobus atropurpureus*
Bitter, Double-blossomed. *Orobus vicioides fl.-pl.*
Bitter, Hairy. *Orobus hirsutus*
Bitter, Lathyrus-like. *Orobus lathyroides*
Bitter, Orange-flowered. *Orobus aurantiacus*
Bitter, Spring. *Orobus vernus*
Bitter, Variegated. *Orobus variegatus*
Bladder. The genus *Phaca*
Bush. *Vicia sepium*

Vetch, Chickling. *Lathyrus sativus*
Common. *Vicia sativa*
Crown. The genus *Coronilla*
Flat-podded. *Lathyrus Cicera*
Four-seeded. *Vicia tetrasperma*
Grass. *Lathyrus Nissolia*
Hairy. *Vicia hirsuta*
Hatchet. *Biserrula Pelecinus* and *Coronilla Securigera*
Horse-shoe. *Hippocrepis comosa*
Indian. *Ervum dispermum*
Liquorice. *Astragalus glycyphyllos*
Medick. The genus *Onobrychis*
Milk. See Milk-Vetch
Narbonne. *Vicia Narbonensis*
Sensitive Joint. *Æschynomene hispida* and other species
Silvery. *Vicia argentea*
Spring. *Vicia lathyroides*
Tufted. *Vicia Cracca*
Tufted Carolina. *Vicia Caroliniana*
Upright. *Vicia Orobus*
Wood. *Vicia sylvatica*
Wood, American. *Vicia americana*
Wood, Great. *Vicia dumetorum*
Yellow-flowered. *Vicia lutea*
Vetchling, Meadow. *Lathyrus Nissolia*
Pea. *Lathyrus pisiformis*
Yellow-flowered. *Lathyrus Aphaca*
Vinatico-wood. See Venatica
Vine, Alleghany. *Adlumia cirrhosa*
American Grape. See Grape
Australian (East). *Vitis hypoglauca*
Australian (North). *Vitis acetosa*
Balloon. *Cardiospermum Halicacabum*
Burdekin. *Vitis (Cissus) opaca*
Climbing. *Psychotria parasitica*
Condor. *Gonolobus Cundurango*
Cross. *Bignonia capreolata*
Cucumber. *Cucumis Dudaim*
Currant. *Vitis vinifera var. corinthiaca*
Eccremocarpus. *Calampelis (Eccremocarpus) scaber*
Elephant's. *Cissus latifolia*
Entire-leaved Ivy. *Vitis indivisa*
Fragrant Wild. *Vitis riparia*
Gippsland Grape. *Vitis (Cissus) hypoglauca*
Glory. The genus *Clianthus*
Golden. *Stigmaphyllon ciliatum*
Gouty-stemmed. *Vitis gongylodes*
Granadilla. *Passiflora quadrangularis*
Grape. Varieties of *Vitis vinifera*
Grape-flower. The genus *Wistaria*
Harvey's. *Sarcopetalum Harveyanum*
Hedge. *Clematis Vitalba*
Hop-leaved. *Vitis humulifolia*
Ice. *Cissampelos Pareira*
India-rubber. *Cryptostegia grandiflora*
Kangaroo, or Kanguru. *Cissus antarctica*
Madeira. *Boussingaultia basselloides*
Maple. *Menispermum canadense*
Matrimony. *Lycium vulgare* and other species
Milk. *Periploca græca*
Mustang. *Vitis candicans*
Mountain. *Viola lutea*
Of Sodom. Supposed to be the Colocynth (*Citrullus Colocynthis*)

Vine, Ornamental-fruited. *Bryonopsis erythrocarpa*
Parsley-leaved. *Vitis laciniosa*
Pea. *Vicia americana*
Pepper. *Ampelopsis bipinnata*
Pipe. *Aristolochia Sipho*
Poison. *Rhus radicans*
Potato. *Ipomœa pandurata*
Red-bead. *Abrus precatorius*
Scrub, of Australia. The genus *Cassytha*
Seven-year. *Ipomœa tuberosa*
Silk. *Periploca græca*
Silver, of the W. Indies. *Pothos argyrea*
Smilax. *Myrsiphyllum asparagoides*
Sorrel. *Cissus acida*
Spanish-Arbour. *Ipomœa tuberosa*
Squaw. *Mitchella repens*
Strainer. *Luffa acutangula*
Turquoise-berried. *Vitis heterophylla humulifolia*
Variegated. *Vitis (Cissus) heterophylla variegata*
Victoria Scrub. *Cassytha melantha*
Water. The genus *Phytocrene*; also applied to *Doliocarpus Calinea*
White Wild. *Bryonia dioica*
Wild Wood. *Ampelopsis quinquefolia*
Wonga-Wonga. *Tecoma australis*
Vine-tie. *Arundo Ampelodesmos (Ampelodesmos tenax)*
Vinegar-plant. *Penicillium glaucum*
Vinegar-tree. *Rhus typhina* and *R. glabra*
Vineyard-Cane, Italian. *Arundo Donax*
Violet. The genus *Viola*
Adder's. *Goodyera pubescens*
Alpine. *Viola montana*
Altaian. *Viola altaica*
American Common Blue. *Viola cucullata*
Arrow-leaved. *Viola sagittata*
Beckwith's. *Viola Beckwithii*
Bird's-foot. *Viola pedata*
Bog. The genus *Pinguicula*
Broad-leaved. *Viola mirabilis*
Bulbous. An old name for the Snow-drop and the genus *Leucojum*
Calathian. *Gentiana Pneumonanthe*
Canada. *Viola canadensis*
Cape. *Ionidium capense*
Corn. *Specularia hybrida*
Dame's. *Hesperis matronalis*
Dog's-tooth. *Erythronium Dens-canis*
Dog's-tooth, Giant. *Erythronium giganteum*
Dog's-tooth, Nuttall's. *Erythronium Nuttallianum*
Dog's-tooth, Oregon. *Erythronium giganteum grandiflorum*
Dog's-tooth, White. *Erythronium albidum*
Dog's-tooth, Yellow. *Erythronium americanum*
Downy Yellow. *Viola pubescens*
False. *Dalibarda repens*
Fringed. See Fringed-Lily
Fringed, of Australia. The genus *Thysanotus*
Hairy. *Viola hirta*
Halberd-leaved. *Viola hastata*
Hand-leaf. *Viola palmata*

Violet, Hedge. *Viola sylvatica*
Hollow-leaved. *Viola cucullata*
Hook-spurred. *Viola canina var. adunca*
Horned. *Viola cornuta*
Kidney-leaved. *Viola renifolia*
Lance-leaved. *Viola lanceolata*
Larkspur-leaved. *Viola delphinifolia*
Long-spurred. *Viola rostrata*
Long-stalked. *Viola pedunculata*
Many-flowered. *Viola multiflora*
Marsh. *Viola palustris*
Mercury's. An old name for *Campanula Medium*
Mountain. *Viola lutea*
Munby's. *Viola Munbyana*
Neapolitan. *Viola odorata var. pallida plena*
New Holland. *Erpetion reniforme (Viola hederacea)*
Olympian. *Viola gracilis*
Pale. *Viola striata*
Palma. *Viola palmaensis*
Pennsylvanian. *Viola pennsylvanica*
Pinnate-leaved. *Viola pinnata*
Primrose-leaved. *Viola primulæfolia*
Pyrenean. *Viola cornuta*
Rock. *Chroolepus Jolithus*
Rouen. *Viola rothomagensis*
Round-leaved. *Viola rotundifolia*
Russian. *Viola suavis*
Shelton's. *Viola Sheltonii*
Siberian. *Viola uniflora*
Silvery-flowered. *Viola argentiflora*
Spear-leaved. *Viola lanceolata*
Spurless. The genus *Erpetion*
Spurred. *Viola calcarata*
Striped-flowered. *Viola striata*
Sweet-scented. *Viola odorata*
Sweet White. *Viola blanda*
Tongue. The genus *Schweiggeria*
Tooth. *Dentaria bulbifera*
Tree. *Viola arborescens*
Twin-flowered. *Viola biflora*
Water. *Hottonia palustris*
Water, American. *Hottonia inflata*
Winter-flowering Double White. *Viola Belle de Chatenay*
Violet-wood, or Myall-wood, Australian. *Acacia homalophylla*
Brazil. Supposed to be a species of *Triptolomœa*
Guiana. *Andira violacea*
Viper's Bugloss. *Echium vulgare*
Viper's-grass. The genus *Scorzonera*
Virgin-tree. *Sassafras Parthenoxylon*
Virgin's-Bower. *Clematis Vitalba*
American. *Clematis virginiana*
Australian. *Clematis Mossmana*
Blue-flowered. *Clematis Viorna*
Purple-flowered. *Clematis Viticella*
Sweet-scented. *Clematis Flammula*
Virginian-Creeper. *Ampelopsis hederacea (A. quinquefolia)*
Evergreen. *Ampelopsis sempervirens*
Heart-leaved. *Ampelopsis cordata*
Variegated. *Ampelopsis heterophylla marmorata*
Veitch's. *Ampelopsis Veitchi*
Virginian-Silk. *Periploca græca*

Vitus's (St.)-wood. *Rhus divica*
"Vouvan"-tree. *Laurelia serrata*
Vutu-tree. The genus *Barringtonia*
Waahoo, of N. America. *Euonymus atropurpureus*
Wafer-Ash. See Ash
Wahahe-tree. See Kohé-tree
Wake-Robin. *Arum maculatum*
American. *Arum dracontium*, *Trillium grandiflorum*, and *T. cernuum*
Walking-fern, of N. America. *Lycopodium alopecuroides*
Walking-leaf, or Walking-Fern. *Camptosorus rhizophyllus*
Wallaba-tree. *Eperua falcata*
Wallaby-bush. *Beyeria viscosa* and other species
Wall-Cress. The genus *Arabis*
Wall-Fern. *Polypodium vulgare*
Wall-flower, Alpine. *Erysimum ochroleucum* (*Cheiranthus alpinus*)
Common. *Cheiranthus Cheiri* and vars.
Fairy. *Erysimum pumilum*
Long-leaved. *Cheiranthus longifolius*
Himalayan. *Erysimum pachycarpum*
Marshall's. *Cheiranthus Marshalli*
Native Victorian. *Pultenæa daphnoides*
Rhætian Dwarf. *Erysimum Rhæticum*
Rock. *Erysimum pulchellum*
Teneriffe. *Cheiranthus scoparius*
Western. *Erysimum arkansanum*
Walnut-tree. Various species of *Juglans*
American. *Juglans nigra*
Belgaum. See Walnut, Indian
Caucasian. *Juglans pterocarpa* (*Pterocarya caucasica*)
Common. *Juglans regia*
Country. See Walnut, Indian
Cut-leaved. *Juglans regia var. laciniata*
Indian, Belgaum, or Country. *Aleurites triloba*
Jamaica. *Picrodendron Juglans*
Otaheite. *Aleurites moluccana* and *A. triloba*
Titmouse. A thin-shelled variety of *Juglans regia*
White. *Juglans cinerea*
Wall-Pennywort. See Penny-wort
Wall-pepper. See Pepper
Wall-Rue. See Rue
Wall-wort. An old name for *Sambucus Ebulus*
Wampee-tree. *Cookia punctata*
Wandering Jew. *Saxifraga sarmentosa*
"Wandoo." *Eucalyptus redunca*
"Wangara"-tree. *Eucalyptus amygdalina*
Wand-plant. *Galax aphylla*
Wapatoo-root, of California. A species of *Sagittaria*
Waratah-tree. *Telopea speciosissima*
Warden. An old name for a variety of Pear used for making "Warden pies"
Ware (Sea-weeds). *Algæ*
Warence. An old name for Madder
Warmot. An old name for Worm-wood
Wart-cress, or Swine's-cress. *Senebiera Coronopus*
American. *Senebiera didyma*

Wart-flower. The genus *Phymatanthus*
Wart-herb. *Rhynchosia minima*
Wart-wort. *Euphorbia Helioscopia*
Washingtonia. The American name for *Wellingtonia* (*Sequoia*) *gigantea*
Water-Agrimony, or Water-Hemp. *Bidens tripartita*
Water-Aloe. *Stratiotes aloides*
Water-Anemone. *Ranunculus aquatilis*
Water-Archer. *Sagittaria sagittæfolia*
Water-Bean. The genus *Nelumbium*
Water-Betony. *Scrophularia aquatica*
Water-Blinks. *Montia fontana*
Water-Cress. See Cress
Water-Crown-Cup. *Sparganophora verticillata*
Water-Crow-foot. *Ranunculus aquatilis*
Water-Dock. See Dock
Water Dropwort. *Œnanthe fistulosa*
Water-Elder. *Viburnum Opulus*
Water Feather-foil. *Hottonia palustris*
Water-Fennel. *Œnanthe Phellandrium*
Water-Fern. *Osmunda regalis*
"Water-fire," of India. *Bergia Ammanoides*
Water-Flag. *Iris Pseud-acorus*
Water-Germander. See Germander
Water-Gladiole. *Butomus umbellatus*
Water-Hemlock. See Hemlock
Water-Hemp. *Bidens tripartita*
Water-Horehound. *Lycopus europæus*
Water-Ivy. See Ivy
Water-leaf, American. The genus *Hydrophyllum*
Sitka. *Romanzoffia sitchensis*
Water-Lemon. *Passiflora laurifolia*
Water-Lentils. The genus *Lemna*
Water-lily, Amazon. *Nymphæa amazonica*
Australian. *Nymphæa gigantea*
Blue Indian. *Nymphæa cyanea*
Chinese. *Nelumbium speciosum* and *Nymphæa* (*Castalia*) *pygmæa*
Dwarf. *Villarsia nymphæoides*
E. Indian Red. *Nymphæa rubra*
Edible-fruited. *Nuphar multisepala*
Egyptian. *Nymphæa Lotus*
Florida. *Nymphæa flava*
Fringed. *Villarsia* (*Limnanthemum*) *nymphæoides*
Hungarian. *Nymphæa thermalis*
"Indian Lotus." *Nymphæa pubescens*
New Zealand, or "of the Shepherds." *Ranunculus Lyallii*
Prickly. *Euryale ferox* (*Annesiea spinosa*)
Pigmy. *Nymphæa pygmæa*
Rose-coloured (hardy). *Nymphæa alba var. rosea*
Rose-coloured, E. Indian. *Nymphæa rosea*
Round-fruited. *Nymphæa sphærocarpa*
Royal. *Victoria regia*
Shield-leaved. *Nymphæa scutifolia*
Striped-flowered. *Nuphar advena*
Sweet-scented. *Nymphæa odorata*
Three-coloured. *Nuphar advena*
Toothed-leaved. *Nymphæa dentata*
Victoria. *Victoria regia*
White. *Nymphæa alba*

English Names of Cultivated, Native,

Water-lily, White, Tuberous-rooted. *Nymphæa tuberosa*
Yellow. *Nuphar lutea*
Yellow, Small. *Nuphar pumila* and *N. kalmiana*; also *Villarsia nymphœoides*
Zanzibar. *Nymphæa stellata zanzibarensis*
Water-Melon. *Cucumis Citrullus*
Water-Milfoil. *Myriophyllum verticillatum*
Water-Moss. See Moss
Water-Nymph, Canadian. *Naias canadensis*
Water-Parsnip. *Sium latifolium*
Water-Pepper. *Polygonum Hydropiper*
Mild. *Polygonum hydropiperoides*
Water-Pimpernel. *Veronica Beccabunga*
Water-Plantain. *Alisma Plantago*
Water-Platter. *Victoria regia*
Water-Purslane. *Peplis Portula*
Water-Rose. An old name for Water-Lily
Water-Shield. *Brasenia peltata*
Water-Snow-cups. *Ranunculus aquatilis*
Water-Soldier. *Stratiotes aloides*
Water-Spike. An old name for the genus *Potamogeton*
Water-Spurge. *Euphorbia Helioscopia*
Water-Star. *Actinocarpus (Alisma) Damasonium*
Water-Star-grass. See Grass
Water-Star-wort. The genus *Callitriche*
Water-Thyme. See Thyme
Water-tree, Red. *Erythrophlœum guineense*
Sierra Leone. *Tetracera alnifolia*
Water-Vine. See Vine
Water-Violet. *Hottonia palustris*
American. *Hottonia inflata*
Water-weed, New Granada. *Murathrium utile*
N. American. *Anacharis Alsinastrum (Elodea canadensis)*
Water-Willow, American. *Dianthera americana*
Water-wort. The genus *Elatine*; also an old name for *Asplenium Trichomanes*
Wattle, African. *Acacia natalitia*
Black. *Acacia decurrens* and *A. mollissima*
Cape, or Crested. *Acacia (Albizzia) lophantha*
Golden, or Green. *Acacia pycnantha*
Gum. *Acacia decurrens*
"Raspberry-jam." *Acacia acuminata*
Savannah. *Citharexylon quadrangulare* and *C. cinereum*
Scrub. *Acacia stipuligera*
Silver. *Acacia decurrens var. dealbata*
Soap-pod. *Acacia concinna*
Wallaby. *Acacia riyens*
Wax-bush, Blue. *Cuphea viscosissima*
Wax-Dammar. *Podocarpus neriifolia*
Singapore. *Podocarpus polystachya*
Wax-flower. The genus *Hoya*
Bell. *Hoya campanulata*
Clustered. *Stephanotis floribunda*
Fleshy-leaved. *Hoya carnosa*
Imperial. *Hoya imperialis*
Manilla. *Hoya coriacea*
N. S. Wales. *Hoya australis*
Wax-Myrtle. *Myrica cerifera*

Wax-plant. *Cerinthe major*
One-flowered. *Monotropa uniflora*
Wax-tree, Brazilian. *Vismia brasiliensis*
Chinese. *Ligustrum lucidum*
Guiana. *Vismia guianensis*
New Granada. *Elæagia utilis*
Otoba. *Myristica Otoba*
Wax-weed, Blue. *Cuphea viscosissima*
Wax-work. *Celastrus scandens*
Wayaka. See Yaka
Way-bread. An old name for *Plantago*
Wayfaring-tree. *Viburnum Lantana*
Indian. *Viburnum cotinifolium*
Weasel-chop. *Mesembryanthemum mustelinum*
Weasel-snout. *Galeobdolon Lamium*
Wedding-flower, Cape. *Dombeya natalensis*
Lord Howe's Island. *Iris (Moræa) Robinsoniana*
Weed-wind, or With-wind. An old name for Bind-weed
Weenong-tree. A species of *Tetrameles*
Weeping Willow, Ash, Beech, &c. See Willow, Ash, Beech, &c.
Weevil-plant. The genus *Curculigo*
Welcome-to-our-house. *Euphorbia Cyparissias*
Weld. *Reseda Luteola*
Wellingtonia. *Sequoia (Washingtonia) gigantea*
Welsh Nut. See Nut
Welsh Onion. See Onion
Welsh Poppy. See Poppy
West Coast Creeper. *Asclepias odoratissima*
Whahoo, or Winged Elm, of N. America. *Ulmus alata*
Whangee, or Wanghee Cane. See Cane
Wheat. The genus *Triticum*
Beech. *Fagopyrum cymosum*
Buck. *Fagopyrum esculentum (Polygonum Fagopyrum)*
Clock. A variety of *Triticum turgidum*
Common, or Soft-grained. *Triticum sativum*
Cow. See Cow-wheat
Dinkel. *Triticum Spelta*
Emmer. *Triticum dicoccum*
Goat's. The genus *Tragopyrum*
Guinea, or Turkey. *Zea Mays*
Hard-grained. *Triticum durum*
Humpy-grained. *Triticum turgidum*
Lammas. See Wheat, Winter
Mummy. *Triticum compositum*
Polish. *Triticum polonicum*
Reputed original of cultivated. *Ægilops ovata*
Single-grained. *Triticum monococcum*
Spelt. *Triticum Spelta*
Spring, or Summer. *Triticum æstivum*
Square-eared. *Triticum sativum var. compactum*
Starch. *Triticum amyleum*
Revet. A variety of *Triticum turgidum*
"Trigo Moro." *Triticum Cevallos*
Turkey. See Wheat, Guinea
Winter, or Lammas. *Triticum hybernum*
Wheel - seed, Tasmanian. *Trochocarpa (Decaspora) thymifolia*
Wheel-tree, of Guiana. *Aspidosperma excelsum*

Whin. *Ulex europæus*
Petty. *Genista anglica* and *Ononis arvensis*
Whin-berry. *Vaccinium Myrtillus*
White Beam-tree. *Pyrus Aria*
White Bothen. An old name for *Chrysanthemum Leucanthemum*
White-bottle. *Silene inflata*
White Goldes. An old name for *Chrysanthemum Leucanthemum*
White Grass, American. See Grass
White Hellebore. The genus *Veratrum*, especially *V. album*
American. *Veratrum viride*
Dark-flowered. *Veratrum nigrum*
White-flowered. *Veratrum album*
White-hoop Wyth, of Jamaica. *Tournefortia bicolor*
White Man's Foot-print. *Plantago major*
White Pot-herb. See Pot-herb
White-root. An old name for Solomon's Seal
American. *Ligusticum actæifolium*
White-rot. *Hydrocotyle vulgaris*
White Runner Bean. See Bean
White-Sapota. *Casimiroa edulis*
White-thorn. See Thorn
White-tree. *Melaleuca Leucodendron*
White-wand Plant. *Galax aphylla*
White-wood. *Tilia americana*; also applied to *Liriodendron tulipiferum*, *Oreodaphne Leucoxylon*, *Nectandra leucantha*, *Tecoma Leucoxylon* and *T. pentaphylla*
Australian. *Lagunaria Patersoni*
Barbadoes. *Tecoma Leucoxylon*
Isle of Bourbon Sweet-scented. *Ruizia cordata*
Java. *Leucoxylon buxifolium*
Of Victoria. *Pittosporum bicolor*
Tasmanian. *Pittosporum bicolor*
Whitlow-grass. The genus *Draba*
Alpine. *Draba alpina*
Danish. *Draba aurea*
Fringed. *Draba ciliata*
Glacier. *Draba glacialis*
Gray. *Draba cinerea*
Hairy-leaved. *Draba aizoides*
Hairy-podded. *Draba Aizoon* and *D. lasiocarpa*
Large. *Draba gigas*
Pigmy. *Draba nivalis*
Pointed-leaved. *Draba cuspidata*
Rock. *Draba rupestris*
Sauter's. *Draba Sauteri*
Sea-green. *Draba aizoides*
Starry. *Draba stellata*
Three-toothed. *Draba tridentata*
Violet-flowered. *Draba violacea*
Woolly-leaved. *Draba tomentosa*
Yellow-flowered. *Draba aizoides*
Yellow-flowered, Large. *Draba ciliaris*
Yellow-flowered, Pale. *Draba bœotica*
Whitlow-wort. The genus *Paronychia*
Forked. *Paronychia dichotoma*
Whorl-leaved. *Paronychia verticillata*
Whitten-tree. *Viburnum Opulus*
Whit-wort. An old name for Feverfew
Whorl-flower, Long-leaved. *Morina longifolia*
Persian. *Morina persica*

Whortle-berry. *Vaccinium Myrtillus*
Bird-Cherry-leaved. *Vaccinium padifolium*
Box-leaved. *Vaccinium buxifolium*
Broad-leaved. *Vaccinium amœnum*
Bushy. *Vaccinium dumosum*
Canadian. *Vaccinium canadense*
Caraccas. *Vaccinium caracasanum*
Corymb-flowered. *Vaccinium corymbosum*
Elongated. *Vaccinium elongatum*
Gale-leaved. *Vaccinium galezans*
Glossy-leaved. *Vaccinium nitidum*
Jamaica. *Vaccinium meridionale*
Madeira. *Vaccinium maderense*
Myrsine-leaved. *Vaccinium Myrsinitis*
Myrtillus-like. *Vaccinium myrtilloides*
Myrtle-leaved. *Vaccinium myrtifolium*
Large-flowered. *Vaccinium grandiflorum*
Leafy. *Vaccinium frondosum*
Narrow-leaved. *Vaccinium angustifolium*
Oriental. *Vaccinium Arctostaphylos*
Ovate-leaved. *Vaccinium ovatum*
Pale-flowered. *Vaccinium pallidum*
Pennsylvanian. *Vaccinium pennsylvanicum*
Privet-leaved. *Vaccinium ligustrinum*
Red, of Mount Ida. *Vaccinium Vitis-idæa*
Red-twigged. *Vaccinium erythrinum*
Resinous. *Vaccinium resinosum*
Rollisson's. *Vaccinium Rollissoni*
Small-flowered. *Vaccinium minutiflorum*
Smooth. *Vaccinium glabrum*
Thick-leaved. *Vaccinium crassifolium*
Trailing. *Vaccinium humifusum*
Tree. *Vaccinium arboreum*
Tufted. *Vaccinium cæspitosum*
Twiggy. *Vaccinium virgatum*
White-berried. *Vaccinium Myrtillus albis-baccis*
White-lipped. *Vaccinium leucostomum*
Willow-leaved. *Vaccinium salicinum*
Whortle-berry-bush, Victorian. *Wittsteinia vacciniacea*
Wicken-tree. See Quick-beam
Widow, Mournful. *Scabiosa atropurpurea*
Mourning. *Geranium phæum*
Widow's-Flower. *Scabiosa atropurpurea*
Widow-wail. The genus *Cneorum*
"Widow-wisse." An old name for *Genista tinctoria*
Wig-tree. *Rhus Cotinus*
Wild-boar's-tree, of San Domingo. *Hedwigia balsamifera*
Wilding (Crab-Apple). *Pyrus Malus*
Wild Irishman, of New Zealand. *Discaria Toumatou*
"Wild Savager." An old name for *Agrostemma Githago*
Wild Snow-ball. *Ceanothus americanus*
Wild Spaniard, of New Zealand. *Aciphylla squarrosa*, *A. Colensoi*, and other species
Wild William. An old name for *Lychnis Flos-cuculi*
Will-o'-the-wisp. *Nostoc commune*
Willow. The genus *Salix*
Almond-leaved. *Salix amygdalina*
American Black. *Salix nigra*
American Dwarf Gray. *Salix tristis*

L

Willow, American Green. *Salix chlorophylla*
American Heart-leaved. *Salix cordata*
American Hoary. *Salix candida*
American Long-leaved. *Salix longifolia*
American Livid. *Salix livida var. occidentalis*
American Prairie. *Salix humilis*
American Shining. *Salix lucida*
American Silky. *Salix sericea*
American Silvery-fruited. *Salix argyrocarpa*
American Water. *Dianthera americana*
Australian. *Geijera parviflora*
Bay. *Salix pentandra*
Bedford. *Salix Russelliana*
Bitter. *Salix purpurea*
Boyton. *Salix purpurea var. Lambertiana*
British Columbia. *Salix lasiandra*
Californian. *Salix lævigata*
Common. *Salix alba*
Crack. *Salix fragilis*
Creeping. *Salix repens*
Cutler's. *Salix Cutleri*
Desert, of California. *Chilopsis saligna*
Dishley. *Salix Russelliana*
Downy. *Salix Lapponum*
Dwarf. *Salix herbacea*
Dwarf Thyme-leaved. *Salix serpyllifolia*
French. *Salix triandra*
French, or Persian. *Epilobium angustifolium*
Goat, or Sallow. *Salix Caprea*
Goat, Weeping. *Salix Caprea var. pendula*
Golden. *Salix vitellina*
Highland. *Salix reticulata*
Huntingdon. *Salix alba*
Kilmarnock. *Salix Caprea*
Kilmarnock, Weeping. *Salix Caprea var. pendula*
Laurel-leaved. *Salix pentandra*
Leicester. *Salix alba var. cærulea*
Netted-leaved. *Salix reticulata*
"Primrose," of the W. Indies. The genus *Œnothera*
Purple. *Salix purpurea*
Rose. *Salix purpurea var. Helix*
Rosemary-leaved. *Salix rosmarinifolia*
Round-eared. *Salix aurita*
Rusty-branched. *Salix Doniana*
Silvery. *Salix Reginæ*
Silvery Woolly-leaved. *Salix lanata*
Swallow-tailed. *Salix alba*
Sweet. *Salix pentandra;* also applied to *Myrica Gale*
Tea-leaved. *Salix phylicifolia*
Violet. *Salix daphnoides*
Virginian. *Itea virginica*
Weeping. *Salix Babylonica*
Weeping Kilmarnock. *Salix Caprea var. pendula*
White. *Salix alba*
White Welsh. *Salix fragilis var. decipiens*
Whortle. *Salix Myrsinites*
Woolly. *Salix lanata*
Willow-herb. The genus *Epilobium*
American. *Epilobium molle* and *E. coloratum*
Broad-leaved. *Epilobium montanum*

Willow-herb, Californian Dwarf. *Epilobium obcordatum*
Californian Yellow-flowered. *Epilobium luteum*
Chickweed-leaved. *Epilobium alsinæfolium*
Dodoen's. *Epilobium Dodonæi*
Hoary. *Epilobium parviflorum*
Hooded. The genus *Scutellaria*
Marsh. *Epilobium palustre*
Mountain. *Epilobium alpinum*
Pale-flowered. *Epilobium roseum*
Rose-bay. *Epilobium angustifolium*
Rosemary-leaved. *Epilobium rosmarinifolium* (*E. angustissimum*)
Square-stemmed. *Epilobium tetragonum*
White-flowered. *Epilobium angustifolium album*
Willow-Thorn. *Hippophaë rhamnoides*
Willow-weed. *Lythrum Salicaria* and *Polygonum lapathifolium*
Wind-flower. The genus *Anemone*
Alpine. *Anemone alpina*
Apennine. *Anemone apennina*
Bastard. *Gentiana Pseudo-Pneumonanthe*
Cape. *Anemone capensis*
Carolina. *Anemone caroliniana*
Cyclamen-leaved. *Anemone palmata*
Forked. *Anemone dichotoma*
Japanese. *Anemone japonica*
Japanese, Pale. *Anemone elegans*
Japanese, Vine-leaved. *Anemone japonica vitifolia*
Japanese, White. *Anemone japonica alba*
Long-fruited. *Anemone cylindrica*
Narcissus-flowered. *Anemone narcissiflora*
Oregon. *Anemone deltoidea*
Pau. *Anemone fulgens*
Peacock. *Anemone pavonina*
Pennsylvanian. *Anemone pennsylvanica*
Poppy. *Anemone coronaria*
River-side. *Anemone rivularis*
Robinson's. *Anemone Robinsoni*
Rocky Mountain. *Anemone multifida*
Scarlet. *Anemone fulgens*
Shaggy. *Anemone vernalis*
Small-flowered. *Anemone parviflora*
Snowdrop. *Anemone sylvestris*
Star. *Anemone stellata*
Sulphur-coloured. *Anemone sulphurea*
Three-leaved. *Anemone trifolia*
Virginian. *Anemone virginica*
White. *Anemone alba*
Winter. *Anemone blanda*
Wood. *Anemone nemorosa*
Wood, Double. *Anemone nemorosa plena*
Windle-straw. *Agrostis Spica-venti* and *Cynosurus cristatus*
Wind-root. *Asclepias tuberosa*
Wind-Rose. *Papaver Argemone* and *Ræmeria hybrida*
Wine-berries. An old name for Whortleberries
Wine-berry-shrub, of New Zealand. *Coriaria sarmentosa*
Wine-cellar Fungus. See Fungus
Wing-Moss. See Moss
Wing-seed. *Ptelea trifoliata*
Winter Aconite. See Aconite

Winter-berry, Atom-bearing. *Prinos atomarius*
Carolina. *Prinos ambiguus*
Deciduous. *Prinos deciduus*
Leathery-leaved. *Prinos coriaceus*
Plum-leaved. *Prinos prunifolius*
Shining. *Prinos glaber*
Smooth. *Prinos lævigatus*
Spear-leaved. *Prinos lanceolatus*
Swamp. *Ilex verticillata*
Winter-bloom. *Hamamelis virginica*
Winter-Cherry. Various species of *Physalis*
Cluster-flowered. *Physalis somnifera*
Common. *Physalis Alkekengi*
Hairy Annual. *Physalis pruinosa*
Tooth-leaved. *Physalis angulata*
Winter Clover. *Mitchella repens*
Winter-Cress. *Barbarea præcox*
Variegated. *Barbarea præcox variegata*
Winter-green. The genus *Pyrola*
Aromatic. *Gaultheria procumbens*
Common. *Pyrola minor*
Creeping. *Gaultheria procumbens*
False, or **Pear-leaved.** *Pyrola rotundifolia*
Intermediate. *Pyrola media*
Larger. *Pyrola rotundifolia*
Notched-leaved. *Pyrola secunda*
One-flowered. *Pyrola uniflora*
Serrated. *Pyrola secunda*
Spotted-leaved. *Pyrola maculata*
Umbelled. *Pyrola umbellata*
Winter-sweet. *Origanum heracleoticum* and *Toxicophlæa spectabilis*
Winter-weed. *Veronica hederæfolia*
Winter Wolf's-bane. *Eranthis hyemalis*
Winter's-Bark, or **Winter's-Cinnamon.** See Cinnamon, Winter's
Wire-bent. *Nardus stricta*
Wire-grass. See Grass
Wistaria, American. *Wistaria frutescens*
Chinese. *Wistaria sinensis*
Japanese. *Wistaria japonica*
Tuberous-rooted. *Apios tuberosa*
Witch, or **Wych, Elm.** See Elm
Hazel. See Hazel
Witches'-butter. *Exidia glandulosa*
Witches'-thimble. *Silene maritima*
Withe, Basket, of Jamaica. *Heliotropium fruticosum*
Prickly. See Prickly-withe
Withe-rod, American. *Viburnum nudum*
Withwind. See Weedwind
Withy. *Salix viminalis*
"Witloof." See Chicory
Woad. *Isatis tinctoria*
Wild. *Reseda Luteola*
Woad-waxen, or **Wood-waxen.** *Genista tinctoria*
Wodier-tree. *Odina Wodier*
Wold-wolle-fibre-plant. *Pinus sylvestris* and *P. Laricio*
Wolf-berry. *Symphoricarpus occidentalis*
Wolf-chop. *Mesembryanthemum lupinum*
Wolf's-bane. The genus *Aconitum*, especially *A. Lycoctonum*
American. *Aconitum uncinatum*

Wolf's-bane, Mountain. *Ranunculus Thora*
Purple. *Aconitum Cammarum*
Trailing. *Aconitum reclinatum*
Wholesome. *Aconitum Anthora*
Wolf's-claw. *Lycopodium clavatum*
Wolf's Fists. An old name for Puff-balls
Wolf's-milk. The genus *Euphorbia*
Wombat-berry. The genus *Eustrephus*
Wood-bine. *Lonicera Periclymenum*
American. *Lonicera grata* and *Ampelopsis quinquefolia*
Spanish, of the W. Indies. *Ipomæa tuberosa*
Wood-broney. An old name for the Ash-tree
Wood-Laurel. *Daphne Laureola*
Wood-lily. *Pyrola minor*
American. The genus *Trillium*
American, Drooping. *Trillium cernuum*
American, Dwarf White. *Trillium nivale*
American, Large White. *Trillium grandiflorum*
American, Painted. *Trillium erythrocarpum*
American, Purple. *Trillium erectum*
Wood-March. An old name for *Sanicula europæa*
Wood-nep. An old name for *Ammi majus*
Wood-Night-shade. See Nightshade
Wood-nut. *Corylus Avellana*
Wood-of-St.-Martha. *Cæsalpinia echinata*
Wood-Oil-plant. Several species of *Dipterocarpus*
China. *Aleurites (Dryandra) cordata*
Wood-Pea, or **Wood-Vetch.** *Orobus sylvaticus*
Wood-reed. *Calamagrostis Epigeios* and *C. lanceolata*
Woodrowel. An old name for Woodruff
Woodruff, Broad-leaved. *Asperula taurina*
Creeping. *Asperula humifusa*
Sweet. *Asperula odorata*
Three-lobed. *Asperula tinctoria*
Wood-rush. The genus *Luzula*
Curved. *Luzula arcuata*
Field. *Luzula campestris*
Great. *Luzula sylvatica*
Hairy. *Luzula pilosa*
Spiked. *Luzula spicata*
Wood-Sage. See Sage
Wood-Sorrel. The genus *Oxalis*
Bowie's. *Oxalis Bowiei*
Brazilian. *Oxalis brasiliensis*
Chilian. *Oxalis Valdiviana*
Common. *Oxalis Acetosella*
Deppe's. *Oxalis Deppei*
Dwarf. *Oxalis tenella*
Flesh-coloured. *Oxalis incarnata*
Jamaica. *Begonia nitida*
Lobed-leaf. *Oxalis lobata (O. granulata)*
Many-flowered. *Oxalis floribunda*
Nasturtium. *Oxalis tropæoloides*
New Zealand. *Oxalis magellanica*
Nine-leaved. *Oxalis enneaphylla*
Oregon. *Oxalis Oregana*
Plumier's. *Oxalis Plumieri*
Procumbent Yellow flowered. *Oxalis corniculata*

Wood - Sorrel, Purple - leaved. *Oxalis atropurpurea*
Red-flowered Procumbent. *Oxalis corniculata var. rubra*
Rosy-flowered. *Oxalis rosea*
Sand. *Oxalis arenaria*
Scented. *Oxalis odorata*
Sensitive. *Oxalis Biophytum*
Upright Yellow-flowered. *Oxalis stricta*
Violet. *Oxalis violacea*
Woolly-stamened. *Oxalis lasiandra*
Yellowish-flowered. *Oxalis luteola*
"**Wood-Sower.**" An old name for *Oxalis Acetosella*
Wood-tongue. The genus *Drymoglossum*
Wood-Vetch. See Wood-Pea
Wood-vine, Yellow. *Morus Calcar-galli*
Wood-waxen, or Woad-waxen. *Genista tinctoria*
Wood-grass. See Grass
Woollen. An old name for Mullein
"**Woolly-butt,**" of Australia. *Eucalyptus longifolia* and *E. viminalis*
Worm-grass, Maryland. *Spigelia marilandica*
Worm-seed. *Artemisia maritima var. Stechmanniana* (*A. Lercheana*)
Spanish. *Halogeton tamariscifolium*
Treacle. See Treacle-Wormseed
Worm-wood. The genus *Artemisia*
African. *Artemisia afra*
Alpine. *Artemisia alpina* and *A. mutellina*
Biennial. *Artemisia biennis*
Bluish. *Artemisia carulescens*
Branchy. *Artemisia ramosa*
Cliff. *Artemisia rupestris*
Common. *Artemisia Absinthium*
Creeping. *Artemisia repens*
Dill-leaved. *Artemisia anethifolia*
Field. *Artemisia campestris*
Forked. *Artemisia furcata*
French. *Artemisia gallica*
Glacier. *Artemisia glacialis*
Hoary. *Artemisia cana*
Holy. *Artemisia Santonica*
Marschall's. *Artemisia Marschalliana*
Medicinal. *Artemisia Abrotanum*
Norwegian. *Artemisia norvegica*
Oriental. *Artemisia orientalis*
Pale-flowered. *Artemisia lactiflora*
Pallas's. *Artemisia Pallasii*
Potentilla-leaved. *Artemisia potentillæfolia*
Rock. *Artemisia saxatilis*
Roman. *Artemisia pontica*
Roman, of N. America. *Ambrosia artemisiæfolia*
Scallop-leaved. *Artemisia pectinata*
Sea-side. *Artemisia maritima*
Sea-side, Cape. *Stœbe cinerea*
Silky-leaved. *Artemisia sericea*
Silvery. *Artemisia argentea*
Slender-leaved. *Artemisia tenuifolia*
Spiked. *Artemisia spicata*
Steller's. *Artemisia Stelleriana*
Tall. *Artemisia procera*
Tansy-leaved. *Artemisia tanacetifolia*
Trailing Evergreen. *Artemisia aprica*
Tree. *Artemisia arborescens*

Worm-wood, Variegated - leaved. *Artemisia vulgaris variegata*
Wulfen's. *Artemisia Wulfenii*
Wound-wort. *Stachys germanica;* also *Anthyllis vulneraria*
Clown's. *Stachys palustris*
Corsican. *Stachys corsica*
Dorcas'. *Solidago Virgaurea angustifolia*
Hercules'. *Heracleum Panaces*
Scarlet. *Stachys coccinea*
Woolly. *Stachys lanata*
Wourali, or Urari, poison-tree. *Strychnos toxifera*
Wrack, Grass. *Zostera marina*
Wreath, Purple. *Petræa volubilis*
Wreathe-wort, Purple. *Orchis mascula*
Writing-reed. *Calamagrostis Epigeios*
"**Wungee.**" *Cucumis cicatrisatus*
Wych, or Witch, Elm. See Elm
Wych-Hazel. See Hazel
Wymot. An old name for Marsh-Mallow
Wyth, Basket. See Basket-Wyth
Wyth, White-hoop. See White-hoop Wyth

Yacca-wood. The wood of *Podocarpus coriacea* and *P. Purdieana*
Yaka, or Wayaka, plant, of Fiji. *Pachyrhizus angulatus*
Yam, American Wild. *Dioscorea villosa*
Chinese, or Japanese. *Dioscorea Batatas*
Common Cultivated. *Dioscorea sativa*
Granada, or Guinea. *Dioscorea bulbifera*
Indian. *Dioscorea trifida*
Japanese. See Yam, Chinese
Kaawi. *Dioscorea aculeata*
Port Moniz. *Tamus edulis*
Red, or White. *Dioscorea alata*
Tivoli. *Dioscorea Nummularia*
Uvi. *Dioscorea alata*
White. See Yam, Red
Yang-mae-tree. *Myrica Nagi*
Yar-nut, or Yor-nut. See Nut
Yarrow. The genus *Achillea*
Asplenium-leaved. *Achillea asplenifolia*
Common. *Achillea Millefolium*
Egyptian. *Achillea ægyptiaca*
Golden-flowered. *Achillea aurea*
Mongolian. *Achillea mongolica*
Noble. *Achillea Filipendula*
Pale-yellow-flowered. *Achillea decolorans*
Rock. *Achillea rupestris*
Serrate-leaved. *Achillea serrata*
Silvery. *Achillea Clarennæ*
Silvery Dwarf. *Achillea (Ptarmica) umbellata*
Soldier's. *Stratiotes aloides*
Woolly. *Achillea tomentosa*
Yate-tree, or Yeit-tree. *Eucalyptus cornuta*
"**Yaupon.**" *Ilex Cassine*
Yaw-root. *Stillingia sylvatica*
"**Yeara,**" of California. *Rhus diversiloba*
Yeast-plant. *Penicillium glaucum*
Yeit-tree. See Yate-tree
Yellow Archangel. *Lamium Galeobdolon*
Yellow Bird's-nest. *Monotropa Hypopitys*
Yellow Bugle. *Ajuga Chamæpitys*
"**Yellowby.**" *Chrysanthemum segetum*
Yellow Cress. *Barbarea præcox*

Yellow Crow-bells. See Crow-bells
Yellow Helmet-flower. See Helmet-flower
Yellow Loose-strife. *Lysimachia vulgaris*
Yellow Parilla. See Parilla
Yellow Pimpernel. *Lysimachia nemorum*
Yellow Rattle. *Rhinanthus Crista-galli*
Yellow Rocket. *Barbarea vulgaris*
Yellow Sultan. *Centaurea suaveolens*
Yellow-root-Shrub. *Xanthorrhiza apiifolia*
Yellow-tops. *Senecio Jacobæa*
Yellow-weed. *Reseda Luteola*
Yellow-wood, American. *Virgilia lutea* (*Cladrastis tinctoria*)
Australian. *Oxleya Xanthoxylon* (*Flindersia Oxleyana*)
Brazilian. *Ochroxylon punctatum*
Cape. *Podocarpus Thunbergii*
E. Indian. *Podocarpus latifolia* and *Chloroxylon Swietenia*
Isle of Bourbon. *Ochrosia Borbonica*
Moreton Bay. *Acronychia lævis* (*Cyminosma oblongifolium*)
N. S. Wales. *Oxleya Xanthoxylon*
Prickly. See Yellow-wood, W. Indian
S. African. *Podocarpus elongata*
W. Indian, or Prickly. *Xanthoxylon Clava-Herculis*
Yellow-wort. *Chlora perfoliata*
Yerba Buena, of California. *Micromeria Douglasii*
Yerba Mansa, of California. *Anemopsis californica*
Yevering Bells. *Pyrola secunda*
Yew-tree. The genus *Taxus*
American. *Taxus baccata* var. *canadensis*
"Blue John." *Taxus stricta*
Californian. *Taxus brevifolia*
Canadian. *Taxus canadensis*
Cheshunt. *Taxus baccata* var. *Cheshuntensis*
Chinese. *Taxus nucifera* and *Podocarpus chinensis*
Cluster-flowered. The genus *Cephalotaxus*
Cluster-flowered, Fortune's. *Cephalotaxus Fortunei*
Cluster-flowered, Long-stalked. *Cephalotaxus pedunculata*
Cluster-flowered, Plum-fruited. *Cephalotaxus drupacea*
Common. *Taxus baccata*

Yew-tree, Common Erect. *Taxus baccata* var. *erecta*
Dovaston's. *Taxus baccata* var. *Dovastoni*
Florence Court, or Irish. *Taxus baccata* var. *fastigiata*
Florida. *Taxus Floridana*
Irish. See Yew, Florence Court
Japanese. *Podocarpus Koraiana* (*Taxus japonica*)
Japan, Short-leaved. *Taxus adpressa*
Jointed. The genus *Arthrotaxis*
Mexican. *Taxus globosa*
Nidpath Castle. *Taxus baccata* var. *Nidpathensis*
Prince Albert's. *Saxgothea conspicua* (*Taxus patagonica*)
Scattered-leaved. *Taxus baccata* var. *sparsifolia*
Silver-variegated. *Taxus baccata* var. *argentea*
Stinking. The genus *Torreya*
Stinking, Large. *Torreya grandis*
Washington's. *Taxus canadensis* var *Washingtoni*
Yellow-berried. *Taxus baccata* var. *fructu-luteo*
Yoke-Elm. See Elm
Yoke-Moss. See Moss
Yoke-wood. *Catappa longissima*
Yorkshire Sanicle. See Sanicle
Youth-and-Old-Age. The genus *Zinnia*
Youth-wort. *Drosera rotundifolia*
Yulan. *Magnolia Yulan*
Zante-wood. *Rhus Cotinus* and *Chloroxylon Swietenia*
Zebra-plant. *Maranta* (*Calathea*) *zebrina*
Zebra-poison, of S. Africa. *Euphorbia arborea*
Zebra-wood, E. Indian. *Guettarda speciosa*
Guiana. *Omphalobium Lambertii*
Guinea. *Omphalobium africanum*
Jamaica. *Eugenia fragrans*
Zedoary, Long. *Curcuma Zerumbet*
Round. *Curcuma Zedoaria*
"Zeloak"-root. The root of *Iris juncea*
Zelkowa-tree. The genus *Zelkova*
Common. *Zelkova crenata* (*Planera Richardi*)
Zephyr-flower. The genus *Zephyranthes*
Keeled. *Zephyranthes carinata*
Rosy. *Zephyranthes rosea*
Sulphur-coloured. *Zephyranthes sulphurea*
Tube-spathed. *Zephyranthes tubispatha*
White. *Zephyranthes candida*

A DICTIONARY
OF
ENGLISH PLANT NAMES.

PART II.

A DICTIONARY
OF
ENGLISH PLANT NAMES.

PART II.

ABBREVIATIONS.—*Sp.*, *species*; *var.*, *variety*; *fl.-pl.*, *flore-pleno*.

Abelmoschus esculentus. *Gobbo*
Aberia Caffra. *Kai, Kau,* or *Kei Apple-tree*
Abies. *Deal-tree, Fir-tree*
— alba. *Single* or *White Spruce*
— alba var. glauca. *Dimsdale's Silver Spruce*
— alba var. minima. *Hedgehog White Spruce*
— alba var. nana. *Dwarf White Spruce*
— Alcockiana. *Alcock Spruce*
— balsamea (A. balsamifera). *Balm of Gilead Fir, Balsam Fir*
— Brunoniana. *Indian Hemlock Spruce*
— canadensis. *Common Hemlock Spruce*
— commutata. *Engelmann's Spruce*
— concolor. *Western White Fir*
— Cunninghami. *Moreton Bay Pine-tree*
— (Pseudotsuga) Douglasii. *Douglas Spruce, Nootka Fir*
— excelsa var. Clanbrasiliana. *Lord Clanbrasil's Dwarf Spruce*
— excelsa var. communis. *Common Norway Spruce, White Fir*
— excelsa var. Finedonensis. *Finedon Hall Spruce*
— excelsa var. pendula. *Weeping Norway Spruce*
— excelsa var. pygmæa. *Dwarf Spruce*
— Fortunei. *Intermediate Fir*
— Fraseri. *Fraser's Double,* or *Southern Balsam, Fir*
— (Larix) Kæmpferi. *Golden Larch*
— magnifica. *Western Red Fir*
— Menziesii. *Menzies' Spruce*
— Mertensiana. *Californian Hemlock Spruce*
— Morinda (A. Smithiana). *Indian Spruce, Prickly Fir*
— nigra. *Black-leaved* or *Double Spruce, Red Fir*
— (Pinus) nobilis. *Tuck-tuch Pine-tree*
— orientalis. *Eastern Spruce*
— Pattoniana. *Patton's Californian Spruce*
— pectinata. *Silver Fir*
— polita. *Tiger's-tail Spruce*
— religiosa. *Mexican Fir*
— rubra. *Newfoundland Red Pine, Red,* or *Arctic Spruce*
— Smithiana (A. Morinda). *Indian* or *Himalayan Spruce*
— Tsuga. *Japan Hemlock Spruce*
— Webbiana. *King Pine*

Abronia umbellata. *Purple Sand-Verbena*
Abrus precatorius. *Coral-bead plant, Love Pea, Red-bead Vine, Rosary-pea Tree, Wild Liquorice*
Abuta rufescens. *White Pareira Brava*
Abutilon. *Indian Mallow*
— Americanum. *Twelve-o'clock-Flower*
— Avicennæ. *American Velvet-leaf, Lantern-Flower*
Acacia acuminata. *Raspberry-jam Wattle*
— arabica. *" Babool"-Bark, Sunt-wood,* and *Gum Arabic-tree*
— arborea. *Wild Tamarind-tree* of Jamaica
— armata. *Kangaroo-Thorn*
— Catechu. *Catechu-tree, Wadalee-gum-tree, Khair-tree*
— Cavenia. *" Cavan"* of Chili
— concinna (Mimosa abstergens). *Soap-pod-tree, Soap-pod Wattle*
— cornigera. *Cuckold-tree*
— dealbata. *Silver Wattle* of Tasmania
— decurrens. *Black Wattle* or *Green Wattle* of Australia
— discolor. *Australian Sunshine Plant*
— Doratoxylon. *Spear-wood* of Australia
— falcata. *Lignum-vitæ* of N. S. Wales
— Farnesiana. *Cassié-oil-plant, Sponge-tree, West Indian Blackthorn*
— Giraffe. *Camelopard's Tree*
— gummifera. *Barbary* or *Morocco Gum-tree*
— homalophylla. *Myall-wood* or *Violet-wood* of Australia
— horrida. *Cape Gum-tree, " Dorn-boom,"* of S. Africa
— Julibrissin. *Bastard Tamarind, Silk-tree* of Constantinople, *Smooth-tree Acacia*
— juniperina. *Prickly Wattle*
— Koa (A. heterophylla (?)) *Koa-tree* of the Sandwich Islands
— Latronum. *Buffalo-thorn*
— leucophlœa. *Kuteera-gum-tree*
— lophantha. *Two-spiked Acacia*
— Melanoxylon. *Black-wood* of N. S. Wales, *Light-wood* of Tasmania
— microbotrya. *" Bad-jong"* of S. W. Australia
— mollissima. *Black Wattle, Silver Wattle*
— moniliformis. *" Tusca"-tree*

Acacia Natalitia. *African Wattle*
Nilotica. *Gum-Arabic-tree, Shittim-wood (?)*
Oxycedrus. *Sharp Cedar*
pendula. *Myall-tree*
penninervis. *Black-wood* of W. Australia
pycnantha. *Golden* or *Green Wattle*
rigens. *Wallaby Wattle*
Sassa. *Abyssinian Myrrh-tree*
Senegal. *Gum-Arabic-tree*
Serissa. *Shireesh,* of India
Soyal. *Gum-Arabic-tree, Thirsty Thorn*
stenocarpa. *Gum-Arabic-tree*
stipuligera. *Scrub Wattle*
Suma. *Shumeo-tree*
tomentosa. *Elephant-thorn, Jungle-nail*
trichophylloides. *Bastard Tamarind* of Jamaica
vera (Mimosa nilotica). *Egyptian Thorn*
Verek. *Gum-Arabic-tree* of Kordofan
villosa. *Yellow Tamarind-tree*
Acæna. *New Zealand Bur*
microphylla (A. Novæ-Zelandiæ). *Rosy-spined N. Zealand Bur*
myriophylla. *Fern-leaved N. Zealand Bur*
ovina. *Sheep-pest* of Australia and New Zealand
Acalypha. *Three-seeded Mercury, Copper-leaf*
rubra. *String-wood* of St. Helena
Acantholimon glumaceum. *Prickly Thrift*
Acanthus. *Bear's-Breech, Bear's-foot, Branke Ursine, Cutherdole,* or *Cutbertill*
latifolius. *Stately Bear's-Breech*
longifolius. *Long-leaved Bear's-Breech*
mollis. *Common* or *Soft-leaved Bear's-Breech*
spinosus. *Spiny Bear's-Breech*
spinosissimus. *Armed Bear's-Breech*
Acer. *Maple-tree*
acuminatum. *Acute-leaved Maple*
atheniense. *Athenian Maple*
barbatum. *Carolina Maple*
campestre. *Common,* or *Field Maple, Dog Oak, Maser-tree*
campestre foliis variegatis. *Variegated Field Maple*
campestre var. austriacum. *Austrian Maple*
campestre var. hebecarpum. *Downy-fruited Field Maple*
circinatum. *Round-leaved Maple, Vine Maple*
coriaceum. *Thick-leaved Maple*
creticum. *Cretan Maple*
dasycarpum. *Silver, Soft,* or *White Maple*
dissectum. *Cut-leaved Maple*
eriocarpum (A. dasycarpum). *Sir Charles Wager's Maple*
Granatense. *Spanish Maple*
Ginnala. *Crimson-leaved Maple*
glaucum. *Glaucous-leaved Himalayan Maple*
heterophyllum (A. sempervirens). *Ever-green Maple*
Hyrcanum. *Hyrcanian Maple*
Ibericum. *Iberian Maple*
lobatum. *Siberian Maple*
Loudoni. *Bohemian Maple, Loudon's Maple*

Acer Monspessulanum. *Montpelier Maple*
Neapolitanum. *Neapolitan Maple*
Negundo (Negundo aceroides). *Ash-leaved Maple, Box-Elder*
Negundo var. crispum. *Curled-leaved Box-Elder*
oblongum. *Oblong-leaved Maple*
obtusatum. *Blunt-lobed,* or *Hungarian Maple*
Opulus. *Italian Maple*
opulifolium. *French,* or *Guelder-rose-leaved Maple*
palmatum. *Japanese Maple*
palmatum var. dissectum atro-purpureum. *Japanese Purple Maple*
Pensylvanicum. *Moose-wood, Striped Dogwood, Striped Maple*
platanoides. *Norway,* or *Platanus-leaved Maple*
platanoides var. laciniatum. *Eagle's-claw Maple*
platanoides var. Lobelii. *Lobel's Platanus-leaved Maple*
platanoides var. pubescens. *Downy Pla-tanus-leaved Maple*
platanoides variegatum. *Silver-variegated Platanus-leaved Maple*
polymorphum var. atro-purpureum. *Crim-son-leaved Japanese Maple*
Pseudo-Platanus. *Mock-Plane, Sycamore Tree*
Pseudo-Platanus var. albo-variegatum. *White-variegated Sycamore*
Pseudo-Platanus var. flavum variegatum. *Yellow-variegated Sycamore*
Pseudo-Platanus var. purpureum. *Purple-leaved Sycamore*
rubrum. *Red-flowered,* or *Swamp, Maple*
saccharinum. *Sugar Maple, Rock Maple*
saccharinum var. nigrum. *Black Sugar-Maple*
saccharinum var. *"Bird's-eye" Maple, "Curled" Maple*
Sikkimense. *Sikkim Maple*
spicatum. *Mountain Maple, Spike-flowered Maple*
striatum. *Moose-wood, Striped - barked Maple, Striped Dog-wood*
tataricum. *Tartarian Maple*
trifidum. *Trifid-leaved Maple*
truncatum. *Truncate-leaved Maple*
villosum. *Woolly-leaved Nepaul Maple*
Wieri laciniatum. *Wier's Cut-leared Maple*
Aceras anthropophora. *Green-man-Orchis, Man-Orchis*
Acerates viridiflora. *Green Milk-weed*
Achania Malvaviscus. *Bastard Hibiscus*
Achillea. *Milfoil, Yarrow*
ægyptiaca. *Egyptian Yarrow*
Ageratum. *Sweet Maudlin, Sweet Milfoil*
alpina. *Alpine Milfoil*
asplenifolia. *Asplenium-leaved Yarrow*
atrata. *Black,* or *Chamomile-leaved, Mil-foil*
aurea. *Golden-flowered Yarrow*
Clavennæ. *Silvery-leaved Yarrow*
decolorans. *Pale-yellow-flowered Yarrow*
Filipendula. *Noble Yarrow*
macrophylla. *Feverfew-leaved Milfoil*

Achillea magna. *Great Milfoil*
Millefolium. *Common Milfoil, Common Yarrow, Hundred-leaved Grass, Nosebleed, Thousand-Seal*
mongolica. *Mongolian Yarrow*
moschata. *Musk Milfoil, Swiss Genipi*
nobilis. *Showy Milfoil*
Ptarmica. *Bastard Pellitory, Fair-Maids-of-France, Goose-tongue, Sneeze-wort*
Ptarmica var. umbellata. *Dwarf Silvery Yarrow*
pubescens. *Downy Milfoil*
rupestris. *Rock Yarrow*
serrata. *Serrate-leaved Yarrow*
squarrosa. *Rough-headed Milfoil*
tomentosa. *Woolly Yarrow*
Achlys triphylla. *Oregon May-apple*
Achras (Sapota) australis. *Australian Brush-apple, Wild Plum, of N. S. Wales*
mammosa. *Surinam Medlar*
Sapota. *Nase-berry, Neese-berry, or Nisberry, Sapodilla or Sapotilla Plum*
Sideroxylon. *Nase-berry Bully-Tree*
Acidoton urens. *Jamaica Cowitch, Cowage, or Cowhage*
Aciphylla Colensoi and A. squarrosa. *Spear-Grass, or "WildSpaniard" of NewZealand*
Acmene floribunda. *Australian Myrtle, Lilly-pilly-tree*
Acnida cannabina. *Water-Hemp, Willow-Hemp*
Acnistus arborescens. *Deadly Dwale*
Aconitum. *Aconite, Monk's-hood, Wolf's-bane*
Anthora. *Jacquin's Yellow-flowered Monks-hood, Wholesome Wolf's-bane, Yellow Helmet-flower*
Arctophonum. *Bear-bane*
autumnale. *Autumn Monk's-hood*
bicolor. *Two-coloured Monk's-hood*
Cammarum. *Purple Wolf's-bane*
chinense. *Chinese Monk's-hood*
ferox. *Bish, Bis, or Bikh-poison, Indian or Nepaul Aconite*
japonicum. *Japanese Monk's-hood*
Lagoctonum. *Hare's-bane*
Lycoctonum. *Wolf's-bane*
Meloctonum. *Badger's-bane*
Myoctonum. *Mouse-bane*
Napellus. *Bear's-foot, Common Aconite, Friar's, Soldier's, or Turk's Cap, Helmet-flower, Luckie's Mutch, Common Monk's-cowl or Monk's-hood*
paniculatum. *Panicled Monk's-hood*
pyramidale. *Blue Rocket*
pyrenaicum. *Pyrenean Monk's-hood*
reclinatum. *Trailing Wolf's-bane*
septentrionale. *Northern Monk's-hood*
Theriophonum. *Beast's-bane*
Tragoctonum. *Goat's-bane*
uncinatum. *American Wolf's-bane*
variegatum. *Variegated-flowered Monk's-hood*
Vulparia. *Fox-bane*
Acorus Calamus. *Myrtle Flag, Myrtle Grass, Myrtle Sedge, Sweet Flag*
gramineus. *Grass-leaved Sweet-Flag*
gramineus variegatus. *Variegated Sweet-Flag*

Acrocomia fusiformis (A. sclerocarpa). *Gru-Gru, or Macaw, Palm*
lasiospatha. *Great Macaw-tree, Macuja or Mucuja Palm*
mexicana. *Coyoli Palm*
sclerocarpa. *Macuja Palm*
Acrodiclidium Camara. *Ackawai Nutmeg-tree*
jamaicense. *Mountain, or Timber, Sweetwood*
Acronychia lævis. *Yellow-wood, of Moreton Bay*
Acrostichum alcicorne. *Elk's-hornFern*
Actæa alba. *Toad-root, White Bane-berry, White-and-red Cohosh*
racemosa. *Black Snake-root, Bug-bane, Cohosh*
spicata. *Common Bane-berry, Grape-wort, Herb Christopher*
spicata var. rubra. *Red Bane-berry*
Actinella grandiflora. *Pigmy Sunflower*
scaposa. *Tufted Dwarf Sunflower*
Actinocarpus. *Thrum-wort*
(Alisma) Damasonium. *Water-star, Raypod*
Actinostrobus. *Swan River Cypress*
Adansonia digitata. *Baobab-tree, Ethiopian Sour Gourd, Indian Cork-tree, Monkey-bread*
Gregorii. *Cream-of-Tartar-tree, Gouty-stemmed-tree, Sour Gourd, of Australia*
Adenanthera pavonina. *Barbadoes Pride, Barricari-seed-plant, Coral Pea-tree, Peacock Flower-Fence, Red Sandal-wood*
Adenium Honghel. *Honghel-bush*
namaquanum. *Elephant's-trunk*
Adenocarpus. *Gland-pod*
Adenophora. *Gland Bell-flower*
Adenostephanus organensis. *Organ Mountain Shelter-tree*
Adenostoma. *"Chamiso," of California*
Adhatoda cydoniæfolia. *Brazilian Bower-plant*
Vasica. *Malabar Nut-tree*
Adiantum æthiopicum. *"English Maidenhair" Fern, of New Zealand*
Capillus-veneris. *Black Maiden-hair, Capillaire, Dudder Grass, Maiden-hair, or Maiden-hair Fern, Venus's-hair*
dolabriforme. *Miniature Basket-Fern*
Farleyense. *Great Maiden-hair Fern*
formosum. *Branching Maiden-hair Fern*
pedatum. *American Maiden-hair Fern*
rubellum. *Roseate Maiden-hair Fern*
Adlumia cirrhosa. *Alleghany Vine, Climbing Colic-weed, Climbing Fumitory, Mountain Fringe-Vine*
Adonis æstivalis. *Summer Pheasant's-eye*
autumnalis. *Adonis-flower, Autumn Pheasant's-eye, Red Chamomile, Red Morocco, Rose-a-Ruby*
pyrenaica. *Pyrenean Adonis*
vernalis. *Ox-eye*
Adoxa Moschatellina. *Hollow-root, Moschatel, Musk-root, Tuberous Crow-foot, Tuberous Moschatel*
Æcidium Berberidis. *Barberry Fungus*
Ægiceras majus. *Queensland Salt Lake Tree*

Ægilops. *Goat-grass, Hard-grass*
ovata. The reputed original of cultivated Wheat
Ægle Marmelos. *Bael,* or *Bhel-fruit Tree, Bengal Quince, Golden Apple*
Ægopodium Podagraria. *Ach-weed, Ash-weed, Ax-weed, Bishop's Elder, Bishop's Weed, Goat-weed, Gout-weed, Gout-wort, Ground Ash, Herb Gerard*
Podagraria variegatum. *Variegated Gout-weed*
Æonium arboreum. *House-leek Tree*
Aeranthes grandiflorum. *Air-flower, Air-plant*
Aerides odorata. *Fragrant Air-plant*
Æschynanthus. *Blush-wort*
Æschynomene aspera. *Pith-hat plant, Singapore Hat-plant, Shola,* or *Solah-plant, Sponge-wood,* of Bengal
(Agati) grandiflora. *West Indian Pea-tree*
hispida. *Sensitive Joint-Vetch*
montevidensis. *Humming-bird Bush*
(Pavia) californica. *Californian Buck-eye*
(Pavia) carnea. *Red-flowered Buck-eye*
(Pavia) flava. *Big Buck-eye, Sweet Buck-eye*
(Pavia) glabra. *Fetid Buck-eye, Ohio Buck-eye*
Hippocastanum. *Horse-Chestnut, Oblion-ker-tree*
Hippocastanum flore-pleno. *Double-flowered Horse-Chestnut*
Hippocastanum var. pallida. *Pale-flowered Horse-Chestnut*
Hippocastanum var. rubra. *Red-flowered Horse-Chestnut*
Hippocastanum var. rubra fl.-pl. *Double-flowered Red Horse-Chestnut*
Æthionema coridifolium (Iberis jucunda). *Lebanon Candytuft*
saxatile. *Candy Mustard*
Æthusa Cynapium. *False Parsley, Fool's Cicely, Fool's, Asses',* or *Dog's Parsley, Lesser Hemlock*
Fatua. *Fine-leaved Fool's Parsley*
Agalmyla longistyla. *Scarlet Root-blossom*
Agapanthus minor. *Small African Lily*
umbellatus. *Blue African Lily*
umbellatus var. albus. *White African Lily*
Agaricus arvensis. *Champillion, Cheese-room, Flaps, Horse-mushroom, Mushroom Snow-ball*
Cæsareus. *Imperial Mushroom*
campestris. *Common Mushroom*
gambosus. *St. George's Mushroom*
muscarius. *Bug Agaric, Fly-bane*
nebularis. *Clouded Mushroom*
olearius. *Luminous Mushroom*
Orcella. *Plum Mushroom, Vegetable Sweet-bread*
(Marasmius) Oreades. *Champignon, Fairy-ring Mushroom, Scotch Bonnets*
ostreatus. *Oyster Mushroom*
personatus. *Blewits, Blue-hats, Blue-legs*
procerus. *Parasol,* or *Scaly Mushroom*
Prunulus. *Autumn Mousseron, Plum Mushroom*

Agaricus rubescens. *Brown Warty,* or *Red-fleshed Mushroom*
Agarum Turneri. *Sea-Colander*
Agathæa amelloides. *Cape Aster*
cœlestis. *Blue Marguerite*
spathulata. *Grass-Thistle*
Agathophyllum aromaticum. *Clove Nut-meg,* of Madagascar, *Ravensara-nut Tree*
Agathyrus. *Blue-flowered Sow-thistle*
Agave americana. *American Aloe, Century-plant*
foetida. *Fetid Aloe*
lurida. *Vera Cruz Aloe*
mexicana and other species. *Maguey-fibre plant*
saponaria. *Mexican Soap-plant*
Sissalana. *Sisal Hemp-plant*
utahensis. *Utah Aloe*
Victoriæ-Reginæ. *Queen Victoria Century-plant*
Virginica. *False Aloe, Rattlesnake's-Master*
Ageratum. *Floss-flower*
conyzoides. *Bastard Agrimony*
Aglaia odorata. *Chu-lan Tree*
Aglaonema commutata. *Poison-dart,* of the Philippine Islands
Agrimonia agrimonioides. *Three-leaved Agrimony*
Eupatoria. *Cockle-bur, Stickle-wort, Common Agrimony, Liver-wort*
odorata. *Sweet-scented Agrimony*
repens. *Creeping Agrimony*
Agrostemma (Lychnis) coronaria. *Rose Campion*
(Lychnis) Coronaria fl.-pl. *Double-blossomed Rose Campion*
Githago. *Bastard Nigella, Corn-cockle, "Wild Savager"*
Agrostis. *Cloud-grass, Spear-grass*
alba. *Fine-top-grass*
elata. *Thin-grass*
nebulosa and A. pulchella. *Ornamental Cloud-grass*
perennans. *Thin-grass*
scabra. *Fly-away Grass, Hair-grass,* of N. America
setacea. *Deer's-foot Grass*
Spica-venti. *Corn-grass, Wind-grass, Windle-straw*
stolonifera. *Florin Grass, Orcheston Grass*
vulgaris. *Bent Grass, Black Quitch Grass, Herd's-grass,* or *Red-top,* of N. America, *Monkey's Grass*
Ailantus glandulosa. *Ailanto, False Var-nish-tree, Japan Varnish-tree, Tree of Heaven, Tree of the Gods*
Aiphanes corallina. *Grigi Palm*
Aira. *Hair-grass*
cæspitosa. *Bent-grass, Hassock-grass, Tussock-grass*
caryophylla. *Mouse-grass*
flexuosa. *Bent-grass*
Ajuga alpina. *Alpine Bugle*
Chamæpitys. *Field Cypress, Forget-me-not, Gout Ivy, Ground Pine, Yellow Bugle*
genevensis. *Erect Bugle*

Ajuga Iva. *Herb Eve or Ivy, Musky Bugle*
pyramidalis. *Pyramidal Bugle*
reptans. *Brown Bugle, Middle Comfrey, Middle Consound*
reptans variegata. *Variegated - leaved Bugle*
Alaria esculenta. *Badderlocks, Hen-ware, Honey-ware*
Albizzia Lebbek. *Siris-Acacia*
Alchemilla arvensis. *Bowel-hive-grass, Fire-grass, Parsley Piert*
sericea. *Silky Lady's-mantle*
vulgaris. *Bear's-foot, Duck's-foot, Great Sanicle, Lady's Mantle, Lion's-foot, Lion's-paw*
Alchornea latifolia. *Dove-wood of the W. Indies*
Alectoria jubata. *Rock-hair Moss*
sarmentosa. *Jaffna Moss*
Alectra brasiliensis. *Cane-killer*
Alectryon excelsum. *New Zealand Oak*
Aletris aurea. *Golden-tipped Star-grass*
farinosa. *Ague-root, Colic-root, Common Star-grass, Crow-corn, Unicorn-root*
Aleurites cordata. *Chinese Wood - Oil-plant, Tung-oil-plant*
laccifera. *Gum Lac Tree*
moluccana. *Otaheite Walnut*
moluccensis. *Candle-nut-tree*
triloba. *Bancoul, Kekune, Kekui*, or *Lumbang-oil-plant, Candle-nut Tree, Indian, Belgaum*, or *Country Walnut, Otaheite Walnut*
Algæ. A general name for *Sea-weeds*
Alhagi Camelorum. *Alhagi Manna-plant, Camel's-thorn*
Maurorum. *Hebrew* or *Persian Manna-plant*
Alibertia edulis. *" Marmeladinha "*
Alisma natans. *Floating Water-Plantain*
Plantago. *Common Water-Plantain, Devil's-spoons, Mad-dog Weed*
ranunculoides. *Small Water-Plantain*
Alliaria officinalis. *Garlic Mustard, Garlic-wort, Hedge Garlic, Jack-by-the-hedge*
Allium. *Garlic, Leek, Onion*
acuminatum. *Purple-flowered Garlic*
alpinum. *Alpine Garlic*
Ampeloprasum. *Blue Leek, Great-headed Garlic, Levant Garlic, Pearl Onion, Vine Leek, Wild Leek*
Ascalonicum. *Shallot*
Ascalonicum var. majus. *Scallion*
azureum. *Azure-flowered Garlic*
canadense. *Wild Garlic of N. America*
carinatum. *Keeled Garlic*
Cepa. *Common Onion*
Cepa var. aggregatum. *Burn Onion, Potato Onion*
Cepa var. proliferum (bulbiferum). *Canada Onion, Tree Onion*
cernuum. *Drooping-flowered Garlic*
Chamæmoly. *Dwarf Moly*
ciliatum. *Fringed Garlic*
cœruleum. *Blue-flowered Garlic*
falcifolium. *Sickle-leaved Garlic*
fistulosum. *Ciboul, Stone Leek, Welsh Onion*

Allium flavum. *Yellow-flowered Garlic*
(Nothoscordon) fragrans. *Sweet-scented Garlic*
grandiflorum. *Large-flowered Garlic*
inodorum. *Carolina Garlic*
leptophyllum. *Himalayan Onion*
lusitanicum. *Perennial Ciboul*
Moly. *Golden-flowered Garlic, Lily Leek, Moly, Sorcerer's Garlic*
Murrayanum. *Murray's Garlic*
neapolitanum. *Daffodil Garlic*
nigrum. *Black Garlic*
odorum. *Fragrant-flowered Garlic*
oleraceum. *Field Garlic*
Ophioscorodon. *Rocambole*
paradoxum. *Quaint Garlic*
pedemontanum. *Piedmont Garlic*
Porrum. *Common Leek*
roseum. *Rosy-flowered Garlic*
sativum. *Clown's Treacle, Common Garlic, Poor-Man's Treacle*
Schœnoprasum. *Chives*
Scorodoprasum. *Sand Leek, Spanish Garlic, Wild Rocambole*
sphærocephalum. *Round-headed Garlic*
subhirsutum. *Moly, of Dioscorides*
tricoccum. *Wild Leek, of N. America*
triflorum. *American Mountain Leek*
triquetrum. *Triangular-stemmed Garlic*
uniflorum. *One-flowered Garlic*
ursinum. *Bear's* or *Wild Garlic, Buck-rams, Gipsy Onion, Hog's Garlic, Ramsons, Wild Leek*
Victoriale. *Long-rooted Garlic*
vineale. *Crow Garlic, Crow Onion, Stag's Garlic*
Wallichianum. *Wallich's Garlic*
Allosorus acrostichoides. *Oregon Rock Brake*
crispus. *Curled Rock Brake, Parsley Fern*
Alnus. *Alder*
glutinosa. *Aar, Aul, Common Alder*
cordifolia. *Heart-leaved Alder*
imperialis asplenifolia. *Fern-leaved Alder*
incana. *Black, Hoary,* or *Speckled Alder*
integrifolia. *Kokra-wood,* of India
maritima. *Sea-side Alder*
oblongata. *Turkey Alder*
rhombifolia. *Californian Alder*
rubra. *Californian Red Alder*
serrulata. *Smooth Alder*
viridis. *Green,* or *Mountain Alder*
Aloe barbadensis. *Barbadoes Medicinal Aloes-tree*
dichotoma. *Quiver-tree*
ferox. *Cape Aloe*
margaritifera. *Pearl Aloe*
prolifera. *Proliferous Aloe*
socotrina. *Socotrine Aloes-tree*
variegata. *Partridge-breast Aloe*
vulgaris. *Yellow-flowered Aloe*
Aloexylon Agallochum. *Aloes-wood, Calambac-tree, Eagle-wood*
Alonsoa incisifolia. *Cut-leaved Mask-flower*
linearis. *Narrow-leaved Mask-flower*
Alopecurus agrestis. *Black Grass, Hunger Grass, Land Grass, Mouse-tail Grass*
geniculatus. *Black Grass, Elbowit Grass*
pratensis. *Fox-tail Grass*

Alopecurus pratensis variegatus. *Gold-striped Fox-tail Grass*
Aloysia citriodora. *Lemon-scented Verbena, Herb Louisa*
Alphitonia excelsa. *Cooper's-wood, Red Ash*
Alpinia Cardamomum. *Bastard Cardamom*
Galanga. *Galingale* (true)
(Hellenia) cœrulea. *Brush Shell-flower*
nutans. *Indian Shell-flower*
officinarum. The rhizome of this plant forms the *Galangal* of medicine
Alsophila. *Grove Fern*
australis. *Hill Tree-Fern*
contaminans. *Glaucous Tree-Fern*
excelsa. *Norfolk Island Tree-Fern*
Leichardtiana. *Whip-stick Fern*
Rebeccæ. *Slender Tree-Fern*
Alstonia scholaris. *Devil's-tree, Dita-bark-tree*
Alstrœmeria. *Herb Lily*
Flos-Martini. *St. Martin's-flower*
pelegrina. *Lily of the Incas*
Alternanthera. *Joy-weed*
Achyrantha. *Chaff-flower*
Althæa. *Hollyhock*
ficifolia. *Antwerp Hollyhock*
Narbonnensis. *Narbonne Hollyhock*
officinalis. *Guimauve, Marsh - Mallow, White Mallow*
rosea. *Common Hollyhock, Holy Hoke*
Alyssum. *Mad-wort*
alpestre. *Alpine Mad-wort*
calycinum. *Small Yellow Alysson*
deltoideum. *Purple Mad-wort*
denticulatum. *Toothed-leaved Mad-wort*
gemonense. *Hoary German Mad-wort*
maritimum. *Sweet Alysson, or Allison*
montanum. *Mountain Mad-wort*
olympicum. *Dwarf Mad-wort*
orientale. *Eastern Mad-wort*
saxatile. *Gold Basket, Gold-dust, Golden-tuft, Rock Mad-wort*
saxatile variegatum. *Variegated - leaved Mad-wort*
saxatile var. compactum. *Dense-flowered Mad-wort*
spinosum. *Spiny Mad-wort*
utriculatum. *Bladder-podded Mad-wort*
Wiersbeckii. *Wiersbeck's Mad-wort*
Alyxia. *Brushland Box* of Australia
buxifolia. *Tasmanian Scent-wood, Heath-Box-tree* of Australia, *Tonga - bean-wood*
Amarantus. *Amaranth*
Blitum. *Wild Blite*
caudatus. *Floramor*, or *Florimer, Love-lies-bleeding, Thrum-wort, Velvet-flower*
hypochondriacus. *Prince's-Feather*
olcraceus. *Edible Amaranth*
salicifolius. *Fountain-plant, Golden Amaranth*
speciosus. *Showy Amaranth*
spinosus. *Prickly Calalu, Thorny Amaranth*
tricolor. *Floramor*, or *Florimer, Flower Gentle, Joseph's Coat, Three-coloured Amaranth, "Tricolor"*

Amaryllis. *Daffodil-Lily*
Atamasco. *Atamasco Lily*
(Lycoris) aurea. *Golden Lily*
Belladonna. *Belladonna Lily*
(Hippeastrum) equestris. *Barbadoes Lily*
formosissima. *Jacobea Lily, St. James's-Cross Lily*
ornata. *Cape Coast Lily*
Reginæ. *Mexican Lily, Queen's Lily*
Amberboa (Centaurea) moschata. *Purple Sweet-Sultan*
suaveolens. *Yellow Sweet-Sultan*
Ambrina ambrosioides (Ambrosia maritima). *"Cappadocian Oak," Demigods' Food, Hedge Mustard* of the W. Indies
anthelmintica. *Worm-seed-oil-plant*
artemisiæfolia. *Bitter-weed, Hog - weed, Roman Worm-wood* of N. America
maritima (Ambrina ambrosioides). *"Cappadocian Oak"*
trifida. *Great Rag-weed*, of N. America
Amelanchier Botryapium. *Grape-pear, Snowy Mespilus*
canadensis. *Shad-bush, June-berry, Service-berry*
florida. *Flowery Amelanchier*
ovalis. *Medlar-bush, Oval-leaved Amelanchier*
sanguinea. *Red-branched Amelanchier*
vulgaris. *Common Amelanchier*
Amianthium muscætoxicum. *American Fall-poison*, or *Fly-poison*
Ammania humilis. *Tooth-cup*
Ammi copticum. *Prickly-seeded Bishop's-weed, True Bishop's-weed*
majus. *Amcos, Bull-wort, Common Bishop's-weed, Wood-nep, Herb William*
Visnaga. *Spanish Tooth-pick, Tooth-pick Bishop's-weed*
Ammobium alatum. *Winged Everlasting, Winged-stalked Sand-flower*
Ammophila arenaria. *Sea Mat-weed*
Amomum aromaticum. *Bengal Cardamoms*
Cardamomum. *Round*, or *Clustered Cardamoms*
maximum. *Jura Cardamoms*
Melegueta. *Malaguetta*, or *Melegueta, Pepper*. The seeds of this plant are the "Grains of Paradise," of medicine
xanthoides. *Bastard Cardamoms*, of Siam
Amorpha canescens. *Lead-plant, Wild Tea* of N. America
crocco - lanata. *Yellow-woolled False-Indigo*
fragrans. *Fragrant False-Indigo*
fruticosa. *Shrubby False-Indigo*
glabra. *Smooth False-Indigo*
herbacea (A. pubescens). *Downy False-Indigo*
nana. *Dwarf False-Indigo*
Amorphophallus campanulatus. *Telinga Potato*
Ampelopsis bipinnata. *Pepper-vine, Two-winged Virginian Creeper*
cordata. *Heart-leaved Virginian Creeper*
dissecta. *Cut-leaved Virginian Creeper*

Ampelopsis hederacea (A. quinquefolia). *American Ivy, Common Virginian Creeper, False Grape, Five-leaves, Wild Wood-vine, Wood-bine*
heterophylla marmorata. *Variegated Virginian Creeper*
sempervirens. *Evergreen Virginian Creeper*
Veitchi. *Veitch's Virginian Creeper*
Amphicarpæa monoica. *Hog-pea-nut, Wild Bean-Vine*
Amygdalus communis. *Common Almond-tree*
communis var. amara. *Bitter Almond*
communis var. macrocarpa. *Long-fruited Almond*
communis var. pendula. *Weeping Almond*
communis var. persicoides. *Almond Peach*
incana. *Woolly Almond*
nana. *Dwarf Almond*
orientalis (A. argentea). *Silvery-leaved Almond*
persica sinensis camelliæflora. *Camellia-flowered Peach*
persica sinensis caryophylliflora. *Carnation-flowered Peach*
persica sinensis fl.-pl. albo. *Double White Chinese Peach*
persica sinensis fl.-pl. sanguineo. *Double Crimson Chinese Peach*
persica vulgaris. *Common Peach*
persica vulgaris flore-pleno. *Double-blossomed Peach*
persica vulgaris foliis purpureis. *Purple-leaved Peach*
persica vulgaris foliis variegatis. *Variegated-leaved Peach*
persica vulgaris fructû-pleno. *Double-fruited Peach*
persica vulgaris var. alba. *White-flowered Peach*
persica vulgaris var. compressa. *Flat-fruited Peach*
persica vulgaris var. hispanica. *Spanish Peach*
persica vulgaris var. Nectarina (A. p. lævis). *Nectarine-tree*
prostrata. *Prostrate Almond*
pumila. *Double-flowered Dwarf Almond*
sibirica. *Siberian Almond*
Amyris. *Shrubby Sweet-wood*
balsamifera. *Candle-wood, or Rhodes-wood, of the W. Indies, Jamaica Rose-wood, Lignum Rhodium, Mountain Torch-wood*
commiphora (Balsamodendron Roxburghii). *False Myrrh*
elemifera (A. Plumieri). *Gum-Elemi Tree*
Floridana. *Florida Balsam Tree*
toxifera. *Janca-tree*
sylvatica. *Torch-wood Tree*
Anacardium. *Monkey-nut*
occidentale. *"Bean of Malacca," Cashew-nut Tree*
rhinocarpus. *Wild Cashew*, of Guiana
Anacampseros. *Love-plant*
Anacharis Alsinastrum (Elodea canadensis). *American Water-weed, Babington's-curse, Choke Pond-weed, Little Snake-weed, Water-Thyme*

Anacyclus Pyrethrum. *Alexander's-foot, Bertram, Long-wort, Pellitory*, of Spain
Anæctangium. *Branched Beardless-Moss*
Anæctochilus setaceus. *"King of the Woods" Orchid, King-plant*
Anagallis arvensis. *Poor-man's-Weather-glass, Red Pimpernel, Shepherd's-Clock*
grandiflora (A. collina, A. fruticosa). *Large-flowered Pimpernel*
indica. *Indian Pimpernel*
Monelli. *Italian Pimpernel*
tenella. *Bog Pimpernel*
Anagyris. *Bean Trefoil*
fœtida. *Stinking-wood*
Anamirta (Menispermum) Cocculus. *Cocculus-indicus-plant*
Ananassa sativa. *Pine-apple*
Anastatica Hierochuntina. *Resurrection-plant, Rose of Jericho*
Anatherum bicorne. *Ridging-grass, W. Indian Foxtail-grass*
macrurum. *W. Indian Foxtail-grass*
Anchusa. *Alkanet, Sea-Bugloss*
arvensis. *Small Bugloss*
capensis. *Cape Alkanet, Cape Forget-me-not*
hybrida. *Hybrid Alkanet*
italica. *Italian Alkanet*
sempervirens. *Evergreen Alkanet*
Andira inermis. *Bastard, or W. Indian Cabbage-tree, Partridge-wood (?), Worm-bark-tree*
violacea. *Violet-wood,* of Guiana
Andrachne telephioides. *Bastard,* or *False, Orpine*
Andreæa. *Split-Moss*
Andrographis (Justicia) paniculata. *Creyat, or Kariyat-plant*
Andromeda acuminata. *American Pipe-stem-wood*
arborea. *Sorrel-tree, Sour-leaved Elk-tree*
buxifolia. *Box-leaved Andromeda*
calyculata. *Leather-Leaf*
Catesbæi. *Spiny-leaved Andromeda*
coriacea. *Thick-leaved Andromeda*
fasciculata. *Bundle-flowered Andromeda*
(Cassiope) fastigiata. *Himalayan Andromeda, Himalayan Heather*
floribunda. *Lily-of-the-Valley Tree, Free-flowering Andromeda*
hypnoides. *Moss-like Andromeda*
jamaicensis. *Jamaica Andromeda*
japonica. *Japanese Andromeda*
Leschenaulti. *Carbolic-acid-plant*
Mariana. *Lamb-kill Stagger-bush*
ovalifolia. *Oval-leaved Andromeda*
polifolia. *Marsh Holy Rose, Marsh Rosemary, Moor-wort, Wild Rosemary* of N. America
racemosa. *Branching Andromeda, Pepper-bush* of N. America
speciosa. *Large-flowered Andromeda*
spicata. *Spike-flowered Andromeda*
(Cassiope) tetragona. *Square-stemmed Andromeda*
Andropogon. *Beard-grass*
bicornis. *Mountain-grass,* of Jamaica
citratus. *Lemon-grass,* or *Siri-oil-plant*
Gryllus. *Brush-grass*

Andropogon halepensis. *Cuba-grass*
Iwarancusa. *Roosa, or Roussa-grass, of India*
laniger. *Sweet Rush*
Martini. *Roosa, or Roussa-grass, of India*
muricatus. *Cus-cus, Khus-khus, or Vettiver Oil-plant*
Nardus. *Ginger-grass Oil-plant, Indian Geranium*
Schœnanthus. *Camel's-Hay, Geranium Grass, Namur, Nemaur, or Spikenard Oil-plant, Lemon-grass, Palmarosa Oil-plant, "Squinant," Sweet Rush*
scoparius. *Broom Grass*
Androsace Chamæjasme. *Rock-Jasmine*
Anemia. *Flower-Fern*
phyllitidis. *Ash-leaf Flower-Fern*
Anemone. *Wind-flower*
alba. *White Wind-flower*
alpina. *Alpine Wind-flower*
apennina. *Apennine Wind-flower*
blanda. *Winter Wind-flower*
capensis. *Cape Anemone*
caroliniana. *Carolina Wind-flower*
coronaria and vars. *Common Garden Anemone, Poppy Wind-flower*
cylindrica. *Long-fruited Wind-flower*
deltoidea. *Oregon Wind-flower*
dichotoma. *Forked Wind-flower*
elegans. *Pale Japanese Wind-flower*
fulgens. *Pau Anemone, Scarlet Wind-flower*
Halleri. *Haller's Pasque-flower*
Hepatica. *Common Hepatica, Noble Liverwort*
Hepatica var. angulosa. *Large Hepatica*
japonica. *Japanese Wind-flower*
japonica alba. *White Japanese Wind-flower*
japonica vitifolia. *Vine-leaved Japanese Wind-flower*
montana. *Mountain Pasque-flower*
multifida. *Rocky Mountain Wind-flower*
narcissiflora. *Narcissus-flowered Anemone*
nemorosa. *Wood Anemone, Wood Wind-flower*
nemorosa plena. *Double White Wood Wind-flower*
palmata. *Cyclamen-leaved Wind-flower*
parviflora. *Small-flowered Anemone*
patens var. Nuttalliana. *American Pasque-flower*
pavonina. *Peacock Wind-flower*
pennsylvanica. *Pennsylvanian Wind-flower*
Pulsatilla. *Bluemony, Coventry Bells, Danes'-Blood, Danes'-Flower, Flaw-flower, Pasque-flower, Passe-flower*
ranunculoides. *Wood-Ginger, Yellow Wood Anemone*
rivularis. *River-side Wind-flower*
Robinsoni. *Robinson's Wind-flower*
stellata. *Star Wind-flower*
sulphurea. *Sulphur-coloured Wind-flower*
sylvestris. *Snowdrop Anemone*
trifolia. *Three-leaved Wind-flower*
vernalis. *Shaggy Wind-flower*
virginica. *Virginian Wind-flower*

Anemopsis californica. *Yerba Mansa, of California*
Anethum graveolens. *Anet, Dill, or Dill-seed*
Sowa (Peucedanum graveolens). *Sowa-plant, of Bengal*
Angelica sylvestris. *Ground Ash, Holy Ghost, Wild Angelica*
Angiopteris. *Turnip-Fern*
Angophora. *Australian Mahogany, Gum Myrtle*
lanceolata. *Apple-tree, of Victoria*
subvelutina. *Apple-tree, of N. S. Wales*
Angræcum fragrans. *Bourbon, or Fuham Tea-plant*
sesquipedale. *Long-spurred Orchid*
Anigozanthus. *Swan River Funeral-flower*
Manglesii. *Kangaroo's-foot Plant*
flavidus. *Australian Yellow Sword-lily*
Anisomeles malabarica. *Malabar Cut-mint*
Anisophyllum laurinum. *Monkey-apple, of Sierra Leone*
Anoda Dilleniana. *Blue Hibiscus*
Anomatheca cruenta. *Flowering Grass*
Anona. *Custard-apple*
Cherimolia. *Cherimoyer-tree, Peruvian Custard-apple*
hexapetala. *Long-leaved Custard-apple*
muricata. *Prickly Custard-apple, Sour-sop*
palustris. *Alligator Apple, Marsh Cork-wood, Monkey-apple of the West Indies, Shining-leaved Custard-apple*
reticulata. *Bullock's-heart, Netted Custard-apple*
squamosa. *Scaly Custard-apple, Sugar Apple, Sweet-sop*
Anopterus glandulosa. *Tasmanian Laurel*
Anredera scandens. *W. Indian Buck-wheat*
Antennaria. *Cat's-ear*
carpatica. *Carpathian Everlasting*
dioica. *Moor Everlasting, Mountain Cat's-foot, Mountain Cud-weed, Purple Mountain Cotton-weed*
dioica var. minor. *Rosy-flowered Mountain Everlasting*
margaritacea. *Pearl Cud-weed, Pearly Everlasting*
plantagiuifolia. *Mouse-ear Everlasting, Plantain-leaved Everlasting, Pussy's-foot*
tomentosa. *Silvery Cud-weed*
Anthacanthus microphyllus. *Tears of St. Peter*
Anthemis Aizoon. *Silvery Chamomile*
arvensis. *Corn Chamomile*
Chia. *Cut-leaved Chamomile*
Cotula. *Camowyne, Camowyne, Dog's Chamomile, Dog's-fennel, Madders, Mathes, Mawther, May-weed, Stinking Chamomile, Stinking May-weed*
inodora fl.-pl. *Double Scentless May-weed*
nobilis. *Camomile, or Chamomile, Camoryne or Camowyne, Roman Chamomile, Scotch Chamomile*
tinctoria. *Yellow-flowered Chamomile*

Anthericum (Czackia) Liliago. *Branched Spider-wort, St. Bernard's Lily* (Czackia, Paradisia) Liliastrum. *Great Savoy Spider-wort, St. Bruno's Lily*
Liliastrum majus. *Giant St. Bruno's Lily*
Anthistiria ciliata (A. australis). *Kangaroo-grass*
Anthocercis. *Australian Ray-flower*
Antholyza (Gladiolus) æthiopica. *African Corn-Flag*
Anthospermum æthiopicum. *Amber-tree*
moschatum. *Tasmanian Sassafras-tree*
Anthoxanthum odoratum. *Vernal or Sweet Vernal Grass*
Anthriscus. *Beaked Parsley*
Cerefolium. *Chervil*
sylvestris. *Cow-parsley, Cow-weed, Deil's-Meal, Mock Chervil, Orchard-weed, Wild Caraway, Wild Chervil, Wild Cicely*
vulgaris. *Bur-Chervil*
Anthurium. *Banner-plant, Flamingo-plant, Tail-flower*
Andréanum. *André's Flamingo-plant*
Bakeri. *Baker's Flamingo-plant*
candidum. *White-flowered Flamingo-plant*
Lindigi. *Lindig's Flamingo-plant*
magnificum. *Splendid Flamingo-plant*
Palmeri. *Palmer's Flamingo-plant*
Saundersi. *Saunder's Flamingo-plant*
Scherzerianum. *Scherzer's Flamingo-plant*
Veitchii. *Veitch's Flamingo-plant*
Warocqueanum. *Warocque's Flamingo-plant*
Antiaris (Lepurandra) saccidora. *Sack-tree,* of Western India
toxicaria. *Malay Arrow-poison, Upas-tree,* of Java
Antigonon leptopus. *Mountain Rose,* of the W. Indies
Antirrhinum. *Snap-dragon*
Asarina. *Heart-leaved Snap-dragon*
majus. *Common Snap-dragon, Dragon's-mouth, Lion's-mouth*
Orontium. *Calf's-Snout, Small Snap-dragon*
rupestre. (Linaria rupestris). *Rock Snap-dragon*
Anthyllis. *Kidney-Vetch*
Barba-Jovis. *Jove's-beard,* or *Jupiter's-beard, Silver-bush*
erinacea. *Hedge-hog-plant, Rushy Kidney-Vetch*
montana. *Mountain Kidney-Vetch*
Vulneraria. *Common Kidney-Vetch, Lady's Fingers, Lamb's-toe, Wound-wort*
Vulneraria var. rubra. *Pink-flowered Kidney-Vetch*
Antidesma. *Nigger's-cord*
Antigonon. *Beetle-nut plant*
Anychia (Queria) dichotoma. *Forked Chick-weed*
Apargia. *Hawk-bit*
serotina. *Cat's-tongue*
Apera (Agrostis) Spica-venti. *Corn-grass, Wind-grass, Windle-straw*

Aphyllanthes monspeliensis. *Lily-pink,* of Montpelier
Aphyllon. *Naked Broom-rape,* of N. America
uniflorum. *One-flowered Cancer-root*
Apios tuberosa. *American Ground-nut, Mic-Mac Potato, Tuberous-rooted Wistaria,* "*Wild Bean*" of N. America
Apium. *Celery*
australe. *New Zealand Celery*
graveolens. *Ache, Common Celery, Marche, Marsh Parsley, Smallage* or *Smalledge, Wild Celery*
graveolens var. rapaceum. *Celeriac,* or *Turnip-rooted Celery*
graveolens rapaceum var. tricolor. *Three-coloured Celeriac*
prostratum. *Australian Celery*
Aplophyllum. *Simple-leared Rue*
Aplotaxis auriculata. "*Costus,*" of the Ancients
Aplectrum hyemale. *Adam-and-Eve. American Putty-root*
Apocynum. *Dog's-bane*
androsæmifolium. *Fly-trap,* of N. America.
Spreading Dog's-bane
cannabinum. *Canada,* or *Indian, Hemp*
Venetum. *Venetian Dog's-bane*
Aponogeton distachyon. *Cape Asparagus, Cape Pond-weed, Hawthorn-scented Pond-weed*
Aporosa (Lepidostachys) Roxburghi. *Kokrawood*
Aquilaria Agallochum. *Agila-wood, Eagle-wood, Uggur-oil-plant*
ovata. *Eagle-wood,* of Malacca
Aquilegia. *Columbine*
alpina. *Alpine Columbine*
Burgeri. *Burger's Columbine*
cœrulea. *Rocky Mountain Columbine*
californica (A. truncata). *Californian,* or *Large Scarlet, Columbine*
canadensis. *Canadian Columbine*
chrysantha. *Golden-flowered Columbine*
fragrans. *Sweet-scented Columbine*
glandulosa. *Altaian Columbine*
grata. *Mauve-coloured Columbine*
leptocerus lutea. *Yellow Long-spurred Columbine*
longibracteata. *Long-bracted Columbine*
longissima. *Longest-spurred Columbine*
pyrenaica. *Pyrenean Columbine*
Skinneri. *Skinner's Columbine*
truncata (A. californica). *Large Scarlet Columbine*
viridiflora. *Green-flowered Columbine*
vulgaris. *Common Columbine*
vulgaris alba. *White-flowered Columbine*
Arabis. *Rock-cress*
albida. *Early-flowering White Rock-cress, Lady's-Cushion, Mountain Snow*
alpina. *White Alysson* or *Allison*
Androsace. *Rosette Rock-cress*
aubrietioides. *Pink-tinted Rock-cress*
blepharophylla. *Rosy-flowered Rock-cress*
canadensis. *Sickle-pod*
ciliata. *Fringed Rock-cress*
deltoidea. *Blue Rock-cress*
hirsuta. *Hairy Rock-cress*

M

Arabis, lucida. *Shining-leared Rock-cress*
lucida variegata. *Variegated Rock-cress*
perfoliata. *Glabrous Rock-cress*
petræa. *Northern Rock-cress*
procurrens. *Spreading Rock-cress*
rosea. *Rosy-flowered Calabrian Rock-cress*
stricta. *Bristol Rock-cress*
Thaliana. *Thale-cress*, or *Thale Rock-cress*
Turrita. *Tower-cress*, or *Tower Rock-cress*
Arachis hypogaea. *Earth-almond, Earth-nut, Grass-nut, Ground-nut* or *Earth-nut Oil-plant* of Medicine, *Teuss-oil-plant, Kat-jang* or *Manilla Nut, Monkey-nut, Pea-nut, Pindar, Underground Kidney-bean*
Aralia hispida. *Bristly Sarsaparilla, Wild Elder*
nudicaulis. *Rabbit-root, Wild Sarsaparilla*
papyrifera. *Chinese Rice-paper Plant*
racemosa. *American Spikenard*
spinosa. *Hercules'-Club, Shot-bush, Virginian Angelica-tree*
Araucaria Bidwillii. *Bunya-Bunya Pine*, of Queensland
brasiliensis. *Brazilian Pine-tree*
Cookii. *Cook's New Caledonia Pine*
Cunninghamii. *Moreton Bay Pine*
excelsa. *Norfolk Island Pine*
imbricata. *Chilian Pine-tree, Monkey-Puzzle*
Rulei. *Rule's New Caledonia Pine*
Arbutus. *Arbute-tree, Strawberry-tree*
alpina. *Alpine Strawberry-tree, Canadian Strawberry-tree*
Andrachne. *Oriental Strawberry-tree*
densiflora. *Densely-flowered Strawberry-tree*
Menziesii. *Madrona*, or *Madrono*, of California. *Menzies's Strawberry-tree*
procera. *Tall Strawberry tree*
tomentosa. *Woolly-branched Strawberry-tree*
Unedo. *Cane-apple, Common Arbute-tree, Common Strawberry-tree, Dalmatian Strawberry-tree*
Unedo var. Croomi. *Scarlet-barked Straw-berry-tree*
Unedo var. rubra. *Red-flowered Strawberry-tree*
Archangelica atropurpurea. *Great Angelica*, of N. America
officinalis. *Angelica* (cultivated), *Archangel*
Archemora rigida. *American Cow-bane*
Arctium Lappa. *Beggar's-Buttons, Bur, Burr, Burdock, Clot-Bur, Clod-Bur, Cuckle, Cuckold, Harlock*
Arctostaphylos (Arbutus) alpina. *Alpine Bear-berry, Black Bear-berry*
glauca and A. pungens. *"Manzanita,"* of California
Uva-ursi. *Bear-berry*, or *Bear Bilberry, Bear's Grape, Brawlins, Burren Myrtle, Creashak, Kinnikinnick, Mountain Box, Upland Cranberry*
Ardisia. *Spear-flower*
coriacea. *Red Beef-wood*
Arduina grandiflora. *Natal Plum*

Areca Catechu. *Betel - nut Palm-tree Catechu Palm-tree, "Drunken Date-tree," Pinang Palm*
monostachya. *Walking-stick Palm*
oleracea. *Cabbage Palm*
(Kentia) sapida. *New Zealand Palm, Nikau Palm*
Arenaria. *Sand-wort*
balearica. *Balearic* or *Majorca Sand-wort*
cæspitosa. *Tufted Sand-wort*
ciliata. *Fringed Sand-wort*
graminifolia. *Grass-leaved Sand-wort*
grandiflora. *Large-flowered Sand-wort*
Grœnlandica. *Mountain Sand-wort*
laricifolia. *Larch-leaved Sand-wort*
montana. *Mountain Sand-wort*
multicaulis. *Many-stemmed Sand-wort*
peploides. *Sea Chick-weed, Sea-purslane*
purpurascens. *Purplish-flowered Sand-wort*
rubra. *Purple Chick-weed, Red Sand-wort*
segetalis. *Sword-grass*
squarrosa. *Pine-barren Sand-wort*
tetraquetra. *Square stemmed Sand-wort*
uliginosa. *Bog Sand-wort*
verna. *Vernal Sand-wort*
Argania Sideroxylon. *Argan-tree, Morocco Iron-wood*
Argemone hispida. *"Chicalote,"* of California
grandiflora. *White Mexican Poppy*
mexicana. *Corooko*, or *Thistle Oil-plant, Devil's Fig, Infernal Fig, Prickly Poppy, Yellow Mexican Poppy, Yellow Thistle*
Argyreia. *Silver-weed*
Argyrodendron trifoliatum. *Iron-wood*, of N. S. Wales
Arisæma Dracontium. *Dragon-root, Green Dragon*
papillosum. *Ceylon Snake-root*
triphyllum. *Indian Turnip, Jack-in-the-pulpit, Preacher-in-the-pulpit*
Aristea major. *Great Blue Flower-rod*
Aristida dichotoma. *Poverty-grass*
Aristolochia. *Birth-wort*
Clematitis. *Upright Birth-wort*
grandiflora. *Pelican-flower, Poisonous Hog-weed*
Guaco. *"Guaco"-plant*, of S. America
odoratissima. *Serpent - withe, Sweet-scented Birth-wort*
recurvilabra. *"Putchuck"*
reticulata. *Red River* or *Texan Snake-root*
rotunda. *"Apple-of-the-Earth"*
Serpentaria. *Serpentary-root, Virginian Snake-root*
Sipho. *Broad-leaved Birth-wort, Dutch-man's Pipe, Pipe-Vine*
tomentosa. *Woolly Birth-wort*
Aristotelia Colensoi. *Mountain Wine-berry*, of New Zealand
fruticosa. *Mountain Currant*, of New Zealand
peduncularis. *Bleeding Heart*
Armeniaca (Prunus Armeniaca). *Apricot-tree*

Armeniaca, brigantiaca. *Briançon Apricot*
dasycarpa. *Thick-fruited Apricot*
sibirica. *Siberian Apricot*
vulgaris. *Common Apricot*
vulgaris foliis variegatis. *Variegated-leaved Apricot*
vulgaris var. ovalifolia. *Oval-leaved Apricot*
Armeria. *Thrift*
alliacea. *Garlic Thrift*
alpina. *Alpine Thrift*
alpina rosea. *Rosy-flowered Alpine Thrift*
Cephalotes. *Great Thrift*
grandiflora. *Large-flowered Thrift*
maritima. *Cliff-rose, Cushion Pink, Lady's-Cushion, Lady's - Pincushion, Marsh Daisy, Sea Gilly-flower, Sea Pink*
maritima alba. *White-flowered Thrift*
plantaginea. *Plantain-leaved Thrift*
scoparia. *Broom Thrift*
vulgaris. *Common Thrift*
Arnebia echioides. *Prophet-flower, Russian Bugloss, Spotted Golden Borage*
Arnica montana. *Medicinal Leopard's-Bane, Mountain Alkanet, Mountain Tobacco*
Arnoseris pusilla. *Dwarf Nipple-wort, Lamb's Succory, Swine's Succory*
Arnopogon. *Sheep's-beard*
Aronia (Pyrus) arbutifolia. *Red Chokeberry*
Arracacha esculenta (Conium Arracacha). *"Arracacha," Peruvian Carrot*
Arrhenatherum avenaceum. *False Oat*
Artabotrys. *Tail-grape*
Artanema fimbriata. *Fringed Heath-cup,* of Australia
Artanthe adunca. *Spanish Elder,* of the W. Indies
Artemisia. *Worm-wood*
Abrotanum. *Boy's-love, Lad's-love, Maid's-love, Medicinal Worm-wood, Old-Man, Southern-wood*
Absinthium. *Absinth, Common Wormwood, Old-Woman*
afra. *African Worm-wood*
alpina. *Alpine Worm-wood*
anethifolia. *Dill-leaved Worm-wood*
aprica. *Trailing Evergreen Worm-wood*
arborescens. *Tree Worm-wood*
arbuscula. *Sage-bush,* or *Sage-brush,* of N. America
argentea. *Old-Woman, Silvery Worm-wood*
biennis. *Biennial Worm-wood*
cærulescens. *Bluish Worm-wood*
campestris. *Field Worm-wood*
cana. *Hoary Worm-wood*
chinensis (A. Moxa). *Moxa-plant*
Dracunculus. *Tarragon-plant*
frigida. *Silky Worm-wood*
furcata. *Forked Worm-wood*
gallica. *French Worm-wood*
glacialis. *Glacier Worm-wood*
hirsuta. *E. Indian Mug-wort*
judaica. *Judean Worm-wood*
lactiflora. *Pale-flowered Worm-wood*
Ludoviciana. *Western Mug-wort*

Artemisia, maritima. *Garden Cypress, Sea-side Worm-wood*
maritima var. Stechmanniana (A. Lercheana). *Worm-seed*
Marschalliana. *Marschall's Worm-wood*
Moxa. *Chinese* or *Japanese Moxa-plant*
mutellina. *Alpine Worm-wood*
norvegica. *Norwegian Worm-wood*
orientalis. *Oriental Worm-wood*
Pallasii. *Pallas's Worm-wood*
pectinata. *Scalloped-leaved Worm-wood*
pontica. *Roman Worm-wood*
potentillæfolia. *Potentilla-leaved Worm-wood*
procera. *Tall Worm-wood*
ramosa. *Branchy Worm-wood*
repens. *Creeping Worm-wood*
rupestris. *Cliff Worm-wood*
Santonica. *Holy Worm-wood, Tartarian Southern-wood*
saxatilis. *Rock Worm-wood*
sericea. *Silky-leaved Worm-wood*
spicata. *Spiked Worm-wood*
Stelleriana. *Steller's Worm-wood*
tanacetifolia. *Tansy-leaved Worm-wood*
taurica. *Taurian Worm-wood*
tenuifolia. *Slender-leaved Worm-wood*
tridentata and A. trifida. *Sage-bush,* or *Sage-brush,* of N. America
vulgaris. *Fellon-herb, Mother-wort, Mug-wort*
vulgaris variegata. *Variegated - leaved Worm-wood*
Arthropodium cirrhatum. *Fringed Lily,* of New Zealand
paniculatum. *Fringed Lily,* or *Fringed Violet,* of Victoria
Arthrostylidium Schomburgkii. *Blowpipe Bamboo*
Arthrotaxis. *Jointed Yew*
Artocarpus incisa. *Bread-fruit Tree*
integrifolia. *Jaca,* or *Juck-tree*
Arum Arisarum. *Friar's Cowl*
atro-rubens. *Brown Dragons*
crinitum. *Dragon's-Mouth*
divaricatum. *Indian Kale*
Dracontium. *American Wake - Robin, Edder-wort, Green Dragons, Indian Turnip*
Dracunculus. *Brook-leek, Dragon Arum, "Dragon's Female," "Faverole"*
italicum. *Italian Arum*
macrorrhizon. *Roasting Coco*
maculatum. *Aaron, Adam - and - Eve, Adder's Meat, Arrow-root, Bloody-Man's-Finger, Bobbin-Joan, Bobbins, Bulls-and - Cows, Calf's - foot, Cuckoo - pint, Friar's - Cowl, Lamb-in-a-pulpit, Lily Grass, Lords - and - Ladies, Mandrake, Nightingales, Portland Starch - root, Wake-Robin*
Arundinaria falcata. *Ningala Bamboo*
Hookeriana. *Praong Bamboo*
macrosperma. *Large Cane,* of N. America
spathulata. *Hardy Bamboo*
tecta. *Small Cane,* of N. America
Arundo. *Reed, Reed-grass*
Ampelodesmos (Ampelodesmos tenax A. festucoides). *Diss, Vine-tic*

English Names of Cultivated, Native,

Arundo, conspicua. *Silvery Reed-grass*, of New Zealand
Donax. *Distaff Cane, Great Reed, Italian Vineyard Cane*
occidentalis. *Trumpet Reed, W. Indian Reed-grass, Wild Cane*
Phragmites. *Common Reed, Spire Reed*
saccharoides. *Wild Cane*
(Ampelodesmos) tenax (A. festucoides). *Diss, Vine-tie*

Asagræa officinalis (Veratrum officinale). *Cebadilla,* or *Cevadilla-plant* of Medicine

Asarum canadense. *Canadian Snake-root, Indian Ginger*
caudatum. *Tailed Snake-root*
europæum. *Asarabacca, "Cabaret," Hazel-wort, Wild Nard*
virginicum. *Heart Snake-root, Sweet-scented Asarabacca*

Asclepias. *Milk-weed, "Silken Cissy," Silk-weed, Swallow-wort*
Cornuti. *Common Milk-weed* or *Silk-weed*
curassavica. *Bastard Ipecacuanha, Blood-flower* of the W. Indies, *Red-head, Wild Ipecacuanha*
Douglasii. *Douglas's Milk-weed*
incarnata. *Swamp Milk-weed*
odoratissima. *West Coast Creeper*
phytolaccoides. *Poke Milk-weed*
purpurascens. *Purple Milk-weed*
quadrifolia. *Four-leaved Milk-weed*
tuberosa. *Butterfly-weed, Pleurisy-root, Tuber-root, Wind-root*
variegata. *Variegated Milk-weed*
verticillata. *Whorled Milk-weed*

Asclepiodora decumbens. *Green Milk-weed*

Ascophora elegans. *Bread-mould*

Ascyrum Crux-Andreæ. *St. Andrew's Cross*
stans. *St. Peter's-wort*

Asimina (Anona) triloba. *Virginian Papaw*

Aspalathus. *African Broom*

Asparagus acutifolius. *Asparagus*, of S. Europe
æthiopicus. *Ethiopian Asparagus*
albus. *Garden-hedge Asparagus*, of Madeira
Broussoneti. *Giant Asparagus*
decumbens. *Decumbent Asparagus*
falcatus. *Sickle-branched Asparagus*
laricinus. *Asparagus*, of S. Africa
officinalis. *Common Asparagus, Sparrow-grass*
plumosus. *Feathery Asparagus*
racemosus. *Racemose Asparagus*
tenuifolius. *Slender-leaved Asparagus*

Aspergillus glaucus. *Blue-mould* of cheese

Asperugo procumbens. *German Madwort, Madder-wort*

Asperula arvensis. *Quinsy-wort*
cynanchica. *Quinsy-wort, Quinancy-wort,* or *Squinancy-wort*
humifusa. *Creeping Wood-ruff*
odorata. *Mugwet, Sweet Grass, Sweet Wood-ruff*
taurina. *Broad-leaved Wood-ruff*
tinctoria. *Three-lobed Wood-ruff*

Asphodelus albus. *White-flowered Asphodel*
fistulosus. *Onion-Asphodel*
luteus. *Jacob's-rod, King's Spear, Yellow Asphodel*
ramosus. *Branching Asphodel, King's-spear, Silver-rod*
sub-alpinus. *Sub-alpine Asphodel*

Aspidistra. *Shield-flower*
lurida. *Parlour Palm*

Aspidium. *Shield-Fern*
acrostichoides. *Christmas Shield-Fern*, of N. America
(Polystichum) aculeatum. *Prickly Shield-Fern*
capense. *Climbing Shield-Fern*, of New Zealand
Lonchitis. *Holly Fern*
(Polystichum) munitum. *Californian Shield-Fern*
nevadense. *Nevada Wood-Fern, Sierra Shield-Fern*

Aspidosperma excelsum. *Paddle-wood,* or *Wheel-tree*, of Guiana
Quebracho. *White Quebracho-tree*

Asplenium. *Spleen-wort*
Adiantum-nigrum. *Black Oak Fern, Black Spleen-wort*
alternifolium. *Alternate-leaved Spleen-wort*
bulbiferum. *Common Spleen-wort*, of New Zealand
Ceterach. *Finger-Fern, Miltwaste, Scale Fern, Scaly Spleen-wort*
flaccidum. *Hanging-tree Spleen-wort*, of New Zealand
fontanum. *Smooth Rock Spleen-wort*
lanceolatum. *Lanceolate Spleen-wort*
marinum. *Sea Spleen-wort*
Nidus. *Bird's-nest Fern*
obtusatum. *Shore Spleen-wort*, of New Zealand
parvulum. *Smaller Ebony Spleen-wort*
rhizophyllum. *Walking-leaf Fern*
Ruta-muraria. *Tent-wort, Wall-Rue, Wall-Rue Spleen-wort, White Maiden-hair*
septentrionale. *Forked Spleen-wort*
Trichomanes. *Common Spleen-wort, English Maiden-hair, Green Spleen-wort, Maiden-hair Spleen-wort, Water-wort*
viride. *Green Spleen-wort*

Astelia. *Otago Cotton-plant*

Aster. *Star-wort*
albescens. *Shrubby Star-wort*
alpinus. *Blue Alpine Daisy*
altaicus. *Large-flowered Altaian Star-wort*
Amellus. *Italian Star-wort*
(Eurybia) argophylla. *Australian Star-wort*
bessarabicus. *Dwarf Amellus Aster*
cabulicus. *E. Indian Star-wort*
canescens. *Hoary Star-wort*
coccineus. *Crimson-flowered Star-wort*
cordifolius. *Heart-leaved Star-wort*
Datschyi. *Christmas-flowering Star-wort*
(Galatella) dracunculoides. *Tarragon-like Star-wort*
dumosus. *Bushy Star-wort*

Aster, elegans. *Elegant Star-wort*
ericoides. *Heath-leaved Star-wort*
Fortunei. *Fortune's Star-wort*
fragilis. *Fragile Star-wort*
grandiflorus. *Christmas Daisy*
horizontalis. *Horizontal-branched Starwort*
hyssopifolius. *Hyssop-leaved Star-wort*
lævis var. lævigatus. *Smooth Star-wort*
laxus. *Loose-branched Star-wort*
longifolius. *Long-leaved Star-wort*
multiflorus. *Many-flowered Star-wort*
Novæ Angliæ. *New England Star-wort*
Novi Belgii. *New York Star-wort*
obliquus. *Oblique Star-wort*
patens. *Spreading Star-wort*
pendulus. *Pendulous Star-wort*
ptarmicoides. *Bouquet Star-flower, Yarrow-leaved Star-wort*
puniceus. *Purple-stemmed Star-wort*
pyrenæus. *Pyrenean Star-wort*
racemosus. *Branching Star-wort*
Reevesi. *Reeves's Star-wort*
rubricaulis. *Red-stemmed Star-wort*
scorzoneræfolius. *Scorzonera-leaved Starwort*
sericeus. *Silky Star-wort*
Shortii. *Short's Star-wort*
spectabilis. *Showy Star-wort*
tardiflorus. *Late-flowering Star-wort*
tenuifolius. *Slender-leaved Star-wort*
Townshendi. *Townshend's Star-wort*
Tradescanti. *Michaelmas Daisy*
turbinellus. *Mauve-flowered Star-wort*
Tripolium. *Blue Chamomile, Blue Daisy, Purple Chamomile, Sea-star, Sea Starwort, Serapias Turbith, Share-wort, Tripoly*
versicolor. *Various-coloured Star-wort*
Asterocephalus stellatus. *Starry Pincushion-flower, Star-head*
atropurpureus. *Sweet Scabious*
Asterolinum stellatum, *Flax-star*
Astilbe. *False Goat's-beard*
decandra. *Common False Goat's-beard*
japonica. *Japanese False Goat's-beard*
rivularis. *River-side False Goat's-beard*
rubra. *Red-flowered False Goat's-beard*
Astragalus. *Milk-Vetch*
Ægicerns. *Goat's-Horn*
alpinus. *Mountain Milk-Vetch*
bœticus. *Triangular-podded Milk-Vetch, Swedish Coffee*
caryocarpus. *American Ground-Plum, Red-fruited Milk-Vetch*
dasyglottis. *Clover Milk-Vetch*
galegæformis. *Galega Milk-Vetch*
glycyphyllos. *Liquorice-Vetch, Sweet-leaved Milk-Vetch*
Hornii. *"Loco"-plant*
hypoglottis. *Purple-flowered Milk-Vetch*
lentiginosus. *Rattle-weed*
monspessulanus. *Montpelier Milk-Vetch*
Onobrychis. *Saintfoin Milk-Vetch*
pannosus. *Shaggy Milk-Vetch*
ponticus. *Pontic Milk-Vetch*
Poterium. *Small Goat's-Horn*
sesameus. *Starry Milk-Vetch*
Tragacantha. *Goat's-thorn Milk-Vetch, Great Goat's-thorn, Gum-Tragacanth-plant*

Astragalus trimestris. *Egyptian Milk-Vetch*
vaginatus. *Sheathed Milk-Vetch*
vimineus. *Silvery Milk-Vetch*
Astrantia major. *Black Hellebore, Great Black Master-wort*
minor. *Small Black Master-wort*
Astrapæa. *Madagascar Fire-plant*
Astrebla (Danthonia) triticoides. *Mitchell-grass,* of Australia
Astrocaryum acaule. *Tu Palm*
aculeatum. *Acuyuru Palm, Gree-Gree-tree,* of Trinidad
Murumuru. *Murumuru Palm-tree*
Tucuma. *Tucuma Palm*
vulgare. *Cumari Palm, "Gru-gru" Palm*
Astroloma humifusum. *Australian Cranberry*
Astrotricha pterocarpa. *Plum-bush,* of Australia
Atalanthus arboreus. *Tree Sow-thistle*
Atalantia glauca. *Desert Lemon,* of Australia
monophylla. *Wild Lime,* of Coromandel
Athamanta Cervaria. *Broad-leaved Spignel, Mountain Hart-wort, "Much-good"*
cretensis. *Candy Carrot, Cretan Carrot, Fine-leaved Spignel*
Libanotis. *Mountain Spignel*
macedonica. *Macedonian Parsley*
sicula. *Flix-weed-leaved Spignel*
Atherosperma moschata. *Plume-Nutmeg* of Australia, *Tasmanian Sassafras*
Athanasia annua. *African Daisy*
Athyrium Filix-fœmina. *Female Polypody, Lady Fern*
Atractylis acaulis. *Spindle-wort*
Atragene (Clematis) alpina. *Alpine Clematis*
(Clematis) americana. *American Clematis*
(Clematis) occidentalis. *Western Clematis*
(Clematis) ochotensis. *Ochotsk Clematis*
(Clematis) sibirica. *Siberian Clematis*
Atriplex. *Orache*
albicans. *White Orache*
halimoides and other species. *Australian Salt-bush*
Halimus. *Broad-leaved Sea Purslane-tree*
hortensis. *Butter Leaves, Garden Orache, Mountain Spinach*
hortensis rubra. *Red-leaved Orache*
littoralis. *Grass-leaved Orache*
Nummularia. *Salt-bush,* of Australia
patula. *Common Orache, Delt Orache*
portulacoides. *Lesser Shrubby Orache, Purslane Orache, Sea-Purslane*
pedunculata. *Stalked Orache*
rosea. *Frosted Orache*
Atropa Belladonna. *Bane-wort, Black Cherry, Belladonna, Deadly Nightshade, Dwale, Dwayberries, Great Morel, Naughty-Man's-Cherry*
Mandragora. *Autumnal Mandrake*
Attalea Cohune. *Cohune Palm*
compta. *Pindova Palm*
excelsa. *Urucuri Palm*

Attalea funifera. *Broom-Palm, Coquilla-nut-Palm, Monkey-grass* or *Para-grass-Palm, Piassaba Palm*
Aubrietia purpurea. *Purple Alysson, Purple Rock-cress*
Aucuba himalaica. *Himalayan Laurel*
japonica. *Blotched-leaved Laurel, Gold-Leaf-plant, Spotted Laurel, Variegated Laurel*
japonica vera. *Green-leaved Aucuba*
Audibertia polystachya. *White Sage*, of California
Auricula alpina. *Yellow Alpine Auricula*
Avena. *Oat, Oat-grass*
brevis. *Short Oat*
elatior. *Button-Grass, Couch Onion, Haver-Grass, Onion-Grass, Pearl-Grass*
fatua. *Drake, Droke, Flaver, Wild Oat*
flavescens. *Yellow Oat-grass*
nuda. *Naked Oat, Pill-corn*, or *Pil-corn*
orientalis. *Tartarian Oat*
pratensis. *Oat-grass, Perennial Oat*
sativa. *Common Cultivated Oat, Haver*, or *Haver*
Smithii and Δ. striata. *Wild Oat*, of N. America
sterilis. *Animal, Fly*, or *Hygrometric Oat*
Averrhoa Bilimbi. *Bilimbi Tree, Illimbing*
Carambola. *Caramba*, or *Carambola-tree, Coromandel Goose-berry, Country Gooseberry* of India
Avicennia nitida. *White Mangrove*
officinalis. *New Zealand Mangrove*
Azalea. *False Honey-suckle*
amœna. *Bright-flowered Azalea*
arborescens. *Smooth Azalea, Tree Azalea*
bicolor. *Two-coloured Azalea*
calendulacea. *Flame-coloured Azalea, Marigold Azalea*
calendulacea var. chrysolecta. *Fine Golden Azalea*
canescens. *Hoary Azalea*
crispiflora. *Crisp-flowered Azalea*
Danielsiana. *Daniels's Azalea*
glauca. *Dwarf Glaucous Azalea*
hispida. *Tall Glaucous Azalea*
indica. *Indian Azalea*
ledifolia. *Ledum-leaved Azalea*
mollis. *Soft-leaved Azalea*
nitida. *Shining-leaved Azalea*
nudiflora. *Pinxter-flower*
nudiflora var. prolifera. *Proliferous Azalea*
obtusa. *Blunt-leaved Azalea*
ovata. *Ovate-leaved Azalea*
ovata var. alba. *White-flowered Azalea*
pontica. *Common Yellow Azalea*
sinensis. *Chinese Azalea*
speciosa. *Showy Azalea*
squamata. *Scaly Azalea*
viscosa. *White Swamp-Honey-suckle*
viscosa var. odorata. *Sweet-scented Azalea*
Azarolus sativa. *Neapolitan Medlar*
Azorella (Bolax) glebaria. *Balsam Bog*

Babiana. *Baboon-root*
Baccharis. *Ploughman's Spikenard*
angustifolia. *American Ploughman's Spikenard*
cordifolia. "*Mio-Mio*," of S. America

Baccharis halimifolia. *Groundsel-tree, Pencil-tree*
scoparia. *Mountain Broom-tree*
Backhousia australis. *Lance-wood*, of Australia
myrtifolia. *Backhouse's Myrtle*
Bactris Maraja. *Marajah Palm*
minor. *Tobago Cane*
Plumicriana. *Prickly-pole*, of the W. Indies
Badiera diversifolia. *Bastard Lignum-vitæ*
Badula (Anguillaria) Barthesia. *Guinea-fowl-wood*
Balanites ægyptiaca. *Zachun-oil-tree*
Balantium Culcita. *Culcit Fern*
Ballota nigra. *Black*, or *Stinking, Horehound*
Baloghia lucida. *Blood-wood*, of Norfolk Island
Balsamodendron (Amyris) Gileadense. *Balm of Gilead Tree, Balsam of Mecca* or *Roghen*
Mukul. *Bdellium-tree* of the Scriptures, *Googul*, or *Indian Bdellium-tree*
Myrrha. *Myrrh-gum-tree*
Playfairii. Supposed to be the "*Hotai*"-*resin-plant*
pubescens. *Bayee Balsam-tree*
Balsamorrhiza. *Balsam-root*, of California
Bambusa. *Bamboo*
arundinacea. *Common Bamboo*
aurea. *Yellow-stemmed Bamboo*
(Arundinaria) falcata. *Common Hardy Bamboo*
Fortunei. *Fortune's Bamboo*
gracilis. *Slender Bamboo*
japonica (B. Metaké). *Metaké Bamboo*
latifolia. *Orinoko Bamboo*
nigra. *Dark-stemmed Bamboo*
reticulata. *Netted-reined Bamboo*
Simmondsii. *Simmonds's Bamboo*
striata. *Striped Bamboo*
viridi-glaucescens. *Grayish Bamboo*
Banksia australis. *Tasmanian Honeysuckle*
compar. *Beef-wood*, of Queensland
Cunninghamii, B. ericæfolia, and B. serrata. *Australian Honeysuckle-tree*
Baphia nitida. *Shining Bar-wood* or *Cam-wood*
Baptisia alba. *White False-Indigo*
australis. *Blue False-Indigo*
exaltata. *Tall False-Indigo*
tinctoria. *Horse-fly-weed, Wild Indigo* of N. America
Barbarea præcox. *American Cress, Bank Cress, Belleisle Cress, Land Cress, Normandy Cress, Winter Cress, Yellow Cress, Yellow Rocket*
præcox variegata. *Variegated Winter Cress*
vulgaris. *Common Winter Cress, St. Barbara's Herb, Winter* or *Yellow Rocket*
Barnardia scilloides. *Chinese Squill*
Barosma betulina, B. crenulata, and B. serratifolia. *Buchu-leaf-plant*
Baroxylon rufum. *Red Heavy-wood*
Barringtonia speciosa. *Hutu, Vutu*, or *Futu-tree* of Tahiti, *S. Sea Island Bottle-brush-tree*
Bartonia aurea. *Golden Barton's-flower*

and Foreign Plants, Trees, and Shrubs. 167

Bartramia. *Apple-Moss*
pomiformis. *Common Apple-Moss*
Bartramidula. *Dwarf Apple-Moss*
Bartsia. *Eye-bright Cow-wheat*
alpina. *Alpine Eye - bright Cow-wheat, Mountain Poly*
Odontites. *Hen-Gorse, Red Eye-bright, Red Eye bright Cow-wheat*
viscosa. *Viscid Eye-bright Cow-wheat*
Basella alba. *E. Indian Spinach, White Malabar Nightshade*
rubra. *E. Indian Spinach, Red Malabar Nightshade*
Bassia butyracea. *Butter-tree* of Nepaul, *Phoolwa-oil-plant*
latifolia. *Epie, Mahoua,* or *Yallah-oil-plant*
longifolia. *Ilpa, Illipoo,* or *Illupie-oil-plant*
Parkii. *African Butter-tree, Shea tree*
Batatas edulis. *Spanish,* or *Sweet, Potato*
(Ipomœa) Jalapa. *Mechoacan Jalap-plant*
paniculata. *Giant Sweet Potato*
Batis maritima. *Jamaica Samphire, W. Indian Salt-wort*
Bauera rubioides. *River-rose,* of Tasmania
Bauhinia anguina. *Snake-charm,* of the E. Indies
Hookeri. *Tree-bean,* of Australia
racemosa. *Maloo Creeper*
tomentosa. *St. Thomas's Tree*
Vahlii. *Maloo Creeper*
variegata. *Mountain Ebony,* of the E. Indies, *Variegated St. Thomas's Tree*
Beatsonia portulacæfolia. *Tea-plant* of St. Helena
Beaufortia decussata. *Beaufort Myrtle*
Beaumontia grandiflora. *Nepaul Trumpet-flower*
Bedfordia salicina. *Tasmanian Dog-wood*
Begonia. *Elephant's-ear*
albo-coccinea. *White-and-scarlet Begonia*
castaneæfolia. *Chestnut-leaved Begonia*
corallina. *Coral-flowered Begonia*
Davisi. *Dwarf Scarlet Begonia*
discolor. *Two-coloured Begonia*
Evansiana. *Beefsteak-plant*
Frœbeli. *Frœbel's Begonia*
fuchsioides. *Fuchsia Begonia*
geranioides. *Geranium-leaved Begonia*
lucida. *Shining Begonia*
manicata. *Collared Begonia*
multiflora. *Many-flowered Begonia*
nitida. *Jamaica Wood-sorrel*
octopetala. *Eight-petalled Begonia*
peruviana. *Peruvian Begonia*
Rex. *King Begonia*
Roezlii. *Roezl's Begonia*
scandens. *Climbing Sorrel*
semperflorens. *Ever-blooming Begonia*
socotrana. *Socotran Begonia*
suaveolens. *Sweet-scented Begonia*
tuberosa. *Tuberous-rooted Begonia*
Veitchii. *Veitch's Begonia*
Bellevalia. *Roman Squill*
Bellis. *Daisy*
annua. *Annual Daisy*
aucubæfolia. *Aucuba-leaved Daisy*
hybrida. *Italian Daisy*

Bellis integrifolia. *Texan* or *Western Daisy*
perennis. *Bairn-wort, Ban-wort, Ban-wood, Benner-Gowan, Bone-flower, Bruise-wort, Common Daisy, Ewe-Gowan, Herb Margaret, Gowan, Marguerite, May-Gowan*
perennis var. alba plena. *Double White Daisy*
perennis var. hortensis. *Garden Daisy*
perennis var. prolifera. *Apes-on-horse back,* "*Hen-and-Chickens*" *Daisy*
rotundifolia var. cœrulea. *Blue Round-leaved Daisy*
sylvestris. *Portugal Wood-Daisy*
Bellidiastrum. *Daisy-star*
Bellium bellidioides. *Common Small Daisy, False Daisy*
crassifolium. *Thick-leaved Small Daisy*
minutum. *Dwarf Small Daisy*
Benincasa (Cucurbita) cerifera. *Wax-Gourd, White Gourd*
Bentinckia coddapanna. *Lord Bentinck's Palm*
Berberis. *Barberry,* or *Berberry*
Aquifolium. *Holly - leaved Barberry,* "*Oregon Grape*"
aristata. *Indian Barberry Bark-tree*
asiatica. *Asiatic Barberry*
Bealei. *Beale's Barberry*
canadensis. *American Barberry*
cratægina. *Hawthorn-Barberry*
cretica. *Box-leaved Barberry*
Darwini. *Darwin's Barberry*
dealbata (B. glauca). *White-leaved Barberry*
dulcis. *Edible-fruited Barberry*
emarginata. *Notch-petalled Barberry*
empetrifolia. *Fuegian Barberry*
fascicularis. *Bundle-flowered Barberry*
floribunda. *Many-flowered Barberry*
glumacea. *Glumaceous Barberry*
heterophylla. *Various-leaved Barberry*
Hookeri. *Hooker's Barberry*
intermedia. *Intermediate Barberry*
japonica. *Japanese Barberry*
Lycium. *Indian Barberry-Bark-tree*
(Mahonia) nervosa. *Nerved-leaved Barberry*
(Mahonia) repens. *Creeping-rooted Barberry*
sibirica. *Siberian Barberry*
stenophylla. *Golden Barberry, Narrow-leaved Barberry*
vulgaris. *Common Barberry,* or *Berberry, Jaundice-berry, Jaundice-tree, Pipperidge,* or *Piprage*
Wallichii. *Wallich's Barberry*
Berberidopsis corallina. *Coral Barberry*
Berchemia volubilis. *American Supple-jack*
Bergera (Murraya) Kœnigi. *Curry-leaf Tree, Limbric-oil-tree*
Bergia Ammannoides. *Water-fire-tree,* of India
Berrya Ammonilla. *Trincomalee-wood*
mollis. *Pet-wood*
Bertholletia excelsa. *Brazilian-nut Tree, Castanha-oil-tree*
Beta. *Beet*

Beta Cicla. *Chard Beet, Leaf Beet, Sicilian Beet, White Beet, Wild Beet*
maritima. *Beet Spinach, Perpetual Spinach, Sea-side Beet, Sugar Beet*
vulgaris. *Red Beet*
vulgaris var. macrorrhiza. *Mangel, or Mangold-Wurtzel*
Betonica grandiflora. *Large-flowered Betony*
Betula. *Birch-tree*
alba. *Bedewen, Bedwen, Birk-tree or Burk-tree, Common Birch, Lady Birch*
alba var. laciniata. *Cut-leaved or Fern-leaved Birch*
alba var. laciniata pendula. *Weeping Fern-leaved Birch*
alba var. pendula. *Common Weeping Birch*
alba var. populifolia. *American White, Gray, or Old Field Birch*
antarctica. *Fuegian or Antarctic Birch-tree*
Bhojputtra. *E. Indian Birch*
glandulosa. *American Dwarf Birch*
lenta. *Black Birch, Cherry Birch, Mahogany Birch, Mountain Mahogany, Sweet Birch*
lutea. *Yellow, or Gray Birch*
nana. *Dwarf Birch*
nigra. *Red, or River Birch*
papyracea. *Canoe, or Paper Birch*
populifolia laciniata. *Cut-leaved Birch*
pumila. *American Low Birch, Hairy Dwarf Birch*
tristis. *Kamtschatka Weeping Birch*
Beureria havanensis and B. succulenta. *Poison-berry*, of the W. Indies, *W. Indian Currant-tree*
Beyeria viscosa. *Wallaby-bush*
Bigelovia nudata. *Rayless Golden-Rod*
Bidens (Cosmos) atrosanguinea. *Black Dahlia*
Beckii. *American Water-Marigold*
bipinnata. *Spanish Needles*
cernua. *Water Hemp-Agrimony*
connata. *Swamp Beggar-ticks*
chrysanthemoides. *Large Bur-Marigold,* of N. America
frondosa. *Beggar-ticks, Stick-tight, Small Bur-Marigold,* of America
pilosa. *New Zealand Cowage*
tripartita. *Common Bur-Marigold, Water-Agrimony, Water-Hemp*
Bignonia alliacea. *Garlic-shrub,* of Guiana
capreolata. *Cross-Vine, Four-leaved Hardy Trumpet-flower, Tendrilled Trumpet-flower*
Chica. *Chica-plant*
Leucoxylon. *Green Ebony-tree,* of Jamaica, *White Cedar,* of Dominica
procera (Jacaranda Copaia). *Fern-leaved Trumpet-flower*
(Tecoma) radicans. *Ash-leaved Hardy Trumpet-flower*
sempervirens. *Carolina Yellow Jasmine, Evergreen Trumpet-flower*
suberosa (Millingtonia hortensis). *Cork-tree,* of the E. Indies

Bignonia Unguis. *Cat-claw, Four-leaved Trumpet-flower*
Billardiera. *Australian Apple-berry-tree*
longiflora. *Long-flowered Apple-berry*
mutabilis. *Colour-changing Apple-berry*
ovalis. *Oval-leaved Apple-berry*
scandens. *Climbing Apple-berry Tree*
Biophytum sensitivum (Oxalis sensitiva). *"Biophytum," Sensitive Wood-Sorrel*
Biota (Thuja) orientalis. *Common Chinese Arbor-vitæ*
orientalis aurea. *Golden Chinese Arbor-vitæ*
orientalis elegantissima. *Dwarf Chinese Arbor-vitæ*
orientalis falcata. *Sickle-spined Chinese Arbor-vitæ*
orientalis pyramidalis. *Pyramidal Chinese Arbor-vitæ*
orientalis Sieboldii (Thuja japonica). *Japanese Arbor-vitæ*
Biscutella. *Buckler Mustard*
auriculata. *Ear-podded Buckler Mustard*
Bisserula Pelecinus. *Hatchet-Vetch*
Bixa orellana. *Arnotta, or Anotta-tree*
Blakea trinervis. *Jamaica Rose* quinquenervia. *Mussel-tree,* of Guiana
Blechnum brasiliense. *Dwarf Brazilian Tree-Fern*
Spicant (B. boreale). *Hard Fern, Herring-bone Fern, Rusty-back, Snake-Fern*
Blighia (Cupania) sapida. *Akee-tree*
Blitum capitatum. *Strawberry-Blite, Straw-berry-Spinach*
maritimum. *Coast-Blite*
Blumea longifolia. *Blume's Thistle*
Bobartia aurantiaca. *Bobart's Orange Iris*
Bocconia. *Plume-Poppy, Tree-Celandine*
cordata. *Heart-leaved Tree-Celandine*
frutescens. *Parrot-weed, Shrubby Tree-Celandine*
Bœhmeria (Urtica) argentea. *Silver-leaved Grass-cloth-plant*
cylindrica. *American False-Nettle*
nivea. *China-grass, Grass-cloth-plant, Ramee, Ramie,* or *Rheea-grass*
Puya. *Puya-fibre-plant*
Boerhaavia decumbens. *Hog-weed,* or *Hog-meat,* of Jamaica, *Ipecacuanha,* of Guiana
erecta. *American Hog-weed*
scandens. *Climbing Hog-weed*
Boerkhausia rubra. *Red Hawk-weed*
Bolax gummifer. *Bog Balsam*
Boldoa fragrans. *Boldo,* or *Boldu-tree, Chilian Sassafras*
Boletus edulis. *Polish Mushroom*
(Polyporus) fomentarius and B. igniarius. *Amadou,* or *German Tinder*
Boltonia glastifolia. *False Chamomile*
Bombax Ceiba. *Silk Cotton-tree,* of S. America
malabaricum. *Silk-cotton-tree,* of Malabar
Bontia daphnoides. *Wild Olive,* of Barbadoes
Borago officinalis. *Common Borage, Cool-tankard, Tale-wort*
laxiflora. *Bell-flowered Borage*

and Foreign Plants, Trees, and Shrubs. 169

Borago orientalis (B. cordifolia). *Early-flowering Borage*
Borassus æthiopum. *Deleb Palm* (?)
flabelliformis. *Brab-tree, E. Indian Wine-Palm, Great Fan-Palm, Palmyra Palm-tree, Tal* or *Tala Palm*
Boronia serrulata. *Native Rose,* of Australia
Borreria (Spermacoce). *Button-weed*
Borrichia arborescens. *Sea-side Ox-eye*
frutescens. *Jamaica Samphire, Sea-side Ox-eye*
Bosea Yervamora. *Tree Golden-rod*
Bossiæa Scolopendrium. *Australian Plank-plant*
Boswellia Carteri, B. thurifera, and B. serrata. *Frankincense-tree,* or *Olibanum tree*
Botrychium. *Grape-Fern, Moon-wort*
Lunaria. *Lunary, Moon-Fern, Moon-wort*
lunarioides. *American Grape-Fern*
ternatum. *Moon-wort Fern,* of New Zealand
virginicum. *Rattle-snake Fern*
Botryodendron latifolium. *Norfolk Island Shade-leaf*
Boussingaultia baselloides. *Madeira Vine*
Bouteloua. *Grama-grass, Muskit-grass*
Bowenia spectabilis. *Bowen's Fern-Palm*
Brabeium stellatifolium. *African Almond, Kaffir Chestnut, Wild Almond,* or *Wild Chestnut,* of the Cape of Good Hope
Brachychæta cordata. *False Golden-Rod*
Brachychiton acerifolium. *Flame-tree*
luridum. *Sycamore,* of Australia
Brachycome iberidifolia. *Swan River Daisy*
Brachypodium. *False Brome-grass*
Bragantia Wallichii. "*Alpam root,*" of Malabar
Brasenia peltata. *American Water-shield, Deer-food*
Brassaia actinophylla. *Queensland Umbrella-tree*
Brassica. *Cabbage*
campestris. *Bargeman's Cabbage, Navew, Summer Rape, Wild Cabbage*
campestis var. Rutabaga. *Swedish Turnip*
Caulo-rapa (B. oleracea gongylodes). *Hungarian Turnip, Kohl-Rabi, Turnip-rooted Cabbage*
chinensis. *Chinese Cabbage,* or *Pak-choi, Shang-hae-oil-plant*
Eruca. *Salad-Rocket*
Erucastrum. *Bastard Rocket*
muralis. *Wall-Rocket, Wild Rocket*
Napo-brassica. *Turnip Cabbage*
Napus. *Navew, Winter Rape*
Napus var. oleifera. *Carcel-oil-plant, Cole, Colesat, Cole-seed, Colza* or *Coltza*
Napus var. Teltow *Turnip*
oleracea. *Garden Cabbage,* or *Cole-wort*
oleracea var. acephala. *Borecole,* or *Kale*
oleracea var. Botrytis cauliflora. *Cauliflower*
oleracea var. Botrytis asparagoides. *Broccoli*
oleracea var. bullata major. *Savoy Cabbage*

Brassica oleracea var. bullata minor. *Brussels Sprouts*
oleracea var. capitata. *Drum-head Cabbage*
oleracea var. costata. *Coure Tronchuda Cabbage*
oleracea var. sabellica. *Scotch Kale*
Rapa var. depressa. *Common Turnip*
Bravoa geminiflora. *Scarlet Twin-flower*
Brayera anthelmintica. *Abyssinian Tape-worm Plant*
Bremontiera Ammoxylon. *Sand-wood*
Brevoortia (Brodiæa) coccinea. *Crimson Satin-flower*
Briza maxima. *Pearl-Grass*
media. *Dadder* or *Dodder Grass, Dithering* or *Dothering Grass, Doddle-Grass, Earthquakes, Jockey-Grass, Lady's Hair, Maiden-hair-Grass, Pearl-Grass, Quake* or *Quaking Grass, Shaking Grass*
Brizoporum spicatum. *Spike-Grass*
Brodiæa. *Brodie's Lily, Californian* or *Missouri Hyacinth*
coccinea. *Crimson-flowered Californian Hyacinth, Crimson Satin-flower, Vegetable Fire-Cracker*
congesta. *Allium-like Californian Hyacinth*
grandiflora. *Large-flowered Californian Hyacinth*
ixioides (Calliprora lutea). *Ixia-like Californian Hyacinth*
volubilis. *Twining Californian Hyacinth*
Bromelia Karatas. *Curra-Tow-plant*
sphacelata. *Hardy Bromelia*
Pinguin. *Penguin,* or *Pinguin-plant,* of the W. Indies, *Wild Pine-apple*
Bromus. *Brome Grass*
ciliatus. *Prairie Grass*
Kalmii. *American Chess*
mollis. *Blubber-Grass, Bull-Grass, Cock-Grass, Goose Corn, Haver-Grass, Hooded Grass, Lob,* or *Lop-Grass, Out-Grass*
Schraderi. *Prairie-Grass, Rescue-Grass*
secalinus. "*Brullaum,*" *Chess* or *Cheats, Cock-Grass, Drake, Droke, Wild Oat*
sterilis. *Black Grass, Drake, Droke, Haver-Grass*
Brosimum Alicastrum. *Bread-nut Tree, Snake-wood-tree*
Aubletii. *Leopard-wood, Letter-wood, Snake-wood*
Galactodendron. *Cow-tree,* or *Milk-tree,* of Guiana, "*Palo de Vaca*"
spurium. *Milk-tree,* of Jamaica
Broussonetia papyrifera. *Paper-Mulberry, Tapa-cloth-tree,* of Polynesia
Brownea Rosa. *West Indian Mountain Rose*
Brugmansia (Datura) arborea. *Fever-plant*
suaveolens. *Angel's-Trumpet-plant, White Peruvian Trumpet-flower*
Brunellia comocladifolia. *W. Indian Sumach*
Brunsvigia (Amaryllis) Josephinæ. *Chandelier Flower, Royal Brunswick Lily*
toxicaria. *Cape Poison-bulb*
Brya Ebenus. *Cocus-wood, Green Ebony,* Jamaica or *W. Indian Ebony*

English Names of Cultivated, Native,

Bryonia. *Bryony*
dioica. *Devil's Turnip, Grape-wort, Isle-of-Wight-Vine, Mandrake, Murrain-berries, Tetter-berry, Red* or *White Bryony, White Wild Vine, Wild Hop*
Bryonopsis erythrocarpa. *Ornamental-fruited Vine*
Bryophyllum calycinum (B. proliferum). *Life-plant, Sea-Leaf, Sprout-Leaf*
Bryum. *Thread-Moss*
affine. *Many-stalked Thread-Moss*
albicans. *Pale-leaved Thread-Moss*
alpinum. *Red Alpine Thread-Moss*
androgynum. *Narrow-leaved Thread-Moss*
argenteum. *Silvery Thread-Moss*
cæspititium. *Lesser Matted Thread-Moss*
capillare. *Greater Matted Thread-Moss*
carneum. *Pink-fruited Thread-Moss*
cuspidatum. *Wood-Moss*
dealbatum. *Pale-leaved Thread-Moss*
gracile. *Slender Thread-Moss*
hornum. *Swan's-neck Thread-Moss*
jubaceum. *Slender-branched Thread-Moss*
palustre. *Marsh Thread-Moss*
pyriforme. *Golden Thread-Moss*
roseum. *Rosette Thread-Moss*
rurale. *Thatch-Moss*
triquetrum. *Long-stalked Thread-Moss*
ventricosum. *Swelling Bog Thread-Moss*
Bubon macedonicum. *Macedonian Parsley*
Buchanania latifolia. *Cheeroojee,* or *Cheeroonjee-oil-plant*
Buchloe dactyloides. *Buffalo-grass,* of N. America
Buchnera americana. *Blue-hearts*
Bucida Buceras. *Black Olive,* of Jamaica, *French Oak, Olive-bark-tree, Ox-horn*
capitata. *Yellow Sanders-wood*
Buddlea globosa. *Orange-ball Tree*
Buena hexandra. *China-Bark-plant*
Bulbocodium vernum. *Spring Meadow-Saffron*
tenuifolium. *Slender-leaved Meadow-Saffron*
trigynum. *Caucasian Meadow-Saffron*
Bumelia ingens. *Black Bullet-tree*
lycioides. *American Iron-wood, Southern Buck-thorn*
montana. *Mountain Bully-tree,* of Jamaica
nigra. *Black Bully-tree*
retusa. *Ballata-tree, Bastard Bullet-tree*
salicifolia. *Galimeta-wood*
Bunchosia. *W. Indian Cherry*
Bunias orientalis. *Hill Mustard*
Bunium Bulbocastanum. *Large Earth-nut*
flexuosum. *Ar-nut, Cat-nut, Cipper-nut, Devil's Oatmeal, Earth-nut, Earth-chest-nut, Fare-nut* or *Vare-nut, Hare-nut, Hawk-nut, Hog-nut, Jur-nut, Kipper-nut, Knipper-nut, Pig-nut, St. Anthony's Nut*
Buonaparten juncea. *Peruvian Hemp*
Buphane (Hæmanthus) toxicaria. *Poison-bulb,* of the Cape
Buphthalmum. *Ox-eye*
aquaticum. *Sweet-scented Ox-eye*
salicifolium. *Willow-leaved Ox-eye, Yellow Ox-eye*
speciosum. *Heart-leaved Ox-eye*

Bupleurum. *Hare's-ear*
fruticosum. *Shrubby Hare's-ear, Tree Thorough-wax*
rotundifolium. *Common Hare's-ear, " Modesty,"* of N. America, *Thorough-wax* or *Thorow-wax*
Burchellia capensis. *Buffelhorn-wood*
Bursaria spinosa. *Tasmanian Box, Christ-Tree,* or *Native Thorn*
Bursera acuminata. *Carana-resin-plant*
gummifera. *American Gum-tree, Chibou,* or *Cachibou-resin-plant, Jamaica,* or *W. Indian Birch, W. Indian Mastich-tree*
Burtonia scabra. *Burton's Pea-bush*
Busbeckia arborea. *Cape-tree,* of N. S. Wales
Butea frondosa. *Bengal-Kino-tree, Dhak-tree, Keeso-flower-tree, Pulas-tree,* or *Pulus-tree*
superba. *Deccan Bear-tree, Palas-tree,* or *Pulas-tree*
Butomus umbellatus. *Flowering Rush, Lily-grass, Water-Gladiole*
Butyrospermum (Bassia) Parkii. *African,* or *Shea, Butter-tree*
Buxus. *Box-tree*
balearica. *Minorca Box, Giant Box-tree*
chinensis. *Chinese Box*
Fortunei rotundifolia. *Fortune's Round-leaved Box*
sempervirens. *Common Box*
sempervirens var. angustifolia. *Narrow-leaved Box*
sempervirens var. arborescens. *Tree Box*
sempervirens var. argentea. *Silvery-leaved Box*
sempervirens var. aurea. *Gold-edged Box*
sempervirens var. myrtifolia. *Myrtle-leaved Box*
sempervirens var. pyramidata. *Pyramidal Box*
sempervirens var. rosmarinifolia. *Rosemary-leaved Box*
sempervirens var. rotundifolia. *Round-leaved Box*
sempervirens var. suffruticosa. *Dwarf Box*
Byrsonima cinerea. *Golden-spoon,* of the W. Indies
coriacea. *W. Indian Locust-tree*
spicata. *Muruxi-Bark-tree, Shoe-maker's-bark-tree*

Cacalia (Emilia) atriplicifolia. *Pale Indian Plantain, Wild Caraway*
aurantiaca. *Orange Tassel-flower*
coccinea. *Candle-plant, Scarlet Tassel-flower*
Kleinia. *Cabbage-tree, Carnation-tree*
reniformis. *Great Indian Plantain*
suaveolens. *American Sweet Centaury*
tuberosa. *Tuberous Indian Plantain*
Cachrys Libanotis. *Rosemary Frankin-cense*
Cactus curassavicus. *Pin-pillow Cactus*
Echinocactus. *" Hedge-hog-thistle "*
flagelliformis. *Rat's-tail Cactus*
Melocactus. *Turk's-cap Melon-thistle*
Opuntia. *Indian Fig, Prickly Pear*
peruvianus. *" Thorny Reed,"* of Peru

and Foreign Plants, Trees, and Shrubs.   171

Cæsalpinia bijuga. *Sarin-tree*, of Jamaica
Bonducella. *Bonduc Seeds-* or *Grey Nicker Seeds-* or *Nuts-plant*
brasiliensis. *Brazil-wood* of commerce
brevifolia. "*Algoborillo*"
coriaria. *Divi-Divi-tree*
Crista. *Bahama Braziletto*
echinata. *Fernambuc*, or *Red Brazilwood*, *Lima-wood*, *Nicaragua-wood*, *Peach-wood*, *Wood-of-St. Martha*
Pipai. *Pipi-pod-tree*
(Poinciana) pulcherrima. *Peacock Flower Flower-fence* of India, *Pride of Barbadoes*
Sappan. *Bukkum-wood*, *Sappan-wood*
sepiaria. *Mysore Thorn*
Cæsia vittata. *Blue Grass-Lily*
Cajanus indicus. *Dhal*, or *Dhol*, of India, *Catjang*, *Angola Pea*, *Pigeon Pea*
indicus bicolor. *Congo Pea*, of Jamaica
indicus flavus. *No-eye Pea*, of Jamaica
Cakile americana. *American Sea-Rocket*
maritima. *Common Sea-Rocket*
Caladium esculentum. *Indian Kale*, "*Tanga*," of S. Carolina, *Taro*, or *Tara-plant*, of the Sandwich Islands
grandiflorum. *Large-leaved Indian Kale*
nymphæifolium. *Water-lily-leaved Indian Kale*
sagittæfolium. *Arrow-leaved Indian Kale*
seguinum. *Dumb Cane*
Calamagrostis arenaria. *Sea-sand Reed*
argentea. *Silvery Wood-reed*
canadensis. *Blue Joint-grass*
Epigejos. *Bush Grass*, *Common Wood-reed*, *Feather-top Grass*, *Writing-reed*
lanceolata. *Spear-leaved Wood-reed*
Calamintha. *Calamint*
Acinos. *Balm*, *Basil Thyme*, *Clairo*, *Field*, *Stone*, or *Wild Basil*, *Field Calamint*
alpina. *Alpine Calamint*
Clinopodium. *Basil-weed*, *Field*, *Stone*, or *Wild Basil*, *Hedge-Calamint*, *Horse-Thyme*
glabella. *Tom Thumb Calamint*
Nepeta. *Field Balm*, *Field Calamint*, *Small-flowered Calamint*
officinalis. *Cat-mint*, *Medicinal Calamint*
sylvatica. *Wood Calamint*
Calampelis (Eccremocarpus) scaber. *Eccremocarpus Vine*
Calamus (Dæmonorops) Draco. *Dragon's-blood Palm*. Yields the "Dragon's-blood" of commerce
flagellum. *Reem Cane*
Rotang. *Chair-bottom Cane*, *Rattan Cane*
rudentum. *Chair-bottom Cane*, *Great Rattan Cane*
Scipionum. *Malacca Cane*
verus and C. viminalis. *Chair-bottom Cane*
Calandrinia discolor. *Common Rock-Purslane*
nitida. *Shining Rock-Purslane*
umbellata. *Umbel-flowered Rock-Purslane*
Calathea zebrina. *Zebra-plant*
Calceolaria. *Slipper-flower, Slipper-wort*
chelidonioides. *Celandine Slipper-flower* or *Slipper-wort*
Kellyana. *Kelly's Hardy Slipper-flower* or *Slipper-wort*

Calceolaria pinnata. *Wing-leaved Slipper-flower*
Calea lobata. *W. Indian Halbert-weed*
Calectasia cyanea. *Australian Blue Spike-flower*
Calendula (Dimorphotheca) hybrida. *Large Cape Marigold*
(Dimorphotheca) pluvialis. *Small Cape Marigold*
officinalis. *Golds*, *Goldins*, *Mary-bud*, *Pot-Marigold*
Calla (Richardia) æthiopica. *Arum Lily*, *Egyptian Lily*, *Lily-of-the-Nile*
palustris. *Bog Arum*, *Water Dragons*
Calliandra. *W. Indian Horse-wood*
comosa. *Horse-wood*, of Jamaica
purpurea. *Soldier-wood*
trinervia. *Silk-flower*
Callicarpa americana. *French Mulberry*
incana. *Urn-fruit Tree*
Calliopsis bicolor. *Eye-flower*
Callirhoe digitata. *Finger-leaved Poppy-Mallow*
involucrata. *Crimson-flowered Poppy-Mallow*
macrorrhiza. *White-flowered Poppy-Mallow*
pedata. *Long-stalked Poppy-Mallow*
triangulata. *Purple-flowered Poppy-Mallow*
Callistemma hortense (Callistephus hortensis). *China Aster*
Callistemon. *Paper-bark-tree*, of Australia
pallidum and C. salignum. *Tea-tree*, of N. S. Wales
Callistephus chinensis (Callistemma hortense). *China Aster*
Callithamnion floridulum. "*Figs*"
Callitriche. *Star-grass*, *Water Star-wort*
verna. *Water Fennel*
Callitris australis. *Oyster Bay Pine-tree*
quadrivalvis (Thuja articulata). *Jointed Arbor-vitæ*, *Juniper* or *Sandarach Gum-tree*, *Pounce-tree*
Ventenatii. *Camphor-wood*
vulgaris. *Heather*, *Ling*, *Long Heath*
Calocephalus Brownii. *Australian Garland-flower*
Calochortus. *Butterfly-Tulip*, *Butterfly-weed*, *Mariposa Lily*, *Pretty-grass*, *Wild Tulip*, of California
albus (Cyclobothra alba). *White Mariposa Lily*
cœruleus. *Blue Mariposa Lily*
citrinus. *Lemon-coloured Mariposa Lily*
Gunnisoni. *Gunnison's Mariposa Lily*
lilacinus. *Lilac Mariposa Lily*
luteus. *Yellow Mariposa Lily*
luteus var. oculatus. *Large-eyed Mariposa Lily*
Nuttalli. *Sego*
pulchellus (Cyclobothra pulchella). *Pretty-flowered Mariposa Lily*
splendens. *Brilliant Mariposa Lily*
uniflorus. *One-flowered Butterfly Tulip*
venustus. *Elegant-flowered Mariposa Lily*
Calodendron capense. *Buchu-tree*, *Cape Chestnut*
Calophaca wolgarica. *Wing-leaved Calophaca*

172 English Names of Cultivated, Native,

Calophyllum angustifolium. *Piney-tree,* of Penang
Calaba. *Calaba-tree, Keena-nut Tree, Santa-Maria Tree*
Inophyllum. *Alexandrian Laurel, Oondee, Pinnacotty, Poon-seed,* or *Poonay Oil-plant, St. Mary's-wood, Tamanu* or *Dilo Oil-plant*
spurium. *Pootungee Oil-plant*
Calonyction. *Midnight Lily*
Calopogon pulchellus. *Grass-Pink Orchis*
Calothamnus. *Australian Net-bush*
Calotropis gigantea. *Bow-string Hemp,* or *Mudar-plant,* of India
procera. *French Cotton, French Jasmine, Mudar-plant*
Caltha. *Marsh-Marigold*
 grandiflora. *Large-flowered Marsh-Marigold*
 leptosepala. *White-flowered Marsh-Marigold*
 palustris. "*Boots*," *Common Marsh-Marigold, Golds, Goldins, May-blobs, Meadow Bright,* or *Meadow Bout, Meadow Gowan, Water Butter-cup, Water,* or *Open Gowan, Yellow Gowan*
 palustris flore-pleno. *Double Marsh-Marigold*
 purpurascens. *Purplish Marsh-Marigold*
Calycanthus floridus. *Common Carolina Allspice*
 glaucus. *Glaucous-leaved Carolina Allspice*
 lævigatus. *Smooth-leaved Carolina Allspice*
Calycocarpum Lyoni. *Lyon's Cup-seed*
Calypso borealis. *Calypso Orchis*
Calyptranthes. *Lid-flower-tree*
 Chytraculia. *Bastard Green-heart,* of the W. Indies, *White Rod-wood*
 Jambolana. *Jambolan-tree, Java Plum*
Calyptronoma Swartzii. "*Thatch*," of the W. Indies
Calystegia. *Bear-bind, Bind-weed*
 dahurica (Convolvulus dahuricus). *Dahurian Bind-weed*
 pubescens plena. *Double Chinese Bind-weed*
 spithamea. *Dwarf Bind-weed*
Calythrix tetragona. *Hair-cup-flower*
Camassia esculenta. *Bear's Grass, Quamash* or *Camash*
 esculenta var. alba. *White-flowered Quamash*
 (Scilla) Fraseri. *Wild Hyacinth,* of N. America
 Leichtlini. *Californian White-flowered Quamash*
Camelina sativa. "*False-Flax*," of N. America, *Gold-of-pleasure, Oil-seed-plant, Siberian Oil-seed*
Camellia. *Japan Rose*
 albaplena. *Double White Camellia*
 anemonæflora. *Anemone-flowered Camellia*
 Bealii. *Bright Crimson Camellia*
 candidissima. *Creamy-White Camellia*
 caryophylloides. *White, rosy-carmine-striped, Camellia*
 commensa. *Vermilion-red Camellia*

Camellia compacta alba. *Fine White Camellia*
 Dride. *Bright-rose-coloured Camellia*
 euryoides. *Small-flowered Camellia*
 fimbriata. *Pure White, fringed, Camellia*
 Grunelli. *Large White Camellia*
 imbricata. *Carmine-rose Camellia*
 japonica and vars. *Common Camellia*
 japonica Rosa-mundi. *Rose-of-the-World*
 Kissi. *Kissi Camellia*
 Lombarda. *Superb Rosy-flowered Camellia*
 maliflora. *Apple-flowered Camellia*
 Matthotiana. *Bright Crimson Camellia, Matthoti's Camellia*
 Matthotiana alba. *Pure White, Imbricated, Camellia*
 Mrs. Cope. *White, crimson-striped Camellia*
 ochroleuca. *Creamy-white Camellia*
 Optima. *Bright-rose, crimson-rayed, Camellia*
 Queen Victoria. *Red, white-ribboned, Camellia*
 reticulata. *Captain Rawes's Camellia*
 Rubens. *Deep-rose, large-flowered, Camellia*
 Saccoi nova. *Rosy-pink Camellia*
 Sasanqua. *Lady Banks's Camellia, Tea-oil-plant*
 Spinii. *Snow-white Camellia*
 sulcata. *Yellowish-white Camellia*
 tricolor. *Three-coloured Camellia*
 Triumphans. *Large Crimson Camellia*
 Valtevaredo. *Fine Rosy-flowered Camellia*
Cameraria latifolia. *False Manchineel Tree*
Campanula. *Bell-flower*
 aggregata. *Crowded-flowered Bell-flower*
 alliariæfolia. *Alliaria-leaved Bell-flower*
 Allionii. *Allioni's Bell-flower*
 americana. *Tall Bell-flower,* of N. America
 alpina. *Alpine Hare-bell*
 aparinoides. *Marsh Bell-flower,* of N. America
 attica. *Attic Annual Bell-flower*
 autumnalis (Platycodon autumnale). *Autumn Bell-flower*
 barbata. *Bearded Hare-bell*
 Barrelieri. *Barrelier's Bell-flower*
 cæspitosa (C. pumila). *Tufted Hare-bell*
 carpatica. *Carpathian Hare-bell*
 celtidifolia. *Nettle-leaved Bell-flower*
 cenisia. *Mont Cenis Hare-bell*
 Cervicaria. *Throat-wort*
 collina. *Sage-leaved Bell-flower*
 divaricata. *Branching Bell-flower*
 Elatines. *Elatine Hare-bell*
 elegans. *Elegant Bell-flower*
 Erinus. *Forked Bell-flower*
 fragilis. *Brittle Hare-bell*
 garganica. *Mt. Gargano Hare-bell*
 glomerata. *Clustered Bell-flower, Danes'-Blood*
 grandiflora (Platycodon grandiflorum). *Noble Hare-bell*
 grandis. *Great Bell-flower*
 hederacea (Wahlenbergia hederacea). *Ivy-leaved Hare-bell*
 Hendersoni. *Henderson's Bell-flower*
 Hohenhackeri. *Hohenhacker's Bell-flower*

Campanula Hostii. *Host's Bell-flower*
hybrida. *Corn-field Bell-flower, Corn Violet*
isophylla (C. floribunda). *Ligurian Hare-bell*
lactiflora. *Milk-white Bell-flower*
lanceolata. *Spear-leaved Bell-flower*
Langsdorffiana. *Langsdorff's Bell-flower*
latifolia. *Broad-leaved Bell-flower, Hask-wort*
Leutweinii. *Leutwein's Bell-flower*
linifolia. *Flax-leaved Bell-flower*
linnæifolia. *Swamp Bell-flower*, of California
Loefflingi. *Loeffling's Annual Hare-bell*
longifolia. *Long-leaved Bell-flower*
Loreyi. *Lorey's Annual Hare-bell*
macrantha. *Large-flowered Bell-flower*
macrantha alba. *Large-flowered White Bell-flower*
macrostyla. *Long-styled Annual Bell-flower, Candelabrum Bell-flower*
Medium. *Canterbury Bells, Coventry Bells, Coventry Rapes, Mariettes, "Mercury's Violets"*
muralis. *Wall Bell-flower*
nitida. *Shining Hare-bell*
nobilis. *Long-flowered Hare-bell*
patula. *Spreading Bell-flower*
persicifolia. *Peach Bells, Peach-leaved Bell-flower*
petræa. *Woolly-leaved Bell-flower*
Portenschlagiana. *Portenschlag's Bell-flower*
primulæfolia. *Primrose-leaved Bell-flower*
pulla. *Dark-coloured Dwarf Hare-bell, Austrian Hare-bell*
pumila. *Dwarf Bell-flower*
punctata. *Spotted Bell-flower*
pusilla. *Diminutive Bell-flower*
pusilla alba. *Diminutive White Bell-flower*
pyramidalis. *Chimney-plant, Pyramidal Bell-flower, Steeple Bells*
Raineri. *Rainer's Hare-bell*
Rapunculus. *Garden Rampion*
rapunculoides. *Creeping Bell-flower*
rhomboidea. *Diamond-leaved Bell-flower*
rotundifolia. *Blawort, Blue-bell* of Scotland, *Common Hare-bell* or *Hair-bell, Lady's Thimble, Witches' Bells*
sarmatica (C. gummifera). *Gummy Bell-flower*
Sibthorpii. *Sibthorp's Bell-flower*
soldanellæflora. *Soldanella-flowered Hare-bell, Double-flowered Hare-bell*
speciosa. *Showy Hare-bell*
(Specularia) Speculum. *Lady's*, or *Venus's, Looking-glass*
Thompsoni. *Thompson's Bell-flower*
Trachelium. *Blue Fox-glove, Canterbury Bells, Coventry-bells, Great Throat-wort, Hask-wort, Nettle-leaved Bell-flower*
turbinata. *Turban Bell-flower, Vase-flowered Hare-bell*
urticæfolia. *Nettle-leaved Bell-flower*
Van Houttei. *Van Houtte's Bell-flower*
(Symphiandra) Wanneri. *Wanner's Bell-flower*

Campanula Zoysii. *Zoys's Hare-bell*
Campomanesia linearifolia. *Bulillos-tree*, of Peru
Camptosorus rhizophyllus. *Walking-leaf Fern*
Cananga odorata. *Ylang-Ylang-tree*
Canarina Campanula. *Canary Island Bell-flower*
Canarium commune. *Chinese Olive, Java Almond, Kanari-oil-plant*
strictum. *Black Dammar-tree*
Canavalia ensiformis. *Horse-bean,* of Jamaica
gladiata. *Horse-bean*, of the W. Indies.
"Overlook" of Jamaica, *Sword-Bean*
obtusifolia. *Sea-side Bean*, of the W. Indies
Canella alba. *Canella Bark-tree, Wild Cinnamon*
Canna indica. *Flowering Reed, Flowering Shot, Indian Reed, Indian Shot*
peruviana. *Peruvian Arrow-root-plant*
Cannabis sativa. *Carl-hemp, Common Hemp, Fimble, Gallow-grass*
Cantharellus cibarius. *Chantarelle Mushroom*
Cantua buxifolia. *Peruvian Magic-tree*
Capparis aphylla. *Kureel-tree*
cynophallophora. *Bottle-cod-root, Devil's-bean*
ferruginea. *Mustard-shrub*
sinaica. *Mountain Pepper*
sodada. *Timbuctoo Caper-bush*
spinosa. *Common Caper-bush*
Capraria biflora. *Goat-weed, Sweet-weed, Tea-plant,* of Martinique, *W. Indian Tea-tree*
Caprifolium belgicum. *Dutch Honey-suckle*
Periclymenum. *Common Honey-suckle, Wood-bine*
Caproxylon Heddewigii. *West Indian Pig-wood*
Capsella Bursa-pastoris. *Mother's-Heart, Pick-purse, "Poor-man's Parmacetie," Shepherd's-purse* or *Shepherd's-scrip, Tooth-wort, Toy-wort*
Capsicum. *Bird's-eye Pepper, Garden Ginger*
annuum and C. fastigiatum. *Cayenne Pepper, Chillies, Guinea Pepper, Red Pepper*
baccatum. *Bird-Pepper*
cerasiforme. *Cherry-Pepper*
frutescens. *Goat-Pepper, Spur-Pepper*
grossum. *Bell-Pepper*
tetragonum. *Bonnet-Pepper*
Caragana. *Siberian Pea-tree*
Altagana. *Flat-podded Siberian Pea-tree*
arborescens. *Common Siberian Pea-tree*
arenaria. *Sand Siberian Pea-tree*
Chamlaga. *Chinese Pea-tree*
frutescens. *Shrubby Siberian Pea-tree*
grandiflora. *Large-flowered Siberian Pea-tree*
jubata. *Maned Siberian Pea-tree*
mollis. *Soft-leaved Siberian Pea-tree*
microphylla. *Small-leaved Siberian Pea-tree*

Caragana pygmæa. *Pigmy Siberian Pea-tree*
spinosa. *Thorny Siberian Pea-tree*
Redowskii. *Redowski's Siberian Pea-tree*
tragacanthoides. *Goat's-thorn-like Siberian Pea-tree*
Carapa angustifolia. *Pottery-tree*, of French Guiana
fasciculata. *Tamacoari Balsam-tree*
guianensis. *Carap, Crab*, or *Andiroba Oil-plant, Crab-wood*, of Guiana
guineensis (C. Touloucouna). *Coondi, Kundah*, or *Tallincoonah Oil-plant, Kundoo-nut-tree*
Cardamine. *Bitter-cress*
amara. *Large Bitter-cress*
asarifolia. *Asarum-leaved Cuckoo-flower*
hirsuta. *Hairy Bitter-cress, Lamb's Cress*
latifolia. *Broad-leaved Cuckoo-flower*
macrophylla. *Large-leaved Cuckoo-flower*
pratensis. *Bread-and-Milk, Common Cuckoo-flower, Cuckoo's-Bread, Cuckoo-Spit, Lady's Smock, May-flower, Meadow-Bitter-cress, Meadow-cress*
pratensis fl.-pl. *Double Cuckoo-flower*
rhomboidea. *American Spring Cress*
rotundifolia. *American Mountain Water-cress, Round-leaved Cuckoo-flower*
trifolia. *Three-leaved Cuckoo-flower*
Cardiospermum Corindum. *Heart-seed, Indian Heart*
grandiflorum. *Supple-jack*, of Jamaica
Halicacabum. *Balloon-Vine, Heart-Pea*
Cardopatium corymbosum. *Black Chamæleon*
Carduus. *Thistle*
acanthoides. *Welted Thistle*
(Cnicus) acaulis. *Dwarf Thistle*
anglicus. *Gentle Thistle*
arvensis. *Bour Thistle, Corn Thistle, Creeping Thistle, Dog Thistle, Waste Thistle*
benedictus. *Blessed Thistle*
eriophorus. *Friar's Crown, Woolly-headed Thistle*
heterophyllus. *Melancholy Thistle*
lanceolatus. *Bird Thistle, Boar Thistle, Bull Thistle, Bur Thistle, Plume Thistle, Spear Thistle*
Marianus. *Blessed Thistle, Milk Thistle, Our Lady's Thistle*
nutans. *Musk Thistle, Scotch Thistle* (of artists)
(Cnicus) palustris. *Marsh Thistle*
(Cnicus) pratensis. *Meadow Thistle*
pycnocephalus (C. tenuiflorus). *Slender Thistle*
(Cnicus) tuberosus. *Tuberous-rooted Thistle*
Carex. *Blue Grass, Sedge, Shore Grass, Spire Grass*
acuta. *Slender-spiked Sedge*
alpina. *Alpine Sedge*
ampullacea. *Bottle Sedge*
appressa. *Otago Tupak-grass*
arenaria. *Sea-side Sedge*
atrata. *Black Sedge*
axillaris. *Axillary Sedge*
baccans. *Crimson-fruited Sedge*
binervis. *Green-ribbed Sedge*

Carex Buxbaumii. *Hoary Sedge*
cæspitosa. *Hassock-grass, Tufted Sedge*
canescens (C. curta). *White Sedge*
capillaris. *Capillary Sedge*
depauperata. *Starveling Sedge*
digitata. *Fingered Sedge*
dioica. *Diœcious Sedge*
distans. *Loose Sedge*
divisa. *Bracteated Marsh Sedge*
divulsa. *Carpeting Sedge*
elongata. *Elongated Sedge*
extensa. *Bracteated Sea Sedge*
filiformis. *Slender-leaved Sedge*
flava. *Hedge-hog Grass, Yellow Sedge*
fulva. *Tawny Sedge*
glauca. *Carnation Grass, Gilliflower Grass, Glaucous Heath Sedge*
hirta. *Hammer Sedge, Hairy Sedge*
humilis. *Dwarf Sedge*
incurva. *Curved Sedge*
lævigata. *Smooth Sedge*
lagopina. *Hare's-foot Sedge*
leporina (C. ovalis). *Oval-spiked Sedge*
limosa. *Drooping Bog Sedge*
montana. *Mountain Sedge*
muricata. *Prickly Sedge*
pallescens. *Pale Sedge*
paludosa. *Lesser Bank Sedge, Sniddel*
panicea. *Carnation-grass, Gilliflower Grass, Pink-leaved Sedge*
paniculata. *Hassock-grass, Panicled Sedge*
paradoxa. *Paradoxical Sedge*
pauciflora. *Few-flowered Sedge*
pendula. *Great Pendulous Sedge*
pilulifera. *Round-headed Sedge*
præcox. *Spring Sedge*
Pseudo-cyperus. *Cyperus-like Sedge*
pulicaris. *Flea Grass, Flea Sedge*
punctata. *Dotted Sedge*
remota. *Remote-spikeleted Sedge*
rigida. *Stiff Mountain Sedge*
riparia. *Greater Bank Sedge*
rupestris. *Rock Sedge*
saxatilis. *Highland Sedge*
stellulata. *Lesser Prickly Sedge*
stricta. *Greater Tufted Sedge, Tussock Sedge*
strigosa. *Loose-flowered Sedge*
sylvatica. *Wood Sedge*
tomentosa. *Downy Sedge*
vesicaria. *Bladder Sedge*
vulgaris. *Lesser Tufted Sedge*
vulpina. *Great Spiked Sedge*
Careya arborea. *Hindostanee Slow-match Tree*
Cargillia arborea. *Gray Plum-tree*, of Australia
australis. *Black Plum*, of Illawarra
Carica Papaya. *Melon-tree, Papaw-tree*
Carissa Carandas. *Carandas-tree*
Xylopicron. *Bitter-wood*, of Madagascar
Carlina acanthifolia. *Acanthus-leaved Carline-thistle*
gummifera. *White Chamæleon*
vulgaris. *Common Carline-thistle*
Carludovica. *Small-Palmetto Palm*
insignis. *Humble Palmetto Palm*
palmata. *Panama Hat-tree*
Carmichaelia australis. *New Zealand Broom*

and Foreign Plants, Trees, and Shrubs. 175

Carolinea (Pachira) Barrigon. *Barrigon-tree*, of Panama princeps. *Guiana Chestnut*
Carpinus americana. *American Horn-beam, Blue Beech, Iron-wood, Water Beech*
Betulus. *Common Horn-beam, Horn Beech, Horse, Horst,* or *Hurst Beech, White Beech, Yoke Elm*
Betulus var. incisa. *Cut-leaved Horn-beam*
Ostrya. *Hop Horn-beam*
Carpodiscus acidus. *Sour Pishamin-tree*, of Sierra Leone
dulcis. *Sweet Pishamin-tree*
Carthamus lanatus. *Distaff Thistle*
tinctorius. *Bastard* or *False Saffron, Kossumba* or *Safflower Oil-plant, Saf-flower, Saffron-thistle*
Carum Carui. *Common Caraway*
(Bunium) Bulbocastanum. *Pig-nut, Tuberous-rooted Caraway*
Gairdneri and Kelloggii. *Edible-rooted Caraway*, of California
(Sison) verticillatum. *Whorled-leaved Caraway*
Carumbium populifolium (Omalanthus populifolius). *Queensland Poplar*
Carya (Juglans). *Hickory*
alba. *Shell-bark,* or *Shag-bark, Hickory*
amara. *Bitter-nut*, of N. America, *Swamp Hickory, White Hickory*
aquatica. *Water Hickory*
compressa. *Compressed-fruited Hickory*
integrifolia. *Entire-leafleted Hickory*
microcarpa. *Small-fruited Hickory*
(Juglans) myristicæformis. *Nutmeg-Hickory*
olivæformis. *Illinois Nut-tree, Pecan-nut-tree*
porcina. *Broom Hickory, Brown Hickory, Pig-nut*
sulcata. *Springfield Nut, Thick Shell-bark Hickory, Western Shell-bark Hickory*
tomentosa. *Mocker-nut, White-heart Hickory*
Carynocarpus lævigatus. *New Zealand Laurel*
Caryocar braziliense. *Piquia-oil-plant*
nuciferum. *Butter-nut-tree*, of S. America, *Souari* or *Suwarrow-nut Tree*
tomentosum. *Butter-nut*, of Guiana
Caryophyllus aromaticus. *Clove-tree*
Caryota urens. *Jaggery Palm, Bastard Sago-Palm, East Indian Wine-Palm, Toddy-Palm*
Casearia ulmifolia. *Brazilian Snake-root*
Casimiroa edulis. *Mexican Apple, White Sapota*
Cassandra (Andromeda) calyculata. *Leather-leaf*
Cassia acutifolia. *Senna-plant* of commerce
alata. *Ring-worm shrub, Winged Senna-plant*
angustifolia. *Senna-plant* of commerce, *Tinnevelly Senna-plant*
Brewsteri. *Cigar Cassia, Queensland Laburnum*
Chamæcrista. *Partridge-Pea*

Cassia emarginata. *Notch-leaved Senna-tree*
Fistula. *Indian Laburnum, Purging Cassia, Pudding-Pipe-tree*
florida. *Iron-wood*, of Dutch E. Indies
glandulosa. *Dutchman's-butter*
lanceolata. *Alexandrian Senna-plant*, of commerce
marilandica. *Maryland Senna-plant*
nictitans. *Sensitive Pea,* or *Wild Sensitive Plant*, of N. America
obovata. *Aleppo,* or *Italian, Senna-plant*
occidentalis. *Stinking-weed,* or *Stinking-wood*
Cassine capensis. *Cape Phillyrea*
Colpoon. *Cape Ladle-wood, Colpoon-tree*
Maurocenia. *Hottentot Cherry*
Cassinia fulgida. *Golden-bush*
Cassinopsis capensis. *Caffre Broom-tree*
Cassytha. *Scrub-Vine*, of Australia
melantha. *Scrub-Vine*, of Victoria
Castanea. *Chestnut*
chrysophylla. *Golden-leaved Chestnut*
pumila. *Chinquapin*, of N. America
vesca. *Spanish,* or *Sweet, Chestnut*
vesca variegata. *Variegated Chestnut*
vesca var. americana. *American Chestnut*
vesca var. asplenifolia. *Fern-leaved Chestnut*
Castanopsis chrysophylla. *Western Chinquapin*
Castanospermum australe. *Australian Bean-tree, Moreton Bay Chestnut*
Castelea Nicholsoni. *W. Indian Goat-bush*
Castilleia coccinea. *Scarlet Painted Cup*
Castilloa elastica. *Caoutchouc-tree*, of Panama
Markhamiana. *Caoutchouc-tree*, of Panama
Casuarina. *Beef-wood, She-Oak*, of Australia
equisetifolia. *Iron-wood*, of the S. Sea Islands, *Swamp Oak*, of Australia, *Tree-Horse-tail*
glauca. *Desert She-Oak*
leptoclada. *River Oak*, of N. S. Wales
paludosa. *Australian Swamp Oak*
quadrivalvis. *Coast She-Oak, Tasmanian She-Oak*
stricta. *Australian Beef-wood, He-Oak*
suberosa. *Erect She-Oak, Swamp Oak*, of Australia
torulosa. *Forest Oak*, of Botany Bay
Catabrosa (Aira) aquatica. *Water Hair-grass, Water Whorl-grass*
Catananche cœrulea. *Blue Cupidone, Blue Succory, Blue-flowered Cupid's-dart*
lobata. *Lobed-leaved Catananch*
lutea. *Yellow Succory*
Catalpa bignonioides. *Indian Bean*, of N. America
longissima. *French Oak, Jamaica Yoke-wood, St. Domingo Oak*
Catha edulis. *Abyssinian*, or *Arabian, Tea-plant*
Cathartocarpus conspicua. *Drum-stick Tree*
marginatus. *Horse-Cassia*
Cathcartia villosa. *Cathcart's Poppy*
Catesbæa spinosa. *Lily-thorn*, of the Bahamas

Caucalis Anthriscus. *Hedge Parsley*
daucoides. *Bastard* or *Bur Parsley, Hedge-hog Parsley, Hen's-foot*
(Torilis) nodosa. *Knotted Bur-weed*
Caulanthus crassicaulis and C. procerus. *Wild Cabbage*, of California
Caulophyllum thalictroides. *Blue Cohosh, Blue Ginseng, Pappoose-root, Squaw-root*
Ceanothus americanus. *Mountain-sweet, New Jersey Tea-plant, Wild Snow-ball*
azureus. *Blue Bush, Blue-flowered Red-root*
buxifolius. *Box-leaved Red-root*
chloroxylon. *Jamaica Cog-wood*
colubrinus (Colubrina ferruginosa). *Bahama Red-wood, Bahama Snake-wood*
cordulatus. *Californian Snow-bush*
dentatus. *Tooth-leaved Red-root*
floribundus. *Free-flowering Red-root*
integerrimus. *Beauty-of-the-Sierras, Californian Lilac*
intermedius. *Intermediate Red-root*
mexicanus. *Mexican Blue-bush*
microphyllus. *Small-leaved Red-root*
ovatus. *Ovate-leaved Red-root*
papillosus. *Pimpled Blue-bush*
sanguineus. *Crimson-twigged Red-root*
spinosus. *Californian Red-wood, Red-wood* of Santa Barbara
thyrsiflorus. *Californian Lilac*
Cecropia peltata. *Trumpet-tree, W. Indian Snake-wood*
Cedrela australis. *Red Cedar*, of Australia
brasiliensis. *Acajou-wood, Brazilian Cedar*
odorata. *Bastard Barbadoes Cedar, Cigar-box-wood, False Cedar, Honduras, Jamaica,* or *W. Indian Cedar*
sinensis. *Chinese Cedar, Cigar-box-wood*
Toonah. *Chittagong-wood, E. Indian Cedar, Singapore Cedar*
Cedronella cana. *Hoary Balm of Gilead-plant*
cordata. *Heart-leaved Balm of Gilead*
triphylla (Dracocephalum canariense). *Common Balm of Gilead-plant*
Cedrus. *Cedar*
atlantica. *African,* or *Mt. Atlas Cedar, Silver Cedar*
Deodara. *Deodar, East Indian Cedar, Fountain-tree*
Deodara var. nivea. *Silvery-leaved Deodar*
Deodara var. robusta. *Heavy-branched Deodar*
Libani. *Cedar of Lebanon*
Celastrus. *Staff-tree*
bullatus. *Scarlet-fruited Staff-tree*
edulis. *Abyssinian Tea-plant*
scandens. *Climbing Bitter-sweet, David-root, Fever-twig, Staff-tree, Staff-Vine, Wax-work*
Celmisia. *New Zealand Aster*
coriacea. *New Zealand " Cotton-plant,"* or *" Leather-plant "*
Celosia cristata. *Cock's-comb*
Celsia cretica. *Cretan Mullein*
Celtis. *Nettle-tree*
aculeata. *Prickly Nettle-tree*
australis. *European Nettle-tree, Honey-berry,* of Greece, *Lote-tree*

Celtis caucasica. *Caucasian Nettle-tree*
crassifolia. *Hoop Ash, Thick-leaved Nettle-tree*
lævigata. *Smooth Nettle-tree*
Lima. *File-leaved Nettle-tree*
micrantha. *Jamaica Nettle-tree*
occidentalis. *American False Elm, Hack-berry, Nettle-tree, Rim-Ash, Sugar-berry*
orientalis. *Eastern Nettle-tree*
parviflora. *Small-flowered Nettle-tree*
pumila. *Dwarf Nettle-tree*
sinensis. *Chinese Nettle-tree*
Tournéforti. *Tournefort's Nettle-tree*
Cenarrhenes nitida. *Port Arthur* or *Tasmanian Plum*
Cenchrus echinatus. *W. Indian Bur-grass*
tribuloides. *Hedge-hog-grass,* or *Bur-grass,* of N. America
Cenomyce pyxidata. *Cup Lichen, Cup Moss*
rangiferina. *Reindeer Moss*
Centaurea. *Centaury, Knap-weed, Mat-fellon*
argentea. *Silvery Knap-weed, Silvery Scabious*
aspera. *Jersey Centaury*
australis. *Australian Centaury*
babylonica. *Babylonian Centaury*
Calcitrapa. *Caltrops, Star-Thistle*
chironoides. *Californian Centaury*
Cyanus. *Blawort, Blue-blaw, Blue Bon-nets, Blue-bottle, Blue Corn-flower, Corn-bottle, Corn Centaury, Corn-flower, Hurt-sickle*
dealbata. *Mealy Centaury*
gymnocarpa. *Silvery-leaved Knap-weed*
Isnardi. *Jersey Thistle*
macrocephala. *Great-headed Centaury*
montana. *Mountain Centaury*
(Amberboa) moschata. *Purple Sweet Sultan*
(Amberboa) moschata var. rubra. *Red-flowered Sweet Sultan*
Myacantha. *Mouse-thorn*
nigra. *Bell-weed, Black Centaury, Bull-weed, Button-weed, Crop-weed, Hard-heads, Iron-hard, Iron-heads, Iron-weed, Knap-weed, Knob-weed, Knot-weed, Log-ger-heads*
Phrygia. *Phrygian Centaury*
ruthenica. *Russian Knap-weed*
Scabiosa. *Black Top, Greater Centaury*
solstitialis. *Barnabas, St. Barnaby's Thistle, Yellow-flowered Centaury*
stricta. *Erect Centaury*
(Amberboa) suaveolens. *Yellow Sweet Sultan*
uniflora. *One-flowered Centaury*
Cetranthus macrosiphon. *Long-spurred Valerian*
ruber. *Fox's-Brush, German Lilac, Red Valerian, Spur Valerian*
Centrosema virginianum. *Spurred Butter-fly-Pea*
Centunculus minimus. *Bastard Pimpernel, Chaff-weed*
Cephalanthera grandiflora. *White Helle-borine*
rubra. *Red Helleborine*

Cephalanthus occidentalis. *Button-bush, Globe-flower-bush, Little - Snow-balls, Swamp Globe-flower, White Pond-Dogwood,* of Louisiana
Cephalotaxus. *Cluster-flowered Yew*
  drupacea. *Plum-fruited Cluster-flowered Yew*
  Fortunei. *Fortune's Cluster-flowered Yew*
  pedunculata. *Long-stalked Cluster-flowered Yew*
Cephalotus follicularis. *Australian Pitcher-plant*
Cerasus. *Cherry-tree*
  (Prunus) Avium. *Hedge-berry, Wild Gean Cherry*
  borealis. *Northern Choke-Cherry*
  Capollin. *Capollin Bird-Cherry*
  caproniana. *Hautbois Cherry*
  caroliniana. *Evergreen Cherry*
  Chamæcerasus. *Ground-Cherry*
  Chamæcerasus var. fragrantissima. *Sweet-scented Ground-Cherry*
  cornuta. *Horned Cherry*
  depressa. *Sand Cherry*
  duracina. *White-heart Cherry*
  duracina var. cordigera. *Bigarreau Cherry*
  Hixa. *Teneriffe Cherry-Bay*
  hyemalis. *Black Choke-Cherry*
  Juliana. *Gean Cherry, St. Julian's Cherry*
  Juliana var. Heaumiana. *Helmet-fruited Cherry*
  (Prunus) japonica. *Japan Cherry*
  Lauro-cerasus. *Common Laurel,* or *Cherry-Laurel*
  Lauro-cerasus var. latifolia. *Versailles Laurel*
  lusitanica. *Portugal Laurel*
  lusitanica var. myrtifolia. *Miniature Portugal Laurel*
  Mahaleb. *Mahaleb Cherry, Perfumed Cherry*
  Marascha. *Marascha Cherry*
  nepalensis. *Nepaul Bird-Cherry*
  nigra. *Black-fruited Cherry*
  occidentalis. *W. Indian Cherry*
  Padus. *Common Bird-Cherry*
  Padus var. rubra. *Red Cornish Cherry*
  persicæfolia. *Peach-leaved Cherry*
  Pseudo-cerasus. *Bastard-Cherry*
  pubescens. *Downy Cherry*
  pumila. *Canadian Cherry, Dwarf Cherry, Sand Cherry*
  pygmæa. *Pigmy Cherry*
  salicina. *Willow-leaved Cherry*
  semperflorens. *Ever-flowering Cherry*
  serotina. *Late-flowering Cherry, Cherry-oil-plant*
  serrulata. *Saw-leaved Cherry*
  sphærocarpa. *Round-fruited Cherry*
  vulgaris (Prunus Cerasus). *Common Cherry*
  vulgaris fl.-pl. *Double-blossomed Cherry*
Cerastium. *Mouse-ear Chick-weed*
  alpinum. *Alpine Mouse-ear Chick-weed*
  Bieberstcini. *Bieberstein's Mouse-ear Chick-weed*
  Boissieri. *Boissier's Mouse-ear Chick-weed*
  glaciale. *Glacier Mouse-ear Chick-weed*
  grandiflorum. *Large-flowered Mouse-ear Chick-weed*

Cerastium latifolium. *Broad-leaved Mouse-ear Chick-weed*
  tomentosum. *Jerusalem Star, Snow-in-Summer, Snow-plant, Woolly Mouse-ear Chick-weed*
  triviale and C. vulgatum. *Common Mouse-ear Chick-weed*
Ceratiola ericoides. *Sand-hill Rosemary*
Ceratochloa. *Horn-grass*
  unioloides. " *Californian Prairie-grass,*" *Rescue-grass*
Ceratonia Siliqua. *Algaroba Bean, Carob-tree, European Locust-tree, St. John's-Bread-tree*
Ceratopetalum. *Leather-wood,* of Australia
  apetalum. *Light-wood,* of Australia
  gummiferum. *Christmas-tree,* of Australia
Ceratophyllum demersum. *Common Horn-wort,* or *Horn-weed, Morass-weed*
Ceratopteris thalictroides. *Pod-Fern*
Cerbera Ahouai. *Serpent's-bane*
  Tanghin. *Ordeal-tree,* of Madagascar
Cercis canadensis. *American Judas Tree, Red-Bud*
  occidentalis. *Californian Judas Tree, Californian Red-Bud*
  Siliquastrum. *Common Judas Tree, Love-tree*
Cercocarpus. *Mountain Mahogany,* of California
  ledifolius. *Wasatch Mountain Mahogany*
Cereus. *Torch-thistle*
  (Aporocactus) flagelliformis. *Creeping Cereus*
  grandiflorus. *Night-flowering Cereus*
  triangularis. *Prickly-withe,* of the W. Indies, *Strawberry Pear*
Cerinthe. *Honey-wort*
  aspera. *Rough-leaved Honey-wort*
  major. *Wax-plant*
Ceroxylon andicola. *Wax-Palm,* of New Granada
Cestrum. *Bastard Jasmine,* or *Poison-berry,* of the W. Indies
  aurantiacum. *Night-blooming Jasmine*
Ceterach officinarum. *Rusty-back, Scale-Fern, Stone-Fern*
Cetraria islandica. *Iceland Moss*
Chærophyllum. *Chervil*
  bulbosum (Anthriscus bulbosus). *Parsnip Chervil, Tuberous-rooted* or *Turnip-rooted Chervil*
  sylvestre. *Cow-Parsley, Wild Chervil*
  temulentum. *Cow-Parsley*
Chailletia toxicaria. *Rat's-bane,* or *Rat-poison,* of W. Africa
Chamæcyparis. *Swamp Cypress,* or *Water Cedar*
Chamædoris. *Merman's Shaving-brush*
Chamælirium luteum. *Blazing-star, Devil's-bit,* of N. America
Chamæpeuce Casabonæ. *Fish-bone, Herring-bone Thistle*
Chamælaucium. *Fringe-Myrtle*
Chamæranthemum. *Brazilian Flower-of-the-Grove*
Chamærops excelsa. *Fan-Palm,* of Ne-paul, *Hemp-Palm*

N

Chamærops Fortunei. *Chusan Palm, Shang-hae Fan-Palm*
humilis. *"African Hair" Palm, Dwarf Fan Palm, European Palm* (Rhapidophyllum) Hystrix. *Blue Palmetto* (Sabal). *Palmetto, Cabbage Palm-tree, Cabbage Palmetto*
Ritchieana. *Tiger-grass Palm*
serrulata. *Saw-Palmetto*
Chamæscilla (Cæsia) corymbosa. *Dwarf Australian Squill*
Chara. *Stone-wort*
vulgaris. *Feather-beds*
Chavica Betle. *Betle Pepper*
officinarum and C. Roxburghii. *Long Pepper*
Cheilanthes. *Lip-Fern*
gracillima. *Californian Lace-Fern*
odora. *Hay-scented Lip-Fern*
vestita. *Hairy Lip-Fern*
Cheiranthus. *Gilliflower, Wall-flower*
Cheiri. *Bleeding Heart, Bloody Warrior, Common Wall-flower, Wall, Winter, or Yellow Gilliflower*
longifolius. *Long-leaved Wall-flower*
Marshalli. *Marshall's Wall-flower*
scoparius. *Teneriffe Wall-flower*
Cheirostemon platanoides. *Hand-flower-tree, Hand-plant*
Chelidonium. *Swallow-wort*
japonicum (C. grandiflorum). *Japanese Celandine*
majus. *Cock-foot, Great Celandine, Tetter-wort*
Chelone glabra. *Balmony, Turtle-head, White Shell-flower, Salt-rheum-weed, Turtle-bloom*
Lyoni. *Lyon's Shell-flower*
obliqua. *Twisted Shell-flower*
Chenopodium. *Goose-foot*
album. *Bacon-weed, Dirt-weed, Dirty Dick, Fat-Hen, Frost Blite, Lamb's-Quarters, Muck-weed, Pig-weed, White Goose-foot*
(Ambrina) ambrosioides. *Demigod's-food, Mexican Goose-foot, Mexican Tea*
aristatum. *Bearded Goose-foot*
Atriplicis. *Purple Goose-foot*
auricomum. *Australian Spinach*
Bonus-Henricus. *All-good, Blite, English, False* or *Wild Mercury, Flowery Docken, Good Henry, Good King Henry, Perennial Goose-foot, Wild Spinach*
Botrys. *Cut-leaved Goose-foot, Feather Geranium, Hind-heal, Jerusalem Oak, Oak-of-Paradise*
erosum. *Australian Spinach*
glaucum. *Oak-leaved Goose-foot*
guineense. *Guinea Goose-foot*
hybridum. *Maple-leaved Goose-foot*
murale. *Nettle-leaved Goose-foot*
polyspermum. *All-seed, Many-seeded Goose-foot*
Quinoa. *Petty Rice*, of Peru, *Quinoa* or *White Quinoa*
rubrum. *Red Goose-foot, Sow-bane, Swine's-bane*
urbicum. *Upright Goose-foot*
Vulvaria. *Dog's-Orache, Notch-weed, Stinking Goose-foot*

Cherleria sedoides. *Cyphel*
Chickrassia tabularis. *Chittagong-wood*
Chilochloa. *Fodder-grass*
Chilopsis saligna. *Desert-Willow*, of California
Chimaphila maculata. *Spotted Winter-green*
umbellata. *Prince's-Pine, Pipsissewa, Ground Holly*
Chimonanthus fragrans (Calycanthus præcox). *Japan Allspice*
grandiflora. *Large-flowered Japan Allspice*
Chiococca angustifolia. *Brazilian Snake-root*
racemosa. *David's-root, Snowberry-tree*, of Jamaica
Chiogenes hispidula. *Creeping Snow-berry*
Chionanthus virginica. *Fringe-tree, Poison Ash, Virginian Snow-flower*
Chionodoxa Luciliæ. *Snow-glory*
nana. *Dwarf Snow-glory*
Chlora perfoliata. *Yellow Centaury, Yellow-wort*
Chloranthus inconspicuus. *Chu-lan Tree*
Chloris truncata. *Windmill grass*, of Australia
ventricosa. *Warted Grass*, of Australia
Chlorocodon Whitei. *Mundi-root*
Chlorogalum pomeridianum. *Amole*, or *Soap-plant*, of California, *Indian Soap-plant*
Chloroxylon Swietenia. *Satin-wood*, or *Yellow-wood*, of the E. Indies, *Wood-oil-plant, Zante-wood*
Choisya ternata. *Mexican Orange-flower*
Chondrilla. *Gum Succory*
juncea. *Common Gum Succory*
Chondrodendron convolvulaceum. *Wild Grape*, of the Peruvians
tomentosum (Cissampelos Pareira). *Pareira-Brava-plant*
Chondrus (Fucus) crispus. *Alga marina, Carrageen, Carageen*, or *Carrigeen Moss, Irish Moss, Pearl Moss*
Chorda Filum. *Sea-laces*
Chordaria flagelliformis. *Whipcord Sea-weed*
Chorisia speciosa. *Vegetable Silk-tree*
Chroolepus Jolithus. *Rock Violet*
Chrysanthemum alpinum. *Alpine Chrysanthemum*
arcticum. *Northern Chrysanthemum*
atro-coccineum. *Scarlet Chrysanthemum*
Burridgeanum. *Tricoloured Ox-eye Daisy*
carinatum. *Tricoloured Chrysanthemum*
coronarium and vars. *Crown Daisy, Old Garden Chrysanthemum, Sicilian Chrysanthemum*
frutescens. *Marguerite, Paris Daisy*
indicum. *Indian Chrysanthemum*
inodorum. *Scentless Chrysanthemum*
lacustre. *Marsh Fever-few, Marsh Ox-eye Daisy*
Leucanthemum. *Big, Bull, Dog, Horse, Moon*, or *Ox-eye Daisy, Espibawn, Horse Gowan, Large White Gowan, Mathes, Mawther, Midsummer Daisy, Moon-flower, Moon-penny, White Hothen, White Goldes*

Chrysanthemum millefoliatum. *Tansy-leaved Chrysanthemum*
montanum. *Mountain Chrysanthemum*
Parthenium. *Feverfew Chrysanthemum*
segetum. *Bigold, Boodle, Buddle, Corn, Field,* or *Wild Marigold, Gill Gowan, Gule Gowan, Golden* or *Yellow Cornflower, Golds, Goldins, Yellowby*
sinense and vars. *Common Garden Chrysanthemum*
Chrysobactron Hookeri. *Hooker's Goldenwand*
Chrysobalanus ellipticus and C. luteus. *Pigeon Plum,* of Sierra Leone
Icaco. *West Indian Cocoa-plum*
Chrysocoma latifolia. *Broad-leaved Goldilocks*
Linosyris. *Common Goldilocks*
Chrysophyllum Buranheim. *Monesia-bark-tree*
Cainito. *W. Indian Star-apple*
oliviferum. *Damson-plum,* of the West Indies, *Wild Star-apple*
Chrysopsis Mariana. *Golden Aster,* of Pennsylvania, *Maryland Golden-star*
Chrysosplenium. *Golden Saxifrage*
alternifolium. *Alternate-leaved Golden Saxifrage*
oppositifolium. *Opposite-leaved Golden Saxifrage*
Chrysostemma. *Golden-Crown*
Cibotium Barometz. *Scythian Lamb, Tartarian Lamb*
regale. *Chignon Fern*
Cicca disticha. *Otaheite Gooseberry*
Cicer arietinum. *Calavance, Chick-pea,* " *Chola,*" of India, *Egyptian Pea, Gram, Garavance*
Cichorium Endivia. *Endive*
Intybus. *Chicory, Succory, Witloof*
Cicuta maculata. *Beaver-poison, Musquash-root, Spotted Cow-bane*
virosa. *Common Cow-bane, Water Hemlock*
Cimicifuga americana. *American Bugbane*
foetida. *Fetid Bug-bane*
racemosa. *Black Snake-root, Cohosh, Rattle-root, Squaw-root*
Cinchona. *Jesuit's-Bark-tree, Peruvian Bark-tree*
Calisaya. *Calisaya Bark,* or *Yellow Cinchona Bark*
latifolia. *Calisayo-bark,* of Santa Fé, or *Soft Columbian-bark*
officinalis (C. Condaminea). *Loxa Crown Bark*
peruviana. *Gray Bark*
Pitayensis. *Pitayo-bark*
succirubra. *Red Bark*
Cinclidotus. *Lattice-Moss*
fontinaloides. *Fountain Lattice-Moss*
Cineraria acanthifolia. *Silvery-leaved Cineraria*
alpina. *Alpine Cineraria*
(Ligularia) macrophylla. *Large-leaved Cineraria*
maritima. *Sea-side Rag-wort, Silvery-leaved Cineraria*

Cinna arundinacea. *Wood Reed-grass*
Cinnamodendron corticosum. *Mountain Cinnamon,* of the W. Indies
Cinnamomum. Several species yield the *Cassia-Bark* of commerce
Cassia. *Chinese Cinnamon-tree*
Culilawan. *Culilawan Bark-plant*
Malabathrum. *Indian,* or *Malabar, Leaf*
Zeylanicum. *Common Cinnamon-tree*
Circæa Lutetiana. *Bind-weed Night-shade, Enchanter's Nightshade*
Cirrhopetalum Caput-Medusæ. *Medusa's-head Orchid*
Cirsium. *Horse-thistle, Plume-thistle*
arvense. *Canada Thistle, Creeping Thistle,* "*Cursed Thistle,*" of N. America
(Chamæpeuce) Casabonæ. *Herring-bone Thistle*
Douglasii. *Brilliant Thistle*
ferox. " *Cruel Thistle* "
oleraceum. *Meadow Distaff*
Cissampelos Pareira. *False Pareira-Brava, Velvet-leaf, Ice-Vine, Pareira-Brava Velvet-Leaf*
Cissus acida. *Sorrel Vine*
antarctica. *Kangaroo* or *Kanguru Vine*
latifolia. *Elephant's Vine*
sicyoides. *Bastard Bryony, W. Indian China-root*
venatorum. *Sportsman's Climber*
Cistus. *Holly-rose, Rock-rose*
albidus. *Whitish-leaved Rock-rose*
asperifolius. *Rough-leaved Rock-rose*
candidissimus. *Canary Island Rock-rose*
Clusii. *Clusius's Rock-rose*
corbariensis. *Corbières Rock-rose*
creticus. *Cretan Rock-rose*
crispus. *Curled-leaved Rock-rose*
Cupanianus. *Heart-leaved Rock-rose*
cymosus. *Cyme-flowered Rock-rose*
cyprius. *Common Gum-Cistus*
Florentinus. *Florentine Rock-rose*
frutescens. *Shrubby Rock-rose*
heterophyllus. *Various-leaved Rock-rose*
hirsutus. *Hairy Rock-rose*
incanus. *Hoary Rock-rose*
ladaniferus. *Bog-Cistus, Ladanum Gum-Cistus*
latifolius. *Broad-leaved Rock-rose*
laxus. *Loose-flowered Rock-rose*
laurifolius. *Laurel-leaved Rock-rose*
Ledon. *Ledon Gum-Cistus, Many-flowered Rock-rose*
longifolius. *Long-leaved Rock-rose*
lusitanicus. *Dwarf Rock-rose*
monspeliensis. *Montpelier Rock-rose*
oblongifolius. *Oblong-leaved Rock-rose*
obtusifolius. *Blunt-leaved Rock-rose*
populifolius. *Poplar-leaved Rock-rose*
purpureus. *Purple-flowered Rock-rose*
salviæfolius. *Sage-leaved Rock-rose*
sericeus. *Silky-leaved Rock-rose*
undulatus. *Waved-leaved Rock-rose*
vaginatus. *Sheathed-stalked Rock-rose*
villosus. *Shaggy Rock-rose*
vulgaris. *Common Rock-rose*
Citharexylon. *Fiddle-wood, Guitar-wood*
cinereum. *Old-woman's-Bitter*
quadrangulare. *Savannah Wattle*

English Names of Cultivated, Native,

Citriobatus. *Native Orange*, or *Orange Thorn*, of Australia
Citrullus Colocynthis. *Bitter - apple, Bitter Gourd, Colocynth-plant, Vine-of-Sodom* (?)
(Cucumis) Pseudo-colocynthis. *Himalayan Colocynth*
Citrus acida. *Lime-fruit-tree*
Aurantium. *Common*, or *Sweet, Orange-tree*
Aurantium var. amara. *Seville, Bigarade,* or *Bitter Orange*
Aurantium var. melitensis. *Blood*, or *Maltese, Orange*
australis (C. Planchoni). *Native Orange,* of Queensland
australasica. *Queensland Native Lime-fruit Tree*
Bergamia. *Bergamot Orange, Lemon Bergamot Tree*
buxifolia. *Box-leaved Orange*
decumana. *Shaddock-tree*
japonica. *Kumquat*
javanica. *Java Lemon*
Limetta. *Adam's Apple, Sweet Lime*
Limonum. *Common Lemon-tree*
Lumia. *Pear Lemon, Sweet Lemon*
Margarita. *Pearl Lemon*
Medica. *Median Lemon*
Medica var. Cedra (C. acida). *Citron* (true)
nobilis var. major. *Mandarin Orange*
nobilis var. Tangeriana. *Tangerine Orange*
Paradisi. *Forbidden Fruit, Pomello*
sarcodactylis. *Fingered Citron*
Cladium Mariscus. *Marsh Saw-grass, Shear,* or *Shere Grass*
mariscoides. *Twig-rush,* of N. America
Cladrastis tinctoria (Virgilia lutea). *Yellow-wood Tree*
Clathrus cancellatus. *Lattice Stink-horn*
Claviceps purpurea. *Ergot-of-Rye Fungus*
Claytonia cubensis. *Cuban Spinach*
perfoliata. *Cuban Winter-Purslane*
sibirica. *Chinese Chick-weed, Siberian Purslane*
virginica. *Spring Beauty*
Cleistocactus colubrinus. *Snake Cactus*
Clematis æthusæfolia. *Fool's - Parsley-leaved Clematis*
alpina (Atragene austriaca). *Austrian Clematis*
aristata. *Australian Supple-Jack*
calycina (C. Balearica). *Evergreen Clematis, Minorca Clematis, Winter-flowering Clematis*
campanulata. *Bell-flowered Clematis*
chinensis. *Chinese Clematis*
cirrhosa. *Evergreen Clematis, Spanish Traveller's-joy, Spanish Wild Cucumber*
coccinea. *Scarlet-flowered Clematis*
crispa. *Curled-sepaled Clematis*
cylindrica. *Cylindrical-flowered Clematis*
Davidiana. *David's Clematis*
erecta. *Erect Clematis*
Flammula. *Sweet-scented Virgin's-bower*
florida. *Large-flowered Clematis*
glauca. *Glaucous-leaved Clematis*
indivisa lobata. *New Zealand Clematis*

Clematis integrifolia. *Entire-leaved Clematis*
Jackmanni and vars. *Jackman's Clematis*
lanuginosa. *Woolly Clematis*
Mossmana. *Australian Virgin's-bower*
montana. *Mountain Clematis*
orientalis. *Oriental Clematis*
ovata. *Oval-leaved Clematis*
paniculata. *Panicled Clematis*
patens. *Open-flowered Clematis*
reticulata. *Net-veined-leaved Clematis*
Simsii. *Sims's Clematis*
tubulosa. *Tube-flowered Clematis*
Viorna. *Blue Virgin's-Bower, Leather-flower*
Virginiana. *American Virgin's-bower*
Vitalba. *Bindwith, Biting Clematis, Hedge - Vine, Lady's - bower, Maiden's Honesty, Old-Man, Old-Man's-Beard, Smoke-wood, Traveller's Joy, Virgin's-bower*
Viticella. *Purple Virgin's-bower, Vine-bower Clematis*
Viticella venosa. *Veined Vine-bower Clematis*
Cleome. *Spider-flower*
cardinalis. *Mexican Cardinal-flower*
pungens. *W. Indian Spider-flower*
Clerodendron. *Glory-tree*
fragrans. *Glory-tree,* of China
Siphonanthus. *Tube-flower*
Clethra. *White-Alder Bush*
acuminata. *Painted-leaved White-Alder*
alnifolia. *Alder-leaved Sweet Pepper-bush, Common White-Alder*
arborea. *Tree White-Alder*
arborea minor. *Smaller Tree White-Alder*
arborea variegata. *Variegated Tree White-Alder*
ferruginea. *Rusty-leaved White-Alder*
Michauxi. *Michaux's White-Alder*
paniculata. *Panicled White-Alder*
scabra. *Rough-leaved White-Alder*
tinifolia. *Bastard Locust-tree, Jamaica Sweet Pepper-bush, Soap-wood, Wild Pear,* of the W. Indies
tomentosa. *Woolly White-Alder*
Clianthus. *Glory Pea, Glory Vine, Parrot-beak Plant*
carneus. *Glory Pea,* of Norfolk Island
Dampieri. *Dampier's Glory Pea, Sturt's Desert Pea*
puniceus. *Glory Pea,* or *Parrot's-bill,* of New Zealand
Clidemia. *Indian Currant-bush*
Cliffortia ilicifolia. *African Evergreen Oak*
Cliftonia ligustrina. *Buck-wheat Tree*
Clitoria Mariana. *Butterfly-Pea*
Clivia. *Caffre Lily*
Clusia. *Balsam-tree, Card-leaf-tree, Scotch Attorney*
alba. *Balsam-Fig,* of the W. Indies
flava. *Jamaica Balsam-Fig,* or *Balsam-tree, Fat-Pork, Monkey-Apple, Mountain* or *Wild Mango, Wild Fig*
rosea. *Balsam Fig, Star-of-night*
Clypeola Jonthlaspi. *Buckler-Mustard*
Cneorum. *Widow-wail*
tricoccum. *Spurge Olive, Smooth Widow-wail*

Cnicus. *Plume-Thistle*
Acarna. *Yellow Plume-Thistle*
benedictus. *Blessed Thistle*
Cobæa scandens. *Mexican Ivy-plant, Cups-and-saucers*
Coccinea indica. *Scarlet-fruited Gourd*
Coccoloba. *Sea-side Grape-plant* (Polygonum) adpressa. *Macquarie Harbour Grape-plant*
diversifolia. *Long-leaved,* or *Small, Pigeon-wood*
Floridana. *Pigeon Plum*
leoganensis and C. punctata. *Small-leaved Pigeon-wood*
pubescens. *Leather-coat-leaf-tree,* of the W. Indies
uvifera. *Jamaica-Kino-tree*
Cocculus Bakis. *Senegal-root*
Plukenetii and C. tuberosus. *Cocculus-indicus Plant*
Cochlearia Armoracia. *Common Horseradish*
officinalis. *Scurvy-grass, Spoon-wort*
Cochlospermum Gossypium. *Kuteera-gum-tree, Leaf-Bellows Plant*
Cocos (Acrocomia) aculeata. *Prickly-Macaw-tree*
butyracea. *New Granada Oil-Palm* or *Wine-Palm*
(Acrocomia) fusiformis. *Great Macaw-tree*
guineensis. *Small Prickly Cocoa-nut-tree*
nucifera. *Cocoa-nut Palm*
oleracea. *Iraiba Palm*
schizophylla. *Aricuri* or *Aracuri Palm*
Codarium. *Wild Tamarind-tree*
acutifolium. *Black, Brown, Velvet,* or *Wild Tamarind,* of Sierra Leone
Codiæum (Croton). *South Sea Laurel*
Codonorchis. *Bell-Orchis*
Coffea arabica. *Arabian Coffee-tree*
sibirica. *Siberian Coffee-tree*
racemosa. *Peruvian Coffee-tree*
Coix Lachryma. *Gromwell-Reed, Job's Tears*
Cola acuminata. *Cola, Kolla,* or *Goora-nut Tree*
Coleanthus subtilis. *Sheath-flowering Grass*
Coleus. *Flame-Nettle*
aromaticus. *Patchouli Plant*
fruticosus. *Nettle Geranium*
Colchicum. *Meadow-Saffron*
alpinum. *Alpine Meadow-Saffron*
autumnale. *Autumn, Fog, Meadow, Michaelmas,* or *Purple Crocus, Common Meadow-Saffron, Naked-Ladies*
Bivonæ. *Bivona's Meadow-Saffron*
byzantinum. *Byzantine Meadow-Saffron*
chionense. *Chian Meadow-Saffron*
Parkinsonii. *Parkinson's Chequered Meadow-Saffron*
speciosum. *Giant Meadow-Saffron*
tessellatum. *Chequered Meadow-Saffron*
variegatum (C. Agrippinæ). *Variegated Meadow-Saffron*
Colletia bictoniensis. *Anchor-plant*
Collinsia bicolor. *Two-coloured Collins's-flower*

Collinsia grandiflora. *Large Collins's-flower*
multicolor. *Many-coloured Collins's-flower*
Collinsonia canadensis. *Canadian Horse-Mint, Collinson's-flower, Hard-hack, Heal-all, Horse-weed, Ox-balm, Rich-weed, Stone-root*
Collophora. *Cow-tree,* of the Rio Negro
Colocasia. *W. Indian Kale*
antiquorum. *Cocoa-root* or *Coco, "Kolcas,"* of the Arabians and Egyptians
esculenta. *Bleeding-Heart, Egyptian Ginger, Poë-plant* of the Sandwich Islands
macrorrhiza. *Taro,* or *Tara-plant,* of Tahiti
Colophonia Mauritiana. *Elemi-Resin-plant,* of Mauritius
Colubrina asiatica. *W. Indian Hoop-withe*
ferruginosa. *W. Indian Green-heart,* or *Snake-wood*
Colutea arborescens. *Common Bladder-Senna*
cruenta. *Red-flowered Bladder-Senna*
Haleppica (C. Pocockii). *Pocock's Bladder-Senna*
herbacea. *Annual Bladder-Senna*
media. *Smaller Bladder-Senna*
nepalensis. *Nepaul Bladder-Senna*
Comandra. *Bastard-Toad-flax,* of N. America
Comarostaphylis. *Grit-berry*
Comarum palustre. *Bog Strawberry, Cow-berry, Marsh Cinquefoil, Meadow-nut, Purple-wort*
Combretum butyrosum. *Caffre-Butter*
Guagea. *Vegetable-glue-plant*
Comesperma scoparium. *Swan River Broom*
volubile. *Blue Creeper,* or *"Love,"* of Tasmania
Cometes alternifolia. *Comet-plant*
Commelina. *Day-flower*
cœlestis. *Blue Spider-wort*
cayennensis. *French-reed*
Commersonia Fraseri. *Black-fellow's Hemp,* of Australia
platyphylla. *Brown Kurrajong,* of Australia
Commidendron rugosum. *Gum-shrub,* or *Scrub-wood,* of St. Helena
spurium. *Cluster-leaved Gum-tree, Little Bastard Gum-wood,* or *Small-umbelled Cabbage-tree,* of St. Helena
Comocladia. *Maiden Plum*
dentata. *Toothed-leaved Maiden Plum*
integrifolia. *Maiden Plum,* of Jamaica
Comptonia asplenifolia. *Fern-leaved Gale, Shrubby Sweet Fern-bush*
Condalia microphylla. *Piquillin-bush*
obovata. *Blue Wood, Texan Log-wood*
Conferva ægagropila. *Moor-balls*
rivularis. *Crow-silk, Duck-mud*
Conioselinum canadense. *Hemlock-Parsley,* of N. America
Conium maculatum. *Hemlock, Herb Bennet*
Conocarpus erecta. *Button-tree, Zaragosa Mangrove*

Conoclinium coelestinum. *Mist-flower*
Conophallus Titanum. *Giant Arum*, of Sumatra
Conopholis (Orobanche) americana. *Cancer-root, Squaw-root*
Conopodium denudatum. *Earth Chestnut*
Convallaria (Smilacina) bifolia. *Two-leaved Lily-of-the-Valley*
japonica. *Japanese Lily-of-the-Valley*
majalis. *Common Lily-of-the-Valley, Conval Lily, Lirieonfancy, May Lily, Mugget*
Convolvulus althaeoides. *Mallow Bindweed, Riviera Bind-weed*
arvensis. *Bear-bind, Common Bind-weed, Bine or Bines, Corn-bind, Corn-lily*
bryoniaefolius. *Bryony-leaved Bind-weed*
cantabricus. *Cantabrian Bind-weed*
Cneorum. *Shrubby Bind-weed, Silvery Bind-weed*
erubescens. *Blushing Bind-weed*, of Australia
lineatus. *Pigmy Bind-weed*
mauritanicus. *Blue Rock Bind-weed*
(Ipomæa) Nil. *Gay-bine*
panduratus. *Man-in-the-Ground*, or *Man-in-the-earth, Wild Jalap, Wild Potato*
Scammonia. *Scammony-plant*
sepium. *Bear-bind, Bed-wind, Bell-bind, Bell Wood-bind, Devil's-Garter, Hedgebells, Hedge Lily*
sepium var. roseus. *Rosy-flowered Bind-weed*
sericeus. *Silky Bind-weed*
Soldanella. *Scotch Scurvy-grass, Sea-bells, Sea Bind-weed, Sea-coal, Sea Foal-foot*
Conyza squarrosa. *Ploughman's Spikenard*
rugosa. *St. Helena Flea-bane*
Cookia punctata. *"Wampee"-tree*, of China
Copaifera. Several species yield *Balsam of Copaiba* or *Copaiva*
bracteata and C. pubiflora. *Purple-heart Tree*, of Guinea
Gibourtiana (Gibourtia copallifera). *Copaltree*, of Sierra Leone
Mopane. *Iron-wood*, of E. Tropical Africa
officinalis. *Purple-heart Tree*, of the W. Indies
Copernicia cerifera. *Carnaüba Palm, Wax-Palm*, of Brazil
Pumos. *Pumos Palm*
tectorum. *"Thatch-Palm"*, of the W. Indies
Coprinus comatus. *Maned Mushroom*
Coprosma. *Tasmanian Native Currant*
foetidissima. *Sterile-wood*, of Otago
lucida. *Otago Orange-leaf* or *Looking-glass Bush*
Coprosmanthus herbaceus (Smilax herbacea). *Carrion-flower*
Coptis Teeta. *Coptis Root, Mishmi-Bitter*
trifoliata. *Three-leaved Gold-thread, Mouth-root*
Corallorrhiza. *Coral-root Orchid*
innata. *Spurless Coral-root*
odontorrhiza. *Adam-and-Eve, Coral-teeth, Crawley, Dragon's-claw*
Corchorus capsularis. *Gunny-bag Plant, Jute-plant* (true)
olitorius. *Jews'-Mallow*
siliquosus. *Broom-weed*

Cordia Collococca. *Clammy Cherry*, of Jamaica, *Turkey-berry-tree*
cylindrostachya. *Black Sage*
elliptica. *W. Indian "Man-jack"*
geraschanthoides. *Prince-wood*, or *Spanish Elm*, of the W. Indies
Geraschanthus. *Dominica Rose-wood, Panama Laurel*
globosa. *Gout-Tea*, of the W. Indies
latifolia. *"Sebestens Plum"*
macrophylla. *Broad-leaved Cherry*, or *Manjack*, of the W. Indies
Myxa. *"Assyrian Plum," " Sebestens Plum."* This tree is supposed to have supplied the *Mummy-case-wood* of the Egyptians
subcordata. *Tow-tree*, of Tahiti
Cordiceps. *Caterpillar-Fungus*
Cordyline. *Club Palm, Palm-Lily*
(Dracæna) australis and C. indivisa. *Cabbage-Palm*, or *Ti-plant*, of New Zealand
Jacquini var. rosea. *Chinese Fire-leaf*
Corema (Empetrum) Conradii. *Broom Crow-berry*
lusitanicum. *Portugal Crake-berry*
Coreopsis. *Tick-seed*
auriculata. *Ear-leaved Tick-seed*
grandiflora. *Large-flowered Tick-seed*
lanceolata. *Lance-leaved Tick-seed*
philadelphica. *Philadelphian Tick-seed*
præcox. *Early Tick-seed*
senifolia. *Six-leafleted Tick-seed*
tenuifolia. *Slender-leaved Tick-seed*
tinctoria. *Dyers' Tick-seed*
trichosperma. *Tick-seed Sun-flower*
verticillata. *Whorled-leaved Tick-seed*
Coriandrum sativum. *Common Coriander*
Coriaria myrtifolia. *Myrtle-leaved Tanner's-tree*
nepalensis. *Nepaul Tanner's-tree*
ruscifolia. *Tutu-poison-plant*, or *"Tootplant,"* of New Zealand
sarmentosa. *Wine-berry-shrub*, of New Zealand
thymifolia. *New Zealand Ink-plant*
Corispermum hyssopifolium. *Bug-seed*
Cornicularia jubata. *Horse-hair Lichen, Tree-hair*
Cornucopiæ cucullatum. *Horn-of-Plenty grass*
Cornus. *Cornel-tree, Dog-wood*
alba. *White-berried Dog-wood*
alternifolia. *Alternate-leaved Dog-wood*
asperifolia. *Rough-leaved Dog-wood*
canadensis. *Bunch-berry, Dwarf Cornel, Pudding-berry*
capitata. *Headed-flowered Dog-wood*
circinata. *Round-leaved Dog-wood*
florida. *Box-wood*, of N. America, *False Box-wood, Flowering Dog-wood, Green Osier* of N. America
grandis. *Tall Mexican Dog-wood*
macrophylla. *Large-leaved Dog-wood*
mascula. *Cornelian Cherry, Male Dog-wood*
oblonga. *Oblong-leaved Dog-wood*
paniculata. *Panicled-flowered Dog-wood* or *Cornel*

Cornus sanguinea. *Catteridge-tree, Common Cornel, Common Dog-wood, Dog-berry, Dog-cherry, Dog-tree, Gadrise, Gaiter* or *Gatter-tree, Hound-berry Tree, Peg-wood, Prick-wood, Skewer-wood*
sericea. *Blue-berried Dog-wood, Kinnikinnik, Silky Dog-wood* or *Cornel*
sibirica. *Siberian Scarlet Dog-wood*
stolonifera. *Red Osier Dog-wood*
stricta. *Stiff Cornel, Upright-branched Dog-wood*
suecica. *Dwarf Cornel, Dwarf Honeysuckle, Plant-of-Gluttony*
Coronilla. *Crown-Vetch*
Emerus. *Scorpion Senna*
iberica. *Iberian Crown-Vetch*
juncea. *Rush-stemmed Crown-Vetch*
minima. *Least Crown-Vetch*
montana. *Mountain Crown-Vetch*
valentina. *Small Shrubby Crown-Vetch*
varia. *Rosy-flowered Crown-Vetch*
Correa speciosa. *Fuchsia, of Australia*
Corrigiola littoralis. *Bastard Knot-grass, Shore Strap-wort*
Cortusa. *Alpine Sanicle*
Matthioli. *Bear's-ear Sanicle*
pubens. *Downy Sanicle*
Coryanthes. *Helmet-Orchid*
speciosa. *Brazilian Helmet-flower*
Corydalis (Fumaria) aurea. *Golden Fumewort*
bulbosa (C. solida). *Solid-rooted Fumewort*
cava (C. tuberosa). *Hole-wort* or *Hollowwort, "Hollow Leek"*
cava var. albiflora. *White-flowered Fumewort*
(Fumaria) claviculata. *Climbing Fumitory*
formosa. *Stagger-weed, Turkey-Corn, Wild Turkey-pea*
glauca. *Colic-weed, Pale Fume-wort*
Ledebouriana. *Ledebour's Fume-wort*
(Fumaria) lutea. *Yellow-flowered Fumitory*
Marschalliana. *Marschall's Fume-wort*
nobilis. *Great-flowered Fume-wort*
solida (Fumaria bulbosa). *Solid-rooted Fume-wort*
tuberosa (C. cava). *Hollow Leek, Hollow-root* or *Hollow-wort, Hollow-rooted Fume-wort*
Corylus. *Hazel* or *Hazel-nut Tree*
americana. *American Wild Hazel*
algeriensis. *Mount Atlas Hazel*
Avellana. *Hale-nut,* or *Common Hazel-nut-tree, Nut Bush, Wood-nut*
Avellana var. barcelonensis. *Barcelona* or *Spanish Nut-tree*
Avellana var. crispa. *Frizzled Filbert*
Avellana var. glomerata. *Cluster Nut*
Avellana var. grandis. *Cob-nut*
Avellana var. laciniata. *Cut-leaved Filbert*
Avellana var. Lamberti, *Lambert's, Large Bond, Spanish,* or *Toker Nut*
Avellana var. purpurea. *Purple-leaved Filbert*
Avellana var. tenuis. *Cosford* or *Thin-shelled Nut*
Avellana var. tubulosa. *Red Filbert*

Corylus Avellana var. tubulosa alba. *White Filbert*
Colurna. *Constantinople Hazel*
heterophylla. *Japan Hazel*
rostrata (C. cornuta.) *Beaked* or *Cuckold Hazel,* of N. America
Corynephorus. *Club-grass*
Corynocarpus lævigatus. *Church-tree, Karaka-nut,* or *Kopi-tree,* of New Zealand, *New Zealand Laurel*
Corypha australis. *Cabbage-tree,* or *Cabbage-Palm,* of Australia
Gebanga. *Gebang Palm*
Talicra. *Taliora* or *Tara Palm*
umbraculifera. *Great Fan-Palm, Talipot Palm*
Cosmos (Bidens) atropurpureus. *Black Dahlia*
bipinnatus. *Purple Mexican Aster*
Costus. *Wild Ginger,* of the W. Indies
Cotoneaster. *Quince-leaved Medlar, Rose-Box*
acuminata. *Pointed-leaved Cotoneaster*
affinis. *Allied Cotoneaster*
buxifolia. *Box-leaved Cotoneaster*
frigida. *Mountain Cotoneaster*
laxiflora. *Loose-flowered Cotoneaster*
microphylla. *Small-leaved Cotoneaster*
Nummularia. *Money-wort-leaved Cotoneaster*
rotundifolia. *Round-leaved Cotoneaster*
tomentosa. *Woolly Cotoneaster*
vulgaris. *Common Cotoneaster, Dwarf-Quince-leaved Medlar*
vulgaris var. erythrocarpa. *Red-fruited Cotoneaster*
vulgaris var. melanocarpa. *Black-fruited Cotoneaster*
Cotyledon Umbilicus. *Common Navel-wort, Corn-leaves, Hip-wort, Kidney-wort, Wall Penny-grass, Penny-leaf* or *Penny-wort, Venus's Navel-wort*
Couroupita (Lecythis) guianensis. *Cannon-ball-tree*
Cowania mexicana. *Mexican Cliff-Rose*
Crambe. *Cole-wort, Kale*
cordifolia. *Heart-leaved Cole-wort*
fruticosa. *Shrubby Cole-wort*
juncea. *Rushy Cole-wort*
maritima. *Sea Cabbage, Sea Cole-wort, Sea-kale*
palmatifida. *Large-leaved Ornamental Kale*
tatarica. *Tartar-Bread-plant*
Crassula. *Thick-leaf*
Cratægus. *Haw-thorn*
æstivalis. *American May,* or *Apple-Haw*
apiifolia. *American Parsley-leaved Thorn*
Aronia. *Aronia Thorn*
Azarolus. *Azarole Thorn, Neapolitan Medlar, Parsley-leaved Haw-thorn*
cordata. *Maple-leaved Thorn, Washington Thorn*
crenulata. *Scalloped-leaved Haw-thorn*
Douglasii. *Douglas's Haw-thorn*
flava. *Summer Haw, Yellow-fruited Thorn*
glandulosa. *Glandular Thorn*
glauca. *Glaucous Evergreen Thorn*

Cratægus heterophylla. *Various-leaved Thorn*
latifolia. *Fontainebleau Thorn*
lobata. *Lobed-leaved Thorn*
macrantha. *Long-spined Thorn*
maroccana. *Morocco Thorn*
mexicana. *Mexican Thorn*
nigra. *Black-fruited Thorn*
orientalis. *Eastern Thorn*
ovalifolia. *Oval-leaved Thorn*
Oxyacantha. *Azzy-tree, Bird-Eagles, Common Haw-thorn, May or May-bush, Quick or Quick-set Thorn, White-thorn*
Oxyacantha plena. *Double White Hawthorn*
Oxyacantha var. coccinea. *Scarlet Hawthorn*
Oxyacantha var. coccinea plena. *Double Scarlet Haw-thorn*
Oxyacantha var. pendula. *Weeping Hawthorn*
Oxyacantha var. præcox. *Glastonbury Thorn, St. Joseph of Arimathea's Thorn*
Oxyacantha var. rosea. *Pink-flowered Haw-thorn*
Oxyacantha var. rosea plena. *Double Pink Haw-thorn*
parvifolia. *American Dwarf Thorn, Gooseberry-leaved Haw-thorn*
prunifolia. *Plum-leaved Thorn*
punctata. *Dotted-fruited Thorn*
purpurea. *Purple-branched Thorn*
Pyracantha. *Christ's Thorn, Egyptian Thorn, Evergreen Thorn, Fire-bush*
Pyracantha fructû-albo. *White-berried Pyracanth*
pyrifolia. *Pear-leaved Thorn*
spathulata. *Spathulate-leaved Thorn*
tanacetifolia. *Tansy-leaved Haw-thorn*
tomentosa. *American Black-thorn or Pear-thorn*
trilobata. *Three-lobed-leaved Thorn*
virginica. *Virginian Thorn*
Cratæva gynandra and C. Tapia. *Garlic Pear*
Crawfurdia japonica. *Climbing Gentian*
Crepis. *Hawk's-beard*
aurea. *Golden Hawk's-beard*
biennis. *Rough Hawk's-beard*
fœtida. *Fetid Hawk's-beard*
hieracioides. *Hawk-weed, Hawk's-beard*
paludosa. *Marsh Hawk's-beard*
rubra. *Red-flowered Hawk's-beard*
taraxacifolia. *Beaked Hawk's-beard*
virens. *Smooth Hawk's-beard*
Crescentia. *Calabash-tree*
Cujete. *Jamaica or Oval-fruited Calabash-tree*
Crinum angustifolium. *Narrow-leaved Queensland Lily*
asiaticum. *Asiatic Poison-bulb*
australe (C. pedunculatum). *Murray Lily*
capense. *Cape Lily*
giganteum. *Large African Lily*
longifolium. *Bengal Lily*
spectabile. *Cape Coast Lily*
Crithmum maritimum. *Crest-marine, Peter's Cress, Samphire, Sea-Fennel*
Critho ægiceras. *Nepaul Barley*

Crocus Aucheri (C. Olivieri). *Aucher's Crocus*
biflorus. *Scotch Crocus, Two-flowered Crocus*
Boryanus. *White Autumn Crocus*
Byzantinus. *Byzantine Crocus*
cancellatus. *Cross-barred Crocus*
Cartwrightianus. *Cartwright's Crocus*
Clusii. *Clusius's Crocus*
dalmaticus. *Dalmatian Crocus*
Damascenus. *Damascus Crocus*
Elwesi. *Elwes's Crocus*
etruscus. *Tuscan Crocus*
Fleischeri. *Fleischer's Crocus*
Hadriaticus. *Adriatic Crocus*
Imperati. *Imperati's Crocus*
Kotschyanus. *Kotschy's Crocus*
lacteus. *Cream-coloured Crocus*
lagenæflorus. *Bottle-flowered Crocus*
longiflorus. *Long-flowered Crocus*
luteus. *Common Yellow Crocus*
medius. *Intermediate Crocus*
minimus. *Pigmy Crocus*
nudiflorus (C. multifidus). *Naked-flowered Crocus*
obscurus. *Dark-flowered Crocus*
ochroleucus. *Straw-coloured Crocus*
odorus. *Sicilian Saffron, Sweet-scented Crocus*
Orphanidesi. *Orphanides' Crocus*
peloponnesiacus. *Morean Crocus*
pulchellus. *Mount Athos Crocus*
pusillus. *Dwarf Crocus*
reticulatus (C. susianus). *Cloth-of-Gold Crocus*
Saltzmannianus. *Saltzmann's Crocus*
sativus. *Common Saffron-plant, Saffron Crocus*
serotinus (C. autumnalis). *Late-blooming Crocus*
Sieberi (C. nivalis). *Sieber's Crocus*
speciosus. *Large Autumn Crocus*
stellaris. *Star-flowered Crocus*
sulphureus. *Sulphur-flowered Crocus*
vernus. *Spring Crocus*
vernus var. niveus. *White Spring Crocus*
versicolor. *Various-coloured Crocus*
Weldeni. *Welden's Crocus*
Crotalaria. *Castanet-plant, Rattle-wort*
arborescens. *Cape Rattle-snake-plant*
juncea. *Bengal Hemp-plant, Sunn-Hemp*
sagittalis. *Rattle-box*
tenuifolia. *Jubbulpore Hemp-plant*
Croton balsamiferum. *Sea-side Balsam, Sea-side Sage*
Cascarilla (C. Eleuteria). *Cascarilla-Bark Tree, W. Indian Sweet-wood, Wild Rosemary of Jamaica*
flavens. *Yellow Balsam*
gossypifolium. *Blood-tree*
humile. *Pepper-rod*
(Aleurites) lacciferum. *Gum-Lac Tree*
lucidum. *False Bahama Cascarilla*
Malambo. *Malambo, or Matius-bark-tree*
niveum. *Copalchi-Bark-plant*
Pseudo-China. *Mexican Copalchi-plant*
sebiferum. *Tallow-tree*
Slonnei. *Jamaica Cascarilla*
Tiglium. *Croton-oil-plant*

Croton (Chrozophora) tinctorum. *Dyers' Litmus-plant*, or *Turnsole*
variegatum. *Variegated Laurel* of India
Verreauxii. *Queensland Poison-tree*
Crucianella. *Cross-wort, Petty Madder*
maritima. *Sea-side Cross-wort*
stylosa. *Large-styled Cross-wort*
Crupina. *Starry Scabious*
Cryptocarya. *Australian Nutmeg*
(Laurus) australis. *Moreton Bay Laurel*
glaucescens. *Laurel of N. S. Wales*
moschata. *Brazilian Nutmeg-tree*
Cryptostegia grandiflora. *India-rubber Vine*
madagascariensis. *Madagascar Caoutchouc-plant*
Cryptostemma calendulacea. *Cape-weed of Australia*
Cryptotænia canadensis. *American Honewort*
Ctenium americanum. *Tooth-ache-grass*
Cubeba officinalis. *Cubeb or Java Pepper*
Cucubalus bacciferus. *Berry - bearing Campion*
Cucumis. *Cucumber, Melon*
Anguria. *Prickly-fruited Gherkin-Cucumber*
Chate. *Hairy Cucumber*
cicatrisatus. "*Wungee*"-*plant*
Citrullus. *Water-melon*
(Citrullus) Colocynthis. *Bitter Apple, Colocynth*
Conomon. "*Connemon*" of Japan
Dudaim *Apple Cucumber, Cucumber Vine, Sweet-scented Melon*
flexuosus. *Snake-Cucumber*
Melo. *Common Melon*
prophetarum. *Globe-Cucumber*
sativus. *Common Cucumber*
Cucurbita. *Gourd, Pumpkin, Squash*
aurantiaca. *Mock-Orange Gourd*
(Benincasa) cerifera. *Wax Gourd, White Gourd*
grossularioides. *Gooseberry Gourd*
maxima. *Great Pumpkin, Winter Squash*
Melopepo. *Elector's-cap Gourd, Jerusalem-artichoke Gourd, Melon-pumpkin Squash*
ovifera. *Vegetable Marrow*
ovifera var. Succada. *Succade Gourd*
Pepo. *Common Pumpkin, Summer Squash*
perennis. *Perennial Gourd*
verrucosa. *Long Squash*
Culcitium. *Cushion-plant*, or *Lion's-ear*, of the Andes
Cuminum Cyminum. *Cumin*, or *Cummin-plant*
Cunila Mariana. *American Dittany*
Cunonia capensis. *Red Alder*, of the Cape
Cupania. *Loblolly-wood*, of Jamaica
americana. *Chestnut*, of the Antilles
anacardioides. *Brush-Deal*, of Queensland
australis. *Tamarind-tree*, of Australia
xylocarpa. *Queensland Marsh Hickory*
Cuphea eminens. *Cigar-flower*
Jorullensis. *Beetle-Sage*
silenoides. *Catch-fly Loose-strife*
viscosissima. *Blue Wax-Bush* or *Wax-weed*

Cupressus. *Cypress*
disticha. *Deciduous Cypress*
excelsa. *Tall Guatemala Cypress*
funebris. *Funereal Cypress*
Goveniana. *Goven's Californian Cypress*
Lawsoniana (C. fragrans). *Ginger Pine, Oregon White Cedar, Port Orford Cedar*
lusitanica. *Cedar of Goa*
lusitanica var. pendula. *Cedar of Bussaco, Portugal Cypress*
macrocarpa. *Monterey Cypress*
Nutkaënsis. *Nootka Sound Cypress*
sempervirens. *Common Pyramidal Evergreen*, or *Italian Cypress*
sempervirens var. horizontalis. *Horizontal Cypress*
sempervirens var. stricta. *Upright Cypress*
thurifera. *Incense-bearing Mexican Cypress*
thyoides. *White Cedar*
torulosa. *Bhotan Cypress*
Whitleyana. *Upright Indian Cypress*
Curatella americana. *Sand-paper-tree*
Curcas multifidus. *Jatropha*, or *Pinhoën-oil-plant*
purgans. *Barbadoes Nut-plant, Jatropha, Napala*, or *Physic-nut-oil-plant*
Curculigo. *Weevil-plant*
Curcuma Amada. *Amada*, or *Mango, Ginger*
angustifolia. *E. Indian Arrow-root-plant*
aromatica. *Wild Turmeric-plant*
longa. *Common Turmeric-plant*
Zedoaria. *Round Zedoary*
Zerumbet. *Long Zedoary*
Curtisia faginea. *Assagay, Assegay*, or *Hassagay, Tree*
Cuscuta. *Dodder*
capitata. *Bengal Dodder*
Epilinum. *Flax Dodder, Wild Flax*
Epithymum. *Hell-weed, Small Dodder, Thyme Dodder*
europæa. *Greater Dodder, Hell-weed, Strangle-tare*
Trifolii. *Ail-weed, Clover Dodder*
Cyananthus incanus. *Hoary-leaved Cyananth*
lobatus. *Lobed-leaved Cyananth*
Cyathea arborea. *Tasmanian Cup-Fern*
Burkei. *Burke's Tree-Fern*
dealbata. *Silvery Tree-Fern*
medullaris. *Black-stemmed Tree-Fern, Gray Tree-Fern*
Cyathodes Oxycedrus. *Prickly Cedar*, or *Sharp Cedar*, of Van Dieman's Land
Cycadeæ. *Fern-Palms*
Cycas circinalis. *Sago-Palm*, of the E. Indies
revoluta. *Fern-Palm*, or *Sago-Palm*, of Japan
Cyclamen. *Apple-of-the-Earth, Dilnote, Mitre-flower*
africanum (C. algeriense macrophyllum). *Large-leaved Cyclamen*
Atkinsii. *Atkins's Cyclamen*
Coum. *Round-leaved Cyclamen*
Coum album. *White-flowered Round-leaved Cyclamen*
Coum vernum (C. Coum zonale). *Variegated Round-leaved Cyclamen*

Cyclamen europæum. *Bleeding Nun, European* or *Common Cyclamen, Sow-bread*
hederæfolium. *Ivy-leaved Cyclamen*
ibericum. *Iberian Cyclamen*
Neapolitanum. *Neapolitan Cyclamen*
persicum. *Persian Cyclamen*
vernum (C. repandum). *Spring Cyclamen*
Cyclanthera pedata. *Climbing Cucumber*
Cyclobothra pulchella. *Golden Star Tulip*
Cycloloma platyphyllum (Salsola platyphylla). *Winged Pig-weed*
Cyclopia genistoides. *Bush-Tree*, of the Cape
Cycnoches. *Swan-neck, Swan-wort*
Loddigesii. *Swan-flower*, of Surinam
Cydonia vulgaris. *Common Quince-tree*
vulgaris var. lusitanica. *Portugal Quince*
Cyminosma oblongifolium. *Yellow-wood*, of Moreton Bay
Cymopterus montanus. *Gamote*, of New Mexico
Cynanchum acutum. *Montpelier Scammony-plant*
Cynara Cardunculus. *Cardoon, Prickly Artichoke*
Scolymus. *French Artichoke, Globe Artichoke*
Cynodon Dactylon. *Bahama Grass, Bermuda Grass, Dog's-tooth Grass, Doob Grass, Doorda*, or *Doorwa-grass*, of India
Cynoglossum. *Hound's-tongue*
alpinum. *Alpine Hound's-tongue*
cheirifolium. *Stock-leaved Hound's-tongue*
montanum (C. sylvaticum). *Green Hound's-tongue*
Morisoni. *Beggar's-lice*
officinale. *Common Dog's-tongue* or *Hound's-tongue, Gipsy-flower*
virginicum. *Wild Comfrey*, of N. America
Cynomorium coccineum ("Fungus melitensis"). *Scarlet Mushroom*, of Malta
Cynorchis. *Dog-Orchis*
Cynosurus cristatus. *Dog's-tail*, or *Dog's Grass, " Trancen" Grass*
echinatus. *Cock's-comb Grass*
indicus. *East Indian Wire-grass*
corymbosa. *E. Indian Matting-Grass*
elegans. *Knife-leaved Rush-plant*
esculentus. *" Chufa," Earth,* or *Ground Almond, Edible Rush-nut*
longus. *Cypress-root, Galangale, Galingale, Sweet Cypress*
proliferus. *" Tagasaste "-plant*
rotundus var. Hydra. *Nut-grass*, or *Rush-nut*, of N. America
tegetum. *Galingale Rush*
Cypripedium. *Lady's Slipper, Moccason* or *Moccasin Flower*
acaule. *Stemless Lady's-slipper*
arietinum. *Ram's-head Lady's-slipper*
barbatum. *Bearded Lady's-slipper*
candidum. *Small White Lady's-slipper*
caudatum. *Long-tailed Lady's-slipper*
guttatum. *Hardy Spotted Lady's-slipper*
humile. *Noah's Ark*
insigne. *Twin-flowered Lady's-slipper*
Irapeanum. *Mexican Lady's-slipper, Pelican-flower*

Cypripedium japonicum. *Japanese Lady's-slipper*
macranthum. *Siberian Lady's-slipper*
niveum. *Snow-white Lady's-slipper*
parviflorum. *Small Yellow Lady's-slipper*
pubescens. *Large Yellow Lady's-slipper, American Valerian, Nerve-root, Noah's Ark, Umbel, Yellow Moccasin-flower*
spectabile. *Showy Lady's-slipper*
Cystopteris alpina. *Alpine Bladder-Fern*
angustata. *Narrow-fronded Bladder-Fern*
dentata. *Toothed Bladder-Fern*
fragilis. *Brittle Bladder-Fern*
montana. *Mountain Bladder-Fern*
Cytisus. *Milk-Trefoil, Shrub-Trefoil, Tree-Trefoil*
Adami. *Adam's Laburnum*
albidus. *Whitish-flowered Laburnum*
albus. *Portugal Laburnum*
alpinus. *Scotch Laburnum*
Ardoinii. *Pigmy Laburnum*
argenteus. *Silver-leaved Laburnum*
austriacus. *Austrian Laburnum*
biflorus. *Two-flowered Laburnum*
calycinus. *Large-calyzed Laburnum*
capitatus. *Cluster-flowered Laburnum*
ciliatus. *Fringed-podded Laburnum*
falcatus. *Sickle-podded Laburnum*
fragrans. *Sweet-scented Laburnum*
grandiflorus. *Large-flowered Laburnum*
hirsutus. *Hairy Laburnum*
incarnatus. *Flesh-coloured Laburnum*
Laburnum. *Common Bean-Trefoil, Common Laburnum, False Ebony, Golden Chain, Golden-rain, He Broom*
Laburnum var. pendulus. *Weeping Laburnum*
laniger. *Woolly Laburnum*
leucanthus. *White-flowered Laburnum*
mollis. *Soft-leaved Laburnum*
multiflorus. *Many-flowered Laburnum*
nanus. *Dwarf Laburnum*
nigricans. *Black-rooted Broom*
nubigenus. *Teneriffe Broom*
orientalis. *Oriental Laburnum*
patens. *Spreading Laburnum*
polytrichus. *Many-haired Laburnum*
proliferus. *Proliferous Laburnum*
purpureus. *Purple-flowered Laburnum*
racemosus. *Evergreen Laburnum*
(Sarothamnus) scoparius. *Common Broom*
serotinus. *Late-flowering Laburnum*
sessiliflorus. *Stalkless-flowered Laburnum*
spinosus. *Prickly Laburnum*
supinus. *Trailing Broom*
triflorus. *Three-flowered Laburnum*
Cyttaria Darwinii. *Beech-Fungus*
sp. *Edible Tree-fungus*, of Tierra del Fuego

Daboecia (Menziesia) polifolia. *Irish*, or *St. Dabeoc's, Heath*
(Menziesia) var. alba. *White-flowered Irish Heath*
Dacrydium cupressinum. *Imou Pine, Red Pine* or *Spruce, Rimu-tree*, of New Zealand
Franklini. *Huon Pine*
Dacryodes hexandra. *Dominica Gum-tree*

and Foreign Plants, Trees, and Shrubs. 187

Dactylis glomerata. *Cock's-foot Grass, Dew Grass, Hard Grass, Orchard Grass, Sticky Grass*
glomerata var. aurea. *Golden-edged Cock's-foot Grass*
Dactyloctenium. *Comb Finger-grass*
ægyptiacum. *Egyptian Grass*
Dæmonorops. *Rope-Palm*
Dahlia (Georgina) coccinea. *Scarlet Dahlia*
imperialis. *Giant or Tree Dahlia, Imperial Dahlia*
scapigera. *Long-flower-stemmed Dahlia*
superflua (vars. of). *Florists' Dahlia*
Yuarezi. *Cactus-Dahlia*
Dais cotinifolia. *African Button-flower*
Dalbergia Brownei. *W. Indian May-flower*
latifolia. *East Indian Black-wood or Ebony-tree, East Indian Rose-wood*
Melanoxylon. *Senegal Ebony-tree*
nigra. *Brazilian Rose-wood, Cariuna-wood-tree, Rose-wood* of commerce
sissoides. *East Indian Rose-wood*
Sissoo. *Sissoo-wood, or Sissum-wood*
Dalibarda repens (D. violæoides). *False Violet*
Dammara. *Wax Pine*
(**Agathis**) australis. *Kauri, Kawrie, or Kowrie Pine-tree, New Zealand Copal-resin-plant*
(**Agathis**) loranthifolia. *Amboyna Pitch-tree*
orientalis. *Amboyna Pine, Dammar Pine*
ovata. *New Caledonia Kauri Pine*
robusta (D. Brownii). *Dundathe Pine, Queensland Kauri Pine*
Daniellia thurifera. *Bumbo, or Bungo-tree, Frankincense-tree,* of Sierra Leone
Danthonia. *Wild Oat-grass,* of North America
penicillata. *Wallaby-Grass*
strigosa. *Bristle-pointed Oat*
Daphne alpina. *Alpine Daphne*
altaica. *Altaian Daphne*
Aucklandiæ. *Lady Auckland's Daphne*
cannabina (D. papyracea). *Indian-paper-tree*
chinensis. *Chinese Daphne*
Cneorum. *Garland Flower*
collina. *Neapolitan Daphne*
Fortunei. *Fortune's Daphne*
Gnidium. *Flax-leared Daphne*
indica. *Indian Daphne*
indica var. rubra. *Red-flowered Indian Daphne*
japonica. *Japan Daphne*
Lagetta. *Lace-bark Tree*
Laureola. *Copse Laurel, Dwarf Bay, Spurge Laurel, Wood Laurel*
Mezereum. *Dwarf Bay, Mezereon, Mysterious Plant, Spurge-Flax, Spurge-Olive*
neapolitana. *Neapolitan Daphne*
odora. *Sweet-scented Daphne*
odora variegata. *Variegated Daphne*
oleoides. *Olive-leared Daphne*
pontica. *Pontic Daphne*
pubescens. *Downy Daphne*
rupestris. *Rock Daphne*
salicifolia. *Willow-leaved Daphne*
sericea. *Silky Daphne*
striata. *Streaked-barked Daphne*

Daphne Tarton-raira. *" Herb Terrible," Silvery-leaved Daphne*
Thymelæa. *Smooth-leaved Daphne*
tinifolia. *Bonnace,* or *Burn-nose-Tree,* of Jamaica
viridiflora. *Green-flowered Daphne*
Daphnidostaphylis glauca. *Red Bark,* of California
Darlingtonia Californica. *Californian Pitcher-plant*
Dasystoma. *False Fox-glove*
Datisca cannabina. *Cretan Hemp-plant*
hirta. *American False Hemp*
Datura alba. *East Indian Thorn-apple*
arborea. *Tree Thorn-apple*
ceratocaula. *Horn-stalked Thorn-apple*
fastuosa. *Purple-flowered Thorn-apple*
ferox. *Long-spined Thorn-apple*
Metel. *Downy Thorn-apple*
Stramonium. *Apple-peru, Common Thorn-apple, Devil's-trumpet, Dewtry, Jamestown-weed, Stink-weed*
Tatula. *Blue-flowered Thorn-apple*
Daucus Carota. *Bee's-nest, Bird's-nest, Wild Carrot*
Gingidium. *Shining-leared Carrot*
(Caucalis) orientalis. *Candy Carrot*
Visnaga. *Spanish Carrot,* or *Pick-tooth*
Davallia fijiensis. *Fijian Hare's-foot Fern*
Mariesi. *Japan Hare's-foot Fern*
Davidsonia pruriens. *Queensland Itch-tree*
Daviesia. *Native Hop,* of Australia
latifolia. *Native Hop,* of Victoria
Delabechea rupestris. *Bottle-tree,* or *" Gouty-stemmed Tree,"* of Australia
Delphinium. *Dolphin-flower, Larkspur*
Ajacis and vars. *Rocket Larkspur*
albiflorum. *White-flowered Larkspur*
alpinum. *Alpine Bee-Larkspur*
azureum. *Azure-flowered Larkspur*
Beatsoni. *Beatson's Larkspur*
Belladonna. *Belladonna Larkspur*
bicolor. *Blue-and-white-flowered Larkspur*
cashmerianum. *Cashmere Larkspur*
Consolida. *Branching Larkspur, King's Consound, Knight's Spur, Wild Larkspur*
elatum. *Common Bee-Larkspur*
elegans. *Elegant Larkspur*
exaltatum. *American Bee-Larkspur*
formosum. *Showy Larkspur*
grandiflorum. *Large-flowered,* or *Siberian, Larkspur*
grandiflorum fl.-pl. *Double Siberian Larkspur*
grandiflorum var. rubrum. *Large Red-flowered Larkspur*
hyacinthiflorum. *Hyacinth-flowered Larkspur*
intermedium. *Intermediate Larkspur*
Keteleeri. *Keteleer's Larkspur*
moschatum. *Musk-scented Larkspur*
nudicaule. *Dwarf Red Larkspur*
puniceum. *Scarlet-flowered Larkspur*
Staphisagria. *Louse-wort, Stavesacre*
tricorne. *Three-spurred Dwarf Larkspur*
trolliifolium. *" Cow-poison,"* of California
Dendrobium Myosurus. *Mouse-tail Orchid*
speciosum. *Rock-lily,* of N. S. Wales

Dendrocalamus strictus. *Male Bamboo*
Dendromecon rigidum. *Californian Poppy-tree* or *Tree-Poppy*
Dentaria. *American Pepper-root, Tooth-wort*
bulbifera. *Bulb-bearing* or *Common Tooth-wort, Coral-root, Tooth-cress, Tooth-Violet*
digitata. *Showy Tooth-wort*
diphylla. *Two-leaved Tooth-wort*
enneaphylla. *Nine-leaved Tooth-wort*
pentaphylla. *Five-leaved Tooth-wort*
pinnata. *Seven-leaved Tooth-wort*
polyphylla. *Many-leaved Tooth-wort*
triphylla. *Three-leaved Tooth-wort*
Dentella repens. *Australian Tooth-flower*
Desmodium. *Tick-Trefoil, West Indian Honeysuckle*
Aparines. *Bed-straw Tick-Trefoil*
canadense. *Canadian Tick-Trefoil*
Dilleni. *Dillen's Tick-Trefoil*
gyrans. *Moving Plant, Telegraph-plant*
japonicum. *Japanese Tick-Trefoil*
tortuosum. *Cock's-head*, of the W. Indies
Desmoncus macracanthos. *Jacitara Palm*
Detarium senegalense. *"Dattock,"* of W. Tropical Africa
Deutzia gracilis. *Japanese Snow-flower*
Dialium guineense. *Velvet Tamarind*, of Sierra Leone
indicum. *Tamarind-Plum*
Dianella. *Flax-Lily*, of Australia
coerulea. *Paroo Lily*
Dianthera americana. *American Water-Willow*
repens. *Balsam-herb*
Dianthus. *Pink*
alpinus. *Alpine Pink*
arbuscula. *China Tree-Pink*
arenarius. *Sand Pink*
Armeria. *Deptford Pink*
atro-rubens. *Dark-red Italian Pink*
barbatus. *London Tuft, Sweet William*
barbatus var. angustifolius. *Sweet John*
barbatus var. magnificus. *Double Red Sweet William*
caesius. *Cheddar Pink, Clove Pink, Cliff Pink, Mountain Pink*
Carthusianorum. *German Pink*
Caryophyllus. *Carnadine, Carnation, Clove Gilly-flower, Clove Pink, Coronation*
Caryophyllus var. fruticosus. *Shrubby Clove Pink*
caucasicus. *Caucasian Pink*
chinensis and vars. *Chinese Pink, Indian Pink, "Mignonette"* of the French
collinus. *Hungarian Pink*
corsicus. *Corsican Pink*
deltoides. *Maiden Pink*
dentosus. *Amoor Pink, Toothed Pink*
diadematus. *Diadem Pink*
Dunnettii superba. *Maroon-coloured Pink*
Fischeri. *Fischer's Pink*
floribundus. *Free-flowering Pink*
fragrans. *Sweet-scented Pink*
glacialis. *Glacier Pink*
hispanicus. *Spanish Pink*
Heddewigi. *Single-flowered Japan Pink*

Dianthus Heddewigi var. diadematus. *Double-flowered Japan Pink*
Heddewigi var. laciniatus. *Cut-flowered Japan Pink*
neglectus. *Grass Rose Pink*
petraeus. *Rock Pink*
pinifolius. *Pine-leaved Pink*
plumarius and vars. *Common Garden Pink, Feathered Pink, Indian Eye, Pheasant's-eye Pink*
plumarius var. annulatus. *Ringed-flowered Pink*
prolifer. *Proliferous Pink*
pungens. *Fragrant White Pink*
ramosissimus. *Bush Pink*
ruthenicus. *Russian Pink*
sinensis (D. chinensis). *Indian Pink*
suavis. *Sweet Pink*
superbus. *Fringed Pink*
sylvestris. *Wood Pink*
Tymphresteus. *Mt. Tymphrestus Pink*
vaginatus. *Sheathed Pink*
versicolor. *Colour-changing Pink*
viscidus. *Viscid Pink*
Dicentra (Dielytra) canadensis. *Squirrel-corn*
Cucullaria. *Breeches-flower, Dutchman's-Breeches*
formosa. *Common Bleeding-Heart*
eximia. *Plumy Bleeding-Heart*
spectabilis. *"Locks and Keys," Seal-flower, Showy Bleeding-Heart*
Dichelachne crinita. *Australian Mouse Grass*
Dichopsis (Isonandra) Gutta. *Gutta-percha-tree*
Dicksonia antarctica. *Tasmanian Tree-Fern, Woolly Tree-Fern*
arborescens. *St. Helena Tree-Fern*
(Balantium) Culcita. *Cushion-Fern*, of Madeira
dissecta. *Jamaica Tree-Fern*
lanata and squarrosa. *New Zealand Tree-Fern*
pilosiuscula. *Hay-scented-Fern*
Dicranum. *Fork-Moss*
adiantoides. *Maiden-hair Fork-Moss*
bryoides. *Mungo Park's Moss*
cerviculatum. *Red-necked Fork-Moss*
flavescens. *Yellowish Fork-Moss*
fulvellum. *Tawny Fork-Moss*
glaucum. *White Fork-Moss*
heteromallum. *Silky-leaved Fork-Moss*
pellucidum. *Pellucid Fork-Moss*
scoparium. *Broom Fork-Moss*
Dictamnus caucasicus. *Caucasian Fraxinella*
Fraxinella. *Burning Bush, Dittany, Fraxinella, Gas-plant*
Fraxinella var. albus. *White-flowered Fraxinella*
Dicypellium caryophyllatum. *Cayenne Rose, Clove-bark-tree, Clove Cassia, Pepper-wood*
Dieffenbachia (Caladium) seguina. *Dumb Cane*
Dielytra. See Dicentra
Diervilla canadensis (D. trifida). *Bush-Honeysuckle*

Dietes (Iris) bicolor. *Butterfly Flag*
Digitalis. *Fox-glove*
  ambigua. *Great Yellow Fox-glove*
  grandiflora. *Large-flowered Fox-glove*
  lanata. *Woolly Fox-glove*
  lutea. *Small Yellow Fox-glove*
  Mariana. *Sierra Morena Fox-glove*
  obscura. *Willow-leaved Fox-glove*
  ochroleuca. *Cream-coloured Fox-glove*
  purpurea. *Dead-Men's-Bells, Bloody Finger, Common Fox-glove, Fairy Fingers, Finger Flower, Flap Dock, Lusmore*
  Thapsi. *Mullein Fox-glove*
Digitaria sanguinalis. *Red Finger-grass*
Dillenia sarmentosa and D. scabrella. *Sand-paper-tree*
Dimorphotheca pluvialis. *Great Cape Marigold*
Dioidia teres and D. virginica. *Button-weed*
Dionæa muscipula. *Venus's Fly-trap*
Dioscorea aculeata. *Kaawi Yam*
  alata. *Red or White Yam, Uvi Yam*
  Batatas. *Chinese, or Japanese, Yam*
  bulbifera. *Yam, of Granada or Guinea*
  Nummularia. *Money-wort-Yam, Tivoli Yam*
  sativa. *Common Cultivated Yam*
  trifida. *Indian Yam*
  villosa. *American Wild Yam, Colic-root*
Diosma (Coleonema) alba. *African Sleet-bush*
Diospyros Ebenaster. *E. Indian Ebony,*
  Omander-wood, of Ceylon
  Ebenum. *Ceylon Ebony*
  Embryopteris. *Tumika-oil-plant*
  Kaki. *Chinese Date-Plum, Keg-fig of Japan*
  Kurzii. *Marble-wood, of the Andaman Islands*
  Lotus. *European Date-plum*
  Mabola (D. discolor). *Mabolo-tree*
  Melanoxylon. *Coromandel Ebony-tree*
  oppositifolia and D. quæsita. *Calamander, or Coromandel-wood*
  reticulata. *Mauritius Ebony*
  tetrasperma. *Pigeon-wood*
  Texana. *Mexican Persimmon*
  virginiana. *Persimmon-tree, Virginian Date-Plum*
  virginiana var. pubescens. *Downy-leaved Virginian Date-Plum*
Diotis maritima. *Sea Cotton-weed, Sea Cudweed*
Dipholis nigra. *Bastard-tree or Bastard Bully-tree*
  salicifolia. *Pigeon-wood, White Bully-tree*
Diphylleia cymosa. *Umbrella-leaf*
Diplarrhena Moræa. *White Lily, of Tasmania*
Diplopappus. *Double-bristled Aster*
  rigidus. *Rigid Double-bristled Aster*
Diplotaxis muralis. *Stink-weed*
  tenuifolia. *Wall-Rocket*
Diplothemium maritimum. *Brazilian Coast Palm*
Dipsacus. *Teasel, " Venus's Bason "*
  Fullonum. *Draper's Teasel, Fuller's Teasel or Thistle*

Dipsacus pilosus. *Shepherd's-rod or Shepherd's-staff, Small Teasel*
  sylvestris. *Barber's Brushes, Carde Thistle, Church Brooms, Shepherd's-rod, Wild Teasel*
Dipterix eboënsis (D. oleifera). *Eboe Nut-tree*
  odorata. *Tonquin Bean-tree*
Dipterocarpus. Several species yield the "Wood Oil" of Medicine
  turbinatus. *Gurjun-tree, Wood-oil-plant*
Dirca palustris. *Leather-wood, Moose-wood*
Disa grandiflora. *Table Mountain Orchid*
Discaria Toumatou. *New Zealand Hawthorn, " Wild Irishman," of New Zealand*
Discopleura. *Mock Bishop-weed*
Disemma (Passiflora) Herbertiana and D. coccinea (Passiflora Banksii). *Australian Passion-flower*
Diuris sulphurea. *Dragon's-head, of Tasmania*
Dodecatheon. *American Cowslip*
  elegans. *Elegant-flowered American Cowslip*
  integrifolium. *Entire-leaved American Cowslip*
  Jeffreyanum. *Giant American Cowslip*
  Meadia. *Common American Cowslip, Shooting-star*
  Meadia var. splendens. *Deep rose-coloured American Cowslip*
Dodonæa. *Native Hop, of Australia*
  viscosa. *Switch-Sorrel, of Jamaica*
  Waitziana. *Iron-wood, of Dutch E. Indies*
Dolichos. *Hyacinth-Bean*
  albus. *White Hyacinth-Bean*
  bicontortus. *Ram's-horn Bean*
  biflorus. *Horse-Gram*
  Catjang. *Red Gram*
  (Lab-lab) cultratus. *Japan Lab-lab*
  filiformis. *Cat's-claw*
  Lab-lab. *Egyptian Kidney-Bean, Purple Hyacinth-Bean*
  (Lab-lab) perennans. *White China Lab-lab*
  (Mucuna) pruriens. *Horse-eye Bean*
  sesquipedalis. *Asparagus Bean*
  (Vigna) sinensis. *Chowlee-plant*
  Soya. *Chinese Soy-plant*
  sphærospermus. *Black-eyed Pea, of the W. Indies*
  tuberosus. *Yam Bean*
  uniflorus. *Horse-Gram*
  (Mucuna) urens. *W. Indian Cow-itch*
  (Lab-lab) vulgaris. *Common Lab-lab*
Doliocarpus Calinea. *Water-vine, of Guiana*
Dombeya natalensis. *Cape Wedding-flower*
Dondia (Hacquetia, Astrantia) Epipactis. *Dwarf Master-wort*
Doona zeylanica. *Doon-tree, of Ceylon*
Dorema Ammoniacum. *Gum Ammoniacum-plant*
Doronicum austriacum. *Austrian Leopard's-bane*
  caucasicum. *Caucasian Leopard's-bane*
  (Aronicum, Arnica) Clusii. *Clusius's Leopard's-bane*

Doronicum Columnæ. *Columna's Leopard's-bane*
nudicaule. *American Leopard's-bane*
Pardalianches. *Common,* or *Cray-fish, Leopard's-bane*
plantagineum. *Plantain-leaved Leopard's-bane*
Dorstenia Contrayerva. *Contrayerva-root-plant, South American Snake-root*
Doryanthes excelsa. *Australian Giant Lily*
Palmeri. *Spear-Lily,* of Queensland
Dorycnium. *Venomous Trefoil*
fruticosum. *Spear-Pea*
Doryphora Sassafras. *Australian Sassafras*
Draba. *Whitlow-grass*
aizoides. *Sea-green Whitlow-grass*
Aizoon. *Hairy-podded Whitlow-grass*
alpina. *Alpine Whitlow-grass*
aurea. *Danish Whitlow-grass*
bœotica. *Pale-yellow-flowered Whitlow-grass*
ciliaris. *Large Yellow-flowered Whitlow-grass*
ciliata. *Fringed Whitlow-grass*
cinerea. *Gray Whitlow-grass*
cuspidata. *Pointed-leaved Whitlow-grass*
gigas. *Large Whitlow-grass*
glacialis. *Glacier Whitlow-grass*
lasiocarpa. *Woolly-podded Whitlow-grass*
nivalis. *Pigmy Whitlow-grass*
rupestris. *Rock Whitlow-grass*
Sauteri. *Sauter's Whitlow-grass*
stellata. *Starry Whitlow-grass*
tomentosa. *Woolly-leaved Whitlow-grass*
tridentata. *Three-toothed Whitlow-grass*
verna. *Common Nail-wort* or *Whitlow-grass*
violacea. *Violet-flowered Whitlow-grass*
Dracæna. *Dragon-plant*
(Cordyline) australis. *Hardy Dragon-plant, New Zealand Dragon-plant,* or *Ti-plant*
borealis. *Oval-leaved Dragon-plant*
Draco. *Dragon's-blood-tree, Dragon-tree,* of the Canary Islands
Goldieana. *Zebra-striped Dragon-plant* (Cordyline) indivisa. *Hardy Dragon-plant*
Dracocephalum argunense. *Argunsk Dragon's-head*
austriacum. *Austrian Dragon's-head*
grandiflorum. *Betony-leaved Dragon's-head*
moldavicum. *Moldavian Balm*
peregrinum. *Prickly-leaved Dragon's-head, Twin-flowered Dragon's-head*
Ruyschianum. *Hyssop-leaved Dragon's-head*
Dracontium polyphyllum. *Labaria-plant,* of Demerara
Dracophyllum latifolium and D. Traversi. *"Nei-Nei"-plant,* of New Zealand
Drepanocarpus lunatus. *S. American Sickle-pod*
Drimys aromatica. *Tasmanian Pepper-plant*
axillaris. *Pepper-shrub,* of New Zealand
dipetala. *Pepper-shrub,* of N. S. Wales
Winteri. *Winter's Bark,* or *Winter's Cinnamon*

Drosera. *Sun-dew*
anglica. *Great English Sun-dew*
auriculata. *Climbing Sun-dew*
binata. *Old-man's Eye-brow*
dichotoma. *Double-leaved Sun-dew*
filiformis. *Thread-leaved Sun-dew*
linearis. *Slender Sun-dew*
longifolia. *Long-leaved Sun-dew*
rotundifolia. *Common* or *Round-leaved Sun-dew, Lust-wort, Red-rot, Youth-wort*
Drosophyllum lusitanicum. *Portuguese Sun-dew*
Dryas Drummondi. *Yellow-flowered Dryad*
octopetala. *Mountain Avens, White-flowered Dryad*
tenella. *Labrador Mountain Avens*
Drymaria cordata. *W. Indian Chick-weed*
Drymoglossum. *Wood-tongue*
Dryobalanus aromatica. *Camphor-tree,* of Borneo
Drypetes glauca. *Wild Orange,* of the W. Indies
Duboisia Hopwoodii. *"Pitchery-bidgery," Pitury-shrub,* of Australia
myoporoides. *Cork-wood,* of N. S. Wales
Duguetia quitarensis. *Lance-wood,* of Cuba or Guiana
Duranta. *Sky-flower*
Durio zibethinus. *Durian,* or *Duryon-tree*
Dysodia chrysanthemoides. *Fetid Marigold*
Dysoxylon rufum. *Bastard Cedar-pencil-wood,* of N. S. Wales

Ecastaphyllum Brownei. *W. Indian May-flower*
Ecbalium agreste (Momordica Elaterium). *Squirting Cucumber*
Eccremocarpus scaber. *Chilian Glory-flower*
Echinacea. *Black Sampson, Purple Cone-flower*
angustifolia. *Narrow-leaved Purple Cone-flower*
Echinaria capitata. *Hedge-hog Plant*
Echinocactus electracanthus. *Fly Cactus*
myriostigma. *Silvery Hedge-hog Cactus*
polycephalus. *Many-headed Hedge-hog Cactus*
Simpsoni. *Simpson's Hardy Hedge-hog Cactus*
Echinochloa Crus-corvi. *Crow's-foot Grass*
Echinocystis lobata. *Canadian Mock-apple, Wild Balsam-apple*
Echinophora spinosa. *Prickly Samphire, Sea-Parsnip*
Echinops. *Globe-thistle*
bannaticus. *Hungarian Globe-thistle*
exaltatus. *Tall Globe-thistle*
Ritro. *Small Globe-thistle*
ruthenicus. *Russian Globe-thistle*
sphærocephalus. *Great Globe-thistle*
Echinospermum. *Stick-seed*
Echites nutans. *Drooping Savannah-flower*
Echium. *Viper's-Bugloss*
pyrenaicum. *Pyrenean Viper's-Bugloss*
rubrum. *Red-flowered Viper's-Bugloss*
violaceum. *Purple Viper's-Bugloss*

Echium vulgare. *Blue-weed*, of N. America, *Bugloss, Common Viper's-Bugloss, Viper's-grass*
Ecklonia buccinalis. *Cape Trumpet-Seaweed, Horn-plant*
Eclipta brachypoda. *American False Daisy*
Edgeworthia Gardneri. *Indian-paper-tree*
Edraianthus Pumilio. *Silvery Dwarf Hare-bell*
Pumiliorum. *Thrift-leaved Dwarf Hare-bell*
Edwardsia grandiflora and E. microphylla. *New Zealand Laburnum*
Ehretia tinifolia. *Bastard Cherry*, of the W. Indies
Ekebergia capensis. *Cape Ash-tree*
Elæagia utilis. *Varnish-tree*, or *Wax-tree*, of New Granada
Elæagnus. *Oleaster*
angustifolia. *Bohemian* or *Narrow-leaved Oleaster, Wild Olive, Zakkoum-oil-plant*
argentea. *Missouri Silver-berry* or *Silver-tree, Silvery-leaved Oleaster*
parvifolia. *Small-leaved Oleaster*
reflexa variegata. *Variegated Oleaster*
Elæis guineensis. "*Macaw-fat*," of the W. Indies, *Oil-Palm*, of Guinea
melanococca. *Caiané Palm*
Elæocarpus grandis. *Queensland Bracelet-tree*
Hinau. *Hinau*, or *Hino-tree*, of New Zealand
persicæfolius. *New Caledonia Bracelet-tree*
Elæodendron australe. *Olive-wood*, of N. S. Wales
croceum. *Saffron-wood*, of S. Africa
glaucum. *Ceylon Tea-tree*
integrifolium. *Olive-wood*, of Australia
orientale. *E. Indian Olive-wood Tree*
Elæoselinum. *Marsh Parsley*
Elaphomyces. *Deer Balls, Hart's Truffles, Lycoperdon Nuts*
Elatine. *Water-wort*
americana. *American Water-wort*
Hydropiper. *Small Water Pepper-wort*
Eleocharis. *Spike-rush*
cæspitosus. *Deer's-hair*
Elephantopus. *Elephant's-foot*
caroliniensis. *Elephant's-foot*, of Carolina
scaber. *Rough Elephant's-foot*
Elettaria (Alpinia) Cardamomum. *Small*, or *Malabar, Cardamoms*
Eleusine coracana. *Natchnee-plant*, or *Ragee-plant*, of India
indica. *Crab-grass*, or *Dog's-tail-grass*, of N. America, *Wire-grass*
Ellobocarpus oleraceus. *Pod-Fern*
Elymus. *Lyme-grass, Wild Rye*, of N. America
arenarius. *Common Lyme-grass*
condensatus. *Bunch-grass*, of British Columbia
Embothrium spathulatum. *Australian Waratah Tree*
Embryopteris glutinifera. *Wild Mangosteen*
Emmenosperma alphitonioides. *Illawarra Dog-wood*

Empetrum nigrum. *Black-berried Heath, Crake-berry, Crow-berry, Monox, Monnaghs*, or *Moonog Heather*
Encalypta. *Extinguisher Moss*
Encephalartos Caffer. *Caffre-Bread*
Endiandra glauca. *Teak*, of N. S. Wales
virens. *Bat-and-Ball-tree*, of N. S. Wales
Entada. *Sword-Bean*
Gigalobium. *Sea-side Chestnut-plant*
Pursætha. "*Lady-nut*"-tree
(Mimosa) scandens. *Match-box Bean*, of Queensland, *Cacoon* or *Cocoon*, of the W. Indies, *Scimitar-pod-plant*, "W. Indian Filbert*"-tree*
Entelea arborescens. *Mulberry*, or *Cork-tree*, of New Zealand
Epacris. *Australian Heath*
exserta. *Tasmanian Heath*
grandiflora. *Large-flowered Australian Heath*
Eperua falcata. *Wallaba-tree*, of Guiana
Ephedra distachya. *Great Shrubby Horse-tail, Sea-Grape*
monostachya. *Small Shrubby Horse-tail*
Epidendrum conopseum. *Bartram's Tree-Orchid, Florida Tree-Orchid*
macrochilum. *Dragon's-mouth Orchid*
Vanilla (Vanilla planifolia, V. aromatica). *Vanilla-plant*, of commerce
Epigæa repens. *American Ground-Laurel, May-flower* of New England, *Trailing Arbutus*
Epilobium. *Willow-herb*
alpinum. *Mountain Willow-herb*
alsinæfolium. *Chick-weed-leaved Willow-herb*
angustifolium. *Bay-Willow, Blood Vine, Blooming Sally, French* or *Persian Willow, Rose Bay, Rose Elder*
angustifolium var. album. *White-flowered Willow-herb*
coloratum. *American Willow-herb*
Dodonæi. *Dodoen's Willow-herb*
hirsutum. *Apple-pie, Cherry-pie, Codlins-and-Cream, Custard-cups, Fiddle Grass, Gooseberry-pie*
luteum. *Yellow-flowered Californian Willow-herb*
montanum. *Broad-leaved Willow-herb*
obcordatum. *Dwarf Californian Willow-herb*
palustre. *Marsh Willow-herb*
parviflorum. *Hoary Willow-herb*
roseum. *Pale-flowered Willow-herb*
rosmarinifolium (E. angustissimum). *Rosemary-leaved Willow-herb*
tetragonum. *Square-stemmed Willow-herb*
Epimedium alpinum. *Bishop's-Hat, Common Barren-wort*
diphyllum (Aceranthus diphyllus). *Two-leaved Barren-wort*
macranthum (E. grandiflorum). *Large-flowered Barren-wort*
Muschianum. *Muschi's Barren-wort*
pinnatum. *Large-flowered Barren-wort*
purpureum. *Purple-flowered Barren-wort*
violaceum. *Violet-flowered Barren-wort*
Epipactis. *Helleborine*
palustris. *Marsh Helleborine*

Epiphegus virginiana. *Cancer-root, Virginian Beech-drops, Virginian Broom-rape*
Epiphyllum. *Leaf-flowering Cactus*
truncatum. *Common Winter Cactus*
Epipremnum mirabile. *Tonga-plant*
Equisetum. *Horse-tail, Joint Grass, Paddock-pipes, Scrub Grass*
arvense. *Bottle-Brush, False Horse-tail*
hyemale. *Dutch Rush, Pewter-wort, Rough Horse-tail, Scouring Rush, Shave-grass*
limosum. *Smooth Horse-tail*
Mackaii. *Mackay's Horse-tail*
maximum. *Fox-tailed Asparagus*
Moorei. *Moore's Horse-tail*
palustre. *Marsh Horse-tail*
ramosum. *Long-branched Horse-tail*
robustum. *American Horse-tail*
scirpoides. *Dwarf Horse-tail*
sylvaticum. *Bottle-Brush, Wood Horse-tail*
Telmateia. *Great Horse-tail*
umbrosum. *Blunt-topped Horse-tail*
variegatum. *Variegated Horse-tail*
Wilsoni. *Wilson's Horse-tail*
Eragrostis elegans. *Feather-grass, Love-grass*
Eranthis hyemalis. *Winter Aconite, Winter Hellebore, Winter Wolf's-bane*
Erechthites hieracifolia (Senecio hieracifolius). *American Fire-weed*
Eremochloe. *Desert-grass, of California*
Eremophila. *Desert Fuchsia, of Australia*
Mitchelli. *Sandal-wood, of Queensland*
Eremostachys. *Desert-Rod*
Erianthus. *Woolly Beard-grass*
Ravennæ. *Giant Woolly Beard-grass, Ravenna-grass*
Erica. *Heath*
arborea. *"Briar-root," of which pipes are made, Tree Heath*
australis. *Spanish Heath*
baccans. *Berried Heath*
carnea. *Spring or Winter Heath*
carnea var. alba. *White-flowered Spring Heath*
carnea var. hibernica. *Connemara Heath, Black Heath, Carlin Heather, Scotch Heather, She Heather*
ciliaris. *Fringed Heath*
codonodes. *Bell-flowered Heath*
mediterranea (E. carnea). *Mediterranean Heath*
multiflora. *Many-flowered Heath*
polytrichifolia. *Polytrichum-leaved Heath*
sicula. *Sicilian Heath*
stricta. *Upright Heath*
Tetralix. *Bell Heath, or Bell Heather, Cross-leaved Heath*
vagans. *Cornish Heath, Moor Heath*
Erigenia bulbosa. *Harbinger-of-Spring*
Erigeron acris. *Blue-flowered Flea-bane*
alpinus (E. uniflorus). *Mountain Flea-bane*
annuus. *Daisy Flea-bane, Sweet Scabious of N. America*
bellidifolius. *Robin's Plantain*
canadensis. *Canadian Flea-bane, Butter-weed, Horse-weed, Pride-weed, Colt's-tail*

Erigeron glabellus. *Smoothish Flea-bane*
glaucus. *Glaucous-leaved Flea-bane*
grandiflorus. *Large-flowered Flea-bane*
(Leptostelma) maximus. *Mexican Daisy*
mucronatus. *Australian Flea-bane*
multiradiatus. *Many-rayed Flea-bane*
Roylei. *Royle's Flea-bane*
speciosus (Stenactis speciosa). *Showy Flea-bane*
strigosus. *Daisy Flea-bane*
Erinosma. *St. Agnes's-flower*
Eriobotrya (Mespilus) japonica. *Loquat, or Japan Medlar*
Eriocaulon septangulare. *Pipe-wort*
Eriocoma cuspidata. *Silk-grass, of N. America*
Eriodendron. *Silk Cotton-tree*
anfractuosum (Bombax Ceiba). *Cabbage-wood, Ceiba-tree, God-tree, Silk Cotton-tree of the W. Indies*
orientale. *East Indian White Silk Cotton-tree*
Samauma. *Silk Cotton-tree, of the Amazon*
Eriodictyon californicum. *Tar-bush, of California*
Eriogonum fasciculatum. *"Bee-feed," or Wild Buck-wheat, of California*
Eriophorum. *Cotton-grass, Cotton-rush, Moor-pawms, Moss-crops, Wild Cotton*
capitatum (E. Scheuchzeri). *Globe Cotton-grass*
polystachyum. *Common Cotton-sedge, Downy Ling, Tassel Cotton-grass*
vaginatum. *Canna-down, Cannach, Common Moss-crops, Hare's-tail Rush*
Erisma Japura. *Japura Tree, of Brazil*
Erithalis fruticosa. *Jasmine-scented-wood Tree*
Eritrichium nanum. *Fairy Borage*
Ernodea littornlis. *Branched Spurge*
Erodium. *Heron's-bill*
alpinum. *Alpine Heron's-bill*
caruifolium. *Caraway-leaved Heron's-bill*
cheilanthifolium. *Cheilanthes-leaved Heron's bill*
Ciconium. *Long-beaked Heron's-bill*
cicutarium. *Common Heron's-bill, Pin-clover, or Pin-grass, of California, Wild Musk*
hymenodes. *Pelargonium Heron's-bill, Three-leaved Heron's-bill*
macradenum. *Black-eyed Heron's-bill*
Manescavi. *Showy Heron's-bill*
maritimum. *Sea-side Heron's-bill*
moschatum. *Ground-needle, Musk Heron's-bill*
petræum. *Rock Heron's-bill*
Reichardi. *Fairy Heron's-bill*
romanum. *Roman Heron's-bill*
trichomanæfolium. *Fern-leaved Heron's-bill*
Erpetion. *Spurless Violet*
reniforme (Viola hederacea). *Australian Pansy, New Holland Violet*
Eruca sativa. *Edible or Salad Rocket*
Ervum dispermum. *Indian Vetch*
Ervilia. *Black Bitter Vetch, Ers, Pigeon's-pea*
hirsutum. *Rough-podded Tare*

and Foreign Plants, Trees, and Shrubs. 193

Ervum Lens. *Common Lentil, Red Pottage Pea*
monanthos. *One-flowered Tare*
Eryngium alpinum. *Alpine Eryngo*
amethystinum. *Amethystine Eryngo*
Bourgati. *Bourgati's Eryngo*
bromeliæfolium. *Pine-apple-leaved Eryngo*
campestre. *Field Eryngo*
eburneum. *Ivory-leaved Eryngo*
fœtidum. *Fitt-weed*
giganteum. *Giant Eryngo*
maritimum. *Sea Eryngo, Sea Holly*
pandanifolium. *Palm-leaved Eryngo*
planum. *Flat-leaved Eryngo*
pusillum. *Dwarf Eryngo*
spinâ-albâ. *White-spined Eryngo*
yuccæfolium. *Button Snake-root, Rattle-snake's Master*
Erysimum Alliaria. "*Eileber*," *Garlic-wort, Sauce-alone*
arkansanum. *Western Wall-flower*
cheiranthoides. *Treacle-Mustard, Treacle-Worm-seed*
ochroleucum (Cheiranthus alpinus). *Alpine Wall-flower*
odoratum. *Sweet-scented Hedge-Mustard*
pachycarpum. *Himalayan Wall-flower*
Perofskianum. *Perofsky's Dwarf Wall-flower*
pulchellum. *Rock Wall-flower*
pumilum. *Fairy Wall-flower*
Rhæticum. *Rhætian Dwarf Wall-flower*
Erythræa. *Blush-wort,*" *Canchalagua*," of California
aggregata. *Clustered Blush-wort*
Centaurium. *Centaury,* " *Cristaldre*," *Earth-gall, Fever-wort*
littoralis. *Sea-shore Blush-wort*
Erythrina Caffra. *Kaffir's-tree*
Corallodendron. *Red Bean-tree, West Indian Coral-tree*
indica. *East Indian Coral-tree, Mootchie-wood*
monosperma. *Gum-Lac-tree,* " *Pismire-tree*"
Erythrolæna conspicua. *Scarlet Mexican Thistle*
Erythronium albidum. *White Dog's-tooth Violet*
americanum. *Serpent's-tongue, Yellow Adder's-tongue, Yellow Dog's-tooth Violet*
Dens-canis. *Common Dog's-tooth Violet*
giganteum. *Giant Dog's-tooth Violet*
grandiflorum. *Oregon Dog's-tooth Violet*
Nuttallianum. *Nuttall's Dog's-tooth Violet*
Erythrophlæum guineense. *Gree-Gree-tree,* or *Sassy-tree,* of Sierra Leone, *Ordeal-bark-tree*
Icononse. *Red-water Tree,* of Sierra Leone
Erythroxylon areolatum. *Jamaica Iron-wood*
Coca. *Coca-bush,* or *Spadic-bush,* of S. America
laurifolium. *Laurel-leaved Red-wood Tree*
Escallonia macrantha. *Chilian Gum-Box*
Eschscholtzia californica. *Californian Poppy,* " *Chriseis* "
californica var. alba. *White Californian Poppy*

Eschscholtzia californica var. crocea *Orange Californian Poppy*
californica var. crocea rosea. *Pink Californian Poppy*
Espeletia. *Lion's-ear,* of the Andes
Eucalyptus. *Australian Gum-tree*
albens. " *White Box*," of Australia, *White Gum-tree*
amygdalina. *Australian Kino-tree, Giant Gum-tree, Red Gum-tree, Tasmanian Peppermint-tree,* " *Wangara*"-*tree*
botryoides. *Blue Gum-tree, Swamp Mahogany,* of N. S. Wales
calophylla. *Red Gum-tree*
capitellata. *Peppermint-tree, Stringy-bark-tree*
coriacea. *Flooded, Swamp, Weeping,* or *White Gum-tree, Peppermint-tree*
cornuta. *Yate-tree* or *Yeit-tree*
corymbosa. *Blood-wood,* of Victoria
corynocalyx. *Sugar Gum-tree*
crebra. *Australian Iron-bark*
dealbata. " *Gray Box*," of Victoria, *River Gum-tree*
decipiens. *Flooded Gum-tree*
diversicolor. *Blue Gum-tree, Karri-tree* of S. W. Australia
doratoxylon. *Spear-wood,* of Australia
drepanophylla. *Australian Iron-bark*
dumosa. " *Mallee*"-*tree,* of Australia
eximia. *Blood-wood,* of Australia, *Rusty Gum-tree*
ficifolia. *Scarlet-flowered Gum-tree*
gigantea. *Stringy-bark-tree,* of Tasmania
globulus. *Blue Gum-tree, Fever Gum-tree*
gomphocephala. " *Tooart*"-*tree,* of S. W. Australia
goniocalyx. *Spotted* or *White Gum-tree*
Gunni. " *Cider-tree*," of Tasmania
hæmastoma. *Blue* or *Spotted Gum-tree, Mountain Ash,* of N. S. Wales
Laboucherii. *Queensland Iron-bark*
Leucoxylon. *Australian Iron-bark,*" *Bastard Box*," of Victoria, *Black Mountain Ash*
Leucoxylon minor. *White Gum-tree*
longifolia. " *Woolly-butt* "
loxophlebs. *York Gum-tree*
macrorrhynca. *Stringy-bark-tree*
maculata. *Spotted Gum-tree*
maculata var. citriodora. *Lemon-scented Gum-tree*
mannifera. *Manna-tree,* of Australia
marginata. *Jarrah-wood, Mahogany* of S. W. Australia
megacarpa. *Blue Gum-tree*
melanophloia. *Australian Iron-bark*
melliodora. *Red Gum-tree,* " *Yellow Box* "
obliqua. " *Mess-mate*," *Stringy-bark-tree*
odorata. *Peppermint-tree, Red Gum-tree*
oleosa. " *Mallee*"-*tree,* of Victoria
paniculata. *Blood-wood,* of Queensland, *She Iron-bark*
paniculata var. fasciculosa. *White Gum-tree*
pilularis. " *Black-butt*," of S. Queensland and N. S. Wales, *Stringy-bark-tree*
pilularis var. acmenioides. *White Mahogany,* of N. S. Wales

o

Eucalyptus Piperita. *Stringy-bark-tree, Tasmanian Peppermint-tree*
platypus. *Maalok-tree,* of Australia
polyanthemos. *Lignum-ritæ,* of N. S. Wales, "*Red Box,*" of S. E. Australia, *Gippsland* "*Den*"-*tree*
populifolia. "*Bembil,*" *Shining-leaved Box Eucalyptus*
pulverulenta. *Argyle Apple, Silver-leaved Iron-bark Tree*
punctata. *Hickory-Eucalypt,* or *Leather-jacket,* of N. S. Wales
redunca. "*Wandoo*"-*tree*
resinifera. *Botany Bay Kino-tree, Gray* or *Red Gum-tree, Hickory* of N. S. Wales, *Leather-jacket, Red Mahogany* of N. S. Wales
resinifera var. *Forest Mahogany* of N. S. Wales
Risdoni. *Drooping* or *Risdon Gum-tree*
robusta. *Brown Gum-tree, Swamp* or *White Mahogany,* of N. S. Wales
rostrata. *Jarrah-wood, Flooded Gum-tree, Red, River,* or *White Gum-tree*
rudis. *Flooded* or *Swamp Gum-tree*
saligna. *Gray* or *White Gum-tree*
salmonophloia. *Salmon-barked Gum-tree*
salubris. *Gimlet-wood, Fluted Gum-tree*
Sideroxylon. *Red Iron-bark Tree*
Sieberiana (E. virgata). "*Gum-top*"
stellulata. *Green, Olive-green, Lead,* or *White Gum-tree*
Stuartiana. *Apple-scented Gum-tree, But-but-tree,* of Australia, *Hickory,* of N. S. Wales
Stuartiana var. longifolia. *Turpentine Gum-tree*
tereticornis. *Blue* or *Red Gum-tree*
triantha (E. acmenioides). *White Mahogany*
viminalis. *Blue, Drooping,* or *Peppermint Gum-tree, Manna Gum-tree,* "*Woolly-butt*"
virgata. "*Gum-top,*" *Mountain-Ash,* of N. S. Wales
Eucharis amazonica. *Amazon Lily*
Sanderi. *Sander's Lily*
Eucheuma speciosum. *Australian Jelly-plant*
Euchlœna (Recana) luxurians. *Teosinte-grass*
Eucomis. *Pine-apple-flower*
regia. *King's Flower*
Eucryphia pinnata. *Brush-Bush*
Euerigeron Philadelphicus. *Frost-weed, Squaw-weed*
Eugenia. *W. Indian Myrtle*
(Jambos) acida. *Dome Myrtle*
acris. *Wild Clove-tree*
axillaris. *Red Rod-wood*
disticha. *Wild Coffee* of the W. Indies
fragrans. *Zebra-wood* of Jamaica
Jambos (Jambosa vulgaris). *E. Indian Rose-apple, Malabar Plum*
lineata. *Guava-berry*
Michellii. *Cayenne Cherry*
(Jambosa) malaccensis. *Malay Rose-apple*
neurocalyx. *Leba-tree,* of Fiji
pallens. *Black Rod-wood*

Eugenia Pimenta. *Jamaica Allspice-tree, W. Indian Bay-berry-tree*
Smithii. "*Lilly Pillies,*" of Victoria
supra-axillaris. "*Tata,*" of Brazil
Ugni. *Edible-fruited* or "*Fruiting*" *Myrtle, Ugni-shrub,* of Chili
uniflora. *Barbadoes Cherry*
variabilis. *Malay Tea-tree*
Eulalia japonica zebrina. *Zebra-striped Rush*
Eulophia campestris. *E. Indian Salep-plant*
Euonymus. *Spindle-tree*
americanus. *Strawberry-bush,* of N. America
angustifolius. *Narrow-leaved Spindle-tree*
atropurpureus. *Burning-bush, Waahoo,* of N. America
europæus. *Ananbeam, Dog-wood, Cat-tree, Cat-wood, Louse-berry-tree, Peg-wood, Prick-wood, Skewer-wood, Spindle-tree*
europæus fructû-albo. *White-fruited Spindle-tree*
fimbriatus. *Fringed Spindle-tree*
garciniæfolius. *Garcinia-leaved Spindle-tree*
grandiflorus. *Large-flowered Spindle-tree*
Hamiltonianus. *Hamilton's Spindle-tree*
japonicus. *Japanese Spindle-tree*
japonicus var. argenteus. *Silver-edged-leaved Spindle-tree*
japonicus var. aureo-variegatus. *Golden-leaved Spindle-tree*
japonicus var. radicans. *Japanese Box*
japonicus radicans variegatus. *Variegated Dwarf Spindle-tree*
japonicus var. tricolor. *Three-coloured-leaved Spindle-tree*
latifolius. *Broad-leaved Spindle-tree, Large-fruited Spindle-tree*
nanus. *Dwarf Spindle-tree*
obovatus. *Obovate-leaved Spindle-tree*
sarmentosus. *Trailing-stemmed Spindle-tree*
Sieboldianus. *Pai-cha-wood*
verrucosus. *Warty-barked Spindle-tree*
Eupatorium ageratoides. *Nettle-leaved Hemp-Agrimony, White Snake-root*
aromaticum. *Aromatic Hemp-Agrimony*
cannabinum. *Andurion, Common Hemp-Agrimony, Dutch Agrimony, Hemp-weed, Holy Rope, Water Agrimony, Water Hemp*
glutinosum. *Matico,* of the Peruvians
hyssopifolium. *American Bone-set, Thorough-wort*
perfoliatum. *Bone-set, Indian Ague-weed, Vegetable Antimony*
purpureum. *Gravel-root, Joe-Pye-Weed, Purple Hemp-Agrimony, Trumpet-weed*
sessilifolium. *Upland Bone-set*
teucrifolium. *Wild Hore-hound*
triplinerve. "*Ayapana*"
Euphorbia. *Devil's-milk, Gum-thistle, Spurge, Wolf's-milk*
amygdaloides. *Wood-Spurge*
antiquorum. *Cattemundoo,* or *Callemundoo Gum-plant*
arborea. *Zebra-poison,* of S. Africa
balsamifera. *Balsam Spurge*

Euphorbia Caput-Medusæ. *Medusa's-Head*
Cattimandoo. *Caoutchouc-plant*, of Madras
corollata. *White-flowered Spurge*
cucumerina. *Cucumber Spurge*
Cyparissias. *Cypress Spurge*, *Welcome-to-our-house*
dulcis. *Sweet Spurge*
Esula. *Faitour's-Grass, Gromwell-leaved Spurge, Leafy Spurge, Quacksalver's Spurge* or *Turbith*
exigua. *Dwarf Spurge*
Helioscopia. *Cat's-milk, Churn-staff, Irbydale Grass, Little-good* or *Little Goody, Sun-Spurge, Wart Grass, Wart-wort*
hiberna. *Irish Spurge, Mackinboy* or *Makkin-bwee*
Hystrix. *Porcupine Spurge*
Lathyris. *Caper-bush, Caper-plant, Caper-Spurge, Euphorbia-oil-plant, Myrtle Spurge, Wild Caper*
maculata. *Milk Purslane, West Indian Eye-bright*
myrtifolia. *Negro's-slippers*
officinarum. *Poisonous Gum-thistle*
palustris. *Marsh Spurge*
Paralias. *Sea-side Spurge*
Peplis. *Hyssop-Spurge, Purple Spurge*
Peplus. *Petty Spurge*
pilosa. *Hairy Spurge*
platyphyllus. *Broad-leaved Spurge*
Portlandica. *Portland Spurge*
prostrata. *Trailing Red Spurge*
punicea. *Scarlet-flowered Spurge, Jamaica Fire-flower, Vegetable Leather*
resinifera. *Euphorbium Gum-plant*
segetalis (E. Portlandica). *Portland Spurge*
Tirucalli. *Indian Tree-Spurge*
Euphoria Litchi. *Chinese Li-tchi*, or *Leechee*
Euphrasia officinalis. *Euphrasy, Eye-bright*
Euryale ferox (Anneslea spinosa). *Gorgon-plant, Prickly Water-lily*, of the East Indies
Euryangium Sumbul. *Sumbul-Root*
Eurybia. *Daisy-tree, Star-wort*
argophylla (Aster argophyllus). *Tasmanian Musk-tree*
lyrata. *Tasmanian Daisy-tree*
ramulosa. *Shrubby Star-wort*
Eurycles amboinensis. *Amboyna Lily* australasica (E. Cunninghamii). *Brisbane Lily, Moreton Bay Lily*
Euryops speciosissimus. *Resin-bush*, of S. Africa
Eusideroxylon Zuageri. *Iron-wood*, of Dutch E. Indies
Eustrephus. *Wombat-berry*
Euterpe edulis. *Assai Palm*
Euxolus caudatus. *Small-leaved Calalu* viridis. *Green Calalu*
Evolvulus Nummularius. *Monkey-wort*
Exacum guianense. *Guiana Centaury*
Excæcaria Agallocha. *Blinding-tree* glandulosa. *Green Ebony*
Exidia Auricula-Judæ. *"Jew's Ears," "Judas's-Ear"*
glandulosa. *Witches'-butter Fungus*

Exocarpus cupressiformis. *Native Cherry*, of Australia
Exochorda (Spiræa) grandiflora. *Pearl-bush*
Exogonium Purga. *Jalap-plant* (true)
Exostemma caribæum. *"Caribæan Bark"-tree, Sea-side Beech*, of Jamaica¹ *'W. Indian Bark"-tree*
floribundum. *"Caribæan Bark"-tree*
Faba. *Bean*
vulgaris var. chlorosperma. *Green Windsor Bean*
vulgaris var. ensiformis. *Long-pod Bean*
vulgaris var. equina. *Field, Horse, Pigeon*, or *Tick Bean*
vulgaris var. hortensis. *Garden Bean*
vulgaris var. macrosperma. *Broad Windsor Bean*
vulgaris var. præcox. *Mazagan Bean*
Fabiana imbricata. *False-Heath*
Fagara lentiscifolia. *Bastard Iron-wood*, or *Satin-tree*, of the W. Indies
Pterota. *Iron-wood*, of the W. Indies
microphylla. *Ram-goat*
Fagopyrum esculentum. *Brank, Buck-wheat*
cymosum. *Perennial Buck-wheat*, or *Beech-wheat*
Fagus. *Beech-tree*
betuloides. *Evergreen Beech*, of Tierra del Fuego
cliffortioides. *" White Birch,"* of New Zealand
Cunninghamii. *Tasmanian "Myrtle"*
ferruginea. *American Beech, Purple Beech*
fusca. *New Zealand Beech, Towai*, or *Tawhai-tree*
Menziesii. *Otago "Birch," Towai*, or *Tawhai-tree*
obliqua. *Oblique-leaved Beech*
sylvatica. *Common Beech*
sylvatica var. aspleniifolia. *Fern-leaved Beech*
sylvatica var. cristata. *Crested Beech*
sylvatica var. cuprea. *Copper-coloured Beech*
sylvatica var. incisa. *Cut-leaved Beech*
sylvatica var. pendula. *Weeping Beech*
sylvatica var. purpurea. *Purple Beech*
Solandri. *" White Birch,"* of New Zealand
Fallucia paradoxa. *Apache Plume*
Faramea odoratissima. *Wild Coffee*, or *Wild Jasmine*, of the W. Indies
Farfugium grande. *Spotted Colt's-foot*
Fedia Cornucopiæ. *Horn-of-Plenty*
Felicia tenella. *Slender Star-wort*
Feronia elephantum. *Elephant-apple*, or *Wood-apple*, of India
Ferraria. *Black Iris*
Ferula. *Giant Fennel*
asparagifolia. *Asparagus-leaved Giant Fennel*
communis. *Common Giant Fennel*
Ferulago. *Broad-leaved Giant Fennel*
galbaniflua. *Galbanum-plant*
(Thapsia) garganica, or F. glauca. *Supposed to be the "Silphium"* of the ancients

Ferula glauca. *Glaucous Giant Fennel*
nodiflora. *Knotted Giant Fennel*
persica. *Persian Asafœtida*
sulcata. *Furrowed Giant Fennel*
(Euryangium) Sumbul. *Sumbul-plant*
tingitana. *Tangier Giant Fennel*
Festuca. *Fescue Grass*
altissima. *Arab's Miss*
duriuscula. *Hard Fescue*
elatior. *Dover Grass*
glauca. *Ornamental Blue Fescue-grass*
(Triodia) irritans. *Australian Porcupine Grass*
Myurus. *Mouse-tail* or *Rat's-tail Grass*
ovina. *Sheep's Fescue-grass*
pratensis. *Meadow Fescue-grass*
rubra. *Red* or *Creeping Fescue-grass*
scabrella. *Bunch-grass*, of British Columbia
sylvatica. *Ant-hill-grass*
Feuillæa (Fevillea) cordifolia. *Antidote Cacoon* or *Cocoon*, of the W. Indies, "*Seyra-seed*"-*plant*
Ficaria grandiflora (Ranunculus calthæfolius). *Great Pile-wort*
Ficus. *Fig-tree*
aspera. *Tongue Fig*
Benjamines. *Willow Fig-tree*
Carica. *Common Fig-tree*
columnaris. *Banyan tree*, of Lord Howe's Island
eburnea. *Ivory Fig*
elastica. *Caoutchouc-tree*, of the E. Indies, *India-rubber Fig-tree*
glomerata (F. vesca). *Cluster-Fig*, of Australia
indica. *Banyan-tree*, *Bo-tree*, *Grove-tree*, *Pagoda-tree*
laurifolia. *Black Fig*
macrophylla. *Moreton Bay Fig*
ochroleuca. *White Fig*
Parcelii. *Clown Fig*
pedunculata. *Jamaica Cherry*, *Red Fig*, *Willow-leaved Fig-tree*
pertusa. *Laurel-leaved Fig-tree*, of S. America
religiosa. *Peepul* or *Pipul Tree*, *Poplar-leaved* or *Sacred Fig-tree*
repens. *Creeping Fig-tree*
repens var. minima. *Dwarf Creeping Fig-tree*
rubiginosa (F. australis). *Banyan-tree*, of N. S. Wales
Sycomorus. *Sycomore-tree*
tinctoria. *Dyers' Fig-tree*
Filago germanica. *Downweed*, *Clod-weed*, "*Cotton-Rose*" of N. America, *Cud-wort*, *Herb Impious*, *Owl's-crown*
minima. *File-wort*
pygmæa. *Pigmy Cotton-rose*
Filix. *Fern*
Fistulina Hepatica. *Beef-steak Fungus*
Fitzroya Patagonica. *Patagonian Cypress*
Flacourtia cataphracta. *E. Indian Plum-tree*
Ramontchi. *Batoko Plum*, of the Zambesi
sapida. *Ceylon Plum*
Flindersia australis. *Callecdra-wood*, *Queensland Rasp-pod*

Flindersia Bennettiana. *Boyum-Boyum-tree*, of Queensland
maculata. "*Spotted-tree*," of Queensland
Flœrkia proserpinacoides. *False Mermaid*
Fœniculum. *Fennel*
dulce. *Finocchio* or *Finicho*, *Sweet Fennel*
piperitum. *Asses' Finocchio*
vulgare. *Common Fennel*, *Fenkelle*
Fœtida mauritiana. *Stink-wood*
Fontinalis. *Water-Moss*
antipyretica. *Greater Water-Moss*
squamosa. *Alpine Water-Moss*
Fontanesia phillyræoides. *Syrian Privet*
Forsythia suspensa. *Japanese Golden-ball-tree*
viridissima. *Chinese Golden-ball Tree*
Fothergilla alnifolia. *American Witch Alder*
Fouquiera splendens. *Californian Candle-wood*
Fourcroya gigantea. *Giant Mexican Lily*
Fragaria. *Straw-berry*
chilensis. *Chili Straw-berry*
collina. *Alpine*, or *Green Pine Straw-berry*
elatior. *Hautbois Straw-berry*
indica. *Indian Straw-berry*, *Rock Straw-berry*
vesca and vars. *Common Straw-berry*
virginiana. *Old Scarlet* or *Scarlet Virginia Straw-berry*
Franciscea uniflora. *Vegetable Mercury*
Frangula caroliniana. *Alder-Buckthorn*, of N. America
Frankenia lævis. *Common Sea Heath*
pulverulenta. *Mealy Sea Heath*
Frasera verticillata. *American Columbo-root*, *Gold Seal*, *Indian Lettuce*
Fraxinus. *Ash* or *Ash-tree*, *Angelica-tree* of N. America
americana var. alba. *American White Ash*
americana var. aucubæfolia. *Aucuba-leaved Ash*
americana var. cinerea. *Gray Ash*
americana var. elliptica. *Oval-leaved Ash*
americana var. fusca. *Brown-branched Ash*
americana var. longifolia. *Long-leaved Ash*
americana var. mixta. *Mixed Ash*
americana var. ovata. *Ovate-leaved Ash*
americana var. pannosa. *Cloth-leaved Ash*
americana var. pulverulenta. *Powdery Ash*
americana var. Richardi. *Richard's Ash*
americana var. rubicunda. *Reddish-veined Ash*
americana var. Boscii. *Bosc's Ash*
americana var. rufa. *Rufous-haired Ash*
chinensis. *Chinese Ash*
excelsior. *Common Ash-tree*, *Culverkeys*
excelsior var. angustifolia. *Narrow-leaved Ash*
excelsior var. argentea. *Silver-striped Ash*
excelsior var. aurea. *Golden-barked Ash*
excelsior var. aurea pendula. *Variegated Weeping Ash*
excelsior var. fungosa. *Fungous-barked Ash*

and Foreign Plants, Trees, and Shrubs. 197

Fraxinus excelsior var. heterophylla. *Various-leaved Ash*
excelsior var. horizontalis. *Horizontal-branched Ash*
excelsior var. jaspidea. *Striped-barked Ash*
excelsior var. Kincairnie. *Kincairney Ash*
excelsior var. lutea. *Yellow-edged-leafleted Ash*
excelsior var. nana (F. o. humilis). *Dwarf Ash*
excelsior var. pallida. *Pale-barked Ash*
excelsior var. parvifolia. *Small-leaved Ash*
excelsior var. parvifolia argentea. *Silvery-leaved Ash*
excelsior var. parvifolia oxycarpa. *Sharp-fruited Ash*
excelsior var. pendula. *Common Weeping Ash*
excelsior var. purpurascens. *Purple-barked Ash*
excelsior var. verrucosa. *Warted-barked Ash*
excelsior var. verticillata. *Whorled-leaved Ash*
floribunda. *Nepaul Ash*
lentiscifolia. *Lentiscus-leaved Ash*
Oregana. *Oregon Ash*
Ornus. *Flowering Ash*
Ornus var. americana. *American Flowering Ash*
Ornus var. floribunda. *Many-flowered Flowering Ash*
Ornus var. mannifera. *Manna-Ash-tree*
Ornus var. rotundifolia. *Round-leaved Flowering Ash*
Ornus var. striata. *Striped-barked Flowering Ash*
platycarpa. *Carolina Water-Ash*
pubescens. *American Red Ash*
quadraugulata. *American Blue Ash*
sambucifolia. *American Black Ash, Water-Ash*
Schiediana. *Schiede's Ash*
viridis. *American Green Ash*
Fremontia californica. "*Slippery Elm*," of California
Frenela robusta. *Murray Pine-tree*
verrucosa. *Cypress Pine*
Fritillaria. *Fritillary*
imperialis. *Crown Imperial*
latifolia. *Broad-leaved Fritillary*
Meleagris. *Chequered Daffodil, Chequered Lily, Drooping Tulip, Guinea-hen Flower, Snake's-head Fritillary*
Meleagris plena. *Double-flowered Fritillary*
Meleagris var. alba. *White Fritillary*
nigra. *Toad-lily*
persica. *Persian Lily*
pyrenaica. *Pyrenean Fritillary*
recurva. *Scarlet Fritillary*
tulipifolia. *Tulip-leaved Fritillary*
Fuchsia. *Ear-ring Flower, Lady's Ear-drops*
coccinea. *Scarlet Fuchsia*
Colensoi. *New Zealand Fuchsia*
conica. *Conical-tubed Fuchsia*
corymbiflora. *Corymb-flowered Fuchsia*

Fuchsia excorticata. *Kotukutuki-tree*, of New Zealand
fulgens. *Brilliant Fuchsia*
globosa. *Globe-flowered Fuchsia*
gracilis. *Slender Fuchsia*
lycioides. *Box-thorn-leaved Fuchsia*
magellanica. *Fuegian Fuchsia*
microphylla. *Small-leaved Fuchsia*
minima. *Dwarf Trailing Fuchsia*
procumbens. *Basket Fuchsia, Trailing Fuchsia*
pumila. *Tom Thumb Fuchsia*
racemosa. *Edible-fruited Fuchsia*
Riccartoni. *Riccarton's Fuchsia*
serratifolia. *Saw-leaved Fuchsia*
Thompsoni. *Thompson's Fuchsia*
thymifolia. *Thyme-leaved Fuchsia*
Fucus amylaceus. *Ceylon Moss*
nodosus. *Kelp-ware* or *Kelp-wrack, Tang* or *Knob-Tang*
vesiculosus. *Kelp-ware* or *Kelp-wrack, Ore, Ore-weed, Oar-weed, Sea Lettuce, Sea Oak*
Fuirena squarrosa. *Umbrella-grass*
Fumaria densiflora. *Dense-flowered Fumitory*
officinalis. *Common Fumitory, Fumiterre*
vesicaria. *Bladdered Fumitory*
Funaria. *Cord-Moss*
hibernica. *Irish Cord-Moss*
hygrometrica. *Hygrometric Moss*
Funkia. *Japanese Day-lily, Plantain lily*
albo-marginata. *White-margined Plantain-lily*
coerulea. *Blue Plantain-lily*
Fortunei. *Fortune's Plantain-lily*
japonica (F. grandiflora). *Sweet-scented Plantain-lily*
lanceolata. *Spear-leaved Plantain-lily*
ovata. *Common Blue Plantain-lily*
Sieboldi. *Siebold's Blue Plantain-lily*
subcordata. *Corfu Lily, White-flowered Plantain-lily*
undulata foliis argenteo-variegatis. *Silver-variegated Plantain-lily*
undulata foliis aureo-variegatis. *Golden-variegated Plantain-lily*
Fusanus acuminatus. *Quandang-nut*, of Australia
spicatus (Santalum spicatum). *Sandal-wood*, of W. Australia

Gagea fistulosa. *Bright-yellow Star-of-Bethlehem*
Gahnia. *New Zealand Cutting-grass*
Gaillardia amblyodon. *Blunt-toothed Blanket-flower*
aristata. *Bristly Blanket-flower*
grandiflora. *Large-blossomed Blanket-flower*
lanceolata. *Lance-leaved Blanket-flower*
Loiseli. *Loisel' Blanket-flower*
Galactia. *American Milk-Pea*
Galanthus. *Snow-drop*
Elwesii. *Elwes's Snow-drop*
Imperati. *Imperati's Snow-drop*
latifolius. *Broad-leaved Snow-drop*
lutescens. *Yellowish Snow-drop*

198 English Names of Cultivated, Native,

Galanthus nivalis. "*Bulbous Violet,*" *Candlemas Bells, Common Snow-drop, Fair-Maids-of-February, Purification-flower*
plicatus. *Crimean Snow-drop*
poculiformis. *Cup-flowered Snow-drop*
præcox. *Early-flowering Snow-drop*
Redoutéi. *Redouté's Snow-drop*
reflexus. *Reflexed Snow-drop*
Reginæ Olga. *Queen Olya's Snow-drop*
serotinus. *Late-flowering Snow-drop*
Shaylocki. *Shaylock's Snow-drop*
virescens. *Greenish-flowered Snow-drop*
Galax aphylla. *Carpenter's-leaf, White-wand Plant*
Galeandra. *Casque-wort*
Galega. *Goat's-Rue*
frutescens. *Tipsy-wood*
officinalis. *Common Goat's-Rue*
orientalis. *Oriental Goat's-Rue*
Galeobdolon luteum. *Weasel's-snout, Yellow Archangel*
Galeopsis. *Hemp-nettle*
Ladanum. "*Donninethell,*" *Holy Hemp, Red-flowered Hemp-nettle*
ochroleuca. *Downy Hemp-nettle*
Tetrahit. *Bee, Day, or Daye Nettle, Common Hemp-Nettle*
villosa. *Yellow Iron-wort*
Galipea Cusparia. *Angostura Bark, Carony Bark, or Cusparia Bark-tree*
Galium. *Bed-straw*
Aparine. *Bur-weed, Catch-weed, Claver-grass, Cleaver-grass, Cleavers, Geekdor, Goose-bill, Goose-grass, Goose-tongue, Grip-grass, Gull-grass, Harif, Heiriff, Hair-eve,* or *Haritch,* "*Kandlegosses,*" *Love-man, Mutton-chops* or *Mutton-tops, Turkey Grass*
asprellum. *Rough Bed-straw,* of N. America
Bloomeri. *Bloomer's Bed-straw*
boreale. *Northern Bed-straw*
circæzans. *Wild Liquorice,* of N. America
Cruciata. *Cross-wort Bed-straw, Honey-wort, May-wort, Mug-weed*
lanceolatum. *Wild Liquorice,* of N. America
Mollugo. *Great Bed-straw, Hedge-straw, Sticky Grass, Wild Madder*
palustre. *Marsh Bed-straw*
parisiense. *Wall Bed-straw*
saxatile. *Heath Bed-straw*
tricorne. *Corn Bed-straw*
tinctorium. *Small Cleavers*
uliginosum. *Swamp Bed-straw*
verum. *Cheese-rennet, Hundred-fold, Keeslip, Lady's Bed-straw, Maid's Hair, Yellow Bed-straw*
Garcinia Mangostana. *Mangosteen-tree*
Morella. *Ceylon Gamboge-tree*
purpurea. *Cocum,* or *Kokum-oil-plant*
Gardenia edulis. *Bread-fruit-tree,* of N. Australia
florida. *Cape Jasmine*
lucida. *Dikamali,* or *Cambi, Resin-plant*
Thunbergii. *Büffelsball-wood*
Gastonia cutispongia. *Sponge-wood,* of the Isle of France
Gastridium lendigerum. *Nit-grass*

Gastrodia Cunninghamii. *Peri-root,* of New Zealand
sesamoides. *Native Potato,* of Tasmania
Gastrolobium bilobum and G. Callistachys. *Sheep-poison-plant,* of Australia
calycinum. *York Road Poison,* of W. Australia
obovatum, G. spinosum, and G. trilobum. *Cattle-poison-plants,* of West Australia
Gaultheria antipoda. *New Zealand Bilberry*
procumbens. *American Mountain Tea, Aromatic Winter-green, Box-berry, Chequer-berry, Creeping Winter-green, Partridge-berry, Spring Winter-green, Tea-berry*
Shallon. *Salal,* or *Shallon-shrub*
Gaylusaccia brachycera. *Box Huckleberry*
dumosa. *Dwarf Huckle-berry*
frondosa. *Blue Tangle, Dangle-berry*
resinosa. *Black Huckle-berry*
Gazania. *Treasure-flower*
pavonia. *Cape* or *Peacock Treasure-flower*
Geaster. *Earth-star Fungus, Man-Fungus*
Geijera parviflora. *Australian Willow*
Geissois. *Tile-seed,* of Australia
Geissorrhiza. *Tile-root*
juncea. *Rushy Tile-root*
Geitonoplesium. *Australian Shepherd's-joy*
Gelsemium nitidum (Bignonia sempervirens). *Carolina Jasmine, Gelsemin, Wild Jessamine*
Genipa americana. *Genipap-tree, Genip-tree, Marmalade Box*
Genista. *Broom*
ægyptinea. *Egyptian Broom*
ætnensis. *Mt. Etna Broom*
amsantica (G. anxantica). *Amsanto Broom*
anglica. *Carlin-spurs, Cat's Whin, Moor Whin, Moss Whin, Needle Furze, Petty Whin*
aphylla. *Leafless Broom, Violet-flowered Broom*
bracteolata. *Small-bracted Broom*
candicans. *Whitish-leaved Broom*
clavata. *Club-calyxed Broom*
congesta. *Close-branched Broom*
diffusa. *Sprawling Broom*
florida. *Flowery Broom*
germanica. *German Broom*
hispanica. *Spanish Broom, Spanish Furze*
horrida. *Large-spined Broom*
humifusa. *Trailing Broom*
(Spartium) juncea. *Fragrant Spanish Broom*
linifolia. *Flax-leaved Broom*
lusitanica. *Portugal Broom*
mantica. *Mantuan Broom*
monosperma. *White-flowered Broom*
multiflora. *White Portugal Broom*
ovata. *Ovate-leaved Broom*
parviflora. *Small-flowered Broom*
patula. *Spreading Broom*
pilocarpa. *Hairy podded Broom*
pilosa. *Hairy Broom, Hairy Green-weed*

Genista polygalæfolia. *Milk-wort-leaved Broom*
præcox. *Early-flowering Broom*
procumbens. *Procumbent Broom*
prostrata. *Prostrate Broom*
purgans. *Purging Broom*
radiata. *Rayed-branched Broom*
sagittalis. *Arrow-jointed Broom, Hare's-foot Green-weed*
Scorpius. *Scorpion Broom, Scorpion-plant*
sericea. *Silky-leaved Broom*
sibirica. *Siberian Broom*
sphærocarpa. *Round-podded Broom*
sylvestris. *Wood Broom*
tetragona. *Quadrangular-branched Broom*
tinctoria. *Base Broom, Dyer's Broom, Dyer's Green-weed, Green Weed or Greening Weed, Kendal Green, "Widow-wisse," Wood Waxen or Woad Waxen*
tinctoria fl.-pl. *Double-flowered Dyer's-weed*
triacanthos. *Three-spined Broom*
triangularis. *Triangular-stemmed Broom*
umbellata. *Umbel-flowered Broom*
virgata. *Long-twigged Broom*
Gentiana. *Gentian*
acaulis. *Gentianella*
alba. *Whitish Gentian*
Amarella. *Autumn Gentian, Bitter-wort, Fel-wort*
Andrewsii. *Closed-flowered Gentian*
asclepiadea. *Milk-weed Gentian, Willow Gentian*
bavarica. *Bavarian Gentian*
brachyphylla. *Small-leaved Gentian*
Burseri. *Burser's Gentian*
campestris. *Field Gentian*
caucasica. *Caucasian Gentian*
cerina. *New Zealand Sea-shore Gentian*
crinita. *Fringed Gentian*
cruciata. *Cross-wort Gentian*
detonsa. *Smaller Fringed Gentian*
gelida. *Cream-coloured Gentian*
incarnata. *Flesh-coloured Gentian*
lutea. *Yellow-flowered Gentian*
nivalis. *Small Mountain Gentian*
ochroleuca. *Yellowish-white Gentian*
pannonica. *Round-petalled Gentian*
pleurogynoides. *New Zealand Yellow Gentian*
Pneumonanthe. *Autumn Bell-flower, Calathian Violet, Harvest Bells, Lung-flower, Marsh Gentian*
Pseudo-Pneumonanthe. *Bastard Wind-flower*
punctata. *Dotted-flowered Gentian*
purpurea. *Purple-flowered Gentian*
pyrenaica. *Pyrenean Gentian*
quinqueflora. *Five-flowered Gentian*
septemfida. *Crested Gentian*
Saponaria. *Barrel-flowered or Soap-wort Gentian*
saxosa. *New Zealand Mountain Gentian*
verna. *"Lucy of Teesdale," Spring Gentian*
Geodorum. *Earthy-scented Orchid*
Geoffroya. *Bastard Cabbage-tree*, of S. America
superba. *Almendor*, or *Almond-tree*, of Brazil

Geranium. *Crane's-bill*
anemonæfolium. *Anemone-leaved Crane's-bill*
armenum. *Large Rosy-purple Crane's-bill*
argenteum. *Silvery Crane's-bill*
Backhousianum (G. armenum). *Back-house's Crane's-bill*
Carolinianum. *Carolina Crane's-bill*
cinereum. *Gray Crane's-bill*
collinum. *Hill Crane's-bill*
columbinum. *Long-stalked Geranium, Pigeon's-foot Crane's bill*
cristatum. *Crested Crane's-bill*
dissectum. *Australian Geranium, Cut-leaved Crane's-bill*
Endresi. *Endres's Crane's-bill*
Fremontii. *Fremont's Crane's-bill*
ibericum. *Iberian Crane's-bill*
Lamberti. *Lambert's Crane's-bill*
lucidum. *Shining-leaved Crane's-bill*
macrorrhizum. *Long-rooted Crane's-bill*
maculatum. *Crow-foot* or *Wild Crane's-bill* of N. America
molle. *Dove's-foot Crane's-bill* or *Geranium*
odoratissimum. *Geranium-oil-plant*
parviflorum. *Native Carrot*, of Tasmania
phæum. *Dusky-flowered Crane's bill, Mourning-Widow*
platypetalum. *Broad-petalled Crane's-bill*
pratense. *Crow-foot, Crane's-bill, Meadow Geranium, Wild Geranium*
pusillum. *Small flowered Geranium*
pyrenaicum. *Mountain Geranium*
Robertianum. *Dragon's-blood, Fox Geranium, Fox Grass, Herb Robert, Nightin-gales, Red-shanks, Stinking Robert*
rotundifolium. *Round-leaved Crane's-bill*
sanguineum. *Blood-red-flowered Crane's-bill*
sanguineum var. lancastriense. *Walney Crane's-bill*
striatum. *Striped Crane's-bill*
subcaulescens (G. asphodeloides). *Dwarf Crane's-bill*
sylvaticum. *Bassinet Geranium, Wood Crane's-bill, Wood Geranium*
sylvaticum var. album. *White-flowered Crane's-bill*
tuberosum. *Tuberous-rooted Crane's-bill*
Gerardia flava. *Downy False-Fox-glove*
pedicularia. *American Louse-wort, Fever-weed, Fern-leaved False-Fox-glove*
purpurea. *Common False-Fox-glove*
quercifolia. *Smooth False-Fox-glove*
Gerrardanthus macrorrhiza. *Natal Bitter-root*
Gethyllis. *Cape Crocus*
Geum aureum. *Golden Avens*
canadense. *Chocolate-root*
chiloënse. *Chiloë Avens*
coccineum. *Scarlet-flowered Avens*
coccineum fl.-pl. *Double Scarlet Avens*
coccineum grandiflorum. *Large-flowered Scarlet Avens*
montanum. *Yellow-flowered Mountain Avens*

Geum pyrenaicum. *Pyrenean Avens*
reptans. *Creeping Avens*
rivale. *Indian Chocolate, Drooping, Purple,*
    or *Water Avens*
sylvaticum. *Wood Avens*
triflorum. *Three-flowered Avens*
urbanum. *City Avens, Clove-root, Herb*
    Bennet, "Minarta"
virginianum. *Throat-root, White Avens*
Gigarthina helminthocorton. *Corsican An-*
    thelmintic Sea-weed
Gillenia trifoliata. *Bowman's-root, Indian*
    Physic, Western Drop-wort
stipulacea. *American Ipecacuanha*
Gladiolus. *Gladiole, Sword-lily*
angustus. *Narrow-leaved Sword-lily*
Brenchleyensis. *Brenchley Sword-lily*
cardinalis. *Cardinal-flowered Sword-lily*
Colvillei. *Colville's Sword-lily*
communis. *Fox-glove Sword-lily*
Cooperi. *Cooper's Sword-lily*
cruentus. *Blood-red Sword-lily*
dracocephalus. *Dragon's-head Sword-lily*
Gandavensis and vars. *Ghent Hybrid*
    Sword-lily
Libanotis. *Hart-wort Sword-lily*
ochroleucus. *Sulphur-flowered Sword-lily*
papilio. *Butter-fly Sword-lily*
praecox. *Early Sword-lily*
purpureus auratus. *Purple-and-gold Sword-*
    lily
ramosus. *Branching Sword-lily*
Saundersi. *Saunders's Sword-lily*
segetum. *Corn-field Sword-lily, Corn-Flag*
tristis. *Sad-coloured Sword-lily*
Glaphiria nitida. *Bencoolen* or *Malay*
    Tea-plant, Tree-of-Long-Life
Glaucium. *Horned-Poppy*
corniculatum. *Various-coloured Horned-*
    Poppy
fulvum. *Orange-flowered Horned-Poppy*
luteum. *Common Yellow Horned*, or *Sea-*
    Poppy
phoeniceum. *Red-flowered Horned-Poppy*
tricolor. *Three-coloured Horned-Poppy*
Glaux maritima. *Black Salt-wort, Sea*
    Milk-weed, Sea Milkwort
Glechoma hederacea (Nepeta Glechoma).
    Ground Ivy, Hedge-maids
Gleditschia brachycarpa. *Curved-spined*
    Honey-Locust
Bugoti pendula. *Weeping Acacia*
caspica. *Caspian Honey-Locust*
ferox. *Flat-spined Honey-Locust*
indica. *East Indian Honey-Locust*
laevis. *Smooth Honey-Locust*
latisiliqua. *Broad-podded Honey-Locust*
macrospina. *Long-spined Honey-Locust*
microspina. *Small-spined Honey-Locust*
monosperma. *Water-Locust*
sinensis. *Chinese Honey-Locust*
triacanthos. *Common Honey-Locust, Three-*
    thorned Acacia
Gleichenia. *Net-Fern*
Globularia. *Blue Daisy, Globe Daisy*
cordifolia. *Heart-leaved Globe Daisy*
nana. *Thyme-leaved Globe Daisy*
nudicaulis. *Naked-stalked Globe Daisy*
trichosantha. *Hair-flowered Globe Daisy*

Globularia vulgaris. *Common Globe Daisy*
Glochidion australe (Bradleia australis).
    Australian Rivulet-tree
Gloriosa nepalensis. *Nepaul Lily*
simplex. *Senegal Lily*
superba. *Malabar Glory-Lily*
virescens. *Mozambique Lily*
Glossodia. *Australian Tongue-flower*
Glossula tentaculata. *Chinese Tongue-flower*
Glyceria. *Sweet Grass*
aquatica. *Leed, Reed Meadow-grass*
canadensis. *Rattle-snake-grass*
maritima. *Sea Spear-grass*
nervata. *Fowl-meadow-grass* of N. Ame-
    rica
fluitans. *Float* or *Flote Grass, Manna*
    Croup, Manna Grass, Poland Manna
Glycosmis citrifolia. *Jamaica Orange*
Glycyrrhiza glabra. *Liquorice-plant* (cul-
    tivated)
lepidota. *Wild Liquorice* of North Ame-
    rica
Glyptostrobus. *Embossed Cypress*
heterophyllus. *Water-Pine* of China
Gnaphalium. *Cotton-weed, Cud-weed,*
    Everlasting
americanum. *"Life-everlasting," Jamaica*
    Everlasting
arboreum. *Tree-Everlasting*
arvense. *Field Cud-weed*
    (Helichrysum) Colensoi. *"Edelweiss"* of
    New Zealand
decurrens. *Common Everlasting* of N.
    America
dioicum. *Mountain Everlasting*
gallicum. *Narrow-leared Cud-weed*
germanicum. *Common Cud-weed*
grandiceps. *"Edelweiss"* of N. Zealand
Leontopodium. *Bridal Everlasting, "Edel-*
    weiss" of the Alps, *Lion's-paw Cud-*
    weed
luteo-album. *Jersey Cud-weed, Jersey*
    Live-long
margaritaceum. *Pearl Cud-weed*
polycephalum. *Common Everlasting* of N.
    America, *Indian Posy, Sweet-scented*
    Everlasting, Old Field or *White Bal-*
    sam
purpureum. *Purplish Cud-weed* of N.
    America
sanguineum. *Crimson Everlasting*
supinum. *Dwarf Cud-weed*
sylvaticum. *Chafe-weed, Wood Cud-weed*
uliginosum. *Marsh Cud-weed*
Gomphia. *Button-flower-tree*
guianensis. *Candle-wood* of Jamaica
nitida. *Glossy-leaved Button-flower*
Gomphocarpus fruticosus. *Wild Cotton*
    of Australia
Gomphrena globosa. *Globe Amaranth,*
    Red Globe Everlasting
Gomutus saccharifera. *Wine-Palm* of the
    Moluccas
Goniophlebium trilobum. *Lizard's-herb*
Gonolobus. *Angle-pod*
carolinensis. *False Choke-dog*
Cundurango. *Condor Vine*
obliquus. *Choke-dog*
Goodia latifolia. *Salisbury Pea* of Australia

Goodyera. *Rattle-snake Plantain*
pubescens. *Adder's Violet, Net-leaf Plantain, Rattle-snake-leaf, Scrofula-leaf, Scrofula-weed, Silvery Rattle-snake-Plantain*
repens. *Creeping Rattle-snake-Plantain*
Gordonia Hæmatoxylon. *Blood-wood or Red-wood* of Jamaica
lasianthus. *Common Loblolly-Bay*
pubescens. *Downy Loblolly-Bay*
Gossypium. *"Bombast," Cotton-plant*
arboreum. *Tree Cotton-plant*
barbadense. *American or Barbadoes Cotton-plant*
herbaceum. *Common Herbaceous Cotton-plant*
indicum. *Indian Cotton-plant*
peruvianum. *Peruvian Cotton-plant, "Kidney" Cotton-plant*
religiosum. *Nankin Cotton-plant*
Gouania domingensis. *Chaw-stick* of St. Domingo
Gourliæa decorticans. *Chanar-tree*
Gracilaria Helminthocorton. *Corsican Moss*
Gramineæ. A general name for the *Grasses*
Grammatophyllum. *Letter-leaf or Letter-plant*
multiflorum. *Letter-plant* of the Philippine Islands
speciosum. *"Queen of the Orchids"*
Grangea (Cotula) maderaspatana. *Marcella*
Graptophyllum hortense. *Caricature-plant*
Gratiola aurea. *Golden-pert*
officinalis. *Hedge-Hyssop, Poor-man's-herb*
Grayia. *Grease-wood*
Grevillea robusta. *Silk, Silky, or Silk-bark Oak* of Australia
Grewia occidentalis. *African Star-bush*
Grias cauliflora. *Anchovy Pear*
Grindelia. *Gum-plant* of California
Griselinia lucida. *New Zealand Broad-leaf or Puhatea-tree*
Guaiacum officinale. *Guaiacum-gum-tree, Guaiac-resin-plant, Lignum-vitæ-tree*
Guarea grandifolia. *Alligator-wood*
Swartzii. *Mush-wood* of Jamaica
Guatteria caffra. *Caffre's Lance-wood*
longifolia. *East Indian Mast-tree*
virgata. *Jamaica* or *Guiana Lance-wood*
Guazuma tomentosa. *Bastard Cedar* of S. America
ulmifolia. *Bastard Cedar* of Jamaica
Guettarda argentea. *Black Guara, Silver-wood-tree*
elliptica. *Velvet-seed*
speciosa. *Pigeon-wood* of Jamaica, *Zebra-wood* of the E. Indies
Guevina Avellana (Quadria heterophylla). *Evergreen Hazel* of Chili
Guilandina Bonduc. *Bonduc-nut-tree, Sumatra Nicker-tree*
Bonducella. *Common or Yellow-seeded Nicker-tree, Molucca Bean*
Moringa. *Horse-radish Tree*
Guilfoylia (Cadellia) monostylis. *Golden Pyramid-tree*
Guilielma speciosa. *Peach Palm*

Guizotia oleifera. *Niger-seed-plant, Oil-seed, Ram-til, Huts-yellow,* or *Valisaloo-oil-plant*
Gymnadenia Conopsea and G. odoratissima. *Fragrant Orchis*
Gymnema lactiferum. *Ceylon Cow-plant*
Gymnocladus canadensis. *"Chicot," Kentucky Coffee-tree*
Gymnogramma. *Rue-leaved Fern*
leptophylla. *Small Rue-leaved Fern*
Gymnolomia Porteri. *Stone-Mountain Star*
Gymnostachys anceps. *Australian Caterpillar-flower*
Gymnostichum Hystrix. *Bottle-brush-grass*
Gymnostomum. *Beardless Moss, Bladder Moss*
Gynerium argenteum. *Pampas Grass*
roseum. *Rosy-spiked Pampas Grass*
Gynocardia odorata. *"Chaulmugra Seed"-tree*
Gypsophila. *Chalk-plant*
Struthium. *Egyptian Soap-root, Spanish Soap-root*
Gyrocarpus asiaticus. *Catamaran-wood Tree*
Gyrophora murina. *"Velvet Moss"*
vellea. *Rock-tripe, "Tripe de Roche" Lichen*

Habenaria. *Fringed Orchis, Rein-Orchis*
bifolia. *Common Butterfly Orchis*
blephariglottis. *White Fringed Orchis*
chlorantha. *Larger Butterfly Orchis*
ciliaris. *Yellow Fringed-Orchis*
cristata. *Small Yellow Fringed-Orchis*
fimbriata. *Great or Purple Fringed Orchis*
lacera. *Ragged Fringed-Orchis*
leucophæa. *White Prairie-Orchis*
niven. *Surrey Orchis, White Butterfly-Orchis*
viridis. *Frog Orchis*
Habzelia æthiopica. *African, Boulon, Ethiopian, Guinea, Malaghatta, Monkey,* or *Negro Pepper*
Hæmanthus. *Blood-flower*
coccineus. *Red Cape Tulip*
Hæmatostaphis Barteri. *Blood Plum* of Sierra Leone
Hæmatoxylon campechianum. *Campeachy-wood, Log-wood*
Hagenia abyssinica (Bragera anthelmintica). *"Cusso"* of Abyssinia
Hakea. *"Wooden-cherry"-tree* of Australia
laurina (H. eucalyptoides). *Cushion-flower* of W. Australia
ulicina. *Native Furze* of Australia
Halenia deflexa and other species. *Spurred Gentian*
Halesia diptera. *Two-winged-fruited Snow-drop-tree*
parviflora. *Small-flowered Snowdrop-tree*
tetraptera. *Common Snowdrop Tree, Silver-bell Tree*
Halimodendron argenteum. *Silvery-leaved Salt-tree*
subvirescens. *Greenish Salt-tree*
triflorum. *Three-flowered Salt-tree*

English Names of Cultivated, Native,

Halleria lucida. *African Fly-Honeysuckle*
Halogeton tamariscifolium. *Spanish Wormseed*
Haloragis. *Sea-berry*, of Australia
 citriodora. *Piri-jiri shrub*, of New Zealand
Hamamelis virginica. *American Witch-Hazel, Snapping Hazel-nut, Spotted Alder, Winter-bloom*
Hamelia patens. *Rat-poison Plant*, of the W. Indies
 ventricosa. *Prince-wood*, of the W. Indies
Hancornia speciosa. *Mangaba*, or *Mangava-tree*, of Brazil
Hardenbergia monophylla. *Lilac*, or *Sarsaparilla-tree*, of Australia
Harpulia Hillii and H. pendula. *Tulipwood*, of Queensland
Hartighsea spectabilis. *Kohé* or *Wahahétree*, of New Zealand
Hedeoma pulegioides. *American Pennyroyal, Squaw-Mint, Tick-weed*
Hedera (Irvingia) australiana. *Queensland Ivy*
 Helix. *Barren Ivy, Bent-wood, Bind-wood, Common Ivy*
 Helix var. arborea. *Tree-Ivy*
 Helix var. arborea aureo-marginata. *Gold-edged Tree-Ivy*
 Helix var. argenteo-marginata. *Silver-edged Tree-Ivy*
 Helix var. aureo-maculata. *Gold-blotched Ivy*
 Helix var. canariensis (H. H. hibernica). *Irish Ivy*
 Helix var. chrysocarpa. *Yellow-berried Roman Ivy*
 Helix var. conglomerata. *Clustered Ivy*
 Helix var latifolia maculata. *Marbled-leaved Ivy*
 Helix var. palmata. *Palmate-leaved Ivy*
 Helix var. poetica. *Poet's Ivy*
 Helix var. Raegneriana (H. H. colchica). *Giant Ivy*
 Helix var. rhombea variegata. *Variegated Japan Ivy*
 Helix var. taurica. *Crimean Ivy*
 Helix var. tricolor. *Three-coloured Ivy*
 Helix variegata. *Variegated Ivy*
Hedwigia balsamifera. *Pig-wood, Wild Boar's-tree*, of San Domingo
Hedycarpa Pseudo-morus. *Mulberry*, of Australia
 angustifolia. *Native Mulberry*, of Victoria
Hedychium coronarium. *Fragrant Garland-flower*
Hedyosmum nutans. *W. Indian Headache-weed*
Hedyotis umbellata. *E. Indian Madder*, or *Chayroot*
Hedysarum Alhagi. *Camel's-Thorn*
 canadense. *Canadian Bush Clover*
 coronarium. *French Honeysuckle, Maltese Clover, Red Satin-flower, Soola Clover*
Heisteria coccinea. *Partridge-pea Tree*
Helonium. *Helen-flower, Sneeze-weed, Sneeze-wort*
 atropurpureum. *Dark-purple Helen-flower*
 autumnale. *Autumn Helen-flower* or *Sneeze-wort, Smooth Helen-flower*

Helenium Hoopesii. *Hoopes's Sneeze-weed*
 pumilum. *Dwarf Sneeze-wort*
 tenuifolium. *Slender-leaved Sneeze-wort*
Heleocharis tuberosa. *Petsi Rush*
Heliamphora nutans. *S. American Pitcher-plant*
Helianthemum. *Holly-Rose, Sun-Rose*
 alpestre. *Alpine Sun-rose*
 alyssoides. *Alyssum-like Sun-rose*
 Andersoni. *Anderson's Sun-rose*
 angustifolium. *Narrow-leaved Sun-rose*
 apenninum. *Apennine Sun-rose*
 arabicum. *Arabian Sun-rose*
 atriplicifolium. *Orache-leaved Sun-rose*
 Barrelieri. *Barrelier's Sun-rose*
 brasiliense. *Brazilian Sun-rose*
 canadense. *Forest-weed*
 candidum. *White-leaved Sun-rose*
 canescens. *Whitish-leaved Sun-rose*
 canum. *Hoary-leaved Sun-rose*
 cheiranthoides. *Wall-flower Sun-rose*
 corymbosum. *Corymb-flowered Sun-rose*
 crassifolium. *Thick-leaved Sun-rose*
 croceum. *Saffron-flowered Sun-rose*
 cupreum. *Copper-colour-flowered Sun-rose*
 dichotomum. *Forked-branched Sun-rose*
 diversifolium. *Various-leaved Sun-rose*
 eriosepalon. *Woolly-sepalled Sun-rose*
 farinosum. *Mealy-leaved Sun-rose*
 foetidum. *Bryony-scented Sun-rose*
 formosum. *Beautiful Sun-rose*
 Fumana. *Heath-like Sun-rose*
 glaucum. *Glaucous-leaved Sun-rose*
 glomeratum. *Cluster-flowered Sun-rose*
 glutinosum. *Clammy Sun-rose*
 grandiflorum. *Large-flowered Sun-rose*
 (Cistus) guttatum. *Spotted Sun-rose*
 halimifolium. *Sea-Purslane-leaved Sun-rose*
 hirsutum. *Hairy Sun-rose*
 hirtum. *Bristly-calyxed Sun-rose*
 hirtum var. Lagascae. *Lagasca's Sun-rose*
 hispidum. *Bristly Sun-rose*
 hyssopifolium. *Hyssop-leaved Sun-rose*
 involucratum. *Short-pedicelled Sun-rose*
 italicum. *Italian Sun-rose*
 juniperinum. *Juniper-leaved Sun-rose*
 kahiricum. *Cairo Sun-rose*
 læve. *Smooth Sun-rose*
 lævipes. *Cluster-leaved Sun-rose*
 lanceolatum. *Spear-leaved Sun-rose*
 lasianthum. *Hairy-leaved Sun-rose*
 lavandulæfolium. *Lavender-leaved Sun-rose*
 leptophyllum. *Slender-leaved Sun-rose*
 lignosum. *Woody-stemmed Sun-rose*
 lineare. *Linear-leaved Sun-rose*
 Lippii. *Small-petalled Sun-rose*
 lucidum. *Shining-leaved Sun-rose*
 macranthum. *Great-flowered Sun-rose*
 macranthum var. multiplex. *Double-flowered Sun-rose*
 marifolium. *Marum-leaved Sun-rose*
 marjoranifolium. *Marjoram-leaved Sun-rose*
 microphyllum. *Small-leaved Sun-rose*
 Milleri. *Miller's Sun-rose*
 molle. *Soft-leaved Sun-rose*
 mutabile. *Colour-changing Sun-rose*

Helianthemum nudicaule. *Naked-stemmed Sun-rose*
Nummularium. *Money-wort-leaved Sun-rose*
obovatum. *Obovate-leaved Sun-rose*
ocymoides. *Basil-like Sun-rose*
œlandicum. *Œland Sun-rose*
origanifolium. *Marjoram-leaved Sun-rose*
ovatum. *Ovate-leaved Sun-rose*
paniculatum. *Panicle-flowered Sun-rose*
poliifolium. *Polium-leaved or White Sun-rose*
penicillatum. *Pencilled Sun-rose*
Pilosella. *Downy Sun-rose*
pilosum. *Stiff-haired Sun-rose*
procumbens. *Procumbent Sun-rose*
pulchellum. *Neat Sun-rose*
pulverulentum. *Powdered-leaved Sun-rose*
racemosum. *Raceme-flowered Sun-rose*
rhodanthum. *Rose-flowered Sun-rose*
roseum. *Rosy-flowered Sun-rose*
rosmarinifolium. *Rosemary-leaved Sun-rose*
scabrosum. *Rough Sun-rose*
serpyllifolium. *Serpyllum-leaved Sun-rose*
sessiliflorum. *Sessile-leaved Sun-rose*
stœchadifolium. *Woolly-leaved Sun-rose*
stramineum. *Straw-coloured-flowered Sun-rose*
strictum. *Upright-branched Sun-rose*
sulphureum. *Sulphur-flowered Sun-rose*
surrejanum. *Surrey Sun-rose*
tauricum. *Taurian Sun-rose*
thymifolium. *Thyme-leaved Sun-rose*
tomentosum. *Woolly-leaved Sun-rose*
Tuberaria. *Plantain-leaved Sun-rose, Truffle Sun-rose*
umbellatum. *Umbel-flowered Sun-rose*
variegatum. *Striped-flowered Sun-rose*
venustum. *Showy Sun-rose*
versicolor. *Parti-coloured-flowered Sun-rose*
vineale. *Slender Trailing Sun-rose*
violaceum. *Violet-calyxed Sun-rose*
virgatum. *Twiggy Sun-rose*
viride. *Green-leaved Sun-rose*
vulgare (Cistus Helianthemum). *Common Sun-rose*
Helianthus. *Sun-flower*
annuus. *Common Sun-flower, Golden Flower of Peru*
argophyllus. *Silvery-leaved Sun-flower*
atrorubens. *Dark Red Sun-flower*
cucumerifolius. *Cucumber-leaved Sun-flower*
decapetalus. *Ten-petalled Sun-flower*
flexuosus. *Zig-zag Sun-flower*
giganteus. *Giant Sun-flower*
indicus. *Dwarf Annual Sun-flower*
Maximiliani. *Maximilian's Sun-flower*
multiflorus. *Perennial Sun-flower*
Nuttallii. *Nuttall's Sun-flower*
orgyalis. *Graceful Sun-flower*
rigidus (Harpalium rigidum). *Prairie Sun-flower, Rough-leaved Sun-flower*
strumosus. *Carrot-rooted Sun-flower*
tuberosus. *Canada Potato, Jerusalem Artichoke*
Helichrysum. " *Everlasting,*" or " *Immortelle*"-*flower, Golden Moth-wort*

Helichrysum apiculatum. *Tasmanian Everlasting*
(Gnaphalium) arenarium. *Common Yellow Everlasting*
bracteatum. *Bracted Everlasting*
bracteatum var. nanum. *Dwarf Everlasting*
(Astelma) eximium. *Crimson Cape Everlasting*
lucidum (H. bracteatum). *Australian Everlasting*
macranthum. *Large-flowered Everlasting*
orientale. *Yellow " Immortelle"-flower*
serpyllifolium. *Cape Colony Tea*
Stœchas. *Common Shrubby Everlasting, God's-Flower, or Gold-flower, Golden Cassidony, Golden Stœchas, Golden Tufts*
Helicia (Macadamia) ternifolia. *Queensland Nut-tree*
Heliconia Bihai. *False Plantain*
psittacorum. *Parrot's Plantain*
Helicteres. *Screw-tree*
Isora. *Large-fruited Screw-tree, Twisted Stick, Twisted Horn, or " Twisty, of S. India*
jamaicensis. *Screw-tree of Jamaica*
Heliophila. *Cape Stock*
pectinata. *Sun Cress*
Heliophytum indicum. *Indian Heliotrope*
Heliopsis lævis. *Ox-eye, of N. America*
platyglossa. *Ramtil-oil Plant*
Heliotropium. *Heliotrope, Turnsole*
europæum. *Common Heliotrope*
fruticosum. *Basket-withe, of Jamaica*
indicum. *Wild Clary*
peruvianum. *Peruvian Heliotrope or Turnsole, Cherry-pie Flower*
Helipterum. " *Everlasting,*" or " *Immortelle*"-*flower*
Manglesii. *Australian Everlasting*
Helleborus. *Hellebore*
abchasicus. *Green-flowered Christmas-rose*
altifolius (H. niger var. maximus). *Giant Christmas-rose*
angustifolius. *Narrow-leared Christmas-rose*
argutifolius. *Holly-leaved Hellebore*
atro-rubens. *Purplish-red-flowered Christmas-rose*
colchicus. *Plum-coloured Christmas-rose*
fœtidus. *Bar-foot, Bear-foot, Ox-heel, Setter-wort, Stinking Hellebore*
niger. *Bear's-foot, Black Hellebore, Common Christmas Rose, Christ's-wort*
niger var. maximus (H. altifolius). *Great or Giant Christmas-rose*
odorus. *Sweet-scented Hellebore*
officinalis. *Black Hellebore* of the ancients
olympicus. *Olympian Hellebore*
orientalis. *Rose-coloured Christmas-rose*
purpurascens. *Purple-flowered Christmas-rose*
vesicarius. *Syrian Christmas-rose*
viridis. *Bastard Hellebore, Bear's-foot, Bear's-foot, Green Hellebore*
Helminthia echioides. " *Lang-de-beefe,*" *Ox-tongue*
Helonias bullata (H. latifolia). *Stud-flower*

Helonias dioica. *Blazing-star, Devil's-bit, Unicorn's-horn*
Helosciadium inundatum. *Mud-weed*
(Sium) nodiflorum. *Fool's Water-cress, Marsh-wort*
Helvella Mitra. *Bishop's-Mitre Mushroom*
Hemidesmus indicus (Periploca indica). *E. Indian Sarsaparilla,* or *Nunnari-Root*
Hemileia vastatrix. *Coffee Blight Fungus*
Hemionitis palmata. *Mule Fern*
Hemitelia Smithii. *Smith's Tree-Fern*
Hemizonia. *Tar-weed,* of California
Hemerocallis. *Day-lily*
disticha. *Two-rowed Day-lily*
Dumortieri. *Dumortier's Day-lily*
flava. *Yellow Day-lily*
fulva. *Tawny Day-lily*
graminea. *Grass-leaved Day-lily*
Kwanso variegata. *Variegated-leaved Day-lily*
lutea. *Dark-yellow Day-lily*
Hepatica acutiloba. *Sharp-lobed Hepatica* or *Liver-leaf*
angulosa. *Large Blue Hepatica*
triloba. *Golden Trefoil, Noble Liver-wort, Round-lobed Liver-leaf*
Heracleum. *Cow-parsnip, Giant-parsnip*
elegans. *Rough-leaved Cow-parsnip*
eminens. *Blunt-lobed Cow-parsnip*
flavescens (Heracleum austriacum). *Yellowish Cow-parsnip*
giganteum. *Giant Cow-parsnip*
lanatum. *American Cow-parsnip, American Master-wort*
latifolium. *Broad-leaved Cow-parsnip*
Panaces. *Fig-leaved Cow-parsnip, Hercules' All-heal, Hercules' Wound-wort*
persicum. *Persian Cow-parsnip*
pubescens. *Downy Cow-parsnip*
Sphondylium. *Cad-weed, Clog-weed, Common Cow-parsnip, Eltrot, Hog-weed, Madnep, Meadow Parsnip*
Wilhelmsii. *Wilhelm's Cow-parsnip*
Heritiera macrophylla (H. littoralis). *Large-leaved Looking-glass Plant* or *Tree*
minor. *Small-leaved Looking-glass Plant*
Herminiera Elaphroxylon (Edemone mirabilis). *Pith-tree,* of the Nile
Herminium Monorchis. *Musk Orchis*
Hernandia guianensis. *Amadou-wood,* of S. America
ovigera. *"Hernant Seeds"-tree*
sonora. *E. Indian Jack-in-a-box, Jamaica Cog-wood*
Herniaria glabra. *Rupture-wort, "Turk's-herb"*
Herpestis Monnieri. *Water Hyssop,* of S. America
Hesperantha. *Evening-flower*
graminifolia. *Grass-leaved Evening-flower*
Hesperis fragrans. *Evening-scented Rocket*
maritima. *Mediterranean Stock*
matronalis. *"Close Sciences," Dame's, Garden,* or *White Rocket, Dame's Violet, Queen's Gilliflower, Rogue's Gilliflower, Winter Gilliflower*
matronalis fl.-pl. *Double Rocket*

Hesperis matronalis var. alba plena. *Double White Rocket*
matronalis var. purpurea plena. *Double Purple Rocket*
tristis. *Melancholy Gentleman, Night-scented Rocket*
violacea. *Violet-flowered Rocket*
Heteranthera. *Mud-Plantain*
Heteromeles arbutifolia. *Tollon* or *Toyon,* of California
Heterothalamus brunoides. *Romerillo-dye-plant*
Heterotoma lobelioides. *Mexican Bird-plant*
Heterotrichum patens. *"American Gooseberry,"* of the W. Indies
Heuchera americana. *Alum-root, American Sanicle*
glabra. *Smooth Alum-root*
micrantha. *Small-flowered Alum-root*
pubescens. *Downy Alum-root*
ribifolia. *Currant-leaved Alum-root*
Richardsoni (H. americana, H. ribifolia). *Satin-leaf*
Hevea brasiliensis. *Caoutchouc-tree* of Pará
guianensis (Siphonia elastica). *India-rubber-tree* of Guiana
Hibiscus Abelmoschus. *Musk Okro*
africanus. *African Ketmia*
arboreus. *W. Indian Mahoe* or *Mohoe*
cannabinus. *Brown Hemp,* of India, *Hemp mallow*
clypeatus. *Mahoe,* of Congo
coccineus. *American Scarlet Rose-mallow*
(Abelmoschus) esculentus. *Okro,* or *Gombo-plant*
heterophyllus. *Green Kurrajong,* of Australia, *Queensland Sorrel-tree*
militaris. *Halbert-leaved Rose-mallow*
moschatus. *E. Indian Musk-mallow*
Moscheutos. *Swamp Rose-mallow*
palustris. *Purple Marsh Rose-mallow*
Rosa-malabarica. *Malabar Rose, Blacking-plant, Chinese Rose-mallow, Shoe-black-plant, Shoe-flower*
Rosa - malabarica var. schizopetalus. *Fringed-flowered Rose-mallow*
roseus. *Rosy-flowered Rose-mallow*
Sabdariffa. *E. Indian Sorrel-plant, Red Sorrel,* of the W. Indies, *"Rozelle"-plant,* of the E. Indies, *Thorny-mallow*
speciosus. *Hardy Scarlet Hibiscus*
splendens. *Queensland Hollyhock-tree*
syriacus. *Common Althæa-frutex, Rose of Sharon*
syriacus var. elegantissimus. *Double Althæa-frutex*
syriacus var. oculatus. *Painted Lady Althæa-frutex*
syriacus var. purpureus. *Purple Althæa-frutex*
tiliaceus. *Cork-wood*
(Ketmia) Trionum. *Bladder Ketmia, Flower-of-an-hour, Good-night-at-noon, Venice Mallow*
Hieracium. *Hawk's-eye, Hawk-weed*
alpinum. *Mountain Hawk-weed*
aurantiacum. *Golden Mouse-ear, Grim-the-Collier, Orange-flowered Hawk-weed*

Hieracium canadense. *Canada Hawk-weed*
cerinthoides. *Honey-wort Hawk-weed*
Gronovii. *Hairy Hawk-weed*
incarnatum. *Pink-flowered Hawk-weed*
longipilum. *Long-bearded Hawk-weed*
murorum. *Golden or French Lung-wort, Wall Hawk-weed*
paniculatum. *Panicled Hawk-weed*
Pilosella. *Mouse-ear, Mouse-ear Hawk-weed*
prenanthoides. *Wall-lettuce Hawk-weed*
Pseudo-pilosella. *Bastard Mouse-ear*
sabaudum. *Savoy Hawk-weed, Shrubby Hawk-weed*
scabrum. *Rough Hawk-weed*
umbellatum. *Umbellate Hawk-weed*
venosum. *Rattle-snake-weed*
Hierochloë. *Holy-grass*
australis. *Southern Holy-grass*
borealis. *Northern Holy-grass*
macrophylla. *Large-leaved Vanilla-grass*
redolens. *Scented Grass*, of New Zealand
Himanthalia lorea. *Sea-thongs*
Hippeastrum. *Knight's-star Lily*
equestre. *Barbadoes Lily*
Reginæ. *Mexican Lily*
Hippobromus alatus. *Paardepis-tree*, of S. Africa
Hippocratea obcordata. *W. Indian Wild Almond-tree*
Hippocrepis comosa. *Horse-shoe Vetch*
Hippomane Mancinella. *Manchineel-tree*
Hippohaë canadensis. *Canadian Sea-Buckthorn*
rhamnoides. *Common Sea-Buckthorn, Sallow-, or Willow-, thorn*
salicifolia. *Willow-leaved Sea-Buckthorn*
Hippuris. *Mare's-tail*
vulgaris. *Bottle-Brush, Common Mare's-tail*
maritima. *Sea-shore Mare's-tail*
Hirneola (Exidia) Auricula-Judæ. *Jew's-Ear Fungus*
Holcus lanatus. *Dart Grass, Duffel Grass, Hose Grass, Meadow Soft-grass, Midge Grass, Rot Grass, Velvet Grass, Yorkshire Fog*
mollis. *Dart Grass, Duffel Grass, Wood Soft-grass, Rot Grass*
Hoheria populnea. *Otago Ribbon-wood*
Holosteum umbellatum. *Jagged Chick-weed*
Honkenya peploides. *Sea Pimpernel*
Hordeum. *Barley*
deficiens. *Red Sea Barley*
jubatum. *Squirrel-tail Grass*
maritimum. *Sea-side Barley, Squirrel-tail Grass*
murinum. *Mouse Barley, Wall Barley, Wild Barley, Squirrel-tail Grass*
nepalense (H. trifurcatum). *Nepaul Barley*
vulgare var. distichum. *Common Barley, Two-rowed Barley*
vulgare var. hexastichum. *Bear, Bere, Beer, Beir, or Bigg, Six-rowed Barley*
Zeocriton. *Battledore, Fan, Putney, or Sprat Barley*
Horminum pyrenaicum. *Pyrenean Dead-nettle*
Hosta cærulea. *Savannah-wood*
Hottonia inflata. *Water-Violet*, of N. America

Hottonia palustris. *Common Water-Violet, Water Feather-foil, Water Gilliflower*
Hovenia dulcis. *Japanese Raisin-tree*
Hoya. *Honey-plant, Wax-flower Climber, Wax-flower-plant*
australis. *N. S. Wales Wax-flower*
campanulata. *Bell Wax-flower-plant*
carnosa. *Fleshy-leaved Wax-flower-plant*
coriacea. *Manilla Wax-flower*
imperialis. *Imperial Wax-flower-plant*
Hudsonia ericoides. *American False Heath*
Hufelandia pendula. *Slog-wood*
Humea elegans. *Amaranth-Feathers*
Humirium balsamiferum. *Humiri-* or *Umire-Balsam-tree* of French Guiana
floribundum. *Umire-tree*, of Brazil
Humulus Lupulus. *Bine* or *Bines, Common Hop*
Hura crepitans. *Monkey's-Dinner-bell, Sand-box-tree*
Hyacinthus. *Hyacinth*
amethystinus. *Amethyst Hyacinth, Spanish Hyacinth*
(Galtonia) candicans. *Spire-Lily, White Cape Hyacinth*
non-scriptus (Scilla nutans). *Common Blue-bell*
orientalis and vars. *Common Garden Hyacinth*
romanus (Bellevalia romana). *Roman Hyacinth*
serotinus. *Late-flowering Hyacinth*
Hyænanche globosa. *Cape Hyæna-poison*
Hydnocarpus venenata (H. inebrians). *Fish-poison* or *Makooloo-tree* of Ceylon
Hydnora. *Madagascar Christmas-rose*
africana. *Jackal's-Kost*
Hydnum Barba-Jovis. *Jove's-beard*, or *Jupiter's-beard*
repandum. *Spiny* or *Hedge-hog Mushroom*
Hydrangea arborescens. *Wild Hydrangea* of N. America
arborescens var. discolor. *Two-coloured-leaved Hydrangea*
cordata. *Heart-leaved Hydrangea*
heteromalla. *Woolly-leaved Hydrangea*
hortensis. *Common Hydrangea*
hortensis var. cærulea. *Blue-flowered Hydrangea*
japonica. *Japan Hydrangea*
japonica variegata. *Silver-edged Hydrangea*
japonica var. aurea superba. *Golden-edged Hydrangea*
japonica var. cærulea. *Blue-flowered Japan Hydrangea*
japonica var. rosea. *Rosy-flowered Hydrangea*
nivea. *White-leaved Hydrangea*
Otaksa. *Hardy Flesh-coloured Hydrangea*
paniculata grandiflora. *Hardy White-flowered Hydrangea*
quercifolia. *Oak-leaved Hydrangea*
stellata. *Starry-flowered Hydrangea*
Hydrastis canadensis. *Ground Raspberry, Golden-seal, Orange-root, Turmeric-root, Yellow Puccoon*
Hydrocera (Tytonia) natans. *Water Balsam*
Hydrocharis Morsus-ranæ. *Frog-bit*
Hydrocotyle asiatica. *Indian Penny-wort*

English Names of Cultivated, Native,

Hydrocotyle vulgaris. *Flock-wort, Marsh Penny-wort, Penny-rot, Sheep's-bane, White-rot*
Hydrophyllum. *American Water-leaf*
virginicum. *Shawanese Salad*
Hygrophorus virginicus. *Viscid White Mushroom*
Hymenæa Courbaril. *Brazilian Gum-Copal-tree. S. American Locust-tree*
Hymenanthera crassifolia. *Scrub Box-wood*, of Chatham Island
dentata. *Scrub Box-wood of Victoria*
Hymenophyllum. *Filmy Fern*
Tunbridgense. *Tunbridge, or Common, Filmy Fern, Tunbridge Goldilocks*
Wilsoni. *Wilson's Filmy Fern*
Hyoscyamus. *Hen-bane*
albus. *White Hen-bane*
Canariensis. *Canary Island Hen-bane*
niger. *Common Hen-bane*
physaloides. *Kite-flower, Purple-flowered Hen-bane*
pusillus. *Dwarf Hen-bane*
reticulatus. *Egyptian Hen-bane*
Scopolia. *Nightshade-leaved Hen-bane*
Hyoseris minima. *Swine's-Succory*
Hypericum. *St. John's Grass, St. John's-wort, Tutsans*
ægyptiacum. *Egyptian St. John's-wort*
Androsæmum. *Park-leaves, Sweet Amber*
angulosum. *Toothed-flowered St. John's-wort*
Ascyron. *Red-leaved St. John's-wort, Siberian St. John's-wort*
balearicum. *Majorca St. John's-wort*
Burseri. *Burser's St. John's-wort*
calycinum. *Aaron's-Beard, Large-flowered St. John's-wort*
canadense. *Canadian St. John's-wort*
canariense. *Canary Island St. John's-wort*
Caprifolium. *Honeysuckle-leaved St. John's-wort*
chinense. *Chinese St. John's-wort*
ciliatum. *Hair-fringed St. John's-wort*
cordifolium. *Heart-leaved St. John's-wort*
Coris. *Heath-leaved St. John's-wort*
crispum. *Curled-leaved St. John's wort*
cuneatum. *Mount Lebanon St. John's-wort*
decussatum. *Cross-leaved St. John's-wort*
dolabriforme. *Hatchet-like St. John's-wort*
dubium. *Imperforate St. John's-wort*
elatum. *Tall St. John's-wort*
elegans. *Elegant St. John's-wort*
Elodes. *Marsh St. John's-wort*
elodioides. *Close-panicled St. John's-wort*
empetrifolium. *Empetrum-leaved St. John's-wort*
cricoides. *Heath-like St. John's-wort*
fasciculatum. *Aspalathus-like St. John's-wort*
fimbriatum. *Fringed St. John's-wort*
floribundum. *Many-flowered St. John's-wort*
foliosum. *Shining Leafy St. John's-wort*
frondosum. *Dense-foliaged St. John's-wort*
galioides. *Galium-like St. John's-wort*
glandulosum. *Glandular St. John's-wort*
glaucum. *Glaucous-leaved St. John's-wort*
grandiflorum. *Large-flowered St. John's-wort*
heterophyllum. *Various-leaved St. John's-wort*

Hypericum hircinum. *Common, or Goat-scented, St. John's-wort*
hirsutum. *Hairy St. John's-wort*
Hookerianum. *Glossy-flowered St. John's-wort, or Tutsan*
humifusum. *Trailing St. John's-wort*
japonicum. *Japanese St. John's-wort*
Kalmianum. *Kalm's St. John's-wort*
lævigatum. *Smooth St. John's-wort*
lanuginosum. *Downy St. John's-wort*
maculatum. *Spotted-flowered St. John's-wort*
mexicanum. *Mexican St. John's-wort*
montanum. *Mountain St. John's-wort*
nepalense. *Nepaul St. John's-wort*
Nummularium. *Money-wort St. John's-wort*
nudiflorum. *Naked-flowered St. John's-wort*
oblongifolium. *Oblong-leaved St. John's-wort*
olympicum. *Mt. Olympus St. John's-wort*
patulum. *Spreading St. John's-wort*
perfoliatum. *Perfoliate St. John's-wort*
pilosum. *Stiff-haired St. John's-wort*
pulchrum. *Slender St. John's-wort*
pusillum. *Small St. John's-wort*
quadrangulare. *Hard-hay, St. Peter's-wort, Square-stalked St. John's-wort*
quinquenervium. *Small-flowered St. John's-wort*
reflexum. *Reflexed-leaved St. John's-wort*
repens, and Hypericum reptans. *Creeping St. John's-wort*
Sarothra (Sarothra gentianoides). *Bastard Gentian, Orange-grass, Pine-weed*
serpyllifolium. *Thyme-leaved St John's-wort*
simplex. *Simple-stalked St. John's-wort*
sphærocarpon. *Round-capsuled St. John's-wort*
tomentosum. *Woolly St. John's-wort*
triflorum. *Three-flowered St. John's-wort*
triplinerve. *Three-nerved St. John's-wort*
Uralum. *Urala St. John's-wort*
veronense. *Verona St. John's-wort*
verticillatum. *Whorl-leaved St. John's-wort*
virgatum. *Twiggy St. John's-wort*
virginicum (Elodes virginica). *Marsh St. John's-wort, Pink-flowered St. John's-wort*
Hyphæne thebaica. *Doum Palm, Ginger-bread-tree*
Hypnum. *Feather-moss*
abietinum. *Spruce-tree Feather-moss*
albicans. *Whitish Feather-moss*
alopecurum. *Fox-tail Feather-moss*
atro-virens. *Dark-green Feather-moss*
cæspitosum. *Green-patch Feather-moss*
commutatum. *Curled Feather-moss*
complanatum. *Flat Feather-moss*
confertum. *Clustered Feather-moss*
Crista-castrensis. *Ostrich-plume Feather-moss*
cupressiforme. *Cypress-leaved Feather-moss*
dendroides. *Tree-like Feather-moss*
denticulatum. *Fern-like Feather-moss*
filicinum. *Lesser Golden Fern Feather-moss*
fluitans. *Floating Feather-moss*
lutescens. *Yellowish Feather-moss*
medium. *Long-headed Feather-moss*

Hypnum micans. *Sparkling Feather-moss*
molle. *Soft-water Feather-moss*
molluscum. *Plumy-crested Feather-moss*
moniliforme. *Beaded Feather-moss*
murale. *Wall Feather-moss*
myosuroides. *Mouse-tail Feather-moss*
nitens. *Shining Feather-moss*
palustre. *Marsh Feather-moss*
plumosum. *Rusty Feather-moss*
populeum. *Matted Feather-moss*
pulchellum. *Elegant Feather-moss*
pumilum. *Dwarf Feather-moss*
purum. *Neat Meadow Feather-moss*
repens. *Creeping Feather-moss*
rufescens. *Red Mountain Feather-moss*
salebrosum. *Yellow Feather-moss*
scorpioides. *Scorpion Feather-moss*
sericeum. *Silky Feather-moss*
speciosum. *Showy Feather-moss*
splendens. *Glittering Feather-moss*
stellatum. *Yellow Starry Feather-moss*
trichomanoides. *Fern-like Feather-moss*
umbratum. *Shaded Feather-moss*
undulatum. *Waxy Feather-moss*
velutinum. *Velvet Feather-moss*
Hypelate paniculata. *Genip Tree, Honey-berry, Madeira-Wood*
Hypocalymma. *Peach-Myrtle*
robustum. *Swan River Peach-Myrtle*
Hypochæris radicata. *Cape-weed*, of Australia and New Zealand, *Long-rooted Cat's-ear*
Hypoxis erecta. *American Star-grass*
Hyptis suaveolens. *W. Indian Spikenard*
Hyssopus nepetoides. *Square-stalked Hyssop*
officinalis. *Common Hyssop*

Iberis. *Candy-tuft*
alpina. *Alpine Candy-tuft*
amara. *Clown's Mustard, Sciatica-cress, Wild*, or *Bitter, Candy-tuft*
Bubani. *Buban's Candy-tuft*
corifolia. *Coris-leaved Candy-tuft*
correæfolia (I. Coriacea). *Late White Candy-tuft*
Garrexiana. *Garrex's Candy-tuft*
gibraltarica. *Gibraltar Candy-tuft*
jucunda (Æthionema coridifolium). *Glaucous Candy-tuft*
linifolia. *Flax-leaved Candy-tuft*
petræa. *Dwarf Alpine Candy-tuft*
pinnata. *Winged Candy-tuft*
rosmarinifolia. *Rosemary-leaved Candy-tuft*
saxatilis. *Rock Candy-tuft*
semperflorens. *Broad-leaved Candy-tuft*
sempervirens. *Evergreen Candy-tuft*
stylosa. *Alpine Candy-tuft*
Tenoreana. *Tenore's Candy-tuft*
Iberis umbellata. *Common Candy-tuft, Purple Annual Candy-tuft*
Icica. Several species yield the Elemi-Resin of Brazil
Abilo. *Elemi-gum-tree* (?)
altissima. *Carana-wood, Samaria-wood, White Cedar* of British Guiana
Aracouchini. *Balsam-of-Acouchi-tree*
Carana. *Carana-gum-tree, S. American Balm-of-Gilead*

Icica heptaphylla. *Hyawa-tree*, or *Incense-wood*, of Guiana
Tacamahaca. *Coumia-resin-plant*
Ileodictyon cibarium. *"Thunder-dirt" Fungus*, of the New Zealanders
Ilex Aquifolium. *Christmas, Common Holly*, or *Holm*
Aquifolium, var. Altaclarensis. *High-Clere Holly*
Aquifolium, var. angustifolia. *Narrow-leaved Holly*
Aquifolium, var. argenteo-variegata. *Silver-striped Holly*
Aquifolium, var. aureo-variegata. *Gold-striped Holly*
Aquifolium, var. ferox. *Hedge-hog Holly*
Aquifolium, var. ferox albo-pictum. *Milk-maid Holly*
Aquifolium, var. ferox argenteo-variegata. *Striped Hedge-hog Holly*
Aquifolium, var. fructû albo. *White-berried Holly*
Aquifolium, .var. fructû luteo. *Yellow-berried Holly*
Aquifolium, var. fructû nigro. *Black-berried Holly*
Aquifolium, var. heterophylla. *Various-leaved Holly*
Aquifolium, var. myrtifolia. *Myrtle-leaved Holly*
Aquifolium, var. pendula variegata. *Variegated Weeping Holly*
Aquifolium, var. verticellata. *Whorled Holly, Swamp Winter-berry*
balearica. *Minorca Holly*
canariensis. *Canary Island Holly*
Cassine. *Black-drink-tree*, of the N. Carolina Indians, *Cassena, Yaupon*
chinensis. *Chinese Holly*
crenata. *Dwarf Japan Holly*
crenata variegata. *Dwarf Golden Holly*
Dahoon. *Dahoon Holly*
Dahoon, var. laurifolia. *Laurel-leaved Dahoon Holly*
dipyrena. *Two-seeded Holly*
Fortunei. *Fortune's Japanese Holly*
latifolia. *Broad-leaved Holly*
latispina. *Broad-spined Holly*
laxiflora. *Loose-flowered Holly*
opaca. *American Holly*
Paraguayensis. *"Mate," Paraguay Tea-plant, S. American Holly*
sideroxyloides. *Dominica Oak*
verticillata. *Swamp Winter-berry*
vomitoria. *Carolina* or *South Sea Tea-plant, Emetic Holly*
Illecebrum. *Knot-grass*
Illicium anisatum. *Star-Anise-*, or *Star-Aniseed-, tree*
Floridanum. *Aniseed-tree*, of Florida, *Poison-Bay*, of Alabama, *Red-flowered Aniseed-tree*
Ilysanthes gratioloides. *False Pimpernel*
Imantophyllum. *Thong-lily*
miniatum. *Natal Lily*
Imbricaria borbonica. *Mat-wood Tree*
Impatiens. *Balsam, Snap-weed*, of N. America
Balsamina and vars. *Common Garden Balsam*

Impatiens fulva. *Orange-flowered Balsam, Spotted Jewel-weed*
glanduligera. *Hardy Indian Balsam, Nuns Noli-me-tangere. Common Yellow Balsam, Touch-me-not*
pallida. *Pale Touch-me-not*
Sultani. *Sultan's*, or *Zanzibar, Balsam*
Imperata arundinacea. *Lalong-Grass*
sacchariflora. *Amoor River Giant Silvery Grass*
Imperatoria Ostruthium. *False Pellitory-of-Spain, Fellow-Grass, Great Master-wort*
Indigofera australis. *Australian Indigo-plant*
floribunda. *Purple-flowered Indigo-plant*
tinctoria. *Dyer's Indigo-plant*
Inga dulcis. *Sappan-fruit-tree*
edulis. *Sweet Tamarind Tree*
Feuillei. *Pacay-tree*, of Lima
purpurea. *Soldier-bush*
Sassa. *Sassa-gum-tree*
spectabilis. *Guava Real*
vera. *Coco-wood*
Inocarpus edulis. *Fiji Chestnut, Otaheite Chestnut*
Intsia amboinensis. *Iron-wood*, of Dutch E. Indies
Inula candida. *White Flea-bane*
Conyza. *Cinnamon-root, Flea-wort, Ladies' Gloves, Ploughman's Spikenard*
crithmoides. *Golden Samphire*
dysenterica. *Common Flea-bane*
glandulosa. *Georgian Flea-bane*
montana. *Pyrenean Flea-bane*
Helenium. *Elecampane, Elf-Dock, Else-Dock, Horse-Elder*
Oculus-Christi. *Christ's-eye, Hoary Flea-bane*
Pulicaria. *Flea-bane, Flea-bane Mullet*
Ionidium capense. *Cape Green Violet*
Ipecacuanha. *False Brazilian*, or *White, Ipecacuanha*
Ionopsidion acaule. *Carpet-plant, Violet-flowered Cress*
Ipomœa. *American Bind-weed, Morning-Glory*
Bona-nox. *Moon-creeper*
cœrulea. *Blue-flowered Bind-weed*
caroliniana. *Cypress vine*, of Carolina
coccinea. *Scarlet-flowered Bind-weed, Star-glory*
fastigiata. *Wild Potato*, of the W. Indies.
Gerrardi. *Wild Cotton*, of Natal
hederacea coccinea. *Ivy-leaved Cypress-vine*
lacunosa (Convolvulus micranthus). *White-star Bind-weed*
leptophylla. *Colorado Man-root*
Nil. *Smaller Morning-Glory*
Orizabensis (I. batatoides). *Jalap-tops, Male Jalap*
pandurata. *"Man-of-the-Earth," "Mechameek,"* of the N. American Indians, *Wild Potato Vine*
Pes-capræ. *Sea-side Potato*, of India
purpurea. *Common Morning glory*
Quamoclit. *Sweet-William*, of Barbadoes
sidæfolia. *Christmas Gambol*
tuberosa. *Seven-year Vine, Spanish Arbour Vine*, or *Spanish Wood-bine*, of Jamaica, *Trellis Convolvulus*, of the W. Indies
Turpethum. *E. Indian Jalap*

Ipomopsis elegans. *Standing Cypress*
Iriartea exorrhiza. *Pashiuba, Paxiuba, Rasp*, or *Zanora Palm*
setigera. *Blowing-cane Palm*
Iresine. *Blood-leaf*
celosioides. *Juba's-brush*
Iridæa edulis. *Dulse*
Iris. *Flag, Rain-bow-Flower*
alata. *Christmas-flowering Iris, Scorpion Iris*
amœna. *Delicately-tinted Iris*
attica. *Greek Iris*
caucasica. *Caucasian Iris*
cristata. *Dwarf Crested Iris, Lady's Calamus*
DeBerghii. *De Bergh's Iris*
dichotoma. *Blue-curls*
ensata. *Sword-leaved Iris*
erratica. *Erratic Iris*
fœtidissima. *Blue Seggin, Gladden, Gladin, Gladwyn*, or *Glauter, Roast-beef-plant, Stinking Gladwyn*
fœtidissima variegata. *Variegated Gladwyn*
flavescens. *Yellowish Iris*
florentina. *Florentine Iris, Orrice*, or *Orris-root-plant*
germanica. *Common* or *German Iris, Fleur de Luce, Orrice*, or *Orris-root-plant*
graminea. *Grass-leaved Iris*
iberica (Onocyclus ibericus). *Iberian Iris*
italica. *Italian Iris*
juncea (I. lusitanica). *Rush-leaved Iris*
Kæmpferi. *Kæmpfer's Iris*
lacustris. *Dwarf Lake Iris*, of N. America
lævigata. *Smooth Iris*
longipetala. *Long-petalled Iris*
missouriensis. *Missouri Iris*
Monnieri. *Golden Iris, Monnier's Iris*
nudicaulis. *Naked-stemmed Iris*
ochroleuca. *Yellow-banded Iris*
olbiensis. *Sardinian Iris*
pallida. *Pale Blue Iris, Orrice*, or *Orris-root-plant*
Pseud-acorus. *Jacob's-Sword, Water-Flag, Yellow Flag, Yellow Fleur de Luce*
pulchella. *Pretty-flowered Iris*
pumila. *Dwarf Iris*
reticulata. *Early Bulbous Iris*
(Moræa) Robinsoniana. *Wedding-flower of Lord Howe's Island*
rubricaulis. *Red-stemmed Iris*
ruthenica. *Ever-blooming Iris, Russian Iris*
Saari. *Saar's Iris*
sambucina. *Elder-scented Iris*
scorpioides (Xiphium planifolium). *Twice-flowering Iris*
setosa. *Bristle-pointed Iris*
siberica. *Siberian Iris*
spatulata (I. desertorum). *Sweet-scented Iris*
spuria. *Spurious Iris*
squalens. *Brown-flowered Iris*
stenogyna. *Cream-coloured Iris*
stylosa. *Algerian Iris*
susiana. *Great Spotted Iris, Sad-flowered Iris*
Swertii. *Swert's Iris*
tectorum. *Wall Iris*
Telfordi. *Telford's Iris*
tenax. *Tough-leaved Iris*
tingitana. *Tangier Iris*
tridentata. *Labrador Iris*

and Foreign Plants, Trees, and Shrubs. 209

Iris tuberosa (Hermodactylus tuberosus). *Onion Iris, Snake's-head Iris*
Van Houttei. *Van Houtte's Iris*
variegata. *Variegated Iris*
verna. *Dwarf Iris*, of N. America, *Spring-flowering Iris*
versicolor. *Larger Blue Flag*, of N. America
virginica. *Boston Iris, Slender Blue Flag*, of N. America
Xiphion. *Clouded Iris, Small Bulbous Iris, Spanish Flag, Thunderbolt Iris*
xiphioides. *"English" Iris, Great Bulbous Iris, Pyrenean Flag*
Irvingia. *Wild Mango*
Barteri. *Dika*, or *Udika, Bread-tree*, of W. Africa
Isanthus cæruleus. *False Penny-royal*, of N. America
Isatis tinctoria. *Ash-of-Jerusalem, Woad*
Ismene Amancaes. *Peruvian Daffodil*
Isnardia palustris. *Water-Purslane*
Isoëtes. *Quill-wort*
alpinus. *Mountain Quill-wort*
lacustris. *Common Quill-wort, Merlin's Grass*
Isonandra Gutta. *Gutta-Percha Tree*
Itea virginica. *Virginian Willow*
Iva. *High-water Shrub, Marsh Elder*
Ixia. *African Corn-lily*
Ixiolirion. *Ixia-lily*
Ledebouri. *Ledebour's Ixia-lily*
montanum. *Mountain Ixia-lily*
Pallasi. *Pallas's Ixia-lily*
tataricum. *Tartarian Ixia-lily*
Ixora. *Wild Jasmine*, of the W. Indies
ferrea. *Hard-wood-tree*

Jacaranda ovalifolia. *Green Ebony*
Jacksonia scoparia. *Dog-wood*, of N. S. Wales, *Jackson's Broom*
Jacquinia armillaris. *Bracelet-wood Tree, W. Indian Currant-tree*
ruscifolia. *Cross-wood*, of San Domingo
Jambosa (Eugenia) malaccensis. *Malay Apple*
Jasione humilis. *Dwarf Sheep's-bit Scabious*
montana. *Common Sheep's-bit Scabious*
perennis. *Perennial Sheep's-bit Scabious*
Jasminum. *Jasmine*, or *Jessamine*
affine (J. ochroleucum). *Large White-flowered Jasmine*
angustifolium. *Narrow-leaved Jasmine*
arborescens. *Tree-like Jasmine*
aureum. *Golden-leaved Jasmine*
azoricum. *White Azorean Jasmine*
capense. *Cape Jasmine*
fruticans. *Common Yellow-flowered Jasmine, Make-bate*
grandiflorum. *Catalonian* or *Spanish Jasmine, Large-flowered Jasmine*
heterophyllum. *Various-leaved Jasmine*
humile. *Italian Yellow Jasmine*
laurifolium. *Bay-leaved Jasmine*
multiflorum. *Many-flowered Jasmine*
nudiflorum. *Winter-flowering Jasmine*
odoratissimum. *Jonquil-scented Jasmine, Yellow Azorean Jasmine*
officinale. *Common White-flowered Jasmine*
officinale fl.-pl. *Double-flowered Jasmine*
officinale foliis argenteis. *Silver-leaved Jasmine*

Jasminum officinale foliis aureis. *Golden-leared Jasmine*
pubigerum. *Downy Nepaul Jasmine*
Reevesii. *Reeves's Jasmine*
revolutum. *Yellow Nepaul Jasmine*
Sambac. *Arabian Jasmine, White-flowered Indian Jasmine*
Jasonia glutinosa. *Glutinous Flea-bane*
Jateorhiza palmata (Cocculus palmatus). *Calumba*, or *Colombo-root*
Jatropha Curcas (Curcas purgans). *Physic-nut*, of S. America
gossypiifolia. *Belly-ache Bush*
(Janipha) Manihot (Manihot utilissima). *Cassava-plant, Mandioc-plant, Tapioca-plant*
multifida. *Coral-plant, Tyle-berry*
stimulans. *Stinging-bush*
urens var. stimulosa (Cnidoscolus stimulosus). *Spurge-nettle, Virginian "Tread-softly"*
Jeffersonia diphylla. *Ground-squirrel Pea, Rheumatism-root*, of N. America, *Twin-leaf*
Jonesia Asoca. *Ushoka-tree*, of Bengal
Jubæa spectabilis. *Coquito Palm, "Little Coker-nut" Palm*
Juglans. *Walnut-tree*
cinerea. *Butter-nut*, of N. America, *White Walnut*
nigra. *American* or *Black Walnut*
pterocarpa (Pterocarya caucasica). *Caucasian Walnut*
regia. *Ban-nut-tree, Common Walnut-tree, French Nut, Welsh Nut*
regia var. laciniata. *Cut-leaved Walnut*
Juncus. *Rush*
acutus. *Great Sharp Sea-side Rush*
articulatus (J. acutiflorus). *Jointed Rush, Sprit*
balticus. *Baltic Rush*
biglumis. *Two-flowered Rush*
bufonius. *Frog-grass, Toad-grass, Toad Rush*
capitatus. *Cluster-headed Rush, Jersey Rush*
castaneus. *Black-spiked Rush*
compressus. *Round-fruited Rush*
communis. *Candle Rush, Common Soft Rush*
communis var. conglomeratus. *Dense-flowered Rush*
communis var. conglomeratus spiralis. *Cork-screw Rush*
communis, var. effusus. *Candle Rush, Loose-flowered Rush, Pin Rush*
filiformis. *Thread Rush*
glaucus. *Common Hard Rush*
maritimus. *Sea-side Rush*
obtusiflorus. *Blunt-flowered Rush*
squarrosus. *Goose-corn Rush, Heath Rush, Moss Rush*
trifidus. *Highland Rush*
triglumis. *Three-flowered Rush*
Jungermannia. *Scale Moss*
Juniperus. *Juniper*
barbadensis. *Barbadoes Cedar*
bermudiana. *Bermuda Cedar, Pencil-wood Cedar*
californica. *Californian Juniper*
Cedrus. *Canary Island Cedar* or *Juniper*
Cedrus var. brevifolia. *Azores Juniper*

P

Juniperus chinensis. *Chinese Juniper*
chinensis var. aurea. *Golden Chinese Juniper*
communis. *Common Juniper*
communis var. hibernica (J. stricta). *Irish Juniper*
communis var. suecica. *Swedish Juniper*
drupacea. *Plum-fruited Juniper*
echiniformis (J. hemisphærica). *Hedge-hog Juniper*
excelsa var. stricta. *Tall Upright Juniper*
Henryana. *Pencil Cedar*, of British Columbia
japonica. *Japan Juniper*
japonica var. alba. *White-variegated Japan Juniper*
japonica var. aurea. *Golden-variegated Japan Juniper*
macrocarpa. *Large Purple-fruited Juniper*
nana. *Dwarf Juniper*
occidentalis. *Californian Juniper*
Oxycedrus. *Brown-berried Juniper, Cade-oil plant, Prickly Cedar, Sharp Cedar*
pachyphlœa. *Sweet-fruited Juniper*
procera. *Abyssinian Juniper*
prostrata. *Carpet Juniper*
Pseudo-Sabina. *Siberian Savin*
recurva. *Drooping Indian Juniper*
religiosa (J. excelsa). *Incense Juniper*
Sabina. *Common Savin*
Sabina var. cupressifolia. "*Kindly Savin*"
Sabina var. nana. *Dwarf Savin, Green-carpet Juniper*
sabinoides. *Gray-carpet Juniper*
tamariscifolia (J. sabinoides). *Tamarisk-leaved Juniper*
tetragona. *Mexican Juniper*
thurifera. *Spanish Juniper*
virginiana. *Virginian Red Cedar*
virginiana var. argentea. *Silvery-leaved Red Cedar*
virginiana var. humilis. *Dwarf Red Cedar*
virginiana var. pendula. *Weeping Red Cedar*
Jussiæa repens. *Clove-strip*
suffruticosa. W. Indian *Loose-strife*
Justicia comata. *Balsam-herb*, of Jamaica
hyssopifolia. *Snap-tree*
Nasuta. *Nasuta*, or *Tong-pang-chong Shrub*, of India
pectoralis. *Garden Balsam*, of the W. Indies

Kæmpferia Galanga. E. Indian *Galingale*
Kageneckia oblonga. *Lyday-wood*, of Chili
Kalmia. *American Laurel*
angustifolia. *Sheep's-poison Laurel*
glauca. *American Swamp-Laurel*
latifolia. *American Mountain Laurel, Calico-bush, Spoon-wood Tree*
rubra. *Red-flowered American Laurel*
arstenia quinquenervia. *Moss-Apple*
Kennedya. *Native Bean-flower*, of Australia
prostrata. *Coral Creeper*
Kentia (Areca) monostachya. *Whip-stick*, or *Walking-stick, Palm*
Baueri. *Norfolk Island Palm*
Belmoreana. *Curly Palm, Earl Belmore's Palm*
Canterburyana. *Umbrella Palm*
sapida. *Nikau Palm*, of New Zealand

Kernera (Cochlearia) saxatilis. *Rock Scurvy-grass*
Khaya senegalensis. *African Mahogany*
Kibara macrophylla. *Queensland Black Ink-berry*
Kingia australis. *Australian Grass-tree*
Kleinia neriifolia. *Carnation-tree*
Knightia excelsa. *Rewa-Rewa-tree*, of New Zealand, *New Zealand Oak*
Kochia (Chenopodium) Scoparia. "*Belvedere*," *Broom Goose-foot, Broom or Summer Cypress*
villosa. *Cotton-bush*, of Australia
Kokoona zeylanica. *Kokoon-tree*, of Ceylon
Krameria Ixina. *Savanilla Rhatany-plant*
triandra. *Rhatany Root*
Krigia virginica. *Dwarf Dandelion*, of N. America
Kunthia montana. *Snake Cane*, of New Granada.
Kunzea corifolia. *Bottle-green Tea-tree*, of Victoria

Lachenalia. *Cape Cowslip, Leopard-lily*
serotina (Uropetalum serotinum). *Late-flowering Cape Cowslip*
Lachnanthes tinctoria. *Paint-root* or *Red-root*, of Carolina
Lachnocaulon Michauxii. *Hairy Pipe-wort*
Lactarius deliciosus. *Orange-milk Mushroom*
Lactuca. *Lettuce*
elongata (L. canadensis). *American Wild Lettuce, Fire-weed*
intybacea. *Lombard* or *Endive-leaved Lettuce*
(Prenanthes) muralis. *Wall Lettuce*
perennis. *Perennial Lettuce*
pulchella. *Californian Lettuce*
quercina. *Oak-leaved Lettuce*
saligna. *Least Lettuce, Willow Lettuce*
sativa and vars. *Common Garden Lettuce*
sativa var. capitata. *Cabbage*, or *Drum-head, Lettuce*
Scariola. *Prickly Lettuce*
sonchifolia. *Blue Dandelion*
virosa. *Medicinal Prickly Lettuce, Sleep-wort, Strong-scented Lettuce, Wild Lettuce*
Lætia Guidonia. *Jamaica Rod-wood*
Thamnia. *Scarlet-seed*
Lagenaria vulgaris. *Common Bottle-Gourd*
vulgaris var. clavata. *Trumpet-Gourd*
Lagenophora. *New Zealand Daisy*
Lagerstrœmia. *Crape Myrtle, Indian Lilac*
microcarpa. *Ben Teak*
Reginæ. *E. Indian Blood-wood, Jarool-wood, Queen's Flower*
Lagetta linteria. *Gauze-tree, Lace-bark-tree*, of Jamaica
Lagœcia cuminoides. *Wild Cumin*
Lagunaria Patersoni. *White-wood*, of Australia, *Cow-itch-tree*, or *White Oak*, of Norfolk Island
Laguncularia racemosa. *White Button-wood, White Mangrove*
Lagurus ovatus. *Hare's-tail Grass, Turk's-head Grass*
Laminaria. *Fyams, Tangle*
bulbosa. *Sea-furbelows, Sea-hangers*
digitata. *Cairn-tangle, Cuvy, Dead-Man's-Toe, Ore, Ore-weed, Oar-weed, Sea-girdles, Sea-wand*

Laminaria saccharina. *Devil's-apron, Honey-ware, Ribbon-weed, Sea-belt, Sweet Tangle*
Lamium. *Dead-Nettle*
album. *Archangel, Bee Nettle, Blind Nettle, Dumb Nettle, White Dead-Nettle*
amplexicaule. *Hen-bit, Lion's-snap*
aureum. *Golden Dead-Nettle*
Galeobdolon. *Weasel's-snout, Yellow Archangel, Yellow Dead-Nettle*
garganicum. *Mt. Gargano Dead-Nettle*
longiflorum. *Long-flowered Dead-Nettle*
maculatum. *Spotted Dead-Nettle, Variegated Dead-Nettle*
Orvala. *Balm-leaved Red Dead-Nettle*
purpureum. *Common Red Dead-Nettle*
Landolphia. *African-rubber-tree*
Lanosa nivalis. *Snow-mould*
Lansium domesticum. *"Langsat," or Lanseh-tree, of Malacca*
Lantana aculeata. *Jamaica Mountain-Sage*
mixta. *W. Indian Coast Bramble*
(various species). *Surinam Tea-plant*
Laportea canadensis. *Wood-nettle, of N. America*
Lappa. *Burdock*
edulis. *Giant Edible-rooted Burdock,"Gobbo," or "Gobo," of Japan*
Lapsana communis. *Nipple-wort*
Larix. *Larch-tree*
americana. *American Larch, Black Larch, Hackmatack, Tamarack*
europæa. *Common Larch*
Griffithii. *Himalayan or Sikkim Larch*
Ledebourii. *Altaian Larch*
microcarpa. *Red American Larch*
occidentalis. *Oregon Larch*
pendula. *Weeping Larch*
sibirica. *Siberian Larch*
Larrea mexicana. *Creosote-bush, Creosote-plant*
Laserpitium. *Laser-wort*
latifolium. *Herb-Frankincense*
Lasiandra. *Brazilian Spider-flower*
Lastrea. *Buckler-Fern, Shield-Fern*
cristata. *Crested Shield-Fern, Darsham Fern*
dilatata. *Broad Prickly-toothed Fern*
Filix-mas. *Male Fern, Male Polypody*
Fœnisecii (L. æmula). *Hay-scented Fern, Recurved Prickly-toothed Fern*
Oreopteris. *Heath Fern, Mountain Shield-Fern, Scented Fern*
rigida. *Rigid Shield-Fern*
spinulosa. *Narrow Prickly-toothed Fern*
Thelypteris. *Ground Fern, Marsh Shield-Fern*
Latania Borbonica. *Common Bourbon-Palm*
Lathræa clandestina. *Hidden Tooth-wort*
squamaria. *Common Tooth-wort*
Lathyrus. *Everlasting Pea*
amphicarpus. *Earth Pea*
Aphaca. *Yellow-flowered Pea, Yellow-flowered Vetchling*
californicus. *Californian Everlasting Pea*
Cicera. *Flat-podded Vetch*
grandiflorus. *Large-flowered Everlasting Pea*
hirsutus. *Rough Pea*
latifolius. *Common Everlasting Pea*

Lathyrus macrorrhizus. *Bitter Vetch, Cara-meile, Heath Pea, Mouse Pea*
magellanicus. *Lord Anson's Pea*
maritimus. *Sea-side Pea*
montanus. *Mountain Everlasting Pea*
Nissolia. *Grass Pea, Grass Vetch, Meadow Vetchling*
odoratus. *Sweet Pea*
palustris. *Marsh Pea*
pisiformis. *Pea Vetchling, Siberian Everlasting Pea*
pratensis. *Angle-berries, Meadow Pea*
roseus. *Rosy-flowered Everlasting Pea*
rotundifolius. *Round-leaved Everlasting Pea*
sativus. *Chickling Vetch*
Sibthorpei. *Sibthorpe's Everlasting Pea*
sylvestris. *Wood Pea*
tingitanus. *Tangier Pea*
tuberosus. *Dutch Mice, Earth-nut Pea, Tine Tare, Tuberous-rooted Everlasting Pea*
Laurelia aromatica. *Chili Laurel*
Novæ-Zealandiæ. *New Zealand Laurel, New Zealand Sassafras, Pukatea-tree*
sempervirens. *Chili Sassafras, Peruvian Nutmeg*
serrata. *" Vouvan "-tree*
Laurencia ohtusa. *Corsican Moss*
pinnatifida. *Pepper Dulse*
Laurus. *Bay-tree, Laurel*
æstivalis. *Summer Laurel*
aggregata. *Crowded-flowered Laurel*
albida. *White Sassafras, Whitish-leaved Laurel*
azorica. *Azores Laurel*
Benzoin (Benzoin odoriferum). *Benjamin-bush, Spice-bush, Spice-wood*
bullata. *African Oak, Cape Laurel*
Camphora. *Camphor-tree*
canariensis. *Canary Island Laurel*
(Persea) carolinensis. *Carolina Laurel*
Cassia. *Bastard Cinnamon-tree*
Catesbiana. *Catesby's Laurel*
Chloroxylon. *Cog-wood-tree, of the W. Indies, Jamaica Laurel*
coriacea. *Leather-leaved Laurel*
crassifolia. *Thick-leaved Laurel*
Diospyros. *Jove's-fruit Laurel*
exaltata. *Lofty Laurel*
floribunda. *Many-flowered Laurel*
fœtens. *Madeira Bay-tree, Strong-scented Laurel, Til-tree*
geniculata. *Jointed Laurel*
indica. *Indian Bay, Madeira Laurel*
nivea. *Snow-white Laurel*
nobilis. *Bay Laurel, Bay-tree, Poet's or Roman Laurel, Sweet Bay, Victor's Laurel*
patens. *Spreading Laurel*
pendula. *Weeping Laurel*
regalis. *Royal Laurel*
salicifolia. *Willow-leaved Laurel*
Sassafras. *Ague-tree, Sassafras-tree*
splendens. *Shining-leaved Laurel*
thyrsiflora. *Bunch-flowered Laurel*
Lavandula. *Lavender*
dentata. *Sweet-scented Lavender*
Spica. *Common Lavender*
Stœchas. *Cassidony, French Lavender, Sticadoue or Sticados*
vera. *True Lavender*

P 2

Lavatera africana. *African Tree-Mallow*
arborea. *Common Tree-Mallow or Sea-Mallow, Velvet-leaf, Wild Tree-Mallow*
arborea variegata. *Variegated Tree-Mallow*
Olbia. *Garden Tree-Mallow*
subovata. *Suborate-leaved Tree-Mallow*
trimestris. *Common Annual Tree-Mallow*
triloba. *Three-lobed Tree-Mallow*
unguiculata. *Samian Tree-Mallow*
Lawsonia alba (L. inermis). *"Camphire," of Scripture, Egyptian Privet, Henna-plant, Jamaica Mignonette*
Layia platyglossa. *"Tidy-tips," of San Francisco*
Lecanora pallescens. *Crab's-eye Lichen*
tartarea. *Cudbear or Cork Cudbear, Cup-Moss*
Lechea. *American Pin-weed*
Lecidea geographica. *Map-Lichen*
Lecythis Ollaria. *Monkey-pot Tree*
Zabucajo. *Sapucaia-nut Tree*
Ledum latifolium. *Labrador Tea-plant*
palustre. *Marsh Cistus, Marsh Rosemary*
Leersia lenticularis. *American Fly-catch-Grass*
oryzoides. *Cut-Grass, Rice Cut-Grass*
virginica. *American White Grass*
Leiophyllum (Ledum) buxifolium (Ammysine buxifolia). *Box-leaved Sand-Myrtle*
Lemna. *Duck-meat, Duck-weed*
gibba. *Thick-leaved Duck-weed*
minor. *Common Duck-weed, Greens, Greeds, Water Lens, Water Lentils*
polyrrhiza. *Greater Duck-weed*
trisulca. *Ivy-leaved Duck-weed*
Lentinus lepideus. *Cellar Fungus*
Leonia glycycarpa. *Achocon-tree, of Peru*
Leonotis. *Lion's-ear*
Leonurus. *Lion's-tail*
Leontice Leontopetalum. *Lion's-leaf, Lion's Turnip*
Leontodon. *Lion's-tooth*
Taraxacum. *Blow-ball, Canker-wort, Dandelion, Horse-Gowan, Irish Daisy, Milk-Gowan, Yellow Gowan*
Leontopodium alpinum. *Lion's-foot*
Leonurus Cardiaca. *Common Mother-wort*
sibiricus. *Siberian Mother-wort*
Leopoldinia major. *Jara-Assu Palm*
Piassaba. *Piassaba Palm*
pulchra. *Jara Palm*
Lepidium. *Cress, Pepper-wort*
campestre. *Churl's-Mustard, Mithridate Mustard or Pepper-wort, Treacle-Mustard*
Cardamines. *Spanish Cress*
Draba. *Hoary Cress*
latifolium. *Broad-leaved Cress, Dittander, Dittany*
oleraceum. *New Zealand Cress*
Piscidium. *Fish-poison, of the S. Sea Islanders*
ruderale. *Narrow-leaved Cress*
sativum. *Garden Cress, Tongue-grass, Town Cress*
virginicum. *American Wild Pepper-grass*
Lepidosperma gladiatum. *Sword-Sedge, of Australia*
Lepigonum. *Sand Spurrey*
Lepironia mucronata. *Chinese Mat-Rush*

Leptandra virginica. *Black-root*
Leptanthus gramineus. *Water Star-grass*
Leptomeria acerba. *Australian Currant*
Billardieri. *Native Currant, of Victoria*
Leptospermum. *South Sea Myrtle*
flavescens. *New Zealand Tea-plant*
lanigerum. *Tea-tree, of Australia*
lævigatum. *"Sand-stay," of Australia*
scoparium. *Captain Cook's Tea-tree, Manuka-scrub,* of New Zealand
Lepturus incurvatus. *Sea-side Hard-grass, Snake's-tail*
Lespedeza. *Bush Clover, of N. America*
striata. *"Hoop-Koop"-plant, Japan Clover*
Lettsomia Bona-nox. *Clove-scented Creeper*
Leucadendron argenteum. *Cape Silver-tree*
Leucanthemum (Chrysanthemum) arcticum. *Arctic Ox-eye Daisy*
Leucocoryne. *White Club-flower*
Leucocrinum montanum. *Californian Soap-root, Rocky Mountain Dwarf White Lily*
Leucæna (Acacia) glauca. *W. Indian Lead-tree*
Leucojum. *"Bulbous Violet," Snow-flake*
æstivum. *Loddon Lilies, Summer Snow-drop, Summer Snow-flake*
autumnale. *Autumn Snow-flake*
carpaticum. *Carpathian Snow-flake*
Hernandezii (L. pulchellum). *Small Summer Snow-flake*
hyemale. *Winter Snow-flake*
roseum. *Rosy-flowered Snow-flake*
trichophyllum (Acis trichophylla). *Many-flowered Snow-flake*
vernum. *Spring Snow-flake*
Leucopogon. *Australian "Native Currant"*
Fraseri. *Otago Heath*
Richei. *Australian Currant, Australian Carrot-wood*
Leucoxylon buxifolium. *White-wood, of Java*
Levisticum officinale (Ligusticum Levisticum). *Common Lovage, Italian Lovage, Mountain Hemlock*
Lewisia rediviva. *Bitter-root, Spætlum or Spatlum*
Leycesteria formosa. *Flowering Nutmeg, Himalayan Honey-suckle*
Liatris. *Button-snake-root, Gay-feather*
elegans. *Hairy-cupped Button-snake-root*
odoratissima. *Vanilla-plant, of N. America*
pycnostachya. *Dense-spiked Button-snake-root, Kansas Gay-feather*
spicata. *Long-spiked Button-snake-root*
squarrosa. *Devil's-bit, of N. America, Blazing-star*
Libocedrus. *Incense Cedar*
chilensis. *Chilian Arbor-vitæ*
decurrens (Thuja gigantea). *White Cedar, of California*
Doniana. *Kawaka-tree, New Zealand Arbor-vitæ*
tetragona. *Alerse-tree, of Chili*
Licania guianensis. *Cayenne Rose, Cayenne Sassafras, Pepper-wood, Pottery-bark-tree*
Licuala acutifida. *Penang Lawyers*
Ligularia Kæmpferi var. argentea. *Japanese Silver-leaf*

Ligusticum actæifolium. *American White-root, Angelico, Nondo*
Lyalli. *Maori Parsnip*
peregrinum. *Parsley-leaved Lovage*
scoticum. *Scotch Lovage*
Ligustrum. *Privet*
Fortunei. *Fortune's Privet*
japonicum. *Japanese Privet*
japonicum variegatum. *Gold-blotched Privet*
lucidum. *Chinese Wax-tree, Shining-leaved Privet*
ovalifolium. *Oval-leaved Privet*
sinense. *Chinese Privet*
sinense var. nanum. *Dwarf Chinese Privet*
spicatum. *Spiked-flowered Privet*
villosum. *Hairy Privet*
vulgare. *Common Privet, Prim-print or Prim*
vulgare variegatum. *Variegated-leaved Privet*
vulgare var. buxifolium. *Box-leared Privet*
vulgare var. leucocarpum. *White-berried Privet*
vulgare var. sempervirens (L. italicum). *Evergreen Privet*
vulgare var. xanthocarpum. *Yellow-berried Privet*
Lilium. *Lily*
auratum. *Gold-striped Lily*
auratum var. rubro-vittatum. *Scarlet-striped Lily*
avenaceum. *Oat-bulbed Lily*
Brownii. *Brown's Lily*
bulbiferum. *Bulb-bearing Lily*
bulbiferum var. aurantium. *Bulb-bearing Orange Lily*
Buschianum. *Busch's Lily*
californicum. *Californian Lily*
callosum. *Warty Red Japanese Lily*
camtschatcense (Fritillaria camtschatcoensis). *Black Lily*
canadense. *American Wild Yellow Lily*
candidum. *Bourbon Lily, Juno's Rose, Madonna Lily, White Lily*
carniolicum. *Carniola Lily*
carolinianum. *Carolina Lily*
Catesbæi. *Southern Red Lily*
chalcedonicum. *Scarlet Martagon Lily*
colchicum (L. Szovitzianum.) *Crimson-anthered Lily, Tall Sulphur-flowered Lily*
columbianum. *Oregon Lily*
concolor. *Japanese Red-Star Lily*
cordifolium. *Heart-leaved Lily*
cordifolium var. giganteum. *Giant Heart-leaved Lily*
Coridion. *Japanese Yellow-Star Lily*
croceum. *Common Orange Lily, Saffron-coloured Lily*
dalmaticum. *Black Martagon Lily*
davuricum. *Siberian Orange Lily*
eximium. *Transparent Trumpet Lily*
giganteum. *Giant Lily*
Harrisi [L. japonicum floribundum (?)]. *Bermuda Easter Lily*
Humboldtii. *Humboldt's Lily*
japonicum. *Japanese Lily*
Krameri. *Kramer's Lily*
lancifolium. *Spear-leaved Japanese Lily*
Leichtlinii. *Max Leichtlin's Lily*

Lilium longiflorum. *Common Trumpet Lily*
Martagon. *Martagon Lily, Huleh Lily*
medeoloides. *Medeola Lily*
monadelphum. *Caucasian Lily*
neilgherrense. *Neilgherry Lily*
pardalinum. *Panther Lily*
Parryi. *Parry's Lily*
Partheneion. *Brownish-red Japanese Lily*
parvum. *Small Lily*
Philadelphicum. *Whorled-leaved American Lily, Wild Orange-Red Lily*
polyphyllum. *Himalayan White Lily*
Pomponium. *Turban Lily*
pulchellum. *Russian Lily, Siberian Scarlet Lily*
pyrenaicum. *Pyrenean Lily*
Robinsoni. *Robinson's Lily*
speciosum (L. lancifolium). *Spotted Lily*
superbum. *Great American Turk's-cap Lily, Swamp Lily*
Szovitzianum (L. colchicum). *Crimson-anthered Lily, Tall Sulphur-flowered Lily*
tenuifolium. *Tom Thumb Lily*
testaceum. *Buff-coloured Lily, Nankeen Lily*
tigrinum. *Tiger Lily*
tigrinum var. Fortunei. *Fortune's Tiger Lily*
umbellatum. *Umbel-flowered Lily*
venustum (L. Thunbergianum). *Late Orange Lily, Thunberg's Lily*
Wallichianum. *Wallich's Trumpet Lily*
Washingtonianum. *Nevada or Washington Lily*
Washingtonianum var. purpureum (L. rubescens). *Purple Washington Lily*
Wilsoni. *Wilson's Lily*
Limnanthemum. *Marsh-flower*
lacunosum (Villarsia cordata). *Floating-Heart*
(Villarsia) nymphæoides. *Fringed Buck-bean, Fringed Water-Lily, Water-Fringe*
Limnobium Spongia. *American Frog-bit*
Limonia acidissima. *Musk-deer Plant*
carnosa. *Keklam-fruit, of Bengal (?)*
monophylla. *Indian Wild Lime*
Limosella aquatica. *Mud-weed, Mud-wort*
Linaria. *Toad-flax*
alpina. *Alpine Toad-flax*
bipartita. *Cloven-flowered Toad-flax*
canadensis. *American Toad-flax*
cirrhosa. *Tendrilled Toad-flax*
crassifolia. *Thick-leaved Toad-flax*
Cymbalaria. *Coliseum or Kenilworth Ivy, Ivy-leaved Toad-flax, Ivy-wort, Mother-of-Thousands, Oxford Weed, Penny-wort*
dalmatica. *Large Yellow Toad-flax*
Elatine. *Cancer-wort, Pointed Toad-flax*
genistæfolia. *Broom-leaved Toad-flax*
hepaticæfolia. *Hepatica-leaved Toad-flax*
minor. *Small Toad-flax*
origanifolia. *Marjoram-leaved Toad-flax*
Pelisseriana. *Pelisser's Toad-flax*
Perezii. *Perez's Toad-flax*
pilosa. *Hairy Toad-flax*
purpurea. *Purple Toad-flax*
pyrenaica. *Pyrenean Toad-flax*
repens. *Creeping Toad-flax*
reticulata. *Netted-flowered Toad-flax*

Linaria reticulata var. aureo-purpurea. *Gold-and-purple Toad-flax*
spuria. *Cancer-wort, Male Fluellin, Round-leaved Toad-flax*
supina. *Trailing Toad-flax*
triornithophora. *Three-birds Toad-flax*
triphylla. *Three-leaved Toad-flax*
villosa. *Shaggy Toad-flax*
vulgaris. *Butter-and-Eggs, Buttered Haycocks, Common Toad-flax, Dragon-bushes, Eggs-and-Bacon, Eggs-and-Butter, Flax-weed, Gall-wort, 'Ramsted' or 'Ransted,' of N. America, Wild Flax, Yellow Toad-flax*
Lindelofia spectabilis (Cynoglossum longiflorum). *Himalayan Lung-wort*
Lindera Benzoin. *Benjamin-bush, Fever-bush, Spice-bush, Wild All-spice*
melissæfolia. *Jove's-Fruit*
Linnæa borealis. *Twin-flower*
Linociera ligustrina. *Jamaica Rose-wood*
Linosyris vulgaris. *Goldilocks*
Linum. *Flax*
alpinum. *Alpine Flax*
alpinum var. album. *Dwarf White-flowered Flax*
angustifolium. *Pale-flowered Flax*
arboreum. *Evergreen Flax*
Berlandieri. *Berlandier's Flax*
catharticum. *Dwarf, Fairy, Mountain or Purging Flax, Fairy Lint, Mill-mountain*
flavum (L. campanulatum). *Yellow Herbaceous Flax*
grandiflorum. *Crimson-flowered Flax*
Macraei (L. Chamissonis). *Orange-flowered Flax*
monogynum. *Native Flax*, of New Zealand, *White-flowered Flax*
narbonnense. *Narbonne Flax*
perenne. *Perennial Flax*
Provinciale. *Bright Blue Perennial Flax*
rubrum. *Sicilian Red-flowered Flax*
salsoloides. *Heath-leaved Flax*
striatum. *Bog Flax*, of N. America
sulcatum. *Large Yellow-flowered Flax*, of N. America
tauricum. *Crimean Flax.*
trigynum. *E. Indian Flax, Winter-flowering Flax*
usitatissimum. *Common Flax, Lin, Line,* or *Lint*
virginianum. *Virginian Yellow-flowered Flax*
viscosum. *Viscid Flax*
Liparis liliifolia. *American Tway-blade, Purple-flowered Tway-blade*
Loeselii. *Fen-Orchis, Green-flowered Tway-blade*
Lippia (Zapania) lanceolata. *Fog-fruit*
(Aloysia) citriodora. *Lemon-Tree, Lemon-scented " Verbena"*
Liquidambar orientalis. *Liquid-Storax-tree, Lord-wood*
styraciflua. *Bilsted, Copalm Balsam-tree, Sweet Gum-tree*
Liriodendron tulipiferum. *Canoe-wood, Lyre-tree, Saddle-tree, Tulip-tree, Western "Poplar," "Whitewood,"* of N. America, *Yellow Poplar*
Lissanthe sapida. *Australian Cran-berry*

Listera convallarioides. *Common Tway-blade,* of N. America
cordata. *Lesser Tway-blade*
ovata. *Bifoil, Double-leaf, Common Tway-blade*
Lithocarpus javensis. *Stone Oak*
Lithospermum. *Gromwell*
arvense. *Bastard Alkanet, Cornfield Gromwell*
(Batschia) canescens. *Alkanet,* of N. America, *Deep-yellow-flowered Gromwell, Hoary Puccoon.*
fruticosum. *Shrubby Gromwell*
Gastoni. *Gaston's Gromwell*
hirtum. *"Alkanna," Hairy Puccoon*
officinale. *Common Gromwell, Grummel,* or *Graymile, Lichwale* or *Lychwale, Pearl-plant*
petræum. *Rock Gromwell*
prostratum. *Gentian Gromwell, Purple Gromwell*
purpureo-cœruleum. *Creeping Gromwell*
rosmarinifolium. *Rosemary-leaved Gromwell*
Litobrochia Currori. *Royal Fern,* of Calabar
Litsæa dealbata. *Brush-land Mist-tree,* of Australia
Littorella lacustris. *Plantain Shoreweed, Shore Grass*
Livistona (Corypha) australis. *Cabbage-Palm,* of Australia, *Fan Palm,* of N. S. Wales, *Gippsland Palm-tree*
Jenkinsiana. *Toko-pat Palm*
Mariæ. *Palm-tree,* of Central & W. Australia
Lloydia serotina. *Mountain Spider-wort*
Loasa lateritia. *Common Chili Nettle*
Lobelia assurgens. *Tree Lobelia*
azurea nana. *Dwarf Blue Lobelia*
cardinalis. *Cardinal-flower Lobelia, Scarlet Lobelia*
coronopifolia. *Stag's-horn-leaved Lobelia*
Dortmanna. *Water Lobelia*
erinoides. *Trailing Lobelia*
Erinus. *Dwarf Spreading Blue Lobelia*
Erinus var. compacta. *Close-growing Dwarf Blue Lobelia*
fulgens. *Brilliant Lobelia*
gracilis. *Slender Lobelia*
heterophylla major. *Blue-and-white Lobelia*
ilicifolia. *Holly-leaved Lobelia*
inflata. *Asthma-weed, Emetic-weed, Indian Tobacco*
Laurentia. *Italian Lobelia*
littoralis. *Shore Lobelia*
lutea. *Yellow-flowered Lobelia*
Paxtoniana. *Dwarf Blue-and-white Lobelia*
pumila. *Dwarf Lobelia*
ramosa. *Branching Lobelia*
speciosa. *Showy Lobelia*
syphilitica. *Tall Blue Lobelia*
Tupa. *Mullein-leaved Lobelia*
urens. *Acrid Lobelia, Flower-of-the-Axe*
Lodoicea Seychellarum. *Double,* or *Sea-, Cocoa-nut Palm*
Loiseleuria (Azalea) procumbens. *Alpine* or *Trailing Azalea*
Lolium italicum (L. perenne var. multiflorum). *Italian Rye-grass* or *Ray-Grass*

Lolium perenne. *Ever-*, or *Earer-*, *Grass*, *Ray-Grass* or *Rye-Grass*, *Red Darnel*, *White None-such*
perenne var. multiflorum (L. italicum). *Italian Rye-grass*
temulentum. *Cheat, Chess, Darnel, Ivray, Juray*
Lomaria Spicant. *Deer Fern*
Lomatia ilicifolia. *Native Holly*, of Australia
Lonchocarpus (Milletia) Blackii. *Queensland Lance-pod*
Lonicera. *Honey-suckle*
alpigena. *Alpine Honey-suckle*
angustifolia. *Narrow-leaved Honey-suckle*
brachypoda. *Short-stalked Honey-suckle*
brachypoda var. aureo-reticulata. *Gold-netted Honey-suckle*
cærulea. *Blue-berried Honey-suckle*
ciliata. *Fly-Honey-suckle*, of N. America
canadensis. *Canadian Honey-suckle*
Caprifolium. *Caprifole, Perfoliate Honey-suckle, White Italian Honey-suckle*
Caprifolium var. flava. *Yellow Italian Honey-suckle*
confusa. *Confused Honey-suckle*
Diervilla. *Yellow-flowered Upright Honey-suckle*
discolor. *Two-coloured Honey-suckle*
diversifolia. *Various-leaved Honey-suckle*
Douglasii. *Douglas's Honey-suckle*
etrusca. *Tuscan Honey-suckle*
etrusca var. rubra. *Red Italian Honey-suckle*
flava. *Yellow Honey-suckle*
flexuosa. *Chinese Honey-suckle*
fragrantissima. *Winter-flowering Honey-suckle*
grata. *American Wood-bine, Evergreen Honey-suckle*
hirsuta. *Hairy Honey-suckle*
iberica. *Iberian Honey-suckle*
implexa. *Minorca Honey-suckle*
japonica. *Japan Honey-suckle*
Ledebourii. *Ledebour's Honey-suckle*
ligustrina. *Privet-leaved Honey-suckle*
microphylla. *Small-leaved Honey-suckle*
nigra. *Black-berried Honey-suckle*
oblongifolia. *Swamp Fly-Honey-suckle*, of N. America
occidentalis. *Western Honey-suckle*
odoratissima. *Sweetest Honey-suckle*
orientalis. *Eastern Honey-suckle*
parviflora. *Small Honey-suckle*
Periclymenum. *Bear-bind, Common Woodbine* or *Honey-suckle*
Periclymenum var. belgica. *Dutch Honey-suckle*
Periclymenum var. rubra. *Red Dutch Honey-suckle*
pubescens. *Downy American Honey-suckle*
punicea. *Crimson-flowered Honey-suckle*
pyrenaica. *Pyrenean Honey-suckle*
quadrifolia. *Four-leaved Honey-suckle*
sempervirens. *Trumpet Honey-suckle*
sempervirens var. minor. *Small Trumpet Honey-suckle*
Stabiana. *Castellamare Honey-suckle*
tatarica. *Tartarian Honeysuckle*
tomentella. *Downy Honey-suckle*

Lonicera triflora. *Three-flowered Honey-suckle*
villosa. *Shaggy Honey-suckle*
Xylosteum. *Common Fly-Honey-suckle*
Lophanthus. *Giant Hyssop*
anisatus. *Anise-Hyssop*
urticæfolius. *Nettle-leaved Giant Hyssop*
Lophira africana. *Scrubby Oak*, of Africa
alata. *Scrubby Oak*, of Sierra Leone
Lophostemon australis. *Red Box*, of N. S. Wales
macrophyllus. *Gray Box*, of Queensland
Lotus. *Bird's-foot-Trefoil*
arabicus. *Red-flowered Bird's-foot-Trefoil*
argenteus. *Silvery Bird's-foot-Trefoil*
australis. *Sheep-poison-plant*, of Australia
corniculatus. *Bird's-foot, Bird's-foot Clover, Cat-in-clover, Cheese-cake Grass, Common Bird's-foot-Trefoil, Crow's-foot, Eggs-and-Bacon, Fingers-and-Thumbs, Fingers-and-Toes, Ground Honey-suckle, Lamb's-toe*
corniculatus fl.-pl. *Double-flowered Bird's-foot-Trefoil*
creticus. *Silvery Evergreen Bird's-foot-Trefoil*
Dorycnium. *Shrubby Bird's-foot-Trefoil*
edulis. *Edible Bird's-foot-Trefoil*
Jacobæus. *St. James's-Flower*
major. *Greater Bird's-foot-Trefoil*
tetragonolobus. *Winged Pea*
tetraphyllus. *Four-leaved Grass*
thermalis. *Hungarian Lotus* or *Water-lily*
Loxopterygium Lorentzii. *Red Quebracho-tree*
Lucuma (Sapota) mammosa. *Jamaica Bullet-tree, Marmalade-tree, Vegetable-Egg-tree*
multiflora. *Broad-leaved Nase-berry Bully-tree*
Ludwigia. *American False-Loose-strife*
(Isnardia) alternifolia and L. hirtella. *Seed-Box*
palustris. *American Water-Purslane*
Luffa acutangula. *Strainer Vine*
ægyptiaca. *Sooty-Qua Gourd, Towel* or *Washing Gourd*
Lunaria annua. *Annual Honesty*
biennis. *Bolbonac, Common Honesty, Money-flower, Satin-flower, Satin-leaves*
biennis var. albiflora. *White-flowered Honesty*
rediviva. *Perennial Honesty*
Lupinus. *Lupine*, or *Lupin*
arboreus. *Tree-Lupine* of California
densiflorus. *Californian "Sheep-poison"*
Hartwegi. *Hartweg's Lupine*
insignis. *Blue-flowered Lupine*
linifolius. *Flax-leaved Lupine*
macrophyllus. *Large-leaved Lupine*
Moritzianus. *Blue-and-white Lupine*
mutabilis. *Changeable-flowered Lupine*
nanus. *Dwarf Blue-flowered Lupine*
Nutkatensis. *Dwarf Perennial Lupine*
polyphyllus. *Tall Blue-flowered Perennial Lupine*
subcarnosus. *White-and-rose-flowered Lupine*
venustus. *Rosy-flowered Lupine*
Luzula arcuata. *Curved Wood-Rush*
campestris. *Black-head Grass, Chimney-sweeps, Cuckoo-grass, Field Wood-Rush, Glow-worm Grass*, of Australia

Luzula pilosa. *Hairy Wood-Rush*
spicata. *Spiked Wood-Rush*
sylvatica. *Great Wood-Rush, Wood-grass, Shadow-grass*
Lychnis. *Campion, Lamp-flower*
alpina. *Alpine Campion*
Bungeana. *Bunge's Campion*
Chalcedonica. *Campion or Flower of Constantinople, Common Rose-Campion, Flower of Bristow, Gardener's-Delight or Gardener's-Eye, Jerusalem Cross, None-such, Scarlet Lychnis*
Cœli-rosa. *Rose-of-Heaven, Smooth-leaved Rose-Campion*
diurna (L. dioica). *Adder's-Flower, Devil's-Flower, Hare's-eye, Red Campion*
dioica fl.-pl. *Double Red Campion*
Flos-cuculi. *Crow-flower, Crow-soap, Cuckoo-flower, Marsh-Gilliflower, Meadow Lychnis, Meadow Pink, Ragged-Robin, Wild William*
Flos-Jovis. *Jupiter's - Flower, Umbelled Lychnis*
fulgens. *Brilliant Lychnis*
Githago. *Cockle or Corn-flower, Corn-Pink*
grandiflora (L. coronata). *Large-flowered Lychnis*
Haageana. *Shaggy Lychnis*
Lagascæ. *Rock Lychnis*
Nicæensis. *Italian Rose-Campion*
Preslii. *Presl's Lychnis*
pyrenaica. *Pyrenean Lychnis*
Senno. *Senno Campion*
Sieboldii. *Siebold's Lychnis*
vespertina. *White Campion or Lychnis*
Viscaria. *Clammy Lychnis, German Catchfly, Lime-wort*
Lycium. *Asses' Box-tree, Box-Thorn, Prickly Box*
afrum. *African Tea-tree*
australe. *Murray Box-thorn*
barbarum. *Barbary Box-thorn, Duke of Argyll's Tea-tree, Matrimony-vine*
carolinense. *Carolina Box-thorn*
chinense. *Chinese Box-thorn*
cinereum. *Ash-coloured Box-thorn*
europæum. *European Box-thorn*
fuchsioides. *Fuchsia-like Box-thorn*
horridum. *Caffre Box-thorn, Very-prickly Box-thorn*
lanceolatum. *Lance-leaved Box-thorn*
microphyllum. *Small-leaved Box-thorn*
rigidum. *Stiff Box-thorn*
ruthenicum. *Russian Box-thorn*
Shawii. *Shaw's Box-thorn*
tenue. *Slender Box-thorn*
tetrandrum. *Four-stamened Box-thorn*
Trewianum. *Trew's Box-thorn*
turbinatum. *Top-shaped-fruited Box-thorn*
vulgare. *Matrimony-Vine*
Lycoperdon Bovista. *Blind Ball, Devil's-Snuff-box, Fist-ball, Frog-cheese, Fuss-ball, Puff-ball*
giganteum. *Giant Puff-ball*
Lycopersicum esculentum. *Love - apple, Tomato*
Lycopodium. *Club-moss*
alopecuroides. *Walking-fern*, of N. America
alpinum. *Alpine Club-moss, Cypress-moss, Heath-Cypress, Savin-leaved Club-moss*

Lycopodium clavatum. *Buck-grass, Buck-horn Moss, Common Club-moss, Creeping Fox-tail, Bur, Stag's-horn Moss, Wolf's-claw*
dendroideum. *Ground-Pine*
inundatum. *Marsh Club-moss*
lucidulum. *Moon-fruit Pine, Shining Club-moss*
rupestre. *Festoon Pine*
selaginoides. *Small Alpine Club-moss*
Selago. *Fir Club-moss, Fir-moss, Tree-moss*
Lycopsis arvensis. *Small Bugloss*
vesicaria. *Bladder-seeded Wild Bugloss*
Lycopus europæus. *Gipsy-wort, Water Horehound*
virginicus. *Bugle-weed*, of N. America
Lygeum Spartum. *Cord-grass, Hooded Mat-weed*
Lygodium. *Snake's-tongue Fern*
articulatum. *Twining String-Fern*, of New Zealand
palmatum. *Climbing, or Hartford, Fern*
scandens. *Japanese Climbing Fern*
Lyperanthus nigricans. *Australian Flower-of-Sadness*
Lyperia crocea. *African Saffron*
Lysiloma Sabicù. *Sabicù-, Saracù-, or Sa-vicù-wood*, of Cuba.
Lysimachia. *Loose-strife*
angustifolia. *Narrow-leaved Loose-strife*
barystachya. *Heavy-spiked Loose-strife*
brachystachya. *Short-spiked Loose-strife*
ciliata. *Fringed Loose-strife*
clethroides. *Clethra-leaved Loose-strife*
dahurica. *Siberian Loose-strife*
Ephemerum. *Willow-leaved Loose-strife*
grandiflora. *Large-flowered Loose-strife*
Leschenaultii. *Carmine-flowered Loose-strife*
Linum-stellatum. *Flax-star*
nemorum. *Wood Loose-strife, Yellow Pimpernel*
Nummularia. *Creeping Jenny, Herb-Twopence, Money-wort, Money-wort Loose-strife, Two-penny Grass*
Nummularia var. aurea. *Golden Creeping Jenny, Golden Money-wort*
punctata. *Dotted Loose-strife*
quadrifolia. *Four-leaved Loose-strife*
thyrsiflora. *Tufted Loose-strife*
vulgaris. *Common Yellow Loose-strife*
Lythrum alatum. *Winged Loose-strife*
Græfferi. *Græffer's Loose-strife*
hyssopifolium. *Grass Poly, Hyssop-leaved Loose-strife*
Salicaria. *Common Purple Loose-strife, Willow-weed*
Salicaria var. roseum. *Rosy-flowed Loose-strife*
virgatum. *Slender-branched Purple Loose-strife*

Maba. *Ebony-tree*, of Cochin-China and Coromandel
buxifolia. *E. Indian Satin-wood*
guineensis. *Bahamas Satin-wood* (?)
Macadamia ternifolia. *Queensland Nut-tree*
Machæranthera tanacetifolia (Aster tanacetifolius). *Dagger-flower*
Machærium Schomburgkii. *Itaka-wood, Tiger-wood*

Maclura aurantiaca. *Osage Orange*
  tinctoria. *Dyers' Fustic-wood*
Macrochloa. *Long Grass*
  (Stipa) tenacissima. *Alfa-*, or *Halfa-*, *Grass*,
    Esparto-Grass
Macromerium jamaicense. *W. Indian Thorn*
Macropiper excelsum. *Native Pepper*, of New Zealand
  methysticum. *Ava-*, or *Kava-*, *shrub*, of the South Sea Islands
Macrozamia Fraseri. *Swan River Fern-Palm*
  Perowskiana (M. Denisonii). *Giant Fern-Palm*
Madaria corymbosa and M. elegans. *Mignonette-vine*
Madia. *Tar-weed*, of California
  sativa. *Madia-oil-plant*
Maharanga Emodi. *Rutton-root*, of India
Mahonia. *Ash-leaved Barberry*
  repens. *Edging Barberry*
Maianthemum bifolium (Smilacina bifolia).
  One-leaf, Two-leaved Lily of the Valley
Magnolia. *Laurel-leaved Tulip-tree*
  acuminata. *"Blue" Magnolia, Cucumber-tree*
  Campbelli. *Campbell's Magnolia*
  citriodora. *Lemon-scented Magnolia*
  conspicua (M. Yulan). *Lily-flowered Magnolia*, *Yulan Magnolia*
  cordata. *Yellow Cucumber-tree*
  Fraseri. *Ear-leaved Umbrella-tree, Indian Physic, Long-leaved Cucumber-tree*
  fuscata. *Brown-stalked Magnolia, Chinese Tulip-tree*
  glauca. *Beaver-tree* or *Beaver-wood, Castor-wood, Elk-bark, Small Laurel Magnolia, Swamp Sassafras, White Bay, White Laurel*
  gracilis (M. conspicua). *Slender Magnolia*
  grandiflora. *Great Laurel-leaved Magnolia, Large-flowered Magnolia*
  grandiflora var. lanceolata. *"Exmouth" Magnolia*
  Halleana. *Halle's Magnolia*
  Lenné. *Lenne's Magnolia*
  macrophylla. *Great-leaved Magnolia*
  parviflora. *Small-flowered Magnolia*
  purpurea (M. obovata). *Purple-flowered Magnolia*
  tripetala (M. Umbrella). *Common Umbrella-tree*
  Yulan (M. conspicua). *Yulan-tree*
Malaxis paludosa. *Bog Orchis*
Malcolmia maritima. *Virgin*, or *Virginia, Stock*
Malope grandiflora. *Large-flowered Mallow-wort*
  malacoides. *Barbary Mallow-wort*
Malpighia coriacea. *Locust-berry*
  glabra and M. punicifolia. *Barbadoes*, or *W. Indian, Cherry*
  saccharina. *Sugar Plum*, of Sierra Leone
  urens. *Cow-hage-*, or *Cow-itch-, Cherry*
Malva. *Mallow*
  Alcea. *Hollyhock Mallow, Vervain Mallow*
  campanulata. *Bell-flowered Mallow*
  crispa. *Curled-leaved Mallow*
  lateritia. *Buff-coloured Mallow*

Malva limensis. *Blue-flowered Mallow*
  Morenii. *Moren's Mallow*
  moschata. *Musk Mallow*
  moschata var. alba. *White Musk Mallow*
  rotundifolia. *Dwarf Mallow*
  scoparia. *Birch-leaved Mallow*
  sylvestris. *Cheeses, Marsh Mallow, Maul, Mauls,* or *Maws, Round Dock*
Malvastrum. *False Mallow*
Mammea americana. *Mammee-apple Tree, Wild Apricot*
  humilis. *Spanish Plum*, of the Antilles
Mammillaria. *Nipple-Cactus*
  elephantidens. *Elephant's-tooth Cactus*
Mandevilla suaveolens. *Chili Jasmine*
Mandragora autumnalis. *Autumn-flowering Mandrake*
  officinarum. *Common,* or *Medicinal, Mandrake, Devil's-Apple*
Mangifera indica. *Mango-tree*
Manicaria Plukenetii. *Sea-apple*
  saccifera. *Bussu Palm, Troolie Palm, Wine-Palm*, of Guiana
Manihot Aipi. *Sweet Cassava-plant*
  utilissima. *Bitter Cassava-plant, Tapioca-plant*
Mantisia saltatoria. *Dancing-Girls, Opera-Girls*
Maranta arundinacea. *Arrow-root Plant*
  (Calathea) zebrina. *Zebra-plant*
Marasmius orendes and M. urens. *Fairy-ring Champignon* or *Mushroom*
Marathrium utile. *Water-weed*, of New Granada
Marattia fraxinea. *Ash-leaf Fern*, of New Zealand
  salicina. *Para-Fern*, of New Zealand
Marcgravia umbellata. *W. Indian Ivy*
Marchantia conica. *Conical Liver-wort*
  hemisphærica. *Hemispherical Liver-wort*
  polymorpha. *Brook-,* or *Common, Liver-wort, Liver-grass*
Margyricarpus setosus. *Pearl-berry,* or *Pearl-fruit*
Marica cærulea. *Toad-cup Lily*
  irioides. *Bermuda Iris*
Mariscus cylindricus. *American Hedgehog Club-rush*
Marliera (Rubachia) glomerata. *"Cambuca,"* of Brazil
Marrubium vulgare. *Mawroll, White Hore-hound*
Marsdenia suaveolens. *Fragrant Bower-plant*, of Australia
  tinctoria. *Indigo-plant*, of Pegu
  viridiflora. *Native Potato*, of N. S. Wales
Marsilea macropus (M. hirsuta, M. salvatrix). *Nardoo-plant*
Martynia proboscidea. *Elephant's-trunk, Unicorn-plant*
Mascarenhasia Curnowiana. *Rosy-flowered Jasmine*
Masdevallia Chimæra. *Spectral-flowered Orchid*
Matisia cordata. *Sapoté-,* or *Chupa-Chupa-tree*, of New Granada
Matricaria Chamomilla. *Dog's Chamomile, Dog-,* or *Horse-, Gowan, German Chamomile*

Matricaria inodora. *Dog-Gowan, May-weed*
Matthiola. *Stock*, or *Stock-Gilliflower*
annua. *Ten-week Stock*
annua fl.-pl. *Double Ten-week Stock*
fenestralis. *Cluster-leaved Stock*
græca. *Wall-flower-leaved*, or *Smooth-leaved, Stock*
incana. "*Hopes*," *Queen's Stock*
incana var. coccinea. *Brompton Stock*
odoratissima. *Persian Stock*
sinuata. *Great Sea-Stock*
tristis. *Dark-flowered Stock*
Mauritia Carana. *Carana Palm*
flexuosa. *Ita, Chiqui-Chiqui, Miriti*, or *Morichi, Palm*
vinifera. *Brazilian Wine-Palm, Buriti Palm*
Maximiliana caribæa. *Crown Palm*
regia. *Inija Palm, Jaqua Palm*
Meconopsis aculeata. *Himalayan Blue*, or *Prickly, Poppy*
cambrica. *Welsh Poppy*
nepalensis. *Nepaul Poppy*
simplicifolia. *Sikkim Poppy*
Wallichiana. *Satin Poppy-wort, Wallich's Blue Poppy*
Medeola virginica. *Indian Cucumber-root*
Medicago. *Medick, Snail-Clover*
arborea. *Moon-Trefoil, Tree-Medick*
denticulata. *Denticulate Medick*
Echinus. *Calvary Clover, Crown-of-Thorns, Sea-egg*
falcata. *Sickle-podded Medick*
intertexta. *Hedge-hog Medick*
lupulina. *Black Grass, Black Medick, Black Nonesuch, Hop Clover, Melilot Trefoil, Yellow Clover*, (sometimes sold in Ireland for the "Shamrock" (Trifolium repens)
maculata. *Heart-Clover, Heart-Trefoil, St. Mawe's Clover, Spotted Medick*
minima. *Bur Medick*
muricata. *Thorny Buttons*
sativa. "*Alfalfa*," *Common Medick, Holy Hay, Lucerne, Purple Medick*
scutellata. *Barbary-Buttons, Snails*
Megarrhiza californica (Echinocystis fabacea). *Californian Big-root*, or *Bitter-root*
Melaleuca. *Australian Tea-tree*
genistæfolia. *White Tea-tree*, of Australia
hypericifolia. *Bottle-brush flower*
Leucadendron. *Cajeput-,* or *Cajuput-, tree, White-tree*
minor. *Small Cajeput-,* or *Cajuput-, tree*
squarrosa. *Swamp Tea-tree*, of Australia, *Tasmanian Tea-tree*
uncinata. *Tea-tree*, of N. S. Wales
Wilsoni. *Palm-bark-tree*
Melampyrum. *Cow-wheat*
arvense. *Purple Cow-wheat*
cristatum. *Crested Cow-wheat*
pratense. *Common Cow-wheat*
sylvaticum. *Horse-flower, Small-flowered Cow-wheat*
Melanodendron integrifolium. *Bastard,* or *Black, Cabbage-tree,* of St. Helena
Melanogaster variegatus. *Red Truffle*
Melanorrhœa usitatissima. *Black* or *Martaban Varnish-tree, Lignum-vitæ,* of Pegu, *Varnish-tree,* of Burmah

Melanoselinum decipiens. *Black Parsley*
Melanoxylon Braúna. *Braúna-wood*, of Brazil
Melanthium junceum. *Rush-leaved Black Lily*
uniflorum (Tulipa Breyiana). *Cape Tulip*
virginicum. *American Bunch-flower*
Melastoma malabaricum. *Malabar Laurel*
malabathrica (M. macrocarpum). *Malabar Goose-berry*
Melhania Melanoxylon. *Black-wood,* or *Ebony,* of St. Helena
Erythroxylon. *Red-wood,* of St. Helena
Melia. *Bead-tree*
australis. *White Cedar,* of Australia
Melia Azadirachta. *Margosa-tree, Neem-tree*
Azedarach. *False-Sycamore, Hill Margosa, Holy-tree, Indian Lilac, Pride-of-China, Pride-of-India, Syrian Bead-tree*
guineensis. *Guinea Lilac*
sempervirens. *Hoop-tree*
Melianthus major. *Great Cape Honey-flower*
Melica. *Melick Grass*
nutans. *Mountain Melick*
uniflora. *Wood Melick*
Melicocca bijuga. *Genip-tree,* or *Honey-berry,* of the W. Indies, *Jamaica Bullace-plum*
Melicytus ramiflorus. *Mahoe* or "*Cow-leaf,*" of New Zealand
Melilotus. *Melilot, Sweet Clover*
alba. *Bokhara Clover, Cabul Clover, White Melilot*
arvensis. *Field Melilot*
cœrulea. *Sweet Trefoil*
leucantha. *Bokhara Clover*
officinalis. *Common Melilot, Hart's Clover, King's Clover, Plaster Clover, Wild Laburnum*
parviflora. *Scented Trefoil,* of Australia
segetalis. *Sword-grass*
Melissa (Calamintha) Acinos. *Basil Thyme*
officinalis. *Common Balm, Balm-leaf* or *Baum-leaf, Bawme, Baum*
Melittis Melissophyllum. *Balm Melittis, Bastard Balm, Baum-leaf, Honey-Balm*
Melocactus. *Melon-thistle, Turk's-cap Cactus*
communis. *Common Melon-Cactus, Pope's-head*
Melocanna bambusoides. *Berry-bearing Bamboo*
Memecylon ferreum. *Iron-wood* of Dutch E. Indies
Menispermum canadense. *Moon-creeper, Moon-seed, Vine-Maple, Yellow Perilla*
dauricum. *Daurian Moon-seed*
smilacinum. *Smilax-like Moon-seed*
Mentha. *Mint*
aquatica. *Fish Mint*
arvensis. *Corn Mint, Field Mint*
arvensis var. glabrata. *Chinese Pepper-mint*
arvensis var. piperascens. *Japanese Pepper-mint*
australis. *Australian Pepper-mint*
canadensis. *Wild Mint,* of N. America
cervina. *Hyssop-leaved Mint, Stag Mint*
citrata. *Bergamot Mint*
crispa. *Curled, Crisped,* or *Cross Mint*

and Foreign Plants, Trees, and Shrubs. 219

Mentha hirsuta. *Brook Mint*
piperita. *Brandy-Mint, Common Peppermint*
piperita officinalis. *White Mint*
piperita vulgaris. *Black Mint*
Pulegium. *Flea Mint, Organ, Organs, Organy, Penny-royal, Pudding-grass*
Requieni (Thymus corsicus). *Requien's Penny-royal*
rotundifolia. *Apple-mint, Round-leaved Mint*
rotundifolia variegata. *Variegated Round-leaved Mint*
sativa. *Whorled Water Mint*
sylvestris. *Brook or Water Mint, Horse-Mint*
viridis. *Garden or Spear Mint, Mackerel Mint*
Mentzelia ornata. *Prairie-Lily*
Menyanthes nymphæoides. *Fringed Buckbean*
trifoliata. *Beck-bean, Bog-bean, Bog-Hop, Bog-nut, Bog-Trefoil, Brook-bean, Buck-bean, Marsh Trefoil, Water Trefoil*
Menziesia (Dabeocia) polifolia. *Irish, or St. Dabeoc's, Heath*
Mercurialis annua. *Baron's Mercury, Boy's Mercury, French Mercury, Girl's Mercury*
perennis. *Dog's Mercury, Kentish Balsam*
perennis var. aurea. *Golden Mercury*
Merendera Bulbocodium (Colchicum montanum). *Pyrenean Meadow-Saffron*
Meriandra bengalensis. *Bengal Sage*
Meriania. *Jamaica Rose*
Mertensia. *Smooth Lung-wort*
alpina. *Alpine Lung-wort*
maritima. *Oyster-plant, Sea-Bugloss*
(Pulmonaria) dahurica. *Siberian Lung-wort*
oblongifolia. *Blunt-leaved Lung-wort*
(Pulmonaria) virginica. *Virginian Cowslip*
Merulius lachrymans. *Dry-rot Fungus, of Conifers*
Mesembryanthemum. *Fig-Marigold, "Mid-day-Flower," of Australia*
acinaciforme. *Scimetar-leaved Fig-Marigold*
æquilaterale. *Australian Pig's-faces*
agninum. *Lamb's-chop*
aurantiacum. *Orange Fig-Marigold*
aureum. *Golden Fig-Marigold*
australe. *New Zealand Ice-plant, New Zealand Pig's-faces*
caninum. *Dog-chop*
conspicuum. *Bright-red Fig-Marigold*
cordifolium. *Heart-leaved Fig-Marigold*
crystallinum. *Common Ice-plant*
edule. *Hottentot-Fig*
erminium. *Ermine-chop*
felinum. *Cat-chop*
glabrum. *Dew-plant*
lupinum. *Wolf-chop*
minimum. *Small Dumpling, Smallest Fig-Marigold*
murinum. *Mouse-chop*
mustelinum. *Weasel-chop*
obcordellum. *Greater Dumpling*
sessile album. *Small Ice-plant*
testiculare. *Broad White Fig-Marigold*
tigrinum. *Tiger-chop*
tricolor. *Annual, or Three-coloured, Fig-Marigold*
Tripolium. *Resurrection-plant*

Mesembryanthemum truncatellum. *Small Truncated Fig-Marigold*
vulpinum. *Fox-chop*
Mespilus. *Medlar-tree*
germanica. *Common, or Dutch, Medlar, Minshull Crab*
germanica var. diffusa. *Spreading Medlar*
germanica var. stricta. *Upright Medlar*
germanica var. sylvestris. *Wild Medlar*
grandiflora. *Snowy Medlar*
lobata. *Lobed-leared Medlar*
Smithii. *Smith's Medlar*
Mesua ferrea. *E. Indian Iron-wood, "Nagkushur-," or "Nagkesur-" tree, Nahor-oil-plant*
Methonica superba. *Superb Lily*
Metrosideros florida and M. robusta. *Rata-tree, of New Zealand*
scandens. *Aka-tree, or Lignum-vitæ, of New Zealand*
tomentosa. *Fire-tree, of New Zealand, Pohutu-kawa-tree*
vera. *Iron-tree (true)*
Meum athamanticum. *Badmoney, Baldmoney, Bear-wort, Meu, Spignel*
Michauxia campanuloides. *Michaux's Bell-flower*
Michelia Champaca. *Champaca- or Chumpaca-, tree, of India, Champ-wood, Fragrant Champac, Pand-oil-plant*
Miconia. *W. Indian Currant-bush*
Microcachrys tetragona. *Strawberry-fruited Tasmanian Cypress*
Micromeria Douglasii. *"Yerba Buena," of California*
Microstylis. *Adder's-mouth Orchis*
Mikania Guaco. *"Guaco"-plant, or Snake-poison Antidote, of S. America*
scandens. *Bone-set, Climbing Hemp-weed, Climbing Thorough-wort, German or Parlour Ivy*
Milium effusum. *Millet Grass*
Milletia. *Moulmein Rose-wood*
Mimosa abstergens. *Soap-nut Tree*
pudica. *Sensitive Plant, Humble Plant*
sensitiva. *Sensitive Plant (true)*
Mimulus. *Monkey-flower*
cardinalis. *Cardinal, or Scarlet, Monkey-flower*
cupreus. *Copper-coloured Monkey-flower*
glutinosus. *Orange Monkey-flower*
guttatus. *Spotted Monkey-flower*
Jamesii. *James's Monkey-flower*
luteus. *Yellow Monkey-flower*
maculosus. *Blotched Monkey-flower*
moschatus. *Musk, Musk-plant*
moschatus var. Harrisoni. *Large-flowered Musk-plant*
primuloides. *Primrose Monkey-flower*
repens. *Creeping Monkey-flower*
Tilingi. *Tiling's Monkey-flower*
Mimusops. *Monkey's-face*
elata. *Cow-tree, of Park*
Elengi. *Bukul-tree, W. Indian Medlar*
globosa. *Gutta-Percha Tree, of Guiana*
Sieberi. *Nase-berry*
sp. *Bullet-tree, or Bully-tree, of Guiana*
Mirabilis calfornica. *Californian Four-o'clock-flower*

English Names of Cultivated, Native,

Mirabilis dichotoma. *Common Four-o'clock-flower*
Jalapa. *Common Marvel-of-Peru, False Jalap, Garden Jalap-plant*
longiflora. *Sweet-scented Marvel-of-Peru*
multiflora. *Many-flowered Marvel-of-Peru*
Mitchella repens. *Chequer-berry, Partridge-berry, Deer-berry, One-berry, Squaw-vine, Winter Clover*
Mitella. *Mitre-wort*, or *Bishop's-cap*, of N. America
Mitraria coccinea. *Scarlet Mitre-pod*
Mohria caffrorum (M. thurifraga). *Frank-incense-Fern*
Molinia cœrulea. *Indian Grass, Lavender-Grass, Purple Moor-Grass*
Mollugo verticillata. *Carpet-weed, Indian Chick-weed*
Moluccella. *Molucca Balm*
lævis. *Shell-flower, Smooth Molucca-Balm*
Momordica Balsamina. *Apple-of-Jerusalem, Balsam-Apple, "Marvellous Apples"*
echinata. *Gooseberry Gourd*
Elaterium. *Squirting Cucumber*
Monarda Bradburyana. *Bradbury's Horse-Mint*
didyma. *Bee-Balm, Mountain Mint, Oswego Tea*
fistulosa. *American Wild Bergamot*
paniculata. *Panicled Horse-mint*
punctata. *Dotted Horse-Mint*
Monizia edulis. *Carrot-tree*, of Madeira
Monodora Myristica. *American, Calabash, Jamaica, or Mexican Nutmeg*
Monotoca elliptica. *Beech-tree*, of N. S. Wales
Monotropa Hypopitys. *False Beech-drops, Pine-sap, Yellow Bird's-nest*
uniflora. *American Ice-plant, Corpse-plant, Fit-plant, Indian Pipe, One-flowered Wax-plant, Ova-ova*
Montia fontana. *Blinks, Water Chick-weed*
Moquilea utilis. *Pottery-tree*, of Parà
Mora excelsa. *Mora-tree*, or *Mora-wood*, of Guiana
Moræa papilionacea. *Butterfly Iris*
Sisyrinchium. *Spanish-nut Iris*
Morchella esculenta. *Morel*
Morina longifolia. *Long-leaved Whorl-flower*
persica. *Persian Whorl-flower*
Morinda. *Morinda-bark-tree*
citrifolia. *Awl-tree, E. Indian Mulberry*
tinctoria. *Ach-root, Dyers' Indian Mulberry*
Moringa pterygosperma. *Ben-oil-plant, Horse-radish Tree*
Moronobea coccinea (Symphonia globulifera). *Hog-gum-tree*, of Jamaica
Morus. *Mulberry-tree*
alba. *White-fruited Mulberry*
alba var. Columbassa. *Columba Mulberry*
alba var. italica. *Italian Mulberry*
alba var. macrophylla. *Large-leaved Mulberry*
alba var. membranacea. *Membranous Mulberry*
alba var. Morettiana. *Dandolo's Mulberry*
alba var. multicaulis. *Many-stemmed Mulberry*
alba var. pumila. *Dwarf Mulberry*
alba var. romana. *Roman Mulberry*

Morus alba var. rosea. *Rosy-leaved Mulberry*
alba var. sinensis. *White Chinese Mulberry*
Calcar-galli. *Cock-spur Mulberry, Yellow Wood-vine*
Constantinopolitana. *Constantinople Mulberry*
indica. *E. Indian Mulberry*
mauritiana. *Mauritian Mulberry*
nigra. *Black*, or *Common, Mulberry*
rubra. *American Red Mulberry*
scabra (M. canadensis). *Rough-leaved Mulberry*
tatarica. *Tartarian Mulberry*
Moschoxylon Swartzii. *Musk-wood*, of Jamaica, *Pameroon-bark Tree*
Mouriria. *Silver-wood*
myrtilloides. *Small-leaved Iron-wood*
Mucuna (Dolichos) urens. *Cowitch* or *Cowage, Horse-eye Bean, Ox-eye Bean*
gigantea. *Cowitch*, of N. S. Wales
Muhlenbeckia adpressa. *Native Ivy*, of Australia
Muhlenbergia. *American Drop-seed Grass*
capillaris. *Hair-grass*, of N. America
diffusa. *"Nimble Will" Grass*
Mulgedium. *Bastard* or *False Lettuce*
alpinum and M. macrorrhizum. *Blue Sow-thistle*
Plumieri. *Purple Sow-thistle*
Muntingia Calabura. *Calabur-tree, Silk-wood Tree*
Muraltia Heisteria. *African Furze*
Murraya exotica. *Chinese Box*
sumatrana. *Sumatra Orange*
Murucuja ocellata. *Bull-hoof*
Musa Banksii. *Queensland Banana*
Cavendishii. *Chinese Banana, Dwarf Banana*
coccinea. *Scarlet-bracted Banana*
Ensete. *Abyssinian Banana*
Livingstoniana. *Livingstone's Banana*
paradisaica. *Adam's - Fig, Plantain-tree, "Pisang"*
sapientum. *Common Banana-tree*
sumatrana. *Sumatran Banana*
textilis. *Manilla Hemp-plant*
vittata. *Channelled-leaved Banana*
zebrina. *Striped-leaved Banana*
Muscari. *Grape-Hyacinth*
armeniacum. *Armenian Grape-Hyacinth*
botryoides. *Sky-blue Grape-Hyacinth*
botryoides var. album. *"Pearls of Spain"*
commutatum. *Dark Purple Grape-Hyacinth*
comosum. *Fair-haired Hyacinth, Purse-tassels, Tassel Hyacinth*
comosum, var. monstruosum (Hyacinthus monstruosus). *Feathered Grape-Hyacinth*
Heldreichii. *Greek Grape-Hyacinth*
luteum. *Yellow Grape-Hyacinth*
moschatum. *Musk Grape-Hyacinth*
racemosum. *Common Grape-Hyacinth, Starch Hyacinth*
Musci. A general name for the *Mosses*
Myagrum perenne. *Perennial Gold-of-pleasure*
Myanthus (Catasetum). *Fly-wort*
Mylitta australis. *Native Bread*, of Australia
Mylocaryum ligustrinum. *Buck-wheat Tree*

Myoporum. *Native Australian Blue-berry-tree*
lætum. *Guitar-wood*, of New Zealand
platycarpum. *Australian Sugar-tree*
tenuifolium. *False Sandal-wood*
Myoschilos oblongus. *Chili Senna*
Myosotidium nobile. *Chatham Island Forget-me-not*
Myosotis. *Forget-me-not, Scorpion-grass*
alpestris (M. rupicola). *Alpine Forget-me-not*
arvensis. *Field Forget-me-not*
azorica. *Azorean Forget-me-not*
collina. *Early Hill Forget-me-not*
dissitiflora (M. montana). *Early Forget-me-not*
dissitiflora var. alba. *White-flowered Early Forget-me-not*
lingulata. *Tongue-leaved Forget-me-not*
palustris. *Common Forget-me-not, Mouse-ear, Scorpion-grass, Snake-grass*
repens. *Blue Eye-bright, Creeping Forget-me-not*
rupicola. *Mountain Forget-me-not*
semperflorens. *Long-flowering Forget-me-not*
sylvatica. *Wood Forget-me-not*
verna. *American Forget-me-not*
versicolor. *Colour-changing Forget-me-not*
Myosurus minimus. *Blood-strange, Mouse-tail*
Myrcia acris. *Wild Cinnamon*
Myrica arguta. *Sharp-toothed-leaved Candle-berry-Myrtle*
californica. *Bay-berry*, or *Wax-Myrtle*, of California
cerifera. *Bay-berry, Candle-berry Myrtle, Tallow-shrub, Wax-Myrtle*
Myrica Faya. *Azorean Candle-berry Myrtle*
Gale. *Bog Myrtle, Common Candle-berry Myrtle, Dutch Myrtle, Gale* or *Sweet Gale, Golden Osier, Moor Myrtle, Sweet Willow*
integrifolia. *Sophee-shrub*, of Sylhet
Nagi. *Yang-mae-tree*, of China
Myricaria germanica. *German Tamarisk*
Myriophyllum verticillatum. *Common Water-Milfoil*
Myristica castaneæfolia. *Fijian Nutmeg-tree*
fatua. *Long*, or *Wild, Nutmeg*
insipida. *Queensland Nutmeg-tree*
moschata. *Nutmeg-tree (true)*
Otoba. *Otoba-wax-tree, Santa Fé Nutmeg*, (yields "White" Mace)
sebifera. *Tallow-Nutmeg*, of Cayenne, *Yamadou-oil-plant*
surinamensis. *Dolice-wood*
tomentosa. *Male*, or *Wild, Nutmeg*
Myrospermum pubescens. *Myrrh-seed, Quinquino-*, or *White Balsam-, plant*
Myroxylon (Myrospermum) Peruiferum (M. Pereiræ). *Balsam-of-Peru-plant*
(Myrospermum) Toluiferum. *Balsam-of-Tolu-plant*
Myrrhis odorata. *Anise, British Myrrh, Great Chervil, Sweet Chervil, Sweet Cicely, Sweet Fern*
Myrsine africana. *African Myrtle*
læta. *Black Soft-wood, Bully-tree* or *Bullet-tree*
variabilis. *Smooth Beech*, of Victoria

Myrsiphyllum asparagoides. *Boston Smilax, Smilax Vine, Wreath-Lily*
Myrtus communis. *Common Myrtle*
communis var. bætica. *Orange-leaved Myrtle*
communis var. belgica. *Broad-leaved Dutch Myrtle*
communis var. belgica fl.-pl. *Double-flowered Myrtle*
communis var. italica. *Upright Italian Myrtle*
communis var. leucocarpa. *White-berried Myrtle*
communis var. lusitanica. *Portugal Myrtle*
communis var. lusitanica acuta. *Nutmeg Myrtle*
communis var. maculata. *Spotted-leaved Myrtle*
communis var. mucronata. *Rosemary-leaved, Thyme-leaved*, or *Small-leaved Myrtle*
communis var. romana. *Roman*, or *Broad-leaved, Myrtle*
communis var. tarentina. *Box-leaved Myrtle*
communis variegata. *Variegated-leaved Myrtle*
(Myrcianthes) edulis. *Edible-fruited Myrtle*, of Uruguay
melastomoides. *Moreton Bay Myrtle*
mespiloides. *Medlar-wood*
Nummularia. *Cran-berry Myrtle*
tenuifolia. *Slender-leaved Australian Myrtle*
tomentosa. *Woolly Myrtle*
trinervis. *Three-ribbed Australian Myrtle*
Nabalus albus. *American Lion's-foot, Rattle-snake-root, White Lettuce*
Fraseri. *Gall-of-the-earth, Lion's-foot*
virgatus. *Slender Rattle-snake-root*
Nageia. *Catkin-bearing Laurel*
japonica. *Japan Laurel*
Naias canadensis. *Canadian Naiad* or *Water-Nymph*
flexilis. *Slender Naiad*
Namia vera. *Iron-wood*, of Dutch E. Indies
Nandina domestica. *Sacred Bamboo*, of the Chinese
Napæa dioica. *American Glade-Mallow*
Naravelia zeylanica. *Narawael-shrub*, of Ceylon
Narcissus. *Daffodil*
Ajax. *Ajax Daffodil*
albicans. *"Trompet Marin"*
angustifolius. *Narrow-leaved Daffodil*
bicolor. *Two-coloured Daffodil*
biflorus. *Peerless Primrose* or *Primrose Peerless, Two-flowered Daffodil*
biflorus fl.-pl. *Sweet Nancy*
Bulbocodium. *Hoop-petticoat Daffodil*
calathinus. *Sea-shore Daffodil*
Clusii. *Clusius's Daffodil*
crenulatus. *Bazelman minor Daffodil*
gracilis. *Yellow Rush-leaved Daffodil*
Graellsi. *Graells's Daffodil*
heminalis. *Smaller Curled-cup Daffodil*
Horsefieldi. *Horsefield's Daffodil*
incomparabilis. *Incomparable Daffodil*
interjectus. *Larger Curled-cup Daffodil*
Jonquilla. *Common Jonquil*
juncifolius. *Rush-leaved Daffodil*
lobularis. *Tenby Six-lobed Daffodil*

English Names of Cultivated, Native,

Narcissus maximus. *Golden Daffodil*
minimus. *Pigmy Daffodil*
minor. *Dwarf Daffodil*
monophyllus. *One-leaved Daffodil*
montanus. *Mountain Daffodil*
moschatus. *Musk-scented Daffodil*
obvallaris. *Sibthorp's Daffodil*
odorus. *Campernelle, Large Jonquil*
orientalis. *Eastern Daffodil*
pachybulbos. *Thick-bulbed Daffodil*
pallidulus. *Pale-flowered Daffodil*
papyraceus. *Paper-white Daffodil*
patulus. *White Musk-scented Daffodil*
poeticus. *Poet's Narcissus, Whitsun Lily*
Pseudo-Narcissus. *Affadil, Averill, Bell-rose, But-rose, Chalice-flower, Common Daffodil, Daffudowndilly, Lent Lily, Lent Rose, Yellow Crow-bells*
pumilus. *Dwarfish Daffodil*
pusillus. *Small Jonquil*
pusillus plenus. *Queen Anne's Jonquil*
rupicola. *Cliff Daffodil*
Telamonius. *Telamon's Daffodil*
Tazetta. *French Daffodil, Polyanthus Daffodil*
tenuior. *Slender Daffodil*
tortuosus. *Twisted Daffodil*
Trewianus. *Bazelman major Daffodil*
triandrus. *Cyclamen Daffodil, Rush Daffodil*
viridiflorus. *Green-flowered Daffodil*
Nardosma (Tussilago) fragrans. *Fragrant Colt's-foot, Winter Heliotrope*
palmata. *Sweet Colt's-foot, of N. America*
sagittata. *Arrow-leaved Colt's-foot*
Nardus stricta. *Common Nard, Mat-grass, Small Mat-weed, Wire Bent Grass*
Narthecium americanum. *American Bog Asphodel*
ossifragum. *Common Bog Asphodel, Lancashire Asphodel, Maiden-hair, Yellow Grass*
Narthex Asafœtida. *Asafœtida-plant*
Nasturtium amphibium. *Great Water-cress, Water Radish, Yellow Cress*
lacustre. *American Lake-Cress*
officinale. *Common Water-cress, Water-grass*
palustre. *Annual Water Radish, Marsh Water-cress, Yellow Cress*
sylvestre. *Creeping Water-cress, Water Rocket*
Nauclea cordifolia. *Moddl-wood, of India*
Nectandra cinnamomoides. *Santa Fé Cinnamon*
cymbarum. *Brazilian, or Orinoco, Sassafras*
leucantha. *Shingle-wood, White or Timber Sweet-wood*
Puchury. *Pichurim-Bean-tree, Sassafras-nut-tree*
Rodiæi. *Beeberee-, or Bibiri-tree, Green-heart of Guiana, Sipiri-tree*
sanguinea. *Lowland, Pepper, White, or Yellow Sweet-wood*
Nectaroscordum. *Honey-Garlic*
Negundo californicum. *Californian Box-Elder*
Nelitris ingens. *Cherry-tree, of N. S. Wales, Scarlet-berried Scrub Myrtle*
Nelumbium aspericaule. *Rough - stemmed Nelumbo*
luteum. *Water Chinquapin, Yellow Nelumbo, Yellow Water-Bean*

Nelumbium speciosum. *Chinese Water-lily, Egyptian Bean, Sacred Bean*
Nemopanthes canadensis. *American Mountain Holly*
Nemophila. *Californian Blue-bell, Love-grove of N. America*
Neottia Nidus-avis. *Bird's-nest Orchis, Goose-nest*
Nepeta cœrulea. *Blue-flowered Cat-mint*
Cataria. *Common Cat-mint, Cat-nep, Cat-nip, Nep or Neps*
Glechoma. *Ale-hoof, Blue Runner, Cat's-foot, Devil's Candlesticks, Gill or Gill-go-by-ground, Ground Ivy, Robin-run-in-the-hedge, Tun-hoof*
macrantha. *Large-flowered Cat-mint*
Mussini. *Scalloped-leaved Cat-mint*
Nepetella. *Small Cat-mint*
Nepenthes. *E. Indian Pitcher-plant, Monkey-cup*
albo-marginata. *White-margined Singapore Pitcher-plant*
ampullacea. *Bottle-like Pitcher-plant*
bicalcarata. *Two-spurred Pitcher-plant*
distillatoria. *Chinese Pitcher-plant*
Hookeriana. *Dr. Hooker's Pitcher-plant*
intermedia. *Intermediate Pitcher-plant*
lævis. *Smooth Pitcher-plant*
Lawrenciana. *Lawrence's Pitcher-plant*
Lindleyana. *Dr. Lindley's Pitcher-plant*
Loddigesii. *Loddige's Pitcher-plant*
Morganiana. *Mrs. Morgan's Pitcher-plant*
phyllamphora. *Ventricose Pitcher-plant*
Rafflesiana. *Yellow Singapore Pitcher-plant*
Rajah. *Rajah Pitcher-plant*
sanguinea. *Blood-red Pitcher-plant*
superba. *Brilliant-red Pitcher-plant*
villosa. *Hairy Pitcher-plant*
Nephelium lappaceum. *Rambutan, or Rampostan-tree*
Litchi. *Litchi-, or Lee-chee-tree*
Longanum. *Dragon's-eye, Longan-tree*
pinnatum. *Dawa-tree*
Nephrodium (Lastrea). *Boss Fern*
cristatum. *Darsham Fern*
Filix-mas. *Basket-fern, Male Fern*
Oreopteris. *Mountain Fern*
Nephrolepis cordifolia. *Ladder Fern, of New Zealand*
Nereocystis Lutkeana. *Sea-otter's Cabbage*
Nerine Fothergilli. *Scarlet Guernsey Lily*
sarniensis. *Common Guernsey Lily, Narcissus of Japan*
Nerium. *Oleander*
odorum. *E. Indian Oleander, Sweet-scented Oleander*
Oleander. *Common Oleander, Rose-Bay, "South Sea Rose," of Jamaica*
tinctorium. *Dyers' Oleander*
Nertera depressa. *Coral-berried Duck-weed, Fruiting Duck-weed*
Nesæa (Decodon) verticillata. *Swamp Loose-strife*
Nesodaphne Tarairi. *Taraire-tree, of New Zealand*
Tawa. *Tawa-tree, of New Zealand*
Neurachne Mitchelliana. *Mulga-grass, of Australia.*
Neurolœna. *Halbert-weed*

Neurolæna lobata. *Golden-rod*, of the W. Indies
Nicandra physaloides. *Apple-of-Peru*
Nicotiana. *Tobacco-plant*
acutifolia. *Acute-leaved Tobacco-Plant*
affinis. *Tuberose-flowered Tobacco-plant*
fruticosa. *Shrubby Tobacco-plant*
glauca. *Glaucous-leaved* (perennial) *Tobacco-plant*
longiflora. *Long-flowered Tobacco-plant*
macrophylla gigantea. *Giant Tobacco-plant*
noctiflora. *Night-flowering Tobacco-plant*
persica. *Persian* or *Shiraz Tobacco-plant*
repanda. *Havannah Tobacco-plant*
rustica. *Latakia Tobacco-plant, Syrian Tobacco-plant, Wild Tobacco-plant*
suaveolens. *Native Australian Tobacco-plant*
Tabacum. *Virginian Tobacco-plant*
tubiflora. *Tube-flowered Tobacco-plant*
vincæflora. *Vinca-flowered* (perennial) *Tobacco-plant*
Nidularia campanulata. *Corn-bells*
Niemeyera (Lucuma) prunifera. *"Cainito,"* of Australia
Nierembergia frutescens. *Tall Cup-flower*
rivularis. *Trailing White Cup-flower*
Nigella. *Fennel-flower, "Gith"*
arvensis. *Wild Fennel-flower*
damascena. *Devil-in-a-bush, Jack-in-prison, Katherine's,* or *St. Katharine's, Flower, Lady-in-the-Bower, Love-in-a-mist, Love-in-a-puzzle*
hispanica. *Spanish Fennel-flower*
sativa. *Black Cumin, Common Fennel-flower*
Nigritella coccinea. *Dark Scarlet-flowered Orchis*
suaveolens. *Fragrant Dark-flowered Orchis*
Nipa fruticans. *Nipa-tree,* of the Philippine Islands
Niphobolus heteractis. *Climbing Polypody*
Nirbisia Bisma. *Bikh-,* or *Bisk-, Poison,* of Nepaul
Nitraria. *Nitre-bush*
tridentata. Supposed to be the *"Lotus-tree"* of the Ancients
Nolana. *Chilian Bell-flower*
Nonatella officinalis. *Asthma-plant,* of Cayenne
Nopalea (Opuntia) coccinellifera. *Cochineal-Cactus, Nopal-plant* of Mexico
Notelæa ligustrina. *Bastard Olive,* of Victoria, *Iron-wood,* of Tasmania and N. S. Wales
longifolia. *Iron-wood,* of Norfolk Island
ovata. *Dunga-runga Tree,* of N. S. Wales
Nothochlæna. *Cloak-fern*
distans. *Woolly Cloak-fern,* of New Zealand
Marantæ. *Hardy Cloak-fern*
Notobasis syriaca. *Syrian Thistle*
Notospartium Carmichæliæ. *Pink Broom,* of New Zealand
Nostoc commune. *Fairies'-butter, Fallen Stars, Star-jelly, Will-o'-the-wisp*
Nuphar advena. *Three-coloured,* or *Stripe-flowered, Water-Lily*
advena var. lutea. *American Yellow Pond-Lily*
Kalmiana. *Small Yellow American Water-Lily*

Nuphar lutea. *Bobbins, Brandy-bottle, Cambie-leaf, Can-Dock, Clot, Clote, Yellow Water-Lily*
multisepala. *Edible-fruited Water-Lily*
pumila. *Small Yellow Water-Lily*
Nuttallia cerasiformis. *Oso-berry-tree,* of California
Nuytsia floribunda. *Flame-tree* or *Fire-tree,* of S. W. Australia
Nyctanthes Arbor-tristis. *"Hursingar,"* of the E. Indies, *Indian Mourner, Night-Jasmine, Night-scented Tree-of-Sadness, Sad-tree, Sorrowful-tree*
Nymphæa alba. *Bobbins, Cambie-leaf, Can-Dock, Flatter Dock, White Water-Lily*
alba var. rosea. *Hardy Red Water-lily*
Amazonica. *Amazon Water-Lily*
cyanea. *Blue Indian Water-Lily*
dentata. *Toothed-leaved Water-lily*
flava. *Florida Yellow Water-lily*
gigantea. *Australian Water-lily*
Lotus. *Egyptian Lotus, Egyptian Water-lily*
odorata. *American Sweet-scented Water-Lily*
pubescens. *E. Indian "Lotus"*
(Castalia) pygmæa. *Chinese Water-lily, Pigmy Water-lily*
rosea. *E. Indian Rose-coloured Water-lily*
rubra. *E. Indian Red Water-lily*
scutifolia. *Shield-Leaved Water-lily*
sphærocarpa. *Round-fruited Water-lily*
stellata zanzibarensis. *Zanzibar Water-lily*
thermalis. *Hungarian Lotus, Hungarian Water-lily*
tuberosa. *Tuberous-rooted White Water-lily*
Nyssa. *Tupelo-tree*
aquatica. *Water Tupelo-tree*
biflora. *Mountain Tupelo-tree*
candicans (N. capitata), *Ogeechee Lime*
multiflora (N. sylvatica). *Black* or *Sour Gum-tree, Common Tupelo-tree, Pepperidge*
uniflora. *Cotton Gum-tree, Large Tupelo*
Ochna Mauritiana. *Jasmine-wood,* of the Isle of France
Ochroma Lagopus. *W. Indian Cork-wood* or *Down-tree*
Ochrosia Borbonica. *Isle of Bourbon Yellow-wood*
Ochroxylon punctatum. *Brazilian Yellow-wood*
Ocotea (Nectandra) cymbarum. *Orinoco Anise-,* or *Sassafras-, tree*
Ocymum Basilicum. *Sweet Basil*
gratissimum. *E. Indian Basil*
minimum. *Bush-Basil*
sanctum. *Holy Basil, Monks' Basil, Purple-stalked Basil*
viride. *Fever-plant,* of Sierra Leone
Odina Wodier. *Goompany-tree,* or *Wodier-tree,* of India
Odontoglossum madrense. *Almond-scented Orchid*
Warneri. *Violet-scented Orchid*
Œdipodium. *Club-stalked Moss*
Œnanthe crocata. *Belder-Root, Ben-Dock, Dead Tongue, Hemlock-Dropwort, Water Hemlock*
fistulosa. *Hemlock-Dropwort, Water-Dropwort*

Œnanthe Lachenalii. *Marsh-parsley*
Phellandrium. *Edge-weed, Horse-bane, Water-Fennel*
pimpinelloides. *Meadow-parsley*
Œnocarpus Batava. *Bacaba-, or Patava-, Palm-tree, Wine-Palm,* of Guiana
Œnothera. *Evening-Primrose, Tree-Primrose*
biennis. *Common Evening-Primrose, Large Rampion*
bistorta Veitchii. *Orange-flowered Evening-Primrose*
cæspitosa. *Tufted Evening-Primrose*
Fraseri. *Fraser's Evening-Primrose*
fruticosa. *Sun-drops*
glauca. *Glaucous Evening-Primrose*
Jamesii. *James's Evening-Primrose*
Lamarckiana. *Tall Large-flowered Evening-Primrose*
linearis. *Narrow-leaved Evening-Primrose*
marginata. *Dwarf Large-flowered Evening-Primrose*
missouriensis (Œ. macrocarpa). *Missouri Evening-Primrose*
montana. *Mountain Evening-Primrose*
pumila. *Dwarf Evening-Primrose*
riparia. *Swamp Evening-Primrose*
serrulata. *Small-toothed-leaved Evening-Primrose*
sinuata. *Scalloped-leaved Evening-Primrose*
speciosa. *Tall White Evening-Primrose*
taraxacifolia. *Dandelion-leaved Evening-Primrose*
trichocalyx. *Hairy-calyxed Evening-Primrose*
Youngi. *Young's Evening-Primrose*
Oïdium Tuckeri. *Vine-Mildew Fungus*
Olax zeylanica. *Malla-tree,* of Ceylon
Oldenlandia (Hedyotis) umbellata. *Chay-root,* or *Che-root,* of India, *E. Indian Madder*
Oldfieldia africana. *African Oak, African Teak*
Olea. *Olive-tree*
apetala. *Botany Bay Olive, Iron-wood* of Norfolk Island
capensis. *Iron-wood,* of S. Africa
Cunninghamii. *Black Maire,* of New Zealand
dioica. *Wild Olive,* of India
europæa. *European Olive-tree*
(Osmanthus) fragrans. *Sweet-scented Olive*
ilicifolia (Osmanthus ilicifolius). *Holly-leaved Olive*
Oleaster. *Wild Olive-tree*
paniculata. *Queensland Olive-tree*
sativa (O. europæa). *Olive-tree* (cultivated)
undulata. *Iron-wood,* of S. Africa
Oleandra neriiformis. *Oleander-Fern*
Olearia dentata. *N. S. Wales Star-wort*
Haastii and other species. *"Daisy-bush,"* of New Zealand
ilicifolia. *New Zealand Holly*
stellulata (Eurybia Gunniana). *Victorian Snow-bush*
Olivia cyanosa. *Hardpeer,* of S. Africa
Ombrophytum. *Mountain Maize,* of Peru
Omphalea diandra. *Ouabe-oil-plant*
triandra. *Cob-nut,* of Jamaica, *Nut-tree* of the Antilles
Omphalobium africanum. *Guinea Zebra-wood*
Lambertii. *Zebra-wood,* of Guiana

Omphalodes. *Venus's Navel-wort*
linifolia. *Flax-leaved Navel-wort*
Luciliæ. *Rock Forget-me-not, Rock Navel-wort*
verna. *Creeping Forget-me-not*
Oncidium carthaginense. *Spread-eagle Orchid*
Papilio. *Butterfly-plant,* of S. America
Sprucei. *Armadillo's-tail Orchid*
tigrinum. *"Flower-of-the-Dead,"* of the Mexicans
Oncosperma filamentosa. *Nibung Palm*
Onobrychis. *Hen's-bill, Medick-Vetch*
Caput-galli. *Cock's-head*
montana. *Mountain Saintfoin*
sativa. *Cock's-head, Common Saintfoin, Fodder-grass, French Grass*
Onoclea sensibilis. *Sensitive Fern*
Onocarpus vitiensis. *Itch-wood-tree*
Ononis. *Rest-harrow*
alba. *White-flowered Rest-harrow*
angustissima. *Very-narrow-leaved Rest-harrow*
antiquorum. *Tall Rest-harrow*
Apula. *Apulian Rest-harrow*
arborescens. *Tree Rest-harrow*
arenaria. *Sand Rest-harrow*
arragonensis. *Arragon Rest-harrow*
arvensis. *Cammock* or *Cammick, Common Rest-harrow, Fin, Ground Furze, Hen Gorse, Land Whin, Petty Whin, Stay-plough, Wild Liquorice*
brachycarpa. *Short-podded Rest-harrow*
biflora. *Two-flowered Rest-harrow*
breviflora. *Short-flowered Rest-harrow*
capensis. *Cape Rest-harrow*
capitata. *Round-headed Rest-harrow*
cenisia. *Mt. Cenis Rest-harrow*
cuspidata. *Pointed-leaved Rest-harrow*
Denhardtii. *Denhardt's Rest-harrow*
diffusa. *Spreading Rest-harrow*
emarginata. *Notched-leaved Rest-harrow*
falcata. *Sickle-podded Rest-harrow*
foetida. *Strong-scented Rest-harrow*
fruticosa. *Shrubby Rest-harrow*
geminiflora. *Twin-flowered Rest-harrow*
glabra. *Smooth Rest-harrow*
hispanica. *Spanish Rest-harrow*
hispida. *Bristly Rest-harrow*
longifolia. *Long-leaved Rest-harrow*
minutissima. *Pigmy Rest-harrow*
Natrix. *Goat-root, Ram Rest-harrow, Yellow-flowered Shrubby Rest-harrow*
oligophylla. *Few-leaved Rest-harrow*
peduncularis. *Long-flower-stalked Rest-harrow*
pendula. *Drooping Rest-harrow*
picta. *Painted-flowered Rest-harrow*
procurrens. *Rooting-branched Rest-harrow*
ramosissima. *Very-branching Rest-harrow*
reclinata. *Small Rest-harrow*
rotundifolia. *Round-leaved Rest-harrow*
spinosa. *Cammock, Thorny Rest-harrow*
tribracteata. *Three-bracted Rest-harrow*
tridentata. *Three-toothed Rest-harrow*
viscosa. *Clammy Rest-harrow*
Onopordon Acanthium. *"Argentine," Cotton-thistle, Down Thistle, Scotch Thistle* (of gardeners), *"Thistle-upon-thistle"*
acaule. *Stemless Cotton-thistle*

Onosma Emodi. *Maharanga-dye-plant*, of India, *Rutton-root*
tauricum. *Golden-Drop*
Onosmodium. *False-Gromwell*
Opegrapha. *Chink-wort, Letter-Lichen, Scripture-wort*
Ophelia Chirata (Gentiana Chirayita). *Chirayta* or *Chiretta-plant*
Ophiocaryon paradoxum. *Snake-nut Tree*, of Guiana.
Ophioglossum lusitanicum. *Small Adder's-tongue Fern*
vulgatum. *Adder's-Spear, Common Adder's-tongue Fern*
Ophiopogon. *Snake's-beard*
Jaburan aureo-variegatus. *Golden-variegated Snake's-beard*
japonicus. *Japanese Snake's-beard*
spicatus. *Spiked Snake's-beard*
spicatus var. argenteo-marginatus. *Silver-variegated Snake's-beard*
Ophiorrhiza Mungos. *Earth-gall*, of the Malays, E. *Indian Snake-root*
Ophioxylon serpentinum. *E. Indian Serpent-wood*
Ophiurus. *Sea Hard Grass*
Ophrys apifera. *Bee-flower, Bee-Orchis*
arachnites. *Black-Spider Orchis, Late Spider-Orchis*
arachnoides. *Spider-like Orchis*
aranifera. *Common Spider-Orchis, Early Spider-Orchis*
aranifera var. limbata. *Bordered Spider-Orchis*
bombylifera. *Bumble-bee-Orchis*
cornuta. *Horned Orchis*
Ferrum-equinum. *Horse-shoe-Orchis*
fucifera. *Drone-Orchis*
fusca. *Brown Orchis*
Lœselii. *Lœsel's Orchis*
lutea. *Yellow Orchis*
muscifera. *Fly Orchis*
Scolopax. *Woodcock-Orchis*
Speculum. *Looking-glass-Orchis*
tabanifera. *Breeze-fly-Orchis*
tenthredinifera. *Saw-fly-Orchis*
tenthredinifera minor. *Small Saw-fly-Orchis*
vespifera. *Wasp-Orchis*
Opopanax Chironium (Pastinaca Opopanax). *Opopanax-plant*
Opuntia. *Bastard Fig, Indian Fig, Prickly Pear*
cochinillifera. *Cochineal Cactus*
crassa. *Thick-lobed Indian Fig*
curassavica. *Pin-pillow Cactus*
humilis. *Hardy Dwarf Cactus, Hardy Prickly Pear*
missouriensis. *Hardy Prickly Pear*
missouriensis var. leucospina. *White-spined Hardy Cactus*
pubescens. *Downy Indian Fig*
Rafinesquii. *Hardy Dwarf Prickly Pear*
Tuna. *Cochineal-Cactus*
vulgaris. *Barbary Fig, Common Hardy Cactus, Indian Fig, Prickly Pear*
Orbignya phalerata. *Cusi-*, or *Cusich-*, *Palm-tree*
Orchis coriophora. *Bug-Orchis*
foliosa. *Madeira Orchis*

Orchis fusca. *Tawny Orchis*
hircina. *Lizard Orchis*
latifolia. *Cuin-and-Abel, Marsh Orchis, Meadow Rocket, Mount Caper*
latifolia var. Lagotis. *Purple-spotted Broad-leaved Orchis*
laxiflora. *Guernsey Orchis, Loose-flowered Orchis*
longibracteata. *Long-bracted Orchis*
longicalcarata. *Long-spurred Algerian Orchis*
maculata. *Adder's Grass, Bullock Grass, Bloody-Man's-Finger, Crake-feet, Cuckoo-flower, Cuckoo Orchis, Dead-Men's-Fingers, Gethsemane, Hand Orchis, "Long-purples"* of Shakespeare, *May Orchis, Purple Orchis, Purple Wreathe-wort, Ram's-horns, Spotted Orchis*
militaris. *Military Orchis*
Morio. *Bleeding Willow, Crake-feet, Fool Orchis, "Gandergosses," Goose-and-goslings, Green-winged Orchis, "Nuns," Salep Orchis*
nigra (Nigritella angustifolia). *Dark-flowered Orchis*
pallens. *Pale-flowered Orchis*
papilionacea. *Purple Butterfly-Orchis*
purpurea. *Lady-Orchis*
pyramidalis. *Pyramidal Orchis*
Robertiana. *Long-bracted Orchis*
sambucina. *Elder-scented Orchis*
Simia. *Ape-Orchis*
spectabilis. *Preacher-in-the-pulpit, Showy Orchis* of N. America
tephrosanthos. *Ash-coloured Orchis, Monkey Orchis*
ustulata. *Dwarf Meadow Orchis*
Oreodaphne. *Mountain Laurel*
bullata. *Stink-wood, Mountain Laurel*
californica. *Balm-of-Heaven, Californian Cajeput-, Laurel-, Olive-, Sassafras-,* or *Spice, - tree*
cupularis. *Isle of France Cinnamon*
(Nectandra) exaltata. *Sweet-wood,* of Jamaica
(Laurus) fœtens. *Fetid Laurel,* or *Til-tree,* of Madeira
Leucoxylon. *Loblolly-,* or *Rio Grande, Sweet-wood*
Oreodoxa (Areca) oleracea. *Cabbage-Palm,* of the W. Indies
regia. *Royal Wine-Palm*
Origanum. *Marjoram*
Dictamnus. *Dittany of Crete*
heracleoticum. *Winter Marjoram, Winter-sweet*
Marjorana. *Knotted* or *Sweet Marjoram*
Onites. *Pot Marjoram*
Sipyleum (O. pulchellum). *Little Hop-plant, Mt. Sipylos Marjoram*
Tournefortii. *Dittany of Amorgos*
vulgare. *English, Grove,* or *Wild Marjoram, Organ, Organs,* or *Organy*
Ormosia dasycarpa. *Necklace-tree,* of the W. Indies
Ornithocephalus gladiatus. *Bird's-head Orchis*
Ornithogalum. *Star-of-Bethlehem*
aureum. *Golden-flowered Star-of-Bethlehem*

Q

Ornithogalum Boucheanum. *Bouche's Star-of-Bethlehem*
comosum. *Short-spiked Star-of-Bethlehem*
latifolium. *Broad-leaved Star-of-Bethlehem*
narbonnense. *Narbonne Star-of-Bethlehem*
nutans. *Drooping Star-of-Bethlehem*
pannonicum. *Star-of-Hungary*
prasinum. *Green Star-flower*
pyramidale. *Tall Star-of-Bethlehem*
pyrenaicum. *Bath or French Asparagus, Pyrenean Star-of-Bethlehem*
scilloides. *Barbadoes Onion*
umbellatum. *Common Star-of-Bethlehem, Eleven-o'clock Lady, Jack-go-to-bed-at-noon*
Ornithoglossum. *Birds'-tongue*
Ornithopus. *Bird's-foot*
compressus. *Hairy Bird's-foot*
ebracteatus (Arthrolobium ebracteatum). *Channel Islands Bird's-foot*
perpusillus. *Common Bird's-foot*
sativus. *"Serradella"* or *"Serratella"*
Orobanche. *Broom-rape*
cœrulea. *Blue-flowered Broom-rape*
caryophyllacea. *Clove-scented Broom-rape*
elatior. *Tall Broom-rape*
major. *Great Broom-rape, New-Chapel-flower*
minor. *Small Broom-rape*
ramosa. *Branched Broom-rape*
rubra. *Red Broom-rape*
virginica. *Virginian Beech-drops*
Orobus. *Bitter-vetch*
atropurpureus. *Dark Purple Bitter-vetch*
aurantius. *Orange-flowered Bitter-vetch*
canescens. *Hoary Bitter-vetch*
(Platystylis) cyaneus. *Blue Bitter-vetch*
Gmelini. *Gmelin's Bitter-vetch*
hirsutus. *Hairy Bitter-vetch*
lathyroides. *Lathyrus-like Bitter-vetch*
niger. *Black Pea, Black-rooted Bitter-vetch*
sylvaticus. *Wood-pea*
tauricus. *Crimean Bitter-vetch*
tuberosus. *Heath-pea, Tuberous-rooted Wood-pea*
variegatus. *Variegated Bitter-vetch*
vernus. *Spring Bitter-Vetch*
vicioides fl.-pl. *Double-blossomed Bitter-Vetch*
Orontium aquaticum. *Golden-club*
Orthrosanthus multiflorus. *Morning-flower*, of Australia
Orthotrichum. *Bristle-Moss*
diaphanum. *White-tipped Bristle-Moss*
Lugwigii. *Club-fruited Bristle-Moss*
pulchellum. *Elegant Bristle-Moss*
rivulare. *River Bristle-Moss*
rupincola. *Rock Bristle-Moss*
speciosum. *Showy Bristle-Moss*
stramineum. *Straw-like Bristle-Moss*
Oryza. *Rice-plant*
mutica. *Mountain-Rice*
sativa. *Cultivated Rice*
Oryzopsis. *American Mountain-Rice*
Osmanthus americanus (Olea americana). *Devil-wood*
fragrans. *Pragrant Olive*
Osmorrhiza. *Sweet Cicely*, of California
Osmunda. *Flowering Fern*
cinnamomea. *Cinnamon Fern*

Osmunda Claytoniana. *Clayton's Flowering Fern*
regalis. *Bog Onion, Buck-horn Brake, Common Flowering Fern, Ditch Fern, Herb Christopher, King Fern, Osmund Royal, Osmund the Water-man, Royal Fern, Royal Osmund Fern, Water-Fern*
Osteospermum. *Bone-seed*
moniliferum. *Jungle Sun-flower*, of the Cape
Ostrya virginica. *Iron-wood (N. American), Hop-Hornbeam, Lever-wood*
Osyris alba (Cassia poetica). *Gard-robe, Poet's Cassia, Poet's Rosemary*
Othonna. *African Rag-wort*
cheirifolia. *Barbary Rag-wort*
Ouvirandra Berneriana. *Pink-flowered Lace-leaf-*, or *Lattice-leaf-, Plant*
fenestralis. *White-flowered Lace-leaf-*, or *Lattice-leaf-, Plant*
Owenia cerasifera. *Sweet Plum*, of Queensland
venosa. *Sour Plum*, of Queensland
Oxalis. *Wood-Sorrel*
Acetosella. *Alleluia, Common Wood-Sorrel, Cuckoo-bread, Cuckoo-Sorrel, French Sorrel, Gowk-meat, Hallelujah, Stub-wort, "Wood-Sower"*
arenaria. *Sand Wood-Sorrel*
atro-purpurea. *Purple-leaved Wood-Sorrel*
Biophytum. *E. Indian Sensitive Wood-Sorrel*
Bowiei. *Bowie's Wood-Sorrel*
brasiliensis. *Brazilian Wood-Sorrel*
caprina. *Goat's-foot*
corniculata. *Procumbent Yellow-flowered Wood-Sorrel*
corniculata var. rubra. *Red-flowered Procumbent Wood-Sorrel*
crenata. *"Oca,"* of Peru
Deppei. *Deppe's Wood-Sorrel*
enneaphylla. *Nine-leaved Wood-Sorrel*
floribunda. *Many-flowered Wood-Sorrel*
incarnata. *Flesh-coloured Wood-Sorrel*
lasiandra. *Woolly-stamened Wood-Sorrel*
lobata (O. granulata). *Lobed-leaved Wood-Sorrel*
luteola. *Yellowish-flowered Wood-Sorrel*
magellanica. *New Zealand Wood-Sorrel*
odorata. *Scented Wood-Sorrel*
Oregana. *Oregon Wood-Sorrel*
Plumieri. *Plumier's Wood-Sorrel*
rosea. *Rosy-flowered Wood-Sorrel*
stricta. *Upright Yellow-flowered Wood-Sorrel*
tenella. *Dwarf Wood-Sorrel*
tropæoloides. *Nasturtium Wood-Sorrel*
tuberosa. *"Oca,"* of Peru, *Tuberous-rooted Wood-Sorrel*
Valdiviana. *Chilian Wood-Sorrel*
violacea. *Violet Wood-Sorrel*
Oxleya xanthoxyla (Flindersia Oxleyana). *Yellow-wood*, of Queensland and N. S. Wales
Oxybaphus. *Umbrella-wort*
Oxycoccus palustris. *Moss-berry*
Oxyria reniformis. *Mountain Sorrel*
Oxytropis campestris. *Yellow-flowered Oxytrope*
fœtida. *Fetid Oxytrope*
pyrenaica. *Pyrenean Oxytrope*

Oxytropis splendens. *Brilliant - flowered Oxytrope*
uralensis. *Purple-flowered Oxytrope*
Pachira (Carolinea) alba. *Silk-cotton-tree, of New Granada*
macrantha. *Silk-cotton-tree, of Brazil*
Pachyma Cocos. *Tuckahoe Truffle*
Pachyphytum bracteosum. *Silver-bracts*
Pachyrrhizus angulatus. *Yaka, or Wayaka-plant, of Fiji*
Pachysandra procumbens. *Alleghany Mountain Spurge, American Thick-stamen*
Pachystigma Canbyi. *Canby's Mountain-lover*
Padina pavonia. *Peacock's-tail Seaweed, Turkey-feather Laver*
Pæderia fœtida. *Chinese Fever-plant*
Pæonia. *Pæony, Peony, or Piony*
albiflora. *White-flowered Pæony*
anemonæflora. *Anemone-flowered Pæony*
arietina. *Ram's-horn Pæony*
arborea (P. Moutan). *Tree-Pæony*
corallina. *"English," Steep Holmes, or Wild Pæony*
edulis. *Edible-rooted Pæony*
humilis. *Dwarf Pæony*
lobata. *Lobed-leaved Pæony*
mollis. *Soft-leaved Pæony*
Moutan (P. arborea). *Chinese Tree-Pæony*
Moutan var. Banksii. *Banks's Tree-Pæony*
Moutan var. papaveracea. *Poppy-flowered Tree-Pæony*
Moutan var. rosea. *Rosy-flowered Tree-Pæony*
odorata. *Scented Pæony*
odorata var. grandiflora. *Large-flowered Scented Pæony*
officinalis. *Common Garden Pæony*
sibirica. *Siberian Pæony*
tenuifolia. *Fennel-leaved Pæony, Slender-leaved Pæony*
Palicourea speciosa. *Gold-shrub*
Paliurus aculeatus. *Christ's Thorn, Garland-Thorn, Ram-of-Libya*
Palmella cruenta. *Gory Dew*
Panax Colensoi. *Otago Ivy-tree*
crassifolium (Aralia crassifolia). *New Zealand Lance-wood*
dendroides. *Mountain-Ash, of Australia*
quinquefolium (Aralia quinquefolia). *Ginseng-plant of N. America*
sambucifolium (P. dendroides). *Victorian Elder-berry Ash*
(Aralia) trifolium. *Dwarf Ginseng, Ground-*
Pancratium indicum. *Indian Autumn Daffodil*
maritimum. *Mediterranean Lily, Sea-Daffodil*
Pandanus. *Screw-Pine*
Candelabrum. *Chandelier-tree, of Guiana*
Forsteri. *Tent-tree, of Lord Howe's Island*
javanicus. *Javanese Screw-Pine*
odoratissimus. *Fragrant Screw-Pine, Keora-, or Pandang-, oil-plant, Moreton Bay Bread-fruit-tree, Nicobar Bread-fruit-tree*
pedunculatus. *Screw-Pine,* of E. Australia

Panicum. *Panick-Grass*
altissimum. *Tall Panick-grass*
bulbosum. *Bulbous Panick-grass*
cæruleum. *"Bajree" Millet*
capillare. *Old-witch Grass*
colonum. *Millet Rice*
Crus-galli. *Barn-yard,* or *Cock's-shin, Grass*
decompositum. *Australian Millet, Umbrella-Grass*
frumentaceum. *Deccan-grass, Shamalo-grass*
glutinosum. *W. Indian Burr-grass, Ginger-grass*
jumentorum (P. maximum). *Guinea-grass*
leucophæum. *Sour-grass*
miliaceum. *Millet (true)*
molle. *Dutch-grass, Parn-grass, "Scotch-grass,"* of Jamaica
stagninum. *Hedge-hog Grass*
virgatum. *Twiggy Panick-grass*
Papaver. *Poppy*
alpinum. *Alpine Poppy*
Argemone. *Pale Poppy, Wind-Rose*
bracteatum. *Great Scarlet Poppy*
croceum. *Golden Poppy*
dubium. *Blind Eyes, Long-headed Poppy, Smooth-fruited Corn-Poppy*
hybridum. *Rough Poppy*
lateritium. *Orange Poppy*
nudicaule. *Yellow Arctic Poppy*
orientale. *Oriental Poppy.*
pilosum. *Hairy-stemmed Poppy*
Rhœas. *Blind Eyes, Canker* or *Canker-rose, Cock-rose, Cop-rose* or *Copper-rose, Corn-flower, Corn Poppy, Corn Rose, Head-ache, Red-weed*
Rhœas, fl.-pl. *John,* or *Joan, Silver-pin*
somniferum. *Bale-wort, Carnation, Pæony,* or *Opium Poppy, Joan Silver-pin, White Poppy*
somniferum, fl.-pl. *Double Opium Poppy*
somniferum, var. nigrum. *Black Opium Poppy*
spicatum. *Spiked Poppy*
umbrosum. *Caucasian Scarlet Poppy, Dark-spotted,* or *Thompson's, Poppy*
Pappea capensis. *Wild Plum-tree,* of the Cape
Papyrus antiquorum (Cyperus Papyrus). *Paper-reed,* or *Paper-rush,* of the Ancients
corymbosus. *Indian-matting Plant*
odoratus. *Fragrant Paper-reed*
syriacus. *Syrian,* or *Silician, Paper-reed*
Parastranthus. *Inverted-flower*
Pardanthus (Ixia) chinensis. *Chinese Black-berry Lily, Chinese Leopard-flower*
Parietaria debilis. *New Zealand Pellitory*
officinalis. *Hammer-wort, Lich-wort, Pellitory-of-the-Wall*
pennsylvanica. *American Pellitory*
Parinarium excelsum. *Gray Plum,* or *Rough-skinned Plum,* of Sierra Leone
macrophyllum. *Ginger-bread-tree*
Mobola. *Mobola* or *Mola Plum* of the Zambesi
Nonda. *Nonda-tree* of Australia
Paris quadrifolia. *Four-leaved Grass, Herb-Paris, One-berry, True-love*
Paritium elatum. *Common, Blue, Gray,* or *Mountain Mahoe, Cuba Bast-tree*
guineense. *Umbrella-tree,* of Guinea

228   English Names of Cultivated, Native,

Parkia africana. *African Locust-tree, Nitta-tree*
Parkinsonia aculeata. *Jerusalem Thorn, W. Indian Prickly Broom*
Parmelia omphalodes. *Crotal or Crottle Lichen*
perlata. *Canary Moss*
Parmentiera cerifera. *Candle-tree of Panama*
Parnassia. *Grass-of-Parnassus*
asarifolia. *Asarum-leaved Grass-of-Parnassus*
caroliniana. *Large Grass-of-Parnassus*
fimbriata. *Fringed Grass-of-Parnassus*
nubicola. *Himalayan Grass-of-Parnassus*
palustris. *Common Grass-of-Parnassus*
Parochetus communis. *Blue-flowered Shamrock, Shamrock-Pea*
Paronychia. *Nail-wort, Whitlow-wort*
argyrocoma. *Silver Chick-weed, Silver-head*
dichotoma. *Forked Nail-wort or Whitlow-wort*
serpyllifolia. *Thyme-leaved Nail-wort*
verticillata. *Whorled Nail-wort*
Parrotia persica. *Iron-wood, of Persia*
Parthenium Hysterophorus. *Bastard Feverfew, Broom-bush, W. Indian Mug-wort*
Paspalum. *Millet Grass*
conjugatum. *W. Indian Sour-grass*
distichum. *Silt-grass*
exile. *Hungary Rice*
scrobiculatum. *Millet Khoda, of India*
Passerina. *Sparrow-wort*
nivalis. *Alpine Sparrow-wort*
Stelleri. *Ground Jasmine*
Passiflora. *Passion-flower*
(Disemma) aurantia. *Norfolk Island Passion-flower*
cærulea. *Common Blue Passion-flower*
edulis. *Edible-fruited Passion-flower*
fœtida. *W. Indian Love-in-a-mist, or Wild Water-Lemon*
incarnata. *Flesh-coloured Granadilla*
kermesina. *Crimson Passion-flower*
laurifolia. *Jamaica Honey-suckle, Water-Lemon*
macrocarpa. *Pumpkin, or Large-fruited, Passion-flower*
maliformis. *Sweet Calabash-plant, Water Lemon*, of the W. Indies
quadrangularis. *Granadilla, Granadilla Vine, Square-stalked Passion-flower*
racemosa (P. princeps). *Racemed Passion-flower*
tetrandra. *New Zealand Passion-flower*
tiliæfolia. *Lime-tree-leaved Passion-flower*
Vespertilio. *Bat-winged Passion-flower*
vitifolia. *Vine-leaved Passion-flower*
Pastinaca Opopanax. *Rough Parsnip*
sativa. *Common Parsnip*
Patersonia. *Purple Lily, or Native Flag*, of Australia
Paullinia barbadensis. *W. Indian Supple-Jack*
curassavica. *S. American Supple-Jack*
pinnata. *Brazilian Fish-poison-tree*
polyphylla. *W. Indian Supple-Jack*
sorbilis. *Guarana-tree*, of S. America
Pavetta sp. *Wild Jasmine, of Jamaica*

Pavia (Æsculus). *Buck-eye, Smooth-fruited Horse-Chestnut*
discolor. *Red-and-yellow-flowered Dwarf Buck-eye*
flava. *Yellow-flowered Buck-eye*
hybrida. *Variegated-flowered Buck-eye*
macrocarpa. *Long-fruited Buck-eye*
macrostachya. *Long-spike-flowered Buck-eye*
neglecta. *Neglected Buck-eye*
rubra. *Red-flowered Buck-eye*
rubra var. arguta. *Sharp-toothed-leaved Red-flowered Buck-eye*
rubra var. humilis. *Dwarf Red-flowered Buck-eye*.
rubra var. humilis pendula. *Weeping Dwarf Red-flowered Buck-eye*
rubra var. sublaciniata. *Deep-toothed-leaved Red-flowered Buck-eye*
Pavonia coccinea. *Scarlet Mallow, of the Antilles*
Pectis punctata. *W. Indian Marigold*
Pedicularis. *Louse-wort*
canadensis. *Wood-Betony, of N. America*
comosa. *Spiked Louse-wort*
palustris. *Marsh Louse-wort, Red-Rattle Grass*
Sceptrum-Carolinum. *King Charles's Sceptre*
sylvatica. *Common Louse-wort, Red-Rattle*
Pedilanthus. *Slipper Spurge*
tithymaloides. *Jew-bush*
Peganum Harmala. *Syrian Rue*
Pelargonium. *Stork's-bill*
Pellæa. *Cliff-Brake-Fern*
densa. *Oregon Cliff-Brake*
flexuosa. *Zig-zag Cliff Brake*
ornithopus. *Bird's-foot Rock-Brake*
Peltandra virginica. *Arrow-Arum*
Peltidea canina. *Ground-Liver-wort*
Peltogyne paniculata. *Purple-heart Tree*, of Trinidad
Peltophorum Linnæi (Cæsalpinia brasiliensis). *Brazilletto-wood, of Jamaica*
Vogelianum. *Brazilletto-wood, of Brazil*
Penea Sarcocolla. *Sarcocolla-tree*
Penicillaria spicata. "*Bajree*," *Gero Corn*
Penicillium glaucum. *Vinegar-plant, Yeast-plant*
Penicillus. *Merman's Shaving-brush.*
Pentaceras australis (Ailantus punctatus). *Moreton Bay Varnish-tree*
australis. *White Cedar*, of Queensland
Pentaclethra filamentosa. *Wild Tamarind-tree*, of Trinidad
macrophylla. *Owala-tree*, of W. Africa
Pentadesma butyracea. *Butter-and-Tallow-tree*, of Sierra Leone
Penthorum sedoides. *Ditch Stone-Crop*, of N. America.
Pentstemon. *Beard-tongue*
acuminatus. *Pointed-leaved Pentstemon*
argutus. *Graceful Pentstemon*
barbatus. *Bearded Pentstemon*.
campanulatus. *Bell-flowered Pentstemon*
centranthifolius. *Centranthus-leaved Pentstemon*
Cobæa. *Cobæa-flowered Pentstemon*
confertus (P. procerus). *Whorled Pentstemon*
crassifolius. *Thick-leaved Pentstemon*
diffusus. *Spreading Pentstemon*

and Foreign Plants, Trees, and Shrubs.

Pentstemon Digitalis. *Fox-glove Pentstemon*
gentianoides (P. Hartwegi). *Common Pentstemon*
glaber. *Dwarf Blue Pentstemon*
glaucus. *Glaucous Pentstemon*
humilis. *Dwarf Pentstemon*
Jaffrayanus. *Gentian-blue Pentstemon*
lætus. *Flame-coloured Pentstemon*
Lewisii. *Lewis's Pentstemon*
Murrayanus. *Murray's Pentstemon*
ovatus. *Oval-leaved Pentstemon*
Palmeri. *Palmer's Pentstemon*
puniceus. *Scarlet-flowered Pentstemon*
Scouleri. *Scouler's Pentstemon*
secundiflorus. *Side-flowered Pentstemon*
speciosus. *Showy Pentstemon*
Torreyi. *Torrey's Pentstemon*
Wrighti. *Wright's Pentstemon*
Pentzia virgata. *Sheep-fodder-bush,* of S. Africa
Peperoma. *Pepper-Elder*
Peplis Portula. *Water-Purslane*
Pereskia aculeata. *American, Barbadoes, or W. Indian Goose-berry*
Pergularia odoratissima. *West Coast Creeper*
Pericampylos incanus (Cocculus Moorei). *Grass Ivy,* of Australia
Periploca græca. *Climbing Dog's-bane, Milk-vine, Silk-vine, Syrian Silk-plant, Virginian Silk*
Mauritiana. *Coffee-Climber*
Periptera punicea (Sida Periptera). *Shuttle-cock-plant*
Peristeria cerina. *Waxen Dove-flower*
elata. *Common Dove-flower* or *Dove-Orchid, Holy-Ghost-flower*
Peristylus (Habenaria) viridis. *Frog-Orchis*
Pernettya mucronata and other species. *Prickly Heath*
Peronospora infestans. *Potato-Fungus*
Persea carolinensis. *Alligator Pear, Isabella-wood of Carolina, Red Bay, Red Laurel*
caryophyllata. *Pink-wood Tree*
gratissima. *Alligator* or *Avocado Pear, Midshipman's Butter, Vegetable Marrow*
indica. *Canary-wood,* E. *Indian Bay-tree, Venatica-,* or *Vinatico-, wood*
Persica canariensis. *Canary-wood*
Petalostemon. *Prairie Clover*
Petasites vulgaris. *Butter Dock, Bog Rhubarb, Butter-bur, Cleats, Water Docken*
Petitia domingensis. *Spur-tree*
Petiveria alliacea. *Common Guinea-henweed, Garlic-shrub, Strong-man's-weed*
octandra. *Dwarf Guinea-hen-weed*
Petræa volubalis. *Purple Wreath*
Petrobium arboreum. *Rock-plant,* of St. Helena
Petrocallis pyrenaica. *Rock-Beauty*
Petroselinum sativum. *Common Garden Parsley*
segetum. *Corn Hone-wort, Corn Parsley*
Petunia nyctaginiflora. *Common White Petunia*
Peucedanum ammoniacum. *Gum-Ammoniac Plant*
Cervaria. *Mountain Hart-wort*
officinale. *Harstrong,* or *Horestrang, Hog's-fennel, Sulphur-weed* or *Sulphur-wort*

Peucedanum Oreoselinum. *Mountain Parsley*
Ostruthium. *Master-wort*
palustre. *Brimstone-wort, Milk-Parsley*
sativum. *"Tanke," Wild Parsnip*
Peziza. *Cup Mushroom*
coccinea. *Blood Cups, Jew's Ears*
cochleata. *Flaps* or *Flats, Jew's Ears*
venosa. *Jew's Ears*
Phaca. *Bastard-Vetch*
australis. *Trailing Bastard-Vetch*
Bætica. *Spanish Bastard-Vetch*
Phædranassa. *Queen-Lily*
Phalænopsis. *Moth-plant, Moth Orchid*
amabilis. *E. Indian Butterfly-plant*
Phalaris (Digraphis) arundinacea. *Canary Reed, Sword Grass*
arundinacea variegata. *Bride's Laces, French Grass, Gardeners' Garters, Lady Grass, Lady's Garters, Lady's Laces, Painted Grass, Ribbon-grass, Silver Grass*
canariensis. *Canary Grass*
Phallus impudicus. *Devil's Horn, Stinkhorn, Stinking-Polecat Fungus*
Pharbitis (Convolvulus) Nil. *Kaladana Resin-plant*
Pharnaceum acidum. *Longwood Samphire,* of St. Helena
Pharus latifolius. *Oat,* of the W. Indies.
Phascum. *Earth-Moss*
bryoides. *Tall Earth Moss*
muticum. *Common Dwarf Earth-Moss*
Phaseolus aconitifolius. *Turkish Gram*
Caracalla. *Climbing Snail-flower*
diversifolius. *Diverse-leaved Kidney-bean*
Hernandezii. *Frijol-bean,* of Mexico
lunatus. *Hibbert Bean, Lima Kidney-bean, Sugar-bean*
multiflorus var. albiflorus. *White Runner Bean*
multiflorus var. coccineus. *Scarlet Runner Bean*
Mungo (P. Max). *Hairy-podded Kidney-bean*
Mungo var. chlorospermus. *Green Gram*
Mungo var. melanospermus. *Black Gram*
saccharatus. *Sugar Bean*
truxillensis. *Wrinkled Bean*
vulgaris. *Common Kidney-bean, French Bean, Year Bean*
Phebalium montanum. *Tasmanian Mountain Myrtle*
Phelipæa Ludoviciana. *Broom-rape,* of N. America
Phellandrium aquaticum. *Water-Hemlock*
Phellodendron amurense. *Siberian Cork-tree*
Philadelphus. *Mock-Orange, Syringa*
aromaticus. *Sweet-scented New Zealand Tea-tree*
coronarius. *Common Syringa* or *Mock-Orange*
coronarius, var. grandiflorus. *Large-flowered Mock-Orange*
floribundus. *Many-flowered Mock-Orange*
Gordonianus. *Gordon's Mock-Orange*
hirsutus. *Hairy-leaved Mock-Orange*
inodorus. *Scentless Mock-Orange*
latifolius. *Broad-leaved Mock-Orange*

Philadelphus laxus. *Loose-branched Mock-Orange*
Lewisii. *Lewis's Mock-Orange*
mexicanus. *Mexican Mock-Orange*
nanus. *Dwarf Mock-Orange*
speciosus. *Showy Mock-Orange*
tomentosus. *Woolly-leaved Mock-Orange*
verrucosus. *Warted Mock-Orange*
Zeyheri. *Zeyher's Mock-Orange*
Phillyrea. *Jasmine-Box, Mock-Privet*
Phlebodium aureum. *Golden Polypody*
Phleum pratense. *Cat's-tail Grass, Timothy Grass*
Phlomis fruticosa and other species. *Jerusalem Sage*
Lychnites. *Lamp-wick*
Phlox cæspitosa. *Tufted Phlox*
canescens. *Gray-leaved Phlox*
carolina. *Carolina Phlox*
bifida. *Cleft-petalled Phlox*
bryoides. *Moss-like Phlox*
decussata. *Cross-leaved Phlox*
divaricata (P. canadensis). *Straggling Phlox*
Drummondi. *Texan Pride*
frondosa. *Leafy Phlox*
glaberrima. *Very Smooth Phlox*
longiflora. *Long-flowered Phlox*
maculata. *Wild Sweet William, of N. America*
Nelsoni. *Nelson's Phlox*
nivalis. *White-flowered Dwarf Phlox*
ovata. *Ovate-leaved Phlox*
paniculata and vars. *Tall Garden Phlox*
pilosa. *Hairy Phlox*
procumbens. *Procumbent Phlox*
reptans (P. verna, P. stolonifera). *Creeping Phlox*
speciosa. *Pride of Columbia*
subulata. *American Moss Pink*
Phœnix dactylifera. *Date Palm*
sylvestris. *E. Indian Wine Palm, Khujjoor or Khurjurah Palm, Wild Date Palm*
tenuis. *Slender Date Palm*
Pholidota. *Rattle-snake Orchid*
Phoradendron flavescens. *N. American Mistletoe*
Phormium Colensoi. *Small Flax-lily*
tenax. *Common Flax-lily, New Zealand Flax*
tenax variegatum. *Variegated New Zealand Flax*
Photinia arbutifolia. *Californian Maybush*
serrulata (Cratægus glabra) and other species. *Chinese Haw-thorn*
Phragmites communis. *Bennels, Common Reed, Ditch Reed*
Phryma leptostachya. *Lop-seed*
Phrynium villosum (Calathea villosa). *Demerara Frog-plant*
Phygelius capensis. *Cape Fig-wort*
Phyllanthus Conami. *Tipsy-wood, of Brazil*
subereulatus. *Queensland Rock-Broom*
Phyllis Nobla. *Bastard Hare's-ear*
Phyllocladus. *Celery-leaved Pine-tree*
rhomboidalis. *Adventure Bay Pine, Celery Pine of Tasmania*

Phyllocladus trichomanoides. *Celery Pine, or Tanekaha-bark-tree, of New Zealand*
Phyllocoryne jamaicensis. *John-Crow's-nose, or Jim-Crow's-nose, of the W. Indies*
Phyllostachys nigra. *Whangee or Wanghee Cane*
Phymatanthus. *Wart-flower*
Physalis. *Ground-cherry, Winter-cherry*
Alkekengi. *Alkekeny, Bladder Herb, Red Nightshade, Red Winter-cherry, Straw-berry-Tomato*
angulata. *Toothed-leaved Winter-Cherry*
edulis. *Edible Cape-Goose-berry*
peruviana. *Peruvian Cape-Goose-berry*
pruinosa. *Hairy Annual Winter-cherry*
pubescens. *Barbadoes Cape-Goose-berry, Straw-berry Tomato*
somnifera. *Cluster-flowered Winter-cherry*
Physianthus (Aranja) albens. *White Bladder-flower*
Physocalymma floribundum. *Tulip-wood Tree, Brazilian Pink-wood*
Physospermum cornubiense. *Bladder-seed*
Physostegia denticulata. *Toothed False-Dragon-head*
imbricata. *Imbricated False Dragon-head*
virginiana. *Virginian False-Dragon-head*
Physostigma venenosum. *Calabar-Bean or Chop-nut*
Phytelephas macrocarpa. *Ivory-nut Palm, Negro's-head, Vegetable-Ivory-plant*
Phyteuma. *Horned-Rampion*
Charmelii. *Long-bracted Horned-Rampion*
comosum. *Tufted Horned-Rampion*
hemisphæricum. *Linear-leaved Horned-Rampion*
limonifolium. *Limonium-leaved Horned-Rampion*
orbiculare. *Round-headed Horned-Rampion*
pauciflorum. *Few-flowered Horned-Rampion*
spicatum. *Spiked Horned-Rampion*
Phytocrena. *Water-vine*
gigantea. *E. Indian Fountain-tree*
Phytolacca acinosa. *Indian Poke*
decandra. *Coakum, Crimson-berry-plant, Dyers'-grapes, Garget, Pigeon-berry, Red-ink-plant, Virginian Poke-weed, Poke or Scoke*
dioica. *Tree Poke, Umbra-tree of S. America*
electrica. *Electrical Poke*
icosandra. *Hydrangea-leaved Poke, Red-stemmed Poke*
octandra. *Spanish Calalu, W. Indian Fox-glove*
purpurea. *Purple-flowered Poke*
Picea. *Silver Fir*
amabilis. *Woolly-coned Silver Fir*
bracteata. *Leafy-bracted Silver Fir*
cephalonica. *Black Mountain Fir, Mount Enos Fir*
firma. *Japan Silver Fir*
(Pinus) grandis. *Great Californian Silver Fir*
holophylla. *Mandschurian Silver Fir*
nobilis. *Noble Silver Fir*
Nordmanniana. *Crimean Silver Fir*
numidica. *Algerian Silver Fir*

Picea obovata. *Siberian Spruce*
pectinata. *Common Silver Fir*
Pichta. *Pitch, or Siberian, Silver Fir*
Pindrow. *Upright Indian Silver Fir*
Pinsapo. *Spanish Silver Fir*
pungens. *Rocky Mountain Spruce*
religiosa. *Sacred Silver Fir*
religiosa var. glaucescens. *Mexican Silver Fir*
sibirica. *Siberian Spruce*
Picræna excelsa. *Bitter Ash, Jamaica Quassia* or *Picræna-wood*
Picramnia. *Bitter-wood*
Antidesma. *Macary-bitter*, of the W. Indies, *Majo-bitter-tree, Old-Woman's-bitter*
Picria Fel-terræ. *Chinese Earth-gall plant*
Picrodendron Juglans. *Jamaica Walnut*
Pierardia sapida. *Ramleh-tree*, of Rangoon
Pilea. *Stingless Nettle*
grandis. *Dwarf Elder*
herniariæfolia and P. serpyllifolia. *Artilleryplant*
pumila. *Clear-weed, Rich-weed*
Pilocarpus pinnatus. *Jaboranli-plant*, of the W. Indies
Pilocereus Houlletti (P. fossulatus). *Houllett's Woolly-Cactus*
senilis. *Old-man's-head Cactus*
Pilularia globulifera. *Common Pill-wort, Pepper-grass, Pepper-moss.*
Pimelea. *Rice-flower*
drupacea. *Victorian Bird-Cherry*
Pimenta acris. *Bay-berry-tree, Black Cinnamon, Wild Clove*, of the W. Indies
Pimpinella Anisum. *Aniseed-plant* of commerce, *Anise, Anny, Sweet Cumin*
Saxifraga. *Burnet Saxifrage, Pimpernel*
Pinckneya pubens. *Bitter-bark-tree, Fevertree* of Georgia, *Georgia-Bark-tree*
Pinguicula. *Butter-wort*
alpina. *Mountain Butter-wort, White-flowered Butter-wort*
caudata. *Mexican Long-spurred Butterwort*
grandiflora. *Large-flowered, or Irish, Butterwort*
longifolia. *Long-leaved Butter-wort*
lusitanica. *Pale-flowered Butter-wort*
lutea. *Yellow-flowered Butter-wort*
vallisneriæfolia. *Vallisneria-leaved Butterwort*
vulgaris. *Bog Violet, Butter-root, Common Butter-wort, Earning Grass, Eccle Grass, Rot-grass, Steep Grass, Yorkshire Sanicle*
Pinus. *Deal-tree, Pine-tree*
aristata. *Awned-cone Pine*
Arizonica. *Arizona Yellow Pine*
australis. *Broom, Brown, Georgia Pitch-, Hard, Long-leaved, Southern, Yellow,* or *Virginia Pine*
austriaca. *Austrian Pine, Black Pine*
bahamensis. *Pitch Pine*, of the Bahamas
Balfouriana. *Fox-tail Pine, Hickory Pine*
Banksiana. *Gray, or Northern, Scrub-Pinetree, Hudson's Bay,* or *Labrador, Pinetree*
Brutia. *Calabrian Cluster Pine.*
Bungeana. *Chinese Lace-bark Pine*
Cembra. *Russian Cedar, Swiss Stone Pine*
Cembra var. sibirica. *Siberian Pine*

Pinus contorta. *Twisted-branched Pine*
contorta var. Murrayana. *Tamarack Pine*
(Abies) Douglasii. *Oregon Pine, Yellow Pine* of Puget Sound
edulis. *American Nut-Pine, Piñon*
excelsa. *Bhotan Pine*
filifolia. *Thread-leaved Pine*
flexilis. *American White Pine*
Fremontiana (P. monophylla). *Californian Nut-Pine*
Gerardiana. *Neoza Pine, Nepaul Nut-Pine*
glabra. *American Spruce Pine*
Halepensis. *Aleppo* or *Jerusalem Pine-tree*
inops. *New Jersey Scrub Pine*
insignis. *Monterey Pine*
insularis. *Timor Pine*
Kæmpferi. *Golden Pine*
koraiensis. *Corean Pine*
Lambertiana. *Gigantic Californian Pine, Shake-Pine, Sugar-Pine*
Laricio. *Corsican, Calabrian,* or *Larch Pine*
Laricio var. pygmæa. *Dwarf Corsican Pine*
longifolia. *Cheer* or *Emodi Pine*
maritima. *Sea-side Pine*
mitis. *Short-leaved Pine, Spruce Pine, Yellow Pine*
monophylla (P. Fremontiana). *Californian Nut Pine*
monticola. *Mountain Pine*
Mugho. *Drooping-coned Pine*
Mugho var. nana. *Knee Pine*
muricata. *Bishop's Pine*
occidentalis. *W. Indian Pine*
orientalis. *Sapindus Fir*
Pallasiana. *Crimean Pine, Tartarian Pine*
Picea. *Silver Fir*
Pinaster. *Cluster,* or *Star, Pine*
Pinaster var. Hamiltonii. *Lord Aberdeen's Pine*
Pinaster var. minor. *Cortean Pine*
Pinea. *Italian Stone Pine*
Pinea var. fragilis. *Tarentina Pine*
ponderosa. *"Bull," Trucker,* or *Yellow Pine*
Pumilio. *Upright-coned Mountain Pine*
Pumilio var. nana. *Knee Pine*
pungens. *Table Mountain Pine*
religiosa. *Oyamel Fir*
resinosa. *"Norway"* or *Red Pine*, of N. America
rigida. *Pitch Pine, Sap Pine*
Sabiniana. *Digger Pine*
serotina. *Fox-tail* or *Pond Pine*
sinensis. *Chinese Pine*
sitchensis (P. Douglasii). *Blue* or *Tideland Spruce*, of California
Strobus. *Pumpkin-Pine*, of Canada, *Weymouth Pine, White Pine* of N. America
Strobus var. nivea. *White Weymouth,* or *Snow, Pine*
sylvestris. *Birk Apples, Scotch Fir*
sylvestris var. horizontalis. *Highland,* or *Spey-side, Pine*
sylvestris var. nana. *Pigmy Scotch Fir*
Tæda. *Frankincense Pine, Loblolly Pine, Old Field Pine, Rosemary Pine, Slash Pine, Swamp Pine*
Webbiana. *Dye-,* or *King, Pine*
Zeocote. *Candle-wood Pine, Twisted Pine*

Piper. *Pepper*
augustifolium (Artanthe elongata). *Matico-plant*, "*Mohomoho*," of Peru, *Soldier's-herb*
Betle. *Betle-leaf*, or *Betle Pepper*
Cubeba. *Cubeb Pepper*, or *Cubebs* (Cubeba) Clusii. *African Cubebs*, W. *African Black Pepper*
excelsum. *Native Pepper*, of New Zealand
methysticum. *Ava-*, or *Kava-*, *shrub*, of the South Sea Islands
nigrum. *Common Black Pepper-plant* (Chavica) officinarum. *Long Pepper*
peltatum. *Lizard's-tail Pepper*
Piptadenia peregrina. *Niopo-tree*, of Brazil
rigida. *Angico-gum-tree*
Piptanthus nepalensis. *Nepaul Laburnum*
Pipturus argenteus. *Grass-cloth-plant*, of Queensland
Pircunia dioica. *Ombu-tree*
Piscidia carthaginensis. *Black Dog-wood, Jamaica Bitch-wood*
Erythrina. *Jamaica White Dog-wood*, or *Fish-poison-tree*
Pisonia aculeata. *Cocks'-spur*, of the W. Indies
cordata. *Loblolly-wood*
morindifolia. *Lettuce-tree*, of India
Pistacia Lentiscus. *Common Mastich-tree*
Terebinthus (P. atlantica). *Algerian* or *Barbary Mastich-tree*, *Terebinth*, or *Turpentine-tree*
vera. *Pistachio-nut-tree*
Pistia Stratiotes. *Tropical Duck-weed*, *Water Lettuce*, of the W. Indies.
Pisum maritimum (Lathyrus maritimus). *Sea-side Pea*
Ochrus. *Yellow-flowered Pea*
sativum. *Common Garden Pea*
sativum var. arvense. *Field Pea, Gray Pea*
Pithecolobium. *Curl Brush Bean*
bigeminum. *Soap-bark-tree*, of Venezuela
dulce. *Manilla-Tamarind-tree*
filicifolium. *Wild Tamarind-tree*
micradenium. *Shay-bark, Saronette-tree*
Saman. *Genisaro-tree*, *Rain-tree*, *Saman-*, or *Zamang-tree*, of Venezuela
Unguis-cati. *Black-bead-shrub*, *Cat's-claw*
Pittosporum bicolor. *Cheese-wood* or *White-wood*, of Victoria, *Tolosa-wood*, *White-wood*, of Tasmania
crassifolium. *Parchment-bark*
eugenioides. *New Zealand Hedge-Laurel*
undulatum. "*White Box*," of N. S. Wales
revolutum. *Yellow-flowered Brisbane Laurel*
rhombifolium. *Diamond-leaved Queensland Laurel*
undulatum. *Victorian Laurel*
Plagianthus betulinus. *Akaroa-tree, Cotton-tree*, or *Lace-bark*, of New Zealand, *Otago Ribbon-tree*
Lyalli. *New Zealand Ribbon-wood*
pulchellus (Sida pulchella). *Victorian Hemp-bush*
sidoides. *Kurrajong*, of Moreton Bay and Tasmania
Planera aquatica and other species. *Planer-tree*
Richardi (Zelkova crenata). *Common Zelkonea-tree*

Plantago. *Plantain*, "*Way-bread*"
Coronopus. *Buck's-horn Plantain*, *Hart's-horn Plantain*, *Herb Eve*, *Herb Ivy*, *Star-of-the-earth*
Cynops. *Shrubby Plantain*
decumbens. "*Spogel-seed*"*-plant*
lanceolata. *Chimney-sweeps*, *Cock Grass, Cocks, Jack-straws, Rib-Grass, Rib-wort Plantain, Ripple Grass, Hen-plant*
lanceolata major. *Great Hen-plant*
major. *Greater Plantain*, "*White-Man's-Foot-print*"
major var. rosea. *Rose* or *Rose-bracted Plantain*
maritima. *Sea-side Plantain*
maxima. *Broad-leaved Plantain*
media. *Fire-leaves*, *Fire-weed*, *Kemps*, *Hoary Plantain*, *Lamb's-tongue*
media var. atropurpurea. *Purple-leaved Plantain*
Psyllium. *Flea-wort Plantain*
Plasmodiophora Brassicæ. *Club-root Fungus*
Platanthera. *False Orchis*
Platanus. *Plane-tree*
acerifolia. *Maple-leaved Plane-tree*
acerifolia var. hispanica. *Spanish Plane-tree*
acerifolia var. umbellata. *Canopy Plane-tree*
cuneata. *Wedge-leaved Plane-tree*
digitata. *Caucasian Plane-tree*
heterophylla. *Various-leaved Plane-tree*
mexicana. *Mexican Plane-tree*
occidentalis. *American Plane-tree*, *American Sycamore*, *Button-wood*, *Water Beech*, *Western Plane-tree*
orientalis. *Eastern*, or *Oriental*, *Plane-tree*
racemosa. *Californian · Plane-tree*, *Californian Sycamore*
striata. *Streaked-barked Plane-tree*
Platycerium alcicorne. *Australian Elk's-horn Fern*
biforme. *E. Indian Elk's-horn Fern*
grande. *Queensland Elk's-horn Fern, Stag's-horn Fern*
Stemmaria. *Guinea Elk's-horn Fern*
Wallichii. *Wallich's Elk's-horn Fern*
Platycodon grandiflorum. *Chinese Bell-flower*
Platylobium. *Flat Pea*
formosum. *Large-flowered Flat Pea*
Platylophus trifoliatus. *S. African* "*White Alder*"
Platystemon californicus. *Californian Poppy*, "*Cream-cups*"
Plectranthus. *Cockspur-flower*
nudiflorus. *Chinese Basil*
ternatus. *Onime-root*
Plectronia ventosa. *African Cockspur-Thorn*
Pleione. *Indian Crocus*
Hookeriana. *Hooker's Indian Crocus*
humilis. *Dwarf Indian Crocus*
Lagenaria. *Bottle-gourd Indian Crocus*
maculata. *Spotted Indian Crocus*
præcox. *Early-flowering Indian Crocus*
tricolor. *Three-coloured Indian Crocus*
Wallichii. *Wallich's Indian Crocus*
Pleroma (Lasiandra) sarmentosa. *Peruvian Glory-bush*

Plocaria lichenoides. *Ceylon Moss*
tenax. *Swallow's-nest Sea-weed*
Pluchea. *Marsh Flea-bane*
camphorata. *Salt-marsh Flea-bane*
Plumbago. *Lead-wort*
capensis. *Cape Lead-wort*
europæa. *European Lead-wort*
coccinea. *Scarlet-flowered Lead-wort*
Larpentæ. *Hardy Blue-flowered Lead-wort*
rosea. *Rosy-flowered Lead-wort*
scandens. *Devil's-herb*, or *Tooth-wort*, of the W. Indies
Plumieria alba. *Frangipani - plant, W. Indian Pagoda-tree, White Nosegay-tree*
rubra. *Frangipani-plant, Jasmine Mango, Red Nosegay-tree*
Poa. *Meadow-grass*
abyssinica. "*Teff*," of Abyssinia
amabilis. *Whorled Love-grass*
annua. *Causeway-grass, Common Meadow-grass, Low Spear-grass* of N. America, *Suffolk-grass*
australis. *Australian Ornamental Grass*
cæspitosa (P. australis). *Australian Meadow-grass*
compressa. *Wire-grass*, of N. America
nemoralis. *Wood-Meadow-grass*
pratensis. *Kentucky Blue-grass, Natural Grass, Smooth-stalked Meadow-grass*
ramigera. *Bamboo*, of Australia
trivialis. *Bird-grass, Fowl-grass, Fold Meadow-grass, Rough-stalked Meadow-grass*
Podalyria sericea. *African Satin-bush*
Podocarpus andina. *Plum Fir*
chinensis. *Chinese Yew-tree*
coriacea. *Yacca-wood*, of the W. Indies
cupressina. *Kaw-Tabua-tree*, of Fiji
(Nageia) dacrydioides. *White Pine*, of New Zealand
elongata. *S. African Yellow-wood*
ferruginea. *Black Pine*, or "*Miro*," of New Zealand
Koraiana (Taxus japonica). *Japanese Yew*
latifolia. *E. Indian Yellow-wood*
neriifolia. *Nerium-leaved "Wax-Dammar"*
polystachya. "*Wax-Dammar*," of Singapore
Purdieana. *Yacca-wood-tree*
spicata. *Black Pine*, of New Zealand, *Otago Black Rue*
spinulosa. *Illawarra Pine, White Pine, Native Damson* or *Plum* of N. S. Wales
Thunbergii. *Yellow-wood*, of the Cape
(Nageia) Totara. *Mahogany Pine* or *Totara Pine*, of New Zealand
Podophyllum Emodi. *Himalayan May-apple*
peltatum. *American Mandrake* or *May-apple, Podophyllin Plant, Raccoon-berry, Wild Duck-foot, Wild Lemon*
Podostemon ceratophyllus. *Thread-foot, River-weed*, of N. America
Pogonia ophioglossoides. *Snake's-mouth Orchis*
(Triphora) pendula. *Three-Bird's Orchis*
Pogostemon Patchouli. *Patchouli-oil-plant, Pucha-pat*

Poinciana (Cæsalpinia) Gilliesii. *Crimson Thread-flower*
pulcherrima. *Barbadoes-Pride*, "*Spanish Carnation*"
regia. *Royal Peacock-flower*
Poinsettia pulcherrima. *Easter-flower*, of Mexico, *Lobster-flower, Mexican Flame-leaf*
Polemonium cæruleum. *Charity, Common Jacob's-ladder, Greek Valerian*
cæruleum var. Richardsoni. *Richardson's Jacob's-ladder*
confertum. *Dense-clustered-flowered Jacob's-ladder*
humile. *Dwarf Jacob's-ladder*
reptans. *Creeping Jacob's-ladder*
Polianthes tuberosa. *Garden Tuberose*
Polycarpon tetraphyllum. *Four-leaved All-seed*
Polygala. *Milk-wort* or *Milk-weed*
amara. *Bitter Milk-wort*
calcarea. *Chalk Milk-wort*
Chamæbuxus. *Box-leaved Milk-wort*
paucifolia. *Fringed Milk-wort*
polygama. *Ground-flowering Milk-wort*
Senega. *Seneca Snake-root*
vulgaris. *Common Milk-wort, Cross-flower, Gang-flower, Procession-flower, Rogation-flower*
Polygonatum. *Solomon's-seal*
giganteum. *Giant Solomon's-seal*
multiflorum. *David's Harp, Fraxinell, Many-flowered Solomon's-seal, Ladder-to-Heaven, Lady's Seal, Lily-of-the-Mountain*
officinale. *Common Solomon's-seal*
roseum (Convallaria rosea). *Rosy-flowered Solomon's-seal*
verticillatum. *Whorled Solomon's-seal*
Polygonum. *Knot-grass, Knot-weed,* "*Tear-thumb*," of N. America
acre. *Water Smart-weed*
alpinum. *Alpine Knot-weed*
amphibium. *Amphibious Knot-weed, Willow-grass*
arifolium. *Halbert-leaved Tear-thumb*
articulatum. *American Joint-weed*
aviculare. *Armstrong, Beggar-weed, Common Knot-grass, Crab-grass, Crab-weed,* "*Door-weed*," of N. America, *Iron Grass, Ninety-knot, Pink-weed, Sparrow-tongue, Swine's-grass, Wire-grass*
Bistorta. *Adder-wort, Bistort, Dragon-wort, Easter Giant, Easter Ledges, Easter Mu-giants, Easter Mangiants, Gentle Dock, Great Bistort,* "*Oysterloit*," *Patient Dock, Patience Dock, Red-legs, Snake-weed*
Brunonis. *E. Indian Knot-weed*
Convolvulus. *Bear-bind, Bind-corn, Climbing Buck-wheat, Corn-bind, Ivy Bind-weed*
cuspidatum (P. Sieboldi). *Giant Knot-weed*
dumetorum. *Copse Knot-weed*
emarginatum. *E. Indian Buck-wheat*
Fagopyrum. *Beech-wheat, Bock-wheat,* or *Buck-wheat, French Wheat*
Hydropiper. *Ciderage, Culrage,* or *Curage, Lake-weed, Red-Knees, Smart-weed, Water-pepper*

Polygonum hydropiperoides. *Mild Water-pepper*
lapathifolium. *Pale-flowered Knot-weed, Willow-weed*
maritimum. *Sea-side Knot-weed*
minus. *Slender Knot-grass*
orientale. *E. Indian Knot-weed*
Persicaria. *Crab's Claw, Lady's-Thumb, Peach-wort, "Persicaria," Red-shanks*
punctatum. *American Smart-weed*
sachalinense. *Giant Sachalin Knot-weed*
sagittatum. *Arrow-leaved Tear-thumb*
tinctorium. *Indigo-plant*, of China and Japan
vaccinifolium. *Rock Knot-weed, Whortleberry-leaved Knot-weed*
virginianum. *Virginian Knot-grass*
viviparum. *Serpent-grass, Small Bistort*
Polymnia. *Leaf-cup*, of N. America
Polypodium. *Polypody*
alpestre. *Alpine Polypody*
(Phlebodium) aureum. *Golden Polypody*
calcareum. *Lime-stone Fern, Rigid Three-branched Polypody*
cambricum. *Welsh Polypody*
Dryopteris. *Moss Fern, Oak Fern*
falcatum. *Oregon Wild Liquorice*
hibernicum. *Irish Polypody*
incanum. *Hoary Polypody*, of N. America
Phegopteris. *Beech Fern*
pustulatum. *Scented Polypody*, of New Zealand
serpens. *Twining Polypody*, of New Zealand
vulgare. *Adder's Fern, Common Polypody, Brake-root, Golden Locks, Golden Maidenhair, Wall Fern, Wood Fern*
Polypogon. *Beard-grass*
littoralis. *Perennial Beard-grass*
monspeliensis. *Annual*, or *Montpelier, Beard-grass*
Polyporus (Boletus) fomentarius, and P. igniarius. *Amadou, German* or *Vegetable Tinder, Moxa, Punk, Spunk, Touchwood*
hybridus. *Dry-rot Fungus*, of the Oak
(Boletus) igniarius. *Hard Amadou* or *German Tinder*
ulmarius. *Dry-rot Fungus*, of the Elm
Polystichum angulare. *Soft Prickly Shield-Fern*
lobatum. *Lobed Prickly Shield-Fern*
Polytrichum. *Hair-Moss*
aloides. *Dwarf Long-headed Hair-Moss*
alpinum. *Alpine Hair-Moss*
commune. *Common Hair Moss, Moor Silk, Golden Maiden-hair, Goldilocks*
gracile. *Slender Hair-Moss*
juniperinum. *Bear's-bed, Ground Moss, Hair-cap Moss, Juniper-leaved Hair-Moss, Robin's-Rye*
nanum. *Dwarf Round-headed Hair-Moss*
urnigerum. *Urn-bearing Hair-Moss*
Pomaderris apetala. *Cooper's-wood*, of N. S. Wales. *Victorian Hazel*
elliptica. *Kumarahou-tree*, of New Zealand
lanigera. *N. S. Wales Hazel*
Pongamia glabra. *Karrunj-, Kurrung-*, or *Poonga-, oil-plant*

Pontederia. *Pickerel-weed*
azurea. *Sky-blue Pickerel-weed, Water-Plantain* of Jamaica
cordata. *Heart-leaved Pickerel-weed*
crassipes. *Bladder-stalked Pickerel-weed*
Populus. *Poplar*
alba. *Abbey, Abele-tree, Dutch Beech, White Asp, White Poplar*
Acladesca. *Black Italian Poplar*
angulata. *Missouri Cotton-wood*
angustifolia. *Rocky Mountain Poplar*
balsamifera. *Balm of Gilead* or *Balsam-Poplar, Tacamahac*
balsamifera var. caudicans (P. macrophylla). *Ontario Poplar*
canescens. *Gray Poplar*
canescens var. pendula. *Weeping Gray Poplar*
fastigiata. *Lady Poplar, Lombardy Poplar*
Fremontii. *Californian Cotton-wood, Californian Poplar*
græca. *Athenian Poplar*
grandidentata. *Large-toothed Aspen, Soft* or *Paper Poplar*
heterophylla. *Cotton Tree, Downy Poplar*
laurifolia. *Laurel-leaved Poplar*
monilifera. *Berry-bearing Poplar, Carolina Poplar, Cotton-wood, Necklace Poplar*
nigra. *Black Poplar*
nigra var. salicifolia. *Willow-leaved Poplar*
pendula. *Weeping Poplar*
suaveolens. *Fragrant Poplar*
tremula. *Aps, Asp*, or *Aspen, Haps-tree*
tremula var. pendula. *Weeping Aspen*
tremuloides. *American Aspen, Quaking Asp*
trichocarpa. *Californian Poplar*
Porphyra laciniata. *Laver*
vulgaris. *Cultivated Sea-weed* (Japan).
Portlandia hexandra. *French Guiana Bark-tree*
Portulaca. *Purslane*
aurea. *Yellow-flowered Purslane*
grandiflora. *Large-flowered Purslane, Sun-plant*
oleracea. *Garden Purslane*
Thellusoni. *Crimson-flowered Purslane*
splendens. *Red-flowered Purslane*
Portulacaria afra. *Purslane-tree, "Speck-boom,"* of S. Africa
Potamogeton. *Pond-weed*
compressus. *Flat-stalked Pond-weed*
crispus. *Curly Pond-weed, Water Caltrops*
densus. *Close-leaved Pond-weed, Frog's Lettuce, Water Caltrops*
gramineus. *Grass-leaved Pond-weed*
heterophyllus. *Various-leaved Pond-weed*
lucens. *Shining Pond-weed*
natans. *Broad Pond-weed, Devil's Spoons, Tench-weed*
pectinatus. *Fennel-leaved Pond-weed*
perfoliatus. *Perfoliate Pond-weed*
prælongus. *Long Pond-weed*
pusillus. *Slender Pond-weed*
rufescens. *Reddish Pond-weed*
Potentilla. *Cinquefoil*
alba. *White-flowered Cinquefoil*
alchemilloides. *Lady's-mantle Cinquefoil*
alpestris. *Alpine Cinquefoil*

and Foreign Plants, Trees, and Shrubs. 235

Potentilla anserina. *Argemone, Argentine, Argentina, Dog's Tansy, Fair Grass, Fair Days, Goose-grass, Goose-Tansy, Marsh Corn Silver-weed, Wild Tansy*
argentea. *Silvery Cinquefoil*
atrosanguinea. *Dark Crimson Cinquefoil*
cæspitosa. *Tufted Cinquefoil*
calabrica. *Calabrian Cinquefoil*
canadensis. *Canadian Fire-fingers*
Clusiana. *Clusius's Cinquefoil*
Comarum (Comarum palustre). *Marsh Cinquefoil, Marsh Potentil*
Fragariastrum. *Barren Strawberry, Strawberry-leaved Potentil*
frutescens. *Tree Cinquefoil*
fruticosa. *Shrubby Cinquefoil*
Hippinna. *Colorado Silvery Cinquefoil*
hybrida. Garden varieties of *Cinquefoil*
minima. *Dwarfest Cinquefoil*
nana. *Dwarf Cinquefoil*
nitida. *Shining Cinquefoil*
nivalis. *Snowy Cinquefoil*
norvegica. *Norwegian Cinquefoil*
palustris. *Marsh Five-fingers*
pyrenaica. *Pyrenean Cinquefoil*
reptans. *Creeping Cinquefoil, Five-finger Grass, Five-leaf, "Sink-field"*
rupestris. *Rock Cinquefoil*
speciosa. *Showy Cinquefoil*
splendens. *Brilliant Cinquefoil*
Tormentilla. *Blood-root, Tormentil, Ewe-daisy, Septfoil, Tormentil, Tormentil-root*
tridentata. *Three-toothed leaved Cinquefoil*
verna. *Spring Cinquefoil*
Poterium canadense. *Canadian Burnet*
Sanguisorba. *Salad Burnet.*
Pothomorpha. *W. Indian Colt's-foot*
Pothos argyrea. *Silver Vine*, of the W. Indies
Poupartia dulcis. *Otaheite Apple*
Prangos pabularia. *Prangos or Hay-plant, of Tibet*
Prasium majus. *Great Spanish Hedge-Nettle*
minus. *Small Spanish Hedge-Nettle*
Premna integrifolia. *Head-ache-tree*, of the E. Indies
Primula. *Primrose*
acaulis (P. vulgaris). *Common Primrose*
Allioni. *Allioni's Primrose*
altaica. *Altaian Primrose*
amœna. *Caucasian Primrose*
Auricula. *Baziers, Bear's Ears, Boar's Ears, Common Auricula, Dusty Miller, French Cowslip, Mountain Cowslip, Tanner's Apron*
auriculata. *Ear-leaved Primrose*
calycina. *Large-calyxed Primrose*
Candolleana. *De Candolle's Primrose*
capitata. *Round-headed Himalayan Primrose*
carniolica. *Carniolic Primrose*
cashmeriana. *Cashmere Primrose*
ciliata. *Fringed Primrose*
commutata. *Changeable Primrose*
cortusoides. *Cortusa-leaved Primrose*
cortusoides var. amœna. *Veitch's Primrose*
denticulata. *Toothed-leaved Primrose*
elatior. *Bardfield or True Ox-lip, Great Cowslip*

Primula elatior var. cærulea. *Blue Ox-lip*
elatior (varieties of). *Common Polyanthus*
farinosa. *Bird's-eye Primrose*
farinosa var. superba. *Brilliant Bird's-eye Primrose*
Floerkiana. *Floerk's Primrose*
Fortunei (P. erosa). *Fortune's Primrose*
glaucescens. *Glaucous-leaved Primrose*
glutinosa. *Glutinous Primrose*
graveolens. *Strong-scented Primrose*
Henryi. *Henry's Primrose*
hirsuta. *Hairy-leaved Primrose*
imperialis. *Java Primrose*
integrifolia. *Entire-leaved Primrose*
intricata. *Pyrenean Cowslip*
involucrata. *Creamy-flowered Primrose*
japonica. *Japanese Primrose*
Kitaibelii. *Kitaibel's Primrose*
luteola. *Yellowish Primrose*
marginata. *Silver-edged Primrose*
minima. *Fairy Primrose, Snow Rosette*
mollis. *Soft-leaved Primrose*
Munroi. *Munro's Primrose*
nivalis and P. nivea. *Snow-white Primrose*
Palinuri. *Large-leaved Primrose, Neapolitan Primrose*
Parryi. *Parry's Primrose*
Pedemontana. *Piedmont Primrose*
pulcherrima. *Yellow-eyed Globe-headed Purple Primrose*
purpurea. *Globe-headed Purple Primrose*
rosea. *Rosy Primrose*
scotica. *Scotch Bird's-eye Primrose*
Sieboldi. *Siebold's Primrose*
sikkimensis. *Sikkim Cowslip*
sinensis. *Chinese Primrose*
sinensis var. fimbriata. *Fringed Chinese Primrose*
spectabilis. *Showy Primrose*
Steini. *Stein's Primrose*
Stuartii. *Stuart's Primrose*
suffrutescens. *Californian Cowslip*
Tyrolensis. *Tyrolese Primrose*
variabilis. *Culverkeys, Common Ox-lip*
veris (P. officinalis). *Bedlam Cowslip, Common Cowslip, Herb Peter, Palsy-wort, St. Peter's-wort*
veris "flore et calyce crispo." *Gallegaskins or Gaskins*
verticillata. *Abyssinian Primrose*
villosa. *Shaggy-leaved Primrose*
viscosa. *Clammy Primrose*
vulgaris var. acaulis. *Common Primrose*
vulgaris var. caulescens. *Ox-lip*
Wulfeniana. *Wulfen's Primrose*
Pringlea antiscorbutica. *Kerguelen's-land Cabbage or Horse-radish*
Prinos ambiguus. *Carolina Winter-berry*
atomarius. *Atom-bearing Winter-berry*
coriaceus. *Leathery-leaved Winter-berry*
deciduus. *Deciduous Winter-berry*
glaber. *Apalachian or Winter-berry Tea-plant, Ink-berry, Shining Winter-berry*
lævigatus. *Smooth Winter-berry*
lanceolatus. *Spear-leaved Winter-berry*
verticillatus. *American Black Alder, Winter-berry*
Prionium Palmita. *Palmite-Rush*, of S. Africa

Priva echinata. *Styptic* or *Velvet Bur*
Proserpinaca palustris. *Mermaid-weed*
Prosopis dulcis. *Cashaw-tree*
glandulosa. *Mezquit-bean*, of Texas
juliflora. *Algaroba* or *Honey-Mesquit*, *Cashaw-tree*, *July-flower* of Jamaica, S. Western *Honey-Locust*
pubescens. *Screw-bean*, or *Screw-pod Mesquit*, of California, *Tornillo-bean*
Prostanthera. *Australian Mint-bush* or *Mint-tree*
lasianthos. *Australian Lilac*, *Mint-tree*, *Victorian Dog-wood* or *Mint-tree*
violacea. *Violet-flowered Lilac*, of Australia
Protea cynaroides. *Cape Artichoke-flower*
mellifera. *Cape Honey-flower* or *Sugar-bush*
Protococcus nivalis. *Red Snow Lichen*
Prumnopitys elegans. *Plum Fir*
Prunella (Brunella). *Self-heal*
grandiflora. *Large-flowered Self-heal*
grandiflora var. laciniata. *Cut-leaved Large-flowered Self-heal*
hyssopifolia. *Hyssop-leaved Self-heal*
pyrenaica. *Pyrenean Self-heal*
vulgaris. *All-heal*, *Brunel*, *Caravaun-beg*, *Carpenter Grass*, *Common Self-heal*, *Heart-of-the-Earth*, *Herb Carpenter*, *Hook-heal*, *Panay*, *Sickle-heal*, *Sickle-wort*
Prunus. *Cherry-tree*, *Plum-tree*
americana. *American Wild Yellow* or *Red Plum*, *Canada Plum*
Avium. *Crab Cherry*, *Gean-tree*, *Mazard* or *Mazzards*, "*Merries*" of Suffolk, "*Merry Tree*" of Cheshire, *Wild Cherry*
Brigantiaca. *Marmottes-oil-plant*
candicans. *Whitish-leaved Plum*
caroliniana. *Carolina Plum* or *Mock-Orange*
cerasifera. *Weeping Plum*
Cerasus. *Dwarf Wild Cherry*
Chicasa. *Chicasaw Plum*
Claudiana. *Green-Gage Plum-tree*
Cocomilla. *Cocomilla Plum*
demissa. *Californian Wild Cherry*
divaricata. *Spreading Plum*
domestica. *Cultivated Plum-tree*
domestica flore pleno. *Double-blossomed Plum*
domestica variegata. *Variegated-leaved Plum*
domestica var. damascena. *Damson Plum*
domestica var. heterophylla. *Various-leaved Plum*
domestica var. myrobalana. *Cherry Plum*, *Myrobella Plum*
domestica var. Turonensis. *Premier Swiss Plum*
ilicifolia. "*Islay Plum*," of California
insititia. *Bullace*, *Bullas*, or *Bollas*, *Wild Damson*
insititia flore pleno. *Double-blossomed Bullace*
insititia fructu luteo-albo. *Yellowish-white-fruited Bullace*
insititia fructu rubro. *Red-fruited Bullace*
japonica (P. sinensis). *Japanese Plum*
Lauro-cerasus. *Common Laurel*
lusitanica. *Portugal Laurel*

Prunus maritima. *Sand*, or *Beach*, *Plum*, of N. America
Mume. *Mume Plum*, of Japan
occidentalis. *Prune-tree*, of the W. Indies
Padus. *Bird Cherry*, *Black Dog-wood*, *Egg-berry*, *Hack-berry*, *Hag-berry*, *Hedge-berry*, *Hog-cherry*
pennsylvanica. *American Bird-Cherry* or *Wild Red Cherry*
pubescens. *Downy Plum*
pumila. *Dwarf Cherry*, or *Sand Cherry*, of N. America
serotina. *American Wild Black Cherry*
sinensis. *Chinese Plum*
sinensis fl.-pl. *Double-flowered Chinese Plum*
spinosa. *Black-thorn*, *Buck-thorn*, *Bullace*, *Sloe-tree*
subcordata. *Californian Wild Plum*
triloba. *Three-lobed-leared Plum*
Virginiana. *American Choke-Cherry*
Psamma arenaria (Ammophila arundinacea). *Bent-grass*, *Maram* or *Marram*, *Marrem-grass*, *Mat-grass*, *Mat-weed*, *Sea-Reed*, *Sea-shore Bent-grass*
Pseudo-Larix. *Chinese*, or *False Larch*
Pseudolmeria spuria. *Bastard Bread-nut*
Psidium. *Guava-tree*
Cattleyanum. *Purple Guava*
cordatum. *Spice Guava*
indicum. *E. Indian Guara*
montanum. *Mountain Guava*
pomiferum. *Apple-shaped Guava*
pyriferum. *Pear-shaped Guava*
Psophocarpus tetragonolobus. *Goa-bean-plant*
Psoralea. *Scurfy Pea*
bituminosa. *Bitumen Trefoil*
corylifolia. "*Buwchee Seed*"-plant
esculenta. *Missouri Bread-root*, "*Pomme de Prairie*," *Prairie Turnip*
glandulosa. "*Jesuit's Tea*," of Chili, *Mexican Tea-plant*
Lupinella. *Small Lupine*
Psychotria daphnoides. *Brush-land Sage-tree*, of Australia
emetica. *Black*, *Peruvian*, or *Striated Ipecacuhana*
parasitica. *Climbing Vine*, of the W. Indies
Pteroxylon utile. *Sneeze-wood*, of the Cape
Ptelea trifoliata. *Hop-tree*, *Shrubby Trefoil*, *Swamp Dog-wood*, *Wafer Ash*, *Wing-seed*.
trifoliata var. aurea. *Yellow-leaved Hop-tree*
Pteris aquilina. *Adder-spit*, *Bracken*, *Brake-fern*, *Eagle Fern*
esculenta. *Edible Fern* of Tasmania and New Zealand, *Tara-fern* of Tasmania
incisa. *Bat's-wing Fern*
serrulata. *Spider Fern*
Pterocarpus dalbergioides. *Andaman Red-wood-tree*, *Burmese Kino-tree*
Draco. *Dragon-gum-tree*
erinaceus (P. echinatus). *African* or *Gambia Gum Kino-tree*, *African Rose-wood*, *Corn-wood* of W. Africa, *Kino-gum-tree*, *Molompi-wood-tree*
indicus. *Burmese Rose-wood*, "*Lingo-tree*"

Pterocarpus Marsupium. *E. Indian or Amboyna Kino-tree*
santalinus. *E. Indian Red-wood, Red Sandal-wood, Red Sanders Wood, or Ruby Wood*
Pterocarya caucasica (Juglans fraxinifolia). *Caucasian Walnut*
Pterocaulon pycnostachyum. *American Black-root*
virgatum. *Golden Cud-weed*
Pterocephalus Parnassi. *Mt. Parnassus Scabious*
Pterogonium. *Wing-moss*
Pterospermum indicum. *Amboyna-wood*
Pterospora Andromedea. *American Dragon-claw or Dragon-root, Fever-root, Pine-drops*
Pterygodium. *Monk's-cowl Orchid*
Ptilostephium trifidum. *"Feather-crown," of Mexico*
Ptychosperma. *Australian Feather-palm*
Alexandræ. *Princess Alexandra's Palm*
Cunninghamii. *Illawarra Palm*
Ptychotis Ajowan. *Adjowan, Adjouan, or Javance-plant, of Bengal*
heterophylla. *Square Parsley*
Puccinia Anemones. *"Conjuror of Chalgrave's Fern"*
graminis. *Wheat Mildew*
Pulmonaria. *Lung-wort*
angustifolia. *Blue Cowslip, Narrow-leaved Lung-wort*
azurea. *Azure-flowered Lung-wort*
dahurica. *Siberian Lung-wort*
maritima. *Sea-Bugloss*
officinalis. *Bedlam Cowslip, Beggar's Basket, Bugloss Cowslip, Common Lung-wort, Jerusalem Cowslip, "Sage of Bethlehem," Virgin Mary's Honeysuckle*
(Mertensia) sibirica. *Siberian Lung-wort*
Pultenæa daphnoides. *Native Victorian Wall-flower*
Punica Granatum. *Carthaginian Apple, Pomegranate-tree*
Granatum fl.-pl. *Double-flowered Pomegranate*
Granatum var. nana. *Dwarf Pomegranate*
Puschkinia scilloides. *Striped Squill*
Putranjiva Roxburghii. *Wild Olive, of India*
Pycnanthemum. *American Mountain Mint, American Wild Basil*
linifolium. *Virginian Thyme*
Monardella. *Small Monarda*
virginicum. *Narrow-leaved Virginian Thyme, Prairie Hyssop*
Pycnocoma macrophylla. *Bomah-nut-tree*
Pyrethrum. *Fever-few*
achilleæfolium. *Narrow-leaved Fever-few*
alpinum. *Alpine Fever-few*
carneum (P. roseum). *Rosy-flowered Fever-few*
(Chrysanthemum) lacustre. *Marsh Fever-few*
Parthenium. *Common Fever-few or Featherfew, May-weed, Pellitory*
serotinum. *Late-flowering Fever-few*
Tchihatchewi. *Turfing Daisy*
uliginosum. *Great Ox-eye*
Willemoti. *Willemot's Fever-few*

Pyrola. *Winter-green*
elliptica. *American Shin-leaf*
maculata. *Spotted-leaved Winter-green*
media. *Intermediate Winter-green*
minor. *Common Winter-green, Wood Lily*
rotundifolia. *Canker-Lettuce, False or Pear-leaved Winter-green, Larger Winter-green*
secunda. *Notched-leaved Winter-green*
(Chimaphila) umbellata. *Umbel-flowered Winter-green*
uniflora. *One-flowered Winter-green*
Pyrolirion. *Fire-Lily, Flame-Lily*
Pyrrhopappus carolinianus. *False Dandelion*
Pyrrhosa tingens. *" Red " Mace-tree*
Pyrularia (Hamiltonia) oleifera. *Buffalo-nut, Elk-nut, or Oil-nut, of N. America*
Pyrus. *Apple-tree, Pear-tree, Quince-tree, Service-tree*
americana. *Mountain-Ash, of N. America*
augustifolia. *American Narrow-leaved Crab-Apple*
arbutifolia. *Arbutus-leaved Aronia, Chokeberry*
Aria. *Chess Apple, Hen Apple, Lot-tree, White Beam-tree*
Aria var. salicifolia. *Willow-leaved Beam-tree*
Aucuparia. *Care, Field Ash, Mountain-Ash, Quick-beam or Quicken-tree, Rantry, Rowan-tree or Roan-tree, Wicken*
Aucuparia var. pendula. *Weeping Mountain-Ash*
baccata. *Cherry-Apple, Siberian Crab*
betulæfolia. *Birch-leaved Pear-tree*
Bollwylleriana. *Woolly-leaved Pear*
(Mespilus) Chamæmespilus. *Bastard Quince, Dwarf Medlar*
communis. *Choke Pear*
communis sativa and vars. *Common Cultivated Pear*
coronaria. *American Crab-Apple, Garland-Crab, Sweet-scented Crab*
crenata. *Notched-leaved Pear*
Cydonia. *Common Quince-tree*
elæagnifolia. *Oleaster-leaved Pear*
floribunda. *Many-flowered Aronia*
grandifolia. *Large-leaved Aronia*
intermedia (P. scandica). *Swedish Beam-tree*
Malus var. acerba. *Bittersgall, Common Crab, Wilding (Crab-Apple)*
Malus var. astracanica. *Astrachan Apple*
Malus var. baccata. *Cherry-Crab, Scarlet-flowered Crab*
Malus var. dioica. *Diæcious Apple*
Malus var. floribunda. *Coral-flowered Apple*
Malus var. præcox. *Paradise Apple*
Maulei. *Maule's Japanese Quince*
melanocarpa. *Black-fruited Aronia*
nivalis. *Alpine Pear-tree, Snow Pear, White-leaved Pear*
pinnatifida. *Bastard Service-tree*
prunifolia. *Siberian Crab*
pubescens. *Downy-branched Aronia*
rivularis. *Oregon Crab-Apple, Pow-itch-tree of India*
salicifolia. *Willow-leaved Pear*
salviæfolia. *Aurelian or Sage-leaved Pear*

Pyrus sambucifolia. *Western Mountain-Ash*
scandica (P. intermedia). *Swedish Beam-tree* or *Service-tree*
sinaica. *Mt. Sinai Pear*
sinensis. *Chinese Pear*
Sorbus (P. domestica). *Common Service-tree*
Sorbus var. auriculata. *Auricled Service-tree*
spectabilis. *Chinese Apple*
spectabilis fl.-pl. *Double-flowered Chinese Apple*
spectabilis var. floribunda. *Profuse-flowering Chinese Apple*
spuria var. pendula. *Weeping Pear*
torminalis. *Chequer-tree, Maple Service, Wild Service-tree*
trilobata. *Three-lobed-leaved Pear*
variolosa. *Variable-leaved Pear*
Pyxidanthera barbulata. *Pine-Barren Beauty*

Quamoclit vulgaris. *American Bell-flower, China Creeper, Cupid's-flower, Cypress-Vine, Indian Forget-me-not, Indian Pink*
Quassia amara. *Quassia-tree*, of Surinam
(Picræna) excelsa. *Quassia-tree*, of Jamaica, *Simaruba-bark-tree*
Quelania lætioides. *Silver-wood*
Queltia. *Mock Narcissus*
Quillaia Saponaria. *Cullay-*, or *Quillai-, tree, Soap-bark-tree*, of Chili
Quercus. *Oak-tree*
abelicea. *False Sandal-wood*, of Crete
Ægilops. *Vallonea* or *Velani Oak*
agnostifolia. *Agnostus-leaved Oak*
agrifolia. *Encено Oak*
alba. *American White Oak, "Quebec" Oak*
alnifolia. *Golden Oak*, of Cyprus
Andersoni. *Anderson's Oak*
aquatica. *Water Oak*
Ballota. *Barbary Oak, Belote Oak, Sweet-Acorn-Oak*
bicolor. *Swamp White Oak*
Breweri. *"Chuppural" Oak*, of California
Buergeri. *Bueryer's Oak*
Castanea. *Yellow Chestnut*
Catesbæi. *American Turkey Oak, Barren Scrub Oak, Forked-leaved Black Jack Oak*
Cerris. *Bitter, Iron, Mossy-cupped, Turkey,* or *Wainscot Oak*
Cerris var. Fulhamensis. *Fulham Oak*
Cerris var. Lucombeana (Q. exoniensis). *Devonshire, Exeter,* or *Lucombe Oak*
Cerris var. Lucombeana crispa. *New Lucombe Oak*
Cerris var. Ragnal. *Ragnal Oak*
chrysolepis. *Golden-cup Oak, Live Oak* of California
cinerea. *Blue Jack, Upland Willow Oak*
coccifera. *Kermes Oak*
coccinea. *Scarlet Oak*
coccinea var. ambigua. *Gray Oak*
coccinea var. tinctoria. *Bartram's Oak*
Concordia. *Golden Oak*
conferta. *Hungarian Oak*
dentata. *Mongolia Oak*
Douglasii. *Californian Blue Oak, Mountain White Oak*
esculus. *Italian Oak*

Quercus falcata. *American "Spanish" Oak*
fastigiata. *Cypress Oak*
glabra. *Japanese Oak*
Gramuntia. *Holly-leaved Oak*
Ilex. *Evergreen* or *Holm Oak, Holly-Oak*
ilicifolia (Q. Banisteri). *Bear* or *Black Scrub-Oak*
imbricaria. *Shingle Oak, Small Laurel Oak*
lanata. *Woolly,* or *Himalayan, Oak*
lanuginosa. *Nepaul Oak, Truffle Oak*
laurifolia. *Large Laurel Oak*
Leana. *Lea's Oak*
Libani. *Mt. Lebanon Oak*
lobata. *"Roble"-tree* of Mexico, *Sacramento White Oak*
lusitanica. *Portugal Oak*
lusitanica var. infectoria. *Aleppo Gall Oak, Nut-gall Oak*
lyrata. *Over-cup Oak, Swamp Post Oak, Water White Oak*
macrocarpa. *Burr Oak, Mossy-cup White Oak, Over-cup Oak*
Mirbecki. *Mirbeck's Oak*
mongolica. *Mongolia Oak*
nigra. *Barren Oak,* or *Black Jack,* of N. America
oblongifolia. *Californian Evergreen White Oak, Live Oak*
obtusiloba (Q. stellata). *Iron Oak, Post Oak, Rough* or *Box White Oak*
palustris. *Pin Oak, Swamp Spanish Oak*
persica. *Oak Manna-tree*
Phellos. *Willow Oak*
Prinus. *American Chestnut Oak*
Prinus var. acuminata. *Yellow Chestnut Oak*
Prinus var. humilis. *Chinquapin Oak, Dwarf Chestnut Oak*
Prinus var. monticola. *Rock Chestnut Oak*
Pseudo-coccifera. *Abram's Oak of Mamre*
pubescens. *Durmast Oak, Truffle Oak*
Robur. *Black Oak, Common Oak, Truffle Oak*
Robur var. pedunculata. *Common Oak (Long-flower-stalked), Female Oak*
Robur var. sessiliflora. *Bay* or *Chestnut Oak, Common Oak (Stalkless-flowered), Maiden Oak, Male Oak, White Oak*
rubra. *Red Oak*
salicina. *Willow Oak*
sericea. *Running Oak*
serrata. *Japanese Silk-worm Oak*
striata. *Striped Oak*
Suber. *Cork Oak-tree, Cork-tree*
tinctoria. *American Black Oak, Dyer's Oak, Quercitron, Yellow-barked Oak*
undulata. *Rocky Mountain Scrub Oak*
virens. *American Live Oak*
Wislizeni var. frutescens. *"Desert Oak,"* of California
Quisqualis indica. *Rangoon Creeper*

Radiola millegrana. *All-seed, Flax-seed*
Ramondia pyrenaica. *Rosette Mullein*
Ramulina furfuracea. *Angola-weed*
Randia aculeata and R. latifolia. *Indigoplant, Ink-berry,* or *Ink-plant,* of the W. Indies
Ranunculus. *Crow-foot, Frog-flower*
aconitifolius. *Fair-Maids-of-France, Fair-Maids-of-Kent*

Ranunculus acris. *Blister-plant, Butter-cup, Butter Daisy, Crow-flower, Meadow Ranunculus, Upright Crow-foot, Yellow Gowan*
alismæfolius. *Water-Plantain Spear-wort*
alpestris. *White Alpine Crow-foot*
alpinus. *Yellow-tinted Alpine Crow-foot*
amplexicaulis. *Snowy-flowered Crow-foot, White Butter-cup*
anemonoides. *Anemone-flowered Crow-foot, Wind-flower Crow-foot*
aquatilis. *"Ram's-foot," Lode-wort, Water-Anemone, Water Crow-foot, Water Snow-cups*
arvensis. *Corn-field Crow-foot, Devil's Claws, Devil's Coach-wheel, Devil's Curry-comb, Hedge-hog, Hunger-weed*
asiaticus and vars. *Common Garden Ranunculus, Double Persian Ranunculus, " Turkey Crow-foot"*
asiaticus var. sanguineus. *Red,* or *Tripoly, Crow-foot*
auricomus. *Goldilocks, Wood Crow-foot*
bulbosus. *Bulbous-rooted Crow-foot, Butter-cup, Butter Daisy, Crow-flower, St. Anthony's-turn'p*
bullatus. *Blistered-leaved Crow-foot, " Portugal " Crow-foot*
chærophyllus. *Chervil-leaved Butter-cup*
cortusæfolius. *Cortusa-leaved Crow-foot*
Cymbalaria. *Sea-side Ivy-leaved Crow-foot*
fascicularis. *American Early Ranunculus*
Ficaria. *Butter-cup, Fig-wort, Crow-foot, Marsh Pile-wort, Small Celandine*
Flammula. *Small Spear-wort*
floribundus. *Profuse-flowering Crow-foot*
fluitans. *Eel-beds, Eel-ware*
glacialis. *Glacier Crow-foot*
Gouani. *Gouan's Butter-cup*
gramineus. *Grass-leaved Butter-cup*
hederaceus. *Frog-wort, Ivy-leaved Crow-foot, Water-Ivy*
hirsutus. *Hairy Crow-foot*
Lingua. *Great Crow-foot, Great Spear-wort*
Lyallii. *New Zealand Water-lily, Rockwood Lily, " Water-lily " of the New Zealand shepherds*
megaphyllus (R. grandifolius). *Madeira Crow-foot*
millefoliatus. *Thousand-leaved Crow-foot.*
monspeliacus. *Montpelier Butter-cup.*
montanus. *Mountain Butter-cup*
multifidus. *Yellow Water Crow-foot*
ophioglossifolius. *Adder's-tongue Spear-wort*
parnassiæfolius. *Parnassia-leaved Crow-foot*
parviflorus. *Small-flowered Crow-foot*
pennsylvanicus. *Bristly Crow-foot*
platanifolius. *Plane-tree-leaved Crow-foot*
pyrenæus. *Pyrenean Crow-foot*
recurvatus. *Hooked Crow-foot*
repens. *Butter-cup, Butter Daisy, Creeping Crow-foot, Crow-flower, Yellow Gowan*
rutæfolius (Callianthemum rutæfolium). *Rue-leaved Crow-foot*
sceleratus. *Celery-leaved Crow-foot*
speciosus (R. grandiflorus fl.-pl., R. bullatus fl.-pl.) *Large Double-flowered Crow-foot*
spicatus. *Spiked Butter-cup*
Stevenii. *Steven's Crow-foot*

Ranunculus Thora. *Kidney-leaved Crow-foot, Mountain Wolf's-bane*
uniflorus. *One-flowered Crow-foot*
Ranwolfia canescens. *Hoary-leaved Milk-wood*
Raoulia eximia. *Vegetable-sheep*, of New Zealand
Raphanus caudatus. *Rat-tailed Radish*
Landra. *Italian Radish, " Landra "*
maritimus. *Sea-side Radish.*
Raphanistrum. *Jointed Charlock, Runch, White Charlock, Wild Radish*
sativus. *Common Cultivated Radish*
Raphia Ruffia and R. tædigera. *Raffia* or *Roffia Palm*
tædigera. *Jupati Palm*
vinifera. *Bamboo Palm, Wine Palm* of W. Africa
Raphiolepis (Cratægus) indica. *E. Indian Haw-thorn*
Ratonia apetala. *Bastard Locust*, of Jamaica, *Bastard Mahogany*
(Cupania) pyriformis. *Australian Brush* or *Bush Apple*
Ravenala madagascariensis (Urania speciosa). *Traveller's-tree*, of Madagascar.
Renanthera arachnitis. *Scorpion-plant.*
Renealmia. *Wild Ginger*, of the W. Indies
Reseda. *Mignonette*
alba. *White Mignonette*
lutea. *Base Rocket, Crambling Rocket, " Italian Rocket "*
Luteola. *Ash-of-Jerusalem, Base Rocket, Dyer's-Rocket, Dyer's-weed, Dyer's Yellow-weed, Weld, Wild Mignonette, Wild Woad, Yellow-weed*
odorata. *Common Garden Mignonette*
odorata (grown under a special mode of treatment). *Tree Mignonette*
odorata var. pyramidalis. *Pyramidal Mignonette*
odorata var. rosea. *Red-flowered Mignonette*
Restio. *Rope-grass*
australis. *Tasmanian Rope-grass*
chondropetalus. *Cape Thatch-grass*
Retinospora. *Japan Cypress* or *White Cedar*
filicoides. *Fern-like Japan Cypress*
filifera. *Thread-bearing Japan Cypress*
lycopodioides. *Club - Moss - leaved Japan Cypress*
obtusa. *Common Japan Cypress, Tree-of-the-Sun*
obtusa var. argentea. *Silver-variegated Japan Cypress*
obtusa var. aurea. *Golden variegated Japan Cypress*
obtusa var. compacta. *Compact Japan Cypress*
obtusa var. pygmæa. *Pigmy Japan Cypress*
pisifera. *Pea-fruited Japan Cypress*
plumosa. *Plume-like Japan Cypress*
plumosa var. argentea. *Silvery Plume-like Cypress*
plumosa var. aurea. *Golden Plume-like Japan Cypress*
Rhabdochloa. *Twig-grass*
Rhacicallis rupestris. *Ear-wort*

Rhagodia. *Australian Red-berry or Seaberry*
 hastata. *Saloop-bush* of *Australia*
Rhamnus. *Buck-thorn, Hart's-thorn, Ram*
 Alaternus. *Barren Privet*
 alpinus. *Alpine Buck-thorn*
 alnifolius. *American Alder-leaved Buckthorn*
 buxifolius. *Box-leaved Buck-thorn*
 californica. *Californian Coffee-tree*
 catharticus. *Common Buck-thorn, French Berries, Hart's-thorn, Rain-berry-, or Rhine-berry-, thorn, Way-thorn*
 chlorophorus. *Chinese "Green Indigo"-plant*
 dahuricus. *Dahurian, or Siberian, Buckthorn*
 Erythroxylon. *Red-wooded Buck-thorn*
 Frangula. *Alder-Buck-thorn, Berry-bearing or Black Alder, Dog-wood*
 franguloides. *Frangula-like Buck-thorn*
 hybridus. *Hybrid Alaternus*
 infectorius. *Yellow-berried Buck-thorn. Yields the "Yellow" or "Persian" Berries of commerce*
 latifolius. *Azorean, or Broad-leaved, Buckthorn*
 longifolius. *Long-leaved Buck-thorn*
 lycioides. *Lycium-like Buck-thorn*
 oleoides. *Olive-leaved Buck-thorn*
 pubescens. *Downy Buck-thorn*
 pumilus. *Dwarf Buck-thorn*
 Purshianus. *Californian Bear-berry*
 pusillus. *Small Buck-thorn*
 rupestris. *Dwarf Rock Buck-thorn*
 saxatilis. *Rock Buck-thorn*
 Theezans. *Theezan Tea-tree*
 tinctorius. *Dyer's Buck-thorn*
 utilis. *Chinese "Green Indigo"-plant, Lokao Dye-plant*
 valentinus. *Valencia Buck-thorn*
 virgatus. *Slender-branched Buck-thorn*
 Wulfenii. *Wulfen's Buck-thorn*
Rhapis flabelliformis. *Ground Rattan Cane*
Rhaponticum cynaroides. *Swiss Centaury*
Rheedia laterifolia. *Wild Mammee-tree*, of Jamaica
Rheum. *Rhubarb*
 australe. *Nepaul Rhubarb*
 compactum. *Thick-leaved Rhubarb*
 crispum. *Curled-leaved Rhubarb*
 Emodi. *Red-veined Himalayan Rhubarb*
 nobile. *Sikkim Rhubarb*
 officinale. *Medicinal Rhubarb*
 palmatum. *Palmate-leaved, or "Turkey," Rhubarb*
 Rhaponticum. *Garden, or Tart, Rhubarb*
 Ribes. *Currant-fruited, or Warted-leaved Rhubarb*
 undulatum. *Bucharian, or Wavy-leaved Rhubarb*
Rhexia virginica. *Deer-grass, Meadow-Beauty*
Rhinacanthus communis (*Justicia nasuta*). *Ring-worm-root*
Rhinanthus Crista-galli. *Cock's-comb, Henpenny Grass, Penny-grass, Rattle Grass, Yellow Rattle* or *Rattle-box*
Rhipogonum scandens. *New Zealand Sarsaparilla*

Rhizophora Mangle. *Mangrove-tree*
 mucronata. *Kunro-bark-tree*
 racemosa. *Red Mangrove*, of Upper Guinea
Rhodanthe Manglesii. *Pink-rosette Everlasting, Swan River Everlasting*
 Manglesii var. maculata. *Rosy-purple Everlasting*
Rhodiola rosea *Heal-all, Rose-root*
Rhododendron albiflorum. *White-flowered Rhododendron*
 anthopogon. *Bearded-flowered Rhododendron*
 arboreum. *Tree-Rhododendron*
 argenteum. *Silvery Rhododendron*
 azaleoides. *Azalea-like Rhododendron*
 barbatum. *Bearded-leaf-stalked Rhododendron*
 blandfordiæflorum. *Blandfordia-flowered Rhododendron*
 Brookeanum. *Rajah Brooke's Rhododendron*
 californicum. *Californian Rhododendron*
 campanulatum. *Bell-flowered Rhododendron*
 campylocarpum. *Curved-podded Rhododendron*
 Camtchaticum. *Kamtschatkan Rhododendron*
 Catawbiense. *Catawba Rhododendron*
 Catesbæi. *Catesby's Rhododendron*
 Caucasicum. *Caucasian Rhododendron*
 Chamæcistus. *Thyme-leaved Rhododendron*
 Championæ. *Mrs. Champion's Rhododendron*
 chrysanthum. *Golden-flowered Rhododendron*
 ciliatum. *Fringed Rhododendron*
 cinnabarinum. *Vermilion-flowered Rhododendron*
 citrinum. *Citron-flowered Rhododendron*
 Dalhousiæ. *Epiphytal Rhododendron, Lady Dalhousie's Rhododendron*
 dahuricum. *Dahurian or Siberian Rhododendron*
 Farreræ. *Mrs. Farrer's Rhododendron*
 ferrugineum. *"Alpine Rose," Rusty-leaved Rhododendron*
 Gibsonii. *Gibson's Rhododendron*
 glaucum. *Glaucous-leaved Rhododendron*
 gracile. *Slender Rhododendron*
 hybridum. *Herbert's Hybrid Rhododendron*
 hirsutum. *"Alpine Rose," Hairy-leaved Rhododendron*
 jasminiflorum. *Jessamine-flowered Rhododendron*
 javanicum. *Javanese Rhododendron*
 Lapponicum. *Lapland Rose-Bay*
 lepidotum. *Scaly Rhododendron*
 longiflorum. *Long-flowered Rhododendron*
 Maddeni. *Major Madden's Rhododendron*
 maximum. *American Great Laurel or Rose-Bay*
 Metternichii. *Metternich's Rhododendron*
 myrtifolium. *Myrtle-leaved Rhododendron*
 Nilagiricum. *Neilgherry Rhododendron*
 niveum. *Snowy-leaved Rhododendron*
 ponticum. *Pontic, or Purple-flowered, Rhododendron*
 ponticum var. odoratum. *Sweet-scented Rhododendron*
 præcox. *Early-blooming Rhododendron*
 punctatum. *Dotted-leaved Rhododendron*
 purpureum. *American Purple-flowered Rhododendron*
 Purshii. *Pursh's Rhododendron*

Rhododendron retusum. *Blunt-leaved Rhododendron*
setosum. *Bristly Rhododendron*
striatum. *"Alpine Rose"*
verticillatum. *Whorled-leaved Rhododendron*
virgatum. *Slender-branched Rhododendron*
Rhodomyrtus tomentosus. *Hill Guava, or Hill Goose-berry*, of India
Rhodora canadensis. *Canadian Rhododendron*
Rhodorrhiza florida and R. scoparia. *Rhodium-oil-plant*
scoparia. *Canary Rose-wood, Lignum Rhodium*
Rhodotypus kerrioides. *"Jamabuki,"* of Japan, *White Kerria*
Rhodymenia palmata. *Dillisk, Dulse*
Rhus. *Sumach*
arborea. *Poison-tree,* of Jamaica
aromatica. *Fragrant Sumach*
caustica. *Lithi-tree,* of Chili
copallina. *Dwarf Sumach*
Coriaria. *Elm-leaved Sumach, Tanner's Sumach*
Cotinus. *Purple Fringed Sumach, "Scotino," Smoke-plant* or *Smoke-tree, Venetian* or *Venus's Sumach, Wig-tree, Zante-wood.* Yields *"Young Fustic"* of commerce
cotinoides. *False Venetian Sumach*
dioica. *St. Vitus's-wood*
diversiloba. *Poison-Oak,* or *"Yeara,"* of California
elegans. *Carolina Sumach*
glabra. *Scarlet Sumach, Vinegar-tree*
glabra var. laciniata. *Fern-leaved Sumach*
juglandifolia (R. vernicifera). *Walnut-leaved Sumach*
lucida. *African Bower-tree*
Metopium. *Burn-wood,* of the W. Indies, *Coral Sumach, Doctor's-Gum-tree, Jamaica Sumach, Mountain Manchineel*
Osbecki. *Osbeck's Sumach*
pentaphylla. *Five-leaved Sumach*
pumila. *Dwarf Sumach*
radicans. *Poison Vine, Rooting-branched Sumach*
semialata. *Chinese,* or *Japanese, Nut-gall-tree*
suaveolens. *Sweet-scented Sumach*
succedanea. *Red Lac Sumach, Tallow-Sumach,* of Japan
Toxicodendron. *American Poison-Ivy* or *Poison-Oak*
typhina. *Stag's-horn Sumach, Vinegar-tree, Virginian Sumach*
venenata. *Poison Ash, Poison Elder, Poison Sumach, Swamp Sumach*
vernicifera. *Japan Lacquer-,* or *Varnish-, Tree*
viridiflora. *Green-flowered Sumach*
zizyphina. *Zizyphus-like Sumach*
Rhynchosia minima. *Wart-herb*
precatoria. *Mexican Rosary-plant*
Rhynchospermum jasminoides (Parechites Thunbergii). *Chinese Ivy, Chinese Jessamine*
Rhynchospora. *Beak-rush,* or *Beak-sedge,* of N. America
alba. *White Beak-sedge*

Rhynchospora corniculata. *Horned Rush*
fusca. *Brown Beak-sedge*
Ribes. *Currant, Goose-berry*
aciculare. *Needle-spined Goose-berry*
albinervium. *White-ribbed-leaved Red Currant*
alpinum. *Alpine Red Currant, Tasteless Mountain Currant*
atropurpureum. *Dark-purple-flowered Black Currant*
aureum. *Buffalo,* or *Missouri, Currant, Golden-flowered Currant*
aureum var. serotinum. *Late-flowering Golden Currant*
bracteosum. *Californian Black Currant*
carpaticum. *Carpathian Red Currant*
caucasicum. *Caucasian Goose-berry*
cereum. *Waxy-leaved Black Currant*
Cynosbati. *Dog-Bramble, Dog-bramble Goose-berry*
diacantha. *Two-spined Goose-berry*
divaricatum. *Spreading-branched Goose-berry*
flavum. *Yellow-flowered Black Currant*
floridum. *Pennsylvania Black Currant*
fragrans. *Fragrant-flowered Goose-berry*
glaciale. *Nepaul Black Currant*
glandulosum. *Glanded-calyxed Red Currant*
gracile. *Slender-branched-Goose-berry*
Grossularia. *"Berries," Cat-berries, Common Goose-berry*
hirtellum. *Hairy-branched Goose-berry*
Hudsonianum. *Hudson's Bay Black Currant*
inebrians. *Intoxicating Black Currant*
lacustre. *Swamp Goose-berry,* of N. America
lacustre var. echinatum. *Hedge-hog Goose-berry*
leptanthum. *Yellow-flowered Goose-berry*
Lobbii (R. subvestitum). *Lobb's Goose-berry, Vancouver Island Goose-berry*
longiflorum. *Long-flowered Black Currant*
macracanthum. *Large-spined Goose-berry*
Menziesii. *Menzies' Goose-berry*
microphyllum. *Small-leaved Goose-berry*
multiflorum. *Many-flowered Red Currant*
nigrum. *Common Black Currant, Gazels* or *Gazles, Quinsy-berry*
niveum. *White-flowered Goose-berry*
opulifolium. *Guelder-Rose-leaved Black Currant*
orientale. *Syrian Goose-berry*
oxyacanthoides. *Hawthorn-leaved Goose-berry*
petræum. *Rock Red Currant*
procumbens. *Procumbent Red Currant*
prostratum. *Fetid Currant,* of N. America
punctatum. *Dotted-leaved Red Currant*
resinosum. *Resinous Red Currant*
rigens. *Stiff-racemed Red Currant*
rotundifolium. *Round-leaved Goose-berry*
rubrum. *Common Red Currant, Garnet-berry, Raisin-tree*
rubrum var. album. *White-fruited Currant*
sanguineum. *Red-flowered Currant*
sanguineum var. album. *White-flowered Currant*

R

Ribes sanguineum var. plenum. *Double-flowered Red Currant*
saxatile. *Rock Goose-berry*
setosum. *Bristly Goose-berry*
speciosum. *Fuchsia-flowered Goose-berry*
spicatum. *Spiked-flowered Red Currant*
subvestitum (R. Lobbii). *Vancouver's Island Goose-berry*
tenellum. *Yellow-flowered Currant*
tenuiflorum. *Slender-flowered Black Currant*
trifidum. *Trifid-calyxed Red Currant*
triflorum. *Three-flowered Goose-berry*
triste. *Dark-flowered Black Currant*
viscosissimum. *Very-clammy Black Currant*
Richardia (Calla) æthiopica. *Lily-of-the-Nile, Trumpet Lily, White Arum-Lily*
hastata. *Yellow Lily-of-the-Nile, Yellow Arum-Lily*
maculata. *Spotted-leaved Lily-of-the-Nile*
Richardsonia scabra. *Mexican Coca-plant, Undulated, or White, Ipecacuanha*
Richea dracophylla. *Australian Grass-tree*
pandanifolia. *Palm Heath*
Ricinus communis. *Castor-bean, Castor-oil-plant, Palma-Christi*
sanguineus. *Red Castor-bean*
Rigidella flammea. *Mexican Stiff-stalk*
Ripogonum parviflorum. *Sarsaparilla, of New Zealand*
Rivea Bona-nox. *Midnapore Creeper*
Rivina. *W. Indian Hoop-withe* or *Hoop-withy*
humilis. *Blood-berry, Carpenters'-herb, Rouge-berry* or *Rouge-plant,* of the W. Indies
octandra. *Hoop-withy,* of Jamaica
Robinia dubia. *Doubtful Locust-tree*
hispida. *Bristly Locust-tree, Common Rose-Acacia*
hispida var. macrophylla. *Large-leafleted Rose-Acacia*
Pseud-Acacia. *Fragrant White-flowered Locust-tree*
Pseud-Acacia var. Decaisneana. *Decaisne's Locust-tree*
umbraculifera. *Parasol Acacia*
viscosa. *Clammy Acacia*
Robinsonia thurifera. *"Resino,"* of Juan Fernandez
Roccella fuciformis. *Angola-weed, Mauritius-weed*
tinctoria. *Archal, Orchal, Orchil,* or *Orchil, Cape-weed, Orchal "Cork," Litmus-plant, Rock Moss*
Rochea (Crassula) coccinea. *Garden Coral*
Rodgersia podophylla. *Rodgers's Bronze-leaf*
Roella ciliata. *African Hare-bell*
Rœmeria hybrida (Glaucium violaceum). *Violet-flowered Horned-Poppy, Wind-rose*
Rollinia longifolia and R. multiflora. *Guiana Lance-wood*
Sieberi. *Sugar-apple*
Romanzoffia sitchensis. *Sitka Water-leaf*
Rosa alpina. *Common Alpine Rose*
alpina var. pyrenaica. *Small Pyrenean Rose*
anemoneflora. *Anemone-flowered Rose*
arvensis. *Briar, Briar Bush, Field Rose*
arvensis var. scandens. *Ayrshire Rose*
Banksiæ. *Lady Banks's Rose*

Rosa Bengalensis. *Bengal Rose*
berberidifolia. *Barberry-leaved Rose*
blanda. *Early Wild Rose,* of N. America, *Hudson's Bay Rose*
Boursaulti. *Boursault Rose*
bracteata. *Macartney Rose*
Brunonii. *Brown's Rose*
canina. *Bird Briar, Briar-rose, Buckie-berries, Canker, Canker-flower, Cat-Whin, Choop-tree, Dog-briar, Dog-rose, Hep-Briar, Hep-Rose,* or *Hep-Tree*
caryophyllacea. *Clove-scented Rose*
centifolia. *Cabbage Rose, Hundred-leaved Rose*
centifolia var. muscosa. *Common Moss Rose*
centifolia var. muscosa cristata. *Mossy-crested Rose*
centifolia var. pomponia. *Provins Rose*
cinnamomea. *Single Cinnamon Rose*
cinnamomea var. fœcundissima. *Double Cinnamon Rose*
damascena. *Damask Rose*
diversifolia. *Diverse-leaved Rose*
Eglanteria. *Eglantine, Sweet-Briar*
fraxinifolia. *Ash-leaved Rose*
gallica. *French Rose*
gallica var. inaperta. *Vilmorin Rose*
glutinosa. *Cretan Rose*
Grevillei. *Seven-sisters Rose*
hibernica. *Irish Rose*
hyacinthina. *Hyacinth-scented Rose*
indica. *China,* or *Monthly, Rose*
indica var. Noisettiana. *Noisette Rose*
indica var. Thea. *Tea-Rose*
lævigata. *Cherokee Rose*
Lawranceana. *Fairy Rose, Miss Lawrance's Rose*
longifolia. *Willow-leaved Rose*
lucida. *Dwarf Wild Rose,* of N. America, *Shining-leaved Rose*
lutea. *Single Yellow Rose*
lutea var. punicea. *Austrian Rose*
majalis. *Dwarf Cinnamon Rose*
Montezumæ. *Mexican Rose*
moschata. *Musk-scented Rose*
mundi (or Gloria-mundi). *York-and-Lancaster Rose*
pimpinellifolia var. pyrenaica. *Pyrenean White Burnet Rose*
pisocarpa. *Pea-fruited Rose*
Pissarti. *Guiland Rose*
polyantha. *Bramble Rose, Many-flowered Rose*
pumila. *Dwarf Austrian Rose*
pyrenaica. *Pyrenean Dwarf Rose*
Regeliana. *Dr. Regel's Rose*
rubifolia. *Bramble-leaved Rose*
rubiginosa. *Eglantine, Sweet-Briar*
rubrifolia. *Red-leaved Rose*
rubrifolia var. Redoutéa. *Redouté's Rose*
rugosa. *Turkestan Dwarf Rose*
rugosa var. alba. *Ramanas Rose*
Sabini. *Sabine's Rose*
semperflorens. *Red China Rose*
sempervirens. *Evergreen Rose*
sericea. *Silky-leaved Rose*
setigera. *Prairie Rose, Wild Climbing Rose* of N. America
sinica. *Three-leaved Rose*

Rosa spinosissima. *Barrow Rose, Fox Rose, Scotch Rose, White Burnet Rose*
spinosissima var. carnea. *Pink-shaded Burnet Rose*
sulphurea. *Double Yellow Rose, Sulphur-flowered Rose*
taurica. *Crimean Rose*
terebinthinacea. *Turpentine Rose*
turbinata var. Francofurtana. *Frankfort Rose*
versicolor. *York-and-Lancaster Rose*
villosa. *Shaggy-fruited Rose*
villosa var. pomifera. *Apple-, or Apple-bearing, Rose*
viridiflora. *Green-flowered Rose*
Yvara. *Japanese Rose*
Rosmarinus officinalis. *Common Rosemary, Old Man*
Rottlera tinctoria (Mallotus philippinensis). *Kamala-tree*, of India. Yields the "Kamala" or "Kamela" of commerce
Roupellia grata. *Cream-fruit-tree*, of Sierra Leone
Royena lucida. *African Bladder-nut, African Snow-drop-tree*
Rubia. *Madder*
cordifolia. *Bengal Madder, Munjeet*
peregrina. *Evergreen Cliver, Wild Madder*
tinctorum. *Dyers' Madder*
Rubus. *Black-berry, Bramble*
arcticus. *Arctic Bramble, Dwarf Crimson-flowered Bramble*
australis. *"Bush Lawyer,"* or *Supple-Jack*, of New Zealand, *New Zealand Bramble*
cæsius. *Blue Bramble, Dew-berry*
canadensis. *Canadian Bramble* or *Dew-berry, Low Black-berry* of N. America
Chamæmorus. *Cloud-berry, Cnout-berry, Knot-berry, Mountain Bramble*
cratægifolius. *Hawthorn-leaved Bramble*
cuneifolius. *Sand Black-berry*, of N. America
deliciosus. *Showy White-flowered Bramble*
discolor fl.-pl. *Double-flowered Bramble*
echinatus. *Hedge-hog Bramble*
fruticosus. *Common Black-berry* or *Bramble, Garten-berries*
fruticosus fl.-pl. *Double-flowered Bramble*
fruticosus var. laciniatus. *Cut-leaved Bramble*
hispidus. *Running Swamp Black-berry*
Idæus. *Common Rasp-berry, Framboise, Framboys, Hind-berry* or *Hine-berry*
leucodermis. *White-skinned Rasp-berry*
nepalensis. *Nepaul Bramble*
Nutkanus. *Nootka-Sound Rasp-berry, Salmon-berry*
occidentalis. *Black American Rasp-berry, Thimble-berry*
odoratus. *Purple-flowered American Rasp-berry, Sweet-scented Bramble*
parvifolius (R. ribesifolius). *Native Australian Bramble*
phœnicolasius. *Japanese Climbing Bramble*
rosæfolius. *Rose-leaved Bramble, Victorian Rasp-berry*
rotundifolius. *Indian Bramble*
rugosus. *Himalayan Rasp-berry*
saxatilis. *Bunch-berry, Roebuck-berry, Stone Bramble*

Rubus spectabilis. *Rose-flowered Bramble, Salmon-berry*
strigosus. *Wild Red Rasp-berry*, of N. America
triflorus. *Dwarf American Rasp-berry*
trivialis. *Low Bush Black-berry*, of N. America
ursinus. *Oregon Black-berry*
villosus. *Common*, or *High, Black-berry*, of N. America
Rudbeckia. *Yellow Cone-flower*
californica. *Californian Cone-flower*
Drummondi (R. columnaris, Obeliscaria pulcherrima). *Drummond's Cone-flower*
fulgida. *Glowing Cone-flower*
hirta. *Hairy Cone-flower*
laciniata. *Cut-leaved Cone-flower*
maxima. *Large Cone-flower*
Newmanniana. *Newman's Cone-flower*
nitida. *Shining Cone-flower*
purpurea. *Purple Cone-flower*
speciosa. *Showy Cone-flower*
triloba. *Three-lobed Cone-flower*
virginiana. *Virginian Cone-flower*
Ruellia paniculata. *Christmas-Pride*, of Jamaica
tinctoria. *Room-*, or *Roum-*, *plant*, of India
tuberosa (Cryphiacanthus barbadensis). *Jamaica Snap-dragon, Many-root* of the W. Indies, *Menow-weed, Spirit-leaf* or *Spirit-weed*
Ruizia cordata. *Isle of Bourbon Sweet-scented White-wood*
Rulingia pannosa. *Black Kurrajong*, of Illawarra
Rumex. *Dock, Docken*
Acetosa. *Green-sauce, Sharp* or *Sour Dock, Sheep's-Sorrel, Sorrel-Dock, Sour Grass, Sour Leek*
alpinus. *Monk's Rhubarb*
aquaticus. *Eldin Docken, Grainless Dock*
Britannica. *Pale Dock*, of N. America
conglomeratus. *Clustered Dock*
crispus. *Curled Dock*
flexuosus. *New Zealand Dock*
Hydrolapathum. *Horse-Sorrel, Water Dock*
hymenosepalus. *"Wild Pie-plant,"* of California
Lunaria. *Tree-Sorrel*
maritimus. *Golden Dock*
montanus. *Maiden Sorrel*
Nemolapathum. *Grove Dock*
obtusifolius. *Butter Dock, Butter Dock;-Broad-leaved Dock, Keddle* or *Kettle Dock*
orbiculatus. *Great Water Dock*, of N. America
Patientia. *Monk's Rhubarb, Patience* or *Patient Dock*
pulcher. *Fiddle Dock*
salicifolius. *White Dock*, of N. America, *Willow-leaved Dock*, of California
sanguineus. *Bloody Dock, Blood-wort, Red-veined Dock*
scutatus. *Buckler-shaped* or *French Sorrel*
tuberosus. *Tuberous-rooted Dock*
verticillatus. *Swamp Dock*, of N. America
vesicarius. *Bladder Dock*
vesicarius var. roseus. *African Spinach*

Ruppia maritima. *Ditch-grass,* of N. America, *Sea-fennel, Sea-grass, Tassel-grass, Tassel Pond-weed*
Ruscus aculeatus. *Box-Holly, Butcher's Broom, Jew's Myrtle, Knee Holly, Knee Holm, Pettigree* or *Pettigrue, Prickly Box, Shepherd's Myrtle, Wild Myrtle*
— androgynus. *Climbing Butcher's Broom.*
— Hypoglossum. *Double-leaved Butcher's Broom, Double-tongue, Horse-tongue, Tongue-blade*
— Hypophyllum. *Thick-leaved Butcher's Broom*
— Hypophyllum var. latifolius. *Broad-leaved Butcher's Broom*
— racemosus. *Alexandrian Laurel*
Ruta. *Rue*
— albiflora. *White-flowered Rue*
— bracteosa. *Sicilian Rue*
— graveolens. *Ave Grace, Common Rue, Countryman's Treacle, Herb-of-Grace, Herb-of-Repentance*
— montana. *Mountain Rue*
— patavina. *Padua Rue*

Sabal (Chamærops) Blackburniana. *Fan-Palm,* or *Thatch-Palm,* of Jamaica
— Palmetto. *Palmetto Palm*
— umbraculifera. *Jamaica Fan-Palm*
Sabbatia. *American Centaury*
— angularis. *Rose Pink*
Saccharum Bengalense. *Bengal Sugar-cane*
— officinarum. *Common Sugar-cane*
— Sara. *Pen-reed-grass*
Sagina. *Pearl-weed* or *Pearl-wort*
— glabra var. corsica (Spergula pilifera). *Lawn Pearl-wort*
— nodosa. *Knotted Spurrey*
Sagittaria chinensis. *Chinese Arrow-head* or *Arrow-leaf*
— montevidensis. *Chilian Arrow-head*
— sagittæfolia. *Common Arrow-head* or *Arrow-leaf, Water-Archer*
— sagittæfolia fl.-pl. *Double-flowered Arrow-head*
— simplex. *Yellow-flowered Arrow-head*
— variabilis. *Various-leaved Arrow-head*
Saguerus saccharifera. *Areng Palm, Gomuti Palm*
Sagus. *Sago Palm*
— amicarum. *Vegetable Ivory-tree*
— farinifera. *Sago Palm,* of the Moluccas
— lævis and S. Rumphii. *Common Sago Palm*
Salicornia herbacea. *Crab-grass, English Sea-grape, Frog-grass, Grass-wort, Marsh Samphire*
Salisburia adiantifolia. *Ginkgo-tree, Maiden-hair-tree*
Salix. *Willow, Osier, Sallow*
— alba. *Common White, Huntingdon,* or *Swallow-tailed Willow*
— alba var. cærulea. *Leicester Willow*
— amygdalina. *Almond-leaved Willow*
— aquatica. *Water Sallow*
— argyrocarpa. *American Silvery-fruited Willow*
— aurita. *Round-eared Willow*
— Babylonica. *Weeping Willow*
— candida. *American Hoary Willow*
— Caprea. *Goat Willow, Great Sallow, Kilmarnock Willow, Northamptonshire "Palm"*

Salix Caprea var. pendula. *Kilmarnock Weeping Willow, Weeping Goat Willow*
— chlorophylla. *American Green Willow*
— cinerea. *Gray Sallow*
— cordata. *American Heart-leaved Willow*
— Cutleri. *Cutler's Willow*
— daphnoides. *Violet Willow*
— discolor. *American Glaucous Willow*
— Doniana. *Rusty-branched Willow*
— fragilis. *Crack Willow*
— fragilis var. decipiens. *White Welsh Willow*
— herbacea. *Dwarf Willow*
— humilis. *Prairie Willow*
— lævigata. *Californian Willow*
— lanata. *Woolly Willow, Silvery Woolly-leaved Willow*
— Lapponum. *Downy Willow*
— lasiandra. *British Columbia Willow*
— livida var. occidentalis. *American Livid Willow*
— longifolia. *American Long-leaved Willow*
— lucida. *American Shining Willow*
— Myrsinites. *Whortle Willow*
— nigra. *American Black Willow*
— pentandra. *Bay-leaved, Laurel-leaved,* or *Sweet Willow, Willow Bays*
— phylicifolia. *Tea-leaved Willow*
— purpurea. *Bitter Willow, Purple Osier, Purple Willow*
— purpurea var. Forbyana. *Fine Basket Osier*
— purpurea var. Helix. *Rose Willow*
— purpurea var. Lambertiana. *Boyton Willow*
— Reginæ. *Silvery Willow*
— repens. *Creeping Willow*
— reticulata. *Highland* or *Netted-leaved Willow*
— rosmarinifolia. *Rosemary-leaved Willow*
— rubra. *Red Osier*
— Russelliana. *Bedford* or *Dishley Willow*
— sericea. *American Silky Willow*
— serpyllifolia. *Thyme-leaved Willow*
— triandra. *French Willow*
— tristis. *American Dwarf Gray Willow*
— viminalis. *"Augers," Common Osier* or *Ozier, Common Withy, Velvet Osier*
— vitellina. *Golden Osier*
Salmalia malabarica. *Silk-cotton-tree* or *Simool-tree,* of Malabar
Salpiglosis sinuata. *Scalloped Tube-tongue*
Salsola fruticosa. *Shrubby Salt-wort*
— Kali. *British Barilla-plant, Prickly Glass-wort, Prickly Salt-wort*
— Soda. *Barilla-plant* of S. Europe
Salvadora persica. *Kikuel-oil-plant, Tooth-brush-tree* of Abyssinia
Salvia. *Sage*
— ægyptiaca. *Egyptian Sage*
— argentea (S. patula). *Silvery Clary, Silvery-leaved Sage*
— arborea. *Tree Sage*
— azurea. *Southern Blue Sage,* of N. America
— bicolor. *Red-and-white-flowered Sage*
— Camertoni. *Camerton's Sage*
— campestris. *Sibthorp's Blue-flowered Garden Sage*
— canariensis. *Paper Sage*
— chionantha. *White-flowered Sage*
— coccinea. *Scarlet-flowered Sage*
— colorata. *Orange-flowered Sage*

Salvia columbaria. *"Chia"-plant*, of California
fulgens. *Scarlet Sage*, of Mexico
gesneræfolia. *Gesnera-leaved Sage*
glutinosa. *Jupiter's-Distaff, Yellow-flowered Hardy Sage*
Grahami. *Graham's Sage*
hians. *Gaping-flowered*, or *Cashmere, Sage*
Horminum. *Red-topped Sage*
interrupta. *Ash-leaved Sage*
lyrata. *Lyre-leaved Sage*
napifolia. *Rape-leaved Sage*
nubicola. *Yellow-flowered Green-house Sage*
officinalis. *Common Garden Sage*
officinalis var. alba. *White-flowered Garden Sage*
officinalis var. tricolor. *Variegated Sage*
patens. *Spreading Blue-flowered Sage*
patens var. alba. *Spreading White-flowered Sage*
patula (S. argentea). *Silvery-leaved Sage, White-flowered Sage*
pomifera. *Apple-bearing Sage*
pratensis. *Meadow Sage, Meadow Clary*
pratensis var. lupinoides. *Lupine Sage*
Rœmeriana. *Dwarf Crimson-flowered Sage*
rutilans. *Pine-apple-scented Sage*
sanguinea. *Red-flowered Sage*
sanguinea var. grandiflora. *Large Red-flowered Sage*
Sclarea. *Clary*
splendens. *Scarlet-flowered Sage*
sylvestris. *Wood Sage*
Texana. *Texan Sage*
urticæfolia. *Nettle-leaved Sage*
Verbenaca. *Eye-seed-plant, Vervain Sage, Wild Clary, Wild Sage*
verticillata. *Whorled Blue-flowered Sage*
violacea. *Violet-flowered Sage*
virgata. *Long-branched Sage*
viridis. *Green-topped Sage*
Samadera indica. *Niepa-Bark-tree*
Sambucus. *Elder*
canadensis. *American Elder, Autumn-flowering Elder*
chinensis. *Chinese Elder*
Ebulus. *Blood Hilder, Dane-ball, Dane's-blood, Dane-wort, Dane's-weed, Dead-wort, Dwarf Herbaceous Elder, Ground-Elder, Wall-wort*
glauca. *Californian Elder*
humilis. *Dwarf Elder*
nigra. *Arn-tree, Boon-tree, Boor-tree, Boutry-tree, Bur-tree, Common Elder* or *Elder-berry*
nigra var. laciniata. *Cut-leaved* or *Parsley-leaved Elder*
nigra var. macrophylla. *Cauliflower Elder*
variegata. *Variegated Cut-leaved Elder*
variegata var. roseflora. *Rosy-flowered Elder*
pubens. *American Red-berried Elder*
racemosa. *Hart's Elder, Scarlet-berried Elder*
xanthocarpa. *N. S. Wales Elder*
Samolus littoralis. *Tasmanian Water Pimpernel* or *Brook-weed, Wild Thyme* of Otago
Valerandi. *Common Brook-weed, Water Pimpernel*
Sandoricum. *Sandal-tree*
Sanguinaria canadensis. *Blood-root, Red Puccoon*
Sanguisorba officinalis. *Great Burnet*
Sanicula canadensis. *Black Snake-root*
europæa. *Common Sanicle, "Wood-March" marilandica. Black Snake-root*
Sanseviera angolensis. *Angola Hemp*
fasciata. *Banded Bow-string Hemp*
guineensis. *Bow-string Hemp*, of Africa
Roxburgii. *Moorva-*, or *Marool-, plant*, of Africa
Zeylanica. *E. Indian Bow-string Hemp*
Santalum acuminatum. *Peach*, or *"Quandong"-tree*, of Australia
album. *E. Indian*, or *White, Sandal-wood*
cygnorum. *Nut-tree*, of W. Australia
Freycinetianum. *Citron* or *Yellow Sandal-wood*, of the Sandwich Islands
latifolium. *Sandal-Wood*, of W. Australia
paniculatum. *Sandal-wood*, of the Sandwich Islands
Preissianum (S. acuminatum). *Australian Peach* or *"Quandong"-tree*
Santolina. *Lavender-Cotton*
alpina. *Alpine Lavender-Cotton*
anthemoides. *Chamomile-leaved Lavender-Cotton*
canescens. *Hoary Lavender-Cotton*
Chamæcyparissus. *Common Lavender-Cotton, Ground-Cypress*
crithmoides. *Samphire-leaved Lavender-Cotton*
ericoides. *Heath-like Lavender-Cotton*
incana. *Woolly Lavender-Cotton*
pectinata. *Scalloped-leaved Lavender-Cotton*
pinnata. *Pinnate-leaved Lavender-Cotton*
rosmarinifolia. *Rosemary-leaved Lavender-Cotton*
squarrosa. *Spreading Lavender-Cotton*
viridis. *Green Lavender-Cotton*
viscosa. *Clammy Lavender-Cotton*
Sapindus emarginatus. *Poongum-*, or *Soap-nut-, oil-plant*
marginatus. *Carolina Soap-berry-tree*
Pappea. *Wild Prune*, of the Cape
Saponaria. *W. Indian Soap-berry-*, or *Soap-nut-tree*
spinosus (Xanthoxylon sapindoides). *Licea-tree*, of Jamaica
Sapium laurifolium. *Milk-wood*, of Jamaica, *W. Indian Gum-tree*
Saponaria. *Soap-wort*
bellidifolia. *Daisy-leaved Soap-wort*
cæspitosa. *Tufted Soap-wort*
caucasica fl.-pl. *Double-flowered Soap-wort*
ocymoides. *Rock Soap-wort*
ocymoides var. splendens. *Brilliant-flowered Soap-wort*
officinalis. *"Bouncing Bet"* of N. America, *Common Soap-wort, Crow-soap, Fuller's Herb, Hedge Pink, Soap-wort Gentian*
Vaccaria. *Cow-Basil, Cow-fat, Cow-herb, Cow-herb Soap-wort, "Glond"*
Sapota Achras. *Maze-berry, Nase-berry-, Nees-berry-*, or *Nis-berry-tree, Sapodilla Plum*
mammosa. *Vegetable Egg*
rugosa. *Beef Apple*
Sideroxylon. *Bully-tree* or *Bullet-tree*

Sarcobatus vermiculatus. *Grease-wood*, of California
Sarcocephalus esculentus. *Guinea, Negro*, or *Sierra Leone Peach*
Sarcodes sanguinea. *Snow-plant*, of California
Sarcopetalum Harveyanum. *Harvey's Vine*
Sarcostemma glaucum. *Venezuelan Ipecacuanha*
Sarcostigma Kleinii. *Adul-*, or *Odal-, oil-plant*
Sargassum bacciferum. *Gulf Sea-weed, Gulf-weed, Sea-Lentils*
Sarothamnus scoparius. *Basam, Basom, Bassam, Bisom*, or *Beesom, Common Broom*
Sarracenia. *Indian Cup, N. American Pitcher-plant, Side-saddle-flower, Trumpet-leaf*
atrosanguinea. *Dark-red Side-saddle-flower*
Drummondi. *Drummond's Side-saddle-flower*
flava. "*Trumpets,*" *Yellow-flowered Huntsman's Horn, Yellow Side-saddle-flower*
formosa. *Handsome Side-saddle-flower*
minor. *Small Side-saddle-flower*
purpurea. *Huntsman's Cup, Purple Side-saddle-flower*
rubra (S. psittacina). *Red Side-saddle-flower*
variolaris. *Hook-leaved Side-saddle-flower*
Sassafras Parthenoxylon. *Oriental Sassafras, Virgin-tree*
Satureia hortensis. *Summer Savory*
montana. *Winter Savory*
Thymbra. *Candian Savory*
viminea. *Penny-royal-tree*
Sauroglossum. *Lizard's-tongue*
Saururus cernuus. *American Swamp-Lily, Lizard's-tail*
Saussurea alpina. *Alpine Saw-wort*
macrophylla. *Large-leaved Saw-wort*
Sauvagesia erecta. *Iron-shrub, St. Martin's-herb*
Saxgothea conspicua (Taxus patagonica). *Prince Albert's Yew*
Saxifraga. *Breakstone, Rock-foil, Saxifrage*
affinis. *Kerry Saxifrage*
aizoides. *Small Yellow-flowered Mountain Saxifrage*
aizoides var. autumnalis. *Larger Yellow-flowered Mountain Saxifrage*
Aizoon. *Marginal Pyramidal Saxifrage*
ajugæfolia. *Bugle-leaved Saxifrage*
ambigua. *Ambiguous Saxifrage*
Andrewsii. *Andrews's Saxifrage*
arctioides. *Arctia Saxifrage*
aromatica. *Aromatic Saxifrage*
aspera. *Rough Saxifrage*
biflora. *Two-flowered Purple Saxifrage*
bryoides. *Bryum-leaved Saxifrage*
Bucklandii. *Buckland's Saxifrage*
Burseriana. *Early White-flowered Saxifrage*
cæsia. *Gray*, or *Silver Moss, Saxifrage*
cæspitosa. *Tufted Saxifrage*
calycifloru. *Rosy-calyxed Saxifrage*
canaliculata. *Channelled-leaved Saxifrage*
carinthiaca. *Carinthian Saxifrage*
ceratophylla. *Horn-leaved Saxifrage, Stag's-horn Saxifrage*

Saxifraga cernua. *Drooping Saxifrage*
Churchillii. *Churchill's Saxifrage*
(Megasea) cordifolia. *Heart-leaved Saxifrage*
coriophylla. *Early Silver Saxifrage*
Cotyledon. *Pyramidal Saxifrage*
(Megasea) crassifolia. *Thick-leaved Saxifrage*
crustata. *Crusted-leaved Saxifrage*
cuneifolia. *Wedge-leaved Saxifrage*
Cymbalaria. *Ivy-*, or *Ivy-leaved, Saxifrage*
diapensioides. *Diapensia Saxifrage*
diversifolia. *Various-leaved Saxifrage*
elegans. *Elegant Saxifrage*
erosa. *Lettuce Saxifrage*
exarata. *Furrowed Saxifrage*
flagellaris. *Whip-cord Saxifrage*
Geum. *Kidney-leaved Saxifrage*
Geum var. serrata. "*Prattling Parnell*"
gibraltarica. *Gibraltar Saxifrage*
granulata. *Fair-Maids-of-France, First-of-May, Meadow Saxifrage*
granulata fl.-pl. *Mountain Rocket*
groenlandica. *Greenland Saxifrage*
Guthrieana. *Guthrie's Saxifrage*
Hausmanniana. *Hausmann's Saxifrage*
Hirculus. *Marsh Saxifrage*
hirsuta. *Hairy Saxifrage*
hypnoides. "*Dovedale Moss,*" *Eve's Cushion, Indian Moss, Mossy Saxifrage, Queen's-Cushion*
incurvifolia. *Incurved-leaved Saxifrage*
juniperina. *Juniper Saxifrage*
lanceolata. *Lance-leaved Saxifrage*
lantoscana. *Fox's-brush Saxifrage*
ligulata. *Great Strap-leaved Saxifrage*
longifolia. *Long-leaved Saxifrage, Pyrenean Saxifrage*
longifolia var. elatior. *Tall Long-leaved Saxifrage*
Maweana. *Mawe's Saxifrage*
Maylii. *Mayl's Saxifrage*
mutata. *House-leek-leaved Pyramidal Saxifrage, Saffron-flowered Saxifrage*
nepalensis. *Nepaul Saxifrage*
nivalis. *Sengreen, Snowy Saxifrage, White Mountain Saxifrage*
oppositifolia. *Purple-flowered Saxifrage*
oppositifolia var. pyrenaica maxima. *Large Rosy-flowered Mountain Saxifrage*
palmata. *Palmate-leaved Saxifrage*
pectinata. *Scalloped-leaved Saxifrage*
pedemontana. *Piedmont Saxifrage*
peltata. *Great Californian Saxifrage, Shield-leaved Saxifrage, Umbrella-plant*
pennsylvanica. *Swamp Saxifrage*
pentadactylis. *Five-fingered Saxifrage*
polita. *Polished Saxifrage*
pubescens. *Downy Saxifrage*
pulchella. *Pretty Saxifrage*
purpurascens. *Himalayan Purple Saxifrage*
pyramidalis. *Pyramidal Saxifrage*
repanda. *Spreading Saxifrage*
retusa. *Pyrenean Purple-flowered Saxifrage*
rivularis. *Brook*, or *Water, Saxifrage*
Rocheliana. *Rochel's Saxifrage*
rosularis. *Rosette Saxifrage*
rotundifolia. *Round-leaved Saxifrage*
sancta. *Yellow-flowered Dwarf Saxifrage*

and Foreign Plants, Trees, and Shrubs. 247

Saxifraga sarmentosa. *"Aaron's-beard"*
  Saxifrage, Creeping Sailor, Creeping Saxifrage, Mother-of-Thousands, Old-Man's-Beard, Straw-berry Geranium, Straw-berry Saxifrage, Wandering Jew
Schraderi. *Schrader's Saxifrage*
spatulata. *Spoon-leaved Saxifrage*
stellaris. *Kidney-wort, Starry Saxifrage*
Sternbergii. *Sternbery's Saxifrage*
Sturmiana. *Sturm's Saxifrage*
tenella. *Yellowish-leaved Saxifrage*
Tombeana. *Tombe's Saxifrage*
tricuspidata. *Three-spined Saxifrage*
tridactylites. *Nail-wort, Rue-leaved Saxifrage, Whitlow-Grass*
umbrosa. *King's-Feather, Leaf-of-St. Patrick, London Pride, "Mignonette" of the French, Nancy-pretty, None-so-pretty, St. Patrick's Cabbage*
valdeusis. *Vaudois Saxifrage*
Vandelii. *Vandel's Saxifrage*
virginiensis. *Early Saxifrage*, of N. America
virginiensis fl.-pl. *Double Virginian Saxifrage*
Wallacei. *Wallace's Saxifrage*
Wilkommiana. *Wilkomm's Saxifrage*

Scabiosa. *Scabious, Pincushion-flower*
arvensis. *Clod-weeed, Egyptian Rose, Field Scabious, Gipsies' Rose*
atropurpurea. *Egyptian Rose, Mournful Widow, Sweet Scabious, Widow's-Flower*
caucasica. *Caucasian Scabious*
Columbaria. *Small or Lilac-flowered Scabious*
Fischeri. *Fischer's Scabious*
graminifolia. *Grass-leaved Scabious*
ochroleuca. *Pale Yellow Scabious*
silenifolia. *Catch-fly-leaved Scabious*
succisa. *Blue-ball, Blue Bonnets, Blue Buttons, Blue Kiss, Blue Scabious, Devil's-bit, Fore-bit, Forebitten More*
Webbiana. *Webb's Scabious*

Scævola cuneiformis. *Fan-flower*, of Tasmania
Lobelia (S. Taccada). *Malay Rice-paperplant, "Taccada"-plant* of India

Scandix Pecten-Veneris. *Adam's, Beggar's, Clock-, Crake-, Crow-, Poke-, Puck-, Pink, Shepherd's, Tailor's, Venus's, or Witches' Needle, Devil's Darning-needle, Ground-Enell, Lady's, Shepherd's, or Venus's Comb, Needle Chervil, Wild Chervil*

Scaphyglottis. *Boat-lip Orchid*
Sceptranthus. *Sceptre-flower*
Schæfferia frutescens. *Crab-wood Tree, False Box*

Schinus molle. *Australian Pepper-tree, Californian Pepper-tree, False Pepper, Peruvian Mastich-tree*

Schistostega. *Cavern-Moss*
pennata. *Feather Cavern-Moss, Iridescent Moss*
Schivereckia podolica. *Russian Mad-wort*
Schizæa. *Comb-Fern or Rush-Fern*, of New Zealand
Schizanthus. *Butterfly-flower, Fringe-flower*
Grahami. *Graham's Fringe-flower*
pinnatus. *Pinnate-leaved Butterfly-flower*
retusus. *Notched Fringe-flower*

Schizophragma hydrangeoides. *Climbing Hydrangea*
Schizostylis coccinea. *Caffre Lily, Crimson Flag*
Schleichera trijuga. *Gum-Lac-tree*
Schoberia fruticosa. *Sea Rosemary, Shrubby Sea-Blite*
Schœnus brevifolius. *Cord-Rush*, of Victoria
nigricans. *Bog-Rush*
Schœpfia chrysophylloides. *White Beef-wood*
Schollera graminea. *Water Star-grass*, of N. America
Schomburgkia tibicinis. *Cow-horn Orchid*
Schotia. *Caffre Bean-tree*
latifolia. *Elephant Hedge Bean-tree*
Schrankia uncinata. *Sensitive Briar*, of N. America
Schrebera swietenoides. *Muccaady-tree*, of India
Schwalbea americana. *Chaff-seed*
Schweiggeria. *Tongue-Violet*
Schweinitzia odorata. *Sweet Pine-sap*
Sciadophyllum Brownei. *"Galapee"-tree*, of the W. Indies
capitatum. *Broad-leaved Balsam, Candlewood-tree*, of S. America
Jacquinii. *Loblolly Sweet-wood*
Sciadopitys. *Umbrella-, or Parasol-, Pine or Fir*
verticillata. *Whorled-leaved Umbrella-Pine*
Scilla. *Squill, Wild Hyacinth*
amœna. *Star-flowered Squill, Star Hyacinth*
amethystina. *Amethyst Squill*
autumnalis. *Autumn-flowering Squill, Winter Hyacinth*
bifolia. *Early Spring Squill*
brachyphylla. *Cape Hyacinth*
campanulata. *Large or Spanish Blue-bell, Spanish Squill*
corymbosa. *Cape Hyacinth*
(Camassia) Fraseri. *Californian Squill, Eastern Quamash, Wild Hyacinth of N. America*
Lilio-Hyacinthus. *Lily-Hyacinth*
nutans. *Bell-Bottle, Common Blue-bell* (of England), *Crake-feet, Crow-bells, Crow-leck, Culverkeys, Dog's Leek, Hare-bell, Wild Hyacinth*
patula. *Spreading Blue-bell*
peruviana. *Cuban Lily, Pyramidal-flowered Squill*
sibirica. *Siberian Squill*
taurica. *Crimean Squill*
umbellata. *Umbel-flowered Squill*
verna. *Sea Onion, Spring-flowering Squill*
Scindapsus pertusus (Monstera deliciosa). *Indian Ivy*
Scirpus. *Club-grass, Club-Rush*
acicularis. *Needle Club-Rush*
cæspitosus. *Deer-hair, Tufted Club-Rush*
Eriophorum. *Wool-grass*, of N. America
fluitans. *Floating Club-Rush*
fluviatilis. *River Bul-Rush*, of N. America
Holoschœnus. *Clustered Club-Rush*
lacustris. *Bass, Bast, Common Bul-Rush, Frail-Rush, Lake Club-Rush, Mat-Rush, Pool-Rush, Spurt-grass*
lacustris var. occidentalis. *"Tule,"* of California

Scirpus maritimus. *Sea-side Club-Rush, Spart-grass*
multicaulis. *Many-stalked Club-Rush*
palustris. *Creeping Club-Rush*
pauciflorus. *Few-flowered Club-Rush*
pungens. *Sharp Club-Rush*
Savii. *Savi's Club-Rush*
setaceus. *Bristly Club-Rush*
sylvaticus. *Wood Club-Rush*
Tabernæmontani var. zebrinus. *Banded Rush*
triqueter. *Triangular-stemmed Club-Rush*
validus. *Great Bul-Rush*, of N. America
Scleranthus annuus. *Knawel*
Scleria. *Nut-Rush*, of N. America
flagellum. *Cutting-grass*
latifolia. *Knife-grass*
scindens. *Razor-grass*
Scleroderma Cepa. *False-Puff-ball*
verrucosum. *Warty False-Puff-ball*
Scolopendrium. *Hart's-tongue Fern*
crispum. *Crisped Hart's-tongue Fern*
erectum. *Tall Hart's-tongue Fern*
Hemionitis. *Mule's Fern*
vulgare. *Burnt-weed, Button-hole, Christ's Hair, Common Hart's-tongue Fern, Horsetongue*
Scolymus hispanicus. *Golden Thistle, Spanish Oyster-plant*
maculatus. *Spotted Golden-Thistle*
Scoparia australis (Teucrium corymbosum). *Australian Wild Liquorice*
dulcis. *Sweet Broom*
Scorodosma fœtidum. *Asafœtida-plant*
Scorpiurus. *Caterpillar-plant*
muricatus. *Prickly Caterpillar-plant*
subvillosus. *Hairy Caterpillar-plant*
sulcatus. *Furrowed Caterpillar-plant*
vermiculatus. *Common Caterpillar-plant*
Scorzonera hispanica. *"Scorzonera," Viper's-grass*
Scrophularia. *Fig-wort*
aquatica. *Bishop's-leaves, Brook-* or *Water-Betony, Brown-wort, Stinking Christopher, Water Fig-wort*
mellifera. *Barbary Fig-wort*
nodosa. *Great Pile-wort, Kernel-wort, Knotted-rooted Fig-wort, Murrain-grass, Stinking Christopher*
nodosa variegata. *Variegated Fig-wort*
Scorodonia. *Balm-leaved Fig-wort*
vernalis. *Yellow-flowered Fig-wort*
Scutellaria. *Helmet-flower, Skull-cap*
alpina. *Alpine Skull-cap*
antirrhinoides. *Snap-dragon Skull-cap*
caucasica. *Caucasian Skull-cap*
galericulata. *Common Skull-cap, Hooded Willow-herb*
integrifolia. *Entire-leaved Skull-cap*
japonica. *Japanese Skull-cap*
lateriflora. *Hood wort, Mad-dog Skull-cap, Mad-dog Weed, Side-flowering* or *Blue Skull-cap*
macrantha. *Large-flowered Skull-cap*
minor. *Hedge Hyssop, Small Skull-cap*
Mocciniana. *Scarlet-flowered Skull-cap*
orientalis *Yellow-flowered Skull-cap, Yellow-Helmet-flower*
purpurascens. *Purplish-flowered Skull-cap*

Scutellaria scordifolia. *Scordium-leaved Skull-cap*
Wrightii. *Wright's Skull-cap*
Scyphanthus elegans. *Chili Cup-flower*
Scyphophorus pyxidatus. *Cup-Lichen, Cup-Moss*
Seaforthia (Ptychosperma) elegans. *Bungalow Palm, Cabbage-Palm* of N. S. Wales
Sebastiania lucida. *W. Indian Poison-wood*
Secale cereale. *Common Rye*
cornutum. *Spurred Rye*
fragile var. *Perennial Rye*
Sechium edule. *Chayota-plant, Choko-, Choco-,* or *Chaka-, plant* of the W. Indies
Securidaca longipedunculata (Lophostylis pallida). *Buaze fibre-plant*, of the Zambesi
Securigera Coronilla. *Ax-weed, Ax-wort,* or *Ax-sitch, Hatchet-Vetch*
Securinega nitida (Lithoxylon nitidum). *Myrtle*, of Otaheite
Sedum. *Stone-crop*
acre. *Common* or *Biting Stone-crop, Country-pepper, Creeping Jack, Gold Dust, Golden Moss, Jack-of-the-Buttery, Wall-Moss, Wall-pepper*
album. *Tall White Stone-crop, Worm-grass*
Anacampseros. *Evergreen Orpine, Herb-of-Friendship*
anglicum. *"English"* or *Dwarf White Stone-crop*
asiaticum. *Asiatic Stone-crop*
atropurpureum. *Dark-purple Stone-crop*
brevifolium. *Short-leaved Stone-crop*
Brownii. *Brown's Stone-crop*
camtschaticum. *Orange Stone-crop*
ciliare (S. oppositifolium). *Fringed Stone-crop*
corsicum. *Corsican Stone-crop*
Crista-galli. *Cock's-comb Stone-crop*
cruentum. *Blood-red Stone-crop*
cyaneum. *Azure Stone-crop*
dasyphyllum. *Thick-leaved Stone-crop*
elongatum. *Elongated Stone-crop*
Ewersii. *Ewers's Stone-crop*
Fabarium (S. spectabile). *Brilliant Stone-crop*
farinosum. *Mealy Stone-crop*
Forsterianum. *Welsh Stone-crop*
glaucum. *Glaucous Stone-crop*
ibericum. *Iberian Stone-crop*
japonicum. *Japanese Stone-crop*
japonicum variegatum. *Variegated Japanese Stone-crop*
Lydium. *Lydian Stone-crop*
Mechani. *Mechan's Stone-crop*
Nævii. *Nævius's Stone-crop*
obtusatum. *Blunt-leaved Stone-crop*
pallidum roseum. *Pale Rose-coloured Stone-crop*
populifolium. *Poplar-leaved Stone-crop*
pulchellum. *Bird's-foot Stone-crop, Purple American Stone-crop*
purpurascens. *Purplish Stone-crop*
purpureum. *Purple Stone-crop*
recurvatum. *Recurved-leaved Stone-crop*
reflexum. *Stone-hore* or *Stone-Orpine, Trip-madam*
rhodanthum. *Rose-flowered Stone-crop*
roseum. *Rosy-flowered Stone-crop*

and Foreign Plants, Trees, and Shrubs. 249

Sedum rotundifolium. *Round-leaved Stone-crop*
rupestre. *St. Vincent's Rocks Stone-crop*
sempervivoides. *Scarlet Stone-crop*
sexangulare. *Six-angled or Six-rowed Stone-crop*
Sieboldi. *Siebold's Stone-crop*
spatulæfolium. *Spoon-leaved Stone-crop*
spectabile (S. Fabarium). *Brilliant Stone-crop*
spirale. *Spiral Stone-crop*
spurium. *Crimson Stone-crop, Large-fringed Stone-crop*
stenopetalum. *Narrow-petalled Stone-crop*
Telephium. *"Alpine" Live-long, Midsummer-men, Orphan-John, Orpine*
telephioides. *Orpine-like Stone-crop*
turgidum. *Thick-leaved Stone-crop*
villosum. *Hairy Stone-crop*
virens. *Deep-green Stone-crop*
virescens. *Greenish-flowered Stone-crop*
Wightmannianum. *Wightmann's Stone-crop*
Selaginella apus. *Creeping-moss, of N. America*
cæsia arborea. *Tree Club-moss*
convoluta. *Rock Lily*
Kraussiana aurea. *Golden Club-moss*
lepidophylla. *Resurrection Plant*
rupestris. *Dwarf Club-moss, of N. America*
serpens. *Serpent-Moss*
Selenipedium. *S. American Lady's-Slipper*
Selinum carvifolium. *Caraway-leaved Milk-parsley*
palustre. *Common Milk-parsley*
Selliera radicans. *Victorian Swamp-weed*
Semecarpus Anacardium. *"Kidney-bean" of Malacca, Malacca-bean-tree, Marking-nut or Marking-fruit-tree, Marsh-nut, Varnish-tree of Sylhet*
Sempervivum. *House-leek*
anomalum. *Anomalous House-leek*
arachnoideum. *Cob-web House-leek*
(Æonium) arboreum. *Tree House-leek*
arenarium. *Sand House-leek*
barbulatum. *Bearded House-leek*
Boutignianum. *Boutigni's House-leek*
calcareum (S. californicum). *Purple-tipped House-leek*
ciliatum. *Teneriffe House-leek*
fimbriatum. *Fringed House-leek*
flagelliforme. *Long-runnered House-leek*
Funckii. *Funck's House-leek*
globiferum. *Hen-and-chickens House-leek*
heterotrichum. *Hair-tipped House-leek*
Heuffeli. *Heuffel's House-leek*
hirtum. *Hairy House-leek*
Laggeri. *Lagger's House-leek*
Mettenianum. *Metten's House-leek*
montanum. *Mountain House-leek*
piliferum. *Hairy-tufted House-leek*
Pittoni. *Pitton's House-leek*
soboliferum. *Hen-and-chickens House-leek*
spinosum. *Spiny House-leek*
tabulæforme. *Table-shaped House-leek*
tectorum. *Bullock's-eye, Common House-leek,* "*Fuet*," *Home-wort, Jupiter's-Beard, Jupiter's-Eye, Sengreen*
tomentosum. *Woolly House-leek*

Sempervivum tortuosum. *Gouty-stalked House-leek*
triste. *Red-leaved House-leek*
ruthenicum. *Russian House-leek*
Wulfeni. *Wulfen's House-leek*
Senebiera Coronopus. *Buck's-horn, Hog-grass, Swine's-Cress, Wart-Cress*
didyma. *American Swine's-Cress or Wart-Cress*
Senecio. *Groundsel, Rag-weed*
abrotanifolius. *Orange-flowered Groundsel*
adonidifolius. *Adonis-leaved Groundsel*
argenteus. *Silvery Groundsel*
artemisiæfolius. *Worm-wood-leaved Groundsel*
aquaticus. *Oak-leaved Rag-wort, Water Rag-wort*
aureus. *American Golden Rag-wort, False Valerian, Life-root, Squaw-weed*
Bolanderi. *Bolander's Groundsel*
campestris. *Field Groundsel, Woolly Groundsel*
Cineraria. *"Dusty Miller"*
cruentus. *Purple-flowered Groundsel*
Doronicum. *Large-flowered or Leopard's-bane Groundsel*
elegans. *Jacoby, Purple Jacobæa*
erucæfolius. *Narrow-leaved Groundsel*
incanus. *Hoary Groundsel*
Jacobæa. *Ben-weed or Bin-weed, Boliaun, Canker-weed, Common Rag-weed or Rag-wort, Fellon-weed, Keddle Dock, James's-, or St. James's-, weed, St. James's-wort, Stagger-wort or Staver-wort, "Yellow-tops"*
japonicus. *Japanese Groundsel*
lobatus. *Butter-weed*
Lyallii. *Mountain Marigold, of New Zealand*
macroglossus. *Cape Ivy, Ivy-leaved Groundsel*
mikanoides. *Climbing Groundsel, German Ivy, Parlour Ivy*
paludosus. *Bird's-tongue*
paludosus and S. palustris. *Fen, or Marsh, Groundsel*
Petasites. *Plume-leaved Groundsel*
pulcher. *Tyerman's Groundsel*
saracenicus. *Broad-leaved Groundsel, "Saracen's Consound"*
scandens. *Yellow German Ivy*
spatulæfolius. *Spoon-leaved Groundsel*
speciosus. *Showy Groundsel*
squalidus. *Oxford Rag-wort*
sylvaticus. *Wood Groundsel*
tomentosus. *Woolly American Rag-wort*
uniflorus. *One-flowered Groundsel*
viscosus. *Clammy Groundsel*
vulgaris. *Common Groundsel, Flower of St. Macarius, Simson*
Sequoia gigantea. *"Big Tree" or Mammoth Tree, of California, Washingtonia, Wellingtonia*
sempervirens. *Californian Evergreen Red-wood*
Serapias cordigera. *Heart-flowered Orchis*
Lingua. *Tongue-flowered Orchis*
Seriocarpus. *White-topped Aster*
conyzoides. *Silk-fruit or White-topped Aster, of Maryland*

Serjania lethalis. *Timboe Fish-poison, of Brazil*
Serratula. *Saw-wort*
  arvensis. *Corn-thistle, Way-thistle*
  coronata. *Crowned Saw-wort*
  tinctoria. *Common Saw-wort, Dyer's Savory*
Sesamum indicum. *E. Indian Sesame or Oily-grain, Gingelly-, or Gingilie-, Oil-plant, Til-, or Teel-, Oil-plant*
  indicum and S. orientale. *Benne-oil-plant*
Sesbania. *Pea-tree*
  aculeata. *"Dunchi-," or "Dhunchi-," plant, of India*
  ægyptiaca. *Sesban*
Seseli. *Meadow-Saxifrage*
  Hippomarathrum. *Horse-fennel*
  cærulea. *Moor Grass*
Sesuvium Portulacastrum. *Samphire or Sea-side Purslane, of the W. Indies*
Setaria germanica. *Dew-grass, German Millet*
  italica. *Bengal Grass, Chinese Corn, Italian Millet*
  viridis. *Bottle-grass, Green Fox-tail-grass*
Seymeria macrophylla. *Mullein-Fox-glove*
Shepherdia argentea. *Beef-suet Tree, Missouri Buffalo-berry, Rabbit-berry*
Sherardia arvensis. *Field Madder, Spurwort*
Shorea robusta. *Sál-, or Saul-, tree, of India*
Sibthorpia europæa. *Cornish Money-wort*
  europæa variegata. *Variegated Cornish Money-wort*
Sicyos angulatus. *One-seeded Star-cucumber*
Sida. *Indian Mallow*
  canariensis. *Canary Island Tea-plant*
  l'eriptera. *Shuttle-cock-plant*
Sideritis. *Iron-wort*
  canariensis. *Iron-wort, of the Canary Islands*
  hyssopifolia. *Hyssop-leaved Iron-wort*
  syriaca. *Syrian Iron-wort*
Siderodendron triflorum. *W. Indian Iron-tree*
Sideroxylon capense. *Iron-wood, of S. Africa*
  dulcificum. *"Miraculous-berry-" tree, of W. Africa*
Silaus pratensis. *Meadow Saxifrage, Pepper Saxifrage*
Silene. *Campion, Catch-fly*
  acaulis. *Cushion-Pink, Moss Campion*
  alpestris. *Alpine Catch-fly*
  anglica. *Small-flowered Catch-fly*
  antirrhina. *Sleepy or Snap-dragon Catch-fly*
  Armeria. *Lobel's Catch-fly*
  Bolanderi. *Bolander's Catch-fly*
  caucasica (S. Zawadskii). *Caucasian Catch-fly*
  conica. *Striated Catch-fly*
  Elisabethæ. *Elizabeth's Catch-fly*
  falcata. *Sickle-leaved Catch-fly*
  gigantea. *Giant Catch-fly*
  Greigi. *Greig's Catch-fly*
  Hookeri. *Hooker's Catch-fly*
  inflata. *Ben or White Ben, Bladder-Campion, Cow-bell, Spatling or Frothy Poppy, White Bottle*

Silene maritima. *Sea-side Catch-fly, Witches'-thimble*
  maritima fl.-pl. *Double-flowered Sea-side Catch-fly*
  noctiflora. *Night-flowering Catch-fly*
  nutans. *Nottingham Catch-fly*
  orchidea. *Orchis-flowered Catch-fly*
  orientalis. *Umbel-flowered Catch-fly*
  Otites. *Spanish Campion*
  paradoxa. *Dover Catch-fly*
  pendula. *Drooping Catch-fly, Italian Catch-fly*
  pennsylvanica. *American Wild Pink, Pennsylvanian Catch-fly*
  Pumilio. *Pigmy Rosy-flowered Catch-fly*
  quadridentata. *Four-toothed Catch-fly*
  quadrifida. *Four-cleft Catch-fly*
  regia. *Royal Catch-fly*
  Requieni. *Requien's Catch-fly*
  rotundifolia. *Large Scarlet-flowered Catch-fly, Round-leaved Catch-fly*
  saxatilis. *Twisted-petalled Catch-fly*
  Saxifraga. *Saxifrage Catch-fly*
  Schafta. *Autumn Catch-fly*
  stellata. *Starry Campion*
  tenella. *Pigmy White-flowered Catch-fly*
  Vallesii. *Swiss Catch-fly*
  virginica. *Fire-Pink*
  viridiflora. *Green-flowered Catch-fly*
Silphium laciniatum. *Compass-plant, Pilot-weed, Polar-plant, Rosin-weed*
  perforatum. *American Cup-plant*
  terebinthinaceum. *Prairie Dock*
Silybum Marianum. *Blessed, Holy, or Milk Thistle, Our Lady's Thistle*
  eburneum. *Elephant or Ivory Thistle, Ivory Milk-thistle*
Simaba Cedron. *Cedron-tree*
Simaruba. *Bitter-wood*
  amara. *Bitter or Mountain Damson, Stave-wood*
  excelsa. *Bitter Ash*
  glauca. *American Bitter-wood*
Sinapis. *Mustard*
  alba. *Charlock, White or Salad Mustard*
  arvensis. *Brassica, Brassock, Carlock, Charlock, Corn Mustard, Field Kale, Kedlock or Kerlock, Wild Kale, Wild Mustard*
  nigra. *Black, Brown, or Grocer's Mustard*
Siphonanthus hastatus. *Elephant's-ear*
Siphonia elastica. *India-rubber-plant, Seringa-oil-plant*
Sison Amomum. *Hedge Hone-wort, Stone-Parsley*
Sisymbrium canescens. *Tansy Mustard*
  Irio. *London Rocket*
  officinale. *Bank Cress, Crambling Rocket, Hedge Mustard*
  Sophia. *Flix-weed or Flux-weed*
Sisyrinchium. *Pig-root, Rush-Lily, Satin-flower*
  anceps. *Grass-leaved Satin flower*
  aureum. *Golden Satin-flower*
  Bermudianum. *Bermuda Satin-flower, Blue-eyed Grass*
  californicum. *Yellow Californian Satin-flower*
  grandiflorum. *Purple Rush-Lily, Spring-bell, Spring Satin-flower*

Sisyrinchium grandiflorum var. album. *White Rush-Lily*
latifolium. *Broad-leaved Satin-flower*
odoratissimum. *Sweet-scented Satin-flower*
striatum. *Yellow Mexican Satin-flower*
Sium. *Water-Parsnip*
angustifolium. *Narrow-leaved Water-Parsnip*
helenianum. *"Jellico," of St. Helena*
latifolium. *Broad-leaved Water-Parsnip*
Sisarum. *Skirret*
Sloanea jamaicensis. *"Break-axe" or Iron-wood*, of the W. Indies
Sloetia Sideroxylon. *Iron-wood*, of Dutch E. Indies
Smilacina. *False-Solomon's-seal*
bifolia. *One-blade or One-leaf*
racemosa. *False-Spikenard*
stellata. *Star-flowered Lily-of-the-Valley*
Smilax. Various species yield the *Sarsaparilla* of commerce
aspera. *Prickly Ivy*
China (S. ferox). *China-Root*
glycyphylla. *Botany Bay or Sweet Tea-plant*, *Sarsaparilla* of Australia
herbacea. *Carrion-flower*
officinalis. *Jamaica Sarsaparilla*
rotundifolia. *Green-Briar*, of N. America
Smyrnium apiifolium. *Candy Alexanders*
aureum. *Golden Alexanders*
Olusatrum. *"Black Pot-herb," Common Alexanders* or *Alisanders*, *Horse-Parsley*
Soja hispida. *Soja-Bean*, of China, *Soy-plant*, *White Gram*
Solandra grandiflora. *Peach-coloured Trumpet-flower*
Solanum. *Nightshade*
Anguivi. *Madagascar Potato*
anthropophagorum. *Cannibal's Tomato*
arboreum. *Tree-Nightshade*
bahamense. *Canker-berry*, of the W. Indies
betaceum. *Beet-leaved Nightshade*
campanulatum. *Bell-flowered Nightshade*
campechiense. *Purple-spined Nightshade*
Capsicastrum. *Star-Capsicum Nightshade*
carolinense. *American Horse-Nettle*
ciliatum. *Fringed Nightshade*
coccineum. *Scarlet-berried Nightshade*
crispum. *Potato-tree*
discolor. *Two-coloured Nightshade*
Dulcamara. *Bittersweet* or *Woody Nightshade*, *Fellon-wood* or *Fellon-wort*, *Felon-wort Mortal*
esculentum. *Jew's-Apple*, *Mad-Apple*
flavum. *Yellow-berried Nightshade*
guineense. *Large Black-berried Nightshade*
hirsutum. *Hairy Nightshade*
hybridum. *Hybrid Guinea Nightshade*
igneum. *Red-spined Nightshade*
jasminoides. *Jasmine Nightshade*
laciniatum. *Kangaroo-Apple*
Lycopersicum. *Love-apple, Tomato*
Lycopersicum var. cerasiforme. *Cherry Tomato*
mammosum. *Macaw-bush, Nipple Nightshade, Turkey-berry* of the W. Indies
marginatum. *White-edged Nightshade*
Melongena (S. ovigerum). *Bringall or Brinjal*, *"Brown Jolly"* of the W. Indies, *Egg-plant, Jew's-Apple*

Melongena fructû-albo. *White-fruited Egg-plant*
Melongena fructû-luteo. *Yellow-fruited Egg-plant*
Melongena fructû-rubro. *Red-fruited Egg-plant*
Melongena fructû-violaceo. *Violet-fruited Egg-plant*
nigrum. *Black-berried Nightshade, Garden Nightshade, Hound-*, or *Hounds-*, *Berry*, *Petty Morel*
nodiflorum. *Branched Calalu*
ovigerum (S. Melongena). *Egg-plant*
Pseudo-Capsicum. *Jerusalem Cherry*, *Winter-Cherry Capsicum*
scandens. *Climbing Nightshade*
sodomeum. *Apple-of-Sodom* or *Dead-Sea-Apple*, *Black-spined Nightshade*
stoloniferum. *Creeping-stemmed Potato*
stramonifolium. *Broad-leaved Nightshade*
Texanum. *Texan Nightshade*
torvum. *Turkey-berry*, of the W. Indies
tuberosum. *Common Potato*
vescum. *"Gunyang,"* of S. E. Australia
Vespertilio. *Canary Nightshade*
villosum and S. xanthocarpum. *Yellow-berried Nightshade*
Soldanella alpina. *Blue Moon-wort*
Solenostemma Argel. *Argel* or *Arghel*
Solidago. *Golden-rod*
canadensis. *Canadian Golden-rod*
grandiflora. *Large-flowered Golden-rod*
multiflora. *Many-flowered Golden-rod*
multiradiata. *Many-rayed Golden-rod*
nutans. *Drooping-flowered Golden-rod*
stricta. *Upright-branched Golden-rod*
Virgaurea. *Common Golden-rod*
Virgaurea var. angustifolia. *Dorcas' Wound-wort*
Virgaurea var. cambrica. *Dwarf Golden-rod*
Sollya heterophylla. *Australian Blue Bell Creeper*
Sonchus. *Sow-thistle*
alpinus. *Mountain Sow-thistle*
arvensis. *Corn Sow-thistle*
cæruleus. *Blue-flowered Sow-thistle*
elegantissimus. *Finely-cut-leaved Sow-thistle*
fruticosus. *Shrubby Sow-thistle*
laciniatus. *Cut-leaved Sow-thistle*
oleraceus. *Common Sow-thistle, Hare's-Lettuce, Hare's-palace, Milk-thistle, Milk-weed*
palustris. *Marsh Sow-thistle*
(Mulgedium) Plumieri. *Panicle-flowered Sow-thistle*
Sonneratia apetala. *Kambala-tree*
Sophora japonica. *Chinese*, or *Japanese, Pagoda-tree*
tetraptera. *Pelu-tree*
tetraptera var. microphylla (Edwardsia microphylla). *New Zealand Laburnum*
Sophronitis grandiflora. *Scarlet-flowered Orchid*
Sorghum. *Millet-grass*
cernuum. *Texas Millet*
halepense. *Aleppo Millet-grass*
nutans. *Broom-corn, Indian-grass, Wood-grass*

Sorghum saccharatum. *Broom-corn, Chinese Sugar-cane, Imphee Cane, Caffre* or *Kaffir Corn, Sugar-Millet, Sweet Reed*
vulgare. *Dhourra, Dhurra, Doura,* or *Durra, E. Indian Millet, Guinea-corn*
Sorindeia trimera. *Balsam of St. Thomas,* W. Africa
Soulamea amara. *Bitter-king*
Soymida (Swietenia) febrifuga. *E. Indian Mahogany, Red-wood* of Coromandel, *Rohun-Bark-tree*
Sparaxis. *African Harlequin-flower*
Sparganium natans. *Floating Bur-reed*
ramosum. *Bede-Sedge, Common Bur-reed, Knop-Sedge*
simplex. *Unbranched Bur-reed*
Sparganophora verticillata. *American Water Crown-cup*
Sparmannia africana. *African Hemp*
Spartina cynosuroides. *Fresh-water Cordgrass, Prairie-grass*
polystachya. *Salt-water Reed-grass*
stricta. *Common Cord-grass, Mat-weed, Spart-grass, Twin-spiked Cord-grass*
Spartium junceum. *Rush-Broom, Spanish Broom*
Spatalanthus speciosus. *Cape Ribbon-flower*
Spathelia simplex. *"May-pole"* of Jamaica, *Mountain-pride,* or *Mountain-green,* of the W. Indies
Specularia hybrida. *Corn Violet*
perfoliata. *Venus's-Looking-glass,* of N. America
Speculum. *Common Venus's-Looking-glass*
Spergula arvensis. *Corn Spurrey, Toad-flax*
pilifera. *Lawn Pearl-wort, Lawn Spurrey*
sativa. *Spurrey-oil-plant*
Spergularia rubra. *Red Sand-wort, Sand Spurrey*
Spermacoce glabra. *Button-weed,* of N. America
Sphæralcea. *Globe-Mallow*
Sphæranthus hirtus. *E. Indian Globe-Thistle*
Sphæria mortosa. *"Black-knot" Fungus*
Sphærococcus (Gracillaria) lichenoides. *Ceylon,* or *Jaffna, Moss*
Sphagnum. *Bog-Moss, Gold Heath*
acutifolium. *Slender Bog-Moss*
cuspidatum. *Long-leaved Floating Bog-Moss*
cymbifolium. *Blunt-leaved Bog-Moss*
squarrosum. *Spreading-leaved Bog-Moss*
Spigelia Anthelmia. *Pink-root,* of Demerara
marilandica. *Carolina Pink, Maryland Pink-root, Worm-grass*
Spilanthes Acmella. *Alphabet-plant*
oleracea. *Para Cress*
Spinacia oleracea. *Garden Spinach*
oleracea var. glabra. *Round-seeded Spinach*
oleracea var. spinosa. *Prickly-seeded Spinach*
Spiræa. *Meadow-sweet*
ariæfolia. *Shrubby Meadow-sweet*
corymbosa. *Alleghany Meadow-sweet, "May"* of N. S. Wales
Filipendula. *Drop-wort, Italian "May"*
Filipendula plena. *Double-flowered Drop-wort*

Spiræa flagelliformis. *Long-sprayed Meadow-sweet*
hypericifolia. *Italian "May"*
hypericifolia var. *American "May"*
lobata. *Queen-of-the-Prairie*
opulifolia. *Nine-bark Snow-ball-tree, Virginian Guelder-Rose*
pachystachya. *Plumy Meadow-sweet*
salicifolia. *American Meadow-sweet, Queen-of-the-Meadow*
tomentosa. *Hard-Hack, Steeple-bush*
Ulmaria. *Bitter-sweet, Bride-wort, Common Meadow-sweet, Courtship-and-Matrimony, Lady-of-the-Meadow, Maid-of-the-Meadow*
Ulmaria variegata. *Variegated Meadow-sweet*
Spiranthes æstivalis. *Summer Lady's-tracess* or *Lady's-tresses*
autumnalis. *Autumn Lady's-traces* or *tresses*
cernua. *Drooping Lady's-tresses*
gemmipara. *Irish Lady's-tresses*
Splachnum. *Collar Moss*
rubrum. *Bon-grace Moss*
Spodiopogon angustifolius. *Bunkuss-grass*
Spondias. *Hog-Plum*
dulcis. *Otaheite Apple*
lutea. *Golden Apple, Jamaica Plum*
Mombin (S. purpurea). *Purple Hog-plum,* or *Spanish Plum,* of the W. Indies
Sporobolus. *American Drop-seed Grass*
Sprekelia (Amaryllis) formosissima. *Jacobea Lily*
Stachys. *Hedge-Nettle*
arvensis. *Field Betony*
Betonica. *Betony, Bishop's-wort, Wood Betony*
coccinea. *Scarlet Wound-wort*
corsica. *Corsican Wound-wort*
germanica. *Common Wound-wort, Lamb's-ear*
lanata. *Woolly Wound-wort*
palustris. *Clown's All-heal* or *Wound-wort, Marsh Betony*
sylvatica. *Archangel, Common Hedge-Nettle, Deye-nettle, Wild Nettle-grass, Wood Betony*
Stachytarpheta. *Bastard Vervain*
indica. *E. Indian False-Vervain*
jamaicensis. *Brazilian Tea-tree*
Stadtmannia (Cupania) Sideroxylon. *Bourbon Iron-wood*
Stangeria paradoxa. *Hottentot's-head*
Stapelia. *Carrion-flower*
Asterias. *Star-fish-flower*
bufonia. *Toad-flower*
Staphylea colchica. *Colchican* or *Ivory-flowered Bladder-nut*
pinnata. *Anthony-nut,* or *St. Anthony's-nut, Common,* or *European, Bladder-nut*
trifoliata. *American Bladder-nut*
Statice. *Sea-Lavender*
alpina. *Alpine Thrift*
angustifolia. *Narrow-leaved Sea-Lavender*
Armeria. *Common Thrift, Sea-pink*
auriculæfolia. *Rock Sea-Lavender*
bellidifolia. *Daisy-leaved Sea-Lavender*
Bonducelli. *Bonducelle's Sea-Lavender*
caspia. *Caspian Sea-Lavender*

and Foreign Plants, Trees, and Shrubs. 253

Statice elata (Goniolimon elatum). *Tall Sea-Lavender*
eximia. *Rosy-flowered Sea-Lavender*
globulariæfolia. *Globularia-leaved Sea-Lavender*
Gmelini. *Gmelin's Sea-Lavender*
Holfordi. *Holford's Sea-Lavender*
incana (Goniolimon callicomum). *Hoary Sea-Lavender*
Kaufmanniana. *Kaufmann's Sea-Lavender*
latifolia. *Great Sea-Lavender*
Limonium. *Common Sea-Lavender, Wild Marsh-Beet*
minuta. *Pigmy Sea-Lavender*
nana. *Dwarf Sea-Lavender*
oleæfolia (S. virgata). *Olive-leaved Sea-Lavender*
profusa. *Profuse Sea-Lavender*
puberula. *Downy Sea-Lavender*
reticulata. *Matted Sea-Lavender*
speciosa. *Showy Sea-Lavender*
spicata. *Spiked Sea-Lavender*
tatarica. *Tartarian Sea-Lavender*
taurica. *Crimean Sea-Lavender*
Stauracanthus aphyllus. *Cross-spine*
Stellaria. *Star-wort, Stitch-grass,* or *Stitch-wort*
aquatica. *Water Star-wort*
borealis. *Northern Star-wort*
graminea. *Small Star-wort* or *Stitch-wort*
graminea aureo-variegata. *Golden Chickweed* or *Stitch-wort*
Holostea. *Agworm-flower, All-bone, Bird's-tongue, Break-bones, Easter Bell, Great Star-wort* or *Stitch-wort, May-grass, Miller's-star, Snap-stalks, Tongue-grass*
media. *Common Chick-weed*
nemorum. *Wood Star-wort*
pubera. *Great Chick-weed,* of N. America
uliginosa. *Bog Star-wort*
Stemodia durantæfolia. *Goat-weed*
maritima. *Bastard,* or *Sea-side, Germander*
Stenocarpus Cunninghamii. *Tulip-tree,* of Queensland
salignus. *Beef-wood,* of N. S. Wales
sinnatus (S. Cunninghamii). *Queensland Tulip-tree* or *Fire-tree*
Stenostomum bifurcatum. *White* or *Wild Mahogany,* of the W. Indies
Stenotaphrum americanum (S. glabrum). *Australian Buffalo-grass*
Stephanotis floribunda. *Clustered Wax-flower, Madagascar Chaplet-flower, Madagasear Jasmine*
Sterculia alata. *Buddha's Cocoa-nut*
caribæa. *"Mahoe"* of the W. Indies
diversifolia. *Bottle-tree,* of Victoria
platanifolia. *Chinese,* or *Sultan's, Parasol*
quadrifida. *"Calool"* of Australia
(Delabechea) rupestris. *Barrel-tree, Bottle-tree, Gouty-stemmed-tree,* of Australia
Tragacantha. *Tragacanth-gum-tree,* of Sierra Leone
urens. *Kuteera-gum-tree*
villosa. *Oadal-,* or *Oo'dhall-,tree,* of India
Sternbergia ætnensis. *Mount Etna Lily*
clusiana. *Turkish Star-flower*
lutea. *Winter Daffodil.* Supposed to be the *"Lily of the Field,"* of Scripture
lutea. *Yellow Star-flower*

Sticta pulmonaria. *Hazel-Crottles, Hazel-Raw, Lung Lichen, Lungs-of-the-Oak, Tree-Lung-wort*
Stictina scrobiculata. *Aikraw*
Stigmaphyllon ciliatum. *Golden Vine*
Stilago Bunius. *Chinese Laurel*
Stillingia sebifera. *Tallow-tree,* of China
sylvatica. *Queen's-delight, Queen's-root, Silver-leaf, Yaw-root*
Stipa avenacea. *Black Oat-grass*
pennata. *Feather-grass*
spartea. *Porcupine-grass*
(Macrochloa) tenacissima. *Esparto-grass*
Stœbe cinerea. *Cape Sea Worm-wood*
Stokesia cyanea. *Stokes's Aster*
Stratiotes aloides. *Crab's-claw, Fresh-water-Soldier, Knight's Pond-wort, Soldier's Yarrow, Wading Pond-weed, Water-Aloe, Water-House-leek, Water-Sengreen, Water-Soldier*
Strelitzia. *Bird's-tongue-Flower, Bird-of-Paradise-Flower*
Streptanthus obtusifolius. *Arkansas Cabbage*
Streptocarpus. *Cape Primrose*
Streptopus. *Twisted-stalk*
Strobilanthes. *Cone-head*
Strophanthus Kombe (S. hispidus). *Tropical African,* or *Gaboon, Arrow-poison*
Struthiopteris germanica (S. pennsylvanica) *Ostrich-Fern,* of N. America
japonica. *Japanese Ostrich-Fern*
Strychnodaphne floribunda. *Black Sweet-wood*
Strychnos Colubrina. *E. Indian Snake-wood*
Ignatii. *St. Ignatius's Bean*
Nux-vomica. *Nux-vomica-tree.* Yields *False-Angostura Bark* and *Strychnine*
potatorum. *Water-filter Nut*
Pseudo-quina. *Brazilian Copalchi-plant*
toxifera. *Curari, Urari, Ourali,* or *Wourali Arrow-poison-plant,* of Guiana
Stylidium graminifolium. *Grass - leaved Trigger-plant*
Stylophorum diphyllum. *Celandine Poppy*
Stylosanthes elatior. *Pencil-flower*
Styrax. *Storax*
Benzoin. *Gum-Benzoin-,* or *Gum-Benjamin-, shrub*
grandifolia. *Large-leaved Storax*
lævigata. *Smooth-leaved Storax*
officinalis. *Officinal Storax*
pulverulenta. *Powdery Storax*
Suæda maritima. *Sea Blite, Sea-side Goose-foot, White Glass-wort*
Subularia aquatica. *Awl-wort*
Sutherlandia frutescens. *Bladder-Senna,* of the Cape
Swainsona Greyana. *Darling River Pea, Horse-poison-plant* of Australia, *Swainson* or *Poison Pea,* of Australia
Swartzia tomentosa. *Palo-Santo-tree,* of Guiana. Yields *Panocoeco Bark*
Swertia perennis. *Marsh Fel-wort*
Swietenia Chloroxylon. *Yellow Satin-wood,* of the E. Indies
Mahogani. *American,* or *Spanish, Mahogany-tree*
senegalensis. *African,* or *False, Mahogany*

Sycomorus antiquorum. *Pharaoh's Fig*
Symphiandra (Campanula) pendula. *Pendulous Bell-flower*
Symphoricarpus. *St. Peter's-wort, Snow-berry-tree*
  microphyllus. *Small-leaved Snow-berry-tree*
  montana. *Mountain Snow-berry*
  occidentalis. *Wolf-berry*
  racemosus. *Common Snow-berry-tree, St. Peter's-wort*
  vulgaris. *Coral-berry, Indian Currant*
Symphytum. *Comfrey*
  asperrimum (S. peregrinum). *Prickly or Forage Comfrey, "Trottles"*
  bohemicum. *Bohemian Comfrey*
  caucasicum. *Caucasian Comfrey*
  ibericum. *Iberian Comfrey*
  officinale. *Alum, Back-wort, Black Root, Black-wort, Bone-set, Comfrey Consound, Common Comfrey, Knit-back*
  tauricum. *Crimean Comfrey*
  tuberosum. *Tuberous-rooted Comfrey*
Symplocarpus fœtidus. *Meadow-cabbage, Pole-cat Weed, Skunk Cabbage, Skunk Weed*
Symplocos racemosa. *Lodh-bark-tree*
  (Hopea) tinctoria. *Horse-sugar, or Sweet-leaf*, of Carolina
Synadenium. *African Milk-bush*
Syncarpia latifolia. *Turpentine-tree*, of N. S Wales
  laurifolia. *Turpentine-tree*, of Queensland
Syngonium auritum. *Five-fingers*, of the W. Indies
Synœum glandulosum (Trichilia glandulosa). *Australian Rose-wood*
Syringa. *Lilac, Pipe-Privet, Pipe-tree*
  chinensis. *Chinese Lilac*
  Emodi. *Himalayan Lilac*
  Josikæa. *Lady Josika's Lilac*
  persica. *Persian Lilac*
  persica var. alba. *White Persian Lilac*
  persica var. laciniata. *Cut-leaved Persian Lilac*
  persica var. salviæfolia. *Sage-leaved Persian Lilac*
  rothomagensis. *Rouen Lilac* (hybrid)
  vulgaris. *Common Lilac, Pipe-tree*
  vulgaris var. alba. *Common White Lilac*
  vulgaris var. alba-major. *Large White Common Lilac*
  vulgaris var. alba-plena. *Double White Common Lilac*
  vulgaris var. cœrulea. *Blue Common Lilac*
  vulgaris var. grandiflora. *Charles X. Lilac*
  vulgaris var. nana. *Dwarf Lilac*
  vulgaris var. rubra. *Red Common Lilac*
  vulgaris var. rubra-major. *Large Red Common Lilac*
  vulgaris var. rubra-plena. *Double Red Common Lilac*
  vulgaris var. violacea. *Common Purple, or Scotch, Lilac*

Tabernæmontana coronaria. *Adam's Apple, E. Indian Rose-Bay*
  utilis. *Cow-tree*, of British Guiana
Tacca pinnatifida. *Otaheite Salep-plant, Pi-plant, South-Sea-Arrow-root-plant*

Tacsonia sanguinea. *Blood-red Passion-flower*
  Van Volxemi. *Van Volxem's Passion-flower*
Tagetes erecta. *African Marigold*
  lucida. *Sweet-scented Mexican Marigold*
  l'arryi. *Parry's Marigold*
  patula. *French Marigold*
  patula var. pumila. *Dwarf Double French Marigold*
  signata. *Striped Mexican Marigold*
  signata var. pumila. *Dwarf Striped Marigold*
  tenuifolia. *Slender-leaved Dwarf Marigold*
Talinum patens. *"Puchero,"* of Mexico
  teretifolium. *Fame-flower*
Tamarix gallica. *Common, or French, Tamarisk*
  gallica var. mannifera. *Tamarisk Manna-plant*
  indica. *E. Indian Tamarisk*
  mannifera. *Manna-tree*, of Mt. Sinai
  orientalis. *"Atlee-Gall"-tree, Tamarisk Salt-tree*
  parviflora. *Small-flowered Tamarisk*
  plumosa. *Feathery Tamarisk*
  spectabilis. *Showy Tamarisk*
Tamus communis. *Black Bryony, Isle-of-Wight Vine, Murrain-berries, Lady's Seal, Mandrake, Ox berry*
  edulis. *Port Moniz Yam*
Tanacetum. *Tansy*
  Balsamita. *Ale-cost or Ale-coast, Cost, Cost-mary*
  suffruticosum. *Shrubby Tansy*
  vulgare. *Buttons, Common Tansy*
Tanghinia (Cerbera) lactaria. *Milk-tree*, of Madagascar
  venenifera. *Ordeal-tree*, of Madagascar
Taraxacum Dens-leonis (Leontodon Taraxacum). *Dandelion, Priest's-crown, Swine's-snout*
Tarchonanthus. *African Flea-bane*
  camphoratus. *Shrubby African Flea-bane, Wild Sage*, of the Cape
Tarrietia Argyrodendron (Argyrodendron trifoliatum). *Queensland Silver-tree*
Tasmannia aromatica. *Pepper-tree*, of Tasmania
Taverniera Nummularia. *E. Indian Money-wort*
Taxodium capense. *Cypress Broom*
  distichum. *Bald, Black, or Deciduous Cypress, Sabino-tree*
  distichum var. mexicanum (T. mucronatum). *Montezuma Cypress*
  distichum var. pendulum. *Weeping Deciduous Cypress*
  sinense. *Chinese Deciduous Cypress*
Taxus. *Yew-tree*
  adpressa. *Short-leaved Japan Yew*
  baccata. *Common Yew, "Palm"*
  baccata var. argentea. *Silver-variegated Yew*
  baccata var. aurea. *Golden Yew*
  baccata var. canadensis. *American Yew, Ground Hemlock*
  baccata var. canadensis Washingtoni. *Washington's Yew*

and Foreign Plants, Trees, and Shrubs. 255

Taxus baccata var. Cheshuntensis. *Cheshunt Yew*
baccata var. Dovastoni. *Dovaston's Yew*
baccata var. erecta. *Erect Common Yew*
baccata var. fastigiata. *Florence Court, or Irish, Yew*
baccata var. fructu-luteo. *Yellow-berried Yew*
baccata var. Nidpathensis. *Nidpath Castle Yew*
baccata var. sparsifolia. *Scattered-leaved Yew*
brevifolia. *Californian Yew*
Floridana. *Florida Yew*
globosa. *Mexican Yew*
nucifera. *Chinese Yew*
stricta. *" Blue John " Yew,* of N. America
Tecoma. *Trumpet-flower*
australis. *Churchill Island Jasmine or Creeper, Wonga-Wonga-Vine*
capensis. *W. Indian Honey-suckle*
(Bignonia) grandiflora. *Large Hardy Trumpet-flower*
jasminoides. *Moreton Bay Trumpet Jasmine or Bower-plant*
Leucoxylon. *White-wood Cedar, White-wood,* of Barbadoes
pentaphylla. *W. Indian, or Jamaica, Box-wood*
radicans. *Rooting-branched,* or *Virginian, Trumpet-flower, Trumpet Creeper*
serratifolia. *" Pony," Saw-leaved Trumpet-flower*
stans. *Shrubby Trumpet-flower, Yellow Elder*
Tecophylæa cyanocrocus. *Chilian Crocus*
Tectona australis. *Beech,* of Australia
grandis. *E. Indian Oak* or *Teak-tree*
Teesdalia Iberis. *Shepherd's-cress*
nudicaulis. *Pepper-cress*
Telephium Imperati. *Tree-Orpine*
Telopea speciosissima. *" Waratah,"* of Australia
Templetonia retusa (T. glauca). *Coral-bush,* of W. Australia
Tephrosia. *Hoary Pea,* of N. America
Apollinea. *Egyptian Indigo-plant*
cinerea. *Goat's-Rue,* of the W. Indies
(Galega) toxicaria. *Indigo-plant,* of the Niger, *W. Indian Fish-poison-plant, Surinam-poison*
virginiana. *Cat-gut, Virginian Goat's-Rue*
Terfezia. *African Truffle*
Terminalia. *Myrobalan-tree*
Bellerica. *Balda-nut-tree*
Catappa. *" Country " Almond-tree,* of the E. Indies, *Malabar Almond-tree*
Chebula. *Negroes' Olive-tree*
citrina. *Hara-nut Tree*
mauritiana. *False Benzoin*
Ternstrœmia obovalis. *Scarlet-seed*
Testudinaria elephantipes. *Elephant's-foot, Hottentot-bread, Tortoise-plant*
Tetracera alnifolia. *Water-tree,* of Sierra Leone
Tigarea. *Red Creeper,* of Cayenne, *Tigarea-tree,* of Guiana
Tetragonia expansa. *New Zealand Ice-plant, New Zealand Spinach*

Tetragonia implexicoma (T. trigyna). *Australian* and *New Zealand Spinach, Tasmanian Ice-plant, Victorian Bower Spinach*
Tetragonolobus. *Winged Pea*
purpureus. *Purple-flowered Winged-Pea*
Tetrameles. *Jungle-bendy-tree,* of India, *Weenong-tree,* of Java
Tetranema mexicana (Pentstemon mexicanum). *Mexican Fox-glove*
Tetranthera geniculata. *Pond-spice*
Teucrium. *Germander*
aureum. *Golden Germander, Golden Poly*
betonicum. *Madeira Germander*
Botrys. *Jerusalem Oak*
campanulatum. *Small-flowered Germander*
canadense. *American Germander* or *Wood-Sage*
Chamædrys. *Wall-,* or *Wild-, Germander*
flavescens. *Yellow Poly*
fruticans. *Tree-Germander*
lucidum. *Shining Germander*
Marum. *Cat Thyme*
Massiliense. *Sweet-scented Germander*
Polium. *Cat-Thyme, Hul-wort, Poly Germander*
Pseudo-Chamæpitys. *Bastard Ground-Pine*
Pseudo-Hyssopus. *Bastard Hyssop*
pyrenaicum. *Pyrenean Germander*
Scordium. *Water Germander*
Scorodonia. *Garlic Sage, Hind-heal, Mountain Sage, Wood-Germander, Wood-Sage*
Thea. *Cochin China Tea*
Thalassia testudinum. *Manatu-grass, Turtle-grass*
Thalictrum. *Meadow-Rue*
alpinum. *Alpine Meadow-Rue*
anemonoides. *Rue-Anemone, Wind-flower Meadow-Rue*
aquilegifolium. *Columbine Meadow-Rue, Feathered* or *Tufted Columbine, Spanish-Tuft*
atropurpureum. *Dark-flowered Meadow-Rue*
Cornuti. *Canadian Tall Meadow-Rue*
dioicum. *Early Meadow-Rue*
elatum. *Hungarian Tall Meadow-Rue*
fœtidum. *Fetid Meadow-Rue*
flavum. *False Rhubarb, Fen Rue, Poor Man's Rhubarb, Yellow - flowered Meadow-Rue, Maiden-hair or Small Meadow-Rue*
purpurascens. *Purplish Meadow-Rue*
tuberosum. *Tuberous-rooted Meadow-Rue*
Thamnochortus. *Shrubby Grass*
Thapsia. *Deadly Carrot*
fœtida. *Stinking Carrot*
garganica. *" Drias "-plant,* of Algeria
Thapsus Laserpitii. *Laser-wort*
Thaspium. *Meadow-parsnip,* of N. America
Thea. *Tea-tree*
Assamensis (T. Assamica). *Assam Tea-tree*
Bohea. *Bohea Tea-tree*
cochin-chinensis. *Cochin-China Tea-tree*
oleosa. *Oily Tea-tree*
viridis. *Green Tea-tree*
Thelygonum Cynocrambe. *Dog's-Cabbage*
Thelymitra. *Woman's-cap Orchid*
nuda. *Tasmanian Hyacinth*
Theobroma Cacao. *" Cocoa "-tree, Chocolate-nut-tree*
Thesium linophyllum. *Bastard Toad-flax*

Thespesia populnea. *Mahoe*, of Demarara, *Portia-nut-oil-plant, Umbrella-tree*
Thevetia neriifolia. *Exile-oil-plant*
Thlaspi. *Bastard Cress, Besom-weed*
arvense. *Boor's Mustard, Dish Mustard, Penny-Cress, Wild-Cress*
latifolium. *Showy Bastard-Cress*
Thrinax argentea. *Broom Palm, "Chip"-tree, Silver-leaved Palmetto Palm, Silver Thatch*
parviflora. *Palmetto Thatch, Royal Palmetto Palm, Small-flowered Jamaica Fan-Palm*
Thuja. *Arbor-vitæ*
aurea. *Golden Arbor-vitæ*
dumosa. *Bush Arbor-vitæ*
gigantea. *Cedar*, of British Columbia, *Western Arbor-vitæ, White Cedar* of California
Lobbii. *Yellow Cypress*
occidentalis. *American Arbor-vitæ, White Cedar*
occidentalis var. argentea. *Silvery-leaved Arbor-vitæ*
occidentalis var. densa or compacta. *Bag-shot Park Arbor-vitæ*
occidentalis var. ericoides. *American Tom Thumb Arbor-vitæ*
occidentalis var. Vervaeneana. *Belgian Variegated Arbor-vitæ*
plicata. *Nootka Sound Arbor-vitæ*
semper-aurea. *Ever-golden Arbor-vitæ*
Thujopsis. *Broad-leaved Arbor-vitæ*
borealis (Cupressus nutkaënsis). *Yellow Cedar, Yellow Cypress*
dolabrata. *Hatchet-leaved Arbor-vitæ*
Thymus. *Thyme*
alpinus. *Alpine Thyme*
Azoricus. *Azorean Thyme*
azureus. *Azure Thyme*
citriodorus. *Lemon-Thyme*
citriodorus aureo-variegatus. *Variegated Lemon-Thyme*
corsicus. *Corsican Thyme*
lanuginosus. *Downy Thyme*
Mastichina. *Herb-Mastick*
melissoides. *Balm-leaved Thyme*
Nummularius. *Money-wort-Thyme*
(Acinos) patavinus. *Marjoram-leaved Thyme*
Piperella. *Small Peppermint Thyme*
rotundifolius. *Round-leaved Thyme*
Serpyllum. *Brother-wort, Hill-wort, Mother-of-Thyme, Pellamountain, Serpolet-oil-plant, Wild Thyme*
Serpyllum var. album. *White-flowered Thyme*
Serpyllum var. citratus. *Lemon-scented Thyme*
thuriferus. *Incense Thyme*
Tragoriganum. *Goat's-Marjoram*
vulgaris. *Common*, or *Pot-herb, Thyme*
Thyrsacanthus. *Thyrse-flower*
Thysanocarpus curvipes. *Lace-pod*
laciniatus var. crenatus. *Fringe-pod*
Thysanotus. *Fringe-Lily, Fringed Violet*, of Australia
Tiarella cordifolia. *False Mitre-wort*
Tiedemannia teretifolia. *False Water-Drop-wort*
Tigridia. *Tiger-Iris, Tiger-flower*
Pavonia. *"Flower of Tigris," Peacock Tiger-flower, Tiger-flower* of Mexico
speciosa alba. *White Tiger-flower*

Tillæa muscosa. *"Mossy Red-shanks"*
Tillandsia usneoides. *Long* or *Black Moss, New Orleans Moss, Old-man's-beard, Spanish Moss, Vegetable Hair*
Tilia. *Lime, Lime-tree*
americana. *American Bass-wood* or *White-wood*
americana var. laxiflora. *Loose-flowered American Lime-tree*
americana var. pubescens. *Downy-leaved American Lime-tree*
americana var. pubescens leptophylla. *Downy Narrow leaved American Lime tree*
europæa. *European Lime-tree, Lin, Line, Linde*, or *Linden-tree, Russian-Bast-tree, Teil-, Teyl-*, or *Til-, tree*
europæa var. alba. *White-leaved Lime-tree*
europæa var. argentea. *Silver-leaved Lime-tree*
europæa var. aurea. *Golden-twigged Lime-tree*
europæa var. dasystyla. *Hairy-styled Lime-tree*
europæa var. intermedia. *Common Lime-tree*
europæa var. laciniata. *Cut-leaved Lime-tree*
europæa var. microphylla. *"Bass,"* or *"Bast," Small-leaved Lime-tree*
europæa var. platyphylla. *Broad-leaved Lime-tree*
europæa var. platyphylla aurea. *Golden-twigged Broad-leaved Lime-tree*
europæa var. rubra or corallina. *Red-twigged Lime-tree*
heterophylla. *American White Bass-wood*
parvifolia. *Bass, Bast, Small-leaved Lime-tree*
Tilletia Caries. *"Bunt"- Fungus*
Tipuana heteroptera. *Angelim - wood*, of Brazil
Tipularia discolor. *Crane-fly-Orchis*
Tobinia coriacea. *Yellow Mast-wood*
Toddalia aculeata. *Lopez Root*
Todea. *Crape-Fern*
Tofieldia. *False-Asphodel*
calyculata. *Small-calyxed False-Asphodel*
palustris. *Marsh False-Asphodel, Scotch Asphodel*
pubens. *Downy False-Asphodel*
Tolpis barbata. *Yellow Garden Hawk-weed*
Tontelea (Salacia) pyriformis. *Tontel-tree*
Tordylium. *Hart-wort*
apulum. *Small Hart-wort*
maximum. *Great Hart-wort*
officinale. *Officinal Hart-wort*
syriacum. *Broad Tooth-pick Chervil*
Torilis Anthriscus. *Hedge-parsley*
Tormentilla officinalis. *Shepherd's-knot*
Torreya. *Stinking-Yew*
grandis. *Large Stinking-Yew*
Myristica. *Californian Nutmeg-tree*
taxifolia. *Stinking Cedar*
Tortula. *Screw-Moss*
canescens. *Hoary Screw-Moss*
muralis. *Wall Screw-Moss*
rigida. *Aloe-like Screw-Moss*
ruralis. *Great Hairy Screw-Moss*
tortuosa. *Frizzled Mountain Screw-Moss*
unguiculata. *Bird's claw Screw-Moss*

and Foreign Plants, Trees, and Shrubs. 257

Tournefortia argentea. *E. Indian Velvet-leaf*
bicolor. *Basket-Wyth*, or *White-hoop-Wyth*, of Jamaica
heliotropioides. *Summer Heliotrope*
Toxicophlæa spectabilis. *Winter-sweet*
(Acokanthera) Thunbergii. *Hottentot's Ordeal-tree*
Trachelium. *Throat-wort*
cæruleum. *Blue Throat-wort*
Trachylobium (Hymenæa) Hornemannianum. *Zanzibar Copal-resin-plant*
Trachymene australis. *Victorian Parsnip*
Tradescantia. *Spider-wort*
congesta. *Crowded-flowered Spider-wort*
discolor. *Purple-leaved Spider-wort*
malabarica. *Grass-leaved Spider-wort*
pilosa. *Dwarf Spider-wort*
virginica. *Flower-of-a-day, Virginian Spider-wort*
virginica var. alba. *White Spider-wort*
Zanonia. *Gentian-leaved Spider-wort*
Tragia volubilis. *Twining Cow-itch*
Tragopogon porrifolius. *Jerusalem Star, Salsify, Vegetable-Oyster, Star-of-Jerusalem*
pratensis. *Goat's-beard, Go-to-bed-at-noon, Joseph's Flower, Nap-at-noon, Noon-tide, Shepherd's-clock, Star-of-Jerusalem*
Tragopyrum. *Goat's-wheat*
buxifolium. *Box-leaved Goat's-wheat*
lanceolatum. *Lance-leaved Goat's-wheat*
polygamum. *Polygamous Goat's-wheat*
Trapa. *Water-Caltrops*
bicornis and T. bispinosa. *Singhara-nut-plant*
natans. *Jesuit's Nut*, of Venice, "*Ling*," of the Chinese, *Water-Caltrops, Water Chestnut*
Trautvetteria palmata. *False Bug-bane*
Treculia africana. *African Bread-fruit-tree*
Tremella albida. *Fairy Butter*
Auricula. *Earth-jelly*
Tribulus. *Caltrops*
cistoides. *Turkey-blossom*, of the W. Indies
terrestris. *Land Caltrops*
Trichadenia zeylanica. *Tettigaha-*, or *Tettigass-, tree*, of Ceylon
Trichilia emetica. *Roka-tree*, of Arabia
glandulosa. *Rose-wood*, of N. S. Wales
hirta. *Bastard Iron-wood*, of the W. Indies
moschata. *Musk-wood*
spondioides. *White Butter-wood*
Trichobasis Rubigo vera. "*Rust*" *Fungus*
Trichocladus crinitus. *Hair-branch Tree*
Trichodesma zeylanica. *Ceylon Borage*
Trichodium. *Winter-green-grass*
Trichomanes. *Bristle-Fern*
radicans. *Common Bristle-Fern, Cup Goldilocks, Killarney Fern*
Trichonema speciosa. *African Crocus*
Trichosanthes anguina. *Snake Gourd*, of India
colubrina. *Serpent Cucumber, Viper Gourd*, of Central America
Trichosma (Eria) suavis. *Hair Orchid*
Trichostema dichotomum. *Bastard Pennyroyal,* "*Blue Curls*"
lanatum. *Black Sage*, of California

Trichostomum. *Fringe-Moss*
aciculare. *Dark Mountain Fringe-Moss*
canescens. *Hoary Fringe-Moss*
funale. *Cord-like Fringe-Moss*
lanuginosum. *Woolly Fringe-Moss*
patens. *Spreading Fringe-Moss*
Tricuspis purpurea. *Sand-grass*, of N. America
seslerioides. *Tall Red-top-grass*, of N. America
Tricyrtis hirta. *Japanese Toad-Lily*
Trientalis americana. *American Chick-weed-Winter-Green, Star-flower*
europea. *European Chick-weed-Winter green, Star-flower*
Trifolium. *Clover, Trefoil*
agrarium. *Hop Clover, Yellow Clover*
Alexandrinum. *Bersin* or *Egyptian Clover*
alpinum. *Alpine Trefoil*
alpestre. *Oval-headed Clover*
arvense. *Hare's-foot Clover* or *Trefoil, Rabbit-foot*, of N. America, *Stone Clover*
Bocconi. *Boccone's Clover* or *Trefoil*
carolinianum. *Carolina Clover*
filiforme. *Yellow Suckling Clover*
fragiferum. *Straw-berry-*, or *Straw-berry-headed, Clover*
glomeratum. *Clustered Clover*
hybridum. *Alsike*, or *Bastard Clover*
incarnatum. *Crimson Clover* or *Trefoil*
involucratum. *Striped-flowered Clover* or *Trefoil*
Lupinaster. *Bastard Lupine*
maritimum. *Sea-side Clover* or *Trefoil*
medium. *Cow Grass, Marl Grass, Trefoil Clover, Zig-zag Clover* or *Trefoil*
minus. *Yellow Trefoil*
ochroleucum. *Sulphur Clover* or *Trefoil*
pratense. "*Bee-bread*," *Common Clover, Cow Grass, Honey-suckle, Marl Grass, Purple Clover* or *Trefoil*
procumbens. *Lesser Clover, Low Hop Clover, Yellow Clover*
reflexum. *Buffalo Clover*
repens. *Dutch* or *White Clover, Honey-suckle Clover, Honey-suckle Grass*
repens var. hibernicum. *Shamrock*, of Ireland
repens var. purpureum. *Four-leaved Shamrock*
resupinatum. *Reversed Clover*
rubens. *Red Clover* or *Trefoil*
scabrum. *Rough Clover* or *Trefoil*
spadiceum. *Brown Clover*
spumosum. *Bladder-podded Clover* or *Trefoil*
stellatum. *Starry Clover* or *Trefoil*
stoloniferum. *Running Buffalo Clover*
striatum. *Knotted Clover* or *Trefoil*
strictum. *Upright Clover*
subrotundum. *Mayad Clover*
subterraneum. *Subterranean Clover*
suffocatum. *Sand Clover* or *Trefoil*
uniflorum. *One-flowered Trefoil*
Triglochin maritimum. *Sea Arrow-grass*
palustre. *Marsh Arrow-grass*
Trigonella Fœnum-græcum. *Common Fenugreek*
ornithopodioides. *Bird's-foot Fenugreek*
ornithorrhyncus. *Bird's-bill Fenugreek*

s

Trillium. *American Wood-Lily, Three-leaved Night-shade*
cernuum. *Drooping Wake-robin* or *Drooping Wood-lily*, of N. America
erectum. *Beth-root* or *Birth-root, Indian Balm, Lamb's-Quarters, Purple-flowered Wood-Lily*
erythrocarpum. *Painted Wood-Lily*
latifolium. *Ground-Lily, Indian Shamrock, Rattle-snake-root*
grandiflorum. *Large White Wood-Lily, Wake-Robin* of N. America
nivale. *Dwarf White Wood-Lily*
pendulum. *Indian Balm*
Trinia vulgaris. *Hone-wort*
Triodia decumbens. *Heath-grass*
Triosteum. *Fever-wort,* of N. America, *Horse-Gentian*
perfoliatum. *Common Fever-root*
Triphasia trifoliata. *Lime-berry-tree,* of Manilla
Triplaris Bonplandiana. *Ant-tree*
Tripsacum dactyloides. *Buffalo-grass, Gama-grass, Sesame-grass*
Triptolomæa sp. *Violet-wood,* of Brazil (?)
Tristania albicans and T. conferta. *Turpentine-tree,* of Australia
conferta (Lophostemon arborescens). *Queensland Box*
neriifolia. *Water Gum-tree*
Triteleia. *Triplet-Lily*
laxa. *Ithuriel's Spear*
lilacina. *Lilac Star-flower*
Murrayana. *Murray's Star-flower*
uniflora. *Spring Star-flower*
Triticum. *Wheat, Wheat-grass*
æstivum. *Spring* or *Summer Wheat*
amyleum. *Amel Corn, Starch Wheat*
caninum. *Awned Wheat-grass, Dog's-tooth-grass*
Cevallos. *Trigo Moro Wheat*
compositum. *Mummy Wheat, Pharaoh's Corn*
dicoccum. *Emmer Wheat, Two-grained Wheat*
durum. *Hard-grained Wheat*
hybernum. *Winter* or *Lammas Wheat*
monococcum. *Single-grained Wheat, St. Peter's Corn*
polonicum. *Polish Wheat*
repens. *Couch, Couch-grass, Couch-wheat, Dog-grass,* "*Felt*," *Lagoon-grass,* of N. America, *Quack-grass, Scutch-grass, Shelly-grass, Skally-grass, Squitch-grass*
sativum and vars. *Common* or *Soft-grained Wheat*
sativum var. compactum. *Square-eared Wheat*
Spelta. *Dinkel-,* or *Spelt-, Wheat*
turgidum. *Humpy-grained Wheat*
Tritoma. *Club-lily, Torch-lily*
Burchelli. *Burchell's Torch-lily*
grandis. *Tall Torch-lily*
media. *Intermediate Torch-lily*
pumila. *Dwarf Torch-lily*
Rooperi. *Rooper's Torch-lily*
Uvaria. *Common Torch-lily, Flame-flower, Red-hot-poker Plant*
Uvaria var. glaucescens. *Glaucous-leaved Torch-lily*

Tritoma Uvaria var. grandiflora. *Large-flowered Torch-lily*
Triumfetta. *Paroquet-Bur,* of Jamaica, *W. Indian Bur-weed*
Lappula. *Great-wort*
semitriloba. *Bur-bark-tree*
Trochocarpa (Decaspora) laurina. *Australian Beech-cherry* or *Brush-cherry*
thymifolia. *Tasmanian Wheel-seed*
Trollius. *Globe-flower, Globe Ranunculus, Troll-flower*
asiaticus. *Asiatic Globe-flower*
europæus. *Bolts, Common Globe-flower, Golden Ball, Lapper, Lopper, Lockin,* or *Luckin Gowan, Troll-flower*
Fortunei. *Fortune's Globe-flower*
japonicus. *Japanese Globe-flower*
japonicus fl.-pl. *Double Japanese Globe-flower*
laxus. *American Globe-flower, Spreading Globe-flower*
Loddigesi. *Giant Globe-flower*
napellifolius. *Napellus-leaved Globe-flower*
Tropæolum. *Garden Nasturtium, Indian Cress, Yellow Larkspur*
aduncum (T. peregrinum). *Canary-bird Nasturtium, Canary Creeper*
atrosanguineum. *Common Garden Nasturtium*
majus. *Tall Nasturtium*
minus. *Dwarf Nasturtium*
pentaphyllum. *Five-leaved Indian Cress*
polyphyllum. *Yellow Rock Indian Cress*
speciosum. *Flame-flowered Nasturtium* or *Indian Cress*
tuberosum. *Peruvian Nasturtium, Tuberous-rooted Nasturtium*
Trophis. *Ramoon-tree*
aspera. *Paper-tree,* of Siam
Tsuga (Abies) Mertensiana and T. Pattoniana. *Hemlock Spruce,* of California
Tuber æstivum. *English Truffle*
album (Choiromyces meandriformis). *White* or *False Truffle*
cibarium. *Earth-ball, Truffle* (true)
magnatum. *Piedmontese Truffle*
melanosporum. *French Truffle*
rufum (Melanogaster variegatus). *Red Truffle*
Tulipa. *Tulip, Dalmatian Cap*
biflora. *Two-flowered Tulip*
bithynica. *Bithynian Tulip*
Celsiana. *Cels's Tulip, Dwarf Yellow Tulip*
Clusiana. *Clusius's Tulip*
elegans. *Elegant-flowered Tulip*
erythronioides. *Dog's-tooth-Violet Tulip*
fragrans. *Sweet-scented Tulip*
fulgens. *Brilliant Tulip*
Gesneriana and vars. *Common Garden Tulip*
Gesneriana var. laciniata. *Parrot Tulip*
Greigi. *Greig's Tulip, Turkestan Tulip*
Hageri. *Hager's Tulip*
iliensis. *Cowslip-scented Tulip*
Kolpakowskyana. *Kolpakowsky's Tulip*
Oculus-solis. *Sun's-eye Tulip*
Orphanidesi. *Orphanides's Tulip*
persica. *Persian Tulip*
præcox. *Large Sun's-eye Tulip*
pulchella. *Dwarf Rosy-purple Tulip*

and Foreign Plants, Trees, and Shrubs.

Tulipa stellata. *Star-flowered Tulip*
suaveolens. *Van Thol Tulip*
sylvestris. *Wild Tulip*
tetraphylla. *Four-leaved Tulip*
triphylla. *Three-leaved Tulip*
turcica. *Florentine Parrot Tulip*
turkestanica. *Turkestan Tulip*
Tupistra. *Mallet-flower*
Turnera ulmifolia. *Holly-rose* or *Sage-rose,* of the W. Indies
Turpinia occidentalis. *Cassava-wood*
Tussilago. *Colt's-foot*
alpina. *Alpine Colt's-foot*
Farfara. *Asses-foot, Bull's-foot, Clay-weed, Cleats, Common Colt's-foot, Cough-wort, Dove Dock, Foal-foot* or *Fole-foot, Hoofs, Horse-hoof*
Farfara variegata. *Variegated Colt's-foot*
Tylophora (Asclepias) asthmatica. *Country* or *E. Indian Ipecacuanha*
Typha (angustifolia). *Reree-plant,* of India, *Small Bul-Rush*
elephantum. *Elephant's-grass*
latifolia. *Baccobolts, Blackcap, Blackheads, Bul-Rush, Cat-o'-nine-tails, Dod, Flax-tail, March, Marish,* or *Marsh Beetle, Marsh Pestle, Mat-reed, Reed-mace*
Tytonia natans. *Water Balsam*

Ulex. *Furze*
europæus. *Common Furze, French Furze, Gorse, Goss, Thorn-Broom, Whin*
europæus plenus. *Double-blossomed Furze*
nanus. *Autumn-flowering Furze, Cut-whin, Dwarf Furze, Tam Furze*
provincialis. *Provence Furze*
strictus (U. hibernicus). *Irish Furze*
Ulmus. *Elm-tree*
alata. *American Small-leaved Elm, Whahoo, Winged Elm*
americana. *American White Elm*
campestris. *Alme, Aum-tree, Common Elm, English Elm*
campestris var. acutifolia. *Acute-leaved Elm*
campestris var. betulæfolia. *Birch-leaved Elm*
campestris var. chinensis. *Chinese Elm*
campestris var. concavafolia. *Concave-leaved Elm*
campestris var. cornubiensis. *Cornish Elm*
campestris var. cucullata. *Hooded-leaved Elm*
campestris var. foliis aureis. *Golden-variegated Elm*
campestris var. foliis variegatis. *White-variegated Elm*
campestris var. latifolia. *Broad-leaved Elm*
campestris var. nana. *Dwarf Elm*
campestris var. parvifolia. *Small-leaved Elm*
campestris var. purpurea. *Purple-leaved Elm*
campestris var. sarniensis. *Jersey Elm*
campestris var. stricta. *Red English Elm*
campestris var. tortuosa. *Twisted Elm*
campestris var. viminalis. *Twiggy Elm*
campestris var. viminalis variegata. *Variegated Elm*

Ulmus campestris var. virens. *Kidbrook Elm*
campestris var. virgata. *Twiggy Elm*
campestris var. viscosa. *Clammy Elm*
carpinifolia. *Hornbeam-leaved Elm*
Dampieri aurea. *Golden Elm*
effusa. *Spreading-flowered Elm*
fulva. *Moose Elm, Red Elm, Slippery Elm*
glabra. *Smooth-leaved,* or *Feathered, Elm*
glabra var. glandulosa. *Glandular-leaved Elm*
glabra var. major. *Canterbury Seedling Elm*
glabra var. pendula. *Downton Elm, Weeping Elm*
glabra var. ramulosa. *Floetbeck Elm*
glabra var. vegeta. *Chichester Elm, Huntingdon Elm*
hollandica (U. major). *Dutch Elm*
(Holoptelea) integrifolia. *E. Indian Elm*
Kaki. *Japanese Elm*
major. *Declining-branched Elm*
montana. *Mountain Elm, Scotch Elm, Witch* or *Wych Elm* or *Hazel*
montana var. cebennensis. *Cevennes Elm*
montana var. crispa. *Curled-leaved Elm*
montana var. fastigiata. *Exeter Elm*
montana var. glabra. *Smooth-leaved Wych Elm*
montana var. major. *Greater Wych Elm*
montana var. minor. *Small Wych Elm*
montana var. nigra. *Black Irish Elm*
montana var. pendula. *Camperdown Weeping Elm, Weeping Wych Elm*
montana var. rugosa. *Rough-leaved Wych Elm*
monumentalis. *Monumental Elm*
nemoralis. *Hornbeam-leaved Elm*
pumila. *Dwarf Elm*
racemosa. *American Cork Elm, Rock Elm, White Elm*
suberosa. *Cork-barked Elm, Dutch* or *Sand Elm*
suberosa var. alba. *White-barked Elm*
suberosa var. angustifolia and var. latifolia. *Hertfordshire Elm*
suberosa var. erecta. *Upright-branched Elm*
Wallichiana. *Himalayan Elm*
Ulospermum. *Broad-seed*
Ulva Lactuca. *Oyster-green, Sea Lettuce, Sloke*
latissima. *Broad-leaved Oyster-green, Green Laver, Green Sloke, Sea Moss*
Umbilicaria. *" Tripe de Roche " Lichen*
Umbilicus chrysanthus. *House-leek Penny-wort*
pendulinus. *Kidney-wort*
spinosus. *Spiny Penny-wort*
Uncaria acida and U. Gambier. *Gambier, Gambier Catechu, Pale Catechu,* or *Terra japonica-plant,* of Medicine
procumbens. *Grapple-plant, Hook-thorn,* or *" Wait-a-bit "-thorn,* of S. Africa
Uniola paniculata. *" Sea Oats," Sea-side Oat, Spike-grass* of N. America
Urceola elastica. *Caoutchouc-tree,* of Borneo
Urceolina aurea. *Golden Urn-flower*
pendula. *Drooping Urn-flower*
Uredo segetum. *Smut of Corn*

Urena. *Indian Mallow*
Urginea (Scilla) maritima. *Medicinal Squill, Sea-Onion*
Urtica. *Nettle*
(Laportea) canadensis. *Kentucky Hemp*
cannabina. *Hemp-leaved Nettle, Kentucky Hemp*
chamædryoides. *American Germander-leaved Nettle*
dioica. *Common Stinging Nettle*
(Laportea) gigas. *Giant Nettle, of N. S. Wales*
gracilis. *Hudson's Bay Nettle*
moroides and U. photiniæphylla. *Nettle-tree, of Australia*
pilulifera. *Roman Nettle*
pumila. *Cool-weed, Rich-weed, Stingless Nettle*
(Bœhmeria) nivea. "*China-grass*"*-plant, Chinese Cotton-nettle, Chinese Cambric-grass-plant*
urens. *Small British Nettle*
urentissima. *Devil's-leaf*, of Timor
Usnea barbata. *Neck-lace Moss*
jubata. *Tree-hair Lichen*
plicata. *Tree Moss*
Ustilago fœtida. *Bunt or Pepper-brand Fungus*
segetum. "*Brancorn*," "*Brawn*," *Smut Fungus*
Utricularia. *Bladder-wort, Hooded Water-Milfoil*
Endresi. *Endres's Bladder-wort*
inflata. *Inflated Bladder-wort*
minor. *Small Bladder-wort*
vulgaris. *Common Bladder-wort or Hooded Water-Milfoil*
Uvaria (Unona) aromatica. *Negro Pepper*
lanceolata. *Jamaica Lance-wood*
Uvularia grandiflora. *Large-flowered Bell-wort*
sessilifolia. *Sessile-leaved Bell-wort*

Vaccaria vulgaris (Saponaria Vaccaria). *Cow-herb, Soap-wort*
Vaccinium. *Whortle-berry*
albiforum. *White-flowered Whortle-berry*
amœnum. *Broad-leaved Whortle-berry*
angustifolium. *Narrow-leaved Whortle-berry*
arboreum. *Farkle-berry, Tree Whortle-berry*
Arctostaphylos. *Oriental Whortle-berry*
buxifolium. *Box-leaved Whortle-berry*
cæspitosum. *Tufted Whortle-berry*
canadense. *Canada Blue-berry*
caracasanum. *Caraccas Whortle-berry*
corymbosum. *Common or Swamp Blue-berry*, of N. America
crassifolium. *Thick-leaved Whortle-berry*
dumosum. *Bushy Whortle-berry*
elongatum. *Elongated Whortle-berry*
erythrinum. *Red-twigged Whortle-berry*
frondosum. *Blue Tangles, Blue Whortle-berry*, of N. America, *Leafy Whortle-berry*
galezans. *Gale-leaved Whortle-berry*
glabrum. *Smooth Whortle-berry*
grandiflorum. *Large-flowered Whortle-berry*
humifusum. *Trailing Whortle-berry*

Vaccinium leucostomum. *White-lipped Whortle-berry*
ligustrinum. *Privet-leaved Whortle-berry*
macrocarpon. *American Cran-berry, Large Cran-berry*
maderense. *Madeira Whortle-berry*
meridionale. *Jamaica Bilberry*
minutiflorum. *Small-flowered Whortle-berry*
Myrsinites. *Myrsine-leaved Whortle-berry*
myrtifolium. *Myrtle-leaved Whortle-berry*
myrtilloides. *Myrtillus-like Whortle-berry*
Myrtillus. *Black-berry, Black-heart, Blae-berry, Blue-berry, Bull-berries, Common Bilberry, Frughan, Frocken, or Frughans, Hart-berries, Horts, Huckle-berry, Hurtle-berry, Whin-berry, Whortle Bilberry*
Myrtillus var. albis-baccis. *White-berried Bilberry*
nitidum. *Glossy-leaved Whortle-berry*
ovatum. *Ovate-leaved Whortle-berry*
Oxycoccos. *Bog-berry, Corn-berries, Cran-berry, Crane-berry, or Craw-berry, Crone-berry, Fen-berry, Fen Grapes, Marsh or Marish Berries, Monox or Moonog Heather, Moor-berries or Moss-berries*
padifolium. *Bird-Cherry-leaved Whortle-berry*
pallidum. *Pale-flowered Whortle-berry*
penduliflorum. "*Ohelo*," of the Sandwich Islands
pennsylvanicum. *Dwarf Blue-berry, Pennsylvanian Whortle-berry*
præstans. *Kamtschatkan Bilberry*
resinosum. *Resinous Whortle-berry*
Rollisoni. *Rollison's Whortle-berry*
salicinum. *Willow-leaved Whortle-berry*
stamineum. *Deer-berry, Squaw Huckle-berry*
uliginosum. *Bog Bilberry*
vacillans. *Low Blue-berry*
virgatum. *Twiggy Whortle-berry*
Vitis-Idæa. *Brawlins, Cow-berry, Flowering Box, Munshock, Red Whortle-berry*, of Mt. Ida
Vaillantia Cruciata. *Cross-wort*
Valeriana. *Valerian*
dioica. *Marsh Valerian*
globulariæfolia. *Globularia-leaved Valerian*
montana. *Mountain Valerian*
officinalis. *All-heal, Cat's Valerian, Cut-finger, Cut-heal, Herb Bennet, Medicinal Valerian, St. George's Herb.*
Phu. *Cretan Spikenard, Garden Valerian*
pyrenaica. *Capon's-tail Grass, Pyrenean Valerian*
sambucifolia. *Elder-leaved Valerian*
tripteris. *Three-winged Valerian*
tuberosa. *Mountain Spikenard*
Valerianella Auricula. *Sharp-fruited Corn-salad*
carinata. *Keeled Corn-salad*
dentata. *Narrow-fruited Corn-salad*
eriocarpa. *Italian Corn-salad*
olitoria. *Common Corn-salad, Lamb's-Lettuce, Milk-grass,* "*White Pot-herb*"
Vallisneria spiralis. *Eel-grass, Tape-grass*
Vallota purpurea. *Scarborough Lily*
purpurea var. eximia. *Scarlet Scarborough Lily*

and Foreign Plants, Trees, and Shrubs. 261

Vancouveria hexandra. *American Barren-wort*
Vanda furva. *Cowslip-scented Orchid*
Vandellia diffusa. *Bitter-blain*, of Guiana
Vanilla claviculata. *" Green-withe,"* of the W. Indies, *Purple-lip Orchid*
Vascoa amplexicaulis. *Liquorice-bush,* of the Cape
Vateria indica. *White Dammar-tree*
Vella annua. *Annual Cress-Rocket, Spanish Cress*
Pseudo-cytisus. *Perennial or Shrubby Cress-Rocket*
Vepris lanceolata. *White Iron-wood*
Veratrum. *White-Hellebore*
album. *Lung-wort* or *Lyng-wort, White-flowered White-Hellebore*
nigrum. *Dark-flowered White-Hellebore*
viride. *Indian Poke, Itch-weed, Swamp Hellebore, White-Hellebore* of N. America
Verbascum. *Mullein*
Blattaria. *Moth Mullein*
Boerhaavii. *Annual Mullein*
Chaixii. *Nettle-leaved Mullein*
grandiflorum. *Great-flowered Mullein*
Lychnitis. *White-flowered Mullein*
macranthum. *Long-flowered Mullein*
Myconi. *Borage-leaved Mullein*
nigrum. *Black-rooted Mullein*
niveum. *Snow-white Mullein*
olympicum. *Olympian Mullein*
phlomoides. *Woolly Mullein*
phœniceum. *Purple-flowered Mullein*
pulverulentum. *Hoary Mullein*
Thapsus. *Aaron's - Rod, Adam's - Flannel, Ag-leaf, Beggar's - Blanket, Blanket-leaf, Bullock's Lung-wort, Candle-wick, Clown's Lung-wort, Cow's Lung-wort, Duffle, Feld-wood* (?), *Felt-wort, Flannel, Flannel-plant, Fluff-weed, Hay-taper, Hedge-taper, Hig-taper* or *High-taper, Hare's-beard, Jacob's-staff, Ladies' Fox-glove, Mullein Dock, Shepherd's-club, Torches, Velvet Dock, White Mullein*
vernale. *Tall Mullein*
virgatum. *Twiggy Mullein*
Verbena Aubletia (V. montana). *Cut-leaved Vervain, Rocky Mountain Vervain, Rose Vervain*
hastata. *Blue Vervain,* or *Simpler's-Joy,* of N. America, *Wild Hyssop*
montana (V. Aubletia). *Rocky Mountain Vervain*
nodiflora. *Creeping Vervain*
officinalis. *Common Vervain, Holy Herb, Juno's Tears, Pigeon's Grass, Simpler's-Joy*
urticæfolia. *Nettled-leaved* or *White Vervain*
venosa. *Hardy Garden Verbena, Large-veined Vervain*
Verbesina. *Crown-beard*
Vernonia. *Iron-weed,* of N. America
anthelmintica. *Khatzum-,* or *Kinka-, oil-plant*
arborescens. *W. Indian Flea-bane*
Noveboracensis. *" Flat-tops,"* of New York, *New York Iron-weed*
Veronica. *Cancer-wort, Speedwell*

Veronica agrestis. *Garden Speedwell, Germander Chick-weed, Procumbent Field Speedwell*
alpina. *Alpine Speedwell*
americana. *American Brook-lime*
amethystina (V. paniculata, V. spuria). *Amethyst Speedwell*
Anagallis. *" Faverell," Great Water Speedwell, Water Pimpernel*
Andersoni. *Anderson's Speedwell*
aphylla. *Naked-stalked Speedwell*
arvensis. *Spiked Field Speedwell, Wall Speedwell*
austriaca. *Austrian Speedwell*
Beccabunga. *Brook-lime, Horse-well-grass, Water Pimpernel*
bellidoides. *Daisy-leaved Speedwell*
Buchanani. *Buchanan's Speedwell*
Buxbaumii. *Buxbaum's Speedwell*
buxifolia. *New Zealand Box*
candida. *Silvery Speedwell*
caucasica. *Caucasian Speedwell*
Chamædrys. *Angel's-eyes, Bird's-eyes, Blewort, Blue-eyes, Blue Stars, Female Fluellin, Germander Speedwell, God's-eye*
corymbosa. *Many-spiked Speedwell*
cuneifolia. *Wedge-leaved Speedwell*
decussata. *Falkland Islands Speedwell*
foliosa. *Leafy Speedwell*
fruticulosa. *Shrubby-stalked Speedwell*
gentianoides. *Gentian Speedwell*
Guthrieana. *Guthrie's Speedwell*
hederæfolia. *Ivy-leaved Chick-weed, Ivy-leaved Speedwell, Mother-of-Wheat, Small Hen-bit, Winter-weed*
hybrida. *Welsh Speedwell*
incarnata. *Flesh-coloured Speedwell*
incisa. *Cut-leaved Speedwell*
laciniata. *Fern-leaved Speedwell*
lactea (V. repens). *White-flowered Speedwell*
longifolia subsessilis. *Japanese Speedwell*
maritima. *Sea-side Speedwell*
montana. *Mountain Speedwell*
multifida. *Narrow-leaved Speedwell*
neglecta. *Gray-leaved Speedwell*
Nummularia. *Money-wort Speedwell*
officinalis. *Common Medicinal-tea-Speedwell, Fluellen, Ground-hele*
officinalis var. rosea. *Pink-flowered Speedwell*
pectinata. *Scalloped-leaved Speedwell*
peduncularis. *White Caucasian Speedwell*
peregrina. *Neck-weed, Purslane Speedwell*
perfoliata. *Digger's Speedwell,* of Australia
pinguifolia. *Thick-leaved Speedwell*
prostrata. *Prostrate Speedwell*
repens. *Creeping Speedwell*
rupestris. *Rock Speedwell*
salicifolia. *Willow-leaved Speedwell*
satureiæfolia. *Savory-leaved Speedwell*
saxatilis. *Deep-blue-flowered Rock Speedwell*
saxatilis var. Grievi. *Rosy-flowered Rock Speedwell*
scutellata. *Bog* or *Marsh Speedwell*
serpyllifolia. *Paul's-,* or *St. Paul's-, Betony, Thyme-leaved Speedwell*
sibirica. *Siberian Speedwell*
spicata. *Spike-flowered Speedwell*
spicata var. corymbosa. *Clustered-flowered Speedwell*

Veronica spuria. *Bastard Speedwell*
taurica. *Taurian Speedwell*
Teucrium. *Hungarian or Saw-leaved Speedwell*
triphyllos. *Variable-leaved Speedwell*
verbenacea. *Verrain Speedwell*
verna. *Vernal Speedwell*
(Leptandra) virginica. *Culver's Physic, Culver's Root, Great Virginian Speedwell*
Verticillaria acuminata. *"Balsam of Maria"-tree*
Verticordia. *Juniper Myrtle*
Vesicaria. *Bladder-pod or Bladder-seed*
græca. *Grecian Bladder-pod*
Shortii. *American Bladder-pod*
utriculata. *Inflated Bladder-pod*
Vestia lycioides. *Box-thorn*, of Chili
Vexillaria virginica. *Virginian Butterflyweed or Flag-flower*
Viburnum acerifolium. *Dockmackie, Maple-leaved Arrow-wood*
cassinoides. *Apalachian Tea-tree*
cotinifolium. *Indian Wayfaring-tree*
Dahuricum. *Dahurian Guelder-Rose*
dentatum. *American-Arrow-wood*
ellipticum. *Arrow-wood*, of California
lævigatum. *Cassioberry-bush*
Lantana. *Common Wayfaring-tree, Cotton-tree, Coven-tree, Lithy-tree, Mealy-tree, The Cottoner*
lantanoides. *American Wayfaring-tree, Hobble-bush*
Lentago. *Sheep-berry, Sweet Viburnum*
macrocephalum. *Large-flowered Snow-ball-tree*
nudum. *American Withe-rod*
Opulus. *Dog-Eller, Dog-Rowan-tree, Cranberry-tree, Guelder-Rose, Marsh, Marish, or Water Elder, Ople-tree, Snow-ball-tree, Whitten-tree*
Opulus var. nanum. *Dwarf Guelder-Rose*
Opulus var. roseum. *Red Guelder-Rose, Rose-Elder*
orientale. *Eastern Guelder-Rose*
Oxycoccos. *High Cran-berry*, of N. America
prunifolium. *American Black Haw*
pubescens. *Downy Arrow-wood*
Tinus. *Laurustinus*
Tinus var. lucidum. *Shining-leaved Laurustinus*
Tinus var. strictum. *Upright-branched Laurustinus*
Vicia. *Vetch*
americana. *American Wood-Vetch*, "*Pea Vine*" of California
argentea. *Silvery Vetch*
bithynica. *Bithynian Vetch*
caroliniana. *Carolina Tufted Vetch*
Cracca. *Common Tufted Vetch, Cow Vetch, Wild Fitches or Fitches*
dumetorum. *Great Wood-Vetch*
Faba. *Straight Bean*
hirsuta. *Common Tare, Hairy Vetch*
lathyroides. *Spring Vetch, Strangle-Tare*
lutea. *Yellow-flowered Vetch*
Narbonnensis. *Narbonne Vetch*
Orobus. *Upright Vetch*
sativa. *Common Vetch, Fetch, Fitch, Fitches, Lints*

Vicia sepium. *Bush-Vetch*
sylvatica. *Wood-Vetch*
sylvestris. *Strangle-Tare, Tine-Tare*
tetrasperma. *Four-seeded Vetch*
Victoria regia. *Queen Victoria's Water-lily, Royal Water-lily, Victoria Water-lily, Water-Maize, Water-platter*
Vieusseuxia. *Peacock Iris*
glaucopis (Iris Pavonia). *Blue-eyed Peacock-Iris*
Vigna luteola. *Sea-side Bean*, of the W. Indies
sinensis. *Chowlee-plant*, of India
unguiculata. *Red Bean*
Vilfa. *Rush-grass*
Villarsia (Limnanthemum) nymphæoides. *Fringed Buck-bean, Fringed Water-lily, Round-leaved Buck-bean, Small Yellow Water-lily*
Vilmorinia multiflora. *Vilmorin's Purple Pea-flower*
Viminaria denudata. *Australian Rush-Broom, Victorian Swamp Oak or Broom*
Vinca. *Periwinkle*
acutiloba. *Italian Periwinkle*
alba. *White-flowered Indian Periwinkle*
herbacea. *Herbaceous Periwinkle*
major. *Band-plant, Cut-finger, Large Periwinkle*
major var. elegantissima. *Variegated-leaved Periwinkle*
minor. *Small Periwinkle*
minor fl.-pl. purpurea. *Double Purple Periwinkle*
minor var. alba. *Small White Periwinkle*
minor var. atropurpurea. *Dark-purple-flowered Periwinkle*
rosea. *Madagascar Periwinkle,* "*Old-maid*" of the W. Indies, *Rosy-flowered Indian Periwinkle*
Vincetoxicum officinale. *Tame Poison*
Viola. *Violet*
altaica. *Altaian Violet*
arborescens. *Tree Violet*
argentiflora. *Silvery-flowered Violet*
Beckwithii. *Beckwith's Violet*
"Belle de Chatenay." *Double White Winter-flowering Violet*
biflora. *Twin-flowered Violet*
blanda. *Sweet White Violet*
calcarata. *Spurred Violet*
canadensis. *Canadian Violet*
canina. *Dog Violet*
canina var. adunca. *Hook-spurred Violet*
cornuta. *Horned, or Pyrenean, Violet*
cucullata. *Common Blue Violet of N. America, Hollow-leaved Violet*
delphinifolia. *Larkspur-leaved Violet*
gracilis. *Olympian Violet*
hastata. *Halberd-leaved Violet*
hirta. *Hairy Violet*
lanceolata. *Spear-leaved Violet*
lutea. *Mountain Violet*
mirabilis. *Broad-leaved Violet*
montana. *Alpine Violet*
multiflora. *Many-flowered Violet*
Munbyana. *Munby's Violet*
odorata. *March, or Sweet-scented, Violet*
odorata var. pallida plena. *Neapolitan Violet*

Viola palmaensis. *Palma Violet*
palmata. *Hand-leaf Violet*
palustris. *Marsh Violet*
pedata. *Bird's-foot Violet*
pedunculata. *Long-stalked Violet*
pennsylvanica. *Pennsylvanian Violet*
pinnata. *Pinnate-leaved Violet*
primulæfolia. *Primrose-leaved Violet*
pubescens. *Downy Yellow Violet*
renifolia. *Kidney-leaved Violet*
rostrata. *Long-spurred Violet*
rothomagensis. *Rouen Violet*
rotundifolia. *Round-leaved Violet*
sagittata. *Arrow-leaved Violet*
Sheltonii. *Shelton's Violet*
striata. *Pale*, or *Striped-flowered, Violet*
suavis. *Russian Violet*
sylvatica. *Dog Violet, Hedge Violet*
tricolor. *Cull-me-to-you, Fancy, Flamy, Garden-gate, Heart's-case, Herb Trinity, Jump-up-and-kiss-me, Kiss-me, Kiss-me-at-the-garden-gate, Live-in-idleness, Love-in-idleness, Pansy, Pink-of-my-John, Three-faces-under-a-hood, Tickle-my-fancy*
uniflora. *Siberian Violet*
Virgilia lutea (Cladrastris tinctoria). *American Yellow-wood*
Virola sebifera. *Candle-nut-tree*, of Guiana
Viscum album. *Common Mistletoe*
Vismia brasiliensis. *Brazilian Wax-tree*
guianensis. *American Gamboge-tree* or *Gutta-Gum-tree, Wax-tree* of Guiana
Mocanera. *Mocan-shrub*
Vitex Agnus-castus. *Chaste Tree, Hemp-tree, Monk's Pepper-tree, Tree-of-Chastity*
Lignum-vitæ. *Queensland Lignum-vitæ*
littoralis. *New Zealand Teak* or *Puriri-tree*
trifolia. *Wild Pepper*, of India
umbrosa. *W. Indian Box-wood*
Vitis. *Vine*
acetosa. *Vine of N. Australia*
æstivalis. *American Summer Grape*
candicans. *Mustang Vine*
(Cissus) antarctica. *Port Jackson Black Grape*, or *Kangaroo Grape*
cordifolia. *Chicken-Grape, Frost-Grape*, or *Winter-Grape*
gongylodes. *Gouty-stemmed Vine*
(Cissus) heterophylla variegata. *Variegated Vine*
heterophylla humulifolia. *Turquoise-berried Vine*
humilifolia. *Hop-leaved Vine*
(Cissus) hypoglauca. *Gipps-land Grape Vine, Vine* of Eastern Australia
indivisa. *Entire-leaved Ivy Grape-Vine*
Labrusca. *American Plum-Grape, Bland's Grape, Isabella-Grape, Northern Fox-Grape*
laciniosa. *Parsley-leaved Vine*
(Cissus) opaca. *Burdekin Vine*
riparia. *Fragrant Wild Vine*, of N. America
rotundifolia. *American Bull-Grape*
vinifera and vars. *Cultivated Grape-Vine*
vinifera var. corinthiaca. *Currant Vine*
vulpina. *American Bullet-Grape, Muscadine*, or *Southern Fox-Grape*
vulpina var. *Scuppernong Grape*
Vittadenia australis (V. triloba). *New Holland Daisy*

Vittaria lineata. *Florida Ribbon-Fern*
Voandzeia subterranea. *Bambarra Ground-nut, Pea-nut* of Madagascar, *Underground Bean*
Vochysia guianensis. *Copai-yé-wood*, of Guiana
Wahlenbergia capillaris. *Australian Bell-flower*
gracilis. *Australian Hare-bell*
(Campanula) hederacea. *Ivy-leaved Hare-bell*
littoralis. *Tasmanian Bell-flower*
lobelioides. *Madeira Bell-flower*
saxicola. *New Zealand Blue-bell*
Waldsteinia (Comaropsis) fragarioides. *Barren Straw-berry*
Walsura Piscidia. *E. Indian Fish-poison*
Washingtonia (Sequoia) gigantea. "*Big Tree*" or "*Mammoth Tree*" of California
Watsonia. *Bugle-Lily*
Wedelia carnosa. *Creeping Ox-eye, W. Indian Marigold*
Weigela rosea and vars. *Bush Honey-Suckle*
Weinmannia racemosa. *Tawai-bark-tree*, of New Zealand
Wellingtonia (Sequoia) gigantea. "*Big Tree*" or "*Mammoth Tree*" of California
Westringia rosmariniformis. *Victorian Rosemary*
Wickstrœmia indica. *Native Daphne*, of Australia
Widdringtonia. *African Cypress*
juniperoides. *Cape Gum-tree*
Wigandia urens (W. Caracasana). *Curaccas Big-leaf*
Wistaria (Glycine). *Grape-flower Vine*
frutescens. *American Kidney-bean Tree*
sinensis. *Chinese Kidney-bean Tree*, "*Fiji*" or "*Fu-ji*," of Japan
Wittsteinia vacciniacea. *Victorian Blac-berry* or *Whortle-berry Bush*
Woodwardia. *Chain-Fern*
angustifolia. *Narrow-fronded* or *Netted Chain-Fern*
radicans. *Californian Chain-Fern*
Wrightia. *Palay*, or *Ivory-tree*, of the E. Indies
antidysenterica. *Conessi-bark Tree*
tinctoria. *Pala Indigo-plant*

Xanthium. *Cockle-bur* or *Clot-bur*
spinosum. *Bathurst Bur, Spiny Clot-bur*
Strumarium. *Ditch-Bur, Louse-Bur, Small Burdock*
Xanthorrhæa. *Black Boy Tree, Grass-tree* or *Grass-Gum-tree*, of Australia
Xanthorrhiza apiifolia. *American Yellow-root Shrub*
Xanthosoma atrovirens. *W. Indian Kale*
sagittæfolia. *Arrow-leaved Spoon-flower*
Xanthostemon (Metrosideros) chrysantha. *Golden Myrtle*, of Queensland
Xanthoxylon. *Tooth-ache-tree*
americanum. *Northern Prickly Ash, Tooth-ache-tree*
capense. *Wild Cardamom*, of S. Africa
Caribæum. *Satin-wood Tree*

Xanthoxylon Clava-Herculis (X. caroliniaum). *Hercules'-Club, Southern Prickly Ash, W. Indian* or *Prickly Yellow-wood*
Daniellii. *Bitter Pepper, Star-Pepper*
fraxineum. *Prickly Ash*
guianense. *Negro's Pepper*
mandschuricum. *Anise Pepper-tree*
mite. *Thornless Tooth-ache-tree*
piperitum. *Chinese,* or *Japanese, Pepper*
Pterota. *Bastard Iron-wood*
tricarpum. *Three-fruited Tooth-ache-tree*
Xeranthemum. "*Everlasting*" or "*Immortelle*"-*flower*
annuum. *Annual Everlasting*
Xerophyllum asphodeloides. *Turkey's-beard*
Xerotes longifolia. *Tussock-grass,* of Australia
Ximenia americana. *False Sandal-wood, Hog-plum, Mountain-plum, Sea-side Plum* of the W. Indies
Xiphopteris serrulata. *Sword-Fern*
Xylia dolabriformis (Inga xylocarpa). *Burmah,* or *E. Indian, Iron-wood*
Xylomelon pyriforme. "*Wooden-Pear*"-*tree,* of Australia
Xylophylla latifolia. *Sea-side Laurel,* of Jamaica
Xylopia frutescens. *Bitter-wood,* of Guiana
glabra. *W. Indian Bitter-wood*
Xylostroma giganteum. "*Oak-leather*" *Fungus*
Xyris. *Yellow-eyed Grass* of N. America

Yucca. *Adam's-needle, Bear-grass, Spanish Bayonet*
acuminata. *Pointed-leaved Adam's-Needle*
aloifolia. *Aloe-leaved Adam's-Needle, Spanish Dagger*
aloifolia var. pendula. *Drooping-leared Adam's-Needle*
angustifolia. *Narrow-leaved Adam's-Needle*
canaliculata (Y. Treculeana). *Channelled-leaved Adam's-Needle*
concava. *Hollow-leaved Adam's-Needle*
conspicua. *Conspicuous Adam's-Needle*
crenulata. *Scalloped-leaved Adam's-Needle*
draconis. *Drooping-leaved Adam's-Needle*
filamentosa. *Adam's - Needle-and - Thread, Thready Adam's-Needle, Silk-grass* of Carolina
flaccida. *Flaccid-leaved Adam's-Needle*
gloriosa. *Common Adam's-Needle, Mound-lily*
gloriosa var. glaucescens. *Glaucous Adam's-Needle*
graminifolia. *Grass-leaved Adam's-Needle*
nivea. *Silvery Adam's-Needle*
obliqua. *Oblique-leaved Adam's-Needle*
recurva. *Recurved-leaved Adam's-Needle*
rufo-cincta. *Reddish-edged Adam's-Needle*
stricta. *Upright-leaved Adam's-Needle*
superba. *Superb Adam's-Needle*

Yucca tenuifolia. *Slender-leaved Adam's-Needle*
undulata. *Wavy-leaved Adam's-Needle*

Zamia furfuracea and Z. integrifolia. *Jamaica Sago-tree*
Zannichellia palustris. *Horned Pond-weed*
Zanonia indica. "*Bandolier-fruit*"-*tree*
Zapania (Lippia) nodiflora. *Creeping Vervain*
Zasmidium cellare. *Wine-cellar Fungus*
Zauschneria californica. *Californian Fuchsia*
Zea Curagua. *Saw-leaved Maize*
japonica. *Japanese Maize*
Mays. *Common Maize, Guinea Wheat, Indian Corn,* "*Mealies*," *Turkey Wheat*
Zelkova crenata. (Planera Richardi). *Common Zelkowa-tree*
Zephyranthes. *Zephyr-flower*
(Amaryllis) Atamasco. *Atamasco Lily*
candida. *Peruvian Swamp-lily, White Zephyr-flower*
carinata. *Keeled Zephyr-flower*
rosea. *Rosy Zephyr-flower*
sulphurea. *Sulphur-coloured Zephyr-flower*
tubispatha. *Tube-spathed Zephyr-flower*
Zieria. *Turmeric-tree,* of Australia
lanceolata. *Stink-wood,* of Tasmania
macrophylla. *Australian Stink-wood*
Smithii. *Sand-fly Bush,* of Australia
Zingiber. *Ginger*
Casumunar. *Bengal Root*
officinale. *Common Ginger, E. Indian, Jamaica,* or *Red Ginger*
Zerumbet. *Broad-leaved Ginger*
Zinnia. *Youth-and-Old-Age*
Zizania aquatica (Hydropyrum esculentum). *Indian* or *Canada Rice, Water-Oats*
Zizia integerrima. *Golden Alexanders*
Zizyphus flexuosa. *Zig-zag Jujube-tree*
incurva. *Incurved-spined Jujube-tree*
Jujuba. *Common Jujube-tree*
Jujuba var. *Chinese Date*
Lotus. *Lotus,* or *Lotos-, tree*
Spina-Christi. *Christ's-Thorn, Nubk-tree* of Palestine
Zostera (Alga) marina. *Bell-ware, Glass Wrack, Glaziers' Sea-weed, Grass Wrack* or *Grass Weed, Turtle-grass, Sea-Hay, Wrack-grass*
mediterranea. *Glaziers' Sea-weed*
nana. *Dwarf Grass-wrack*
Zuelania lætioides. *Wild Coffee,* of the W. Indies
Zygadenus venenosus. "*Death Quamash,*" *Hog's Potato*
Zygodon. *Yoke Moss*
conoideus. *Small Yoke Moss*
Zygophyllum. *Bean-Caper*
Fabago. *Syrian Bean-Caper*